LANDMARKS IN SOIL MECHANICS

THE RANKINE LECTURES 1981 – 1990

Published by Thomas Telford Services Ltd, Thomas Telford House, 1 Heron Quay, London E14 4JD

This is the third in a series of volumes, each consisting of 10 years of Rankine lectures; this volume is reprinted from *Géotechnique* 1981-1990. The first volume *Milestones in soil mechanics* was published in 1975 and the second volume, *Developments in soil mechanics* was published in 1983, both by Thomas Telford Ltd.

ISBN: 0 7277 1908 1

Printed and bound in Great Britain by Galliard (Printers) Ltd, Great Yarmouth

CONTENTS

The Rankine Lectures 1981 – 1990

The British Geotechnical Society annually commemorates the great engineer and physicist, William John Maquorn Rankine (1820–1872), by holding a lecture in London presented by a distinguished soil mechanics specialist. Following the success of the first volume of lectures, which were given in the years 1961–1970, and the second volume of lectures given from 1971–1980, this volume has been made up of the ten lectures, spanning the years 1981–1990. These were published annually in the September or December issues of Géotechnique, the international journal of soil mechanics and foundation engineering. These lectures represent the work of the highest acclaimed authorities in soil mechanics from throughout the world.

The Rankine Lecture

The Twenty-first Rankine Lecture of the British Geotechnical Society was given by Professor N. R. Morgenstern at Imperial College, London on 3 March, 1981. The following introduction was given by Professor A. W. Skempton.

It is with pleasure and perhaps more than a hint of justifiable pride that I am introducing my former student and colleague, Professor Morgenstern, as the 21st Rankine Lecturer.

Norbert Morgenstern, born in May 1935, took his degree in civil engineering at the University of Toronto in 1956. He then came to Imperial College as a graduate student on an Athlone Fellowship. Here he so distinguished himself that we gladly took the opportunity of converting him into a research assistant, and in 1960 he came on the staff as a Lecturer. Certainly to the advantage of the College, and I think to his own benefit, he then stayed with us for a further 8 years.

This was an exciting period in soil mechanics research, associated particularly at Imperial College with the discovery of residual strength and the study of shear zones both in the laboratory and the field. Morgenstern took an active part in this work. His own contributions included an examination of the mechanics and morphology of shear zones, in conjunction with John Tchalenko, and the development, with Dr Price, of an accurate method of analysing stability on non-circular surfaces. But I also remember the delight of having contact with such a keen intellect ready to sustain long, frequent and always stimulating discussions. And I recall with the utmost gratitude his devoted and inspiring assistance in the field investigations at Sevenoaks. Moreover, a few years later, in a brilliant analysis of the consolidation of thawing soils, he provided the key to a quantitative understanding of the Pleistocene solifluction movements which form such a striking feature of that site.

However, in 1968 he received an offer to take the Chair of Civil Engineering at the University of Alberta. Our loss was Canada's gain. There he has built up one of the leading soil mechanics schools in North America, which now consists of 7 staff, a research assistant and 3 technicians, with 35 graduate students. His personal achievements during the past 12 years, since arriving at Edmonton, are formidable and place him securely in the top rank of world authorities on geotechnical engineering science and practice: a position which causes no surprise to his friends in London, but gives them much pleasure to recognize.

Morgenstern's research work, covering an exceptionally wide range of subjects, has resulted in the publication of rather more than 100 papers, while his consulting practice has included work on 5 large earth dams, on slopes in Hong Kong, Brazil and Madagascar, on drilling and oil production in the Beaufort Sea, on Arctic pipelines and on oil sands. It is with geotechnical problems in the two last classes of project that he will chiefly be concerned this evening.

As we are keenly looking forward to hearing what he has to say, I will without further delay ask him to give his Lecture.

Professor N. R. Morgenstern

MORGENSTERN, N. R. (1981). *Géotechnique* **31**, No. 3, 305–365

Geotechnical engineering and frontier resource development

N. R. MORGENSTERN*

The traditional concepts that constitute the framework for geotechnical engineering are often insufficient on their own to provide a basis for solving geotechnical problems associated with frontier resource developments. Studies are reported on the creep of permafrost slopes, the mechanics of heave in freezing soils and the behaviour of frozen soils subjected to thaw to illustrate this. These problems are encountered in the exploration and production of hydrocarbon resources in the Arctic. Considerations of ice rheology, fundamental thermodynamics and heat conduction in soils are additional concepts needed to solve these problems. Other examples are drawn from the geotechnical concerns that enter into the development of the Alberta oil sands. Here the geotechnical engineer must deal with gas-saturated, diagenetically-altered sands and with deformability and strength under high temperatures. Illustrations are given of the novel forms of behaviour encountered under these conditions. Initial results are presented of pore pressures developed under undrained heating and of the theoretical relation between the rate of heating and the dissipation of pore pressures.

Rankine is actually better known for his work on thermodynamics and properties of fluids and gases than for his work on earth pressure and therefore it seems fitting in a Rankine Lecture to draw attention to the significance of the main body of Rankine's work in many new areas of geotechnical endeavour.

Les concepts traditionnels sur lesquels se base le génie géotechnique ne suffisent souvent pas, à eux seuls, à permettre de résoudre les problèmes géotechniques associés au développement des ressources frontalières. Pour illustrer ce point, il est fait mention d'études sur le fluage de pentes à gel permanent, la mécanique du soulèvement dans des sols en train de geler, et le comportement de sols gelés soumis au dégel. Ces problèmes se posent lors de l'exploration et de la production de ressources hydrocarbonées en Arctique. La rhéologie de la glace, la thermodynamique élémentaire ainsi que la transmission de la chaleur dans les sols sont des concepts supplémentaires nécessaires à la résolution de ces problèmes. D'autres exemples sont tirés des préoccupations d'ordre géotechnique relatives au développement des Sables Pérolifères de l'Alberta. Dans ce cas, l'ingénieur géotechnique a affaire à des sables saturés de gaz diagénétiquement modifiés et qui présentent une certaine déformabilité et une résistance à des températures élevées. Les nouveaux types de comportement rencontrés dans ces conditions sont décrits. Des premièrs résultats sont présentés pour les pressions interstitielles engendrées par le chauffage sans drainage, ainsi que pour le rapport théorique entre l'intensité du chauffage et la dissipation des pressions interstitielles. Rankine est, en fait, mieux connu pour ses travaux sur la thermodynamique et les propriétés de fluides et de gaz que pour ses travaux sur la poussée des terres et il semble donc approprié, lors d'une conférence sur Rankine, d'attirer l'attention sur l'essentiel de son oeuvre et son influence dans bien des nouveaux domaines de la recherche géotechnique.

* University of Alberta.

INTRODUCTION

In selecting the subject of this lecture, I have reflected on my activities since my return to Canada in 1968. Since that time I have had the opportunity of working on and conducting research into a variety of problems related to landslides, dams, foundations, etc. But most of all I have been involved in a series of novel geotechnical problems in remote environments and it is from this experience that I have chosen to draw the material for this lecture. I hope that in so doing I will not convey information of only parochial interest, but will be able to convince you that results have emerged that are of wide scientific and engineering interest. These results have been obtained in attending to special problems associated with geotechnical engineering in frontier resource development with particular reference to the Arctic environment and to the exploitation of the Alberta oil sands. Figure 1 indicates the general region of activity, the location of some of the projects and some place names for guidance.

Geotechnical engineering is remarkable in the variety of materials that are encountered in the practice of it. This is indicated in Fig. 2 which contains a classification of geotechnical materials in terms of origin, composition and consistency.[1] Figure 2 is not intended to include all earth materials but is meant merely to be illustrative of the range of materials met in professional practice. It is of interest to attempt to isolate those principles of geotechnical engineering that unify the subject and thereby provide a framework whereby activit-

[1] The first version of this classification was produced by Professor A. W. Skempton in 1964.

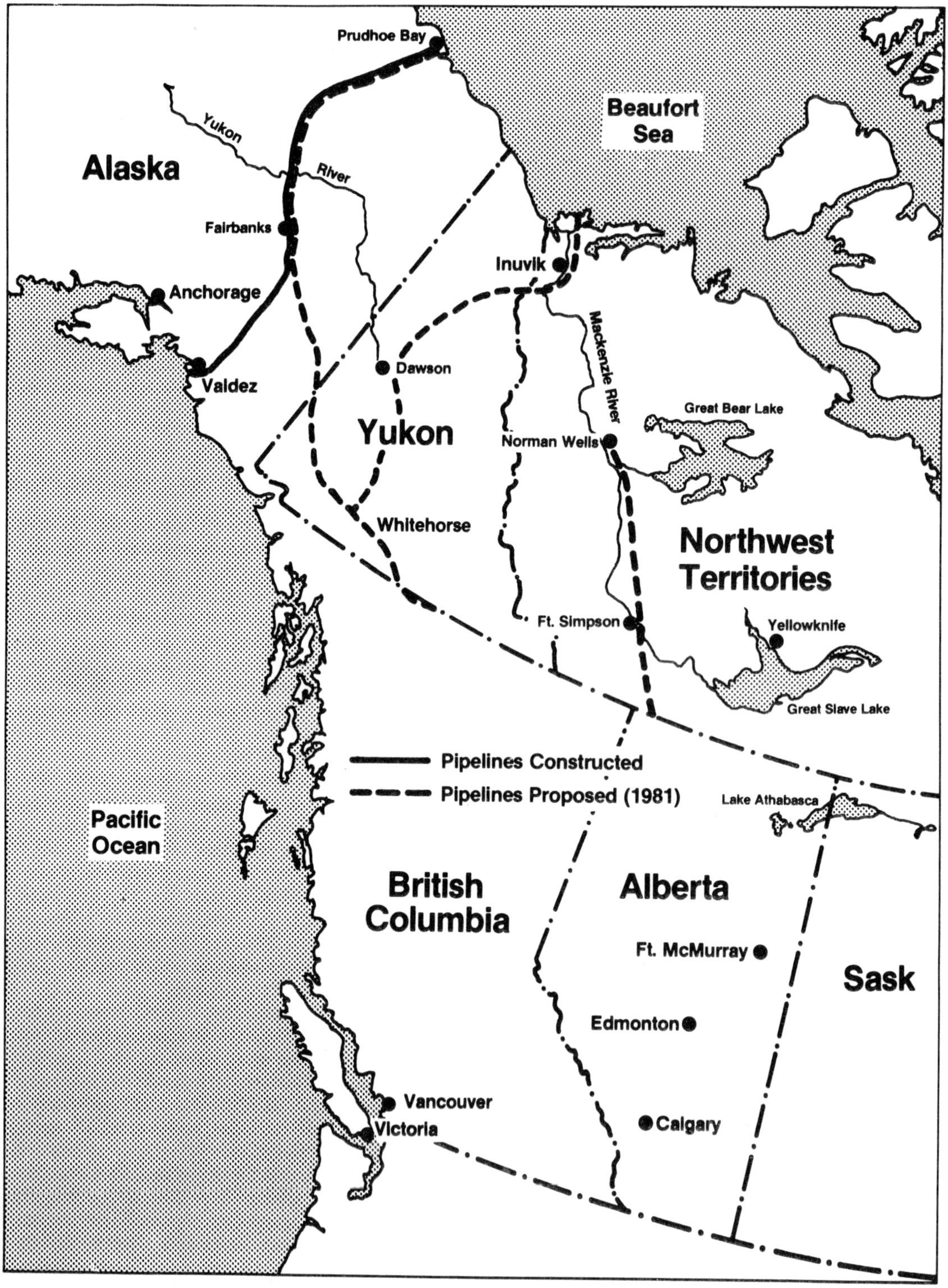

Fig. 1. Region and place names of interest

Consistency \ Origin and Composition		Sedimentary: Clastic: Arenaceous	Sedimentary: Clastic: Argillaceous	Sedimentary: Chemical: Carbonates	Sedimentary: Chemical: Evapourites	Organic	Igneous and Metamorphic
Soil	Cohesionless	Alluvial Sand and Gravel	Rock Flour	Calcareous Sands	Gypsiferous Sands	Topsoil	Talus
Soil	Cohesive	Oil Sand	Clay Clay Shale	Oozes Marl		Peat	Laterite
	Slaking and Softening						
Rock	Soft	Friable Sandstone	Mudstone	Chalk	Gypsum	Lignite	Weathered Granite
Rock	Compressive Strength 500 kPa						
Rock	Hard	Sandstone	Shale	Limestone	Potash	Coal	Granite

Fig. 2. The range of geotechnical materials by origin, composition and consistency

ies over a broad spectrum of earth materials may be undertaken. It seems to me that there are three unifying concepts and they are

(a) the concept of effective stress: a rational explanation of the mechanical behaviour of soils and rocks is best developed in terms of effective stress
(b) the recognition of frictional behaviour: with few exceptions both stiffness and strength of soils and rocks increase with increasing effective normal stress
(c) a continual awareness of the role of structure detail: at one extreme a sample of soil amenable to laboratory testing may adequately characterize the structure of a soil while at the other extreme a discontinuity in otherwise sound rock may be the only element of practical interest; fissured clays and clay shales fall between these two extremes

For an increasing range of problems, these three unifying concepts do not, on their own, provide an adequate basis for the geotechnical engineer to resolve the problems that confront him. He is obliged instead to extend his considerations to additional physical concepts from thermodynamics, heat conduction and other physicochemical phenomena, in order to meet his obligations. Just as the explorer for resources extends the frontiers of technological activity, so the geotechnical engineer working with him expands the range of our activities.

My intent in this lecture is twofold: firstly, to bring to this Society a geotechnical perspective of the nature of these undertakings; and secondly, to encourage particularly our younger colleagues to abandon the view that geotechnical engineering is mature, ready for standardization, but instead to adopt the view that the range of natural materials is so great and the contribution of geotechnical engineering to many technological undertakings is so central, that the limits to our profession expand continually.

By way of presentation, firstly the way a particular problem or class of problems has arisen will be identified. Then the specific research undertaken to solve the problem will be discussed. This will be followed by a summary of the results and some comments on their broader applicability. This procedure will be repeated in a discussion of several issues the have arisen in the development of oil and gas resources in the Arctic and in Alberta.

CREEP IN A NATURAL PERMAFROST SLOPE

The problem

There have been several proposals to bring both oil and gas pipelines down the Mackenzie Valley (Fig. 1). In order to contribute to the orderly design of these projects, as well as for fundamental scientific reasons, a series of research studies were undertaken into the nature of mass movements in permafrost terrain (e.g. McRoberts, 1973; McRoberts & Morgenstern, 1974a, b; Pufahl, 1976). At least for the glaciated terrain of the Mackenzie Valley, it was found that slope failures occurred both through frozen ground and through thawing ground. The latter were far more frequent and were caused by high rates of thaw generating

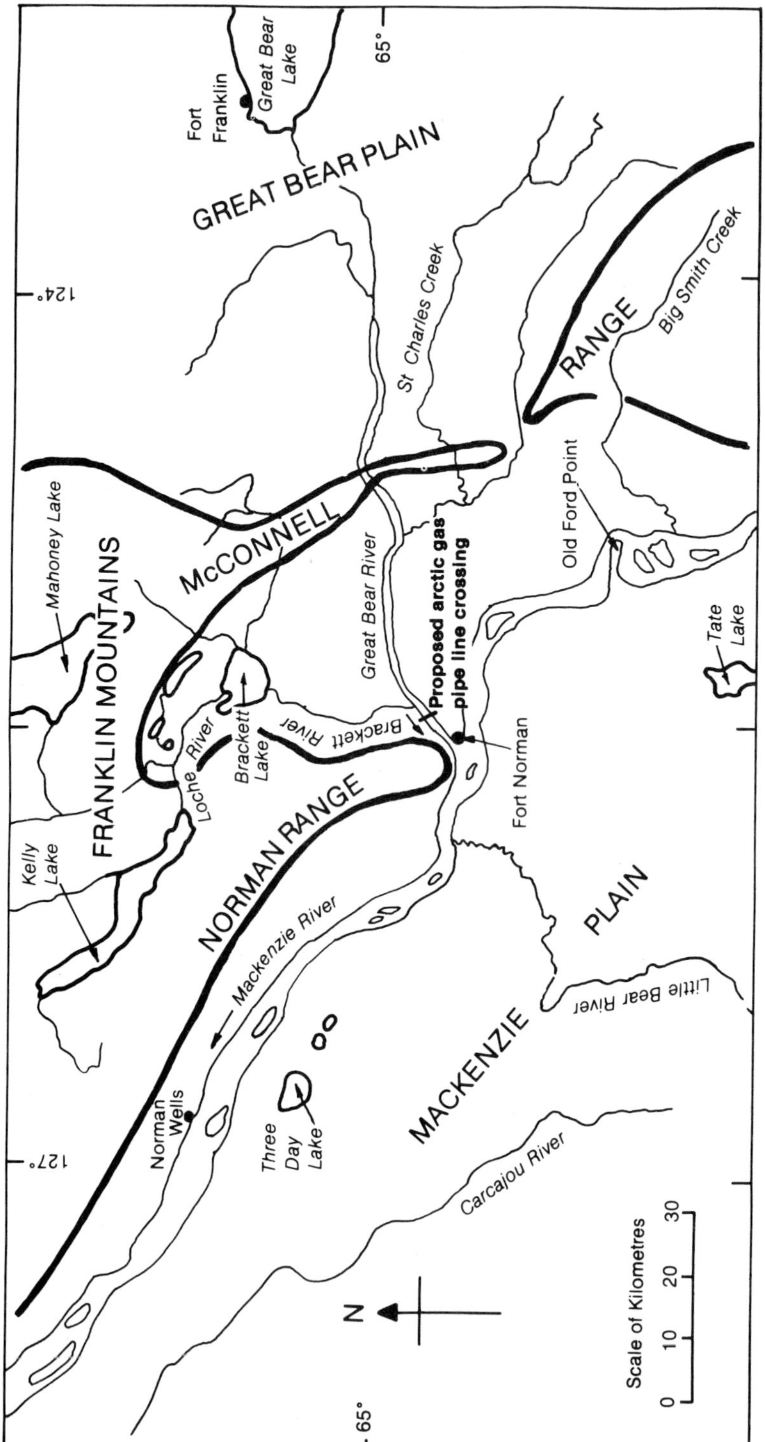

Fig. 3. Physiography of the Great Bear River area

pore pressures, high rates of ablation at ice-rich faces or a variety of more conventional mechanisms in previously thawed material. Failure through frozen ground was a much less frequent occurrence and generally was restricted to large-scale features. The circumstances where failure through frozen ground had occurred or appeared likely could generally be avoided by judicious route location. However, if soil failure had been avoided, the possibility remained for long-term creep deformations, particularly in ice-rich materials, which could result in damage to the pipeline or to any other structure buried in the frozen ground.

Studies of the creep behaviour of frozen ground in the laboratory are not new. The subject is of interest in evaluating the support capacity of artificially frozen ground as well as naturally occurring permafrost. Comprehensive reviews have been published by Andersland & Anderson (1978) and Vyalov, Dokuchayev & Sheynkman (1980). However, most studies utilize artificially prepared specimens and experiments have usually been conducted at relatively cold temperatures and for comparatively short times. This is in contrast with the need to evaluate creep in the relatively warm, fine-grained, ice-rich, structurally non-homogeneous permafrost soils of the Mackenzie Valley. There are serious limitations to relying on laboratory tests under these conditions.

Ice is known to exhibit creep behaviour and the rheology of ice has been investigated extensively in both the laboratory and the field by glaciologists. It seems reasonable to assume that the creep of ice will provide a sensible upper bound to the creep of ice-rich frozen soil.[2] Therefore, using data available at the time that expressed the secondary creep of soil in a power law relation, McRoberts (1975) adopted an infinite slope analysis to calculate the downslope velocities as a function of depth of ice-rich soil and slope inclination. For relatively warm ice (say, warmer than $-4\,^{\circ}\mathrm{C}$) the analysis indicated that surface velocities of about 10 cm/year might be expected on a slope with 10 m of ice-rich soil inclined at 15° to the horizontal. This is a very aggressive geomorphological process and, if true, would be readily discernible in the field. Casual observation was not in accord with these predictions and it was recognized that the available data on creep of ice were probably of limited value in the range of stress, temperature and duration of testing of geotechnical interest.

The evaluation of creep in a natural permafrost slope is best undertaken in detail in the field and it was this phenomenon that was studied. Additional studies were also undertaken to define the flow law of ice in more detail.

Field studies

The site selected for instrumentation is on the southern flank of Great Bear River, a major tributary of Mackenzie River. The site is about 7 km upstream from Fort Norman at the confluence of the two rivers and lies within the widespread discontinuous permafrost zone on the permafrost map of Canada. The site shown on Fig. 3 was selected for several reasons.

(a) It was an intended crossing for a proposed major pipeline.
(b) It was among the highest and steepest slopes in fine-grained soils encountered in the Mackenzie Valley.
(c) The stratigraphy was characteristic of extensive areas of Mackenzie Plain.

A cross-section of the Tertiary and Quaternary stratigraphy along this reach of Great Bear River is given in Fig. 4. The location is near the thalweg of a buried valley. This topographic low was preserved after the Wisconsin glaciation and received an anomalously large thickness of fine-grained sediment when glacial lakes became impounded in the area. The sediments are presently within the zone of discontinuous permafrost and characteristically contain ground ice in a reticulate network. They are overlain by a thick deposit of glaciodeltaic sand in the uplands, but only a thin veneer of organic soil is present on the steep slopes of the Great Bear River valley.

The field studies had four main objectives

(a) the installation of borehole inclinometers to measure in situ creep deformation in the ice-rich soils comprising the slope
(b) the installation of thermistor strings to establish the temperature gradient affecting each inclinometer casing
(c) the installation of piezometers below the base of the permafrost to assess the overall stability of the slope against deep-seated failure
(d) to obtain continuous undisturbed cores from each hole in order to establish the stratigraphy, to determine basic soil properties and to permit detailed laboratory investigation of deformation properties under simulated field conditions

This investigation has been discussed in detail by Savigny (1980) from whom much of this material is drawn.

The logistic difficulties of northern site investigation in remote areas present special problems. Land-use regulations often preclude work in summer months by tracked or wheeled vehicles

[2] Actually a small amount soil will accentuate the creep characteristics of ice but adding additional mineral soil will lead to an attenuation (Hooke *et al.*, 1972).

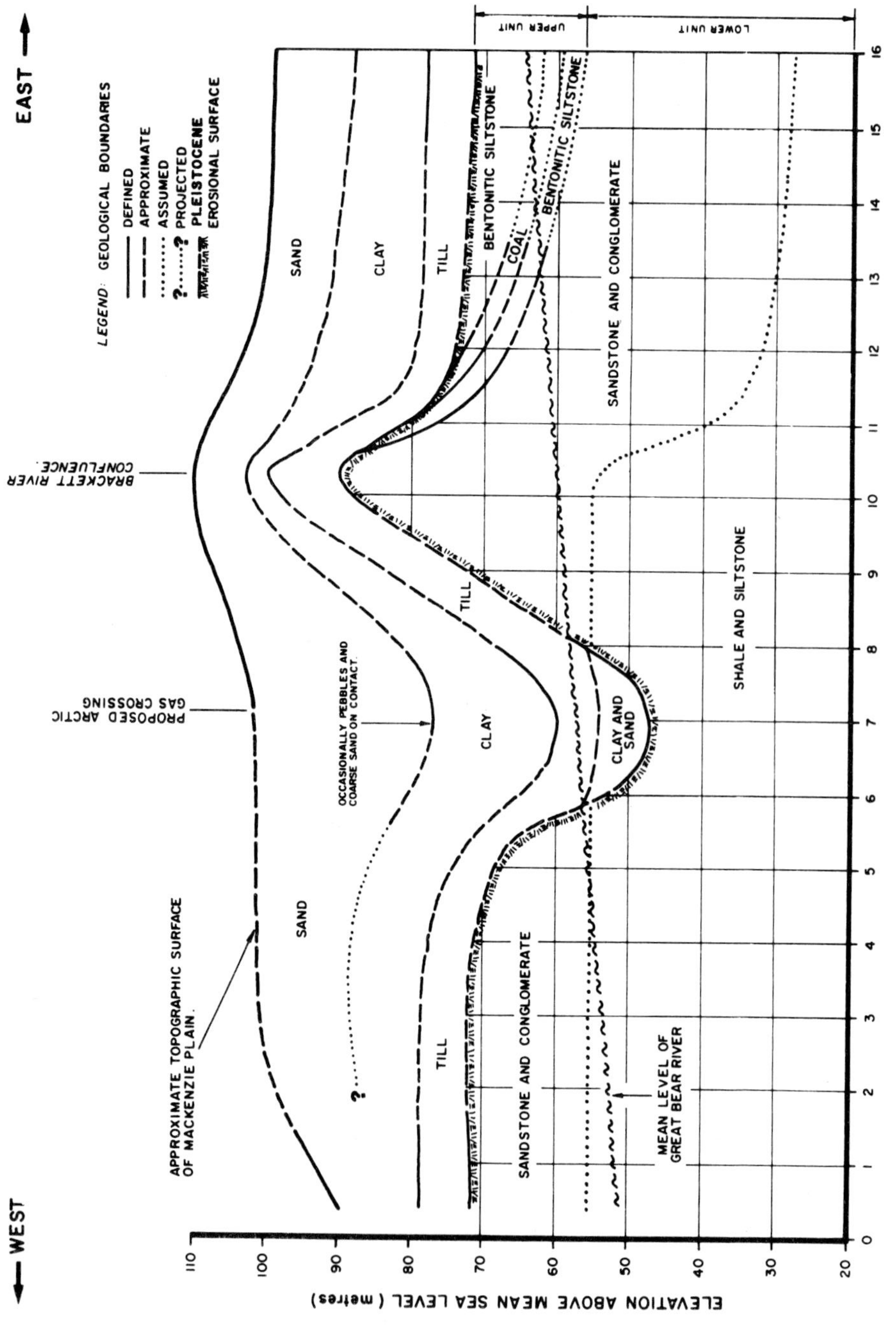

Fig. 4. **Cross-section of tertiary and quaternary stratigraphy between km 0 and km 16 of Great Bear River**

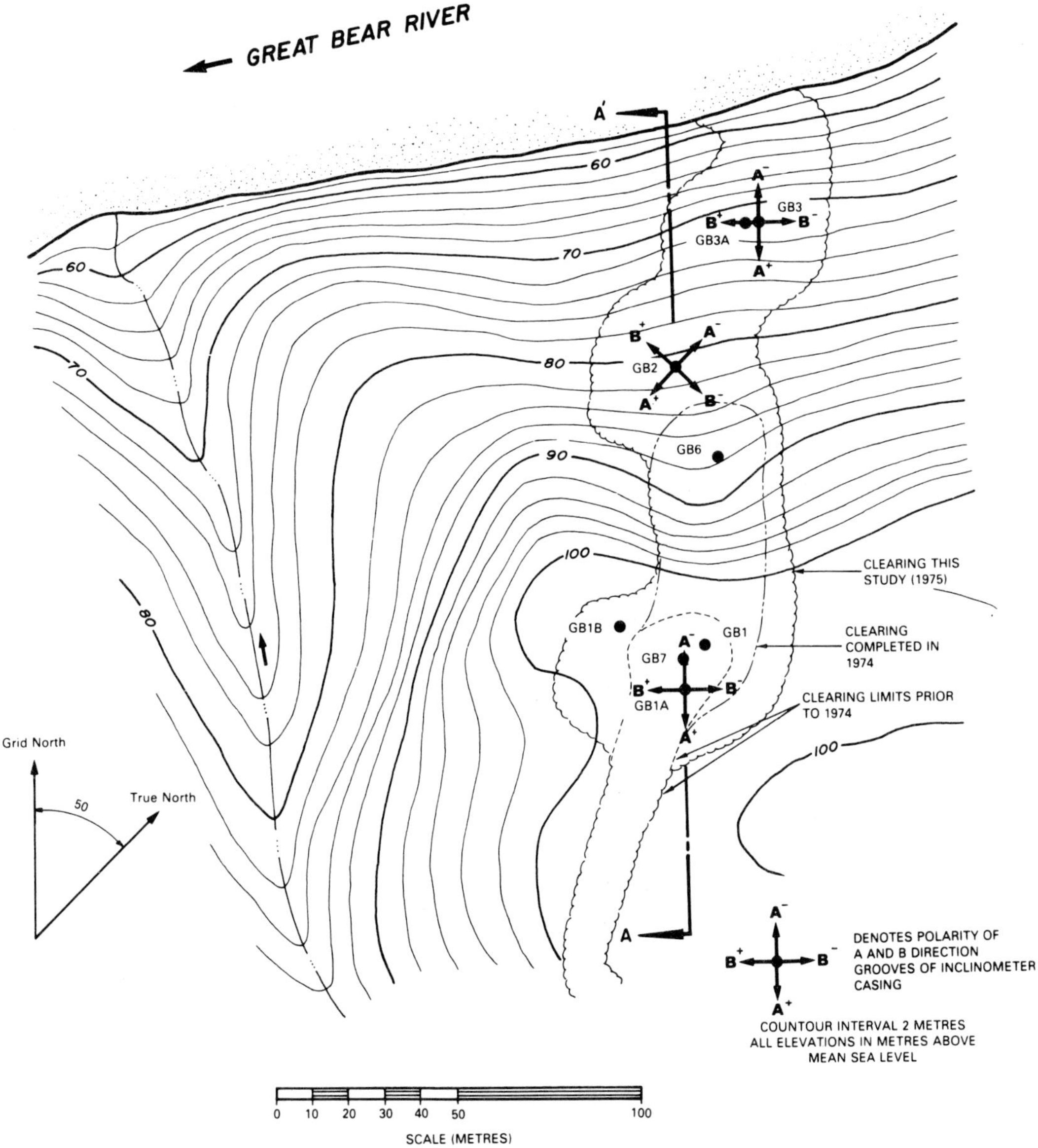

Fig. 5. Site plan, proposed arctic gas crossing of Great Bear River (left bank)

when trafficability is restricted. During parts of the winter, cold is extreme and daylight is limited. Nevertheless we have witnessed a steady stream of innovative solutions to these problems with the development of helicopter-portable drills and self-contained mobile field camps and laboratories for extended route investigations (Roggensack, 1979).

This investigation which required the installation of accurate instrumentaion of very high quality presented its own special requirements. The programme called for a helicopter-portable wet drilling rig with minimum depth capabilities of 60 m. Dry sampling was to be carried out with modified CRREL[3] ice augers at least to the limit of fine-grained sediments. Wet sampling was to commence with a PQ wire-line core barrel, if and when the dry auger reached refusal in stony sediments, and was to continue to the desired depth. Stringent environmental and technical regu-

[3] Cold Regions Research and Engineering Laboratory, US Corps of Engineers, Hanover, New Hampshire.

Fig. 6. Great Bear River instrumented slope

lations required the drilling fluid to be a non-toxic biodegradable water-based mud which was chillec constantly to at least −2 °C. Inclinometers were tc be installed well below the deepest ice-rich zone encountered in Quaternary sediments, and grouted to the surface with a chilled, low heat of hydration grout. Piezometers were to be installed in holes advanced by wet-rotary drilling with sampling being limited to grab samples.

Figure 5 is a site plan indicating the location of the boreholes and the orientation of the inclinometer casings. A photograph of the site is given in Fig. 6. Figure 7 is a stratigraphic cross-section based on the boreholes and outcrop mapping. The siltstone and shale bedrock is Tertiary in age. The rocks are laminated, highly arenaceous, weakly cemented and soften only slightly when soaked in water. The bedrock is overlain unconformably by interbedded clay, sand and coal. These strata are mainly alluvial in origin and represent buried river channel deposits probably of Pleistocene age. They are predominantly grey, highly plastic, intensely fissured and slickensided clays. The bedding structures appear to have been highly contorted by ice-thrusting.

Glacial till deposited by the Wisconsin Laurentide ice sheet rest unconformably on the alluvial deposits. The till is comprised of brown, low to medium plastic, fissured, silty clay and

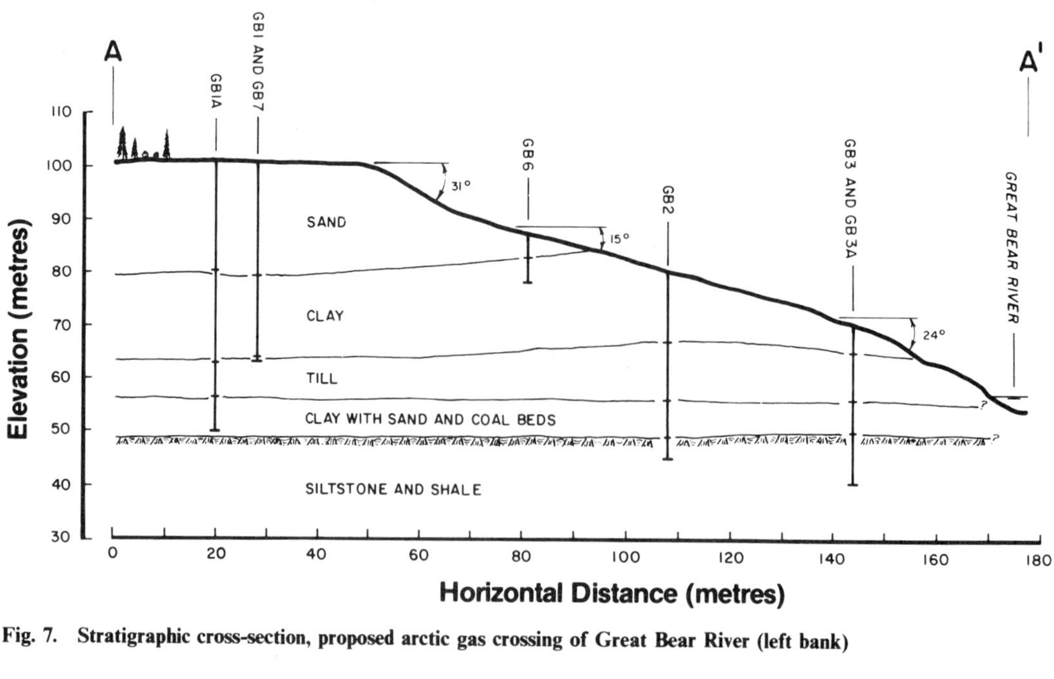

Fig. 7. Stratigraphic cross-section, proposed arctic gas crossing of Great Bear River (left bank)

contains clasts ranging to boulder sizes. Pockets of medium sand are common and reticulate ice occurs near the upper till contact.

Overlying the till with apparent conformity are thick deposits of glaciolacustrine clay. These sediments are dark grey, rhythmically laminated, medium to highly plastic, silty clay. They are fissured throughout and commonly slickensided in association with ice veins. Reticulate ice is the most common ice form but other more tabular forms are also present. Examples are shown in Fig. 8.

Glaciodeltaic sand, the uppermost unit at the site, lies conformably on the clay. A pebble unit at the bottom testifies to the sudden end of the glaciolacustrine phase. The quartzose sands are varicoloured, medium to fine-grained with horizontally bedded and cross-bedded structures. Pore ice is the most common type of ground ice but occasional steeply dipping ice veins were also noted.

An extensive series of classification and strength tests were performed on both thawed and frozen material. The results are summarized in Table 1. These results are unexceptional and generally consistent with experience gained from similar Mackenzie Valley soils. However, excluding visible segregated ground ice, the glaciolacustrine clays exist in situ at a liquidity index of about 0. This is characteristic of heavily overconsolidated clays (Morgenstern, 1967) but there is no evidence that the glaciolacustrine clays have been subjected to greater overburden than exists at this time. It is likely that the clays have been consolidated by the pore water suctions set up during freezing and the formation of reticulate ice (Mackay, 1974). If this clay were to thaw, most of the water liberated would escape through the fissure network, leaving in place a heavily overconsolidated, fissured and slickensided clay. As a result attempts to reconstruct past overburden loads from consolidation behaviour or infer high horizontal stresses due to preconsolidation history would be in error. Caution must be exercised when applying tradi-

Table 1. Properties of glaciolacustrine clay

Liquid limit; %	~50
Plastic moisture content: %	~20
Natural moisture content: %	~22
Bulk density: Mg/m^3	~2·05
c': kPa	10
ϕ'	23°
ϕ' (residual)	14°
c (frozen): kPa	232
ϕ (frozen)	24°

Fig. 8. Ground ice structures

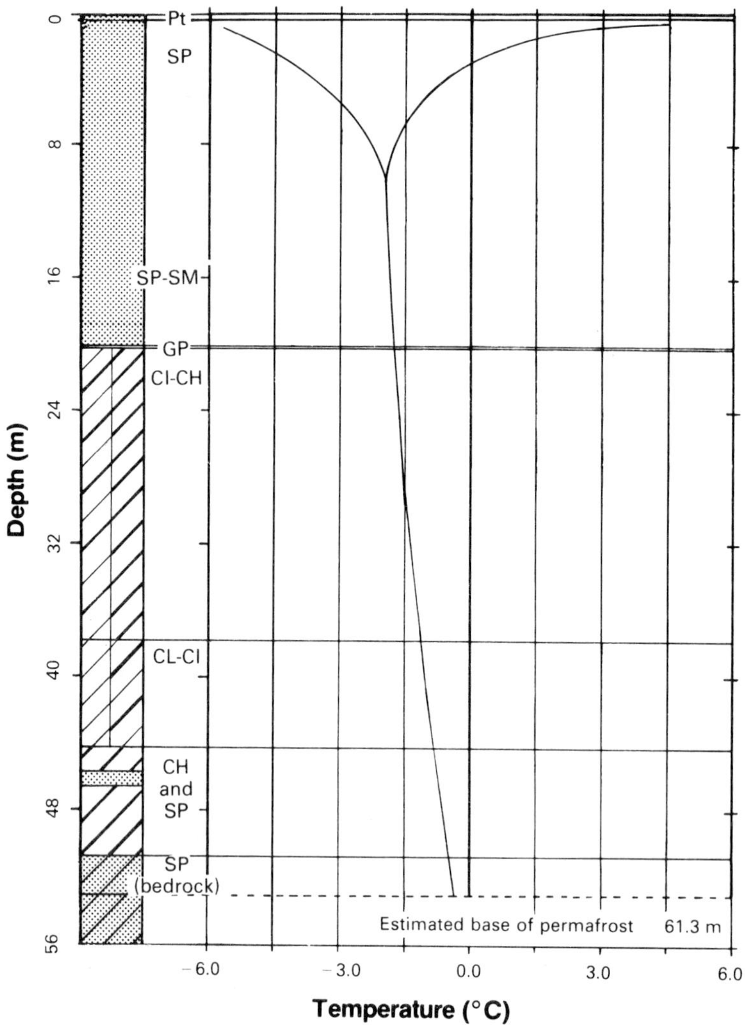

Fig. 9. Temperature gradient for hole GB1A

tional geotechnical concepts to soils that have been frozen in their geological past.

Readings were taken on 12 occasions from April 1975 to June 1977 following completion of the field programme. Most trips were scheduled in March and October of each year to coincide with the periods of coldest and warmest ground temperature respectively. In the following, data from the uppermost hole GB1A will be presented. More complete information is available in Savigny (1980).

The ground temperature profiles or trumpet curve for GB1A are presented in Fig. 9 and a cross-section showing isotherms in the slope is given in Fig. 10. The data on this diagram represent mean annual temperatures below the depth of zero mean annual temperature fluctuation (ZMTF).

In the sandy area at the top of the slope the active layer is 3 m thick and the depth of ZMTF is between 9 and 10 m. The detailed temperature data show that a warming trend has been in progress since monitoring began and was probably initiated by widespread clearing in 1974. This recent adjustment is superimposed on an earlier cooling trend which began in approximately 1950 and is manifest in the steep thermal gradient between 28 and 34 m.[4] Subsurface thermal conditions within the valley slope are slightly different because of the combined effects of aspect, vegetation cover and the micro-climate of the river valley.

The piezometers were a combination

[4] Detailed analysis of the ground temperature data suggests that this cooling trend was initiated by a change in mean annual air temperature of approximately 0·6 °C.

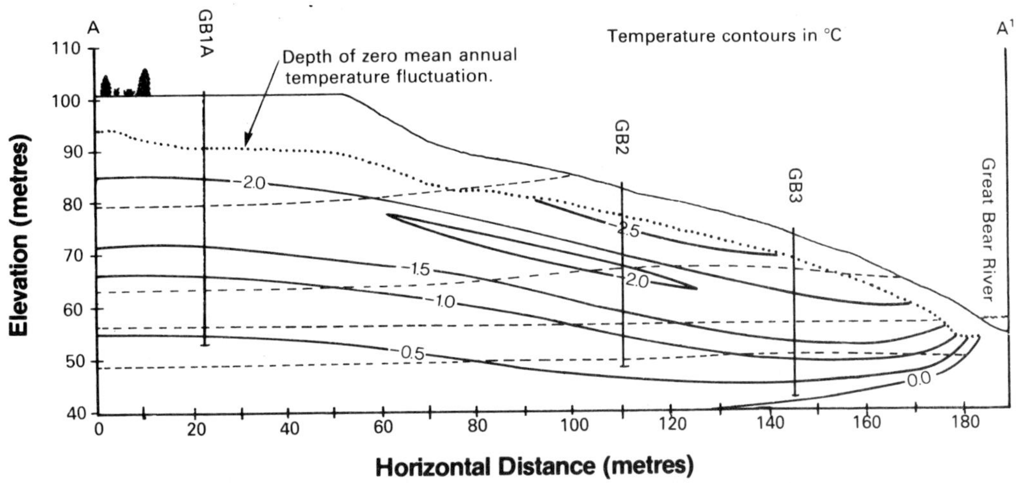

Fig. 10. Thermal cross-section, proposed arctic gas crossing of Great Bear River (left bank)

pneumatic/hydraulic type chosen primarily because of the back-up hydraulic system in which light oil or ethylene glycol could be used in the event that the pneumatic leads became damaged or if verification of the pneumatic reading was required. Only the piezometer at GB3A operated successfully and it indicated that the piezometric elevation at the base of permafrost in the vicinity of Great Bear River corresponded closely with the river level. The presence of sandy zones, joints and thin sandstone laminae in the bedrock provide a means of rapid pore water communication.

It was recognized that if meaningful observations of creep were to be obtained in a reasonable length of time it would be necessary to rely on the limiting accuracy of the inclinometer system. A servo-accelerometer type (SINCO Digitilt Model) was selected as the most suitable for the following reasons

(a) adequate accuracy and precision
(b) negligible non-linearity, hysteresis, temperature stability and zero drift
(c) proven reliability

A variety of special precautions and reading sequences were adopted, particularly after it was established that lateral movements were marginally inside the resolution of the Digitilt system. The parallel-to-slope results from GB1A are shown in Fig. 11 while the transverse-to-slope results are given in Fig. 12. The very complex pattern of movement is a result of the degree to which deformations of the casing approach the limits of resolution of the inclinometer system. A comprehensive testing programme was undertaken to assess the repeatability, resolution and temperature-drift characteristics of the measuring system. In addition, consideration was given to casing spiral and sensor rotation error.

In situ repeatability tests showed that the average performance exceeded by 10 times the manufacturer's specifications. Resolution tests to determine accuracy in a specially constructed calibration frame revealed that errors were negligible. Large temperature changes were found to have an effect on the sensing elements and approximately 20 min were required to achieve stable readings. This gave guidance for field practice. The sensor also displayed a linear temperature drift but it was of no significance to this study because of the small differences in temperature observed throughout the installation profile. An evaluation of casing spiral and sensor axis rotation error due to shifting of sensing elements indicated neither to be of concern.

Several external factors related to the installation procedure and site conditions have affected the readings. These include recovery of thermal equilibrium around the casing, the effect of stratigraphy, and settlement and heave of the casing. They are not peculiar to this study but are particularly important because the magnitude of associated movements is significant in relation to the lateral deflexions measured. A statistical analysis of the inclinometer results revealed that recovery of temperature and stress equilibrium around inclinometer casings cause erratic local deformations, and it was possible to establish an instrument response above which erratic deformations dominate the measurements to the extent that net ground movement at the scale of creep deformations are obscured. In the case of GB1A this occurred for about 75–100 days after the placement of grout.

A correlation exists between deformations and ice-rich zones, especially those with pervasive ice lenses more than 25 mm thick. Where single ice

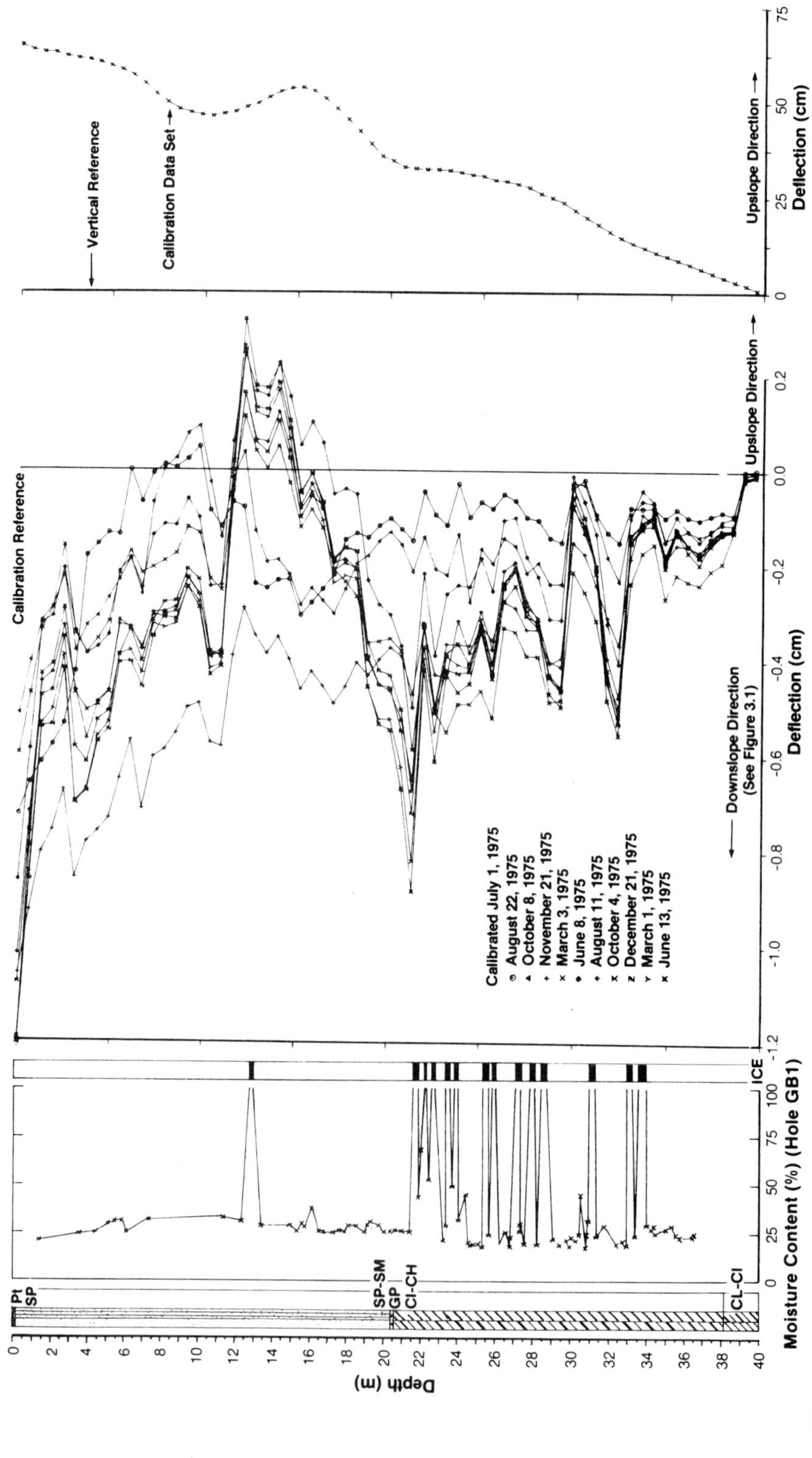

Fig. 11. A-direction (parallel to slope) inclinometer data for hole GB1A

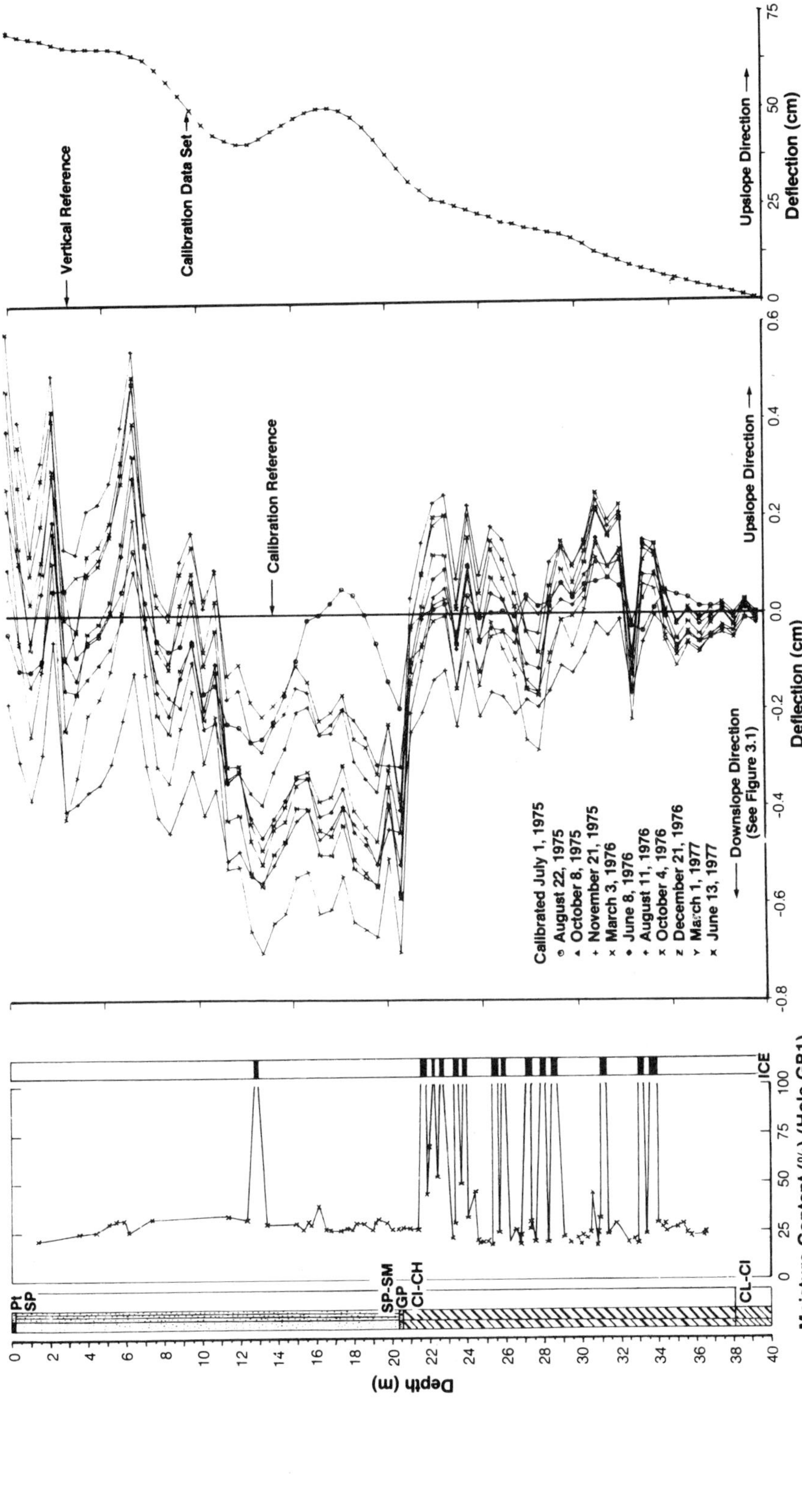

Fig. 12. B-direction (transverse to slope) inclinometer data for hole GB1A

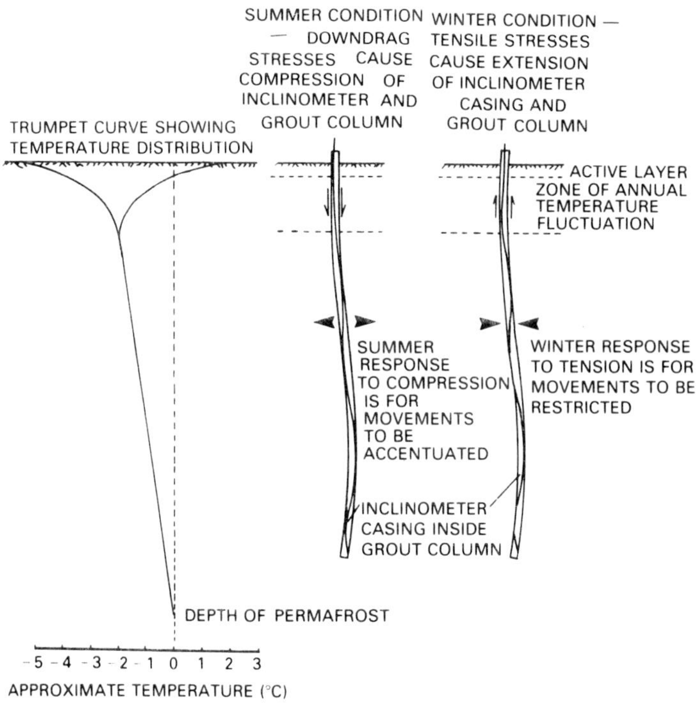

Fig. 13. Schematic representation of heave and settlement of inclinometer casing and grout column

lenses or zones containing closely spaced ice lenses are separated by 2 m or more, relative movements are typically large and cause very sharp deflexions. Examples of this occur at the 15 m depth and between 29 and 34 m in GB1A (see Fig. 11). Where single ice lenses or zones containing closely spaced ice lenses are separated by less than 1 m, and the natural moisture content of soil between the ice lenses is at least 25% to 30%, movements are typically smaller and much less abrupt. These movements are generally progressive with time in the downslope direction, although the pattern is occasionally interrupted by a reversal in the sense of movement. Net downslope deflexion occurs between 20 m and 25 m in GB1A. While the data indicate a correlation between movement and ice lenses, the resulting deflexion pattern approximates simple shear in terms of homogeneous strain through any ice-rich section of the overall soil profile.

The large annual variations in near-surface ground temperatures induce both settlement and heave of the casing as illustrated in Fig. 13. It is probable that compressive and tensile stresses seated in both the active layer and the zone of annual temperature fluctuation are transmitted through the inclinometer casing and grout column. Through the summer season, and up to the approximate culmination of warming, lateral movement outward in response to settlement is progressive, while through the winter season, lateral movements are progressively inward in response to heave. This is supported by Fig. 14 which shows typical plots of deflexion as a function of time for the A (downslope) and B (cross-slope) directions at four discrete measuring depths together with mean velocities determined from least-squares linear regression analysis. In the B direction, which is assumed to be unaffected by downslope net overall ground deformations, each data set has a sinusoidal distribution about its mean velocity with a wave length of approximately 365 days. Lateral movement associated with settlement and heave is progressive, but occurs in the opposite direction during periods of ground warming and cooling respectively, and the net lateral movement after one year is small. In the A direction, conditions are identical, although the sinusoidal distribution is distorted because lateral movements resulting from settlement and heave are superimposed on natural ground deformations associated with creep. This type of plot provides a means for discriminating net ground deformation from seasonal fluctuations.

Velocity data obtained in this way for GB1A are shown in Fig. 15. Although the results are scattered and vary with ice distribution, the velocity at the top of the clay layer is between 0·25 and 0·30

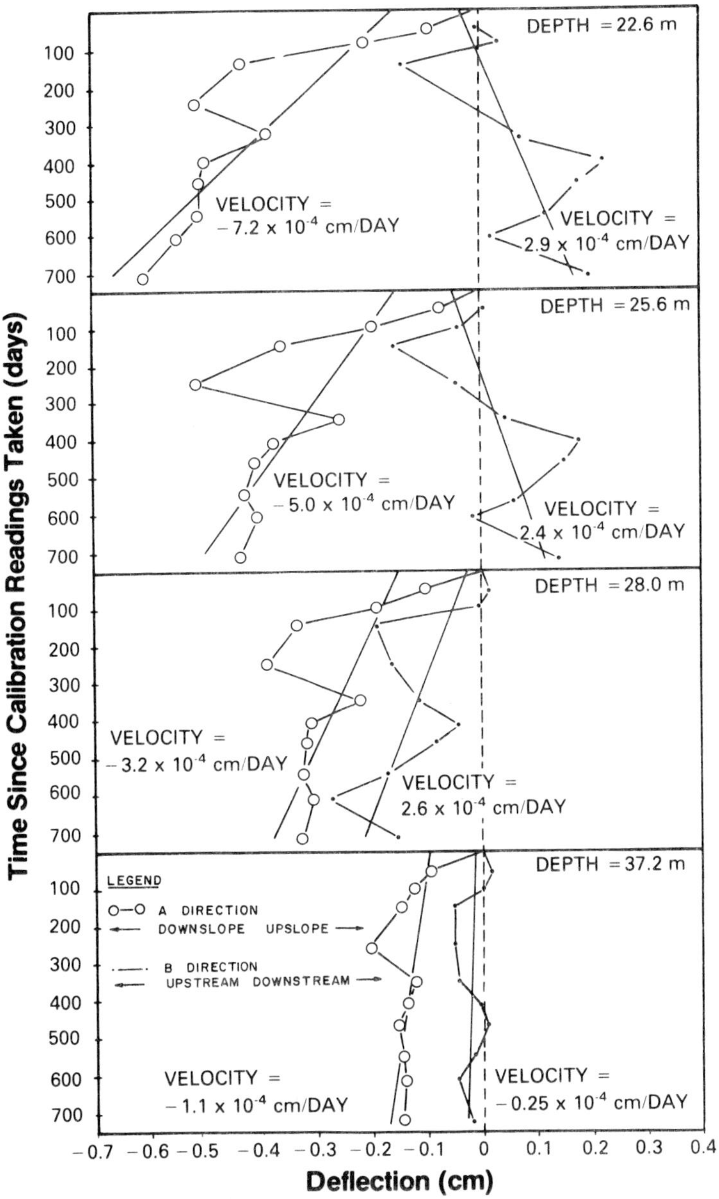

Fig. 14. Typical plots of deflexion against time at different depths in glaciolacustrine clay in hole GB1A

:m/year. Above the 29 m depth where ice lenses are arge and closely spaced, the velocity gradient is ılmost uniform. The shear strain rate through this :one is approximately $2{\cdot}0 \times 10^{-4}$/year. At depths 'rom approximately 29 to 34 m, where large ice enses are more widely separated, the velocity is :rratic, with proportionally more movement ıssociated with the large ice lenses. Below the 34 m lepth, where only small ice lenses are present, the /elocity gradient becomes more uniform with a ;hear strain rate of about $0{\cdot}4 \times 10^{-4}$/year. B-direction deflexions in the clay oscillate about approximately zero net deformation with a small but insignificant downstream velocity. No creep deformations are evident in the sand. This does not preclude the possibility of creep in frozen sand but the data obtained are judged to be less reliable because of more drilling and grouting difficulties experienced during installation.

Laboratory studies

In order to undertake numerical analyses of the

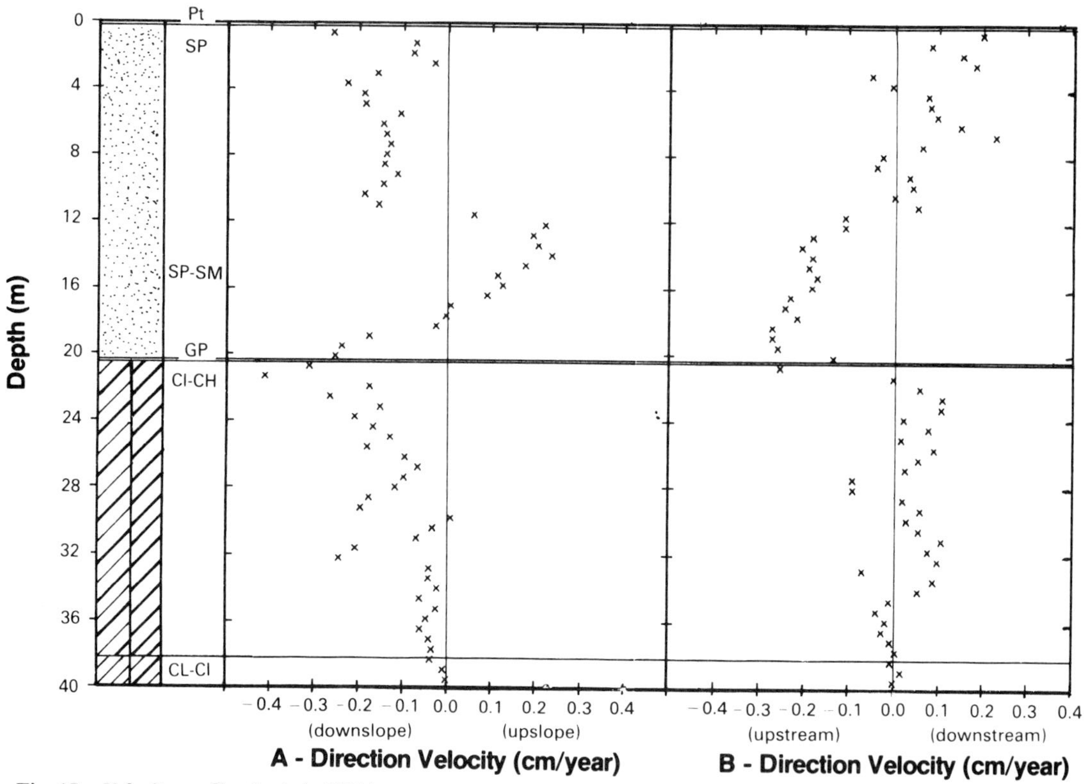

Fig. 15. Velocity profiles for hole GB1A

apparent steady state creep deformation in the slope it is necessary to determine constitutive equations which describe the stress–strain–time behaviour of the materials. There are serious limitations to relying on laboratory tests alone and long-term data on the creep of undisturbed fine-grained permafrost soils are difficult to obtain. Nevertheless, it is still of interest to relate the field creep behaviour to a body of laboratory test data.

The creep of ice is known to follow a power law relation between strain rate and stress at temperatures and stresses of geotechnical interest (Morgenstern, Roggensack & Weaver, 1980; Sego, 1980) and the creep of ice-rich permafrost has been interpreted within the same framework. A plot of the variation of minimum strain-rate with stress observed in creep tests for Great Bear River area glaciolacustrine soils is given in Fig. 16. Recent suggested flow laws for polycrystalline ice and other fine-grained permafrost soils are also shown for purposes of comparison. From the experimental data there is no clear relation between minimum strain rate and stress. Many specimens failed prematurely and the failure mechanism seemed closely related to specific ground ice features (see Fig. 17) where shear developed principally along the soil–ice interface of pervasive primary ice veins. Samples subjected to higher confining pressures generally failed sooner than unconfined specimens. It appears that local stress concentrations are set up at ice–soil interfaces in response to confining pressures and that at least some time should be allowed for creep to dissipate high stress gradients. Systematic procedures are not yet in place to lead to reliable long-term test data on heterogeneous ice-rich soils. However, several tests did display long-term steady state behaviour after about 6 months of sustained loading. The data cluster about the flow law for ice but the scatter is substantial.

Finite element simulation

A visco-elastic finite element analysis of steady state deformation occurring in the slope was undertaken to assess the validity of the power law for describing the creep of ice-rich permafrost. It was assumed that

(a) creep strain causes no volume change
(b) the hydrostatic state of stress has no effect on creep rate
(c) the principal strain rate and stress tensors are coaxial
(d) the stress–strain relation for multiaxial states of stress reduces to the uniaxial power law for uniaxial loading

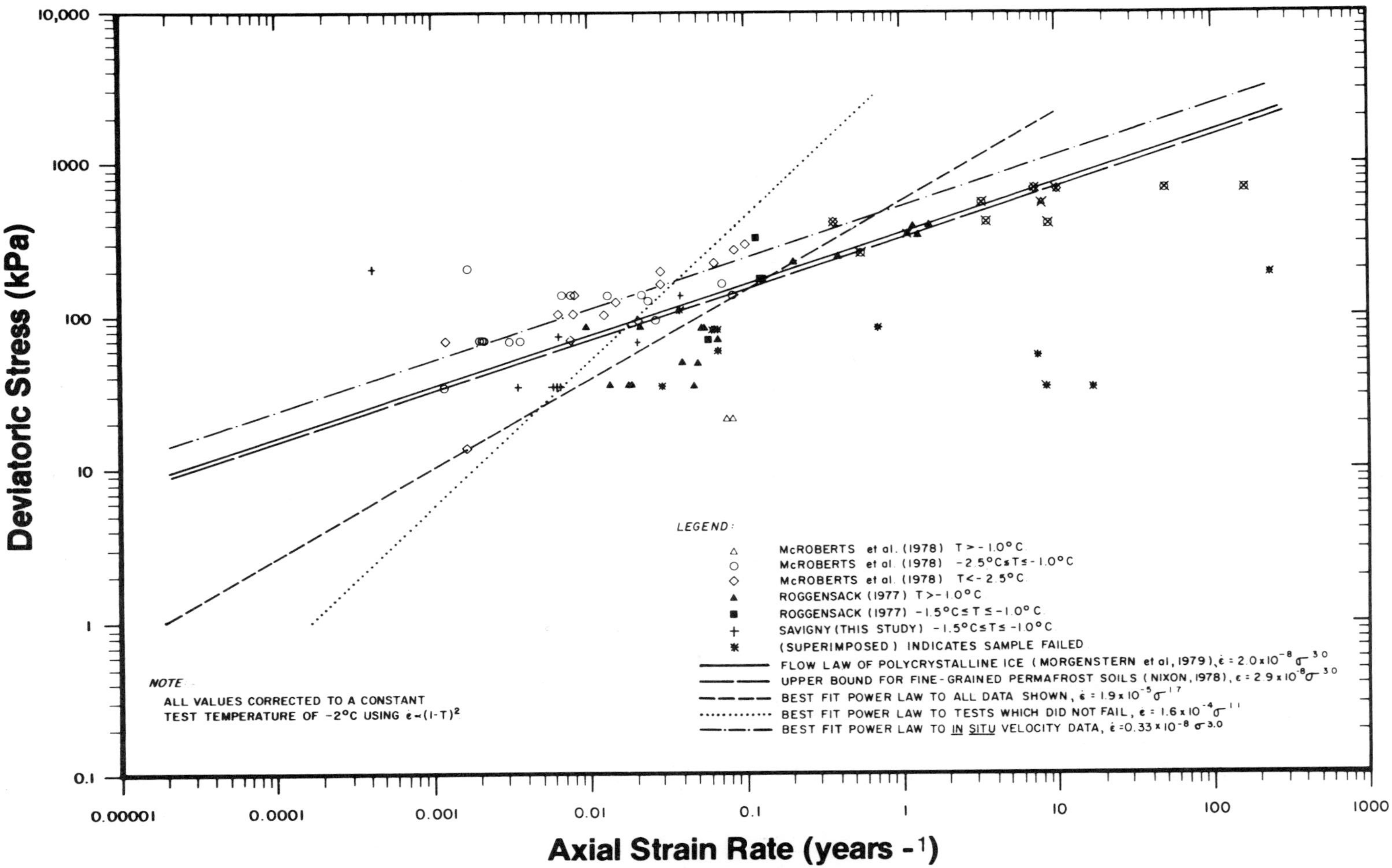

Fig. 16. Comparison of creep relations used in numerical analysis

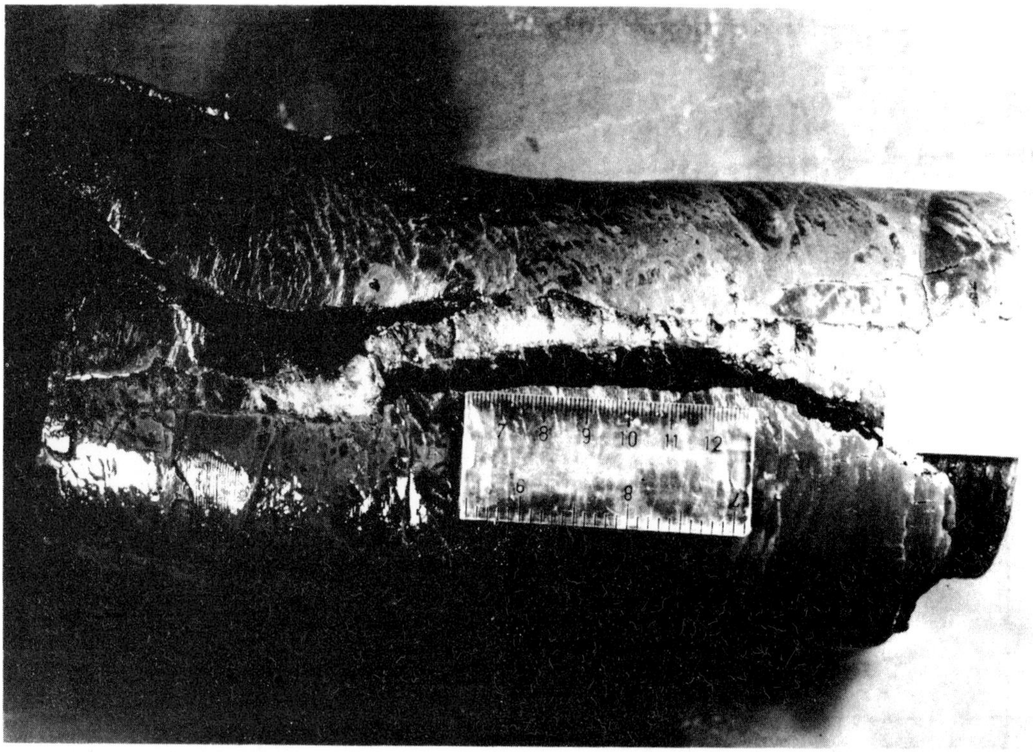

Fig. 17. Failure along ice structure

In the first formulation it was assumed that frozen sand will creep in a manner similar to that exhibited by the clay, particularly if tension develops in the sand. Figure 18 shows the comparison between measured and predicted velocities. If the flow law for ice is used, velocities are grossly overestimated. It is necessary to reduce the modulus in the flow law by 6 times in order to achieve reasonable correspondence with observations in the clay. If the frozen sand is not allowed to creep, the creep of the underlying clay is also restrained but very high horizontal tensile stresses develop in the sand which could not be sustained in the long term. This illustrates the tendency in some instances for tensile cracks to develop in material overlying creeping frozen ground.

Commentary

Despite the remote and hostile conditions, it has been possible to install and monitor instrumentation thereby demonstrating that natural slopes in ice-rich soils do creep. Shear strain rates of the order of 10^{-4}/year have been detected. The movements are in part associated with localized shear in widely separated, pervasive ground ice features. The process is more subdued than predictions based on the flow law of ice alone and the flow law that matches the field behaviour can be used for engineering design in similar soils elsewhere, at least until further data are forthcoming. Special limitations to the use of laboratory tests for evaluating the deformation behaviour of heterogeneous ice-rich permafrost have been indicated.

While the results of the Great Bear study are of direct use for frozen ground engineering in the Mackenzie Valley, they are also of more general interest. Students of the mechanics of periglacial phenomena will have noticed that the creep observed at the slope may be indicative of the process of valley bulging that so far lacks a satisfactory quantitative explanation.

The antecedents to the discovery and description of valley bulging and related phenomena may be found in Horswill & Horton (1976) which now constitutes the definitive description. Salient features are shown in Fig. 19. Briefly, clay has been squeezed upwards into the valley bottom resulting in thinning of the clay layers and forward rotation (cambering) of the overlying strata. The upper portion of the clay is brecciated but the limit of brecciation reflects closely the overlying valley topography. Hence the process which resulted in brecciation must have extended down from an old valley surface.

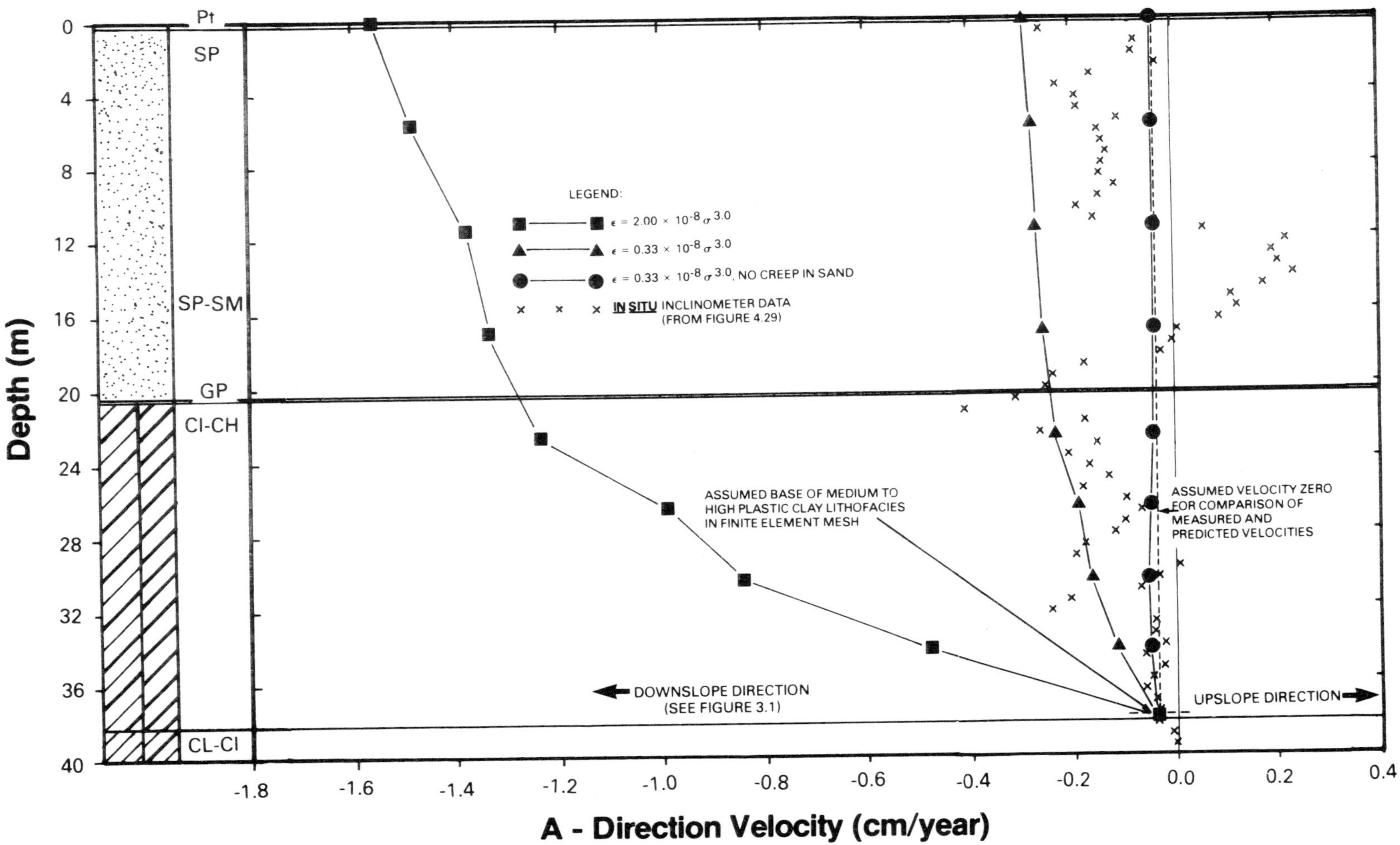

Fig. 18. Comparison of measured and predicted steady state creep velocities in test hole GB1A

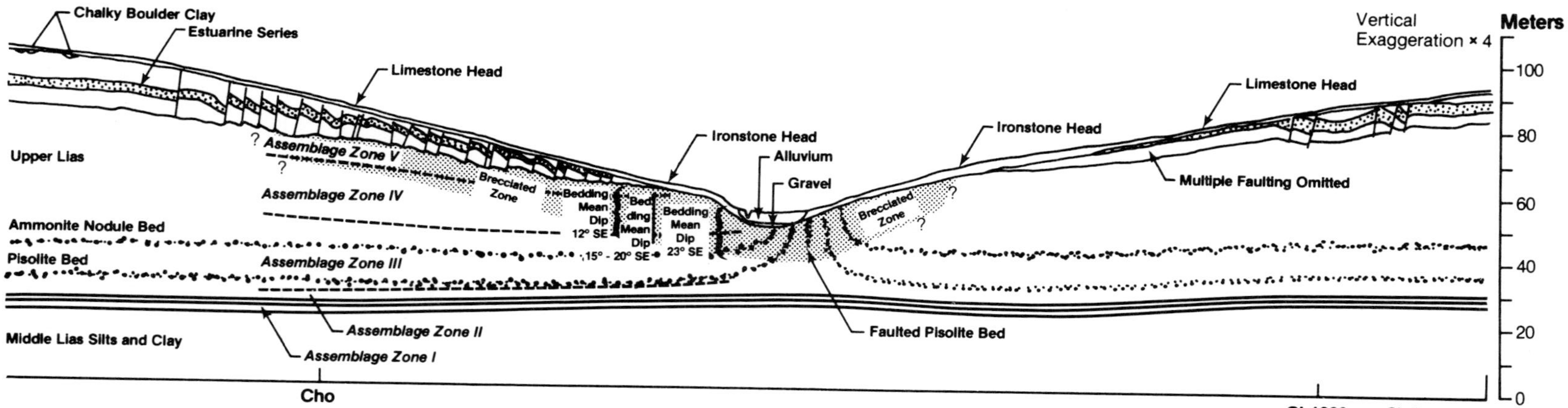

Fig. 19. Gwash valley section showing cambering and valley bulging (after Horswill & Horton, 1976)

Vaughan (1976) has reconstructed the deformation history of the Empingham Valley slope and offered several alternative mechanisms to account for the strains and displacement implied by the present valley slope morphology. He considered lateral movements due to stress relief, vertical loading due to overlying ice and downslope sliding of frozen ground. None are satisfactory in accounting for the magnitude of the movements, the pattern of the deformations and the minor structures within the underlying clay. Various lines of reasoning developed in these recent studies point to the presence of permafrost as a necessary condition for valley bulge formation and the observations at Great Bear River equally support this hypothesis.

In addition to geometrical considerations, the mechanics of valley bulging should account for the flow-like behaviour of the clay, the limits of brecciation and the distinct change in water content displayed by the brecciated clay. A consistent mechanism may be constructed based on the view that valley bulging is due to sustained creep of ice-rich clay following enrichment due to cyclic freezing and thawing. It is unlikely that in situ freezing of the Upper Lias clay alone could lead to significant ice segregation because of the low water content of the undisturbed clay. Instead, cyclic freezing and thawing could disrupt the fabric and permit ingress of water from the overlying sands. If the clays were frozen at depth while free water was available during thaw above, substantial ice enrichment could occur. When the ice content became high enough and the ice structures sufficiently pervasive, creep would be initiated and sustained. Flow of frozen ground toward the valley would cause tensile failure of the overlying material, while erosion in the valley would result in progressive thinning of the mobile members. Vaughan (1976) has deduced valleyward displacements at Empingham of 100 m near the base and 200 m at the top of the Upper Lias. Simple transfer of the Great Bear observations of approximately 0·3 cm/year at the top of the layer indicates some 65 000 years for the bulge process. If ice-rich Upper Lias crept as fast as ice this might be as little as 10 000 years. Finite element modelling is required to explore this explanation in more detail.

FROST HEAVE MECHANICS

The problem

The transfer of oil by pipeline from the Arctic to southern markets has, so far, involved operating at oil temperatures far above 0 °C. When the pipeline is buried in permafrost, thaw results with attendant problems where the ground is ice-rich. These problems are overcome in the delivery of natural gas by pipeline by chilling the gas to below 0 °C. For the major projects that have been considered to date, the extra throughput attainable by chilling the gas compensates in part for the cost of refrigeration. A chilled gas pipeline can therefore be constructed without serious economic penalties. Burying a chilled gas pipeline in permafrost preserves the frozen state and thereby resolves most of the problems associated with pipeline operation in ice-rich ground. However, permafrost is not continuous. The chilled gas pipeline must traverse streams underlain by unfrozen ground and as the pipeline extends further southward even the sub-aerial permafrost becomes increasingly discontinuous. At some point, the gas is no longer chilled below 0 °C and pipeline design beyond this point proceeds on a more or less conventional basis. However, up to the last point of cold flow the pipeline crosses a considerable extent of unfrozen ground which will become frozen if the chilled pipeline is buried in it. The pipeline may then be subjected to frost heave. Two important new design considerations arise. Under these conditions, how much frost heave will occur over the lifetime of the project? In addition, how much differential heave will occur and will it lead to unacceptable strains in the pipe? For example, where the pipeline crosses from frozen to unfrozen and back to frozen ground, it will be restrained from heaving where it is buried in frozen ground but will be subjected to heave across the unfrozen ground. Can this differential heave lead to distress?

The subject of frost action in soils has received considerable attention in the literature. Jessberger (1970) has assembled a bibliography that contains hundreds of citations. Most studies of frost heave have fallen into one of the following classes

(a) index tests to establish the degree of frost susceptibility of various soils
(b) fundamental thermodynamic analyses
(c) empirical studies attempting to relate laboratory investigations to field performance in a quantitative manner

Notwithstanding the considerable research devoted in the past to the frost heave process, there has been no agreement on an engineering theory of frost heave.

It is well known that the propensity of a soil to heave under freezing conditions is affected by grain size distribution, availability of water, rate of heat extraction and applied loads. For a given soil, an engineering theory of frost heave would lead to the predictions of the magnitude and rate of frost heave as a function of certain characteristics of the freezing system and boundary conditions. Prior to freezing, the temperature profile and boundary conditions controlling the availability of water can be established by measurement. A knowledge of the

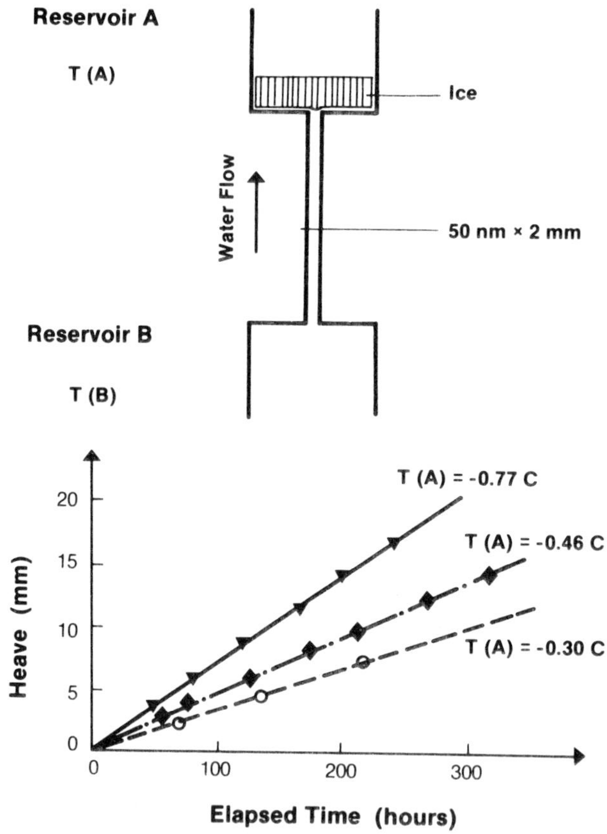

Fig. 20. Experimental results obtained by Vignes & Dijkema (1974)

soil profile can be translated into the moisture content distribution, the thermal conductivity and the permeability of the soil. A change in heat flux or temperature at a boundary must be specified in order to account for the onset of freezing. As a frost front advances into a fine-grained soil, moisture is drawn to the front. It is this coupling of the heat and mass flow that constitutes the complex element in the theory of frost heave. Recently there have been some attempts to embrace heat and mass flux in a coupled theory but predictive results from these studies have not been convincing.

An understanding of why moisture is attracted to a frost front in a fine-grained soil may be obtained in various ways. We have benefited most by considering the thermodynamic equilibrium between ice and water in porous media. If consideration is given initially only to the conditions where no external loads are applied so that the ice will be at atmospheric pressure and temperature close to that at which phase change takes place T^*, the requirement that the free energy of the ice equals that of the water leads to a simple form of the Clausius–Clapeyron equation (e.g. Kay & Groenevelt, 1974)

$$P_w = L(T^* - T_0^*)/V_w T_0^* \quad (1)$$

where

V_w	denotes the specific volume of water
P_w	denotes the water pressure
L	denotes the latent heat of phase change per mole
T^*	denotes the absolute temperature (K)
T_0^*	denotes the temperature at the standard state (273·15 K)

For convenience we can write

$$T = T^* - T_0^* \quad (2)$$

where T denotes the temperature in °C at which ice and water are considered to be in equilibrium.

Equation (1) indicates that if ice is at atmospheric pressure as the temperature decreases below T_0^*, the water pressure becomes negative, and close to 0 °C there is a linear relation between the suction and the temperature. Elegant validations of equation (1) have been provided by Vignes & Dijkema (1974) and Biermans, Dijkema & de Vries (1978). Vignes & Dijkema measured water

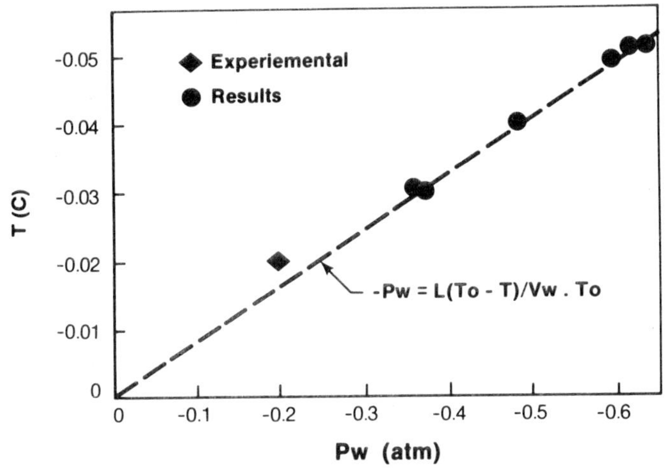

Fig. 21. Experimental results obtained by Biermans *et al.* (1978)

migration rates using the experimental set-up shown in Fig. 20. Two reservoirs, one containing water either above 0 °C or super-cooled, the other containing water and ice, were separated by a narrow slit 50 nm by 2 mm in cross-section and 50 mm long. As predicted by equation (1), water flowed toward the ice regardless of the temperature in reservoir B where the water pressure was maintained at atmospheric pressure. The flow rate was constant for a given temperature in reservoir A. Since the hydraulic conductivity of the slit is constant, equation (1) predicts that the flow-rate should be proportional to the temperature of the ice–water interface. The experimental results were in good accord with this prediction.

Using glass filters in order to increase the flow, Biermans *et al.* (1978) also confirmed the Clausius–Clapeyron relation simplified for atmospheric pressure in the ice. This was achieved by measuring the suction P_w that had to be applied to the water in reservoir B in order to stop the flow to the ice lens and by comparing it with the theoretical prediction. Their results are shown in Fig. 21 and support the theoretical relation to a high degree of accuracy.

Previously Hoekstra (1969) and Radd & Oertle (1973) had measured the pressure P_h necessary to prevent heave as a function of the temperature in soil freezing with access to water. If one assumes that $P_w = 0$ at the ice lens and that the ice pressure is equal to the heaving pressure, the Clausius–Clapeyron relation becomes

$$P_h = -(L/V_i)\ln(T^*/T_0^*) \qquad (3)$$

Their measurements of heaving pressure were in good agreement with this relation, providing further support for the validity of the thermodynamic explanation of the origin of the pore water suction during frost heave.

For frost heave to occur, water must co-exist with ice at temperatures colder than 0 °C. However, if suctions deduced from equation (1) for a possible range of temperatures are applied directly to unfrozen soils of known permeability, flows far in excess of those observed in the laboratory are predicted. Other factors in the frost heave mechanism impede the direct transfer of this suction to the unfrozen soil.

When a fine-grained soil is frozen, not all of the water within the soil pores freezes at 0 °C. In some clay soils up to 50% of the moisture may exist as a liquid at temperatures of −2 °C. This unfrozen water is mobile and can migrate under the action of a potential gradient. The characteristics of unfrozen water have been reviewed by Anderson & Morgenstern (1973) and Tsytovich (1975). Miller (1972) reviewed evidence that water transport to an ice lens takes place through liquid films between ice and mineral matter. This led Miller to propose that an ice lens in a freezing soil grows somewhere in the frozen soil, slightly behind the frost front, i.e. behind the 0 °C isotherm. The temperature at the base of the ice lens is referred to here as the segregational freezing temperature T_s because the segregational heaving process takes place at that temperature. The temperature at which ice can grow in soil pores T_i depends upon pore size and ice–water interfacial energy through the Kelvin equation. This domain between T_i and T_s is referred to as the frozen fringe. In silty soils, the average pore size is relatively large and T_i is close to 0 °C. T_i can also be affected by solute concentration and other factors which are ignored here. Direct evidence for the existence of a frozen fringe has been published by Loch & Kay (1978) and Penner & Goodrich (1980).

In addition to these considerations, Mageau &

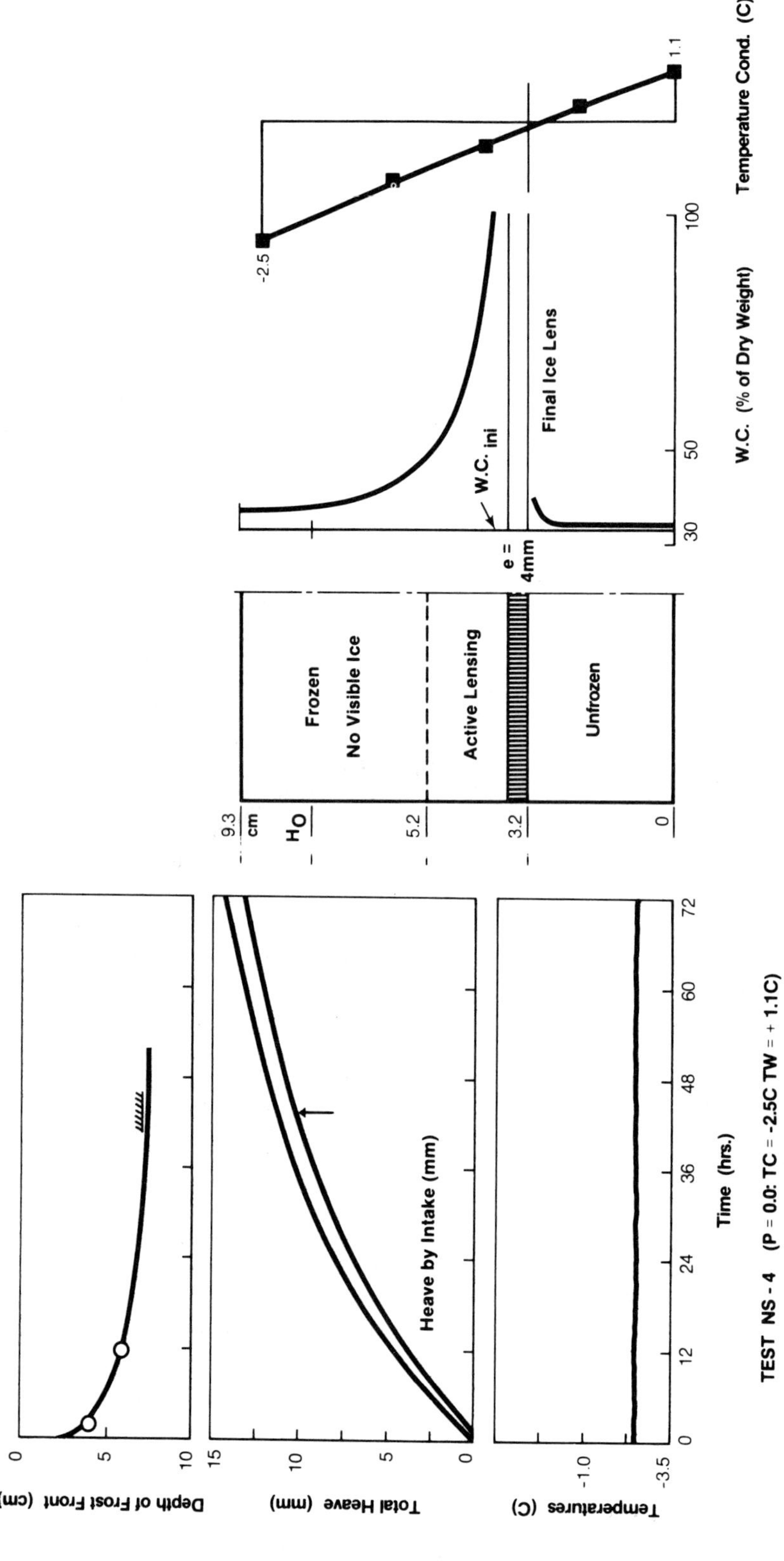

Fig. 22. Experimental results from test NS-4

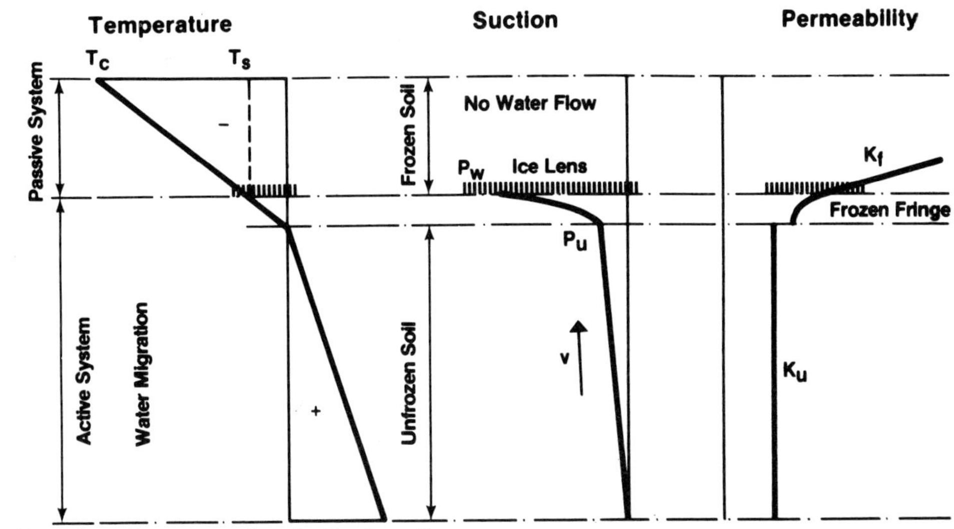

Fig. 23. Schematic representation of a freezing soil

Morgenstern (1979) published experimental results indicating that frozen soil on the cold side of the warmest ice lens had little to no effect on the rate of water intake to that lens. That is, an ice lens acts like an impermeable barrier with regard to water migration in the frozen soil. This is confirmed by field studies. The results from a test pipeline designed to study in situ frost heave showed that all the heave occurred near the frost front since heavy gauges installed throughout the soil profile did not exhibit any further relative movement once the frost front had passed them (Slusarchuk *et al.* 1978).

It appears then that the mechanics of frost heave can be regarded as a problem of impeded drainage to an ice–water interface that exists at the segregation freezing temperature T_s. Substantial suctions are generated at this interface but the reduced permeability of the frozen fringe impedes the flow of water to the ice lens thereby reducing the suction that acts on the unfrozen soil. In order to understand this process in detail it would be necessary to obtain precise knowledge of the distribution of temperature and permeability within the frozen fringe. Rather than pursue this, we have taken the view that precise point measurements of permeability and temperature would not ultimately be of direct value in a comprehensive theory but that instead the coupling of heat and mass transfer should be deducible from an appropriate laboratory test in the same way that Darcy's law relates mass transfer to potential gradient without local measurements of fluid velocity.

Analytical and laboratory studies

One-dimensional freezing tests are conducted conveniently in the type of cell described by Mageau & Morgenstern (1979). Cold- and warm-side temperatures may be controlled and temperature profiles obtained throughout the test. Water inflow and heave may be monitored with time. The test may be performed under a back pressure and, if converted from open flow to a closed system, the pore water suction may be measured.

The results of a typical open system freezing test with constant temperature boundary conditions are shown in Fig. 22. Three distinct phases of frost heave may be recognized

(a) an advancing frost front created by a positive net heat extraction rate
(b) a stationary frost front corresponding to a zero net heat extraction rate
(c) a retreating frost front in which the frozen fringe below the ice lens thaws

It is convenient to analyse first the conditions at the onset of the formation of the final ice lens under zero overburden pressure, which is a simplified case where the effect of frost front advance is almost eliminated (Fig. 23).

At the base of any ice lens, the Clausius–Clapeyron equation (1) relates the pressure in the liquid film to the temperature T_s and can be written

$$P_w = M\,T_s \tag{4}$$

where M is a constant. Neglecting elevation head, equation (4) in terms of total potential becomes

$$H_w = (M/\gamma_w)\,T_s \tag{5}$$

where

H_w denotes the total potential
γ_w denotes the bulk density of water

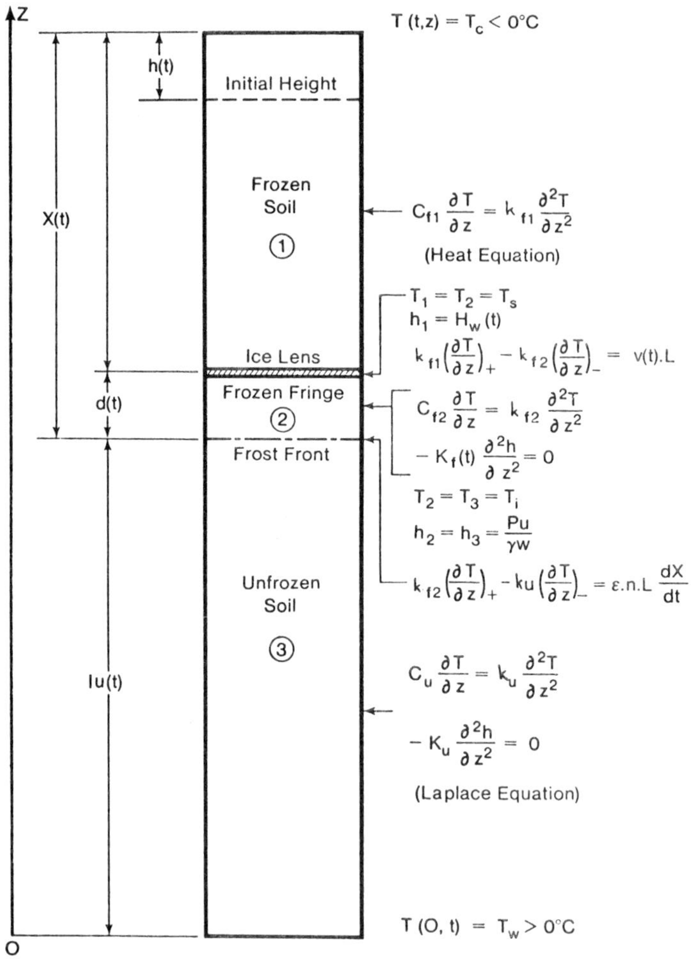

Fig. 24. Equations for the one-dimensional frost heave model, no externally applied load

The soil beneath the ice lens may be treated as a two-layered incompressible system in which there is no accumulation of water or ice and Darcy's law holds. Assuming zero pressure at the base of the system, the velocity of water movement $v(t)$ is given by

$$v(t) = \frac{|H_w(t)|}{(l_u(t)/K_u) + [d(t)/\bar{K}_f(t)]} \tag{6}$$

where

$l_u(t)$ denotes the thickness of the unfrozen soil
$d(t)$ denotes the thickness of the frozen fringe
K_u denotes the permeability of the unfrozen soil
$\bar{K}_f(t)$ denotes the overall permeability of the frozen fringe

The heave due to segregational processes $h_s(t)$ is found directly from equation (6) by

$$h_s(t) = 1{\cdot}09 \int_0^t v(t)\, d(t) \tag{7}$$

Routine considerations of heat conduction lead to the equations for temperature T shown in Fig. 24. For one-dimensional heat flow

$$\frac{\partial}{\partial z}\left(\lambda \frac{\partial T}{\partial z}\right) + Q = C\frac{\partial T}{\partial t} \tag{8}$$

where

C is the volumetric heat capacity
λ is the thermal conductivity
Q is an internal heat generation term per unit area and per unit time

The internal heat is liberated at two different locations: the segregation-freezing temperature T_s and the in situ freezing temperature T_i. At T_s

$$Q = v(t)\,L \tag{9}$$

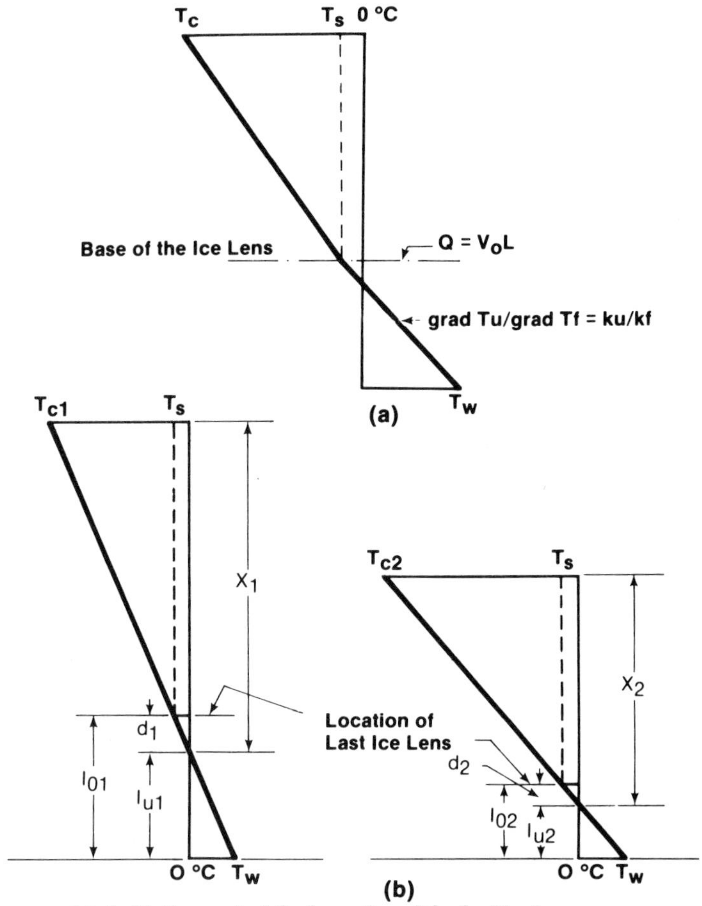

Fig. 25. Conditions associated with the onset of the formation of the final ice lens

and at T_i

$$Q = \varepsilon n L (\mathrm{d}X/\mathrm{d}t) \qquad (10)$$

where

ε	is a dimensionless factor taking into account the unfrozen water remaining in the sample and lumped at T_i
n	is the porosity of the soil
$\mathrm{d}X/\mathrm{d}t$	is the rate of advance of the frost front

Konrad & Morgenstern (1980) have developed a model that avoids the requirement for local measurements of T_s and $\bar{K}_f$ needed to solve the equations given in Fig. 24. They have argued that, since the permeability of frozen soil is influenced by temperature, it is expected that for a given soil the final ice lens should be initiated around the same segregation-freezing temperature T_s, independent of the temperature gradient across the frozen zone.

Freezing two identical samples with different heights under different cold side temperatures T_c and the same warm-side temperature T_w leads to temperature profiles at the beginning of steady state as shown in Fig. 25. From considerations of both geometric similarity and Darcy's law, it can be shown that regardless of T_c

$$v = \mathrm{SP} \times \mathrm{grad}\, T \qquad (11)$$

where

$$\mathrm{SP} = \left| \frac{H_w - h_u}{T_s} \right| \bar{K}_f \qquad (12)$$

$$\mathrm{grad}\, T = \frac{T_w + |T_s|}{lt} = \frac{T_w}{l_u} \qquad (13)$$

h_u	denotes the suction at the frozen–unfrozen interface
SP	denotes the segregation potential

Equation (11) states that if the segregation freezing temperature of a soil is unique, the water intake velocity will be proportional to the temperature gradient on the warm side of the ice lens. The constant of proportionality is called the segregation potential , SP; and the prediction of equation (11)

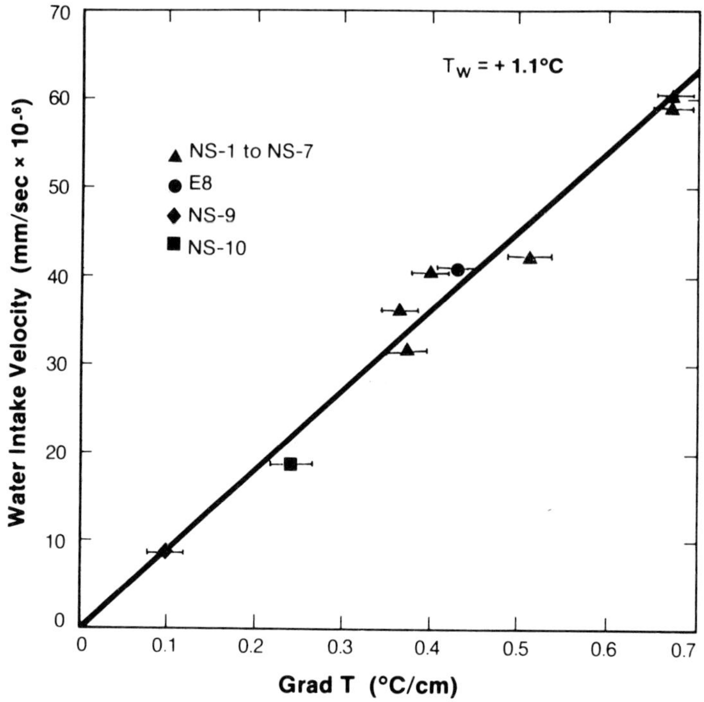

Fig. 26. Relation between water intake velocity and temperature gradient across the frozen fringe at the initiation of the final ice lens

can be tested directly by experiment.

In order to investigate the validity of equation (11) a series of freezing tests on replicate specimens of silt has been conducted at a constant warm-side temperature T_w and different cold-side temperature T_c. These tests were conducted in such a manner that both v and grad T could be identified at the onset of the last ice lens. Details are given in Konrad (1980). The results are shown in Fig. 26 and support the conceptual development reviewed here.

The segregation potential is itself explicable in terms of the detailed characteristics of the frozen fringe. However, from an engineering point of view it is more important to recognize that equation (11) constitutes the necessary coupling between heat and mass flow required to predict frost heave and that the parameter characterizing the freezing system, SP, is readily found from well-defined laboratory tests. The system of equations summarized in Fig. 24 are readily recast in terms of SP and can be solved by numerical means to predict heave under the specified boundary conditions.

The development of the segregation potential has so far been restricted to conditions of constant T_w, almost equilibrium cooling and zero external pressure. To be of general value each of these restrictions must be removed.

Konrad (1980) argued on thermodynamic grounds that when water flows through frozen soil, the suction in the frozen medium is no longer related solely to temperature and the unfrozen water content becomes a function of both temperature and suction. Since the unfrozen water content distribution directly affects the permeability of the frozen soil, different average suctions within the frozen fringe will yield different freezing characteristics for a given soil, although the average temperature in the fringe may remain constant.

By recognizing the effect of different temperature boundary conditions on the location of the final ice lens in a laboratory freezing test, and bearing in mind that changes in cold-side step temperature alone do not affect SP, it can readily be shown that the warm-side temperature alone affects the value of the suction at the frost front. Figure 27 presents simplified temperature distributions across a sample for different boundary conditions. The temperature profiles with identical numbers result in identical characteristics of the frozen fringe whereas different warm-end temperatures give different suction profiles in the fringe. From geometrical considerations and considering Darcy's law in the unfrozen soil, assuming for example, a given value of water intake flux for a fringe of thickness unity, it can readily be shown

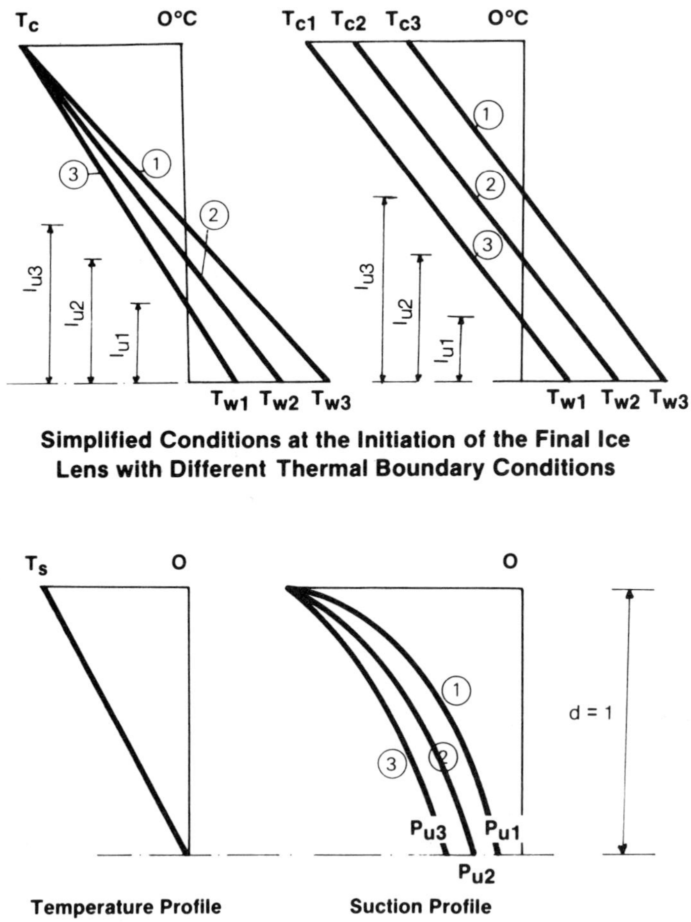

Fig. 27. Effect of warm-plate temperature on suction profile in the frozen fringe

that

$$\frac{|l_{u_1}|}{T_{w_1}} = \frac{|l_{u_2}|}{T_{w_2}} = \frac{|l_{u_3}|}{T_{w_3}} \tag{14}$$

For $T_{w_1} < T_{w_2} < T_{w_3}$ it follows that

$$|l_{u_1}| < |l_{u_2}| < |l_{u_3}|$$

Further, assuming that the segregation temperature T_s does not change drastically with T_w, the shape of the suction profile can be drawn schematically as shown in the Fig. 27. Since the average suction is strongly related to the shape of the suction profile which in turn depends on the actual shape of the permeability profile it is impossible to determine with any degree of accuracy the value of that average suction. For a given warm-side temperature the suction profile in the frozen fringe and particularly the suction at the frost front P_u is unique for a given soil. Therefore, P_u has been adopted as a reflection of the average suction of the frozen fringe. The advantage of using P_u lies in the ease with which it can be determined by applying Darcy's law to the unfrozen soil alone.

A variety of tests, including layered systems, were performed to induce different magnitudes of P_u and to measure SP. The relation between SP and P_u is illustrated in Figure 28. SP decreases with increasing suction at the frost front. This might be viewed at one level as an experimental finding, but in the Author's view it supports the concept that the average suction in the frozen fringe is a fundamental parameter of a freezing soil. The decrease in SP with increasing suction is accounted for primarily by a reduction in frozen permeability with increasing suction. Both SP and P_u can be determined from simple laboratory freezing tests.

Characteristic freezing surface

The first test of the frost heave theory developed here is the recovery of laboratory freezing test data. The theory has been developed and parameters deduced for conditions of near-stationary frost

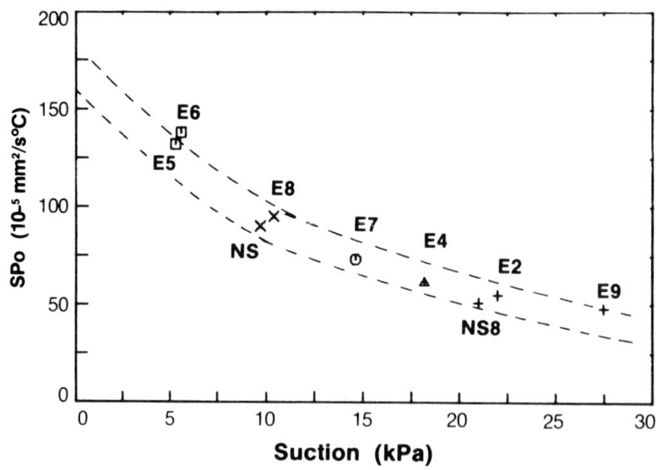

Fig. 28. Segregation potential against suction at the frost front for Devon silt

fronts and additional parameters may be needed to characterize freezing with an advancing frost front. This has proved to be the case (Konrad, 1980).

The governing equations for one-dimensional frost heave summarized in Fig. 24 may be solved numerically using established techniques. Figure 29 compares the total and segregational heave measured in two tests with the predicted values. It appears that good agreement is obtained at the beginning of freezing for about 12 h, after which a substantial difference arises. However, the computed rate of heave compares well with the measured value as steady state conditions are approached. This is not surprising since the input parameters characterizing the freezing system are representative of quasi-steady-state conditions associated with the growth of the final ice lens. Although the predicted heave is about 85% of the observed value at the onset of the formation of the final ice lens, the simulation is not all that satisfactory. This is illustrated by comparing computed and observed water intake velocities for a particular test (see Fig. 30). Substantial differences are apparent. These differences can be accounted for by the influence of changing suction profiles on the characteristics of the frozen fringe.

During a laboratory freezing test, the suction at the frost front changes continually. Initially, relatively long flow paths in the unfrozen soil associated with high flow velocities indicate quite high suctions at the frozen–unfrozen interface. With time, both flow path and water velocity decrease with a concomitant decrease in suction. While it is possible to account for the changing freezing characteristics in terms of variation in T_s and K_f during rapid freezing, a direct evaluation in terms of SP leads to results that are more readily applicable in practice. However, the relation between SP and P_u obtained at quasi-steady-state conditions cannot be applied to the unsteady heat flow condition with an advancing frost front. This is evident from observations that for a given suction P_u, different values of SP can be obtained depending on the degree of thermal imbalance in the test.

Many studies have explored the relation between rates of cooling and frost heave but no clear picture has emerged. It is tempting to relate SP to the suction and rate of frost front advance. However, since the frozen fringe is the seat of segregational process, it can be shown that, under certain circumstances, a given frost front penetration over a given time interval does not necessarily induce identical changes in the anatomy of the frozen fringe. This is illustrated in Fig. 31. If two identical samples are subjected to different geometrical and thermal boundary conditions and compared upon reaching a specified rate of frost penetration, there will be differences in temperature gradients in the frozen and unfrozen soil. This in turn affects the thickness of the frozen fringe. If, for simplicity, it is assumed that T_s is the same in both specimens, the dimensions of the frozen fringe are then fully defined at time t in both samples. If the frost front advances in both cases an identical length dX during an interval dt, the result is a change in temperature distribution in both samples and this is shown in Fig. 31. The ratio of the hatched area and the area defined by the frozen fringe at time t can be interpreted as a measure of the degree of cooling of the fringe. The frozen fringe cooled by a different amount in each case. Therefore the degree of thermal imbalance has been related to the rate of cooling of the frozen fringe during freezing. Hence, a freezing soil subjected to an advancing frost front may be characterized by the segregation

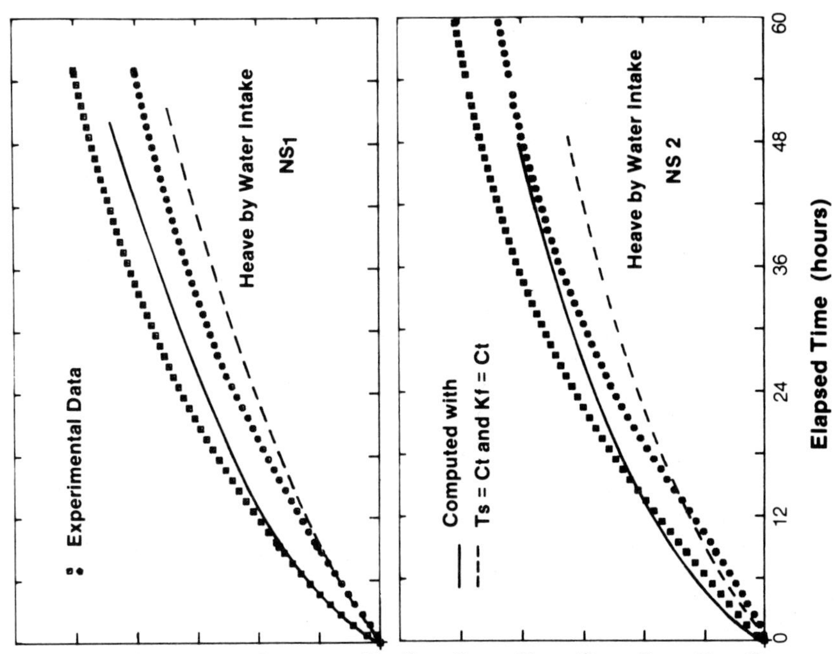

Fig. 29. Comparison of prediction using constant freezing parameters with actual data

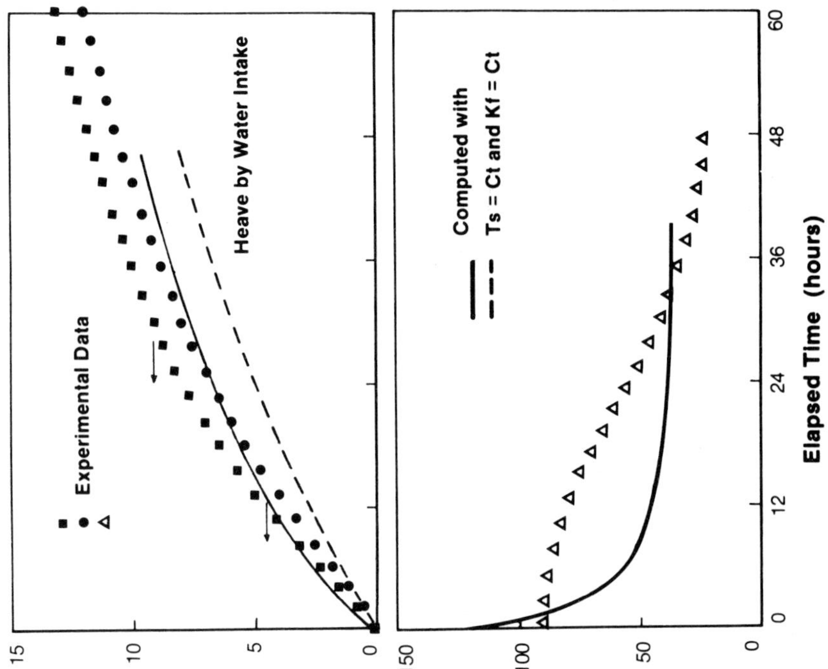

Fig. 30. Comparison of prediction using constant freezing parameters with actual data

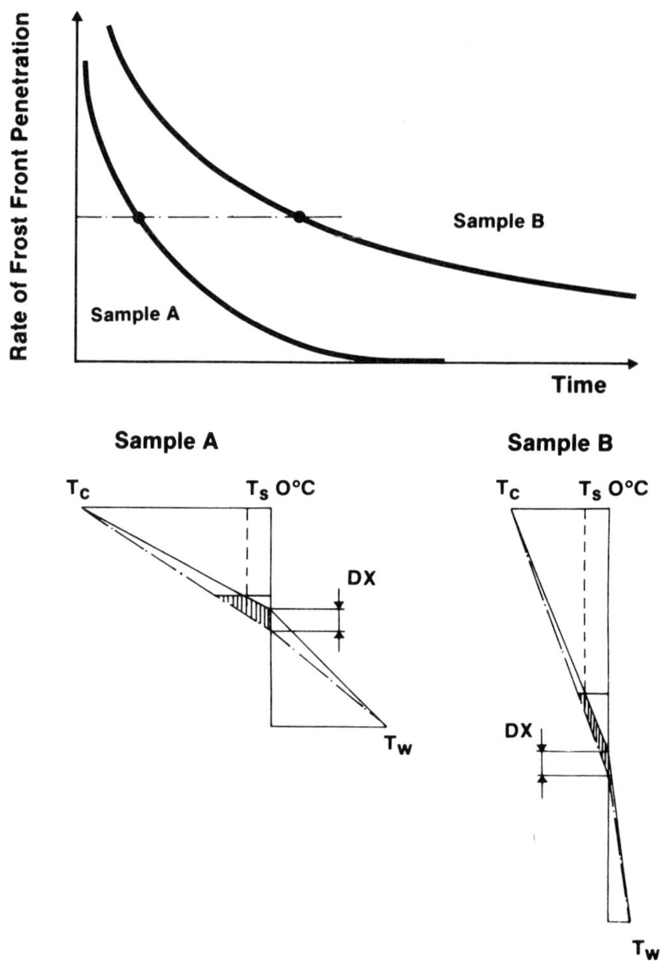

Fig. 31. Changes in frozen fringe at a given rate of frost front advance

potential, which is a function of two independent parameters: the suction at the frost front P_u, and the rate of cooling of the fringe $\mathrm{d}T_f/\mathrm{d}t$. This results in acceptable input for frost heave prediction in the more general heat and mass transfer formulation.

The frost heave characteristic surface (SP, P_u, $\mathrm{d}T_f/\mathrm{d}t$) can be determined from controlled freezing tests in which the variation of the length of unfrozen soil at any time is known from temperature measurements. Details of tests and their interpretation are given by Konrad (1980). Figure 32 summarizes results from several different tests and shows that a unique relation between SP and P_u exists for a particular value of $\mathrm{d}T_f/\mathrm{d}t$. Such a relation has already been established at the onset of the formation of the final ice lens. Results like these can be combined to form the surface shown in Fig. 33. The transients are extreme at high rates of cooling and the surface may not be well defined for these conditions, particularly if the unfrozen soil is compressible. Cavitation also limits the suction. However, this is only of concern for the early stages of laboratory tests and will not be a restriction when applied to field conditions.

By fitting functions to the experimental relations between SP and P_u at different rates of cooling and providing interpolation procedures, the surface can be used to characterize mass transfer in the formulation presented in Fig. 24. Unsteady heat flow is first solved across the whole specimen. The resulting temperature profile can then be used to determine the rate of cooling of frozen fringe. From the current rate of cooling, SP can be fixed as a function of P_u. Knowing SP determines the water intake velocity as a function of suction at the frost line. However, for a given length of unfrozen soil the velocity of water flow is related by Darcy's law to the difference in total potential across the unfrozen length. This requirement thus fixes the particular value of SP and P_u at the time under consideration

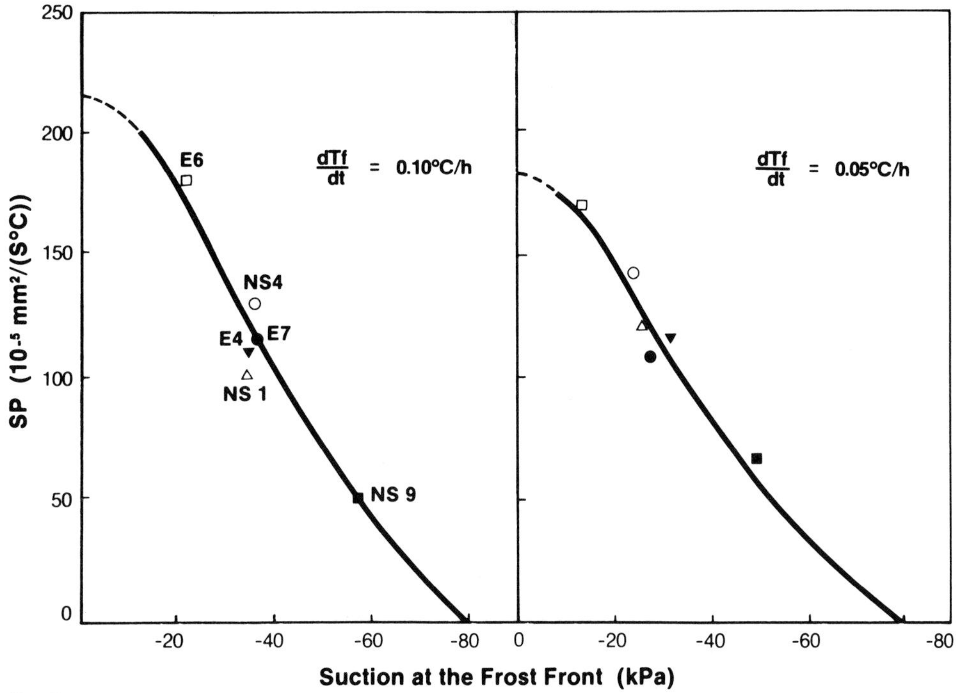

Fig. 32. Freezing characteristics for Devon silt

and the solution process can march forward in time. Comparisons between predicted and measured heaves in a variety of laboratory tests are given in Fig. 34.

All the simulated freezing tests discussed so far have been conducted with fixed temperature boundary conditions during the whole freezing period. It is tempting to conclude that the validity of the proposed characterization of a freezing soil is therefore restricted to those specific thermal conditions. In order to demonstrate that the characteristic freezing surface is independent of freezing path one sample was frozen in two stages. During the first stage, the temperatures at the top and bottom of specimen were maintained constant for 24 h. During that period, the frost front penetrated approximately to the middle of the sample. Then the second stage was initiated by changing the temperatures at both ends in order to force further penetration of the frost front. During the second phase the temperatures were also maintained constant with time. The warm-plate temperature was lowered from 3·5 °C to 1 °C. This results, in the early stage of the phase, in heat flow to both ends of the specimen since the temperature distribution is at a maximum somewhere within the unfrozen soil. Figure 35 shows the comparison between computed and measured results. The model predicts remarkably well the change in the rate of heaving that occurred as the temperature boundary conditions were changed. Furthermore, the computed frost front penetration is also in agreement with the measured profile and visual observations after the test was completed. In addition, Fig. 36 demonstrates that the model predicts extremely well the actual increase in water content in the frozen soil.

The final parameter that needs consideration in the development of a comprehensive theory for frost heave is applied pressure. It has been known for a long time that applied pressure inhibits frost heave and this can also be illustrated in terms of the SP (see Fig. 37). The influence of applied pressure can be explained in terms of stress-induced changes in unfrozen water content, frozen fringe permeability and segregation freezing temperature; but these are not necessary in order to accept data like Fig. 37 as an experimetal finding of value in predicting the influence of applied stress on frost heave.

Applications

In order to understand more clearly the chilled gas pipeline problem, both laboratory model and full-scale field studies have been carried out. A model box utilized in one study (Northern Engineering Services Ltd, 1975) is shown in Fig. 38. The tests were intended only to obtain qualitative information; temperature data were not sufficiently complete for analytical purposes. Boundary

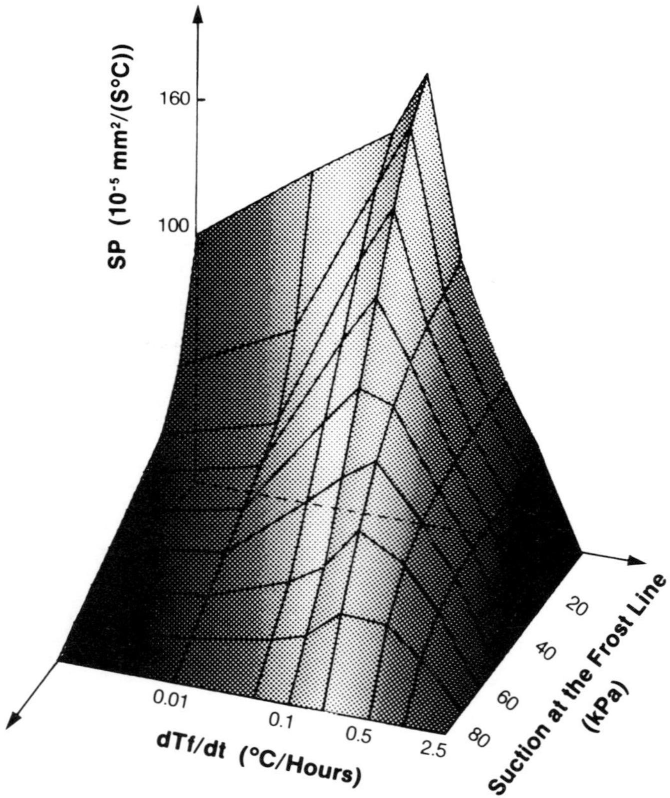

Fig. 33. Characteristic freezing surface for Devon silt used for simulation

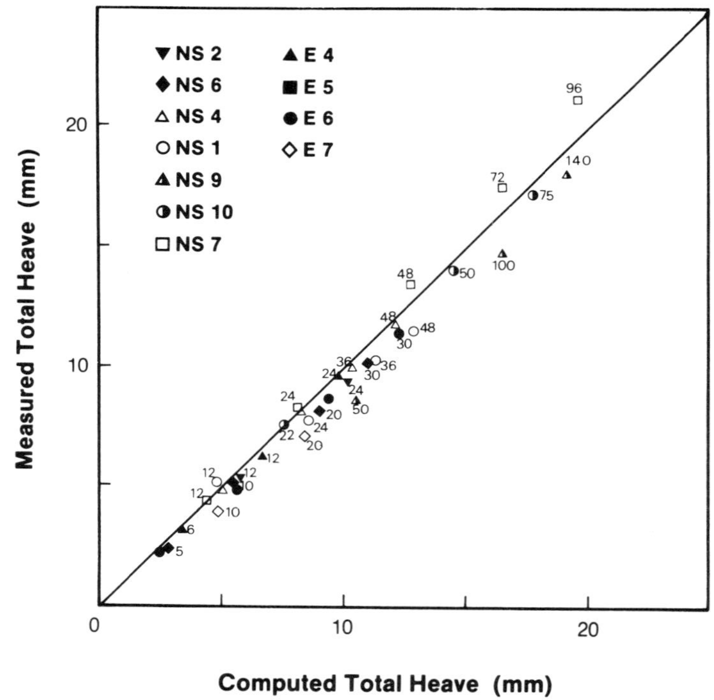

Fig. 34. Comparison of predicted total heave with actual laboratory data using SP, P_u, $dT_f/d(t)$

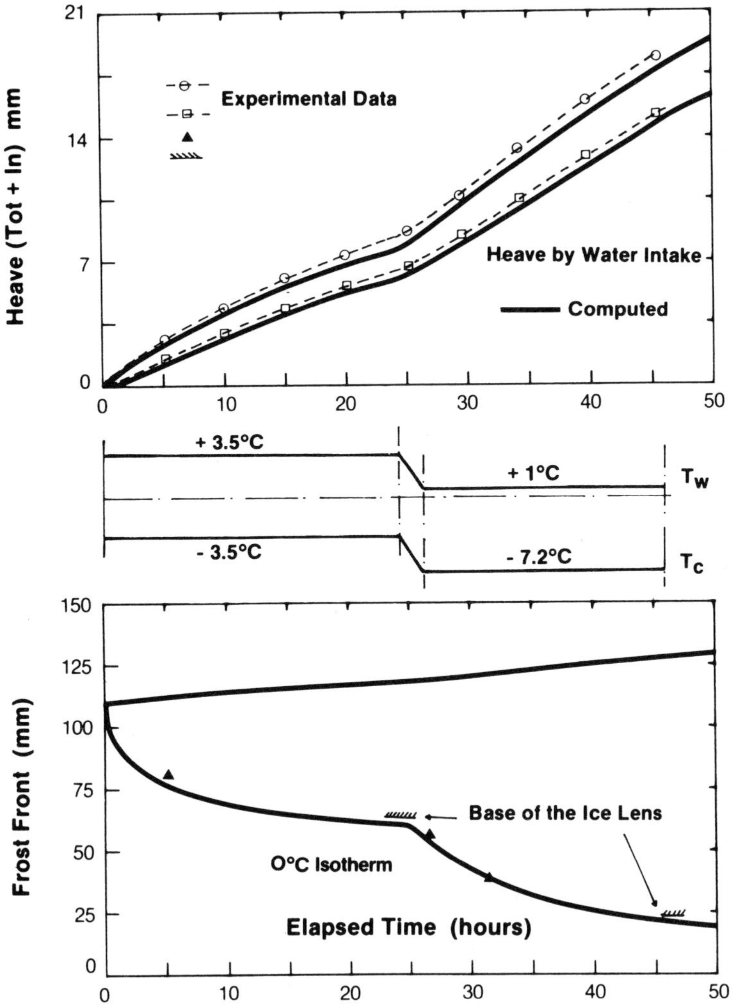

Fig. 35. Comparison of prediction with actual data for test E-1

conditions had to be deduced by back-analysis. Tests U-1, U-2 and U-3 were run consecutively to assess the effect of alternate freezing and thawing on pipeline performance. For tests U-2 and U-3 the initial conditions corresponded to the final conditions at the end of the thawing cycle for the previous test. The initial ground temperatures below the pipe were therefore warmer than otherwise expected thereby accounting for shallower frost penetration. Tests U-5 and U-6 were essentially duplicate tests and the soil had not been frozen previously in either case. The results of these tests are compared with theoretical predictions in Fig. 39.

The analysis of the model tests reveals that the best fit for tests U-2 and U-3 is obtained with a permeability of the frozen fringe of $1{\cdot}4 \times 10^{-9}$ cm/s and that tests U-5 and U-6 are fitted best with 1×10^{-9} cm/s. These permeabilities are in the range deduced from laboratory freezing tests. The prediction of heave with the matched data is encouraging. It appears that the segregation potential of a soil is increased after a freeze–thaw cycle. This increase, reflected in the permeability of the frozen fringe, is thought to be associated with changes in soil structure. Tests U-5 and U-6 demonstrate that for the same freezing temperature in the model pipe, the deeper the frost front, the smaller the resulting heave. This result, which is not intuitively obvious, confirms that colder ground temperatures which lead to deeper frost penetration are actually more favourable conditions with regard to pipeline heaving than warmer ground temperatures.

A field test facility was constructed in Calgary, Alberta in 1973. Four test sections using 1·22 m dia.

pipe were buried in a frost-susceptible silt and have been maintained at a temperature of $-8{\cdot}5\,°C$ since that time. Many detailed results have been reported by Slusarchuk *et al.* (1978). Laboratory freezing tests were performed on undisturbed samples but in a less controlled manner than would be specified today. However, a reasonable fit to the laboratory data provides a relation between SP and applied pressure which can be used in the field prediction to give the results shown in Fig. 40. The good correspondence is encouraging.

In many field freezing conditions P_u will be small enough to ignore. In a laboratory test this would correspond to a warm-plate temperature close enough to 0 °C to ensure small values of P_u. Under these circumstances it is possible to predict natural heaving if the relation between surface freezing temperature and time is known or alternatively to invert the process and deduce SP from observations of natural freeze-back and associated heave. An illustration of this applied to the interpretation of natural freezing in Fairbanks silt (Aitken, 1974) is shown in Fig. 41. By measuring the penetration of the frost front and heave, the in situ magnitude of SP is readily found. If a surcharge is applied to the ground the relation between SP and applied stress can also be determined. This obviates the need to extract samples and conduct laboratory tests to determine frost heave characteristics of many natural soils.

Commentary

The initial objective of the research programme described here was to develop a procedure for forecasting the heave of a chilled buried gas pipeline both unrestrained and under restrained conditions. The examples cited previously demonstrate that unrestrained heave is predicted in a reasonable manner. Restrained heave may also be predicted by calculating the normal stress required to deform the pipeline encased in frozen soil in a differential manner. This stress can be used as an externally

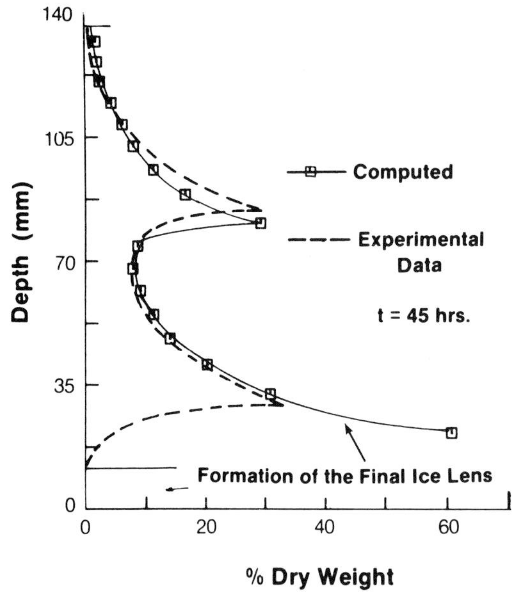

Fig. 36. Water content profile at the end of freezing for test E-1

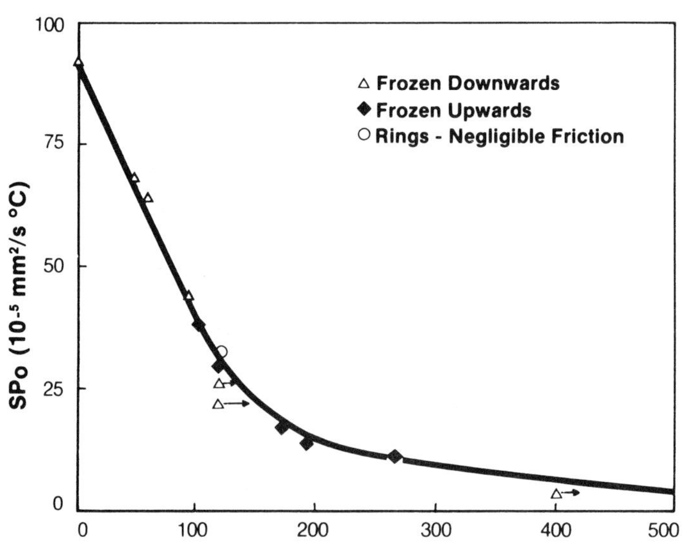

Fig. 37. Segregation potential for Devon silt under different applied loads (series C)

applied stress in the frost heave calculation to moderate the predicted local heave. In this way, an iterative solution can be developed for soil–structure interaction analyses of differential frost heave.

The segregation potential SP provides a new basis for frost heave classification. Existing procedures are not very effective in discriminating among differing degrees of frost-susceptibility. A standard test can be devised to determine SP under representative boundary conditions and laboratory-based parameters readily correlated with field values of SP deduced from natural freeze-back tests. Existing classification work has been hampered by a lack of a clear transfer to field conditions.

Experimental studies in terms of the segregation potential or the equivalent parameters T_s and k_f provide a means for exploring in a fundamental manner the influence of mineralogy, pore water solutes and other compositional factors known to influence frost heave susceptibility.

Finally the theory sheds new light on both engineering and geological freezing processes by showing that only ice at less than 0 °C in a fine-grained soil is needed to attract water whether the ground surface is cooling or not. Segregational processes can even occur under summer conditions as has been observed in the field (Mackay, 1980). Climatic conditions necessary for the accumulation of ground ice do not require sustained neat extraction at the ground surface, although if the ice warms to 0 °C the segregational process stops.

MECHANICS OF THAWING GROUND

The problem

While it had been recognized for a long time that thawing of ice-rich frozen ground results in large settlements and reduced bearing capacity, procedures for including the effects of thawing permafrost in geotechnical design were virtually non-existent in North American practice prior to the late 1960s. Possible exceptions to this were the development of hydro-electric facilities along the Nelson River (MacPherson, Watson & Koropatrick, 1970) and some highway and railroad construction in Alaska and the Canadian north. After the discovery of oil at Prudhoe Bay on the Alaskan North Slope it was finally concluded that the transport of oil from the Arctic coast to an ice-

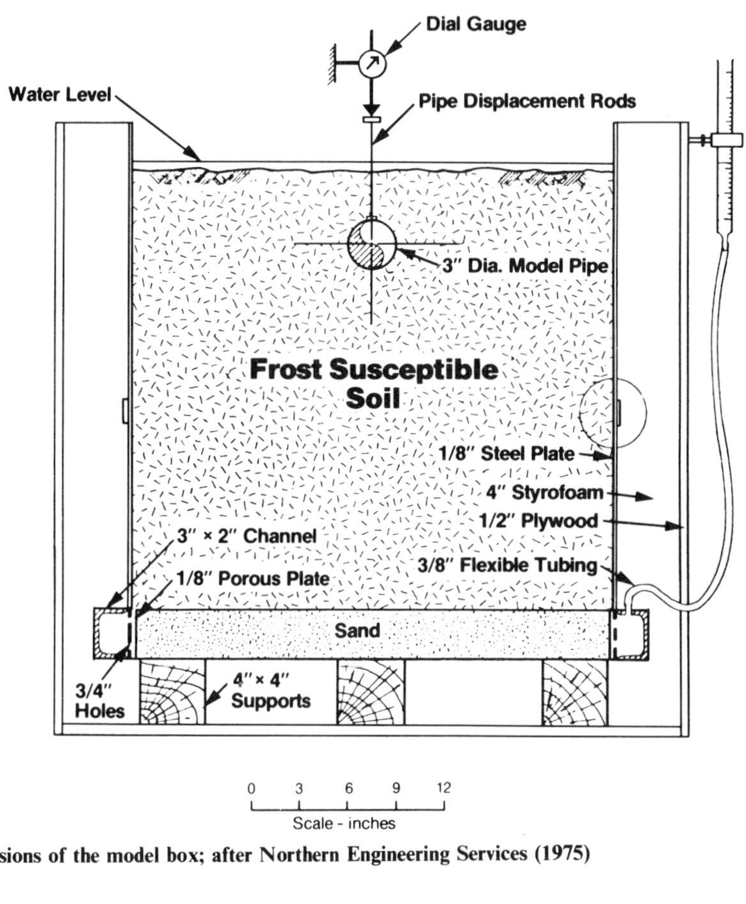

Fig. 38. Dimensions of the model box; after Northern Engineering Services (1975)

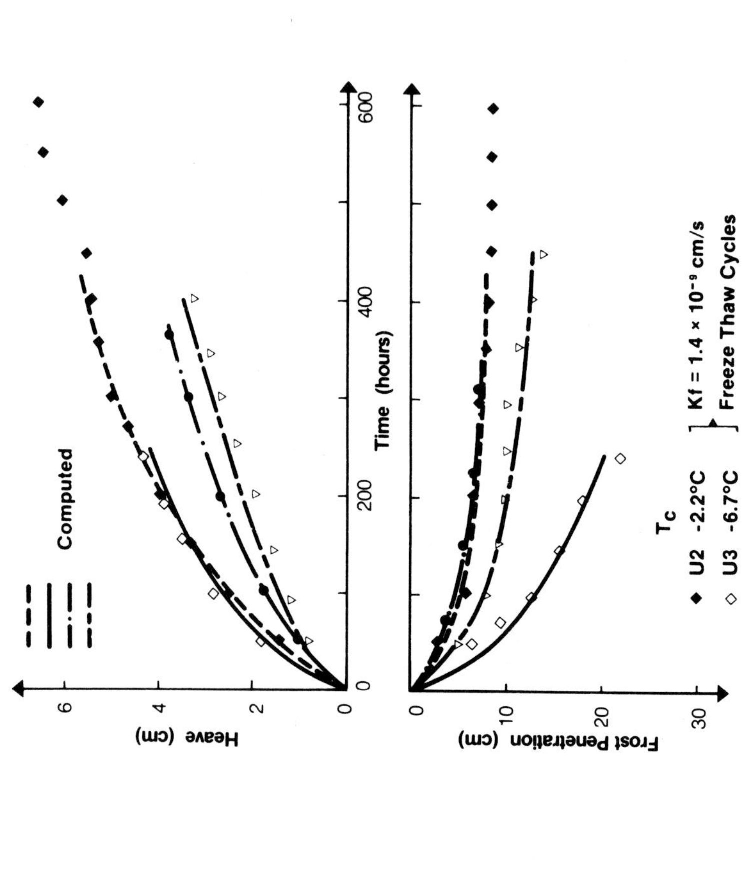

Fig. 39. Comparison between prediction and actual laboratory data for the model pipeline test

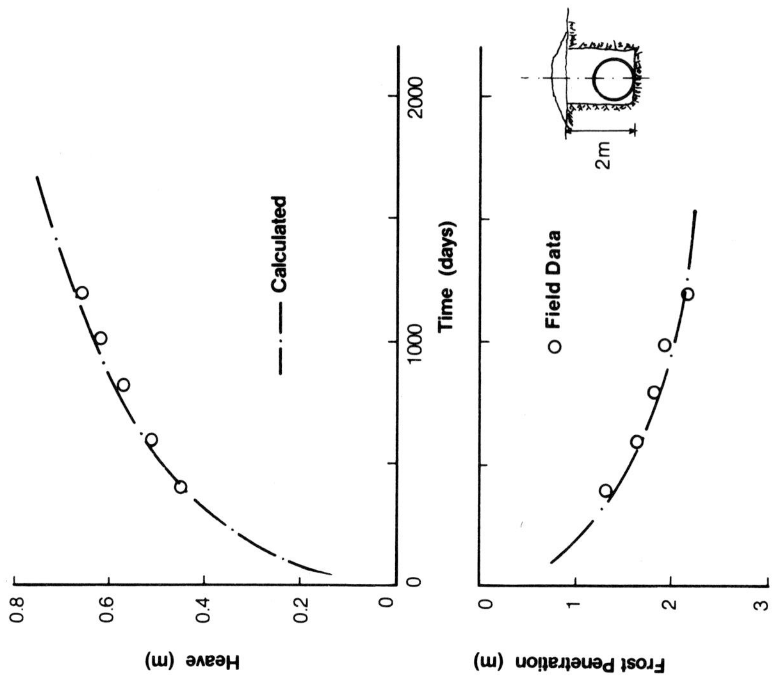

Fig. 40. Best fit for the results at the control section

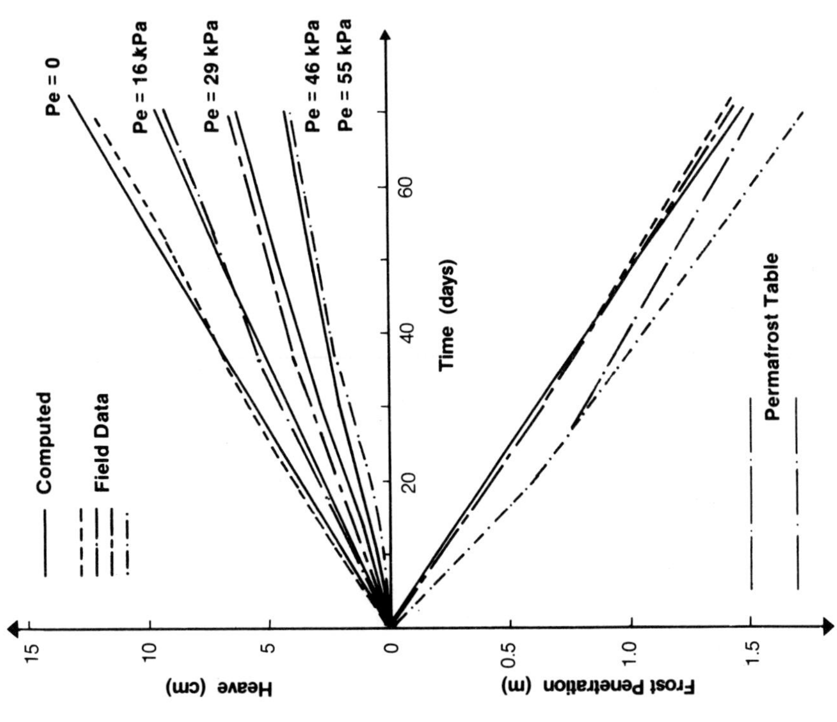

Fig. 41. Comparison of prediction with actual field data with different surcharges; freezing season 1961/62 (Aitken, 1974)

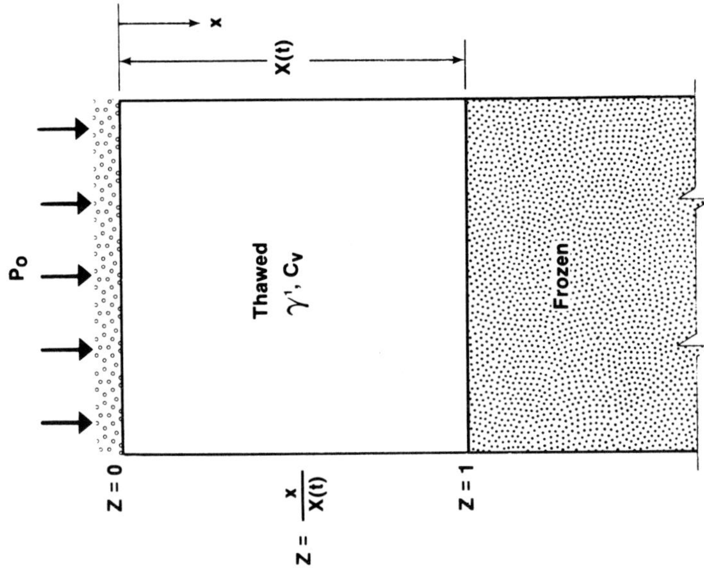

Fig. 42. One-dimensional thaw consolidation

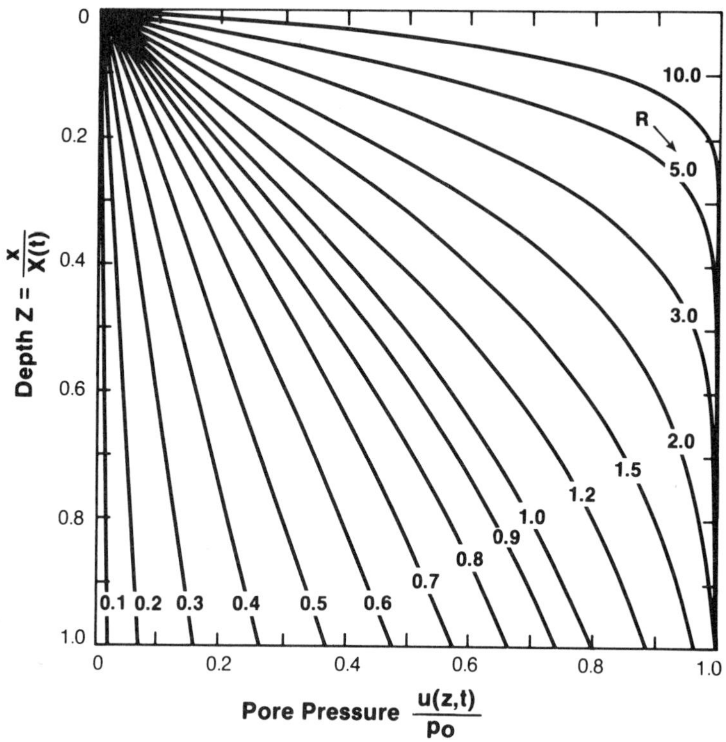

Fig. 43. Excess pore pressures (weightless material)

free port should be accomplished with a 48 in (1·22 m) dia. pipeline that was originally intended to be buried along most of its route. Since it was necessary to maintain oil temperatures at about 70 °C, this would result in thawing of the surrounding ground wherever the pipeline was buried in permafrost. Lachenbruch (1970) drew attention to the potential problems created by the presence of a hot-oil pipeline in permafrost. Depending upon boundary conditions, it was shown that a thaw bulb some 10–12 m in diameter might develop over the design life of the pipeline and if the melted soil were considered a viscous fluid, catastrophic slope instability could result. While the conclusions of this study were based on a limited perspective of the mechanical properties of thawed soils, they did serve to draw attentions to the importance of geotechnical aspects of pipeline design in permafrost.

In the early 1970s investigations into the design and construction of hot-oil pipelines from the Mackenzie delta to southerly markets were initiated and the same geotechnical concerns that had arisen over the Alaskan project became applicable to the developments proposed in Canada. In order to design in a rational manner it was essential to establish the effective stress changes in a soil consequent upon thaw. If thaw led to low effective stresses, the thawed soil would indeed be unstable on slopes and disposed to large settlements as pore pressures dissipated. However, if during and after thaw substantial effective stresses resulted, design could proceed in a more or less conventional manner without undue concern for the presence of permafrost. The problem of thaw-consolidation was therefore systematically attacked.

Determining the actual settlement of soil subjected to thaw is conceptually a straight-forward matter. When thawed under fully drained conditions components of total settlement arise from phase change considerations, settlement under self-weight, and settlement due to additional applied load. The parameters characterizing this behaviour can be studied in the laboratory and used in conventional procedures to estimate one-dimensional settlement. The difficulty in practice arises from the extreme variability of the magnitude of thaw-strain parameters over short distances (Speer, Watson & Rowley, 1973). If thawing proceeds under fully drained conditions the time–settlement relation is simply proportional to the progress of the thaw front with time. However, if thaw proceeds too quickly for the pore pressures generated to dissipate, excess pore pressures are set up, settlement is impeded and the shear strength is reduced accordingly. In order to determine the

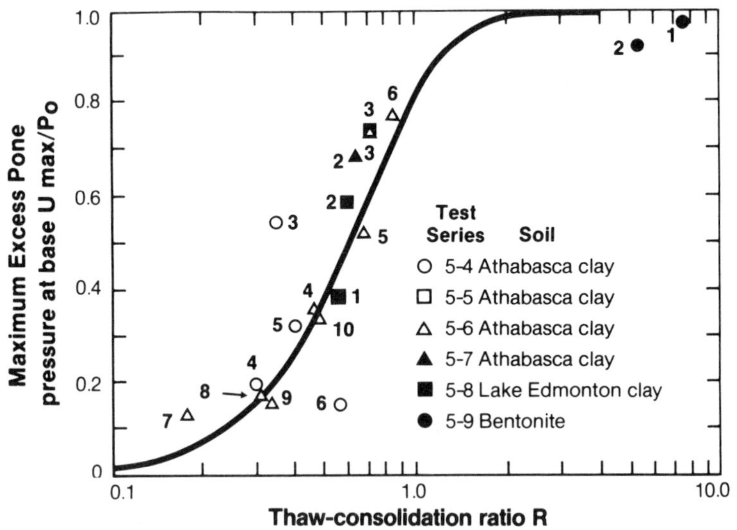

Fig. 44. Maximum excess pore pressures measured for reconstituted soils

extent of drainage that occurs during thaw it is necessary to couple the analysis of the process of thaw with the process of consolidation.

Theoretical and experimental studies

Figure 42 illustrates a uniform layer of frozen soil of semi-infinite extent subjected to a step increase in temperature at the surface. The solution to this type of heat conduction problem is well known and the movement of the thaw plane is given by

$$X(t) = \alpha t^{1/2} \tag{15}$$

where

- α is a constant that depends upon the thermal properties of the soil, its water content and the thermal boundary conditions
- X is the distance from the thaw plane to the surface
- t is time

The thawed soil is compressible and in the simplest development the Terzaghi theory of consolidation is assumed to hold. For a saturated soil a continuity condition can be written at the thaw front by noting that any flow from the thaw front is accommodated by a change in volume of the soil. Details of the solution to this moving boundary problem have been given by Morgenstern & Nixon (1971) and need not be repeated here. Similar but not identical results were obtained by Zaretskii (1968).

This solution permits calculation of pore pressure distributions in thawing layers loaded by both externally applied stresses and self-weight. It emerges from the analysis that the excess pore pressure distributions and the degree of consolidation in thawing soils are dependent upon the thaw–consolidation ratio R

$$R = \alpha/2\sqrt{C_v} \tag{16}$$

where C_v is the coefficient of consolidation. This ratio expresses the relative rate at which water is generated and dissipated at the thaw front. Drainage is enhanced at low values of R, while at very high values of R the process is essentially undrained. It was also assumed in the simplest theoretical development that, if the soil were to thaw under undrained conditions, the initial effective stress would be zero. This assumption is appropriate for the more fine-grained soils.

The thaw–consolidation ratio R served to clarify the accuracy with which soil thermal properties had to be known for geotechnical purposes. Experiments showed that even for natural soils (excluding organic soils) the published data for conductivity and specific heat were adequate to predict α within about 10% which is far superior to the accuracy with which C_v is generally known. The confidence with which R can be evaluated is therefore dominated by traditional geotechnical concerns.

The pore pressure distribution anticipated in an oedometer is shown in Fig. 43. Morgenstern & Smith (1973) described the development of a permafrost oedometer suitable for remoulded soils to assess the validity of the one-dimensional thaw-consolidation theory. As shown in Fig. 44, the observed dependence of the maximum excess pore pressure upon R is in good agreement with the theoretical relation.

The linear theory of thaw-consolidation can be extended to layered systems, other temperature

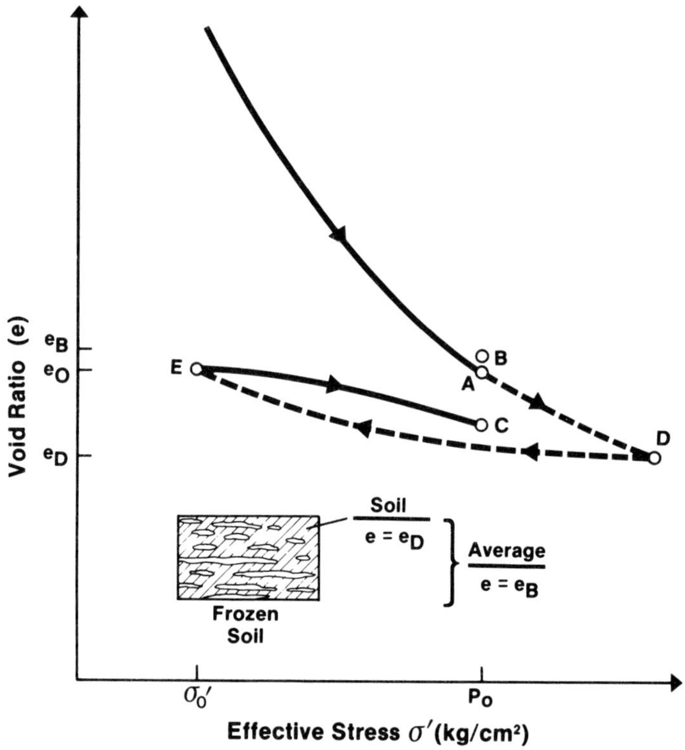

Fig. 45. Stress path in a close-system freeze–thaw cycle (schematic)

boundary conditions and non-linear material formulations. Nixon & Ladanyi (1978) provide a convenient summary of these extensions.

The non-linear theories require a starting point on the relation between void ratio and effective stress. This led Nixon & Morgenstern (1973) to the recognition of the significance of the residual stress which is the initial effective stress in soil thawed under undrained conditions. In addition to phase change effects, it is the departure from the residual stress that results in volume change. While it is reasonable to set the residual stress equal to zero in ice-rich soils with high void ratios, this will not necessarily be the case when the stress and thermal histories associated with the formation of a permafrost soil have caused the void ratio of the soil to be reduced prior to thawing.

The origin of the residual stress can be explained by referring to the experiment illustrated in Fig. 45. A sample of unfrozen soil was normally consolidated to an effective stress P_0 at A. The sample was then frozen with zero drainage and the void ratio increases to B in order to accommodate the volume change associated with phase change as most of the water in the pores turns to ice. If the sample is now allowed to thaw with no drainage, the void ratio returns to A. However, this is accompanied by an increase in pore water pressure which, in the limit, may reduce the effective stress to zero. Now, if drainage is permitted under P_0, the specimen will consolidate to C.

Externally, the freeze–thaw cycle under constant external stress has brought about a net decrease in volume represented by AC. Internally the stress path has been different. Suction develops when fine-grained soils freeze and this can result in an internal redistribution of moisture even under conditions of no overall drainage. At some locations the soil will become highly stressed as water is extracted from it while at other locations segregated ice will form.

Upon thawing, the overconsolidated elements in the soil may sustain effective stresses greater than P_0 but free water is made available locally from the thaw of segregated ice. The soil will swell by absorbing this water and the local stress path taken by freezing and thawing may then follow ADE. If the soil can absorb all of the free water it will come to equilibrium at the residual stress σ_0'. If not, excess free water will remain with the residual stress being zero. When drainage is permitted the soil will reconsolidate to P_0 along EC in a manner characteristic of an overconsolidated soil and exhibit thaw-strain.

Permafrost, when thawed, is influenced by the

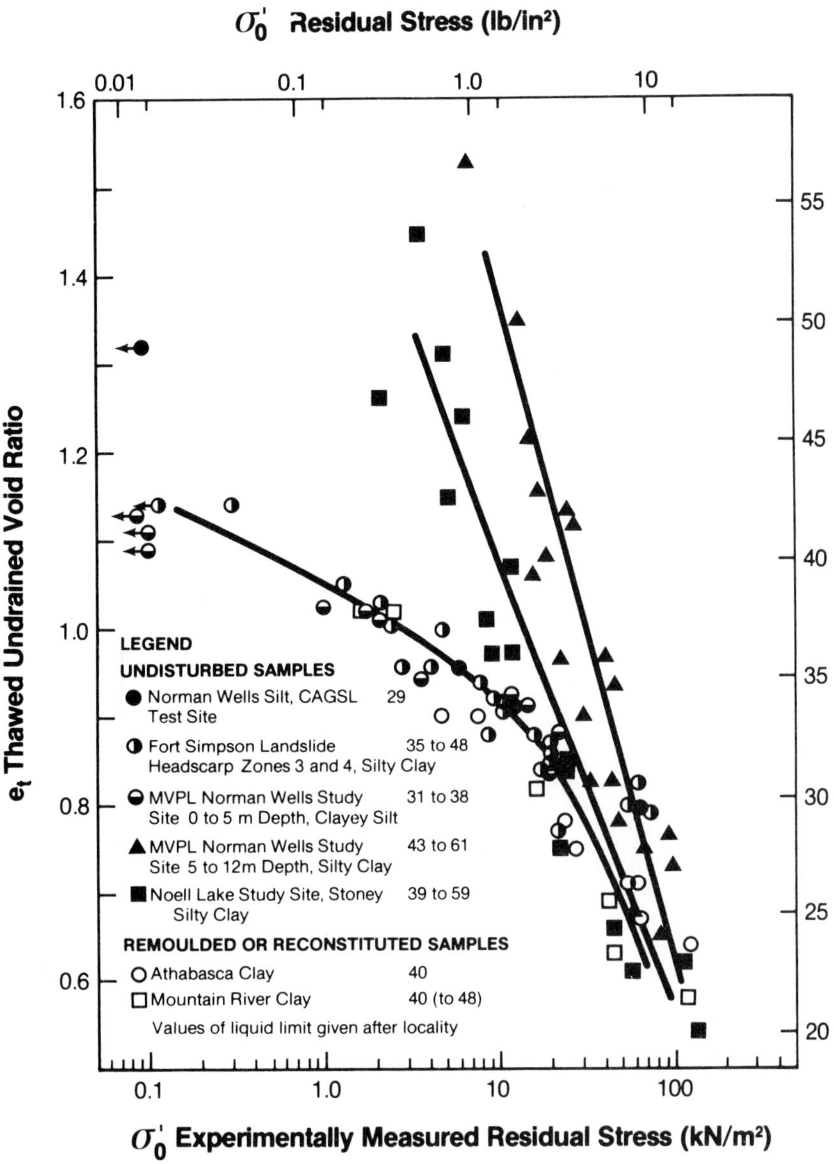

Fig. 46. Relation between residual stress and thawed, undrained void ratio

stress history and thermal history, as well as the hydrogeologic conditions that prevailed prior to the onset of freezing. In some instances σ_0' will be greater than P_0 and frozen ground might even swell when thawed (Crory, 1973). The residual stress will affect pore pressures, settlements associated with thaw and the undrained strength of the soil mass. For example, if a permafrost were thawed under undrained conditions, the undrained shear strength C_u would be given by

$$\frac{C_u}{\sigma_0'} = \frac{[K_0 + A(1-K_0)]\sin\phi'}{1+(2A-1)\sin\phi'} \qquad (17)$$

where

- K_0 denotes the ratio between lateral and vertical effective stress under conditions of zero lateral yield
- A denotes the pore pressure parameter
- ϕ' denotes the effective angle of shearing resistance

The first measurements of residual stress were reported by Nixon & Morgenstern (1973) who tested reconstituted specimens in an oedometer modified for freezing, thawing and pore pressure

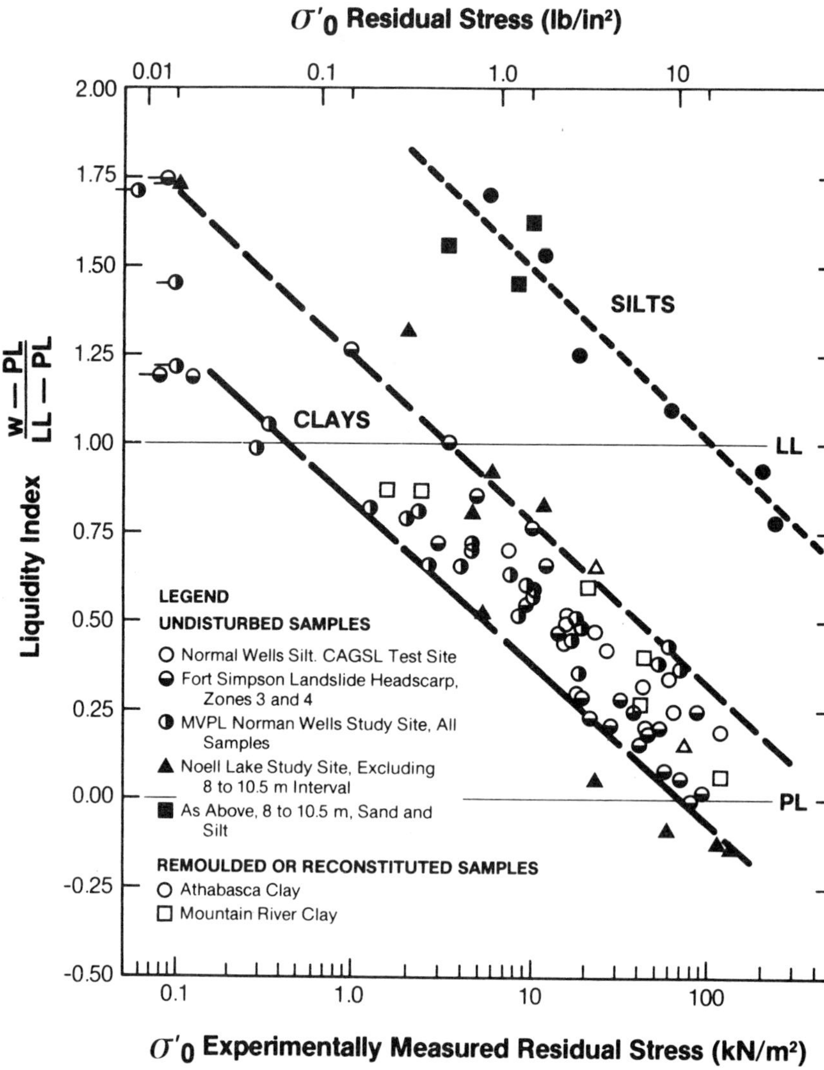

Fig. 47. Relation between liquidity index and residual stress

measurements. The tests revealed a linear relation between thawed undrained void ratio *et* and the logarithm of the effective stress, that is essentially independent of stress path, at least for a limited exploration. Nixon & Morgenstern (1974) also measured residual stress in a number of undisturbed samples of silt and showed that the non-linear theory of thaw consolidation accounting for σ_0' correctly predicted measured pore pressures, that again the thawed undrained void ratio *et* varied linearly with the logarithm of the residual stress, and that there was a tendency for σ_0' to increase with depth.

The study of the behaviour of undisturbed fine-grained permafrost soils from a variety of locations has been pursued in more detail by Roggensack (1977). As illustrated in Fig. 46, the existence of linear relation between *et* and $\log \sigma_0'$ was confirmed. The slope of this relation appears to be related to soil plasticity. Higher plasticity is usually associated with an increased clay content which, in turn, creates greater compressibility and a potential for larger negative pore pressures during freezing. The combination of these two features produces a steeper curve for the *et*–$\log \sigma_0'$ relation.

When experimentally determined residual stresses are replotted in terms of the liquidity index instead of void ratio, as shown in Fig. 47, the points for clay soils fall in a distinct band. This correlation is useful to extrapolate to the conditions when frozen ground is thawed at great depth such as arises in some oil well casing stability problems.

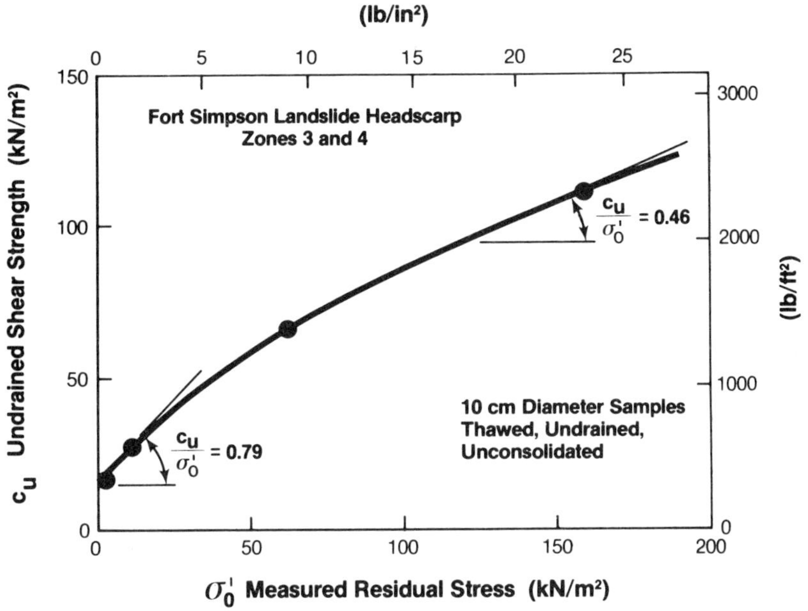

Fig. 48. Undrained strength as a function of residual stress; Fort Simpson site

The slope of the band is essentially identical to that for the sedimentation compression of clays reported by Skempton (1970). However, at any particular liquidity index residual stresses fall significantly below corresponding effective overburden pressures usually anticipated for normal consolidation. The difference between the two is related to the stress path followed to reach each condition. Freeze–thaw action brings about a large decrease in void ratio under conditions of constant applied stress. To obtain the same void ratio or liquidity index along the virgin compression line would require much larger effective stresses. This emphasizes once again the dominant effect that freezing history can have on stress history and the caution that should be exercised before attributing apparent overconsolidation to ice-loading, erosion or drying.

Roggensack (1977) also performed undrained compression tests to investigate the applicability of equation (17). In terms of effective stress, thawed clays display curved strength envelopes and A values that increase with increasing σ_0'. Both features can be attributed to the cryogenic texture found in thawed fine-grained permafrost soils. Experimentally measured undrained strengths (Fig. 48) compare well with C_u/σ_0' values computed by substituting appropriate values for ϕ', A and K as found in the laboratory into equation (17). In situ values will likely be less unless K_0 is equal to unity. Neither the in situ value for A nor K_0 for soils subjected to freezing and thawing have been studied.

Applications

Recognition of the consolidation of fine-grained soils during thaw and of the presence of a residual stress even if thawed under undrained conditions provides two mechanisms to account for the strength of thawing ground. As a result, thawed soil will not generally behave like a viscous fluid and problems such as the stability of the thaw bulb around a buried warm-oil pipeline will therefore be less acute than might otherwise be anticipated. The opportunity for validation was provided by the study of an instrumented test section installed near Inuvik, NWT.

The test section consisted of a 27 m length of 610 mm buried pipe through which hot oil at 71 °C was circulated. The field test was started on 22 July, 1971, and the ice-rich permafrost in which the pipeline segment was founded began to thaw. The soil around the pipe was instrumented to measure settlements, temperatures and pore water pressures. Undisturbed samples of the permafrost were collected in advance for laboratory testing. The field instrumentation has been described by Slusarchuk, Watson & Speer (1973), the experimental data have been presented by Watson, Rowley & Slusarchuk (1973) and a comparison between observed and predicted results has been given by Morgenstern & Nixon (1975).

The test section was overlain by 1·4 m of gravel fill. The first 0·6 m of the soil profile was comprised of compressed organic soil, silty clay and pure ice. The base of the pipe was placed in this layer. Ice-rich clayey silt extended for about 2 m below the

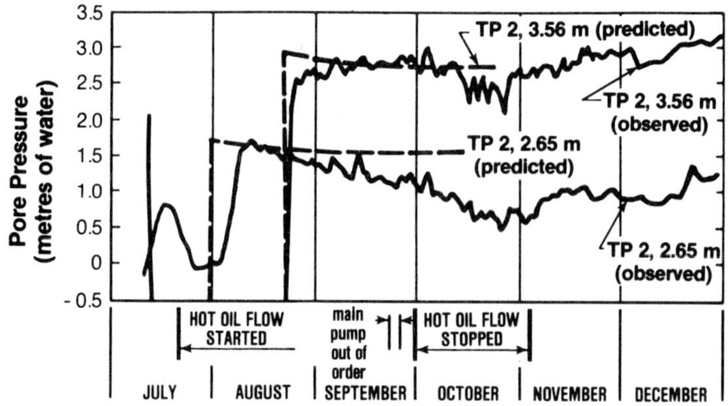

Fig. 49. Comparison between measured and predicted pore pressure; Inuvik test site

pipe and was underlain by a relatively incompressible gravelly till.

The piezometers installed in frozen ground at the site probably provided the first measurements of thaw-induced pore pressures. From a knowledge of the thaw-consolidation ratio R predictions could be made and the comparison with some of the observations is shown in Fig. 49. The values of excess pore pressure predicted at ten locations were about 25% of the ultimate value of the effective stress at each location. The observed values lay between 15 and 39% with an average of 24%. This agreement has been extremely encouraging and supports the more routine use of thaw-consolidation theory in practice.

Field studies reported by McRoberts (1973) and McRoberts & Morgenstern (1974a) indicate a widespread propensity for slope instability when fine-grained permafrost is subjected to thaw. Moreover, the gentle inclination of many solifluction slopes has long been paradoxical to the geotechnical engineer.

Thaw-consolidation theory can be introduced into slope stability analysis to account for these features. For example, if infinite slope analysis is extended to consider thawing conditions the factor of safety F of a slope inclined at α to the horizontal becomes

$$F = \frac{\gamma'}{\gamma}\left(1 - \frac{1}{1+(1/2)R^2}\,\frac{\tan\phi'}{\tan\alpha}\right) \qquad (18)$$

where

- γ denotes the bulk density of the soil
- γ' denotes the submerged density of the soil
- ϕ' denotes the effective angle of shearing resistance
- R denotes the thaw-consolidation ratio

Support for the development of excess pore pressures during thawing of slopes in fine-grained soils has been provided by McRoberts, Fletcher & Nixon (1978). Two sites adjacent to the Mackenzie River Valley, NWT, that had been exposed to long-term degradation of permafrost, were studied; at both sites, situated on modest slopes, excess pore pressures were measured. It was further shown that the highest excess pore water pressures measured were consistent with predictions from thaw-consolidation theory. While not conclusive, due to a variety of site complications, the general correspondence between prediction and measurement is again encouraging. As a result of integrating thaw-consolidation with stability analysis it has been possible to evaluate rational stabilization measures for thawing slopes. Pufahl & Morgenstern (1979) have shown how substantial increases in factor of safety could be obtained if surcharge loading were combined with only modest amounts of insulation to increase the effective stress across a potential slip surface. Skempton & Weeks (1976) have also found these considerations of value in an analysis of instability of gently inclined fossil periglacial slopes.

Thawing of permafrost is also encountered adjacent to production casing of oil wells and well stability must be evaluated. Experience at Prudhoe Bay (Mitchell & Goodman, 1978) indicated that thaw occurred under drained conditions in gravelly dense soils and no operational problems have been encountered. However, more recently oil has been discovered offshore in the Beaufort Sea and the possibility exists that wells will be developed through a considerable thickness (approximately 500 m) of fine-grained, sub-sea permafrost soils. Under these conditions undrained thaw must be anticipated. Arching of the soil about the well will affect the stresses transferred to the casing tending to make it buckle. As anticipated by Palmer (1972), the existence of high residual stresses will exercise considerable influence on the arching mechanism and attendant stress transfer. However, even if the undrained strength of the thawed soil is very high, strain can still develop in a casing placed in thawing

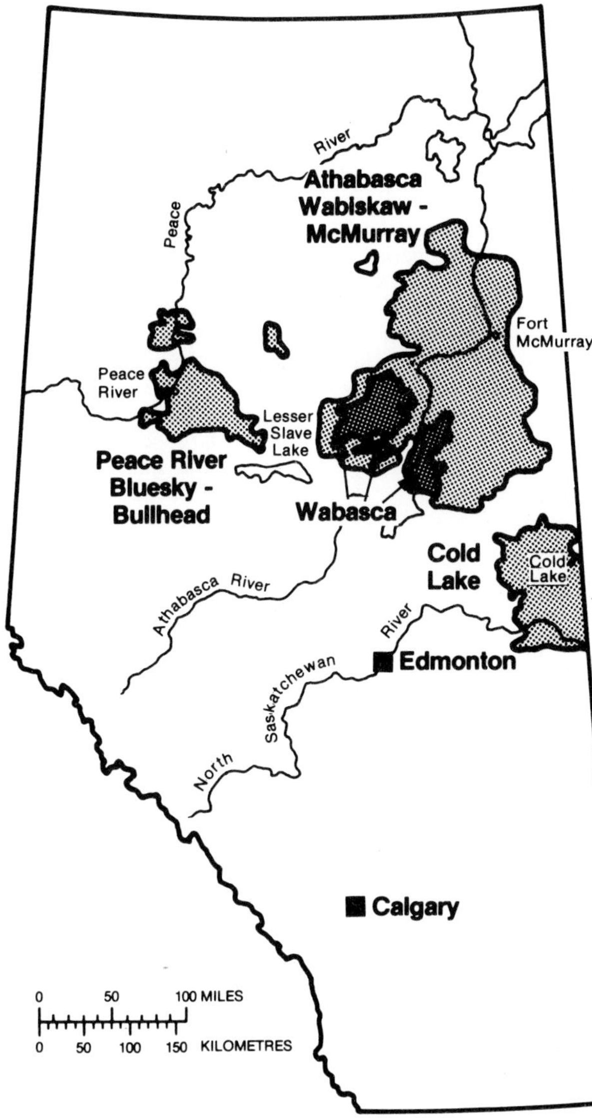

Fig. 50. Location in Alberta of the four major Cretaceous oil sands deposits

permafrost as a result of the volume changes and stiffness changes associated with undrained thaw. Studies of these soil-structure interaction problems are not yet well developed, and testing to obtain the appropriate deformation parameters is in its infancy.

OIL SAND GEOTECHNICS

Introduction

Oil sands may be defined as sands which contain heavy hydrocarbons that are chemically similar to conventional oils but which have higher densities and viscosities. The hydrocarbons range from heavy crudes to natural bitumen. While oil sand deposits are widespread, by far the largest occur in Canada and Venezuela. The Canadian deposits are situated primarily in the Province of Alberta (see Fig. 50). The magnitude of these deposits can be appreciated when one realizes that they are comparable in size, as is the Venezuelan Orinoco Oil Belt, to the in-place volumes of conventional crude oil for the entire Middle East. Demaison (1977) has stated that Alberta's Athabasca deposit is the world's largest known accumulation of hydrocarbons, and is at least four times as large as the largest of all giant oil fields, Ghawar, in Saudi Arabia.

Although abundant, the bitumen has physical

properties such that it cannot be pumped out like light crude and alternate extraction procedures have had to be developed. The bitumen occurs in beds of sand, and more recently has also been discovered in underlying porous carbonate rocks. The sand grains are usually covered with a film of water, and bitumen occupies most of the remaining pore space, along with minor amounts of fine clay particles, other mineral matter and occasionally some natural gas. Following a long period of research, it was eventually shown that crude bitumen could be separated from the oil sands with hot water. This method eventually became the basis of commercial production in 1967 by the integration of the hot-water extraction process with a mining operation.

About 0·3 million hectares of the Athabasca deposit is overlain by 50 m or less of overburden and is potentially capable of being mined from the surface. The remaining 6·7 million hectares of the major Alberta deposits vary considerably in bitumen content and are buried at such depths that the crude bitumen can only be recovered by in situ extraction methods. Generally these methods involve heating the extremely viscous bitumen with steam so that it will flow and can be pumped to the surface. Experience exists for in situ extraction of bitumen where overburden thickness is greater than 150 m but the extraction techniques for those reserves lying between 50 and 150 m is uncertain at present. The rapid rise in the cost of conventional light crude since 1973 has made it economic to begin large-scale development of the Alberta oil sands and this has provided a very substantial incentive for resolving the technological problems associated with extraction.

Figure 51 illustrates in a general manner the various ways that have been either adopted or proposed to extract the bitumen. The geology is characteristic in a schematic manner of the Athabasca setting, where the oil sands outcrop in a river valley and dip gently to the west. At shallow depth, open-cast mining provides a means of extracting the oil sand after which it must be delivered to a plant for processing. Where the overburden is deep, steam injection is utilized in a cycle of injection and production to reduce viscosity and recover bitumen. Underground combustion is also under active investigation at the pilot stage. In situ steaming without combustion is more advanced and a commercial-scale operation is currently being designed. At intermediate depths, extraction by mining has been advocated on

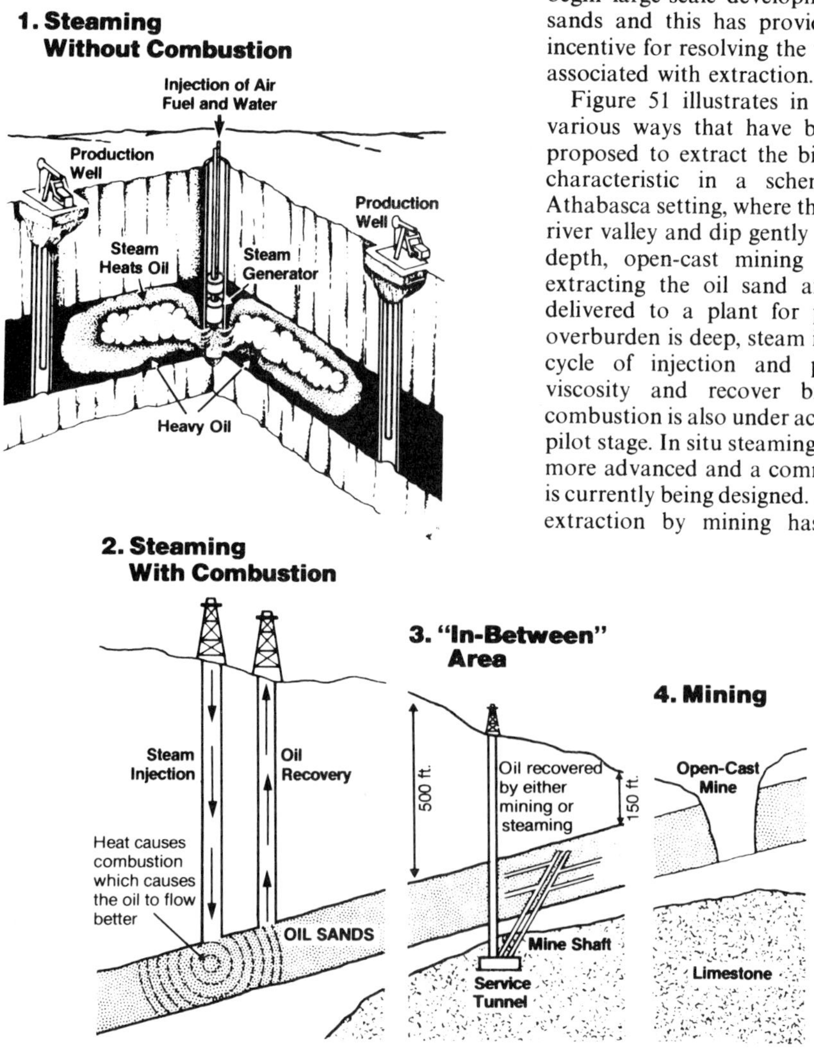

Fig. 51. Extraction of heavy oil

occasion but economic evaluation is not supportive. However, an interesting hybrid technology is under active investigation. This is mine-assisted in situ processing (MAISP) where an underground mine system is to be developed by means of vertical shafts and tunnels in or adjacent to the oil sands in order to provide access for the installation of horizontal steam injection and recovery wells. This layout is intended to result in more cost-effective production through a higher density of wells per unit cost in the formation. The concept has already been adopted in the USSR to recover bitumen in north-eastern Siberia near Yarega.

In each of the circumstances listed above, extraction by surface mining, extraction by MAISP and extraction by steam injection, novel geotechnical problems arise because of the peculiar properties of the oil sands, the scale of the extractive undertakings, and the pressure–temperature environment of some of the in situ processes.

Mining oil sand

The mining of oil sand involves earth-moving on a grand scale. The first commercial operation owned by Suncor (Sun Oil Co. Ltd) moves about 25 000 m^3 of overburden and 60 000 m^3 of oil sand per day in order to produce 7200 m^3 (45 000 barrels) of oil per day. The leases are covered by organic soil (muskeg) which, following a period of gravity drainage, is removed by front-end loaders and a fleet of dump trucks. This takes place during the winter when the surface is frozen. Both the remaining overburden and the usable oil sand are then mined by means of bucket wheel excavators on a three bench-mining configuration. The overburden wheel has a 12 m dia. digging head and a theoretical peak digging rate of 13 000 t/h. Under normal operations it has achieved a consistent average of 6800 t/h. The bench-mining machines, which have a 10 m dia. head, have produced peak quantities of 9000 t/h for short periods, but each has an average output closer to 4500 t/h (Supple, 1980). Bench height has only been about 20 m and slope instability has not proven to be a particular hazard to the bucket-wheel mining scheme. Trafficability and abrasion of digging teeth have proved troublesome but these difficulties have been reduced with experience.

The outstanding geotechnical challenge of this project has been associated with tailings disposal. As a result of the hot-water separation process about 250 000 t of tailings are handled daily including 100 000 t of solids. All of this material must be stored permanently in a closed system. Until space was available in mined-out areas, a retention dyke was necessary. For economic reasons it was desirable to construct the dyke from the tailings. However, sufficient fines remain in the tailings to preclude their direct use as a construction material without separation and when sluiced into the pond the fines separate and consolidate very slowly. Mittal & Hardy (1977) have described the innovative techniques of materials handling, dyke design and construction that have culminated in the building of a dyke from these tailings. The dyke is some 3·4 km long and has been built by the upstream method of hydraulic construction to a height of over 90 m, being founded in part directly on muskeg and thick normally consolidated alluvial sediments.

The next commercial operation was the Syncrude project which began construction in 1973 with the intent of producing 20 700 m^3 (130 000 barrels) per day which entails moving about 250 000 m^3 of overburden and oil sand every day. After a period of intensive study, this project adopted draglines for primary mining. As reviewed by Adam & Regensburg (1980), dragline mining appeared to have advantages over bucket-wheel excavation by minimizing the transportation distances for waste disposal even though ore grade oil sands had to be handled twice. Other contributions to the cost advantage were the relatively rapid opening of the mine and the potential for selective mining. Ultimately draglines of 60 m^3 bucket capacity and 110 m boom length were selected, each costing about $30 million. Since they were obliged to sit on a steep highwall some 50–60 m high, confidence in slope stability was central to the approval of the dragline mining scheme.

At the Syncrude site overburden is composed of Holocene, Pleistocene and Cretaceous sediments. The oil-bearing McMurray formation is comprised of both sand-dominated and clay-dominated facies resulting in non-uniform oil saturation. It lies unconformably over Devonian carbonates and is the result of a more or less continuous transgressive sequence. Stable slopes observed in natural outcrops provided a high level of confidence that draglines could be supported safely on oil sand slopes provided weak overburden had been removed. For example, Dusseault & Morgenstern (1978a) undertook a survey of natural slopes along river valleys where the McMurray formation outcrops and encountered no massive rotational or planar failures. They found that bitumen-rich oil sands may form very steep slopes (50°–55°) up to 70 m in height. Over limited sections, inclinations as steep as 75° were found. Bitumen-free portions of the McMurray formation were also steep with high slopes indicating substantial natural strengths. These observations were supported by the excavation of a 55 m deep test pit having a highwall slope of 60°. This trial was extensively instrumented and led to agreement in principle by a Board of Consultants to the application of

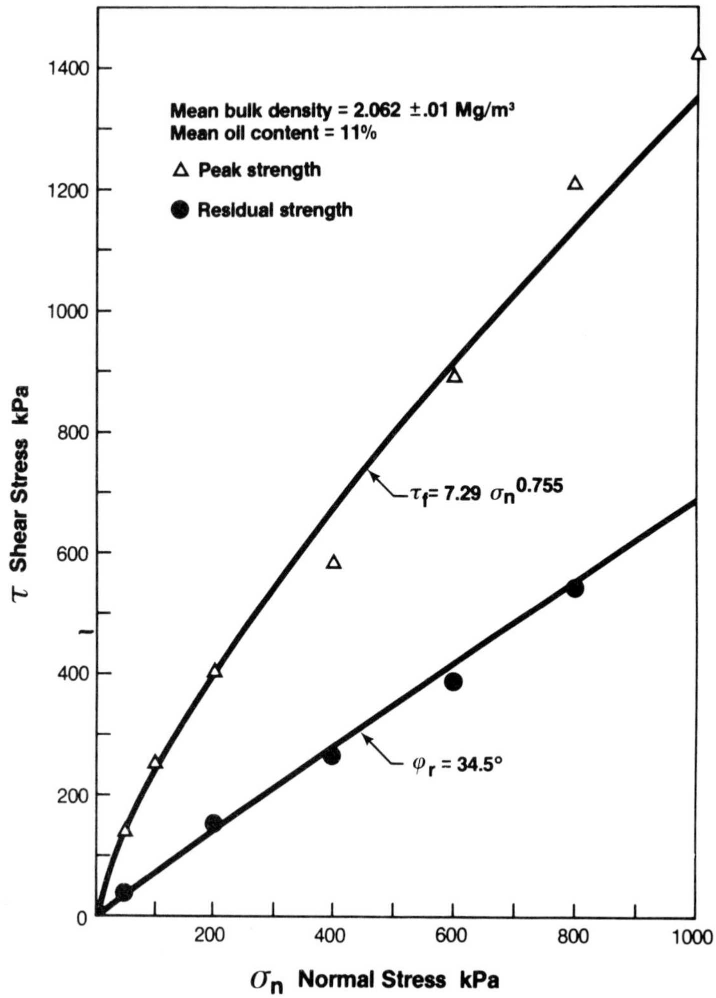

Fig. 52. Failure envelope for oil sands, shear box tests

draglines, subject to certain operating restrictions. It was clear that in practice instability might be controlled by such minor geological details as intraformational lenses of silt and clay, basal clays beneath the oil sands, joints and other defects. Notwithstanding the apparent strength of the oil sand in mass, the stability of the slopes from a conventional geotechnical perspective remained paradoxical.

The evaluation of the shear strength of oil sands is made difficult by the presence of dissolved gas that comes out of solution causing serious sample disturbance due to expansion. The first strength tests performed by Hardy & Hemstock (1963) gave low values which were correctly attributed to this effect. Brooker (1975) provided the first detailed assessment of the shear strength of the McMurray oil sands, and found dilatant behaviour with an angle of shearing resistance slightly below 45° and a cohesion intercept of about 80 kPa. However, these data were limited in quantity, not yet consistent with field observations, and were based on specimens with void ratios that exceeded typical in situ values.

Fresh oil sand can be remoulded readily in the hand suggesting a lack of cohesion or abnormally high negative pore pressures. Mineral or clay cementation is absent from the greater proportion of most profiles although cemented stringers are encountered. The interstitial bitumen is thought to behave as a fluid and therefore does not contribute directly to the stability of slopes. In addition, the specific surface of oil sand is low, so it is unlikely that interfacial tensions in the quartz–oil–water–gas system contribute to strength in any significant manner.

Fig. 53. Locked sand fabric

Dusseault & Morgenstern (1978b) obtained high quality samples of oil-rich sand by using downhole freezing techniques to inhibit gas expansion. When tested in both triaxial and shear box equipment, these samples produced remarkably high angles of shearing resistance accompanied by high rates of dilatation, particularly at low normal stress. As shown in Fig. 52, the envelope passes through the origin for all practical purposes but is markedly curved as normal stress is increased. The residual strength and the strength of remoulded oil sand is characteristic of values reported elsewhere for quartzose sands and is unexceptional. This suggests that the origin of the remarkable strength of natural oil sands should reside in their structure.

Microscope studies revealed an unusual integranular fabric. Mineral cement is absent, grain-to-grain contact area is large, many contacts are characterized by an interpenetrative structure with grain surfaces displaying a rugose solution–recrystallization texture. An example is given in Fig. 53. The lack of mineral cement is consistent with zero cohesion at zero normal stress. The interpenetrative fabric results in high rates of dilatation at low stresses. As normal stress levels increase, dilatancy is suppressed in favour of grain shear giving rise to the curved Mohr envelope. The rugose surface texture results in a residual friction angle that is somewhat higher than the value observed from testing smooth Ottawa sand.

These observations explained the stability of slopes in oil sands but they are also of more general interest. As a result of this texture, Dusseault & Morgenstern (1979) suggested that these materials constitute a distinct class of materials separate from loose and dense sands, and that they be called locked sands. Locked sands develop when sands loaded for long periods of time are subjected to diagenetic processes. If the dominant processes are solution and quartz overgrowth formation, the result may be a densified, uncemented aggregate with an interlocked structure. Experience so far indicates that locked sands possess an in situ porosity that is less than the minimum attainable in the laboratory and that they are generally pre-Quaternary in age. Locked sands are not peculiar to Alberta and are probably widespread. The St Peter sandstone in the Minneapolis region has been shown to be a locked sand and it is likely that many of the soft or friable sandstones referred to in the literature are locked sands. Provided care is taken not to disrupt the fabric, locked sands are strong and capable of supporting substantial loads with only small deformations.

Undergound access to oil sands

Underground access to oil sand deposits is an integral part of any MAISP scheme. Devenny & Raisbeck (1980) have illustrated the types of facilities required and indicate that access from underground drilling chambers is being considered because of the following

(a) more of the drilled hole contacts the reservoir
(b) it is probable that horizontal or near horizontal wells can be placed more efficiently with better control of location
(c) with improved location, it will be possible to place wells closer together
(d) with closer well spacing, control of fracture flow paths may become possible, facilitating more rapid and uniform heating and, hence, more efficient extraction
(e) increased resource recovery at lower cost should be possible

The MAISP concept is contingent upon the feasibility of sinking shafts through the oil sands and, in some instances, tunnelling in them at depths of 250–500 m from the ground surface. They are strong but uncemented sands. However, during unloading, gas comes out of solution. This disrupts the interlocked fabric which leads to both swelling and weakening. Therefore in addition to the more routine considerations of deep shaft and tunnel design, it is necessary to have a clear understanding of the geotechnical behaviour of gas-saturated porous media in order to proceed with design and construction.

Early laboratory tests on oil sand reported by Hardy & Hemstock (1963) found that contrary to conventional geotechnical experience, the strength in borings decreased with depth. They correctly attributed this to the exsolution of gas upon release

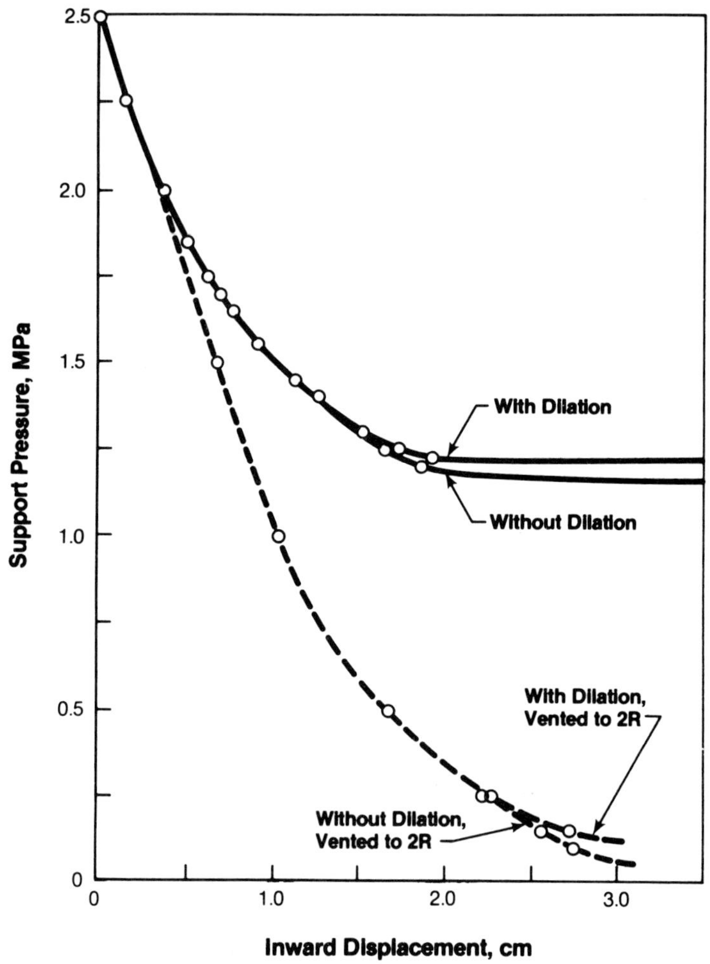

Fig. 54. Convergence of cylindrical shaft in oil sand; radius, $R = 2{\cdot}5$ m (Byrne *et al.*, 1980)

of stress, primarily from the oil phase. In addition, they observed that the gas pressure in the pores will increase with increasing temperature and the oil phase will impede the dissipation of the gas pressure because of its high viscosity. Hence gas-saturated oil sand acts in the short term as a relatively impervious material with respect to excess pore gas pressure because of the immobility of the pore fluids.

The geotechnical implications of undrained gas expansion during unloading are multiple. Undrained gas expansion leads to substantial volume increase. This disrupts the interlocked fabric of the oil sand and thereby reduces its shear strength. Until gas drainage occurs by venting, pore pressures during unloading are higher than in a comparable material that is gas-free, and the available shearing resistance is reduced accordingly. For surface works, gas exsolution can result in heave of excavations which will contribute to increased settlement upon reloading. It can also induce weakening of material within slopes and thereby result in shallow instability. Exfoliation of freshly cut slopes is common and is a major factor affecting efficient dragline operations. For underground works, the expansion of the oil sand around a cavity must be considered in both the design of temporary and permanent support systems. The stand-up time of excavation faces will be affected and the production of exsolved gas must be considered when designing ventilation systems. The exsolution of gas constitutes a major impediment to geotechnical design because it makes undisturbed sampling of at least the oil-rich sands virtually impossible. Dusseault (1980) has recently summarized data that show that oil-rich sands have the greatest potential for expansion and that a reduction in bulk density of about 10% compared with the in situ value determined by geophysical means is not uncommon.

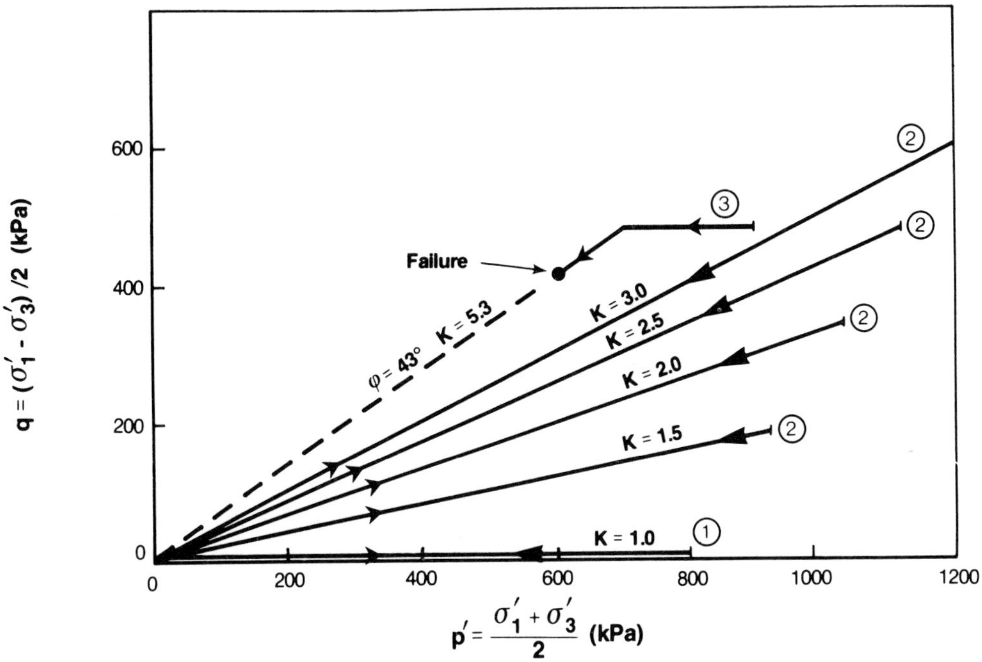

Fig. 55. Stress paths for undrained tests on dense sand with H_2O/CO_2 pore fluid ($Sw_i = 100\%$)

Only two underground excavations in the Alberta oil sands have been described so far and both were completed at relatively shallow depths. The first was a test shaft sunk in 1963 which was abandoned at a depth of 23·5 m because of poor mining methods which did not provide adequate control of seepage. Nevertheless gas was encountered bubbling through the water at the base of the excavation; and below a depth of about 18 m the walls of the shaft deteriorated by progressive slabbing to a depth of 0·3 m or more in less than a few hours (Hardy & Scott, 1978).

The second case was the construction of a short creek diversion tunnel associated with some landslide stabilization works (Chatterji *et al.*, 1979; Harris, Poppen & Morgenstern, 1979). In this instance a 4·4 m dia., 107 m long lined tunnel was constructed through oil-rich sands. While provision was made in the design for considerable swelling of the tunnel face, generally less than 2 cm swell was encountered. It was possible to excavate the tunnel with a point-attack machine and the stand-up time must be rated in days to weeks. Clearly gas exsolution was not a problem and this is attributed to the proximity of the tunnel location to the valley wall. It appears that during the unloading, natural valley formation was sufficiently slow to permit the gas to drain, probably by diffusion, so that little gas existed during the excavation of the tunnel. It is unlikely that these favourable conditions will be encountered at greater depths.

The development of an analytical framework for dealing with the stresses and deformations of gas-saturated oil sand was initiated by Harris & Sobkowicz (1977) who formulated the change of pore pressure and volume due to stress and temperature changes when oil sand is unloaded and gas comes out of solution from both the oil and water phases. These considerations have been incorporated in a series of finite element programs in which the skeletal behaviour of the oil sand is modelled with increasing complexity. A recent example, which treats the soil skeleton in a non-linear manner, includes shear dilation and satisfies strain compatibility between the skeleton and the pore fluid phase (Byrne *et al.*, 1980).

Figure 54 illustrates the application of this analytical capability. A cylindrical shaft in oil sand with radius equal to 2·5 m is treated as a plane problem. The initial stress was 2·5 MPa and the convergence of the shaft wall as the support pressure is relieved is shown. Soil data are specified in Byrne *et al.* (1980). The solid line represents the inward displacement predicted when excavation occurs under undrained conditions and gas venting is prevented. Large inward movements occur when the support pressure drops below about 1·2 MPa. The dashed line indicates the displacements when depressurization due to drainage has occurred to a radius of 5 m and the pore pressure is zero in this zone. The support may be reduced to about 0·2 MPa

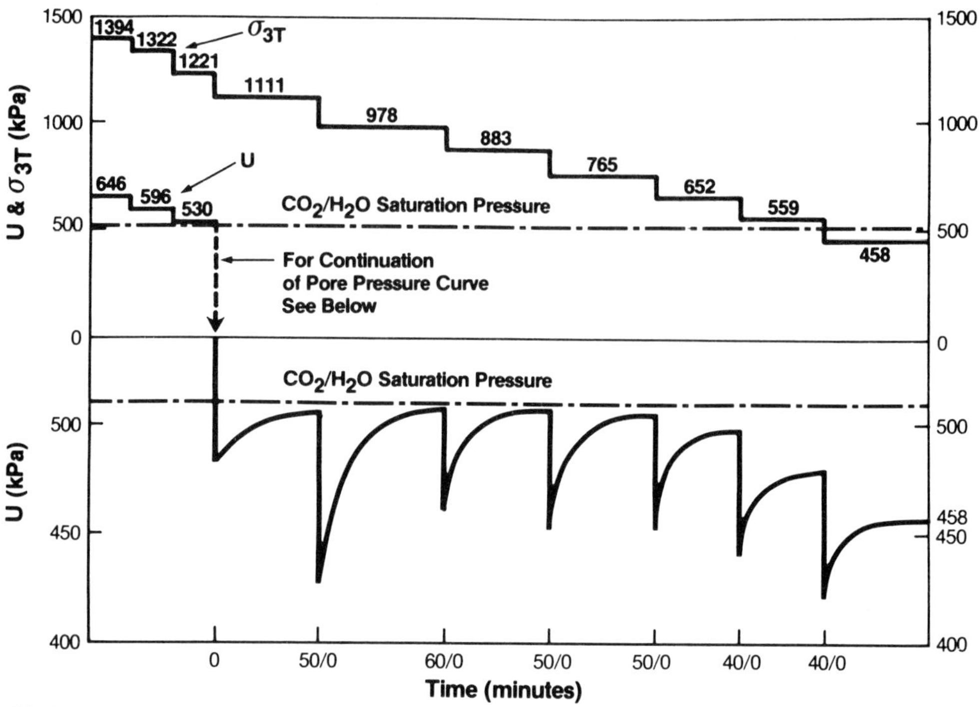

Fig. 56. Isotropic undrained unloading test on dense sand with H_2O/CO_2 pore fluid ($Sw_i = 100\%$)

before large inward movements occur.

Little is known experimentally about the influence of alternate stress paths on the behaviour of gas-saturated porous media and the requirements for gas drainage which are of such practical significance. In order to study these effects a test facility has been assembled in which a pore fluid containing dissolved carbon dioxide can be flooded into sand and the sample then subjected to various tests.

Figure 55 illustrates some of the undrained stress paths imposed on samples of very dense Ottawa sand ($n = 31\%$, $M_v = 2$ to $5 \times 10^{-3}\,\text{MPa}^{-1}$) containing a water–carbon dioxide mixture as the pore fluid. Initial liquid saturations are 100%, so that the CO_2 gas is totally dissolved in the water. The pressure at which gas will just begin to exsolve in the pore fluid is referred to as the liquid–gas saturation pressure. In the tests, time-dependent changes in pore fluid pressure and strain are monitored as total stress changes. Major findings so far are as follows.

(a) For those conditions where the pore fluid pressure remains above the liquid–gas saturation pressure, the soil remains totally liquid-saturated and behaves in a typical undrained fashion. For very dense soils B is slightly less than 1, but for more compressible soils $B = 1$. Behaviour is essentially time-independent.

(b) As soon as the pore fluid pressure decreases below the liquid–gas saturation pressure, gas starts to exsolve and bubbles form in the pore space. This exsolution process is time-dependent and continues until an equilibrium is reached between the liquid and gas pressures and the gas concentrations in the bubble and in the liquid.

(c) The exsolution of gas causes time-dependent pore pressure changes and hence time-dependent changes in effective stresses and strain. However, the stress–strain relations for dense cohesionless soils do not appear to be affected significantly by the presence of small amounts of gas, to gas saturations of about 20%.

(d) For tests on dense cohesionless materials containing large amounts of dissolved gas, the exsolution process proceeded in such a way as to maintain, in the long term, pore fluid pressures nearly equal to the initial liquid–gas saturation pressure. This is illustrated in Fig. 56 which shows the results of an isotropic unloading test.

(e) For stress paths with the minor principal stress decreasing to failure, the stress–strain curves for undrained failure with gas in the pore fluid are almost identical with those of a drained sample with no gas in the pore fluid. However, the rate

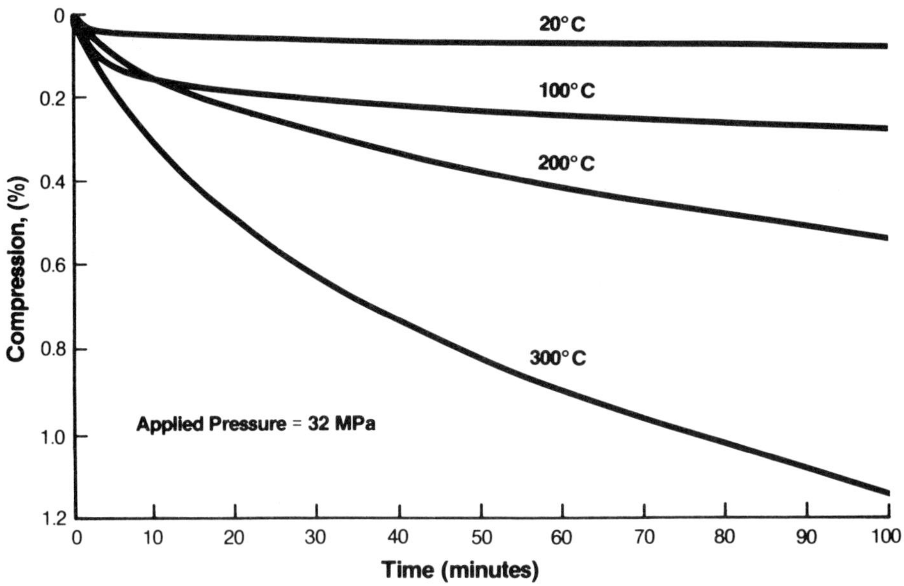

Fig. 57. Compression of Quartz sand under elevated temperature

of failure is governed by the rate of decrease of the minor principal effective stress which is related directly to the rate of gas exsolution.

Interest in the geotechnical behaviour of gas-saturated porous media is not restricted to oil sands. In fact, gas-saturated soils are probably more common than is generally recognized. Okumura (1977) has analysed the implications on strength of dissolved air in deep-sea samples and shown that the expansion of the pore fluid upon isothermal stress release greatly influences the effective stress of the sample in the laboratory. As a result, undrained strength measured in the laboratory can be much less than the in situ strength unless compensation is made for the gas expansion effects. Methane-saturated sediments are particularly common in areas of high rates of recent sedimentation.

An interesting example of a gas-saturated soil was encountered in the vicinity of Montalto di Castro, Italy where a nuclear power plant was under construction.[5] The stratigraphy consists of a 35–40 m layer of sand and gravel overlying about 30 m of sandy clay. The clay in turn overlies a layer of silty sand. The deposits are all Pleistocene. Both the clay and underlying sand are virtually saturated with carbon dioxide. Since carbon dioxide is extremely soluble in water considerable volumes of gas can be seen to exsolve upon sampling. The potential existed for undrained gas exsolution in the clay due to excavation and ground water lowering in the overlying sand layer. This gas exsolution can have an important bearing on the prediction of heave of unloaded areas and subsequent settlement upon reloading.

In situ extraction from oil sands

Geotechnical considerations enter only in a limited way in conventional hydrocarbon reservoir engineering. Subsidence effects in compressible reservoirs are calculated in terms of changes in effective stress. In situ stresses and rock strength enter into the mechanics of hydraulic fracture propagation. However, most conventional processes concerned with fluid injection and withdrawal in hydrocarbon reservoirs are not intimately dependent on the deformation and strength properties of the reservoir soil or rock. While still speculative, geotechnical considerations may play a more significant role in the in situ extraction processes associated with oil sands because of the weakness and deformability of these materials.

The most common process for in situ recovery involves massive injection of steam in order to reduce the viscosity of the bitumen. The efficiency of subsequent withdrawal is much influenced by the permeability and compressibility induced by the massive injection. In addition, it is important to ensure that the injected steam stays within the stratum intended for stimulation. An understanding of the mechanics of the injection process requires knowledge of how hot pressurized fractures extend in a deformable cohensionless

[5] The geotechnical implications of this clay were investigated by the Author in conjunction with D'Appolonia Consulting Engineers, Inc., Brussels.

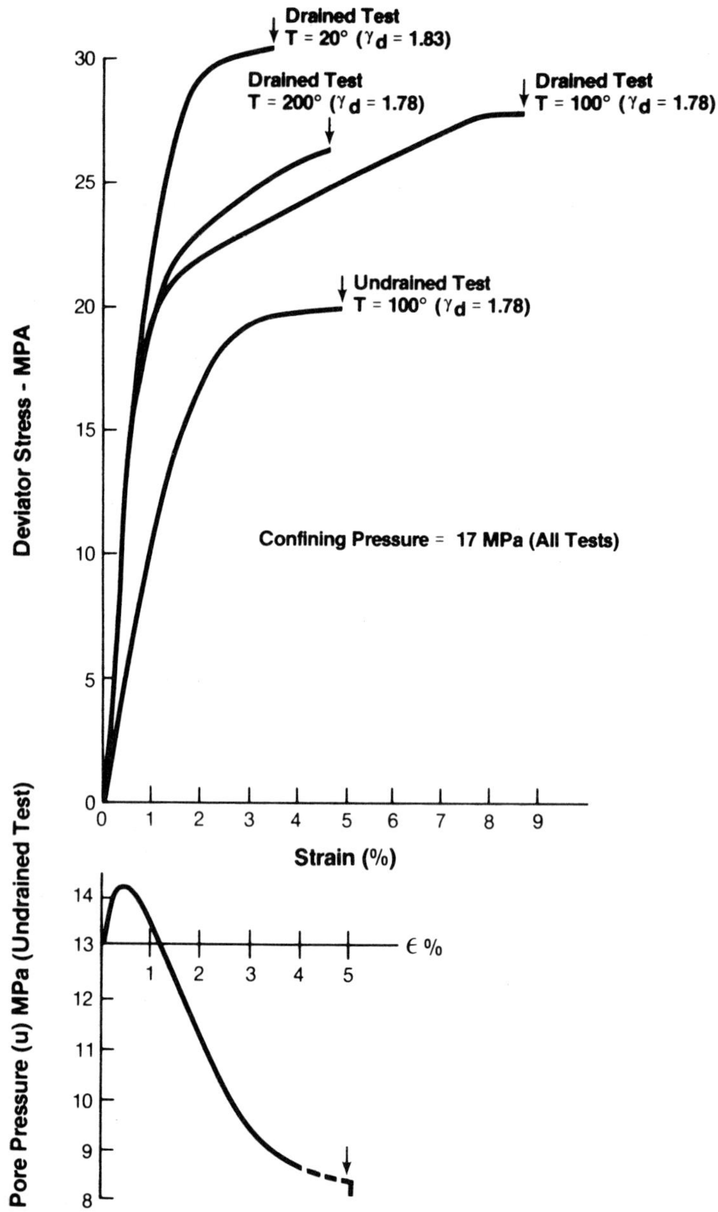

Fig. 58. Temperature effects on strength of dense sand

medium. Undoubtedly fractures can extend by both parting and by shear. Massive injection can also have surface effects and potential heave could be a factor in movement of surface facilities.

The elastic analysis of pressurized fractures (e.g. Hungr & Morgenstern, 1980) may provide an adequate basis for the prediction of far field effects. However, these theories or alternate theories derived from linear fracture mechanics constitute an excessive simplification of the actual process of fracture extension. Studies of the mechanics of fracture extension in cohesionless media are needed and these studies will ultimately have to embrace both fluid and heat transfer processes in order to make a realistic contribution to process simulation.

Heating of oil sands due to injection or in situ combustion raises novel geotechnical considerations. The first has to do with the effect of heat on geotechnical properties. The strength and compressibility of oil sands at elevated temperatures has a bearing on a variety of both short-term and long-term considerations related to underground access and the mechanics of the in situ recovery processes. In order to investigate these properties a test facility has been assembled capable of performing compressibility, triaxial shear and permeability tests at confining stresses to 27 MPa and temperatures to 320 °C. Steam generation facilities allow injection to be simulated and experiments to be undertaken with liquid or vapour back pressures. Only limited test data have been produced so far.

Even conducting experiments on Ottawa sand has revealed major changes in geotechnical properties at elevated temperatures. Figure 57 illustrates the influence that pressure and temperature have on one-dimensional compressibility. The compressibility at room temperature is similar to other data in the literature. A small amount of comminution occurs at high pressures. Repeating the test at 100 °C intervals reveals a dramatic change in compressibility and a marked increase in its time dependence. This is due to weakening of the particles as evidenced by reduction in grain size measured after the test. Preliminary shear strength tests on Ottawa sand also show temperature effects, but the effects are not marked for dry sand.

An equally important aspect of heating oil sands is the change in pore pressures that can arise. If heating is rapid, the expansion of the pore fluid may occur under conditions of impeded drainage, and in the limit conditions might even be totally undrained. As a result pore pressures can increase during heating with the consequences of swelling and a reduction in shearing resistance. This class of problems has already been identified by Campanella & Mitchell (1968), and Mitchell (1976) provides an excellent summary of the interaction between undrained heating and induced pore pressure changes.

Just as the effect of an isothermal total stress change generates a pore pressure reaction expressed in terms of B which is reducible to the amounts and compressibilities of the phases composing the soil, so an undrained temperature change induces a pore pressure change that can be expressed by B_t. This coefficient is reducible in a similar manner to the stress and temperature-dependent volume changes of the components of the soil. Typical values may be deduced from Mitchell (1976). While the effects on pore pressure changes of removing samples from the ground at 5–10 °C and placing them in the laboratory at 20 °C are small, this is probably not the case when the ground is subjected rapidly to temperature changes of 250 °C by the injection of pressurized steam. Figure 58 presents the Author's first measurements of the pore pressure changes in dense Ottawa sand heated to 100 °C and then sheared under undrained conditions. It is evident that the pore pressure reaction to heating is substantial ($B_t \sim 0{\cdot}77$) and its effect on available shear strength is significant.

In order to assess whether significant pore pressures develop it is necessary to evaluate whether heating occurs in an undrained manner. This introduces the class of problems of heat-consolidation. If heating occurs slowly, there will be time for drainage and the ground can expand due to thermally induced strain alone. Hence the magnitude of the pore pressures that arise at a point during heating will depend upon the relation between the rate of increase of the temperature and the propensity for the pore pressures to dissipate at that point. To determine the pore pressures requires coupling the heat transfer problem and consolidation problem through the thermal pore pressure coefficient B_t.

Figure 59 presents results from a simple example intended to illustrate rapid heating of oil sands but neglecting the temperature dependence of permeability and any convective effects. For a step temperature applied to the boundary, the governing solution for heat conduction is known. This can be used as input to a moving boundary problem in the theory of consolidation with pore pressure generation due to local temperature changes. Results have been calculated numerically using an explicit procedure. As might be anticipated, the maximum pore pressure that develops depends upon the ratio of the diffusivity of the medium to its coefficient of consolidation. More complex formulations will be needed to simulate in situ conditions realistically. However, this case does serve to draw attention to the major factors influencing thermally induced pore pressure changes.

The Author has been drawn to the investigation of geotechnical behaviour at elevated temperatures through his involvement in oil sand development, but there is increasing interest in high temperature effects for other reasons. In situ retorting of oil shales and in situ gasification of coal will both utilize underground cavities that must remain stable at elevated temperatures. Underground storage of nuclear waste generates heat and the long-term implications on security of containment provides another reason for interest in elevated temperature studies. Both temperature dependence of pore pressures and shearing resistance also have a bearing on the mechanics of faulting. Sibson (1973) has pointed out that the heating of confined water can reduce the effective stress and thereby facilitate fault movement. More recently, Lachenbruch (1980) has analysed in a comprehensive manner the interaction between fault movement and heat consolidation. His study

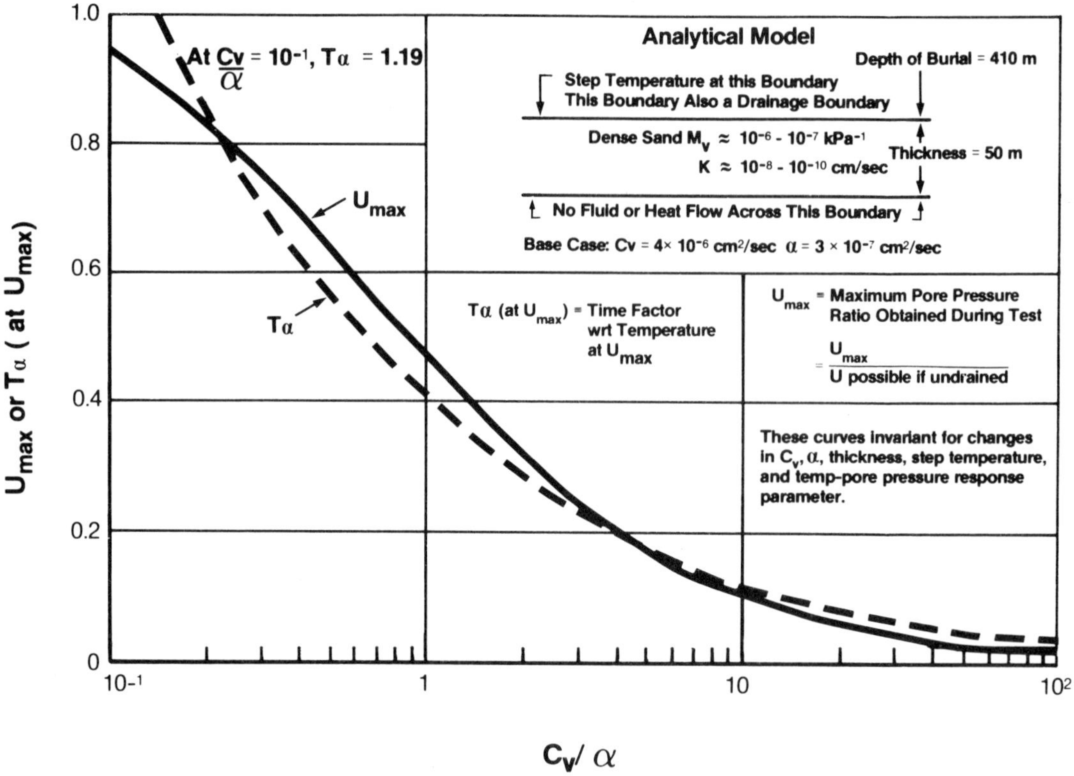

Fig. 59. Heat-consolidation: plot of U_{max} and $T\alpha$ (at U_{max}) against C_v/α

reveals that there are several plausible mechanisms that can dramatically affect frictional resistance during an earthquake, but that present knowledge of the controlling parameters makes it difficult to determine which, if any, play a significant role.

SUMMARY

My selection of examples of geotechnical problems presented by frontier resource development is not intended to be restricted in a geographical sense and I fully recognize that other problems, in particular those associated with recent activities in the North Sea, are equally challenging to the geotechnical community. However, one feature of the problems reviewed that guided my selection is that in each case it has been necessary to reach beyond conventional concepts in order to contribute to their resolution in a rational manner. Moreover, by doing so, the potential of geotechnical engineering is extended to a broader range of activities.

In the case of creep in naturally frozen soils, it is possible to quantify the process occurring in nature in rheological terms, and this is of value both for solving immediate problems in permafrost engineering and for shedding light on the mechanics of some periglacial processes. However, the composite nature of natural permafrost appears to create anomalies when conventional sampling and testing is used to obtain design data, and these procedures need re-evaluation. In the case of frost heave, it has been necessary to absorb certain thermodynamic considerations in order to develop a predictive theory suitable for engineering needs. Both thaw-consolidation and heat-consolidation theories are concerned with the interaction of heat transfer and volume change of soils and both theories have broad application. Concepts from physical chemistry are necessary to account for the behaviour of gas-saturated porous media and the novel problems that they present.

Rankine is honoured in geotechnical engineering primarily for his work on earth pressure theory. However, this was really a very small portion of his contribution to engineering and science and in fact he is far better known for his contributions to thermodynamics and for his studies of the behaviour of gases and fluids. While assembling the material for this lecture it has given me some comfort to believe that illustrating the expanded range of geotechnical concerns draws even more from the work of this great engineer and scientist and thereby enhances his role in geotechnical engineering. We should be encouraged by his example to survey the diversity of geotechnical problems around us.

I have in my personal library a volume of Rankine's miscellaneous scientific papers published posthumously (Rankine, 1881). In a memoir of the author within the volume there is a quotation from an article by Clerk-Maxwell on Rankine.

> 'The scientific career of Rankine was marked by the gradual development of a singular power of bringing the most difficult investigations within the range of elementary methods. In his earlier papers, indeed, he appears as if battling with chaos, as he swims, or sinks, or wades, or creeps, or flies, . . . but he soon begins to pave a broad and beaten way over the dark abyss . . . .'

Geotechnical engineering has important contributions to make to many frontier resource developments. The problems are complex, but one hopes that some future commentator will be able to speak of geotechnical engineering in this area of endeavour as Clerk-Maxwell did of Rankine.

ACKNOWLEDGEMENTS

In assembling the material presented here I have drawn on the efforts of a large number of people not only within the University of Alberta but also associated with us in professional practice. My colleagues at the University of Alberta have always been supportive in every way and we have enjoyed the collaboration of a remarkably talented group of graduate students. The research on creep of a permafrost slope was undertaken by Dr K. W. Savigny who drew on earlier studies by Dr E. C. McRoberts and Dr W. D. Roggensack. Dr J. F. Nixon, Mr L. B. Smith and Dr Roggensack contributed much of the material on thaw-consolidation behaviour. The investigations into frost heave mechanics have been brought to fruition by Dr J.-M. Konrad. Our research into oil sand behaviour has been conducted mainly by Dr M. B. Dusseault and Mr J. C. Sobkowicz.

I would like to thank my colleague Dr J. D. Scott for providing data from our high-temperature test facility for inclusion and Mr Sobkowicz for performing the heat-consolidation calculations. Both Dr W. Roggensack and Dr S. Thomson gave valuable assistance by critically reading drafts of the text but they bear no responsibility for the final version.

Finally, I would like to acknowledge the contribution of Dr R. M. Hardy, past-Dean of Engineering at the University of Alberta. Dr Hardy was the first in a Canadian university to initiate research into permafrost engineering and was first to study the geotechnical behaviour of oil sands. His pioneering efforts made it much easier for those who followed.

REFERENCES

Adam, D. G. & Regensburg, B. O. (1980). Dragline mining at Syncrude. *Proceedings of international mining conference, Calgary*, session 1. Calgary: Alberta Chamber of Resources.

Aitken, G. (1974). *Reduction of frost heave by surcharge stress.* Technical report no. 184. Hanover, New Hampshire: Cold Regions Research and Engineering Laboratory.

Andersland, O. B. & Anderson, D. M. (1978). *Geotechnical engineering for cold regions.* New York: McGraw-Hill.

Anderson, D. M. & Morgenstern, N. R. (1973). Physics, chemistry and mechanics of frozen ground: a review. In *Permafrost: the North American contribution to the 2nd international conference, Yakutsk*, 257–288. Washington: National Academy of Sciences.

Biermans, M. K., Dijkema, K. & de Vries, D. A. (1978). Water movement in porous media towards an ice front. *J. Hydrol.* **37**, 137–148.

Brooker, E. W. (1975). Tar sand mechanics and slope evaluation. *Proc. 10th Can. Rock Mech. Symp.* **1**, 409–446.

Byrne, P. M., Smith, B. L., Grigg, R. F. & Stewart, W. P. (1980). A computer model for stress–strain and deformation analysis of oil sands. *Proceedings of applied oil sands geoscience conference*, Edmonton: University of Alberta. In press.

Campanella, R. G. & Mitchell, J. K. (1968). Influence of temperature variations on soil behaviour. *J. Soil Mech. Fdns Div., Am. Soc. Civ. Engrs* **94**, 709–734.

Chatterji, P. K., Smith, L. B., Insley, A. E. & Sharma, L. (1979). Construction of saline creek tunnel in Athabasca oil sand. *Can. Geotech. Jl* **16**, 90–107.

Crory, F. E. (1973). Settlement associated with the thawing of permafrost. In *Permafrost: the North American contribution to the 2nd international conference, Yakutsk*, 599–607. Washington: National Academy of Sciences.

Demaison, G. J. (1977). Tar sands and super giant oil fields. In *The oil sands of Canada–Venezuela*, Redford, D. A. and Winestock, A. G. (eds), 9–16. Special volume 17. Montreal: Canadian Institute of Mining and Metallurgy.

Devenny, D. W. & Raisbeck, J. M. (1980). Rock mechanics considerations for in-situ development of oil sands. In *Underground rock engineering*, 90–96. Montreal: Canadian Institute for Mining and Metallurgy.

Dusseault, M. B. (1980). Sample disturbance in Athabasca oil sand. *Jl Can. Petrol. Tech.* **19**, 85–92.

Dusseault, M. B. & Morgenstern, N. R. (1978a). Characteristics of natural slopes in the Athabasca oil sands. *Can. Geotech. Jl* **15**, 202–215.

Dusseault, M. B. & Morgenstern, N. R. (1978b). Shear strength of Athabasca oil sands. *Can. Geotech. Jl* **15**, 216–238.

Dusseault, M. B. & Morgenstern, N. R. (1979). Locked sands. *Q. Jl Engng Geol.* **12**, 117–132.

Hardy, R. M. & Hemstock, R. A. (1963). Shearing strength characteristics of Athabasca oil sands. In *Karl A. Clark Volume*, Carrigy, M. A. (ed.), 109–122. Information series no. 45. Edmonton: Research Council of Alberta.

Hardy, R. M. & Scott, J. D. (1978). The 1963 GCOS test shaft. *Proceedings of seminar on underground excav-*

ation in oil sands, paper no. 13. Edmonton: Alberta Oil Sands Technology and Research Authority.

Harris, M. C., Poppen, S. & Morgenstern, N. R. (1979). Tunnels in oil sand. *Jl Can. Petrol. Tech.* **18**, 1–7.

Harris, M. C. & Sobkowicz, J. C. (1977). Engineering behaviour of oil sand. In *The oil sands of Canada–Venezuela*, Redford, D. A., and Winestock, A. G. (eds), 270–281. Special volume 17. Montreal: Canadian Institute of Mining and Metallurgy.

Hoekstra, P. (1969). Water movement and freezing pressures. *Proc. Soil Sci. Soc. Am.* **33**, 512–518.

Hooke, R. L., Dahlin, B. B. & Kauper, M. T. (1972). Creep of ice containing dispersed fine sand. *J. Glaciology* **11**, 327–336.

Horswill, P. & Horton, A. (1976). Cambering and valley bulging in the Gwash Valley at Empingham, Rutland. *Phil. Trans. R. Soc.*, series A, **283**, 427–451.

Hungr, O. & Morgenstern, N. R. (1980). A numerical approach to predicting stresses and displacements around a three-dimensional pressurized fracture. *Int. Jl Rock Mech. Mining Sci.* **17**, 333–338.

Jessberger, H. L. (1970). *Ground frost: a listing and evaluation of more recent literature dealing with the effect of frost on the soil.* Document no. AD 865 128. Springfield, Virginia: National Technical Information Service.

Kay, B. D. & Groenevelt, P. H. (1974). On the interaction of water and heat transport in frozen and unfrozen soils: I. Basic theory; the vapour phase. *Proc. Soil Sci. Soc. Am.* **38**, 395–400.

Konrad, J. M. (1980). *Frost heave mechanics.* PhD thesis, University of Alberta, Edmonton.

Konrad, J. M. & Morgenstern, N. R. (1980). A mechanistic theory of ice lens formation in fine grained soils. *Can. Geotech. Jl* **17**, 473–483.

Lachenbruch, A. H. (1970). *Some estimates of the thermal effects of a heated pipeline in permafrost.* Circular 632. Washington, DC: US Geological Survey.

Lachenbruch, A. H. (1980). Frictional heating, fluid pressure, and the resistance to fault motion. *J. Geophys. Res.* **85**, 6097–6112.

Loch, J. P. G. & Kay, B. D. (1978). Water redistribution in partially frozen, saturated silt under temperature gradients and overburden loads. *Proc. Soil Sci. Soc. Am.* **42**, 400–406.

Mackay, J. R. (1974). Reticulate ice veins in permafrost, northern Canada. *Can. Geotech. Jl* **11**, 230–237.

Mackay, J. R. (1980). The origin of hummocks, western Arctic coast, Canada. *Can. Jl Earth Sci.* **17**, 996–1006.

MacPherson, J. G., Watson, G. H. & Koropatrick, A. (1970). Dykes on permafrost foundations in northern Manitoba. *Can. Geotech. Jl* **7**, 356–364.

McRoberts, E. C. (1973). *Stability of slopes in permafrost.* PhD thesis, University of Alberta, Edmonton.

McRoberts, E. C. (1975). Some aspects of a simple secondary creep model for deformations in permafrost slopes. *Can. Geotech. Jl* **12**, 98–105.

McRoberts, E. C., Fletcher, E. B. & Nixon, J. F. (1978). Thaw consolidation effects in degrading permafrost. *Proc. 3rd Int. Conf. Permafrost, Edmonton* **1**, 693–699.

McRoberts, E. C., Law, T. C. & Murray, T. K. (1978). Creep tests on undisturbed ice-rich silt. *Proc. 3rd Int. Conf. Permafrost, Edmonton* **1**, 539–545.

McRoberts, E. C. & Morgenstern, N. R. (1974a). The stability of thawing slopes. *Can. Geotech. Jl* **11**, 447–469.

McRoberts, E. C. & Morgenstern, N. R. (1974b). Stability of slopes in frozen soil, Mackenzie Valley, NWT. *Can. Geotech. Jl* **11**, 554–573.

Mageau, D. & Morgenstern, N. R. (1979). Observations on moisture migration in frozen soils. *Can. Geotech. Jl* **17**, 54–60.

Miller, R. D. (1972). Freezing and heaving of saturated and unsaturated soils. *Highw. Res. Rec.*, no. 393, 1–11.

Mitchell, J. K. (1976). *Fundamentals of soil behaviour.* New York: J. Wiley.

Mitchell, R. F. & Goodman, M. A. (1978). Permafrost thaw-subsidence casing design. *J. Petrol. Tech.* **30**, 455–460.

Mittal, H. K. & Hardy, R. M. (1977). Geotechnical aspects of a tar sand tailings dyke. *Proceedings of conference on geotechnical practice for disposal of solid waste materials*, 327–347. New York: American Society of Civil Engineers.

Morgenstern, N. R. (1967). Shear strength of stiff clay. *Proceedings of geotechnical conference, Oslo* **2**, 59–72.

Morgenstern, N. R. & Nixon, J. F. (1971). One-dimensional consolidation of thawing soils. *Can. Geotech. Jl* **8**, 558–565.

Morgenstern, N. R. & Nixon, J. F. (1975). An analysis of the performance of a warm-oil pipeline in permafrost, Inuvik, NWT. *Can. Geotech. Jl* **12**, 199–208.

Morgenstern, N. R., Roggensack, W. D. & Weaver, J. S. (1980). The behaviour of friction piles in ice and ice-rich soils. *Can. Geotech. Jl* **17**, 405–415.

Morgenstern, N. R. & Smith, L. B. (1973). Thaw-consolidation tests on remoulded clays. *Can. Geotech. Jl* **10**, 25–40.

Nixon, J. F. (1978). First Canadian geotechnical colloquium: foundation design approaches in permafrost areas. *Can. Geotech. Jl* **15**, 96–112.

Nixon, J. F. & Ladanyi, B. (1978). Thaw consolidation. In *Geotechnical engineering for cold regions*, Andersland, O. B. and Anderson M. (eds), chapter 4. New York: McGraw-Hill.

Nixon, J. F. & Morgenstern, N. R. (1973). The residual stress in thawing soils. *Can. Geotech. Jl* **10**, 571–580.

Nixon, J. F. & Morgenstern, N. R. (1974). Thaw-consolidation tests on undisturbed fine-grained permafrost. *Can. Geotech. Jl* **11**, 202–214.

Northern Engineering Service Ltd, Calgary (1975). *Interim report on results from frost effects study.* Unpublished.

Okumura, T. (1977). Stress change of soil sample taken from sea floor. *Proc. 9th Int. Conf. Soil Mech., Tokyo.* Soil sampling, speciality session 2, 141–146.

Palmer, A. C. (1972). *Thawing and differential settlement close to oil wells through permafrost.* Division of Engineering report ARPA E-83. Providence, RI: Brown University.

Penner, E. & Goodrich, L. E. (1980). Location of segregated ice in frost susceptible soil. *Proc. 2nd Int. Symp. Ground Freezing, Trondheim*, 626–639.

Pufahl, D. (1976). *The stability of thawing slopes.* PhD thesis, University of Alberta, Edmonton.

Pufahl, D. E. & Morgenstern, N. R. (1979). Stabilization of planar landslides in permafrost. *Can. Geotech. Jl* **16**, 734–747.

Radd, F. J. & Oertle, D. H. (1973). Experimental pressure studies of frost heave mechanisms and the growth-fusion behaviour of ice. In *Permafrost: the North*

American contribution to the 2nd international conference, Yakutsk, 257–288. Washington: National Academy of Sciences.

Rankine, W. J. M. (1881). *Miscellaneous scientific papers.* London: Charles Griffin.

Roggensack, W. D. (1977). *Geotechnical properties of fine-grained permafrost soils.* PhD thesis, University of Alberta, Edmonton.

Roggensack, W. D. (1979). Techniques for core drilling in frozen soils. *Proceedings of symposium on permafrost field methods and permafrost geophysics*, technical memorandum no. 124. Ottawa: Associate Committee for Geotechnical Research, National Research Council of Canada.

Savigny, K. W. (1980). *In situ analysis of naturally occurring creep in ice-rich permafrost soil.* PhD thesis, University of Alberta, Edmonton.

Sego, D. C. (1980). *Deformation of ice under low stresses.* PhD thesis, University of Alberta, Edmonton.

Sibson, R. H. (1973). Interactions between temperature and pore-fluid pressure during earthquake faulting and a mechanism for partial or total stress relief. *Nature, Lond.* **243**, 66–68.

Skempton, A. W. (1970). The consolidation of clays by gravitational compaction. *Q. Jl Geol. Soc. Lond.* **125**, 373–411.

Skempton, A. W. & Weeks, A. G. (1976). The Quaternary history of the Lower Greensand escarpment and Weald clay vale near Sevenoaks, Kent. *Phil. Trans. R. Soc.*, Series A, **283**, 493–525.

Slusarchuk, W., Clark, J., Nixon, J. F., Morgenstern, N. R. & Gaskin, P. (1978). Field test results of a chilled pipeline buried in unfrozen ground. *Proc. 3rd Int. Conf. Permafrost, Edmonton*, 878–890.

Slusarchuk, W. A., Watson, G. H. & Speer, T. L. (1973). Instrumentation around a warm oil pipeline buried in permafrost. *Can. Geotech. J.*, **10**, 227–245.

Speer, T. L., Watson, G. H. & Rowley, R. K. (1973). Effects of ground-ice variability and resulting thaw settlements on buried oil pipelines. In *Permafrost: the North American contribution to the 2nd international conference, Yakutsk*, 746–752. Washington: National Academy of Sciences.

Supple, M. A. (1980). Mining with bucket-wheel excavators. *Proceedings of international mining conference, Calgary*, Session 1. Calgary: Alberta Chamber of Resources.

Tsytovitch, N. A. (1975). *The mechanics of frozen ground.* New York: McGraw-Hill.

Vaughan, P. R. (1976). The deformations in the Empingham Valley slope. *Phil. Trans. R. Soc.*, Series A, **283**, 451–461.

Vignes, M. & Dijkema, K. (1974). A model for the freezing of water in a dispersed medium. *J. Colloid Interface Sci.*, **49**, 165–172.

Vyalov, S. S., Dokuchayev, V. V. & Sheynkman, D. R. (1980). *Ground ice and ice-rich ground as structure foundations.* Draft translation 737. Hanover, New Hampshire: Cold Regions Research and Engineering Laboratory.

Watson, G. H., Rowley, R. K. & Slusarchuk, W. A. (1973). Performance of a warm oil pipeline buried in permafrost. In *Permafrost: the North American contribution to the 2nd International conference, Yakutsk*, 759–766. Washington: National Academy of Sciences.

Zaretskii, Y. K. (1968). Calculations of the settlement of thawing soil. *Soil Mech. Fdn. Engng.*, No. 3, 151–155.

VOTE OF THANKS

In proposing a vote of thanks to Professor Morgenstern, Dr A. C. Meigh said:

'There are always interfaces in engineering. In our everyday geotechnical problems we have an interface between two disciplines—geology on the one hand, and soil and rock mechanics on the other. We frequently complain that some geologists and engineers are unable or unwilling to cross the boundary. It is clear that we need have no such complaint in the case of Professor Morgenstern. He always views engineering within its geological context.

'Again, tonight, he has put all his work properly into its geological framework. But he has done much more than that; he has straddled other boundaries. To investigate frozen soil problems he has had to consider thermodynamics and heat conduction in soils. In connection with the oil sands he has had to face the problems of dissolved gases and the effects of their coming out of solution.

'I am sure that we have all been impressed by both the scale and complexity of the problems which have been described to us and the practical difficulties which have accompanied their resolution. What is also impressive is that Morgenstern and his colleagues have focussed their attention, and their research efforts, on the major problems confronting the community within which they live, to the benefit not only of that community but others elsewhere. Surely this is the hallmark of a centre of engineering excellence. Furthermore they have tackled these problems in a comprehensive way, and have faced up to the necessity of developing new techniques in the laboratory and in the field, and of developing new analytical concepts.

'We have enjoyed a most stimulating Rankine Lecture. We have been shown dramatically that geotechnical problems cannot always be solved by conventional geotechnics. It is with the greatest of pleasure that I now propose a vote of thanks to Professor Morgenstern.'

The vote of thanks was accorded with acclamation.

The Rankine Lecture

The twenty-second Rankine Lecture of the British Geotechnical Society was given by Dr D. J. Henkel at Imperial College of Science and Technology, London, on 3 March 1982. The following introduction was given by Professor C. P. Wroth.

David John Henkel was born and brought up in Southern Rhodesia, and he obtained a BSc degree at the University of Natal in 1941. Following four years' war service in the Royal Corps of Signals, he was appointed Head of the Soil Mechanics Section of the National Building Institute in Pretoria. He worked with the late Professor Jennings on the problem of expansive clay soils and the structural damage caused to houses by inadequate foundations.

In 1949 Dr Henkel made a decision which has been to the lasting benefit of the development of soil mechanics in the UK. This was a result of his concern to leave the South African scene because of his far-sighted doubts about the long-term future there, allied with his wish to enter the mainstream of soil mechanics research. He joined the team being established at Imperial College by Dr Skempton (as he then was) as a lecturer in civil engineering. He became a senior lecturer and stayed there for 14 singularly productive and influential years.

In 1963, in a spirit of adventure and with a desire to do his bit for the developing countries, he accepted the appointment of Professor of Soil Mechanics at the new Indian Institute of Technology in New Delhi, with the intention of building up a centre of geotechnical expertise in India. For reasons beyond his control, this hope was not fulfilled and he moved after two years to Cornell University to succeed Professor Broms as Professor of Civil Engineering and Head of the Department of Geotechnical Engineering there. He held this post for five years until, in 1970, he returned to London as a full-time consultant in the geotechnical group in Ove Arup & Partners. In 1977 he was appointed a Director.

In collecting my thoughts for this introduction, I have asked a number of people, who have had close associations with David Henkel, what they consider to have been his main contribution to the geotechnical community. Without exception the feature that has been singled out has been his role as an educator. Throughout his working career he has influenced many people by his teaching, by his example and by demanding the highest standards from those with whom he works. Much of this role as educator has naturally taken place during his 21 years of university teaching, but he has continued to educate in the widest sense during the past 12 years in full-time practice.

I believe that his success as a teacher is due to special characteristics. In the first place, his style of teaching is not solely in a formal, didactic manner, but rather one of prompting his students, his juniors, his colleagues and even his adversaries into learning for themselves by getting them to ask—and then to answer— the right questions.

Second, Henkel has a very real ability to think clearly and in a simple and penetrating manner. He has a particular knack of unravelling what is the essential core of a problem facing a geotechnical engineer, be it one of soil behaviour, of geology, of hydrology, of chemistry, of basic mechanics—or even of personal relationships or of politics. It is this ability to get to the heart of a problem, and to strip it of all complex irrelevances that is at once so striking, so successful—and at times so humbling!

He has made many specific technical contributions to our discipline, the most notable being in his time at Imperial College. He played a major part in the early 1950s in developing with Professor Bishop the triaxial test and collaborating on the classic book. *The measurement of soil properties in the triaxial test*, which has become a standard reference in any reputable soil testing laboratory. He took a key role in directing and interpreting the remarkable series of tests on Weald clay and London clay, conducted by a sequence of research students and enshrined in their PhD theses. These tests provided the first comprehensive and consistent set of high quality data of the effective stress-strain properties of clay. They set the standard for the conduct of experimental research, and they have been an invaluable data bank still referred to today. Regarding the interpretation of these data, he was involved with Drucker and Gibson in the first serious attempts to apply plasticity theory to the stress–strain behaviour of clays.

At the same time that this basic research was being undertaken at Imperial College, the staff there were providing important advice to consulting civil engineers who had not received any basic education in soil mechanics and foundation engineering. This advice was far from being routine, and some significant developments took place. One notable achievement was the work that Henkel did in conjunction with Skempton on the back-analysis of landslips in overconsolidated clays, which showed conclusively that the failures could only be explained rationally in terms of effective stresses, and by taking the long-term value of the cohesive component of strength as zero.

I must inject a personal anecdote into this public occasion, to illustrate the endearing characteristics of our lecturer. When I was contemplating taking sabbatical leave in 1967–68, my first choice was to go to Cornell University so that I could experience Henkel's special talents for a whole year. I wrote a long letter to him outlining my plans and seeking some financial support. The reply I received epitomized Henkel's way of doing business: swiftly, decisively and briefly. The letter came by return of post and read: 'Dear Peter, Yes, do come. I am sure we can work something out. Yours sincerely, David.' I had an enthralling year and my education was greatly advanced: Henkel persuaded me to take two excellent graduate courses in engineering geology and air photo interpretation, which I would not have done without his firm advice. I owe him a great debt of gratitude.

David Henkel has had an equally remarkable impact in practice during his time with Arups. He has played a central role in tackling a wide variety of geotechnical problems. These have included deep excavations in Hong Kong for the Mass Transit System, with particular regard for the performance of the slurry trench diaphragm walls in the decomposed granite, the slope stability problems of Hong Kong, deep excavations in London for the Barbican development and the British Library, the Dubai dry dock and the Pompidou Arts Centre in Paris. So the list could go on.

Henkel has always relished intellectual challenges, and I feel sure he has responded magnificently to the challenge of giving a Rankine Lecture. It is with very real pleasure and keen anticipation that I ask Dr Henkel to give his lecture.

Dr D. J. Henkel

HENKEL, D. J. (1982). *Géotechnique* **32**, No. 3, 175–194

Geology, geomorphology and geotechnics

D. J. HENKEL*

The importance of collaboration between geologists and geotechnical engineers is emphasized and the common interest in geomorphology is suggested as a useful link to enable both the geological engineering skills to be mobilized. The role of geomorphology in the understanding of soil movements in the Gulf of Mexico during hurricanes is discussed. Attention is drawn to problems of tropical weathering and changes in soil chemistry which need further study. Some of the problems associated with groundwater lowering in an area underlain by dolomite are described together with the effects on stability of minor changes in surface drainage of an inclined rock layer.

L'article souligne l'importance d'une collaboration plus étroite entre les géologues et les ingénieurs géotechniciens, afin que leurs études combinées puissent améliorer simultanement leurs deux disciplines géologie et géotechnique. Puis est discuté le rôle joué par la morphologie dans la compréhension des mouvements du sol pendant les ouragans dans le Golfe du Mexique. Le besoin existe d'une étude approfondie des problèmes causés par la dégradation dans les zones tropicales et des changements dans la chimie du sol. Finalement l'article décrit quelques-uns des problèmes posés par l'abaissement de l'eau souterraine dans une zone sousjacente de dolomie et discute les effets sur la stabilité de changements de faible importance dans le drainage superficiel d'une couche de roche inclinée.

INTRODUCTION

The subject of my lecture this evening reflects my experience over the past 30 years that engineers and geologists have not yet learned to communicate efficiently with each other. We still do not always ensure that essential geological knowledge and experience is applied to the design and construction of projects.

We still come across problems in construction which could and should have been foreseen at an early stage in the design process. Part of the problem arises from the excessively obscure jargon too often used by geologists and part is due to the fact that the engineer may not know what the geologist has to offer. In addition, both are often unclear about their respective roles. I believe that engineers and geologists need to clarify their respective functions and use of language so that they can work together in a more productive manner. The problem is not new. It was considered by Peck in 1973 and by Legget in the 1977 Terzaghi Lecture. I hope my lecture will promote more efficient communication.

Geologists have for many years recognized that they have an important role to play in the construction of civil engineering works. History provides many examples of the outstanding contribution of geologists to the art of civil engineering, particularly in the fields of dam and tunnel construction.

As long ago as 1801, William Smith suggested that a book he proposed to publish, but never did, would provide geological information to enable the canal engineer 'to choose his stratum, find the most appropriate materials, avoid slippery ground, or remedy the evil' (Sheppard, 1917). We still from time to time encounter slippery ground and on some sites come across the evil which we have to remedy. The problems do not seem to have changed much over the past 180 years.

The straightforward and unambiguous role of the geologist in civil engineering became confused when the term 'engineering geology' was introduced into the geological vocabulary. There have been so many conflicting definitions of that term that even today I am not sure what it means.

In 1961 Terzaghi presented a paper entitled 'Engineering geology on the job and in the classroom' to the Boston Society of Civil Engineers. The term 'engineering geology' appeared to have originated as the name of a course of elementary geology taught to civil engineering students. The discussion on his paper produced a wide spectrum of opinion ranging from the idea that the engineering geologist should fulfil the role of both engineer and geologist to the more rational view that engineering geology was geology and no different from any other branch of applied geology. It was also suggested by Dolmage (1962) that, because a little knowledge was a dangerous thing, it might be easier if the engineer knew nothing about geology and the geologist knew nothing about engineering.

Terzaghi was firmly against the engineering geologist assuming any of the responsibilities of the engineer and drew attention to the writings of

* Ove Arup & Partners.

Berkey (1929). Berkey, a geologist by profession, defined the geologist's role as follows: 'It is his duty to discover, warn, explain without assuming the particular responsibility of the engineer who has to design the structure and determine how to meet all the conditions presented and stand forth as the man responsible for the project.' I believe that the role of the geologist has remained unchanged and that his duty is still to discover, warn and explain.

In spite of all the discussion, the confusion about engineering geology remained, and, in an attempt to resolve the problem at a meeting of the engineering Group of the Geological Society in 1970, Professor Dearman (1971) gave the following definition: 'Engineering Geology is the science or discipline of geology applied to Civil Engineering, particularly as applied to the design construction and performance aspects of engineering structures in and on the ground. The extremes of the subject merge into the disciplines of Soil Mechanics, Rock Mechanics and Materials Science and merge also into some aspects of the extractive industries including quarrying, opencast mining and deep mining.' Dearman also made it clear that engineering geology was not a special kind of geology but covered the whole spectrum of the science.

This was a good, clear definition of the function of engineering geology—very similar to that adopted by the Association of Engineering Geologists. Definitions were concerned with the areas of civil engineering activity in which the discipline of geology should be applied but did not face the central question of how the geological involvement was to be achieved.

Ten years later the Engineering Group of the Geological Society held a meeting to discuss the question 'Should engineering geology be taught and if so how?' The discussion produced no common viewpoint but the teaching of engineering subjects to geologists was suggested as a step in the process of teaching engineering geology to geologists. There was, however, still confusion over what the engineering geologist needs to be able to do. In my view engineering geology has its roots in the field and can only be learnt by painstaking field observations of how a site works.

The way to clarify the situation is to leave the arguments of the classroom and the lecture theatre and look at what is needed from the geologist to enable the geotechnical engineer to define and solve his design and construction problems at a particular site.

The geotechnical engineer needs answers to the following questions.

(*a*) What soils and rocks are there on the site, how have they been formed and what are their

Fig. 1. Sky and water (M. C. Escher)

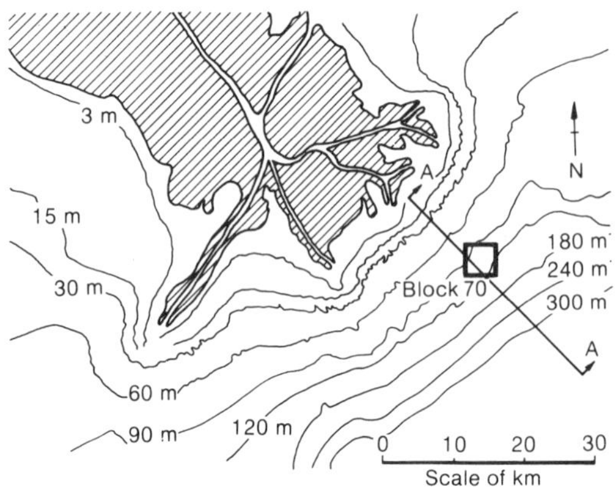

Fig. 2. Birdfoot Delta; contours indicate water depth (Shepard, 1955)

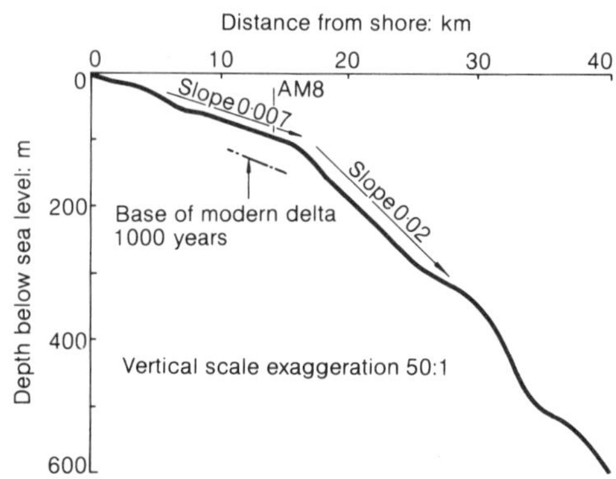

Fig. 3. Section AA, 1940 data

properties?
(*b*) What is the relationship between the shape and form of the site and the geological processes at work?
(*c*) How will the proposed engineering works change the geomorphological environment and what will be the consequences?

The skills and knowledge needed to answer all these questions cover a wide range of the subjects included in the science of geology and indicate the wide background needed by any geologist who wishes to practise in the professional field of engineering geology?

A study of these questions indicates that they are also a partial prescription for the often neglected branch of physical geology known as geomorphology: the study of the origin, evolution and shape of the earth's surface. As well as considering the present land form, it is necessary to take account of earlier land forms which might be buried beneath the present land surface.

If design problems are approached in the light of three basic questions—What is there? Why does it have its present form? What will happen if any of the environmental factors are changed?—a rational framework for the integration of geotechnical and geological skills can be provided. These questions fit in well with Berkey's ideas of discovering, warning and explaining. If the engineering geologist is asked these specific questions rather than asked to produce a geological report both he and the civil engineer will understand more clearly their roles in the design and construction process.

The interface between the separate disciplines of geology and geotechnical engineering is epitomized by the remarkable drawing by Escher entitled 'Sky and water' (Fig. 1). The birds and the fishes retain their separate identities away from the geomorphological boundary between sky and water but, at the interface, they are indistinguishable. In order to understand the nature of the air–water interface we need to view it from above as well as from below. We need the input from both the birds and the fishes. There will, of course, be maverick flying fish and diving birds that can exist fleetingly in another medium but, in the end, they have to return to their native element.

At this geomorphological interface we need to abandon the complex and often unnecessary jargon of geology and geotechnics and communicate in common words to be found in contemporary English dictionaries. The geomorphological approach is particularly important when we are working away from the particular geotechnical conditions with which we are familiar. The engineer's work extends from the permafrost of the polar regions, through the temperate zones and the baking deserts to the tropical rain forest. In all these diverse conditions we need to be aware of the geomorphological processes at work.

Over the years I have been involved in a wide variety of construction projects and those which have proved to be most demanding and stimulating and have contributed most to my education have always been associated with the need to bring together geology, geomorphology and geotechnical engineering. In order to emphasize the prime importance of the interplay between geomorphology and geotechnics I now describe a number of projects which require a multidisciplinary approach so that the engineering problems are understood.

THE MISSISSIPPI DELTA

During the early 1960s there were a number of breakages of offshore oil pipelines in the Mississippi Delta which were associated with the major hurricanes that had swept across the delta, the most important being Carla in 1961, Hilda in 1964 and Betsy in 1965. In addition flare pile had been destroyed during Carla and a small well jacket was lost during Betsy.

In 1967 the Shell Oil Company was planning to install production platforms in the area known as South Pass Block 70 and I became involved with the Shell Development Company in considerations of the geotechnical problems on the site.

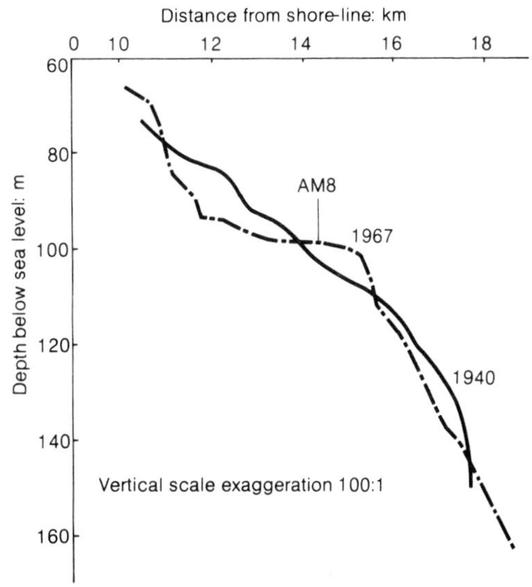

Fig. 4. Section AA, comparison of 1940 and 1967 data (Bea & Arnold, 1973)

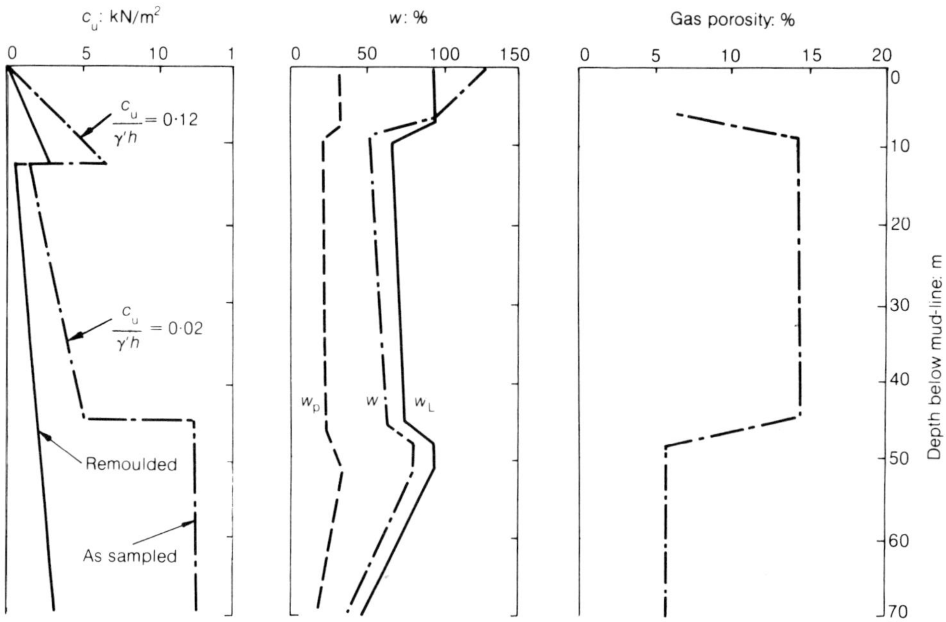

Fig. 5. **Soil properties measured in borehole AM 8 (Bea & Audibert, 1980)**

The Mississippi River is one of the world's largest rivers and carries an enormous sediment load to the sea every year. Shepard (1955) has shown that between 1870 and 1940 the delta advanced into the Mexican Gulf by between 1 km and 3 km, giving an average distance rate of about 30 m/year.

A plan of the Birdfoot Delta area of the Gulf is shown in Fig. 2. The contours of water depth extend to 300 m below sea level. Below 120 m the contours are smooth and evenly spaced but in the shallower waters the complex nature of the contours is apparent. Shepard (1955) considered that this complexity was the result of a series of underwater landslides and his geological interpretation was confirmed by Terzaghi in 1956. Terzaghi showed that the large excess pore-water pressures associated with this high rate of deposition were consistent with slides on the flat delta slopes.

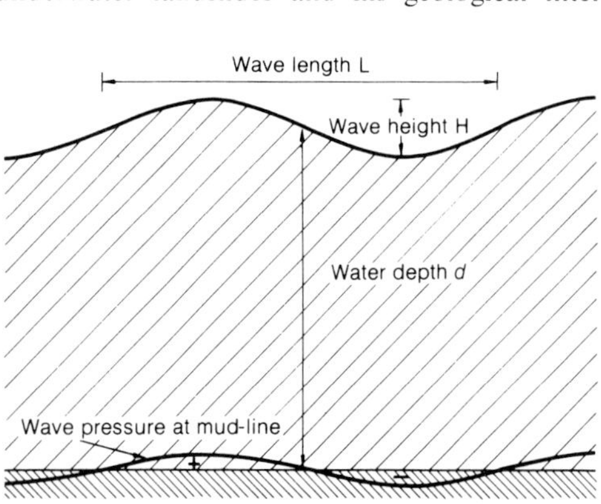

Fig. 6. Effect of waves on mud-line pressures

A section through the delta, on line AA in Fig. 2, based on the United States Coastal and Geodetic Survey of 1940, is shown in Fig. 3. The significant features of the section are the very flat sea bed slopes down to a depth of about 100 m and the abrupt change at this depth from a slope of 0·007 rad to one of 0·02 rad. It has been estimated that the base of the modern delta lies at a depth of about 50 m below the mud-line and that this level represents about 1000 years B.P. (Bea & Audibert, 1980).

A further topographic survey of the delta floor in the vicinity of Block 70 was carried out by Shell in 1967. The two surveys are compared in Fig. 4, again on the section line AA. Even allowing for possible navigational inaccuracies between the 1940 and 1967 surveys there is clear evidence of significant changes in underwater topography. The bulge at about 17 km from the shore-line, shown on the 1940 section, had disappeared by 1967 and there was a substantial reduction in elevation between 11 km and 14 km from the shore-line. A new bulge had developed 15 km from the shore-line.

Borehole AM 8 (Fig. 4) was put down for Shell in 1967; the results of soil tests on the samples are

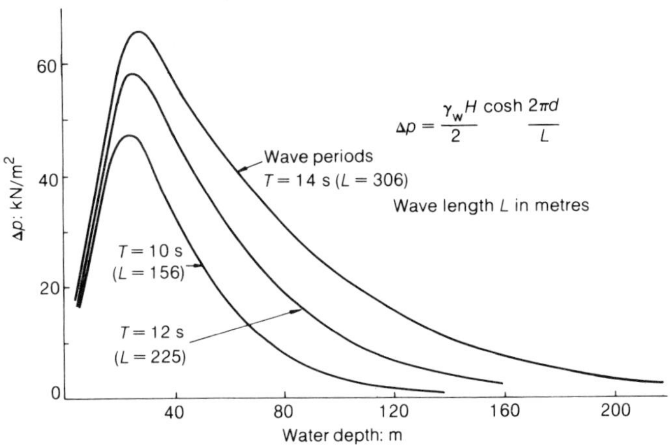

Fig. 7. Wave pressures at mud-line for 20 m wave height

shown in Fig. 5 (Bea & Audibert, 1980). The variation in undrained shear strength with depth is also shown. The major change in strength, which is at a depth of about 45 m, has been identified as the base of the modern delta. Above this elevation the shear strength depth profile is divided into an upper strong crust, with a ratio of $c_u/\gamma' h$ of 0·12, which extends to a depth of about 12 m, and a lower zone, which has a value of $c_u/\gamma' h$ of 0·02 and extends to the base of the modern delta. The value of 0·02 for $c_u/\gamma' h$ is very low and reflects the fact that below the crust the clays are underconsolidated and that there are very large excess pore water pressures. Prior & Suhayda (1979) report cases in other parts of the delta where only about 2% of the submerged overburden pressure is carried by the effective stresses in the clays.

The relationships between the Atterberg limits and the natural water contents are normal for the recently sedimented clays, but an odd feature is the high gas porosities found between depths of 12 m and 45 m.

It can be shown that for gentle slopes equilibrium under gravity forces requires that $c_u/\gamma' h$ is equal to β, the slope angle in radians. The general slope angles in Block 70, on the 1940 section, were about 0·007 rad, while the minimum value of $c_u/\gamma' h$ in borehole AM 8 was 0·02. The changes between 1940 and 1967 could not therefore be explained in terms of gravity slides as the factor of safety against gravity sliding was about 3.

The earlier evidence of the association of pipeline breaks with storms prompted the attempt to find a possible connection between storm waves and sea bottom instability.

When a wave passes over a point on the sea bed there is an increase in pressure beneath the crest of the wave and a decrease in pressure beneath the trough of the wave as shown in Fig. 6. The pressure on the sea bed depends on wave height, wave length and water depth. The real problem is extremely complicated but, as is often the case in engineering, a simplification of the problem, to one in which an analytical solution can be obtained, throws light on the mechanisms at work.

If it is assumed that a sinusoidal wave is travelling across a rigid sea bed, the pressure changes on the sea bed may be calculated easily. The pressure change or wave pressure Δp is given by $\Delta p = (\gamma_w H/2)\cosh(2\pi d/L)$ where γ_w is the unit weight of sea water, H is the wave height, d is the water depth and L is the wave length.

Storm waves with a height of 20 m are not uncommon in the Gulf and as these waves move in towards the shore their height and wavelength are influenced by the water depth. When allowance is made for these factors the wave pressures, as a 20 m high wave moves from deep water into shallow water, change as shown in Fig. 7. Longer waves have a greater influence on wave pressure; the maximum wave pressures occur in water depths of 20–30 m. These maximum wave pressures correspond with the most complex underwater contours and suggest that there is a causative link.

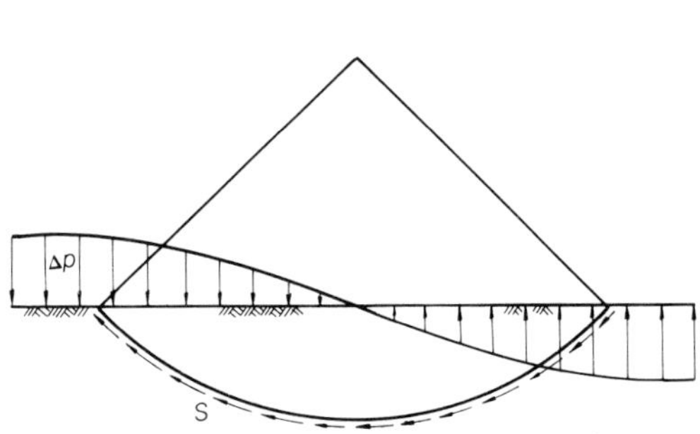

Fig. 8. Limit equilibrium model for stability

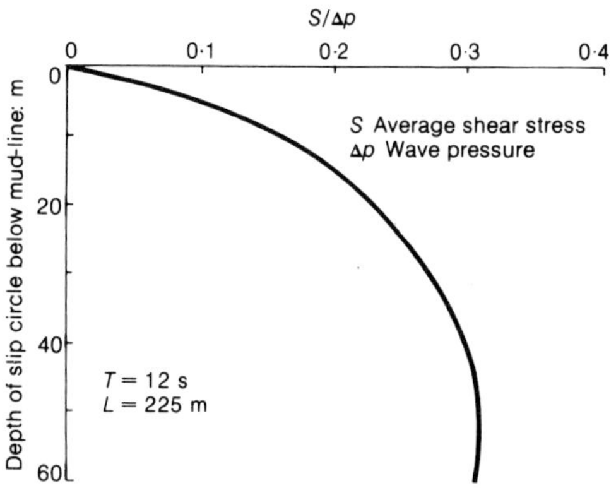

Fig. 9. Ratio of average shear stress to maximum wave pressure

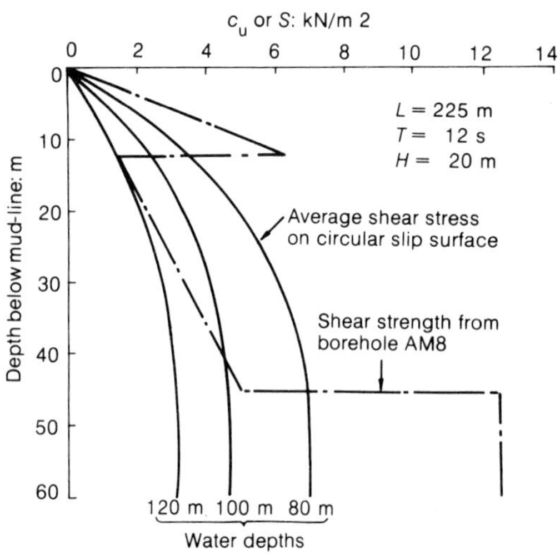

Fig. 10. Variation of average shear stress with depth of slip circle

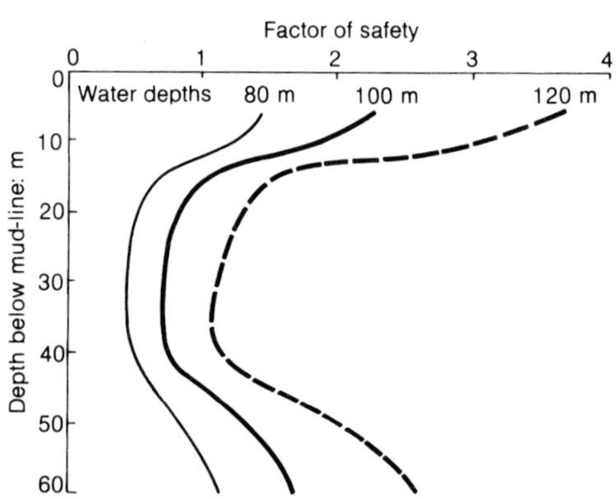

Fig. 11. Variation of factor of safety with depth

The stability of the sea bed may be investigated in a simple way by considering a circular arc failure surface and a sinusoidal wave pressure loading as shown in Fig. 8. For any depth of slip circle below the sea bed, the relationship between the average shear stress on the circular surface and the wave pressure Δp can be calculated. The result of calculations for a wave with a period of 12 s and length of 225 m is shown in Fig. 9. The maximum average shear stress is about 0·3 times the wave pressure and occurs for a depth of slip surface of about 50 m below the mud-line or at about a quarter of the wave length.

In order to compare the shear stresses imposed by the wave and gravity forces with the shear strength of the sediments the 20 m high wave with a period of 12 s and length of 225 m is again used. The ground slope is taken as 0·007 rad. The shear stresses induced in the clay by the wave and gravity forces are plotted in Fig. 10 against the depth of penetration of the slip surface below the mud-line. The shear strengths measured in borehole AM 8 are included and the results for water depths of 80 m, 100 m and 120 m are also shown.

It is difficult to compare the shear stresses and the shear strength directly in Fig. 10 as one needs to compare the average shear strength on the slip surface with the average induced shear stresses. This has been done and the resulting factors of safety are plotted in Fig. 11 against depth for the three water depths. Within the limits of the simplifying assumptions that have been made, it can be seen that, in water depths of less than 100 m, shear failure can be induced by the passage of 20 m high waves with a wavelength of 225 metres.

This simple analysis is concerned with the statics of a dynamic problem which involves the propagation of a stress wave through the sediments as water waves pass across the surface of the sea. The physical consequences of the passage of a wave of shear stresses through the sediment are very difficult to handle analytically and so to help in the understanding of the complex interaction between waves and the sea bed some small-scale experiments were carried out at Cornell University.

A 10% by weight suspension of Bear Paw shale in water with a sodium chloride concentration of 34 g/l was prepared and, after thorough mixing, allowed to sediment. During sedimentation small cracks developed on the surface of the clay, and where these intersected small mud volcanoes were formed. It was not possible to determine how deep the cracks were but the presence of the mud volcanoes showed that the vertical permeability near the surface was rather high.

Gravity slides were initiated at various times after sedimentation started by tilting the tank until

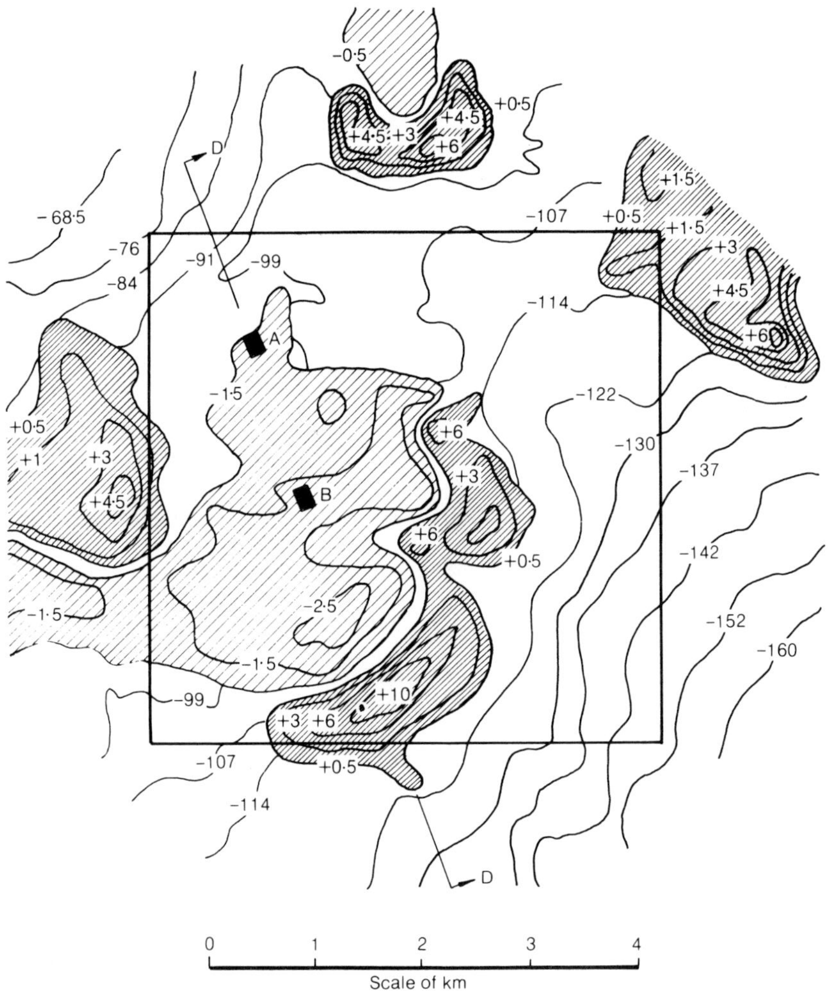

Fig. 12. Changes in bed level in metres due to Camille (Bea et al., 1975)

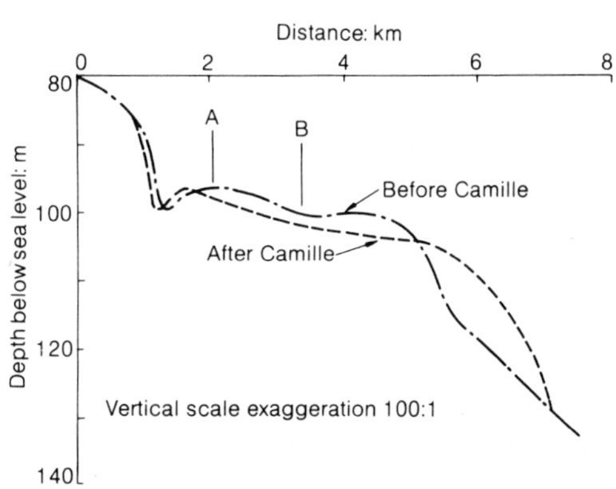

Fig. 13. Change in section DD due to Camille (Bea & Audibert, 1980)

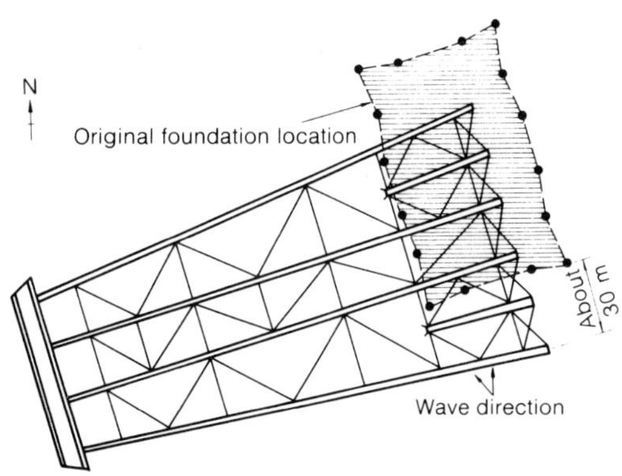

Fig. 14. Platform B after Camille (Bea & Audibert, 1980)

a slide occurred. This procedure, which was essentially a measurement of shear strength against time, suggested that the best way to measure very low shear strengths might well be by using a tilting tank.

After the relationship between consolidation time and shear strength had been established by observing the onset of gravity slides, wave loadings were introduced into the tank. At small wave heights the sediments oscillated in sympathy with the waves. However, when wave heights sufficient to cause shear failure on a sloping bed were generated an unsymmetrical movement in the sediment resulted and a series of mud snouts migrated down the slope. There was good agreement between the calculated shear stresses from the wave loading and the clay shear strength measured in the static tests.

The changes that took place in Block 70 between 1940 and 1967 were very similar to the change in the wave tank as the mud snouts advanced and it seems highly probable that the changes in the profile at Block 70 were due to the effects of storm waves. It also seems probable that the change in slope at a depth of 100 m occurred at the point at which the wave-assisted transport of sediment gave way to the more usual gravity slide.

Studies of gas in recent sediments (Oppenheimer & Kornicker, 1958; Volkmann & Oppenheimer, 1962; Anderson, Harwood & Lovelace, 1971) have shown that gas is formed as bacteria decompose the organic materials available to them. In the absence of any disturbance in the sediment the process of gas generation slows down as the supply

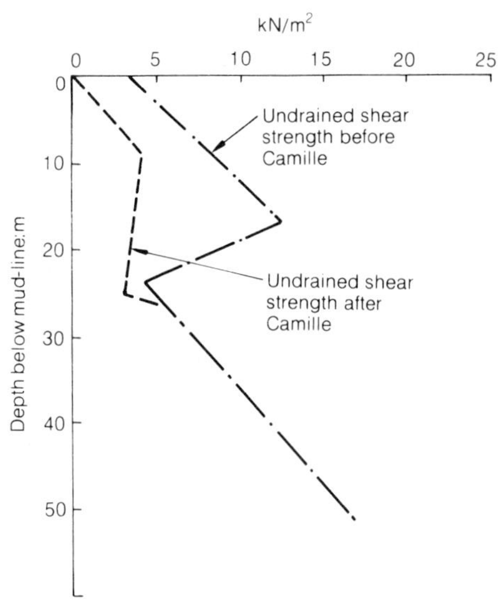

Fig. 15. Changes in shear strength at platform A (Bea & Arnold, 1973)

of organic material is exhausted. If the sediments are disturbed new supplies of organic materials become available to the bacteria and the process of gas generation is renewed. It thus appears (Bea & Arnold, 1973) that the presence of gas in sediment is an indicator that the sediment has been recently disturbed and the field data confirm that the presence of gas correlates well with other evidence of landslide activity. The existence of gassy sediments may be determined by remote sensing because, due to their ability to dissipate acoustic energy, no seismic reflections are obtained from gassy sediments.

The high gas porosities in borehole AM 8 between depths of 12 m and 45 m suggest that underwater landslide movements had extended deep into the recent sediments. The other significant feature in borehole AM 8 was the existence of the stronger crust near the mud-line. A possible explanation of this phenomenon has been supplied by Doyle (1973) as a result of model tests he carried out to investigate the relationship between waves and sediment movement. Doyle also found that, during the consolidation of the sediment in the tank, vertical pore tubes were formed in the soil and that these vertical drains permitted the rapid escape of water from the upper layers of the sediment. Small volcano-like structures were formed at the mud-line as clay particles were ejected from the pore tubes.

When wave loading was initiated the drainage from the pore tubes was reactivated and water and soil spewed out as additional excess pore-water pressures were generated by the wave loading. The wave action combined with the natural vertical drains led to an accelerated consolidation process together with the upward migration of fine particles. The higher shear strengths and Atterberg limits near the mud-line in borehole AM 8 may well be the field expression of this laboratory phenomenon.

Additional field evidence of the effects of waves in Block 70 was provided by the passage of hurricane Camille—the most intense hurricane ever recorded—to the east of the Birdfoot Delta in September 1969. By this time the area had been thoroughly surveyed and production platforms carried on piled foundations were in operation.

The changes in bottom topography which took place during hurricane Camille are shown in Fig. 12 and the positions of production platforms A and B are also shown. An enormous area sunk by up to 2 m and at the south end of the block a massive accumulation of material led to the formation of a mound with a maximum height of 10 m.

The changes on the section line XX in Fig. 12 are shown in Fig. 13. The mound formed is very

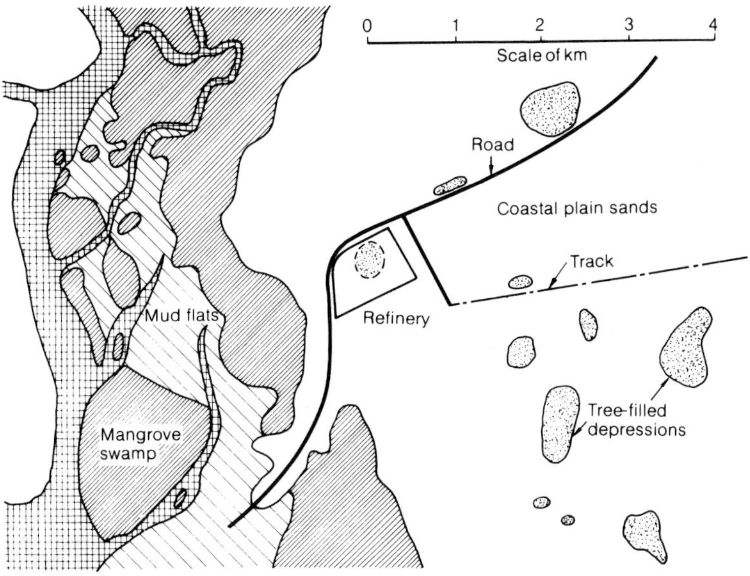

Fig. 16. Plan of refinery area

similar to that seen on the 1940 section, and in the migrating mud waves in the laboratory wave tank. Bea & Audibert (1980) reported that, based on high resolution geophysical data, a nose of soil advanced 1200 m down-slope and that large-scale soil displacements took place to a depth of 30 m.

During the hurricane Camille, platform B disappeared beneath the waves. It was found lying on its side on the sea floor as shown in Fig. 14. The lateral down-slope translation of the platform base was about 30 m. This event provided striking additional evidence in support of the hypothesis that storm waves could lead to massive instability in the weak sediments of the Mexican Gulf.

The sediments at the site of platform A were considerably stronger and precision measurements indicated that the structure had been displaced by about one metre down-slope without its operational functions being impaired. Borehole data obtained before and after Camille showed that a considerable reduction of strength had taken place during the storm and provided field evidence of the increase in excess pore-water pressures and loss of strength associated with repeated loading as shown in Fig. 15.

Since these events, and partly as a result of them, an enormous research effort has gone into the problems associated with rapid accumulation of sediments in the Mississippi Delta. The real problems are probably much more complex than I have indicated (Bea & Audibert, 1980). In the enormous area of the Mississippi Delta there are wide variations in the rate of deposition and the types of material being deposited and many geomorphological processes are at work. However, even a simple examination of one problem shows that we need to know what is there and why it has its present form before we can start to understand what is going on.

TROPICAL WEATHERING AND CHEMICAL CHANGE

Very different types of problem are associated with tropical weathering and the stability of soils subject to changes in groundwater chemistry.

A substantial cavity discovered beneath a concrete slab in the main process area of an oil refinery had no obvious cause and so an investigation was made to find out why the cavity had developed.

The site was on the edge of the Niger Delta, close to one of the discharge mouths on the Bonny River. The general geological conditions at the site are Pleistocene coastal plain sands overlying thick sandy and clayey delta deposits.

The details of the surface features in the vicinity of the refinery were examined using aerial photography. The only visible natural feature, on the otherwise flat coastal plain, was well-defined clumps of trees. When stereo pairs of the area were examined all the trees appeared to be growing in hollows. Aerial photographs taken before the refinery was constructed showed that the process area in which the cavity had been found was located where a clump of trees had been growing. If the problems were to be understood the geomorphological significance of the tree-filled hollows needed to be assessed.

A plan of the ground in the vicinity of the oil

refinery is shown in Fig. 16. In order to establish the possible significance of the tree-filled depressions, the depression closest to the refinery along the track was visited first. The general appearance of the clump of trees from the track was not spectacular but among the trees there was a dank smell of rotting vegetation and a chaotic mass of plant debris, as well as an army of ferocious ants. The ground level of the clump of trees was about 2 m lower than that of the adjacent ground and the surface soils showed signs of intense leaching.

Although the geological description of the surface soils at the site was coastal plain sands, the processes of weathering had produced a matrix of kaolinite holding together the relatively unweathered sand grains. The explanation for the depressions appears to be that the organic acids produced by the rotting vegetation in the depressions had led to an accelerated rate of breakdown of the coastal plain sands with a consequent decrease in volume.

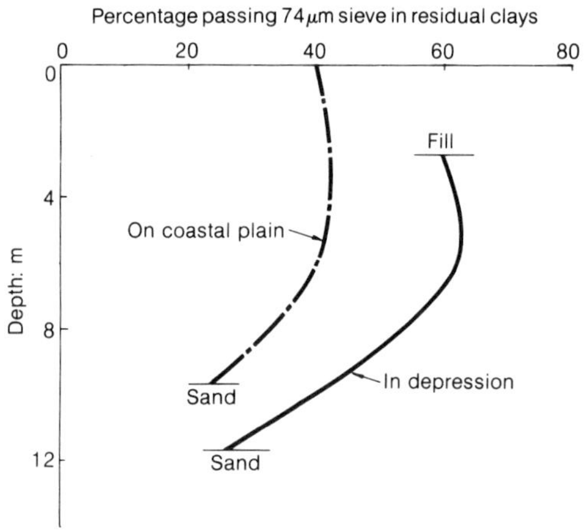

Fig. 17. Comparison of coastal plain and depression soils

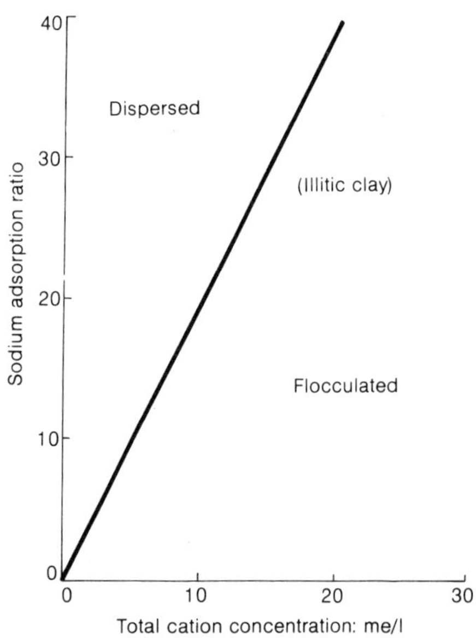

Fig. 18. Boundary between dispersed and flocculated states (after Collis-George & Smiles, 1963)

A simple indicator of the intensity of weathering is the percentage of material passing the 74 μm sieve. In Fig. 17 the soils in the depression and on the flat coastal plains are compared. In the depression the percentage of fine material is much higher and the weathering has proceeded to a greater depth.

In the area of the refinery, a further small cavity in the ground was found in a drain into which water, treated with sodium carbonate, was being discharged. Although there had been some contamination by hydrocarbon wastes it was possible to establish that erosion of the soil along fissures had taken place. The texture of the surface of the natural soil and the fact that the sand grains stood out very clearly suggested that a chemical dispersion process was involved.

The cavity in the process area, which led to the initial concern on the site, was downstream of an ion exchanger used to condition the boiler feed-water. In order to recondition the ion exchanger 14 kg of 98% sulphuric acid and 80 kg of flake caustic soda were passed through the ion exchanger and flushed into the drainage system every eight hours. It appeared that leaks had developed in the drainage system and that some of the chemical waste materials had found their way into the ground. The natural pH of the groundwater at the site is about 5, but in many places near the drains, the pH had increased to about 9 because of contamination from caustic soda.

For many years dam engineers have been concerned about the possibilities of internal erosion in dam foundations and it has been established that internal erosion can take place when the clay particles are in a dispersed rather than a flocculated array. When flocculated the clay particles cling together but when dispersed they are readily removed by flowing water. Dam engineers and soil scientists have a common interest in this problem because whether the soils are dispersed or flocculated has an important effect on their permeability and also the agricultural yield.

The factors which control the flocculation or dispersion of clays are very complex and there are no adequate theories to explain all the phenomena. However, there is strong pragmatic evidence that the presence of sodium ions is one of the more important factors leading to dispersion. As an example the results obtained by Collis-George & Smiles (1963) are shown in Fig. 18. They indicate that the ratio between the sodium adsorption ratio and the total cation concentration control whether the clays are in a dispersed or flocculated state.

I am a believer in simple field tests, and a sample of the alkaline effluent from the ion exchanger was allowed to flow through a small hole in a sample of the soil from the process area. The hole had an initial diameter of about 3 mm which enlarged rapidly, and very dirty water carrying fine particles in suspension emerged from the hole. It was clear that the caustic soda was causing surface erosion of the clay particles.

The effectiveness of dilute caustic soda in causing dispersion is shown in Figs 19 and 20. In Fig. 19 the surface of a sample of the soil from the oil refinery which had been subjected to a flow of distilled water is shown. Although coarse particles are visible they are obscured by a thin layer of clay-sized materials. The grid consists of 2 mm squares. Fig. 20 shows the soil surface after a weak solution of caustic soda had been allowed to flow across the sample. The surfaces of the coarser particles have been washed clean. Again the grid consists of 2 mm squares.

On a larger scale the effects of the caustic soda in the effluent were examined using a scanning electron microscope. Fig. 21 shows the untreated soil; it has the typical appearance of a weathered kaolinite. Fig. 22 was taken after the surface of the sample had been washed with dilute caustic soda solution. The kaolinite sheets had broken down into a mass of small particles which could easily be eroded by flowing water.

Holes into which the particles can be washed are provided by the many fissures which are formed as the soil volume decreases due to the weathering process. The presence of an extensive network of termite burrows provides further routes for the erosive effluents to carry away the dispersed clay particles.

Experience elsewhere in West Africa has shown that the adverse effects of allowing effluents containing caustic soda to come in contact with the well-drained soils produced by tropical weathering of sandy materials are fairly common.

These examples of the effect of changes in

Fig. 19. Surface of natural soil (2 mm grid)

Fig. 20. Surface of soil washed with dilute caustic soda (2 mm grid)

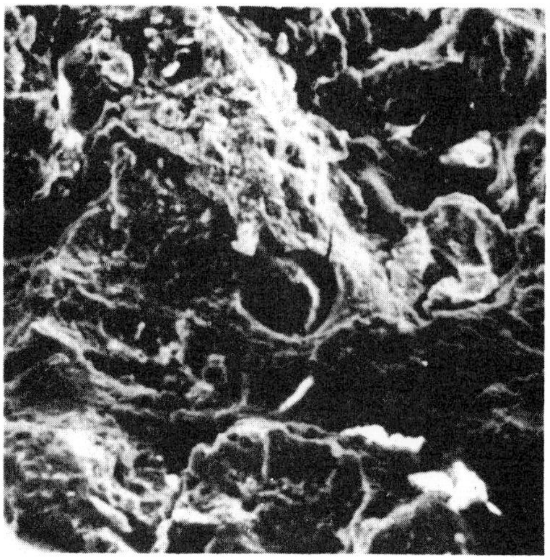

Fig. 21. Electron microscope picture of natural soil; width of sample shown 0·7 mm

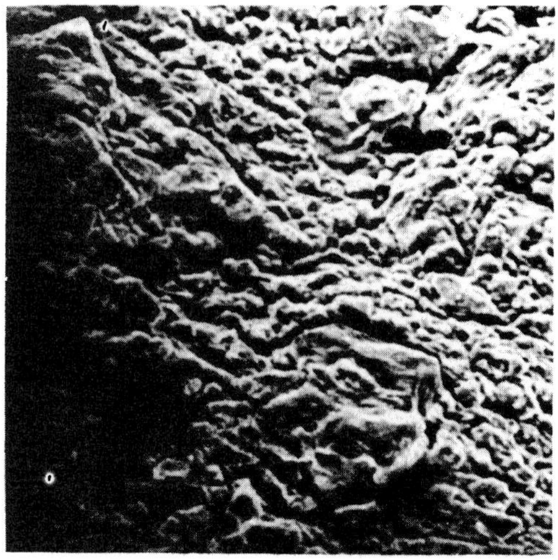

Fig. 22. Electron microscope picture of soil treated with dilute caustic soda; width of sample shown 0·7 mm

chemical environment emphasizes the importance at each site of the three fundamental questions of what is there, why does it have its present form and what will happen if we change anything.

Weathering plays an extremely important role in determining the engineering behaviour of materials and I think that engineering geologists could provide a valuable service to the civil engineering profession by giving more information on the geomorphological consequences of weathering in a variety of climatic and geological circumstances.

In the high rain-fall areas of the Western Ghats in India, where 10 m of rain fall every year, all the silica in the residual soils is dissolved and very porous soils of high permeability are produced. By squeezing the soil in the hand it is possible to squeeze out water as from a sponge.

Weathering processes are very sensitive to the micro-climate at any point and modest changes in surface can produce significant differences in weathering rates and hence soil properties. As an example, Fig. 23 shows a cutting in weathered basalt.

Over the past few years there has been a series of massive construction projects in Hong Kong where many of the engineering problems are associated with the weathering of volcanic or granitic rocks to form residual soils. Provided the in situ weathered rock is not disturbed, the original structure of the granite is retained. The retention of the original coarse-grained structure in spite of intense weathering produces a soil with high permeability and high compressibility. This means that in excavations or in diaphragm wall construction significant swelling can take place during the short period that the stresses on the ground are reduced. Unless large excess bentonite heads are maintained large settlements take place during the construction of diaphragm walls.

EFFECTS OF GROUNDWATER CHANGES

The effects of groundwater lowering in causing settlement are well known from experience in Venice, Mexico City, Long Beach California and London. In most cases groundwater lowering does not lead to sharp discontinuities in the ground surface but in certain geological circumstances the results can be catastrophic.

A massive groundwater lowering operation was carried out in Far West Rand in South Africa, where the land surface is old, the most recent rocks being the Karroo system, which corresponds to the Carboniferous of the Northern Hemisphere.

A plan of the Bank Compartment is shown in Fig. 24. The term compartment is used because of the syenite and diabase dykes which divide the water-bearing dolomite into a number of virtually watertight compartments. Some of these dykes are shown on the plan. Associated with each dyke was a spring which carried the groundwater over the dyke from one compartment to another.

The West Driefontein Gold Mine is situated in the Oberholzer Compartment, immediately to the west of the Bank Compartment, and in October 1968 an unprecedented and unexpected inflow of water occurred in a stope in the eastern part of the mine near the dyke between the Oberholzer and Bank Compartments (Taute & Tress, 1971).

The water inflow was in excess of

Fig. 23. Weathered basalt

360 000 m^3/day or 250 m^3/min and it became apparent that a fissure connecting the two compartments had opened up. Part of the mine became flooded. The only practical way to reinstate the West Driefontein Mine was to drain the Bank Compartment, and permission to do this was obtained from the Department of Water Affairs. The depth of groundwater lowering required was about 1000 m and in the Bank Compartment water levels were lowered during 1969, 1970 and 1971.

A section through the Bank Compartment along the line XX in Fig. 24 is shown in Fig. 25. The Witwatersrand system, which contains the important gold-bearing conglomerates, lies

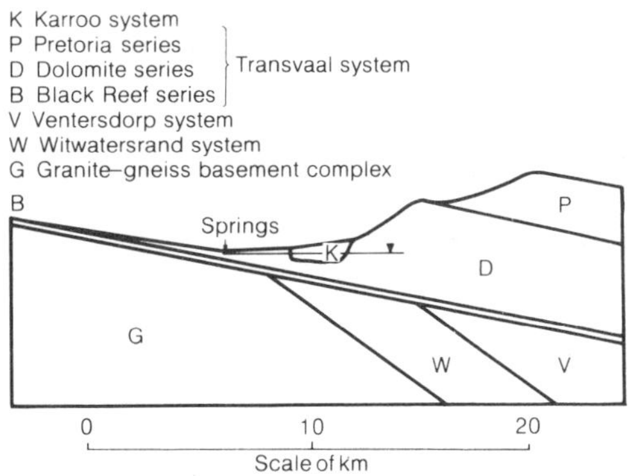

Fig. 25. Section XX through Bank Compartment

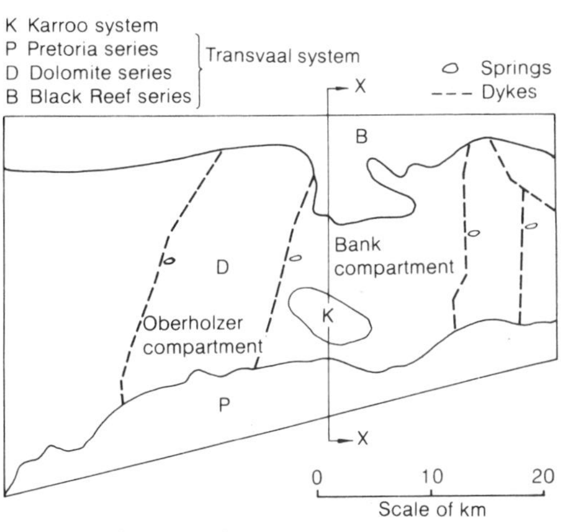

Fig. 24. Plan of Bank Compartment

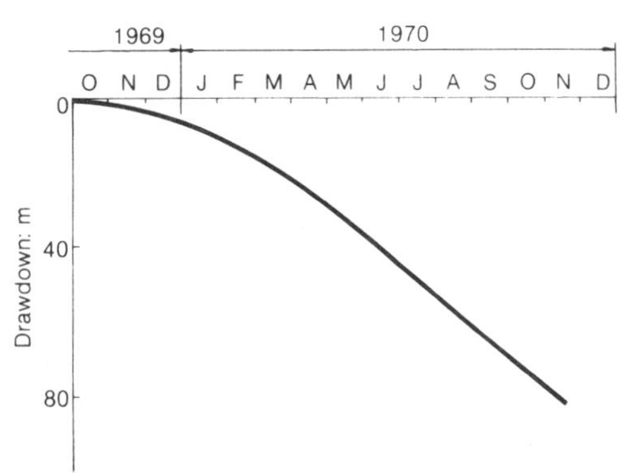

Fig. 26. Water levels in borehole near Brickor

directly on the granite–gneiss basement complex. The Transvaal system, which includes the water-bearing dolomite, rests uncomfortably on an erosion surface which cuts across the older rocks. An isolated pocket of the Karroo system is laid down on the weathered and glaciated surface of the dolomite. The weathered shales of the Karroo system provided admirable raw material for the manufacture of bricks and the Driefontein brick-works, known as Brickor, were established on this outlier.

Water levels recorded in a borehole near the Brickor site are shown in Fig. 26. The readings were ceased in October 1970 because the water level had sunk below the bottom of the borehole.

As the water levels in the Bank Compartment were lowered problems were encountered with the continuous kiln process being used. The cars carrying the bricks became jammed in the kilns and it was decided to measure the settlements of the kilns and adjacent areas. The settlements which took place between July 1970 and April 1972, when movements had effectively stopped, are shown in Fig. 27. During this period the settlements amounted to 180 mm and it had become impossible to operate the kilns which required very tight tolerance on level for their successful working.

The detailed geology of the area was investigated thoroughly by Brink (1979); a section through the area of maximum settlement is shown in Fig. 28. The Karroo sediments were laid down in a solution channel in the dolomite and the settlement profile approximates to the relative thickness of the Karroo sediments which had been dewatered in the pumping operation. The reduction in the hydrostatic uplift led to consolidation of the sediments.

A more alarming aspect of the geological investigations was the discovery of a substantial zone of material known as 'wad'. This is an insoluble and highly compressible material left after the dolomite has been dissolved by percolating waters containing carbonic acid. The solution of the dolomite took place after the deposition of the Karroo sediments. It was fortunate that, at the site of the brickworks, the Karroo sediments, which had infilled a solution feature in the dolomite at the time of their deposition, were able to arch across the very weak and compressible wad produced by additional solution and thus prevent a catastrophic collapse.

Catastrophic collapses had occurred in a number of places adjacent to the brickworks and in a short helicopter trip a number of surface features associated with the collapse of wad were seen. These are shown in Figs 29–31. Fig. 29 shows an early stage in the development of a hole with deformation and cracking of the ground surface. Fig. 30 shows a situation in which most of the disturbed area has collapsed and Fig. 31 shows a view of this hole seen from the ground.

The catastrophic settlements that can result from groundwater lowering in dolomitic or lime-

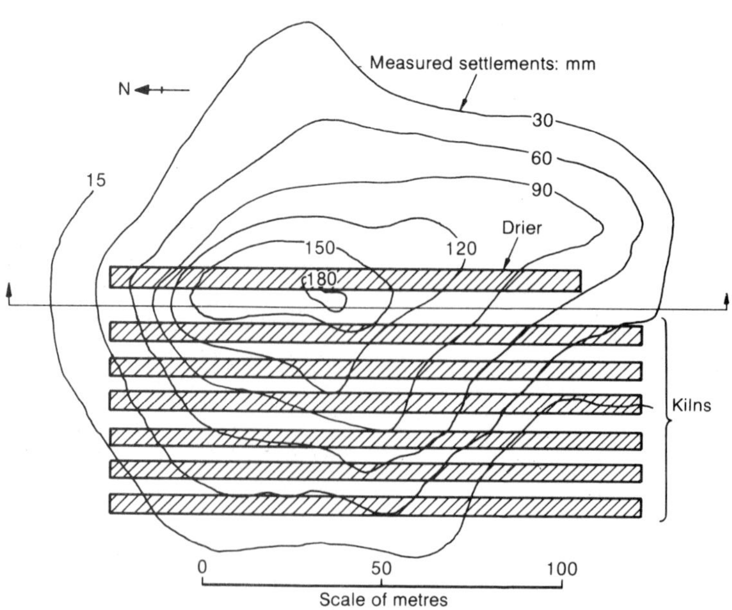

Fig. 27. Settlements at Brickor from July 1970 to April 1972

stone rocks show how important it is for geology and engineering to work together so that situations in which this hazard exists can be identified. It is also clear that where geological unconformities exist it is necessary to examine not only the surface geomorphology but also the geomorphology of the underlying surface.

The solution of limestones and dolomites is continuing in many parts of the world and in order to appreciate the scale of the features that contribute to the problem it is useful to look at areas where the rocks are at the surface. A view of tropical karst in Malaysia is shown in Fig. 32 and the details of its complex, nearly vertical pinnacles are shown in Fig. 33. Where such features exist the difficulties of predicting overall engineering behaviour are formidable and can only be attempted if the processes which led to the formation of the buried topography are understood.

ALTERATIONS IN DRAINAGE

It is not often that in the course of a single job one is able to see the interaction between geology, man and geomorphology. I had this opportunity some years ago while working on the Beas Dam in India. At the site of the dam the river cuts through a plunging anticline. The Siwalik rocks are a sequence of weak sand rocks and shales which have been folded as part of the Himalayan uplift. The surface of one of the sand rock layers on the flank of the anticline is shown in Fig. 34 in which the regional jointing pattern, associated with the folding, can be seen.

The alteration of the surface topography by the construction of access roads and drainage ditches led to a concentrated flow of stormwater over the surface of the rock. The effect on the rock exposure was spectacular as is shown in Fig. 35. The water pressure associated with the surface flooding caused the joints to open up and horizontal displacement of about 13 m took place. Fig. 36 shows the area after the movement had occurred.

This experience was a lesson on the delicate

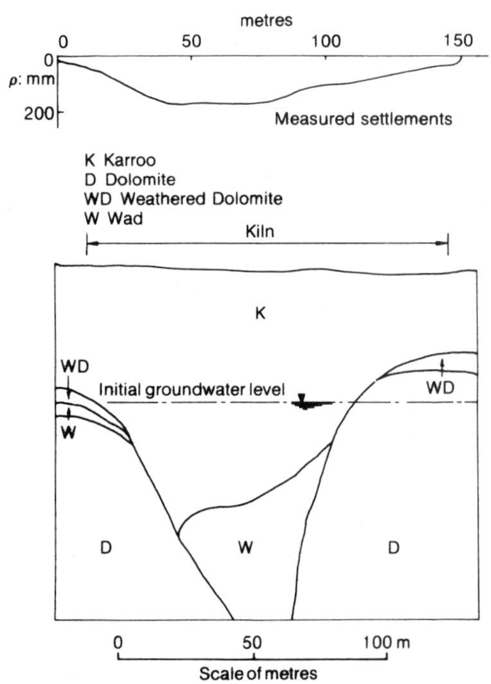

Fig. 28. Detailed geology at Brickor

Fig. 29. Early stage of hole development

Fig. 30. Collapse of disturbed area

Fig. 31. Hole from ground

Fig. 32. General view of Malaysian karst

Fig. 33. (right). Vertical limestone pinnacle

Fig. 34. Rock surface on Beas River

equilibrium that exists in many natural situations, and unless we ask ourselves why the land has its present form and why it is in equilibrium we may not realize what we are doing.

GLACIAL MATERIALS

I cannot leave my subject without a few words about the difficulties of working in some of the glacial material we encounter. The Pleistocene glaciation involved many advances and retreats of the ice and, in the process, complex variations in local erosion and infilling occurred. The one lesson we must learn from experience in glacial materials is that the unexpected should always be expected. In spite of site investigations we will, in many cases, not know what is there until we have opened up the foundation or other excavations. The cliffs in Norfolk near Overstrand (Fig. 37) illustrate the complexity of ice marginal glacial materials. It would be a brave man who would hazard a guess, based on a series of even closely spaced borings, of what could be expected.

An important point emerging from the consideration of the inherent variability of many glacial materials and the variability associated with other geological processes is that in some circumstances it will not be possible to understand what is going on. We must recognize this fact and make it clear at an early stage in the design process so that everyone is aware of the risks involved. However, we should try to define the limiting conditions that might be encountered so that construction strategies to cope with the possible limiting conditions can be developed.

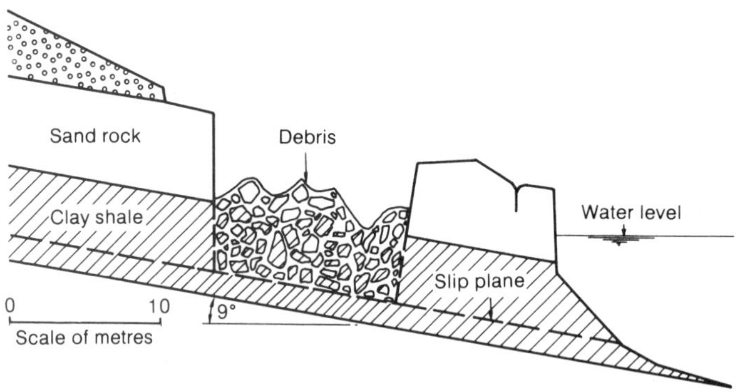

Fig. 35. Section through slip

Fig. 36. Rock surface after slip

Fig. 37. Glacial deposits at Overstrand

CONCLUSION

Over the past few decades we have developed sophisticated analytical and numerical methods for the solution of almost any problem provided we are able to make the correct fundamental assumptions. My message is that, in order to define the fundamental assumptions, there is no substitute for painstaking study in the field of geology and geomorphology. We still need something of the Victorian virtue known as 'an eye for the ground'.

The geological environmental is complex with so many facets that control its behaviour that the only way to achieve a fuller understanding is for there to be an interdisciplinary approach in which engineers and geologists work much more closely together. The meeting ground can, I believe, be found by both the professions concentrating on understanding the geomorphology of construction sites.

The birds and fishes in Fig. 1 remind us that in order to understand our problems we have to approach them from two sides—geology and engineering.

ACKNOWLEDGEMENT

The electron microscope photographs reproduced in Figs 21 and 22 were taken by Dr Tovey.

REFERENCES

Anderson, A. L., Harwood, R. J. & Lovelace, R. T. (1971). *Investigation of gas in bottom sediments.* Final report to Office of Naval Research, Applied Research Laboratories, University of Texas at Austin.

Bea, R. G. & Audibert, J. M. E. (1980). Offshore platforms and pipelines in Mississippi River Delta. *J. Geotech. Engng Div. Am. Soc. Civ. Engrs* **106**, GT8, 853–869.

Bea, R. G. & Arnold, P. (1973). Movements and forces developed by wave-induced slide in soft clays. *Proc. 5th Ann. Offshore Technol. Conf.* **2**, 731–742.

Bea, R. G., Bernard, H. A., Arnold, P. & Doyle, E. H. (1975). Soil movements and forces developed by wave-induced slides in the Mississippi Delta. *J. Petrol. Technol.* **27**, Apr., 500–514.

Berkey, C. P. (1929). Responsibilities of the geologies in engineering projects. *Tech. Publs Am. Inst. Min. Metall. Engrs*, No. 215, 4–9.

Brink, A. B. A. (1979). *Engineering geology of Southern Africa*, vol. 1. Pretoria: Building Publications.

Collis-George, N. & Smiles, D. E. (1963). An examination of catcon balance and moisture characteristic methods of determining the stability of soil aggregates. *J. Soil Sci.* **14**, No. 1, 21–32.

Dearman, W. R. (1971). Introductory statement to regional meeting of the Engineering Group of the Geological Society Dublin. *Q. Jl Engng Geol.* **4**, No. 3, 187–190.

Dolmage, V. (1962). Discussion on Engineering geology on the job and in the classroom, by K. Terzaghi. Contributions to soil mechanics 1954–62. *J. Boston Soc. Civ. Engrs*, 365–369.

Doyle, E. H. (1973). Soil–wave tank studies of marine soil instability. *Proc. 5th Ann. Offshore Technol. Conf.* **2**, 753–766.

Henkel, D. J. (1970). The role of waves in causing submarine landslides. *Géotechnique* **20**, No. 1, 75–80.

Legget, R. F. (1979). Geology and geotechnical engineering. *J. Géotech. Engng Div. Am. Soc. Civ. Engrs* **105**, GT 3, 342–391.

Oppenheimer, C. H. & Kornicker, L. S. (1958). Effect of the microbial production of hydrogen sulfide and carbon dioxide on the pH of recent sediments. *Publs Inst. Mar. Sci. Univ. Tex.* **5**.

Peck, R. B. (1973). Presidential address. *Proc. 8th Int. Conf. Soil Mech., Moscow* **4**, 156.

Prior, D. B. & Suhayda, J. N. (1979). Application of infinite slope analysis to subaqueous sediment stability, Mississippi Delta. *Engng Geol.* **14**, No. 1, 1–10.

Shepard, F. P. (1955). Delta front valleys bordering on the Mississippi distributaries. *Bull. Geol. Soc. Am.* **66**, 1489–1498.

Sheppard, T. (1917). William Smith: his maps and memoirs. *Proc. York. Geol. Soc.* **19**, 75–253.

Taute, A. H. & Tress, P. W. (1971). Dewatering of the flooded underground workings on the Bank Compartment. *S. Afr. Min. Engng J* **82**, Oct, 23–45.

Terzaghi, K. (1956). Varieties of submarine slope failures. *Proc. 8th Conf. Soil Mech., Texas*

Terzaghi, K. (1961). Engineering geology on the job and in the classroom. Contributions to soil mechanics 1954–62. *J. Boston Soc. Civ. Engrs*, 335–347.

Volkmann, C. M. & Oppenheimer, C. H. (1962). The microbial decomposition of organic carbon in surface sediments of Marie Bays of Central Texas Gulf Coast. *Publs Inst. Mar. Sci. Univ. Tex.* **8**.

VOTE OF THANKS

In proposing a vote of thanks to Dr Henkel, Professor J. N. Hutchinson said:

'In his introductory remarks, Professor Wroth highlighted the distinguished and wide-ranging nature of Dr Henkel's career. The lecture which we have just heard has truly reflected these qualities, moving expertly from consideration of soft Holocene deposits in the Gulf of Mexico to the deep geology of ancient rocks in the Transvaal; from the subtle manifestations of tropical weathering in the Niger Delta to the complexities of glacial deposits in East Anglia. With his soundly based, all-round knowledge and his enviable ability, noted by Professor Wroth, to cut through a maze of inessentials to reach the heart of a problem, Dr Henkel bids fair to be a rare exception to Terzaghi's dictum concerning the supposed impossibility of combining, in one person, equal competence in engineering and geology. Whether he achieves this in the form of a flying fish or a diving bird is perhaps less clear.

'I found Dr Henkel's account of his, now classic, work on the generation of submarine landslides in soft clays by differential wave loading to be particularly elegant and satisfying. In this connection, it is interesting to note that, on the bed of the North Sea, similar cyclic wave loading seems to have had the beneficial effect of densifying the widespread sand deposists there, which otherwise may well have been prone to liquefaction.

'Dr Henkel's eminence in both the professional and academic spheres of our subject makes him unusually well equipped to comment on current teaching practices. In the lecture, his views on these were uncharacteristically restrained, but three important points emerge. First, that engineering geology should become more distinct than at present from geotechnical engineering, so that the two disciplines may be truly complementary. Second, that in this context, geomorphology has been seriously neglected and that there is now a pressing need to give this discipline its proper place in the geotechnical spectrum. The claim of neglect is indeed supported by the fact that the term geomorphology has not been mentioned in any of the previous twenty-one Rankine Lectures. Third, Dr Henkel suggests that the discipline of geomorphology can act as a long-needed catalyst to bring about a more effective combination between geology and geotechnics. I believe that these views deserve our serious consideration.

'During his time in the Civil Engineering Department at Imperial College, Dr Henkel was able to put into effect some of his teaching ideas. In particular, he organized and led the first geotechnical field excursion for postgraduate soil mechanics students. Indeed, such was his belief in the educative value of the Norfolk Pleistocene cliffs that we were taken through a field of anti-tank mines to see them! I have never been quite sure how to interpret the fact that this was done with the agreement of the Rector! I would suggest that this episode is symbolic of Dr Henkel's subsequent professional career. In this, he has continued to enter engineering and geological minefields from which, however, because so well armed with the geotechnical virtues, he manages to emerge unscathed.

'There is one matter for which, I believe, there is cause for regret. That is, because of Dr Henkel's intense professional activity over the past decade, he has had little time to record the fruits of this in the technical literature. Tonight's lecture has shown vividly what we have been missing. I hope that, in the future, Dr Henkel will make time to enlarge on, and add to, the wisdom that he has shared with us tonight.

'In conclusion, it gives me great pleasure to propose a warm vote of thanks to Dr Henkel for his outstanding and thought-provoking address'.

The vote of thanks was accorded with acclamation.

The Rankine Lecture

The twenty-third Rankine Lecture of the British Geotechnical Society was given by Dr Evert Hoek at Imperial College of Science and Technology, London, on 1 March 1983. The following introduction was given by Dr P. Hackett, Camborne School of Mines.

I first met Evert Hoek a long time ago. Long enough to belie his youthful appearance, and hopefully mine! In fact, it was in 1961, during the first of many visits to South Africa, that I first encountered the wit, enthusiasm and imaginative thinking of this resourceful man.

I seem to recall that, at the time, the *cognoscenti* were deeply involved in analytical discussions about the effects of 'switching-on gravity' on classical elastic solutions of the hole-in-plate type in order to simulate the effects of mining deep underground workings. (Indeed, I seem to recall contributing something to these erudite discussions myself.) I was therefore rather surprised when, while visiting Hoek in Pretoria to further these abstract thoughts, I first saw a centrifuge of considerable proportions designed to impose body forces on rock mechanic models. I have long since accepted that such a direct and practical approach to problem solving typifies the uncluttered thinking for which he has become internationally renowned.

However, both the technology and the man have moved on since then and, while I doubt that he would wish now to build a bigger and better centrifuge, his continued application to the discipline of rock engineering has resulted in many significant advances to both the science and technology. Not least of these is the elegant and simple design of the triaxial load cell with which his name in synonymous. One of the monuments to an earlier pre-Hoekian triaxial testing period is probably still lurking menacingly in the rock mechanics laboratory at the University of Nottingham; for, just before he and Franklin published their design (Hoek & Franklin, 1968), the University had ordered a cell from a well known supplier of quality soil mechanics equipment. When this device arrived, it was clear that it was not going to be used very often, being of vast proportions, extremely heavy (requiring cranage!) and of doubtful performance even given skilled handling. How things have changed!

During the early 1970s, Evert Hoek's interests developed to encompass the particular problems of rock slopes which led to the publication of that 'best seller' with Bray (Hoek & Bray, 1974). Indeed, his interests have ranged so widely across the discipline that one might have been forgiven for thinking that he had sought to embrace a zoological content in his paper with Roberts and Fish (Roberts, Hoek & Fish, 1972).

However, he was soon back in mainstream rock engineering with interests ranging from the surface to underground, from small to large excavations from mining to storage and waste disposal schemes, all of which led to his second *opus magnum* (Hoek & Brown, 1980).

Hoek's breadth of coverage of the discipline is matched by a great depth of interest and thus, while extending and expanding the rock engineering design criteria for mining and civil construction, he is also still involved in furthering our understanding of the rheology of rock as a material—particularly in respect of its anisotropy.

All of this has taken Evert Hoek through research appointments and through academic life to jet-set consultancy work, collecting on the way such notable awards as the Consolidated Goldfields Medal (IMM 1970) and the Burwell award from the Geological Society of America (1979). He had always been a lecturer of great character and distinction, and it is with great pleasure that I ask him to deliver his lecture.

REFERENCES

Hoek, E. & Bray, J. W. (1974). *Rock slope engineering.* London: Institution of Mining and Metallurgy.

Hoek, E. & Brown, E. T. (1980). *Underground excavations in rock.* London: Institution of Mining and Metallurgy.

Hoek, E. & Franklin, J. A. (1968). A simple triaxial cell *Trans. Instn Min. Metall.*, Section A, Vol. 77.

Roberts, B., Hoek, E. & Fish. (1972). The concept of the Mammoth Quarry. *Quarry Managers Journal* **56**, No. 7.

Dr E. Hoek

HOEK, E. (1983). *Géotechnique* **33**, No. 3, 187—223

Strength of jointed rock masses

E. HOEK*

Jointed rock masses comprise interlocking angular particles or blocks of hard brittle material separated by discontinuity surfaces which may or may not be coated with weaker materials. The strength of such rock masses depends on the strength of the intact pieces and on their freedom of movement which, in turn, depends on the number, orientation, spacing and shear strength of the discontinuities. A complete understanding of this problem presents formidable theoretical and experimental problems and, hence, simplifying assumptions are required in order to provide a reasonable basis for estimating the strength of jointed rock masses for engineering design purposes. This Paper summarizes some of the basic information upon which such simplifying assumptions can be made. A simple empirical failure criterion is presented and its application in engineering design is illustrated by means of a number of practical examples.

Des masses jointives de rochers comprennent des particules angulaires enchevêtrées ou des blocs de matière dure et cassante séparés par des surfaces discontinues enrobées ou non de matières de moindre résistance. La résistance de masses rocheuses de ce genre dépend de la résistance des morceaux intacts et de leur liberté de mouvement, qui sont fonctions elles mêmes du nombre, de l'orientation, de l'écartement et de la résistance à la rupture au cisaillement des discontinuités. La compréhension complète de ce problème presente des difficultés considérables d'aspect théorique et expérimental, de sorte que des hypothèses simplificatrices sont nécessaires pour avoir une base raisonnable sur laquelle on peut estimer la résistance des masses jointives de rochers en vue de la construction. Cet article résume quelques-unes des données de base sur lesquelles de telles hypothèses simplificatrices peuvent être faites. Un critère de rupture empirique de nature très simple est donné, son application à la construction ètant illustrée au moyen d'un certain nombre d'exemples pratiques.

INTRODUCTION

The past twenty years have seen remarkable developments in the field of geotechnical engineering, particularly in the application of computers to the analysis of complex stress distribution and stability problems. There have also been important advances in the field of geotechnical equipment and instrumentation and in the understanding of concepts such as the interaction between a concrete or steel structure and the soil foundation on which it is built or, in the case of a tunnel, the interaction between the rock mass surrounding the tunnel and the support system installed in the tunnel. Similarly, there have been significant advances in our ability to understand and to analyse the role of structural features such as joints, bedding planes and faults in controlling the stability of both surface and underground excavations.

In spite of these impressive advances, the geotechnical engineer is still faced with some areas of major uncertainty and one of these relates to the strength of jointed rock masses. This problem is summed up very well in a paper on rockfill materials by Marachi, Chan & Seed (1972) when they say 'No stability analysis, regardless of how intricate and theoretically exact it may be, can be useful for design if an incorrect estimation of the shearing strength of the construction material has been made'. These authors go on to show that, although laboratory tests on rockfill are difficult and expensive because of the size of the equipment involved, there are techniques available to permit realistic and reliable evaluation of the shear strength of typical rockfill used for dam construction.

Unfortunately, this is not true for jointed rock masses where a realistic evaluation of shear strength presents formidable theoretical and experimental problems. However, since this question is of fundamental importance in almost all major designs involving foundations, slopes or underground excavations in rock, it is essential that such strength estimates be made and that these estimates should be as reliable as possible.

In this Paper the Author has attempted to summarize what is known about the strength of jointed rock masses, to deal with some of the theoretical concepts involved and to explore their limitations and to propose some simple empirical approaches which have been found useful in solving real engineering problems. Examples of such engineering problems are given.

DEFINITION OF THE PROBLEM

Figure 1 summarizes the range of problems

* Golder Associates, Vancouver.

	Description	Strength characteristics	Strength testing	Theoretical considerations
	Hard intact rock	Brittle, elastic and generally isotropic	Triaxial testing of core specimens in laboratory relatively simple and inexpensive and results usually reliable	Theoretical behaviour of isotropic elastic brittle rock adequately understood for most practical applications
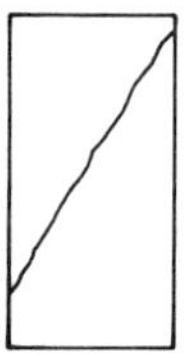	Intact rock with single inclined discontinuity	Highly anisotropic, depending on shear strength and inclination of discontinuity	Triaxial testing of core with inclined joints difficult and expensive but results reliable. Direct shear testing of joints simple and inexpensive but results require careful interpretation	Theoretical behaviour of individual joints and of schistose rock adequately understood for most practical applications
	Massive rock with a few sets of discontinuities	Anisotropic, depending on number, shear strength and continuity of discontinuities	Laboratory testing very difficult because of sample disturbance and equipment size limitations	Behaviour of jointed rock poorly understood because of complex interaction of interlocking blocks
	Heavily jointed rock	Reasonably isotropic. Highly dilatant at low normal stress levels with particle breakage at high normal stress	Triaxial testing of undisturbed core samples extremely difficult due to sample disturbance and preparation problems	Behaviour of heavily jointed rock very poorly understood because of interaction of interlocking angular pieces
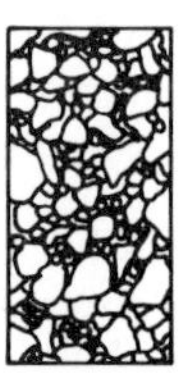	Compacted rockfill	Reasonably isotropic. Less dilatant and lower shear strength than in situ jointed rock but overall behaviour generally similar	Triaxial testing simple but expensive because of large equipment size required to accommodate representative samples	Behaviour of compacted rockfill reasonably well understood from soil mechanics studies on granular materials
	Loose waste rock	Poor compaction and grading allow particle rotation and movement resulting in mobility of waste rock dumps	Triaxial or direct shear testing relatively simple but expensive because of large equipment size required	Behaviour of waste rock adequately understood for most applications

Fig. 1. Summary of range of rock mass characteristics

considered. In order to understand the behaviour of jointed rock masses, it is necessary to start with the components which go together to make up the system—the intact rock material and the individual discontinuity surfaces. Depending on the number, orientation and nature of the discontinuities, the intact rock pieces will translate, rotate or crush in response to stresses imposed on the rock mass. Since there are a large number of possible combinations of block shapes and sizes, it is necessary to find behavioural trends which are common to all of these combinations. The establishment of such common trends is the most important objective of this Paper.

Before embarking upon a study of the individual components and of the system as a whole, it is necessary to set down some basic definitions.

Intact rock refers to the unfractured blocks which occur between structural discontinuities in a typical rock mass. These pieces may range from a few millimetres to several metres in size and their behaviour is generally elastic and isotropic. Their failure can be classified as brittle which implies a sudden reduction in strength when a limiting stress level is exceeded. In general, viscoelastic or time-dependent behaviour such as creep is not considered to be significant unless one is dealing with evaporites such as salt or potash.

Joints are a particular type of geological discontinuity but the term tends to be used generically in rock mechanics and it usually covers all types of structural weakness with the exception of faults. Hence the term jointed rock mass may refer to an assemblage of blocks separated by joints, bedding planes, cleavage or any other type of structural weakness.

Strength, in the context of this Paper, refers to the maximum stress level which can be carried by a specimen. No attempt is made to relate this strength to the amount of strain which the specimen undergoes before failure nor is any consideration given to the post-peak behaviour or the relationship between peak and residual strength. It is recognized that these factors are important in certain engineering applications but such problems are beyond the scope of this Paper.

The presentation of rock strength data and its incorporation into a failure criterion depends on the preference of the individual and on the end use for which the criterion is intended. In dealing with slope stability problems where limit equilibrium methods of analyses are used, the most useful failure criterion is one which expresses the shear strength in terms of the effective normal stress acting across a particular weakness plane or shear zone. The presentation which is most familiar to soil mechanics engineers is the Mohr failure envelope. On the other hand, when analysing the stability of underground excavations, the response of the rock to the principal stresses acting upon each element is of paramount interest. Consequently, a plot of triaxial test data in terms of the major principal stress at failure versus minimum principal stress or confining pressure is the most useful form of failure criterion for the underground excavation engineer. Other forms of failure criterion involving induced tensile strain, octahedral shear stress or energy considerations will not be dealt with.

Most of the discussion on failure criteria will be presented in terms of Mohr failure envelopes. With the Author's background being in underground excavation engineering the starting point for most of his studies is the triaxial test and the presentation of failure criteria in terms of principal stresses rather than shear and normal stresses. This starting point has an important bearing on the form of the empirical failure criterion presented.

STRENGTH OF THE INTACT ROCK

A vast amount of information on the strength of intact rock has been published during the past fifty years, and this was reviewed by the late Professor J. C. Jaeger in the eleventh Rankine lecture (1971).

In this context, one of the most significant steps was a suggestion by Murrell (1958) that the brittle fracture criterion proposed by Griffith (1921, 1925) could be applied to rock. Griffith postulated that, in brittle materials such as glass, fracture initiated when the tensile strength of the material is exceeded by stresses generated at the ends of microscopic flaws in the material. In rock, such flaws could be pre-existing cracks, grain boundaries or other discontinuities. Griffith's theory, summarized for rock mechanics applications by Hoek (1968), predicts a parabolic Mohr failure envelope defined by the equation

$$\tau = 2(|\sigma_t|(|\sigma_t|+\sigma'))^{1/2} \tag{1}$$

where τ is the shear stress, σ' is the effective normal stress and σ_t is the tensile strength of the material (note that tensile stresses are considered negative throughout this Paper).

Griffith's theory was originally derived for predominantly tensile stress fields. In applying this criterion to rock subjected to compressive stress conditions, it soon became obvious that the frictional strength of closed cracks has to be allowed for, and McClintock & Walsh (1962) proposed a modification to Griffith's theory to account for these frictional forces. The Mohr failure envelope for the modified Griffith theory is defined by the equation

$$\tau = 2|\sigma_t| + \sigma' \operatorname{Tan} \phi' \tag{2}$$

where ϕ' is the angle of friction on the crack

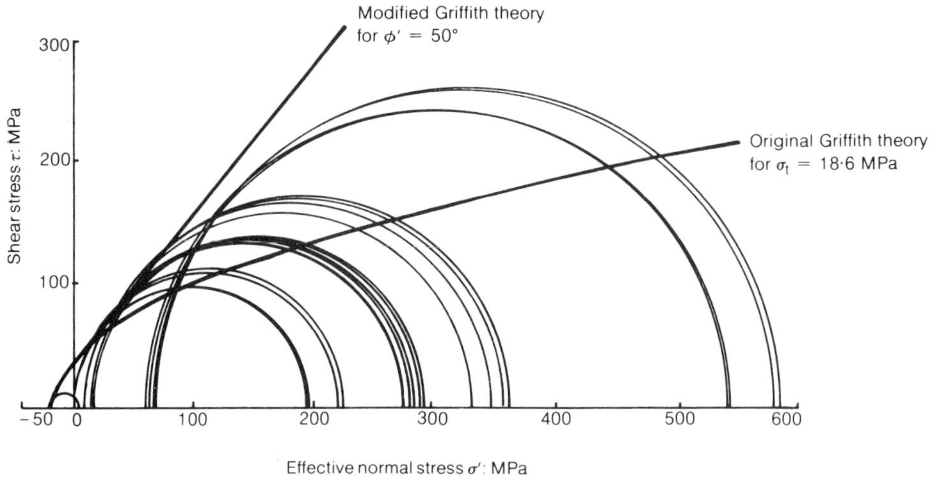

Fig. 2. Mohr circles for failure of specimens of quartzite tested by Hoek (1965). Envelopes included in the figure are calculated by means of the original and modified Griffith theories of brittle fracture initiation

surfaces. (Note, this equation is only valid for $\sigma' > 0$.)

Detailed studies of crack initiation and propagation by Hoek & Bieniawski (1965) and Hoek (1968) showed that the original and modified Griffith theories are adequate for the prediction of fracture initiation in rocks but that they fail to describe fracture propagation and failure of a sample. Fig. 2 gives a set of Mohr circles for failure of specimens of quartzite tested triaxially (Hoek, 1965). Included in this figure are Mohr envelopes calculated by means of equations (1) and (2) for $\sigma_t = 18{\cdot}6$ MPa and $\phi' = 50$ degrees. Neither of these curves can be considered acceptable envelopes to the Mohr circles representing failure of the quartzite under compressive stress conditions. In spite of the inadequacy of the original and modified Griffith theories in predicting the failure of intact rock specimens, a study of the mechanics of fracture initiation and of the shape of the Mohr envelopes predicted by these theories was a useful starting point in deriving the empirical failure criterion.

Jaeger (1971), in discussing failure criteria for rock, comments that 'Griffith theory has proved extraordinarily useful as a mathematical model for studying the effect of cracks on rock, but it is essentially only a mathematical model; on the microscopic scale rocks consist of an aggregate of anisotropic crystals of different mechanical properties and it is these and their grain boundaries which determine the microscopic behaviour'.

Recognition of the difficulty involved in developing a mathematical model which adequately predicts fracture propagation and failure in rock led a number of authors to propose empirical relationships between principal stresses or between shear and normal stresses at failure. Murrell (1965), Hoek (1968), Hobbs (1970) and Bieniawski (1974a) all proposed different forms of empirical criteria. The failure criterion on which the remainder of this Paper is based was presented by Hoek & Brown (1980a, 1980b) and resulted from their efforts to produce an acceptable failure criterion for the design of underground excavations in rock.

AN EMPIRICAL FAILURE CRITERION FOR ROCK

In developing their empirical failure criterion, Hoek & Brown (1980a) attempted to satisfy the following conditions

(*a*) The failure criterion should give good agreement with experimentally determined rock strength values.
(*b*) The failure criterion should be expressed by mathematically simple equations based, to the maximum extent possible, upon dimensionless parameters.
(*c*) The failure criterion should offer the possibility of extension to deal with anisotropic failure and the failure of jointed rock masses.

The studies on fracture initiation and propagation suggested that the parabolic Mohr envelope predicted by the original Griffith theory adequately describes both fracture initiation and failure of brittle materials under conditions where the effective normal stress acting across a pre-existing crack is tensile (negative). This is because fracture propagation follows very quickly upon

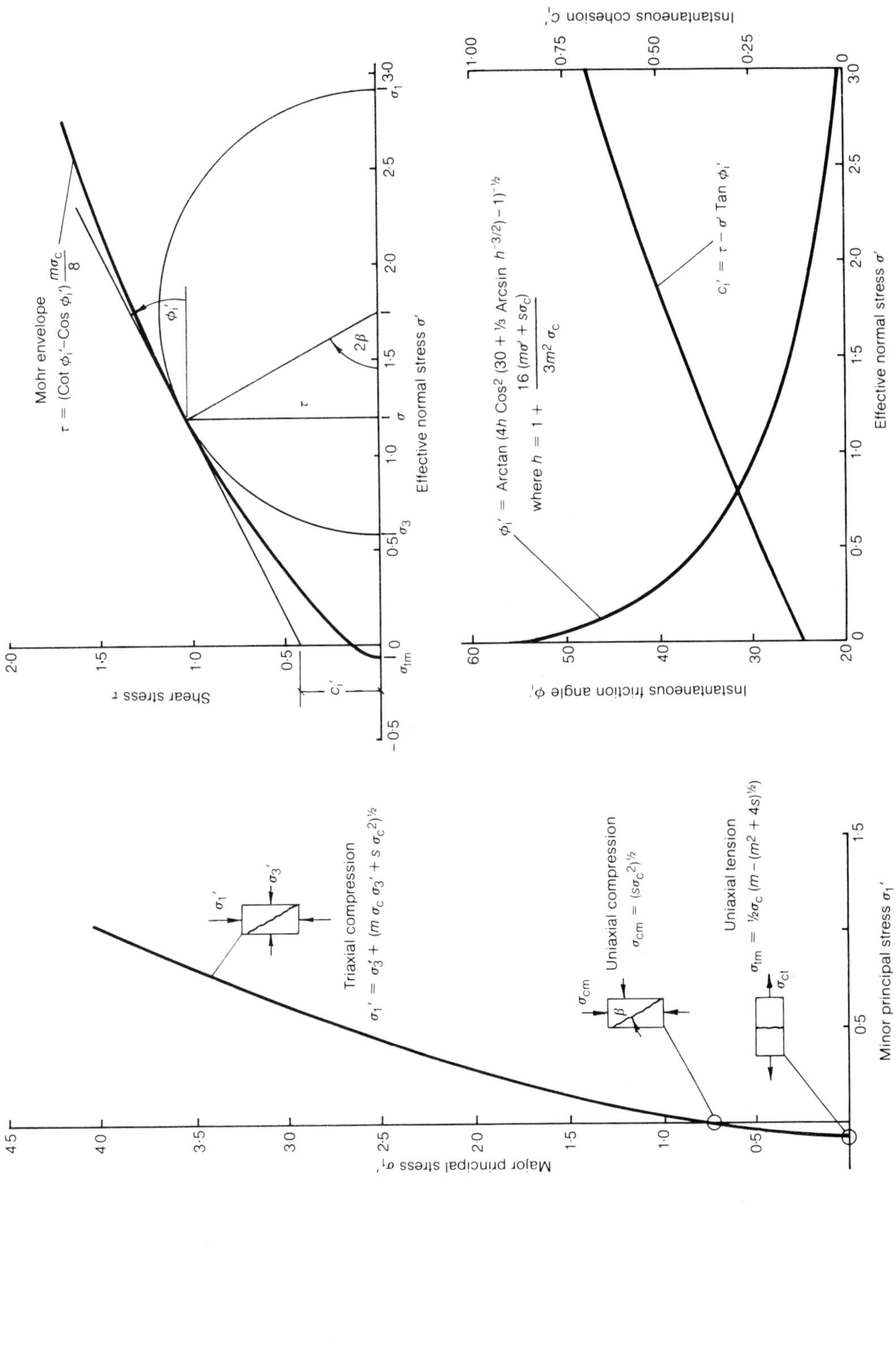

Fig. 3. Summary of equations associated with the non-linear failure criterion proposed by Hoek & Brown (1980b)

fracture initiation under tensile stress conditions, and hence fracture initiation and failure of the specimen are practically indistinguishable.

Figure 2 shows that, when the effective normal stress is compressive (positive), the envelope to the Mohr circles tends to be curvilinear, but not to the extent predicted by the original Griffith theory.

Based upon these observations, Hoek & Brown (1980a) experimented with a number of distorted parabolic curves to find one which gave good coincidence with the original Griffith theory for tensile effective normal stresses, and which fitted the observed failure conditions for brittle rocks subjected to compressive stress conditions.

The process used by Hoek & Brown in deriving their empirical failure criterion was one of pure trial and error. Apart from the conceptual starting point provided by Griffith theory, there is no fundamental relationship between the empirical constants included in the criterion and any physical characteristics of the rock. The justification for choosing this particular criterion over the numerous alternatives lies in the adequacy of its predictions of observed rock fracture behaviour, and the convenience of its application to a range of typical engineering problems.

The Author's background in designing underground excavations in rock resulted in the decision to present the failure criterion in terms of the major and minor principal stresses at failure. The empirical equation defining the relationship between these stresses is

$$\sigma_1' = \sigma_3' + (m\sigma_c\sigma_3' + s\sigma_c^2)^{1/2} \qquad (3)$$

where σ_1' is the major principal effective stress at failure, σ_3' is the minor principal effective stress or, in the case of a triaxial test, the confining pressure, σ_c is the uniaxial compressive strength of the intact rock material from which the rock mass is made up, and m and s are empirical constants.

The constant m always has a finite positive value which ranges from about 0·001 for highly disturbed rock masses, to about 25 for hard intact rock. The value of the constant s ranges from 0 for jointed masses, to 1 for intact rock material.

Substitution of $\sigma_3' = 0$ into equation (3) gives the unconfined compressive strength of a rock mass as

$$\sigma_1' = \sigma_c = (s\sigma_c^2)^{1/2} \qquad (4)$$

Similarly, substitution of $\sigma_1' = 0$ in equation (3), and solution of the resulting quadratic equation for σ_3', gives the uniaxial tensile strength of a rock mass as

$$\sigma_3' = \sigma_t = \tfrac{1}{2}\sigma_c(m - (m^2 + 4s)^{1/2}) \qquad (5)$$

The physical significance of equations (3)–(5) is illustrated in the plot of σ_1' against σ_3' given in Fig. 3.

While equation (3) is very useful in designing underground excavations, where the response of individual rock elements to in situ and induced stresses is important, it is of limited value in designing rock slopes where the shear strength of a failure surface under specified effective normal stress conditions is required. The Mohr failure envelope corresponding to the empirical failure criterion defined by equation (3) was derived by Dr J. Bray of Imperial College and is given by

$$\tau = (\mathrm{Cot}\,\phi_i' - \mathrm{Cos}\,\phi_i')\frac{m\sigma_c}{8} \qquad (6)$$

where τ is the shear stress at failure, ϕ_i' is the instantaneous friction angle at the given values of τ and σ'—i.e. the inclination of the tangent to the Mohr failure envelope at the point (σ', τ) as shown in Fig. 3.

The value of the instantaneous friction angle ϕ_i' is given by

$$\phi_i' = \mathrm{Arctan}\,(4h\,\mathrm{Cos}^2\,(30 + \tfrac{1}{3}\mathrm{Arcsin}\,h^{-3/2}) - 1)^{-1/2} \qquad (7)$$

where

$$h = 1 + \frac{16(m\sigma' + s\sigma_c)}{3m^2\,\sigma_c}$$

and σ' is the effective normal stress.

The instantaneous cohesive strength c_i', shown in Fig. 3, is given by

$$c_i' = \tau - \sigma'\,\mathrm{Tan}\,\phi_i' \qquad (8)$$

From the Mohr circle construction given in Fig. 3, the failure plane inclination β is given by

$$\beta = 45 - \tfrac{1}{2}\phi_i' \qquad (9)$$

An alternative expression for the failure plane inclination, in terms of the principal stresses σ_1' and σ_3', was derived by Hoek & Brown (1980a):

$$\beta = \tfrac{1}{2}\,\mathrm{Arcsin}\frac{\tau_m}{\tau_m + m\sigma_c/8}(1 + m\sigma_c/4\tau_m)^{1/2} \qquad (10)$$

where $\tau_m = \tfrac{1}{2}(\sigma_1' - \sigma_3')$.

CHARACTERISTICS OF EMPIRICAL CRITERION

The empirical failure criterion presented in the preceding section contains three constants; m, s and σ_c. The significance of each of these will be discussed in turn later.

Constants m and s are both dimensionless and are very approximately analogous to the angle of friction, ϕ', and the cohesive strength, c', of the conventional Mohr–Coulomb failure criterion.

Figure 4 illustrates the influence of different values of the constant m on the Mohr failure envelope for intact rock. In plotting these curves,

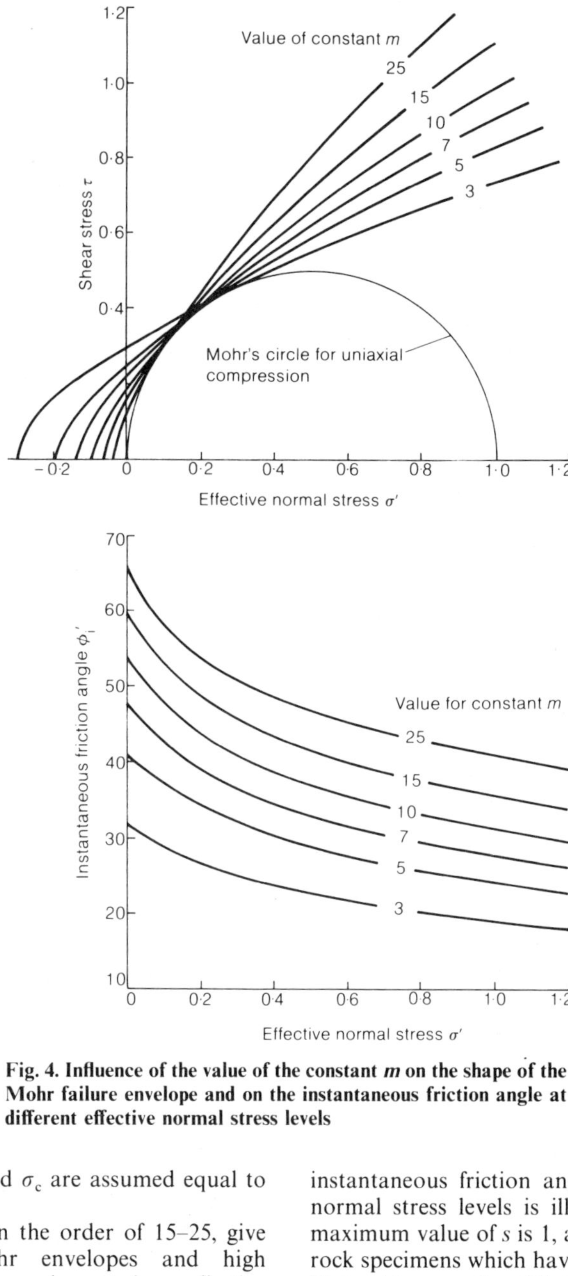

Fig. 4. Influence of the value of the constant *m* on the shape of the Mohr failure envelope and on the instantaneous friction angle at different effective normal stress levels

the values of both s and σ_c are assumed equal to unity.

Large values of m, in the order of 15–25, give steeply inclined Mohr envelopes and high instantaneous friction angles at low effective normal stress levels. These large m values tend to be associated with brittle igneous and metamorphic rocks such as andesites, gneisses and granites. Lower m values, in the order of 3–7, give lower instantaneous friction angles and tend to be associated with more ductile carbonate rocks such as limestone and dolomite.

The influence of the value of the constant s on the shape of the Mohr failure envelope and on the instantaneous friction angle at different effective normal stress levels is illustrated in Fig. 5. The maximum value of s is 1, and this applies to intact rock specimens which have a finite tensile strength (defined by equation (5)). The minimum value of s is zero, and this applies to heavily jointed or broken rock in which the tensile strength has been reduced to zero and where the rock mass has zero cohesive strength when the effective normal stress is zero.

The third constant, σ_c, the uniaxial compressive strength of the intact rock material, has the dimensions of stress. This constant was chosen after very careful consideration of available rock strength data. The unconfined compressive strength is probably the most widely quoted constant in rock mechanics, and it is likely that an estimate of this strength will be available in cases where no other rock strength data are available. Consequently, it was decided that the uniaxial compressive strength σ_c would be adopted as the basic unit of measurement in the empirical failure criterion.

The failure criterion defined by equation (3) can be made entirely dimensionless by dividing both sides by the uniaxial compressive strength

$$\sigma_1'/\sigma_c = \sigma_3'/\sigma_c + (m\sigma_3'/\sigma_c + s)^{1/2} \tag{11}$$

This formulation, which can also be achieved by simply putting $\sigma_c = 1$ in equation (3), is very useful when comparing the shape of Mohr failure envelopes for different rock materials.

A procedure for the statistical determination of the values of the constants m, s and σ_c from experimental data is given in Appendix 1.

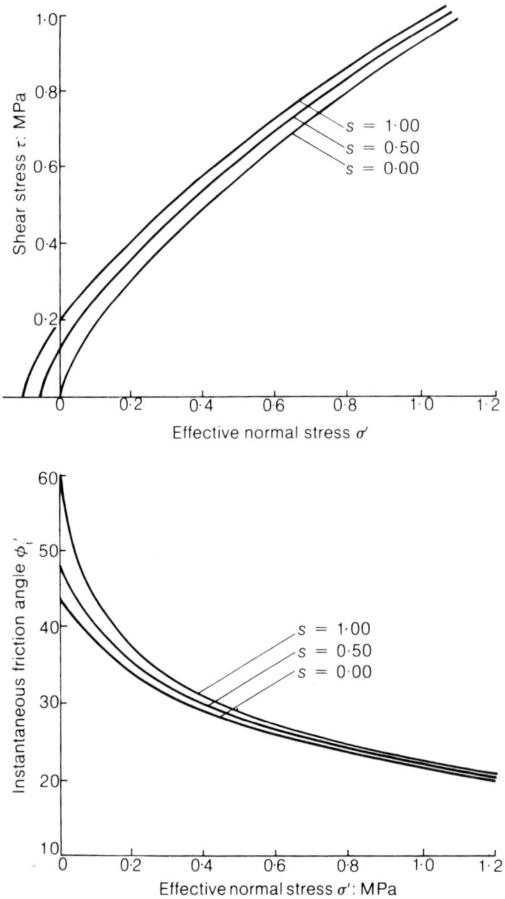

Fig. 5. Influence of the value of the constant s on the shape of the Mohr failure envelope and on the instantaneous friction angle at different effective normal stress levels

TRIAXIAL DATA FOR INTACT ROCK

Hoek & Brown (1980a) analysed published data from several hundred triaxial tests on intact rock specimens and found some useful trends. These trends will be discussed in relationship to the two sets of data plotted as Mohr failure circles in Fig. 6. The sources of the triaxial data plotted in Fig. 6 are given in Table 1.

Figure 6(a) gives Mohr failure envelopes for five different granites from the USA and UK. Tests on these granites were carried out in five different laboratories using different triaxial equipment. In spite of these differences, the failure characteristics of these granites follow a remarkably consistent pattern, and the Mohr failure envelope predicted by equations (6) and (7) for $\sigma_c = 1$, $m = 29{\cdot}2$ and $s = 1$ fits all of these Mohr circles very well. Table 1 shows that a correlation coefficient of 0·99 was obtained by statistically fitting the empirical failure criterion defined by equation (3) to all of the granite strength data.

The term granite defines a group of igneous rocks having very similar mineral composition, grain size and angularity, and hence the failure characteristics exhibited by these rocks are very similar, irrespective of the source of the granite. The trend illustrated in Fig. 6(a) has very important practical implications, since it suggests that it should be possible, given a description of the rock and an estimate of its uniaxial compressive strength, to predict its Mohr failure envelope with a relatively high degree of confidence. This is particularly important in early conceptual or feasibility studies where the amount of reliable laboratory data is very limited.

In contrast to the trends illustrated in Fig. 6(a) for granite, the plot given in Fig. 6(b) for limestone is less convincing. In this case, eleven different limestones, tested in three different laboratories, have been included in the plot. Table 1 shows that the values of the constant m, derived from statistical analyses of the test data, vary from 3·2 to 14·1, and that the correlation coefficient for the complete data set is only 0·68.

The scatter of the data included in Fig. 6(b) is attributed to the fact that the generic term limestone applies to a range of carbonate rocks formed by deposition of a variety of organic and inorganic materials. Consequently, mineral composition, grain size and shape, and the nature of cementing materials between the grains, will vary from one limestone to another.

Comparison of the two plots given in Fig. 6 suggests that the empirical failure criterion presented here gives a useful indication of the general trend of the Mohr failure envelope for

Table 1. Sources of data included in Fig. 6*

Rock type	Location	Reference	Number tested	Uniaxial compressive strength		Each sample		Rock type	
				lb/in^2	MPa	m	r^2†	m	r^2†
Granite	Westerley, USA	Heard *et al.* (1974)	17	31 040	214·0	26·7	1·00		
	Westerley, USA	Wawersik & Brace (1971)	7	43 310	298·6	27·0	1·00		
	Westerley, USA	Brace (1964)	7	49 820	343·5	28·3	0·98		
	Westerley, USA	Mogi (1967)	6	32 440	223·7	32·8	0·99		
	Stone Mountain, USA	Schwartz (1964)	14	16 850	116·2	28·9	0·93	29·2	0·99
	Blackingstone, UK	Franklin & Hoek (1970)	48	30 410	209·7	20·8	0·91		
	Mount Sorrel, UK	Misra (1972)	5	39 910	275·2	26·5	0·99		
	Carnmarth, Redruth, UK	Misra (1972)	5	23 540	162·3	27·7	0·99		
Limestone	Portland, UK	Franklin & Hoek (1970)	30	13 300	91·7	7·5	0·72		
	Indiana, USA	Schwartz (1964)	6	7090	48·9	3·2	0·95		
	Bath, UK	Misra (1972)	7	6830	47·1	5·5	0·97		
	Grindling Stubbs, UK	Misra (1972)	6	19 450	134·1	8·8	0·97		
	Kirbymoorside, UK	Misra (1972)	5	23 830	164·3	12·3	0·98		
	Blackwell, UK	Misra (1972)	5	29 211	201·4	10·0	0·92	5·4	0·68
	Foster Yeoman, UK	Misra (1972)	5	24 265	167·3	14·1	0·95		
	Gigglewick, UK	Misra (1972)	5	22 423	154·6	8·8	0·97		
	Kelmac, UK	Misra (1972)	5	16 897	116·5	7·3	1·00		
	Threshfield, UK	Misra (1972)	5	21 423	147·7	6·9	0·98		
	Swinden Cracoe, UK	Misra (1972)	5	16 027	110·5	8·4	0·96		

* Material constant $s = 1$ for intact rock.
† r^2. is coefficient of derermination or correlation coefficient.

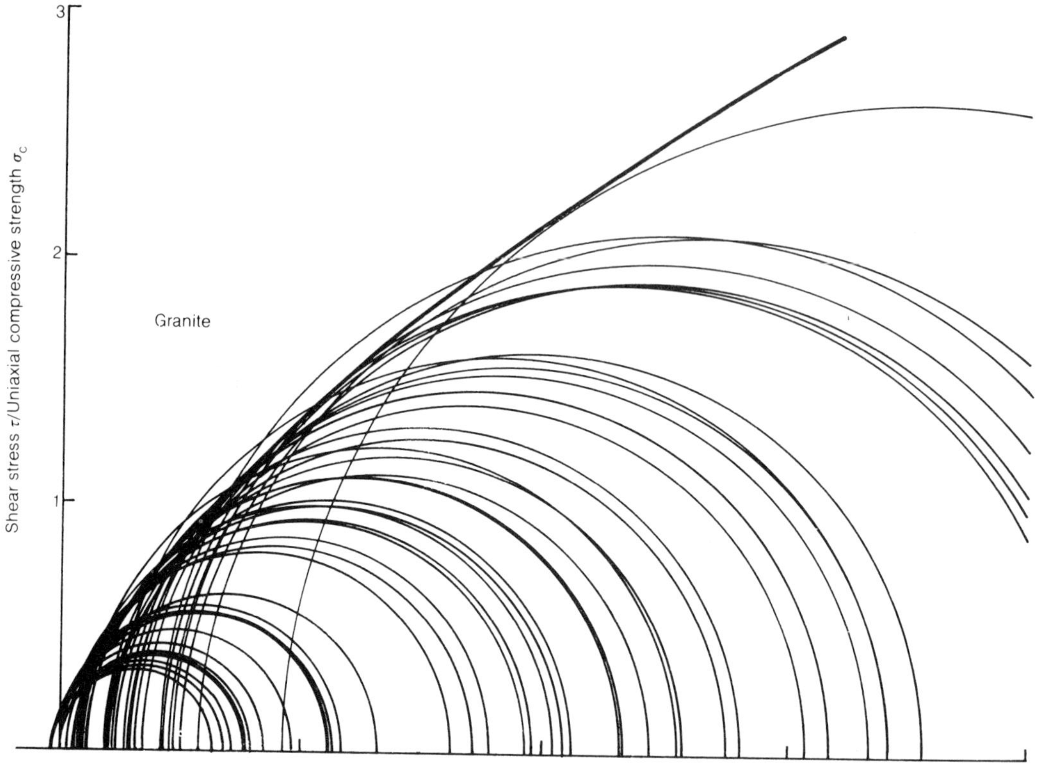

(a)

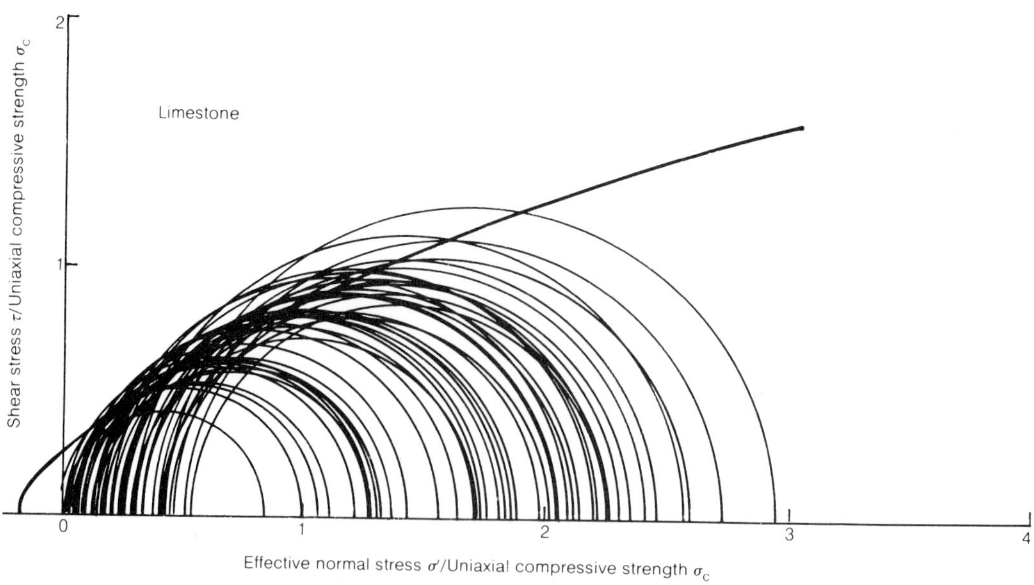

(b)

Fig. 6. Mohr failure circles for published triaxial test data for intact samples of (a) granite and (b) limestone

different rock types. The accuracy of each prediction will depend on the adequacy of the description of the particular rock under consideration. In comparing the granites and limestones included in Fig. 6, there would obviously be a higher priority in carrying out confirmatory laboratory tests on the limestone than on the granite.

Hoek & Brown (1980a) found that there were definite trends which emerged from the statistical fitting of their empirical failure criterion (equation (3)) to published triaxial data. For intact rock (for which $s = 1$), these trends are characterized by the value of the constant m which, as illustrated in Fig. 4, defines the shape of the Mohr failure envelope. The trends suggested by Hoek & Brown (1980a) are

(*a*) Carbonate rocks with well developed crystal cleavage (dolomite, limestone and marble); $m = 7$
(*b*) Lithified argillaceous rocks [mudstone, shale and slate (normal to cleavage)]; $m = 10$
(*c*) Arenaceous rocks with strong crystals and poorly developed crystal cleavage (sandstone and quartzite); $m = 15$
(*d*) Fine grained polymineralic igneous crystalline rocks (andesite, dolerite, diabase and rhyotite); $m = 17$
(*e*) Course grained polymineralic igneous and metamorphic rocks (amphibolite, gabbro, gneiss, granite, norite and granodiorite); $m = 25$

These trends will be utilized later when the estimation of the strength of the jointed rock masses is discussed.

The fitting of the empirical failure criterion defined by equation (3) to a particular set of triaxial data is illustrated in Fig. 7. The Mohr circles plotted were obtained by Bishop & Garga (1969) from a series of carefully performed triaxial tests on undisturbed samples of London clay (Bishop, Webb & Lewin, 1965). The Mohr envelope plotted in Fig. 7 was determined from a statistical analysis of Bishop & Garga's data (using the technique described in Appendix 1), and the values of the constants are $\sigma_c = 211{\cdot}8$ kPa, $m = 6{\cdot}475$ and $s = 1$. The correlation coefficient for the fit of the empirical criterion to the experimental data is 0·98.

This example was chosen for its curiosity value rather than its practical significance, and because of the strong association between the British Geotechnical Society and previous Rankine lecturers and London clay. The example does serve to illustrate the importance of limiting the use of the empirical failure criterion to a low effective normal stress range. Tests on London clay at higher effective normal stress levels by Bishop *et al.* (1965) gave approximately linear Mohr failure envelopes with friction angles of about 11°.

As a rough rule of thumb, when analysing intact rock behaviour, the Author limits the use of the empirical failure criterion to a maximum effective normal stress level equal to the unconfined compressive strength of the material. This question is examined later in a discussion on brittle–ductile transition in intact rock.

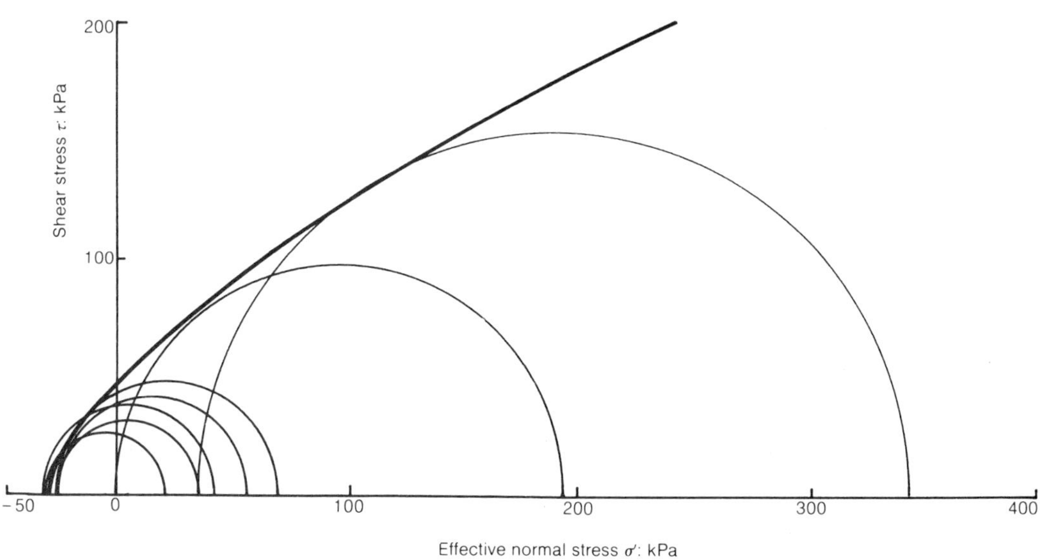

Fig. 7. Mohr failure envelope for drained triaxial tests at very low normal stress levels carried out by Bishop & Garga (1969) on undisturbed samples of London clay

Table 2. Observed and predicted failure plane inclination for Tennessee marble (Wawersik, 1968)

Confining pressure: MPa	Axial strength: MPa	Observed fracture angle	Predicted fracture angle
0	134·48	18·0	26·61
3·45	143·45	23·4	27·0
6·90	160·00	24·8	27·7
13·79	186·21	31·7	28·7
20·69	201·38	35·1	29·1
27·59	220·00	36·3	29·7
34·48	251·03	37·8	30·6
48·28	286·21	38·8	31·4

ASSUMPTIONS INCLUDED IN EMPIRICAL FAILURE CRITERION

A number of simplifying assumptions have been made in deriving the empirical failure criterion, and it is necessary to discuss these assumptions before extending the criterion to deal with jointed rock masses.

Effective stress

Throughout this discussion, it is assumed that the empirical failure criterion is valid for effective stress conditions. In other words, the effective stress σ' used in equations (7) and (8) is obtained from $\sigma' = \sigma - u$, where σ is the applied normal stress and u is the pore or joint water pressure in the rock. In spite of some controversy on this subject, discussed by Jaeger & Cook (1969), Brace & Martin (1968) demonstrate that the effective stress concept appears to be valid in intact rocks of extremely low permeability, provided that loading rates are sufficiently low to permit pore pressures to equalize. For porous rocks such as sandstone, normal laboratory loading rates will generally satisfy effective stress conditions (Handin, Hager, Friedman & Feather, 1963) and there is no reason to suppose that they will not apply in the case of jointed rocks.

Influence of pore fluid on strength

In addition to the influence of pore pressure on strength, it is generally accepted that the pore fluid itself can have a significant influence on rock strength. For example, Colback & Wiid (1965) and Broch (1974) showed that the unconfined compressive strength of quartzitic shale, quartz-diorite, gabbro and gneiss can be reduced by as much as two by saturation in water as compared with oven dried specimens. Analyses of their results suggest that this reduction takes place in the unconfined compressive strength σ_c and not in the constant m of the empirical failure criterion.

It is important, in testing rock materials or in comparing data from rock strength tests, that the moisture content of all specimens be kept within a narrow range. In the Author's own experience in testing samples of shale which had been left standing on the laboratory shelf for varying periods of time, the very large amount of scatter in strength data was almost eliminated by storing the specimens in a concrete curing room to bring them close to saturation before testing. Obviously, in testing rocks for a particular practical application, the specimens should be tested as close to in situ moisture content as possible.

Influence of loading rate

With the exception of effective stress tests on very low porosity materials (e.g. Brace & Martin, 1968), or tests on viscoelastic materials such as salt or potash, it is generally assumed that the influence of loading rate is insignificant when dealing with rock. While this may be an oversimplification, the Author believes that it is sufficiently accurate for most practical applications.

Influence of specimen size

Hoek & Brown (1980a) have analysed the influence of specimen size on the results of strength tests on the intact rock samples. They found that the influence of specimen size can be approximated by the relationship

$$\sigma_c = \sigma_{c50}(50/d)^{0\cdot18} \qquad (12)$$

where σ_c is the unconfined compressive strength, d is the diameter of the specimen in millimetres and σ_{c50} is the unconfined compressive strength of a 50 mm diameter specimen of the same material.

In the case of jointed rocks, the influence of size is controlled by the relationship between the spacing of joints and the size of the sample. This problem is dealt with later in the discussion on jointed rock masses.

Influence of intermediate principal stress

In deriving the empirical failure criterion presented here, Hoek & Brown (1980a) assumed

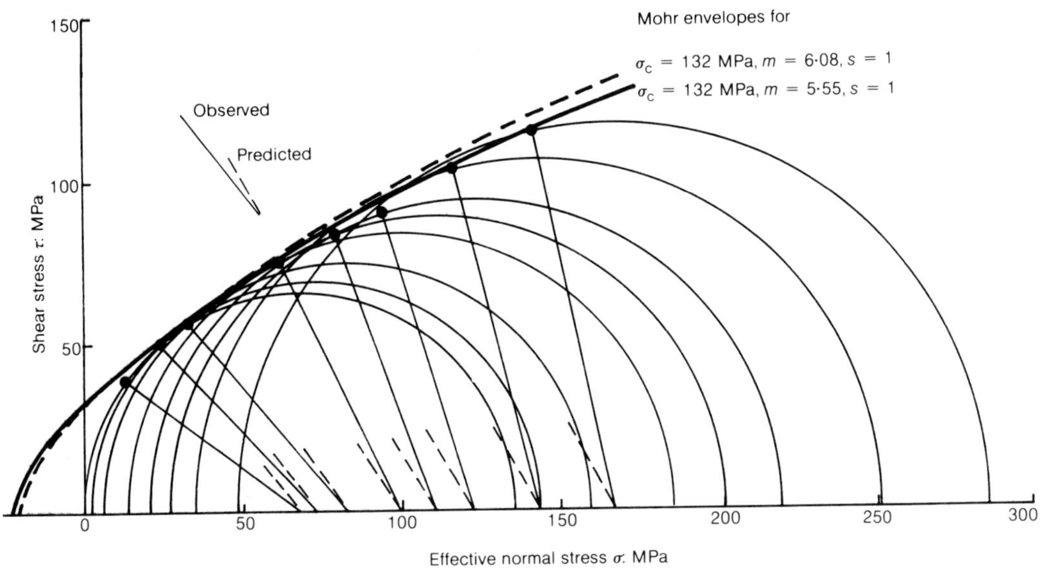

Fig. 8. Plot of Mohr failure circles for Tennessee marble tested by Wawersik (1968) giving comparison between predicted and observed failure plane inclinations

that the failure process is controlled by the major and minor principal stresses σ_1' and σ_3', and that the intermediate principal stress σ_2' has no significant influence upon this process. This is almost certainly an over-simplification, but there appears to be sufficient evidence (reviewed by Jaeger & Cook, 1969) to suggest that the influence of the intermediate principal stress can be ignored without introducing unacceptably large errors.

Failure surface inclination

The inclination of an induced failure plane in an intact rock specimen is given by equations (9) or (10). This inclination is measured from the direction of the maximum principal stress σ_1', as illustrated in Fig. 3.

The results of a series of triaxial tests by Wawersik (1968) on Tennessee marble are listed in Table 2, and plotted as Mohr circles in Fig. 8. Also listed in Table 2 and plotted in Fig. 8, are observed failure plane inclinations.

A statistical analysis of the triaxial test data gives the following constants: $\sigma_c = 132$ MPa, $m = 6{\cdot}08$, $s = 1$, with a correlation coefficient $r^2 = 0{\cdot}99$. The Mohr envelope defined by these constants is plotted as a dashed curve in Fig. 8.

The predicted fracture angles listed in Table 2 have been calculated for $\sigma_c = 132$ MPa and $m = 6{\cdot}08$ by means of equation (10), and there are significant differences between observed and predicted fracture angles.

However, a Mohr envelope fitted through the shear stress (τ) and effective normal stress (σ') points defined by construction (using the Mohr circles), gives a value of $m = 5{\cdot}55$ for $\sigma_c = 132$ MPa and $s = 1$. The resulting Mohr envelope, plotted as a full line in Fig. 8, is not significantly different from the Mohr envelope determined by analysis of the principal stresses.

These findings are consistent with the Author's own experience in rock testing. The fracture angle is usually very difficult to define, and is sometimes obscured altogether. This is because, as discussed earlier, the fracture process is complicated and does not always follow a clearly defined path. When the failure plane is visible, the inclination of this plane cannot be determined to better than $\pm 5°$. In contrast, the failure stresses determined from a carefully conducted set of triaxial tests are usually very clearly grouped, and the pattern of Mohr circles plotted in Fig. 8 is not unusual in intact rock testing.

To conclude, the failure plane inclinations predicted by equations (9) or (10) should be regarded as approximate only, and that, in many rocks, no clearly defined failure surfaces will be visible.

Brittle–ductile transition

The results of a series of triaxial tests carried out by Schwartz (1964) on intact specimens of Indiana limestone are plotted in Fig. 9. A transition from brittle to ductile behaviour appears to occur at a principal stress ratio of approximately $\sigma_1'/\sigma_3' = 4{\cdot}3$.

A study of the failure characteristics of a number

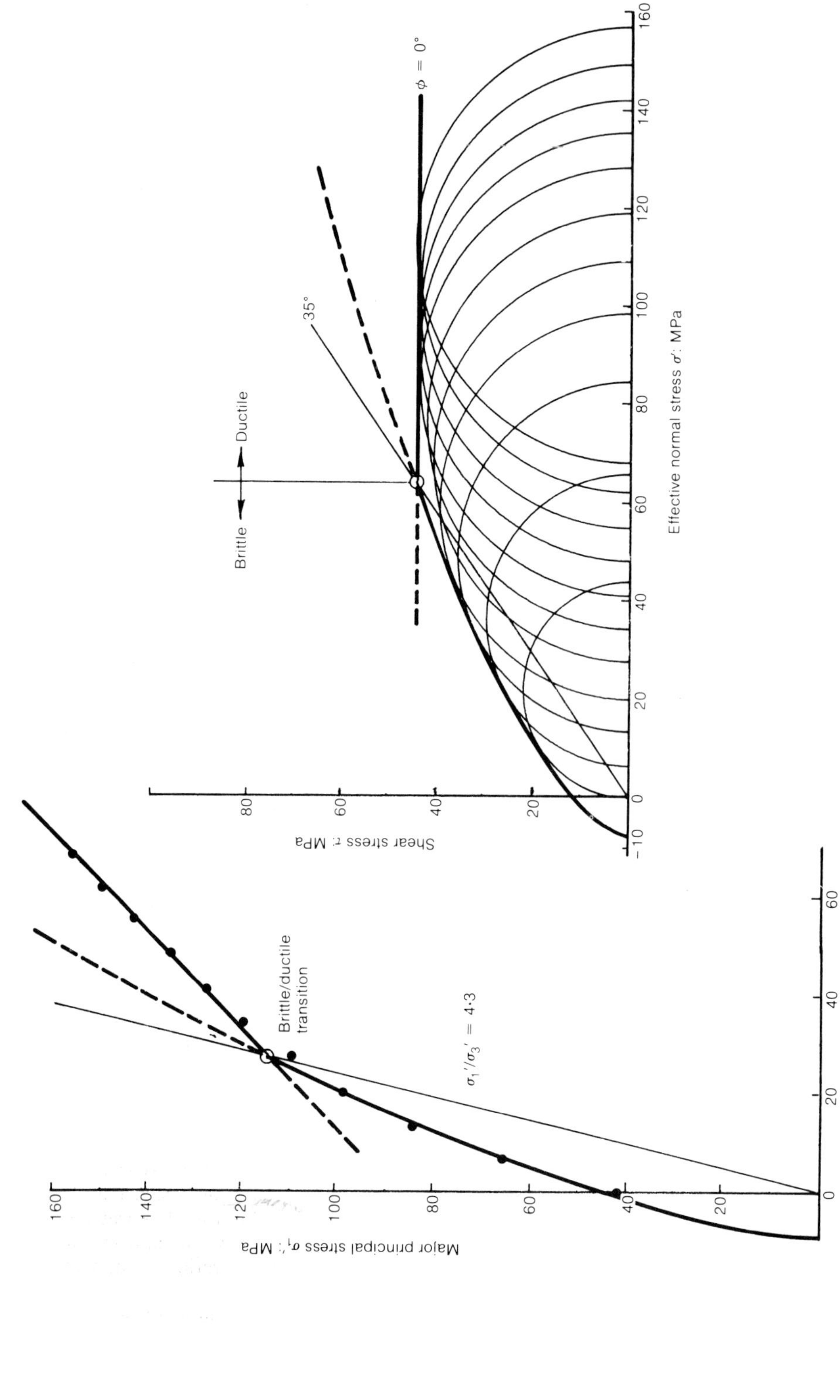

Fig. 9. Results of triaxial tests on Indiana limestone carried out by Schwartz (1964) illustrating brittle–ductile transition

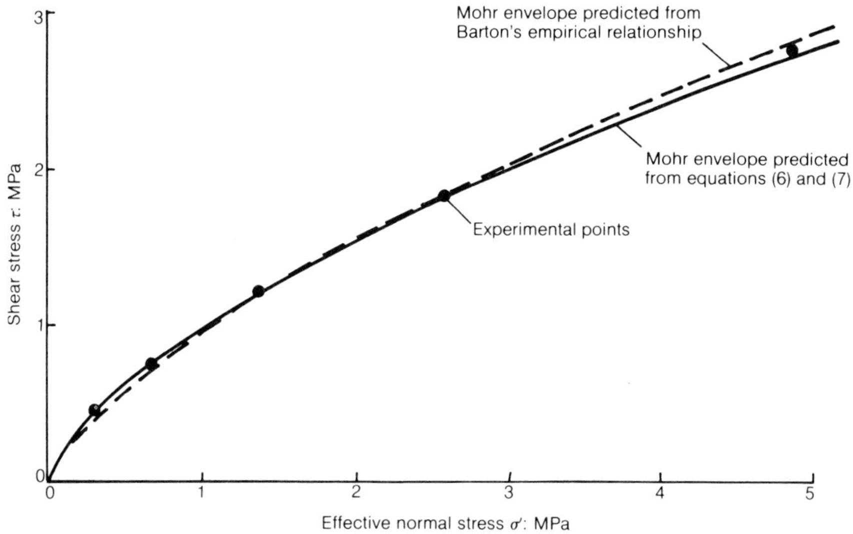

Fig. 10. Results of direct shear tests on moderately weathered greywacke, tested by Martin & Miller (1974), compared with empirical failure envelopes

of rocks by Mogi (1966) led him to conclude that the brittle–ductile transition for most rocks occurs at an average principal stress ratio $\sigma_1'/\sigma_3' = 3{\cdot}4$.

Examination of the results plotted in Fig. 9, and of similar results plotted by Mogi, shows that there is room for a wide variety of interpretations of the critical principal stress ratio, depending on the curve fitting procedure employed and the choice of the actual brittle–ductile transition point. The range of possible values of σ_1'/σ_3' appears to lie between 3 and 5.

A rough rule of thumb used by the Author is that the confining pressure σ' must always be less than the unconfined compressive strength σ_c of the material for the behaviour to be considered brittle. In the case of materials characterized by very low values of the constant m, such as the Indian limestone considered in Fig. 9 ($m = 3{\cdot}2$), the value of $\sigma' = \sigma_c$ may fall beyond the brittle–ductile transition. However, for most rocks encountered in practical engineering applications, this rule of thumb appears to be adequate.

SHEAR STRENGTH OF DISCONTINUITIES

The shear strength of discontinuities in rock has been extensively discussed by a number of authors such as Patton (1966), Goodman (1970), Ladanyi & Archambault (1970), Barton (1971, 1973, 1974), Barton & Choubey (1977) and Richards & Cowland (1982). These discussions have been summarized by Hoek & Bray (1981).

For practical field applications involving the estimation of the shear strength of rough discontinuity surfaces in rock, the Author recommends the following empirical relationship between shear strength (τ) and effective normal stress (σ') proposed by Barton (1971, 1973).

$$\tau = \sigma' \operatorname{Tan}(\phi_b' + \mathrm{JRC}\,\mathrm{Log}_{10}(\mathrm{JCS}/\sigma')) \qquad (13)$$

where ϕ_b' is the 'basic' friction angle of smooth planar discontinuities in the rock under consideration, JRC is a joint roughness coefficient which ranges from 5 for smooth surfaces, to 20 for rough undulating surfaces, and JCS is the joint wall compressive strength which, for clean unweathered discontinuities, equals the uniaxial compressive strength of the intact rock material.

While Barton's equation is very useful for field applications, it is not the only one which can be used for fitting to laboratory shear test data, e.g. Krsmanovic (1967), Martin & Miller (1974) and Hencher & Richards (1982).

Figure 10 gives a plot of direct shear strength data obtained by Martin & Miller (1974) from tests on 150 mm by 150 mm joint surfaces in moderately weathered greywacke (grade 3, test sample number 7). Barton's empirical criterion (equation (13)) was fitted by trial and error, and the dashed curve plotted in Fig. 10 is defined by $\phi_b' = 20°$, JRC = 17 and JCS = 20 MPa.

Also included in Fig. 10 is a Mohr envelope defined by equations (6) and (7) for $\sigma_c = 20$ MPa, $m = 0{\cdot}58$ and $s = 0$ (determined by the method described in Appendix 1). This curve is just as good a fit to the experimental data as Barton's curve.

A number of analyses, such as that presented in Fig. 10, have convinced the Author that equations (6) and (7) provide a reasonably accurate prediction of the shear strength of rough discontinuities in rock under a wide range of effective

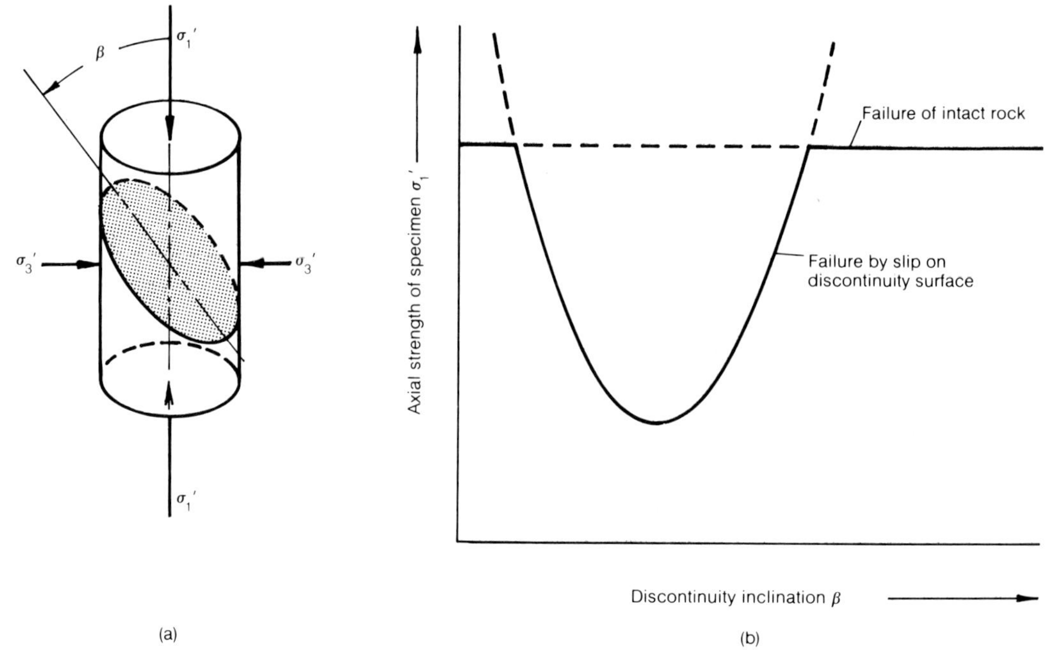

Fig. 11. (a) Configuration of triaxial test specimen containing a pre-existing discontinuity; (b) strength of specimen predicted by means of equations (14) and (3)

normal stress conditions. This fact is useful in the study of schistose and jointed rock mass strength which follows.

STRENGTH OF SCHISTOSE ROCK

In the earlier part of this Paper, the discussion on the strength of intact rock was based on the assumption that the rock was isotropic, i.e. its strength was the same in all directions. A common problem encountered in rock mechanics involves the determination of the strength of schistose or layered rocks such as slates or shales.

If it is assumed that the shear strength of the discontinuity surfaces in such rocks is defined by an instantaneous friction angle ϕ_i' and an instantaneous cohesion c_i' (see Fig. 3), then the axial strength σ_1' of a triaxial specimen containing inclined discontinuities is given by the following equation (see Jaeger & Cook, 1969, pp. 65–68)

$$\sigma_1' = \sigma_3' + \frac{2(c_i' + \sigma_3' \operatorname{Tan} \phi_i')}{(1 - \operatorname{Tan} \phi_i' \operatorname{Tan} \beta) \operatorname{Sin} 2\beta} \quad (14)$$

where σ_3' is the minimum principal stress or confining pressure, and β is the inclination of the discontinuity surfaces to the direction of the major principal stress σ_1 as shown in Fig. 11(a).

Equation (14) can only be solved for values of β within about 25° of the friction angle ϕ'. Very small values of β will give very high values for σ_1', while values of β close to 90° will give negative (and hence meaningless) values for σ_1'. The physical significance of these results is that slip on the discontinuity surfaces is not possible, and failure will occur through the intact material as predicted by equation (3). A typical plot of the axial strength σ_1' against the angle β is given in Fig. 11(b).

If it is assumed that the shear strength of the discontinuity surfaces can be defined by equations (6) and (7), as discussed previously, then in order to determine the values of ϕ_i' and c_i' for substitution into equation (14), the effective normal stress σ' acting across the discontinuity must be known. This is found from

$$\sigma' = \tfrac{1}{2}(\sigma_1' + \sigma_3') - \tfrac{1}{2}(\sigma_1' - \sigma_3') \operatorname{Cos} 2\beta \quad (15)$$

However, since σ_1' is the strength to be determined, the following iterative process can be used

(*a*) Calculate the strength σ_{1i}' of the intact material by means of equation (3), using the appropriate values of σ_c, m and s.

(*b*) Determine values of m_j and s_j for the joint (discontinuity) surfaces from direct shear or triaxial test data. The value of σ_c, the unconfined compressive strength, is the same for the intact material and the discontinuity surfaces in this analysis.

(*c*) Use the value of σ_{1i}', calculated in (*a*), to obtain the first estimate of the effective normal stress σ' from equation (15).

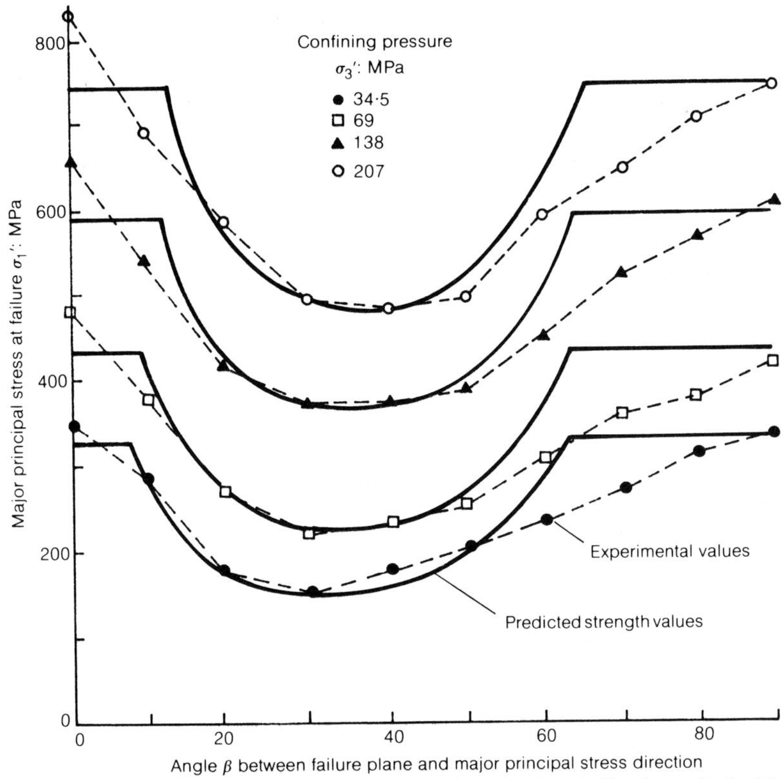

Fig. 12. Triaxial test results for slate with different failure plane inclinations, obtained by McLamore & Gray (1967), compared with strength predictions from equations (3) and (14)

(*d*) Calculate τ, ϕ_i' and c_i' from equations (6)–(8), using the value of m_j and s_j from (*b*), and the value of σ' from (*c*).

(*e*) Calculate the axial strength σ_{1j}' from equation (14).

(*f*) If σ_{1j}' is negative or greater than σ_{1i}', the failure of the intact material occurs in preference to slip on the discontinuity, and the strength of the specimen is defined by equation (3).

(*g*) If σ_{1j}' is less than σ_{1i}' then failure occurs as a result of slip on the discontinuity. In this case, return to (*c*) and use the axial strength calculated in (*e*) to calculate a new value for the effective normal stress σ'.

(*h*) Continue this iteration until the difference between successive values of σ_{1j}' in (*e*) is less than 1%. Only three or four iterations are required to achieve this level of accuracy.

Examples of the analysis described above are given in Figs 12 and 13.

The results of triaxial tests on slate tested by McLamore & Gray (1967) for a range of confining pressures and cleavage orientations are plotted in Fig. 12. The solid curves have been calculated, using the method outlined above, for $\sigma_c = 217$ MPa (unconfined strength of intact rock), $m = 5{\cdot}25$ and $s = 1{\cdot}00$ (constants for intact rock), and $m_j = 1{\cdot}66$ and $s_j = 0{\cdot}006$ (constants for discontinuity surfaces).

The values of the constants m_j and s_j for the discontinuity surfaces were determined by statistical analysis of the minimum axial strength values, using the procedure for broken rock described in Appendix 1.

A similar analysis is presented in Fig. 13, which gives results from triaxial tests on sandstone by Horino & Ellikson (1970). In this case the discontinuity surfaces were created by intentionally fracturing intact sandstone in order to obtain very rough fresh surfaces. The constants used in plotting the solid curves in Fig. 13 were $\sigma_c = 177{\cdot}7$ MPa (intact rock strength), $m = 22{\cdot}87$ and $s = 1{\cdot}00$ (constants for intact rock), and $m_j = 4{\cdot}07$ and $s_j = 0$ (constants for induced fracture planes).

The rougher failure surfaces in the sandstone, as compared to the slate (compare values of m_j), give more sudden changes in axial strength with discontinuity inclination. In both these cases, and in a number of other examples analysed, the agreement between measured and predicted

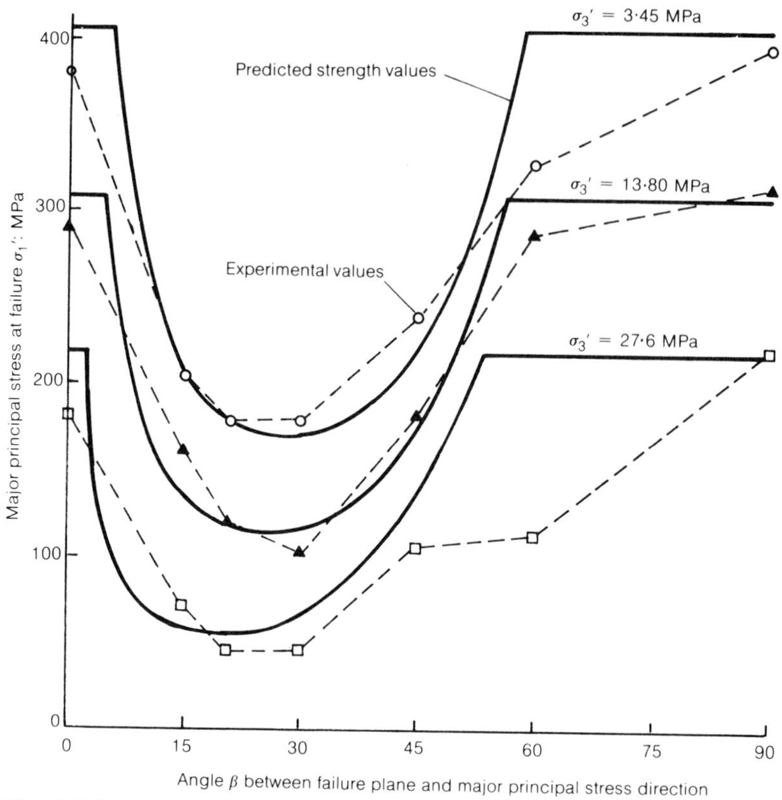

Fig. 13. Triaxial test results for fractured sandstone, tested by Horino & Ellikson (1970), compared with predicted anisotropic strength

strengths is adequate for most practical design purposes.

An example of the application of the analysis of anisotropic failure, is given later. This example involves the determination of the stress distribution and potential failure zones in highly stressed schistose rock surrounding a tunnel.

FAILURE OF JOINTED ROCK MASSES

Having studied the strength of intact rock and of discontinuities in rock, the next logical step is to attempt to predict the behaviour of a jointed rock mass containing several sets of discontinuities. The simplest approach to this problem is to superimpose a number of analyses for individual discontinuity sets, such as those presented in Figs 12 and 13, in the hope that the overall behaviour pattern obtained would be representative of the behaviour of an actual jointed rock mass.

Verification of the results of such predictions presents very complex experimental problems, and many research workers have resorted to the use of physical models in an attempt to minimize these experimental difficulties. Lama & Vutukuri (1978) have summarized the results of model studies carried out by John (1962), Muller & Pacher (1965), Lajtai (1967), Einstein, Nelson, Bruhn & Hirschfield (1969), Ladanyi & Archambault (1970, 1972), Brown (1970), Brown & Trollope (1970), Walker (1971) and others. One of these studies, published by Ladanyi & Archambault (1972), will be considered here.

Ladanyi & Archambault constructed models from rods, with a square cross-section of 12·7 mm by 12·7 mm and a length of 63·5 mm, which had been sawn from commercial compressed concrete bricks. The Mohr failure envelopes for the intact concrete material and for the sawn 'joints' in the model are given in Fig. 14. These curves were derived by statistical analysis of raw test data supplied by Professor Ladanyi.

One of the model configurations used by Ladanyi & Archambault (1972) is illustrated in Fig. 15. Failure of the model in the direction of the 'cross joints' (inclined at an angle α to the major principal stress direction) involves fracture of intact material as well as sliding on the joints. A crude first approximation of the model strength in the α direction is obtained by simple averaging of the Mohr failure envelopes for the intact material

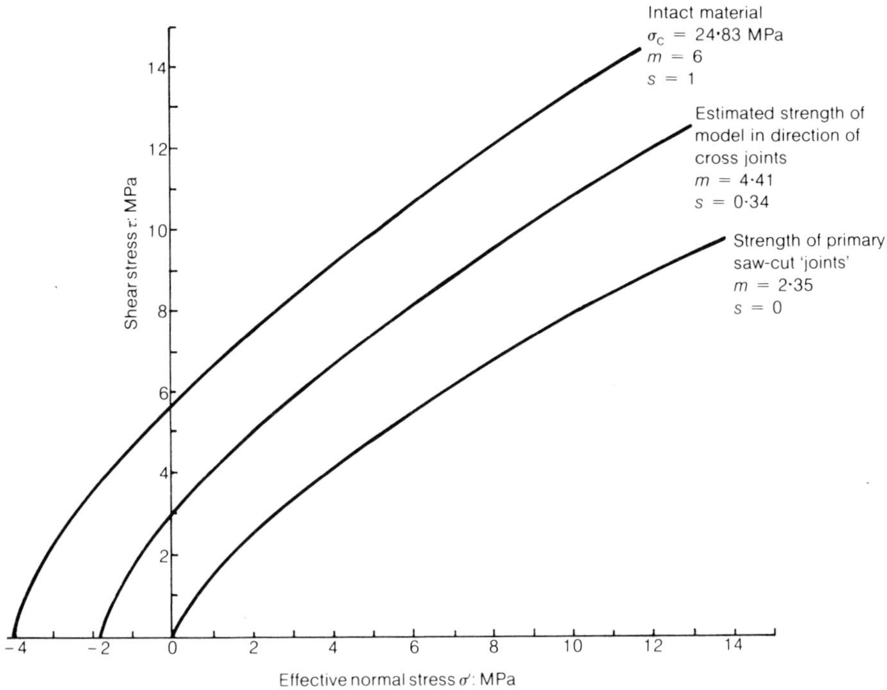

Fig. 14. Mohr failure envelopes for brickwall model tested by Ladanyi & Archambault (1972)

and the through-going joints. The resulting strength estimate is plotted as a Mohr envelope in Fig. 14.

The predicted strength behaviour of Ladanyi & Archambaults' 'brickwall' model, for different joint orientations and lateral stress levels, is given in Fig. 16(a). These curves have been calculated, from the strength values given in Fig. 14, by means of equations (14), (15) and (3). The actual strength values measured by Ladanyi & Archambault are plotted in Fig. 16(b). Comparison between these two figures leads to the following conclusions

(*a*) There is an overall similarity between predicted and observed strength behaviour which suggests that the approach adopted in deriving the curves plotted in Fig. 16(a) is not entirely inappropriate.

(*b*) The observed strengths are generally lower than the predicted strengths. The intact material strength is not achieved, even at the most favourable joint orientations. The sharply defined transitions between different failure modes, predicted in Fig. 16(a), are smoothed out by rotation and crushing of individual blocks. This behaviour is illustrated in the series of photographs reproduced in Fig. 17. In particular, the formation of 'kink bands', as illustrated in Fig. 17(c), imparts a great deal of mobility to the model and results in a

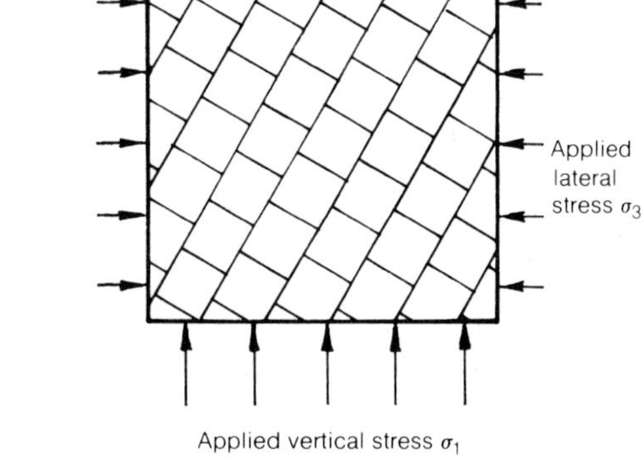

Fig. 15. Configuration of brickwall model tested by Ladanyi & Archambault (1972)

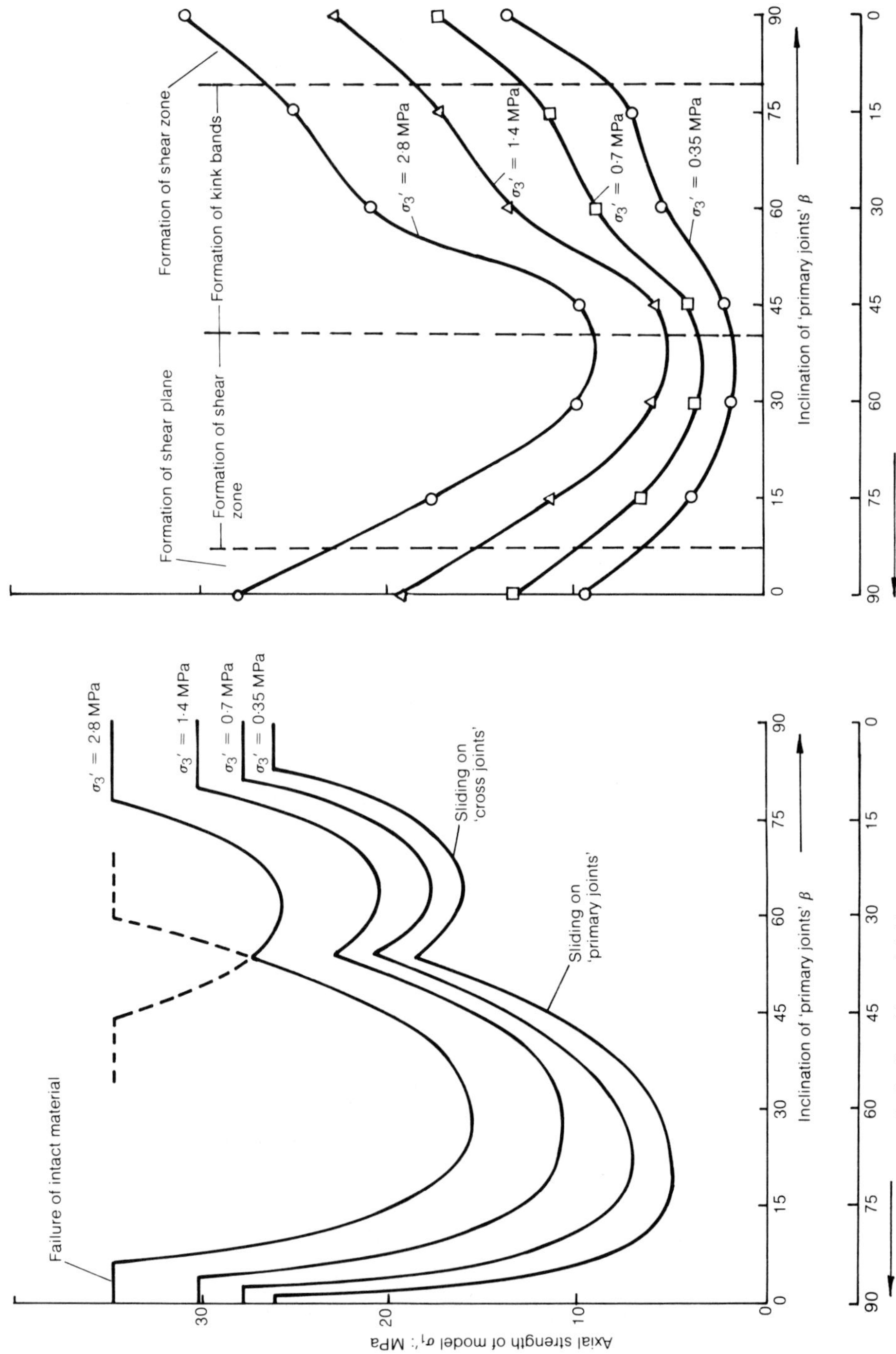

Fig. 16. Comparison between (a) predicted and (b) observed strength of brickwall model tested by Ladanyi & Archambault (1972)

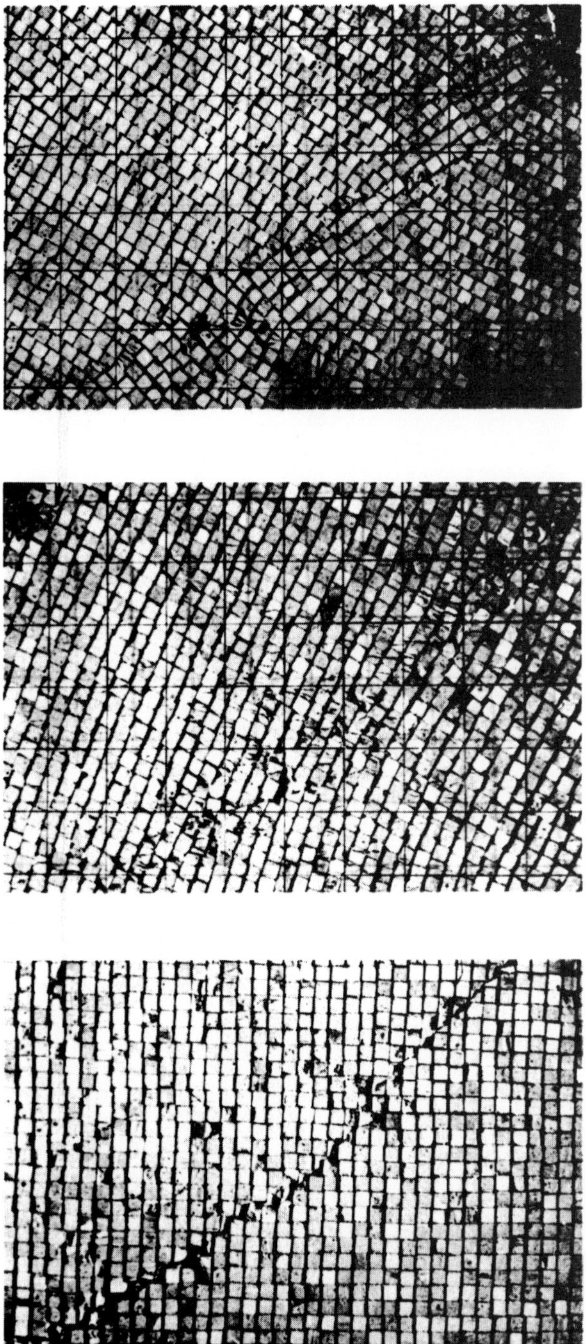

Fig. 17. (a) Shear plane failure; (b) shear zone failure; and (c) kink band failure observed in concrete brick models tested by Ladanyi & Archambault (1972). Photograph reproduced with the permission of Professor B. Ladanyi

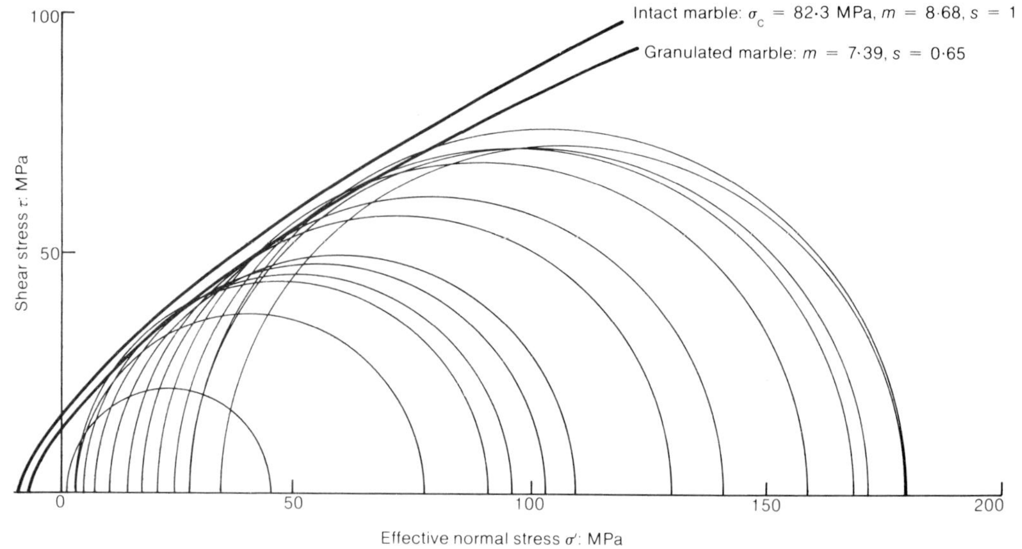

Fig. 18. Comparison between the strengths of intact and granulated Carrara marble tested by Gerogiannopoulos (1979)

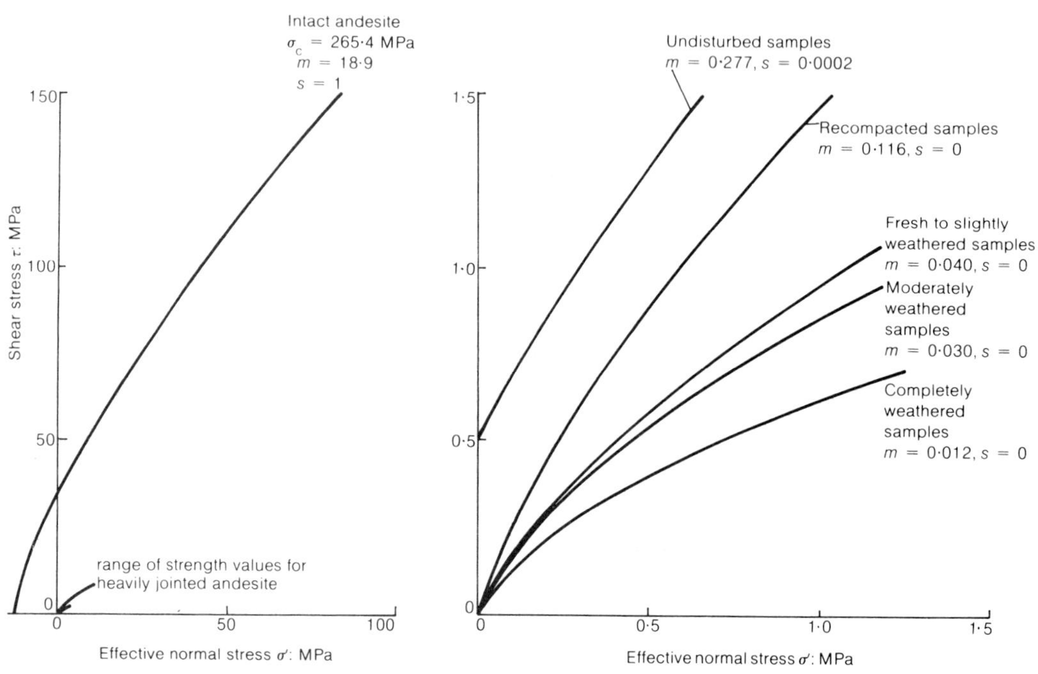

Fig. 19. Mohr failure envelopes for (a) intact and (b) heavily jointed Panguna andesite from Bougainville, Papua New Guinea (see Table 3 for description of materials)

Table 3. Details of materials and test procedures for Panguna andesite

Material	Tested by	Sample diameter: mm	Material constants
Intact Panguna andesite	Jaeger (1970) Golder Associates	25 50	$\sigma_c = 265{\cdot}4$ MPa $m = 18{\cdot}9$ $s = 1$ Correlation coefficient 0·85
Undisturbed core samples of heavily jointed andesite obtained by triple-tube diamond core drilling in exploration adit	Jaeger (1970)	152	$m = 0{\cdot}277$ $s = 0{\cdot}0002$ Correlation coefficient 0·99
Recompacted sample of heavily jointed andesite collected from mine benches (equivalent to compacted fresh rock-fill)	Bougainville Copper	152	$m = 0{\cdot}116$ $s = 0$
Fresh to slightly weathered andesite, lightly recompacted	Snowy Mountains Engineering Corporation	570	$m = 0{\cdot}040$ $s = 0$
Moderately weathered andesite, lightly recompacted	Snowy Mountains Engineering Corporation	570	$m = 0{\cdot}030$ $s = 0$
Completely weathered andesite (equivalent to poor quality waste rock)	Snowy Mountains Engineering Corporation	570	$m = 0{\cdot}012$ $s = 0$

significant strength reduction in the zone defined by $15 > \alpha > 45$, as shown in Fig. 16(b).

(*c*) Intuitive reasoning suggests that the degree of interlocking of the model blocks is of major significance in the behaviour of the model since this will control the freedom of the blocks to rotate. In other words, the freedom of a rock mass to dilate will depend on the interlocking of individual pieces of rock which, in turn, will depend on the particle shape and degree of disturbance to which the mass has been subjected. This reasoning is supported by experience in strength determination of rockfill where particle strength and shape, particle size distribution and degree of compaction are all important factors in the overall strength behaviour.

(*d*) Extension of the principle of strength prediction used in deriving the curves presented in Fig. 16(a) to rock masses, containing three, four or five sets of discontinuities, suggests that the behaviour of such rock masses would approximate to that of a homogeneous isotropic system. In practical terms, this means that, for most rock masses containing a number of joint sets with similar strength characteristics, the overall strength behaviour will be similar to that of a very tightly interlocking rockfill.

The importance of the degree of interlocking between particles in a homogeneous rock mass can be illustrated by considering the results of an ingenious experiment carried out by Rosengren & Jaeger (1968), and repeated by Gerogiannopoulos (1979). By heating specimens of coarse grained marble to about 600°C, the cementing material between grains is fractured by differential thermal expansion of the grains themselves. The material produced by this process is a very low porosity assemblage of extremely tightly interlocking but independent grains. This 'granulated' marble was tested by Rosengren & Jaeger (1968) and Gerogiannopoulos (1979) in an attempt to simulate the behaviour of an undisturbed jointed rock mass.

The results obtained by Gerogiannopoulos from triaxial tests on both intact and granulated Carrara marble are plotted in Fig. 18. In order to avoid confusion, Mohr failure circles for the granulated material only are included in this figure. However, statistical analyses of the data sets for both intact and granulated materials to obtain σ_c, m and s values gave correlation coefficients in excess of 90%.

Figure 18 shows that the strength difference between intact material and a very tightly interlocking assemblage of particles of the same material is relatively small. It is unlikely that this degree of interlocking would exist in an in situ rock mass, except in very massive rock at considerable depth below surface. Consequently, the Mohr failure envelope for granulated marble, presented in Fig. 18, represents the absolute upper bound for jointed rock mass strength.

A more realistic assessment of the strength of heavily jointed rock masses can be made on the basis of triaxial test data obtained in connection with the design of the slopes for the Bougainville open pit copper mine in Papua New Guinea. The results of some of these tests, carried out by Jaeger (1970), the Snowy Mountain Engineering Corporation and in the mine laboratories, have been summarized by Hoek & Brown (1980a).

The results of tests on Panguna andesite are plotted as Mohr envelopes in Fig. 19. Fig. 19(a) has been included to show the large strength difference between the intact material and the jointed rock mass. Fig. 19(b) is an enlargement of the low stress portion of Fig. 19(a), and gives details of the test results on the jointed material. Details of the materials tested are given in Table 3.

Particular mention must be made of the undisturbed 152 mm diameter core samples of jointed Panguna andesite tested by Jaeger (1970). These samples were obtained by careful triple-tube diamond core drilling in an exploration adit in the mine. They were shipped to Canberra, Australia, in the inner tubes of the core barrels, and then carefully transferred onto thin copper sheets which were soldered to form containers for the specimens. These specimens were then rubber sheathed and tested triaxially. This series of tests is, as far as the Author is aware, the most reliable set of tests ever carried out on 'undisturbed' jointed rock.

The entire Bougainville testing programme extended over a ten year period and cost several hundred thousand pounds. This level of effort was justified because of the very large economic and safety considerations involved in designing a final slope of almost 1000 m high for one side of the open pit. Unfortunately, it is seldom possible to justify testing programmes of this magnitude in either mining or civil engineering projects, and hence the results summarized in Fig. 19 represent a very large proportion of the sum total of all published data on this subject.

A similar, although less ambitious, series of tests was carried out on a highly fractured greywacke sandstone by Raphael & Goodman (1979). The results of these tests, plotted in Fig. 20, show a much lower reduction from intact to jointed rock mass strength than for the Panguna andesite (Fig. 19). This is presumably because the intact sandstone tested by Raphael & Goodman is significantly weaker than the andesite tested by Jaeger, and hence there is less possibility of the block rotation mechanism (see Fig. 17(c)) which appears to contribute so much to the weakness of jointed systems in strong materials. This suggestion is highly speculative, and is based on intuitive reasoning rather than experimental facts.

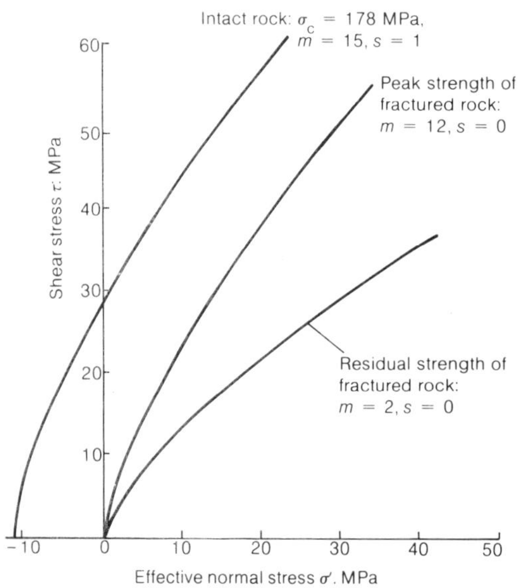

Fig. 20. Mohr failure envelopes estimated from plotted triaxial test data (Raphael & Goodman, 1979) for highly fractured, fresh to slightly altered greywacke sandstone

ESTIMATING THE STRENGTH OF JOINTED ROCK MASSES

Based on their analyses of the results from tests on models, jointed rock masses and rockfill, Hoek & Brown (1980b) proposed an approximate method for estimating the strength of jointed rock masses. This method, summarized in Table 4, involves estimating the values of the empirical constants m and s from a description of the rock mass. These estimates, together with an estimate of the uniaxial compressive strength of the intact pieces of rock, can then be used to construct an approximate Mohr failure envelope for the jointed rock mass.

As a means of assisting the user in describing the rock mass, use is made of the rock mass classification systems proposed by Bieniawski (1974b) and Barton, Lien & Lunde (1974), which has been summarized by Hoek & Brown (1980a).

The Author's experience in using the values listed in Table 4 for practical engineering design suggests that they are somewhat conservative. In other words, the actual rock mass strength is higher than that estimated from the Mohr envelopes plotted from the values listed. It is very difficult to estimate the extent to which the predicted strengths are too low, since reliable field data are almost non-existent. However, based on comparisons between observed and predicted behaviour of rock slopes and underground excavations, the Author tends to regard the strength estimates made from Table 4 as lower bound values for design purposes. Obviously, in

Table 4. Approximate relationship between rock mass quality and material constants

Empirical failure criterion $\sigma_1' = \sigma_3' + (m\sigma_c\sigma_3' + s\sigma_c^2)^{1/2}$ σ_1' = major principal stress σ_3' = minor principal stress σ_c = uniaxial compressive strength of intact rock m, s = empirical constants		Carbonate rocks with well developed crystal cleavage, e.g. dolomite, limestone and marble	Lithified argillaceous rocks, e.g. mudstone, siltstone, shale and slate (tested normal to cleavage)	Arenaceous rocks with strong crystals and poorly developed crystal cleavage, e.g. sandstone and quartzite	Fine grained polymineralic igneous crystalline rocks, e.g. andesite, dolerite, diabase and rhyolite	Coarse grained polymineralic igneous and metamorphic crystalline rocks, e.g. amphibolite, gabbro, gneiss, granite, norite and quartzdiorite
Intact rock samples Laboratory size samples free from pre-existing fractures Bieniawski, 1974b (CSIR)* rating Barton *et al.*, 1974 (NGI)† rating	 100 500	$m = 7$ $s = 1$	$m = 10$ $s = 1$	$m = 15$ $s = 1$	$m = 17$ $s = 1$	$m = 25$ $s = 1$
Very good quality rock mass Tightly interlocking undisturbed rock with rough unweathered joints spaced at 1 to 3 m Bieniawski, 1974b (CSIR) rating Barton *et al.*, 1974 (NGI) rating	 85 100	$m = 3{\cdot}5$ $s = 0{\cdot}1$	$m = 5$ $s = 0{\cdot}1$	$m = 7{\cdot}5$ $s = 0{\cdot}1$	$m = 8{\cdot}5$ $s = 0{\cdot}1$	$m = 12{\cdot}5$ $s = 0{\cdot}1$
Good quality rock mass Fresh to slightly weathered rock, slightly disturbed with joints spaced at 1 to 3 m Bieniawski, 1974b (CSIR) rating Barton *et al.*, 1974 (NGI) rating	 65 10	$m = 0{\cdot}7$ $s = 0{\cdot}004$	$m = 1$ $s = 0{\cdot}004$	$m = 1{\cdot}5$ $s = 0{\cdot}004$	$m = 1{\cdot}7$ $s = 0{\cdot}004$	$m = 2{\cdot}5$ $s = 0{\cdot}004$
Fair quality rock mass Several sets of moderately weathered joints spaced at 0·3 to 1 m, disturbed Bieniawski, 1974b (CSIR) rating Barton *et al.*, 1974 (NGI) rating	 44 1	$m = 0{\cdot}14$ $s = 0{\cdot}0001$	$m = 0{\cdot}20$ $s = 0{\cdot}0001$	$m = 0{\cdot}30$ $s = 0{\cdot}0001$	$m = 0{\cdot}34$ $s = 0{\cdot}0001$	$m = 0{\cdot}50$ $s = 0{\cdot}0001$
Poor quality rock mass Numerous weathered joints at 30 to 500 mm with some gouge. Clean, compacted rockfill Bieniawski, 1974b (CSIR) rating Barton *et al.*, 1974 (NGI) rating	 23 0·1	$m = 0{\cdot}04$ $s = 0{\cdot}00001$	$m = 0{\cdot}05$ $s = 0{\cdot}00001$	$m = 0{\cdot}08$ $s = 0{\cdot}00001$	$m = 0{\cdot}09$ $s = 0{\cdot}00001$	$m = 0{\cdot}13$ $s = 0{\cdot}00001$
Very poor quality rock mass Numerous heavily weathered joints spaced at 50 mm with gouge. Waste rock Bieniawski, 1974b (CSIR) rating Barton *et al.*, 1974 (NGI) rating	 3 0·01	$m = 0{\cdot}007$ $s = 0$	$m = 0{\cdot}010$ $s = 0$	$m = 0{\cdot}015$ $s = 0$	$m = 0{\cdot}017$ $s = 0$	$m = 0{\cdot}025$ $s = 0$

* CSIR Commonwealth Scientific and Industrial Research Organization.
† NGI Norway Geotechnical Institute.

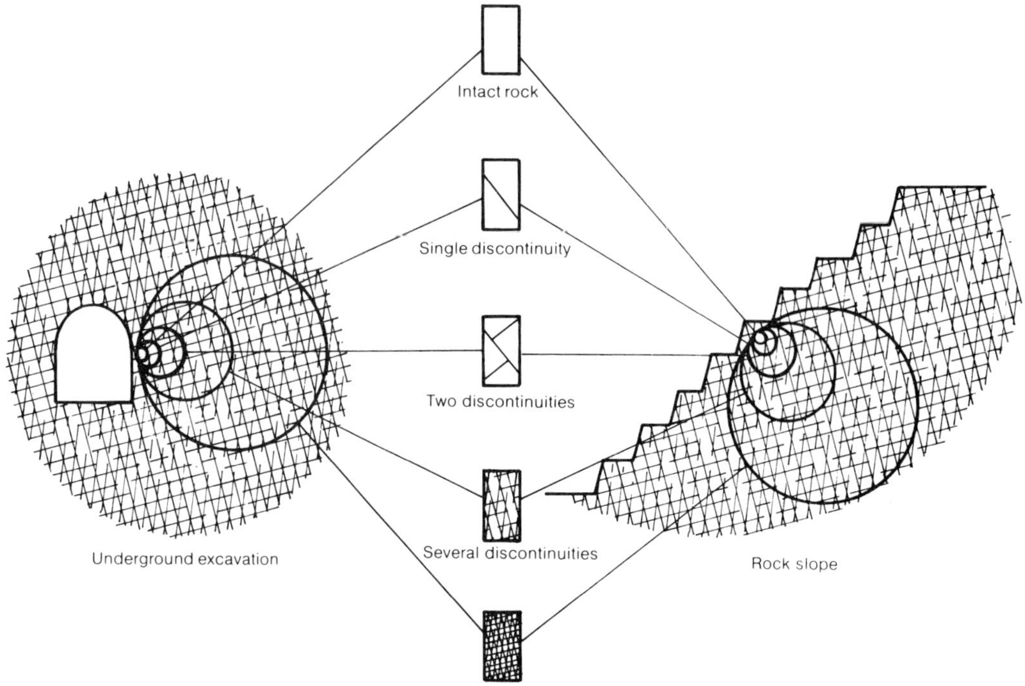

Fig. 21. Simplified representation of the influence of scale on the type of rock mass behaviour model which should be used in designing underground excavations or rock slopes

designing an important structure, the user would be well advised to attempt to obtain his own test data before deciding to use strength values significantly higher than those given by Table 4.

In order to use Table 4 to make estimates of rock mass strength, the following steps are suggested:

(*a*) From a geological description of the rock mass, and from a comparison between the size of the structure being designed and the spacing of discontinuities in the rock mass (see Fig. 21), decide which type of material behaviour model is most appropriate. The values listed in Table 4 should only be used for estimating the strength of intact rock or of heavily jointed rock masses containing several sets of discontinuities of similar type. For schistose rock or for jointed rock masses containing dominant discontinuities such as faults, the behaviour will be anisotropic and the strength should be dealt with in the manner described in Example 1.

(*b*) Estimate the unconfined compressive strength σ_c of the intact rock pieces from laboratory test data, index values or descriptions of rock hardness (see Hoek & Bray, 1981 or Hoek & Brown, 1980a). This strength estimate is important since it establishes the scale of the Mohr failure envelope.

(*c*) From a description of the rock mass or, preferably, from a rock mass classification using the system of Barton *et al.* (1974) or Bieniawski (1974b), determine the appropriate row and column in Table 4.

(*d*) Using equations (6) and (7), calculate and plot a Mohr failure envelope for the estimated values of σ_c, m and s. Draw an approximate average Mohr–Coulomb linear envelope through the plotted points, and estimate the average friction angle and cohesive strength of the rock mass. Compare these values with published data for rockfill (Marachi, *et al.*, 1972; Marsal, 1967, 1973; Charles & Watts, 1980) or with data given in this Paper to ensure that the values are reasonable.

(*e*) Use the estimated strength values for preliminary design purposes and carry out sensitivity studies by varying the values of m and s to determine the importance of rock mass strength in the design.

(*f*) For critical designs which are found to be very sensitive to variations in rock mass strength, establish a site investigation and laboratory testing programme aimed at refining the

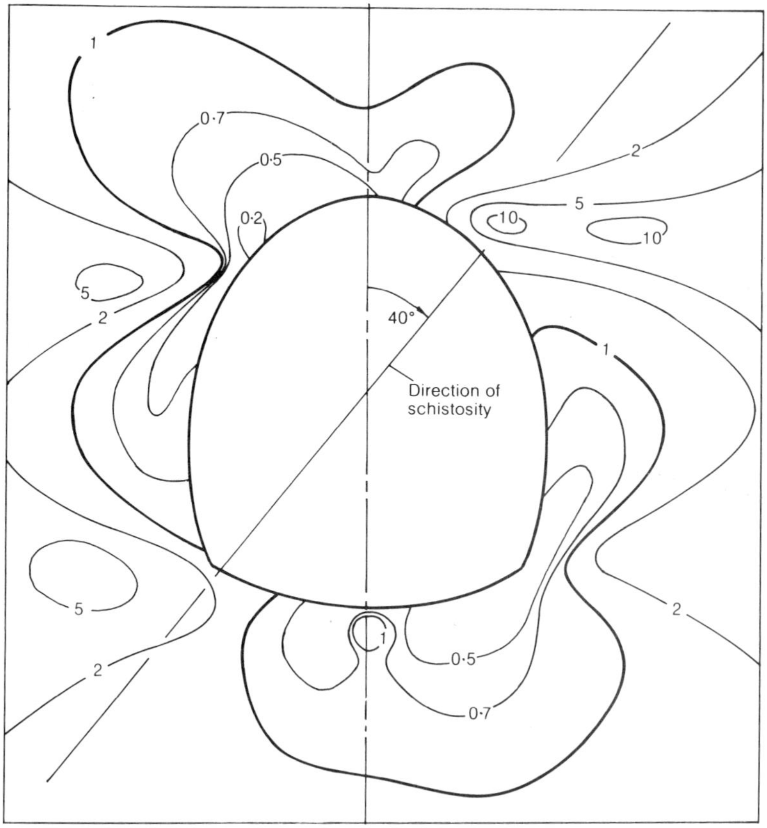

Fig. 22. Contours of ratio of available strength to stress in schistose rock surrounding a highly stressed tunnel

strength estimates made on the basis of the procedure outlined in the preceding steps.

EXAMPLES OF APPLICATION OF ROCK MASS STRENGTH ESTIMATES IN ENGINEERING DESIGN

In order to illustrate the application of the empirical failure criterion presented to practical engineering design problems, three examples are given. These examples have been carefully chosen to illustrate particular points and, although all of the examples are hypothetical, they are based on actual engineering problems studied by the Author.

Example 1

Figure 22 gives a set of contours of the ratio of available strength to induced stress in a schistose gneiss rock mass surrounding a tunnel. The following assumptions were made in calculating these ratios.

The vertical in situ stress in the rock surrounding the tunnel is 40 MPa, corresponding to a depth below surface of about 1500 m. The horizontal in situ stress is 60 MPa or 1·5 times the vertical stress.

The rock strength is defined by the following constants: uniaxial compressive strength of intact rock ($\sigma_c = 150$ MPa), material constants for the isotropic rock mass ($m_i = 12{\cdot}5$, $s_i = 0{\cdot}1$) material constants for joint strength in direction of schistosity ($m_j = 0{\cdot}28$, $s_j = 0{\cdot}0001$).

The direction of schistosity is assumed to be at 40° (measured in a clockwise direction) to the vertical axis of the tunnel.

The rock mass surrounding the tunnel is assumed to be elastic and isotropic. This assumption is generally accurate enough for most practical purposes, provided that the ratio of elastic moduli parallel to and normal to the schistosity does not exceed three. In the case of the example illustrated in Fig. 22, the stress distribution was calculated by means of the two-dimensional boundary element stress analysis technique, using the programme listing published by Hoek & Brown (1980a). A modulus of elasticity

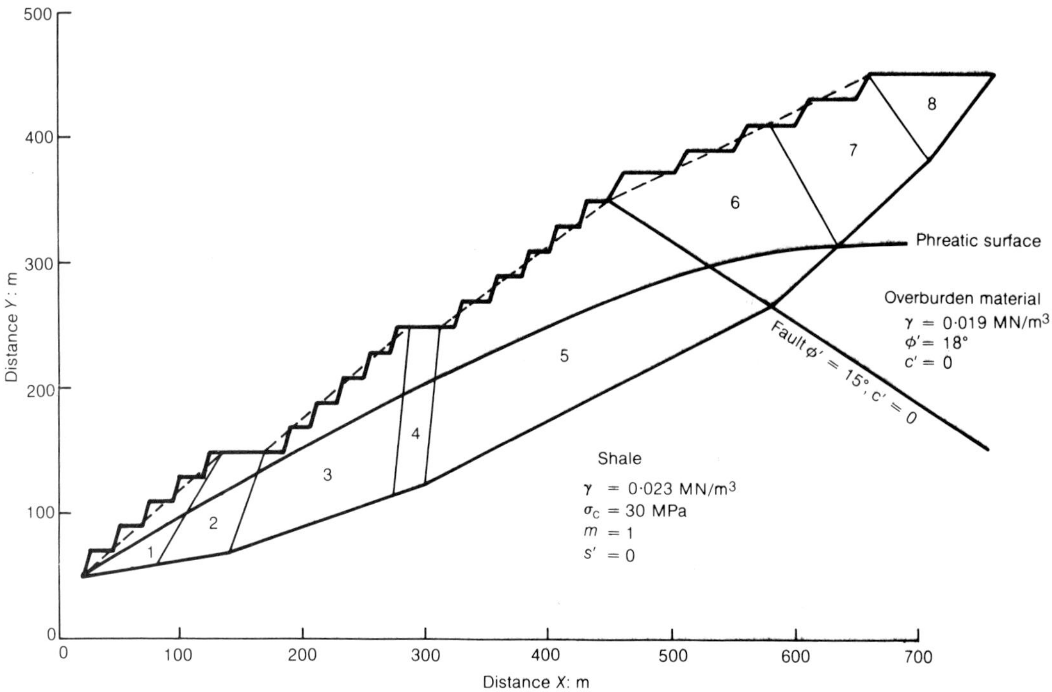

Fig. 23. Rock slope analysed in example 2 (see Table 5 for co-ordinates of slope profile, phreatic surface and failure surface)

of $E = 70$ GPa and a Poisson's ratio $\nu = 0{\cdot}25$ were assumed for this analysis.

The shear and normal stresses τ and σ', acting on a plane inclined at 40° (clockwise) to the vertical axis, were calculated for each point on a grid surrounding the tunnel. The available shear strengths in the direction of this plane, τ_{as}, were calculated by means of equations (7) and (6) (for $\sigma_c = 150$ MPa, $m_j = 0{\cdot}28$ and $s_j = 0{\cdot}0001$). Hence, the ratio of available shear strength τ_{as} to the induced shear stress τ was determined for each grid point.

In addition, the available strength σ_{ai}' of the isotropic rock mass was calculated at each grid point by means of equation (3), using the principal stresses σ_1' and σ_3' and the isotropic rock mass material properties ($\sigma_c = 150$ MPa, $m_i = 12{\cdot}5$ and $s_i = 0{\cdot}1$). This available strength σ_{ai}' was compared with the induced maximum principal stress σ_1' to obtain the ratio σ_{ai}'/σ_1' at each grid point.

In plotting the contours illustrated in Fig. 22, the lower of the two ratios τ_{as}/τ and σ_{ai}'/σ_1' was used to define the strength to stress ratio value.

The zones surrounded by the contours defined by a strength to stress ratio of one contain overstressed rock. The general method used in designing tunnels and caverns in highly stressed rock is to attempt to minimize the extent of such overstressed zones by choice of the excavation shape and orientation in relationship to the in situ stress direction.

When zones of overstressed rock, such as those illustrated in Fig. 22, are unavoidable, appropriate support systems have to be designed in order to restrict the propagation of fracture of rock contained in these zones. Unfortunately, the analysis presented in this example cannot be used to predict the extent and direction of fracture propagation from the zones of overstressed rock and the choice of support systems tends to be based on very crude approximations.

Such approximations involve designing a system of rock bolts with sufficient capacity to support the weight of the rock contained in the overhead overstressed zones and of sufficient length to permit anchoring in the rock outside these zones.

Improved techniques for support design are being developed, but are not yet generally available for complex failure patterns such as that illustrated in Fig. 22. These techniques, discussed by Hoek & Brown (1980a), involve an analysis of the interaction between displacements, induced by fracturing in the rock surrounding the tunnel, and the response of the support system installed to control these displacements. It is hoped that these support–interaction analyses will eventually be developed to the point where they can be used to evaluate the support requirements for tunnels such

Table 5. Stability analysis of slope shown in Fig. 23

Slice	1	2	3	4	5	6	7	8	9
XT	20	135	170	288	312	450	580	660	765
YT	50	150	150	250	250	350	410	450	450
XW	20	106	162	284	308	530	635	710	765
YW	50	100	132	196	210	300	311	380	450
XB	20	82	140	274	300	580	635	710	765
YB	50	60	68	115	123	265	311	380	450
Unit weight γ: MN/m^3	0·023	0·023	0·023	0·023	0·023	0·019	0·019	0·019	Factor of safety
First iteration									
ϕ_B'	30	30	30	30	30	18	18	18	
c_B'	1·0	1·0	1·0	1·0	1·0	0	0	0	
ϕ_S'	0	30	30	30	30	15	18	18	1·69
c_S'	0	1·0	1·0	1·0	1·0	0	0	0	
σ_B'	1·32	0·77	1·40	1·57	1·89	0·58	1·38	0·54	
σ_S'	0	0·09	0·55	0·66	0·75	2·01	1·21	0·52	
Second iteration									
ϕ_B'	40·03	45·08	39·46	38·36	36·58	18	18	18	
c_B'	0·48	0·32	0·51	0·55	0·64	0	0	0	
ϕ_S'	0	62·08	48·11	46·48	45·32	15	18	18	1·57
c_S'	0	0·06	0·25	0·28	0·31	0	0	0	
σ_B'	0·74	1·07	1·31	1·76	1·96	0·57	1·37	0·53	
σ_S'	0	0·16	0·46	0·53	0·62	2·00	1·19	0·51	
Third iteration									
ϕ_B'	45·44	42·02	40·10	37·26	36·23	18	18	18	
c_B'	0·31	0·41	0·48	0·61	0·66	0	0	0	
ϕ_S'	0	58·10	49·67	48·44	47·04	15	18	18	1·57
c_S'	0	0·10	0·21	0·24	0·27	0	0	0	
σ_B'	0·74	1·07	1·31	1·76	1·96	0·57	1·37	0·53	
σ_S'	0	0·15	0·44	0·52	0·61	2·00	1·19	0·51	

as that considered in this example.

Example 2

This example involves a study of the stability of a very large rock slope such as that which would be excavated in an open pit mine. The benched profile of such a slope, having an overall angle of about 30° and a vertical height of 400 m, is shown in Fig. 23.

The upper portion of the slope is in overburden material comprising mixed sands, gravels and clays. Back-analyses of previous failures in this overburden material, assuming a linear Mohr failure envelope, give a friction angle $\phi' = 18°$ and a cohesive strength $c' = 0$. The unit weight of this material averages 0·019 MN/m^3.

The overburden is separated from the shale forming the lower part of the slope by a fault which is assumed to have a shear strength defined by $\phi' = 15°$ and $c' = 0$.

No strength data are available for the shale, but examination of rock exposed in tunnels in this material suggests that the rock mass can be rated as 'good quality'. From Table 4, the material constants $m = 1$ and $s = 0{\cdot}004$ are chosen as representative of this rock. In order to provide a measure of conservatism in the design, the value of s is downgraded to zero to allow for the influence of stress relaxation which may occur as the slope is excavated. The strength of the intact material is estimated from point load tests (see Hoek & Brown, 1980a) as 30 MPa. The unit weight of the shale is 0·023 MN/m^3.

The phreatic surface in the rock mass forming the slope, shown in Fig. 23, is estimated from general knowledge of the hydrogeology of the site and from observations of seepage in tunnels in the slope.

Analysis of the stability of this slope is carried out by means of the non-vertical slice method (Sarma, 1979). This method is ideally suited to many rock slope problems because it permits the incorporation of specific structural features such as the fault illustrated in Fig. 23. This analysis has been slightly modified by the Author, and the equations used in the examples are listed in

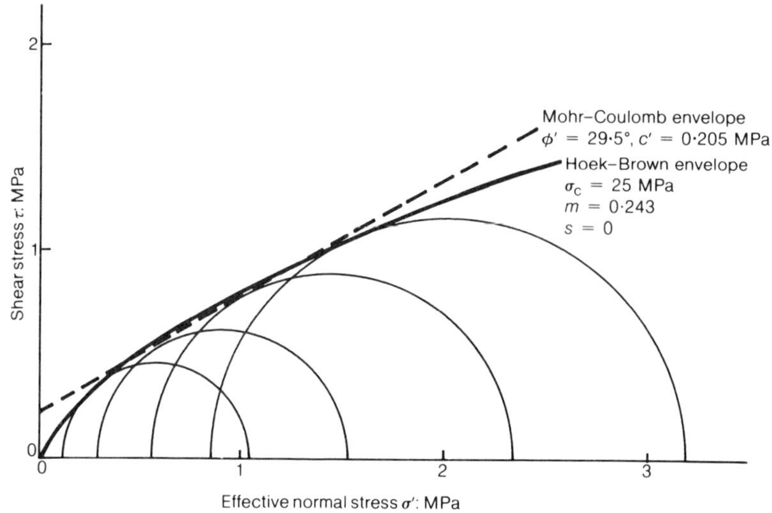

Fig. 24. Mohr circles derived from drained triaxial tests on retorted oil shale waste

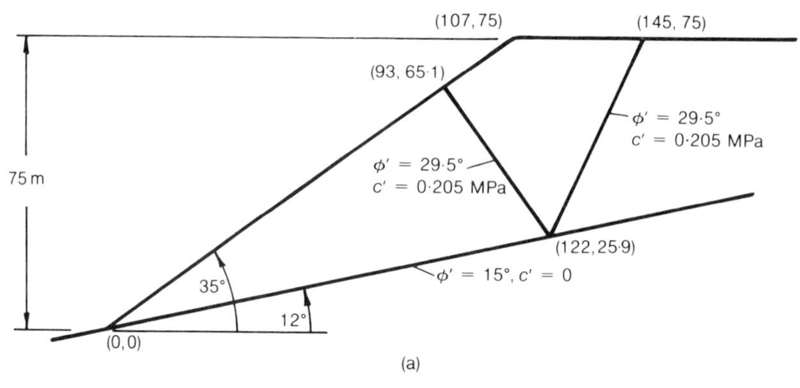

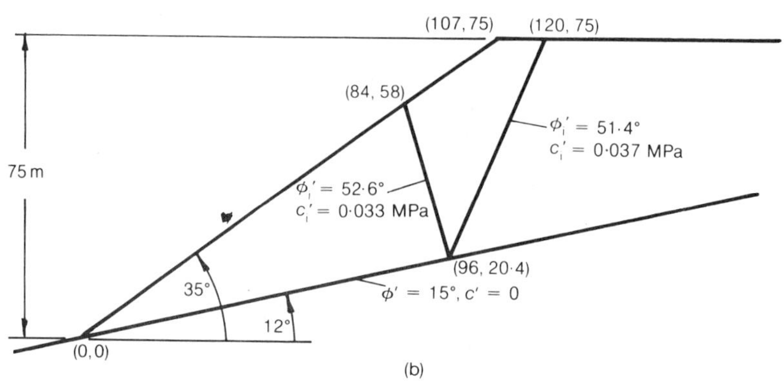

Fig. 25. Analyses of active–passive wedge failure in waste dumps of retorted oil shale resting on weak foundations. (a) Mohr–Coulomb failure criterion, factor of safety = 1·41; (b) Hoek–Brown failure criterion, factor of safety = 1·08

Appendix 2.

Table 5 lists the co-ordinates of the slope profile (XT, YT), the phreatic surface (XW, YW), and the base or failure surface (XB, YB) which was found, from a number of analyses, to give the lowest factor of safety. As a first approximation, the strength of the shale is assumed to be defined by $\phi' = 30°$ and $c' = 1$ MPa. Analysis of the slope, using these values, gives a factor of safety of 1·69.

The effective normal stresses σ_B' and σ_S' on the slice bases and sides, respectively, are calculated during the course of this analysis and these values are listed, for each slice, in Table 5. These values are used to determine appropriate values for the instantaneous friction angle ϕ_i' and instantaneous cohesive strength c_i' for the shale by means of equations (6) and (7) (for $\sigma_c = 30$ MPa, $m = 1$ and $s = 0$). These values of ϕ_i' and c_i' are used in the second iteration of a stability analysis and, as shown in Table 5, the resulting factor of safety is 1·57.

This process is repeated a third time, using values of ϕ_i' and c_i' calculated from the effective normal stresses given by the second iteration. The factor of safety given by the third iteration is 1·57. An additional iteration, not included in Table 5, gave the same factor of safety and no further iterations were necessary.

This example is typical of the type of analysis which would be carried out during the feasibility or the basic design phase for a large open pit mine or excavation for a dam foundation or spillway. Further analyses of this type would normally be carried out at various stages during excavation of the slope as the rock mass is exposed and more reliable information becomes available. In some cases, a testing programme may be set up to attempt to investigate the properties of materials such as the shale forming the base of the slope shown in Fig. 23.

Example 3

A problem which frequently arises in both mining and civil engineering projects is that of the stability of waste dumps on sloping foundations. This problem has been studied extensively by the Commonwealth Scientific and Industrial Research Organization in Australia in relation to spoil pile failures in open cast coal mines (see, for example, Coulthard, 1979). These studies showed that many of these failures involved the same active–passive wedge failure process analysed by Seed & Sultan (1967, 1969) and Horn & Hendron (1968) for the evaluation of dams with sloping clay cores.

In considering similar problems, the Author has found that the non-vertical slice method published by Sarma (1979) is well suited to an analysis of this active–passive wedge failure. Identical results to those obtained by Coulthard (1979) are given by assuming a drained spoil pile with a purely frictional shear strength on the interface between the active and passive wedges. However, Sarma's method also allows the analysis of a material with non-linear failure characteristics and, if necessary, with ground water pressures in the pile.

The example considered here involves a 75 m high spoil pile with a horizontal upper surface and a face angle of 35°. The unit weight of the spoil material is 0·015 MN/m^3. This pile rests on a weak foundation inclined at 12° to the horizontal. The shear strength of the foundation surface is defined by a friction angle of $\phi' = 15°$ and zero cohesion. The pile is assumed to be fully drained.

Triaxial tests on retorted oil shale material forming the soil pile give the Mohr circles plotted in Fig. 24. Regression analysis of the triaxial test data, assuming a linear Mohr failure envelope, gave $\phi' = 29{\cdot}5°$ and $c' = 0{\cdot}205$ MPa with a correlation coefficient of 1. Analysis of the same data, using the 'broken rock' analysis given in Appendix 1, for $\sigma_c = 25$ MPa (determined by point load testing) gave $m = 0{\cdot}243$ and $s = 0$. Both linear and non-linear Mohr failure envelopes are plotted in Fig. 24, and both of these envelopes will be used for the analysis of spoil pile stability.

Figure 25 gives the results of stability analyses for the Mohr–Coulomb and Hoek–Brown failure criteria. These analyses were carried out by optimizing the angle of the interface between the active and passive wedge, followed by the angle of the back scarp followed by the distance of the back scarp behind the crest of the spoil pile. In each case, these angles and distances were varied to find the minimum factor of safety in accordance with the procedure suggested by Sarma (1979).

The factor of safety obtained for the Mohr–Coulomb failure criterion ($\phi' = 29{\cdot}5°$ and $c' = 0{\cdot}205$ MPa) was 1·41, while that obtained for the Hoek–Brown criterion ($\sigma_c = 25$ MPa, $m = 0{\cdot}243$ and $s = 0$) was 1·08. In studies on the reason for the difference between these two factors of safety, it was found that the normal stresses acting across the interface between the active and passive wedges and on the surface forming the back scarp range from 0·06 to 0·11 MPa. As can be seen from Fig. 24, this is the normal stress range in which no test data exists and where the linear Mohr–Coulomb failure envelope, fitted to test data at higher normal stress levels, tends to overestimate the available shear strength.

This example illustrates the importance of carrying out triaxial or direct shear tests at the effective normal stress levels which occur in the actual problem being studied. In the example considered here, it would have been more appropriate to carry out a preliminary stability analysis, based on assumed parameters, before the testing programme was initiated. In this way, the correct range of normal stresses could have been used in the tests. Unfortunately, as frequently happens in the real engineering world, limits of time, budget and available equipment means that it is not always possible to achieve the ideal testing and design sequence.

CONCLUSION

An empirical failure criterion for estimating the strength of jointed rock masses has been presented. The basis for its derivation, the assumptions made in its development, and its advantages and limitations have all been discussed. Three examples have been given to illustrate the application of this failure criterion in practical geotechnical engineering design.

From this discussion and from some of the questions left unanswered in the examples, it will be evident that a great deal more work remains to be done in this field. A better understanding of the mechanics of jointed rock mass behaviour is a problem of major significance in geotechnical engineering, and it is an understanding to which both the traditional disciplines of soil mechanics and rock mechanics can and must contribute. The Author hopes that the ideas presented will contribute toward this understanding and development.

ACKNOWLEDGEMENTS

The Author wishes to acknowledge the encouragement, assistance and guidance provided over many years by Professor E. T. Brown and Dr J. W. Bray of Imperial College. Many of the ideas presented originated from discussions with these colleagues and co-authors.

The stimulating and challenging technical environment which is unique to the group of people who make up Golder Associates is also warmly acknowledged. This environment has provided the impetus and the encouragement required by this Author in searching for realistic solutions to practical engineering problems. Particular thanks are due to Dr R. Hammett, Dr S. Dunbar, Mr M. Adler, Mr B. Stewart, Miss D. Mazurkewich and Miss S. Kerber for their assistance in the preparation of this Paper.

APPENDIX 1. DETERMINATION OF MATERIAL CONSTANTS FOR EMPIRICAL FAILURE CRITERION

Failure criterion

The failure criterion defined by equation (3)

$$\sigma_1' = \sigma_3' + (m\sigma_c\sigma_3' + s\sigma_c^2)^{1/2} \tag{3}$$

can be rewritten as

$$y = m\sigma_c x + s\sigma_c^2 \tag{16}$$

where $y = (\sigma_1' - \sigma_3')^2$ and $x = \sigma_3'$

Intact rock

For intact rock, $s = 1$ and the uniaxial compressive strength σ_c and the material constant m are given by

$$\sigma_c^2 = \frac{\Sigma y}{n} - \left[\frac{\Sigma xy - \dfrac{\Sigma x \Sigma y}{n}}{\Sigma x^2 - \dfrac{(\Sigma x)^2}{n}}\right]\frac{\Sigma x}{n} \tag{17}$$

$$m = \frac{1}{\sigma_c}\left[\frac{\Sigma xy - \dfrac{\Sigma x \Sigma y}{n}}{\Sigma x^2 - \dfrac{(\Sigma x)^2}{n}}\right] \tag{18}$$

The coefficient of determination r^2 is given by

$$r^2 = \frac{(\Sigma xy - \Sigma x \Sigma y/n)^2}{(\Sigma x^2 - (\Sigma x)^2/n)(\Sigma y^2 - (\Sigma y)^2/n)} \tag{19}$$

Broken rock

For broken or heavily jointed rock, the strength of the intact rock pieces is determined by the analysis given above. The value of the constant m for broken or heavily jointed rock is found from equation (18). The value of the constant s is given by

$$s = \frac{1}{\sigma_c^2}\left[\frac{\Sigma y}{n} - m\sigma_c\frac{\Sigma x}{n}\right] \tag{20}$$

The coefficient of determination is found from equation (19).

When the value of s is very close to zero, equation (20) will sometimes give a small negative value. In such cases, put $s = 0$ and calculate the constant m as follows

$$m = \frac{\Sigma y}{\sigma_c \Sigma x} \tag{21}$$

When equation (21) is used, equation (19) is not valid.

Mohr envelope

The Mohr failure envelope is defined by the following equation, derived by Dr J. Bray of Imperial College

$$\tau = (\text{Cot}\,\phi_i' - \text{Cos}\,\phi_i')\frac{m\sigma_c}{8} \tag{22}$$

where the instantaneous friction angle ϕ_i' is given by

$$\phi_i' = \text{Arctan}\,(4h\,\text{Cos}^2\,(30 + \tfrac{1}{3}\text{Arcsin}\,h^{-2/3}) - 1)^{-1/2} \tag{23}$$

where

$$h = 1 + \frac{16(m\sigma' + s\sigma_c)}{3m^2\sigma_c}$$

and the instantaneous cohesive strength c_i' is given by

$$c_i' = \tau - \sigma'\,\text{Tan}\,\phi_i' \tag{24}$$

where σ' is the effective normal stress.

Determination of m and s from direct shear test data

The following method for determination of the material constants m and s from direct shear test data was devised by Dr S. Dunbar (unpublished report) of Golder Associates in Vancouver.

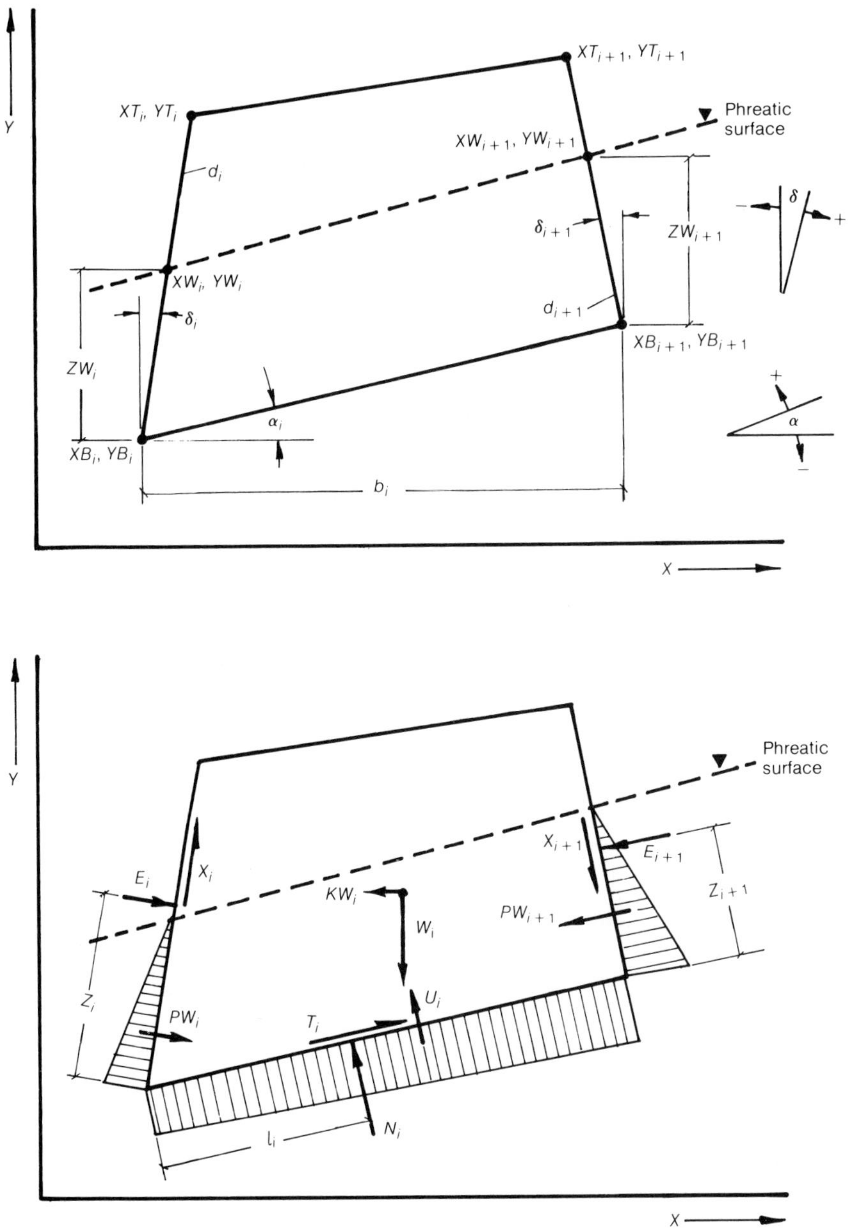

Fig. 26. Geometry of and forces acting on a single slice in the Sarma (1979) non-vertical method for stability analysis

The major and minor principal stresses σ_1' and σ_3' corresponding to each τ, σ' pair can be calculated as follows

$$\sigma_1' = \frac{(\sigma'^2 + (\tau - c')\tau) + \tau(\sigma'^2 + (\tau - c')^2)^{1/2}}{\sigma'} \quad (25)$$

$$\sigma_3' = \frac{(\sigma'^2 + (\tau - c')\tau) - \tau(\sigma'^2 + (\tau - c')^2)^{1/2}}{\sigma'} \quad (26)$$

where c' is an estimate of the cohesion intercept for the entire τ, σ' data set. This estimate can be an assumed value greater than or equal to zero or it can be determined by linear regression analysis of the shear test results.

After the calculation of the values of σ_1' and σ_3' by means of equations (25) and (26), the determination of the material constants m and s is carried out as for broken rock.

An estimate of the uniaxial compressive strength σ_c of the intact rock is required in order to complete the analysis.

APPENDIX 2. SARMA NON-VERTICAL SLICE METHOD FOR THE ANALYSIS OF SLOPE FAILURE ON SURFACES OF GENERAL SHAPE

Introduction

This analysis, published by Sarma (1979), is a general method of limit equilibrium analysis which can be used to determine the stability of slopes of a variety of shapes. Slopes with complex profiles sliding on circular, non-circular or plane surfaces or any combination of such surfaces can usually be analysed by this method. In addition, active–passive wedge failures such as those which occur in spoil piles on sloping foundations or in clay core embankments can also be analysed. The analysis allows different shear strengths (defined by cohesion and angle of friction) to be specified for each slice base and side. The freedom to change the inclination of the sides of the slice also allows the incorporation of specific structural features such as faults. Water pressures acting on the sides and base of each slice are included in the analysis. External forces due to water pressure in tension cracks or to reinforcement installed in the slope can be incorporated but have not been included in this version.

The geometry of the sliding mass is defined by the co-ordinates of the corners of a number of three- or four-sided elements. The phreatic surface is defined by the co-ordinates of its intersections with the slice sides. A closed form solution is then used to calculate the critical horizontal acceleration K_c required to induce a state of limiting equilibrium in the slope. The static factor of safety of the slope is then found by reducing the values of $\text{Tan}\,\phi$ and c to $\text{Tan}\,\phi/F$ and c/F (where F is the factor of safety) until $K_c = 0$.

In order to determine whether the analysis is acceptable, a final check is carried out to assess whether all the effective normal stresses acting across the bases and sides of the slices are positive. If negative stresses are found, the slice geometry must be varied until these negative stresses are eliminated. An additional check on moment equilibrium is also recommended for critical slopes.

The equations listed here have been arranged for programming on a Hewlett Packard 41CV calculator and a full analysis (excluding the moment equilibrium check) can be carried out for ten slices. These equations differ slightly from those published by Sarma (1979) in that a more complete equation is used for the calculation of the effective normal stress on the slice base. This calculation is essential for the analysis of failure of slopes in materials with a non-linear failure criterion.

Geometrical calculations

The geometry of the ith slice is defined in Fig. 26. Assuming that ZW_i, δ_i and d_i are available from the previous slice

$$d_{i+1} = ((XT_{i+1} - XB_{i+1})^2 + (YT_{i+1} - YB_{i+1})^2)^{1/2} \quad (27)$$

$$\delta_{i+1} = \text{Arcsin}\,((XT_{i+1} - XB_{i+1})/d_{i+1}) \quad (28)$$

$$b_i = XB_{i+1} - XB_i \quad (29)$$

$$\alpha_i = \text{Arctan}\,((YB_{i+1} - YB_i)/b_i) \quad (30)$$

$$W_i = \tfrac{1}{2}\gamma((YB_i - YT_{i+1})(XT_i - XB_{i+1}) + (YT_i - YB_{i+1})(XT_{i+1} - XB_i)) \quad (31)$$

$$ZW_{i+1} = (YW_{i+1} - YB_{i+1}) \quad (32)$$

Calculation of water forces

$$U_i = \tfrac{1}{2}\gamma_w(ZW_i + ZW_{i+1})\,b_i.\,\text{Sec}\,\alpha_i \quad (33)$$

$$PW_i = \tfrac{1}{2}\gamma_w\,ZW_i^2\,\text{Sec}\,\delta_i \quad (34)$$

$$PW_{i+1} = \tfrac{1}{2}\gamma_w\,ZW_{i+1}^{\ 2}\,\text{Sec}\,\delta_{i+1} \quad (35)$$

where γ and γ_w are the unit weights of rock and water respectively.

Calculation of critical acceleration K_c

$$K_c = \frac{a_n + a_{n-1}.e_n + a_{n-2}.e_n.e_{n-1} + \ldots + a_1.e_n.e_{n-1}\ldots e_3 e_2}{p_n + p_{n-1}.e_n + p_{n-2}.e_n.e_{n-1} + \ldots + p_1.e_n.e_{n-1}\ldots e_3 e_2} \quad (36)$$

where

$$a_i = Q_i(W_i.\text{Sin}\,(\phi_{Bi} - \alpha_i) + R_i.\text{Cos}\,\phi_{Bi} + S_{i+1}.\text{Sin}\,(\phi_{Bi} - \alpha_i - \delta_{i+1}) - S_i.\text{Sin}\,(\phi_{Bi} - \alpha_i - \delta_i)) \quad (37)$$

$$p_i = Q_i.W_i.\text{Cos}\,(\phi_{Bi} - \alpha_i) \quad (38)$$

$$e_i = Q_i(\text{Cos}\,(\phi_{Bi} - \alpha_i + \phi_{Si} - \delta_i)\,\text{Sec}\,\phi_{Si} \quad (39)$$

$$Q_i = \text{Sec}\,(\phi_{Bi} - \alpha_i + \phi_{Si+1} - \delta_{i+1}).\text{Cos}\,\phi_{Si+1} \quad (40)$$

$$S_i = (c_{Si}.d_i - PW_i.\text{Tan}\,\phi_{Si}) \quad (41)$$

$$S_{i+1} = (c_{Si+1}.d_{i+1} - PW_{i+1}.\text{Tan}\,\phi_{Si+1}) \quad (42)$$

$$R_i = (c_{Bi}.b_i.\text{Sec}\,\alpha_i - U_i.\text{Tan}\,\phi_{Bi}) \quad (43)$$

Calculation of factor of safety F

For slopes where $K_c \neq 0$, the factor of safety is calculated by reducing the shear strength simultaneously on all sliding surfaces until the acceleration K calculated by means of equation (36) is equal to zero. This is achieved by substitution, in equation (37) to (43) of the following shear strength values

$$c_{Bi}/F, \quad \text{Tan}\,\phi_{Bi}/F, \quad c_{Si}/F,$$
$$\text{Tan}\,\phi_{Si}/F, \quad c_{Si+1}/F \quad \text{and} \quad \text{Tan}\,\phi_{Si+1}/F$$

Check on acceptability of solution

Having determined the value of K for a given factor of safety, the forces acting on the sides and bases of the slices are found by progressive solution of the following equations, starting from the known condition that $E_1 = 0$.

$$E_{i+1} = a_i - p_i\,.\,K + E_i\,.\,e_i \tag{44}$$

$$X_i = (E_i - PW_i)\,\text{Tan}\,\phi_{Si} + c_{Si}\,.\,d_i \tag{45}$$

$$\begin{aligned} N_i = (W_i + X_{i+1}\,.\,\text{Cos}\,\delta_{i+1} - X_i\,.\,\text{Cos}\,\delta_i \\ - E_{i+1}\,.\,\text{Sin}\,\delta_{i+1} + E_i\,.\,\text{Sin}\,\delta_i \\ + U_i\,.\,\text{Tan}\,\phi_{Bi}\,.\,\text{Sin}\,\alpha_i - c_{Bi}\,.\,b_i\,.\,\text{Tan}\,\alpha_i) \\ \times \text{Cos}\,\phi_{Bi}\,.\,\text{Sec}(\phi_{Bi} - \alpha_i) \end{aligned} \tag{46}$$

$$T_i = (N_i - U_i)\,\text{Tan}\,\phi_{Bi} + c_{Bi}\,.\,b_i\,.\,\text{Sec}\,\alpha_i \tag{47}$$

The effective normal stresses acting across the base and the sides of a slice are calculated as follows

$$\sigma_{Bi}' = (N_i - U_i)/b_i\,.\,\text{Sec}\,\alpha_i \tag{48}$$

$$\sigma_{Si}' = (E_i - PW_i)/d_i \tag{49}$$

$$\sigma_{Si+1}' = (E_{i+1} - PW_{i+1})/d_{i+1} \tag{50}$$

In order for the solution to be acceptable, all effective normal stresses must be positive.

A final check recommended by Sarma is for moment equilibrium. Referring to Fig. 26 and taking moments about the lower left hand corner of the slice

$$\begin{aligned} N_i l_i - X_{i+1}\,.\,b_i\,.\,\text{Sec}\,\alpha_i\,.\,\text{Cos}\,(\alpha_i + \delta_{i+1}) \\ - E_i Z_i + E_{i+1}(Z_{i+1} + b_i\,.\,\text{Sec}\,\alpha_i\,.\,\text{Sin}\,(\alpha_i + \delta_{i+1})) \\ - W_i(XG_i - X_{Bi}) + K_c W_i(YG_i - Y_{Bi}) = 0 \end{aligned} \tag{51}$$

where XG_i, YG_i are the co-ordinates of the centre of gravity of the slice.

Starting from the first slice, where $Z_1 = 0$, assuming a value for l_i, the moment arm Z_{i+1} can be calculated or vice versa. The values of Z_i and Z_{i+1} should lie within the slice boundary, preferably in the middle third.

REFERENCES

Barton, N. R. (1971). A relationship between joint roughness and joint shear strength. *Proc. Int. Symp. Rock Fracture, Nancy, France*, 1–8.

Barton, N. R. (1973). Review of a new shear strength criterion for rock joints. *Engng Geol.* **7**, 287–332.

Barton, N. R. (1974). *A review of the shear strength of filled discontinuities in rock.* Publication No. 105. Oslo: Norwegian Geotechnical Institute.

Barton, N. R. & Choubey, V. (1977). The shear strength of rock joints in theory and practice. *Rock Mech.* **10**, No. 1, 1–54.

Barton, N. R., Lien, R. & Lunde, J. (1974). Engineering classification of rock masses for the design of tunnel support. *Rock Mech.* **6**, No. 4, 189–236.

Bieniawski, Z. T. (1974a). Estimating the strength of rock materials. *Jl S. Afr. Inst. Min. Metall.* **74**, No. 8, 312–320.

Bieniawski, Z. T. (1974b). Geomechanics classification of rock masses and its application in tunnelling. *Proc. 3rd Int. Congr. Soc. Rock Mech. Denver* **2**, Part A, 27–32.

Bishop, A. W., Webb, D. L. & Lewin, P. I. (1965). Undisturbed samples of London clay from the Ashford Common shaft. *Géotechnique* **15**, No. 1, 1–31.

Bishop, A. W. & Garga, V. K. (1969). Drained tension tests on London clay. *Géotechnique* **19**, No. 2, 309–313.

Brace, W. F. (1964). Brittle fracture of rocks. In *State of stress in the earth's crust* (ed. W. R. Judd) pp. 111–174. New York: Elsevier.

Brace, W. F. & Martin, R. J. (1968). A test of the law of effective stress for crystalline rocks of low porosity. *Int. J. Rock Mech. Min. Sci.* **5**, No. 5, 415–426.

Broch, E. (1974). The influence of water on some rock properties. *Proc. 3rd Int. Congr. Soc. Rock Mech. Denver*, **2**, Part A, 33–38.

Brown, E. T. (1970). Strength of models of rock with intermittent joints. *J. Soil Mech. Fdns Div. Am. Soc. Civ. Engrs* **96**, SM6, 1935–1949.

Brown, E. T. & Trollope, D. H. (1970). Strength of a model of jointed rock. *J. Soil Mech. Fdns Div. Am. Soc. Civ. Engrs* **96**, SM2, 685–704.

Charles, J. A. & Watts, K. S. (1980). The influence of confining pressure on the shear strength of compacted rockfill. *Géotechnique* **30**, No. 4, 353–367.

Colback, P. S. B. & Wiid, B. L. (1965). The influence of moisture content on the compressive strength of rock. *Proc. 3rd Can. Rock Mech. Symp. Toronto*, 57–61.

Coulthard, M. A. (1979). *Back analysis of observed spoil failures using a two-wedge method.* Australian CSIRO Division of Applied Geomechanics. Technical report No. 83. Melbourne: CSIRO.

Einstein, H. H., Nelson, R. A., Bruhn, R. W. & Hirschfeld, R. C. (1969). Model studies of jointed rock behaviour. *Proc. 11th Symp. Rock Mech. Berkeley, Calif.* 83–103.

Franklin, J. A. & Hoek, E. (1970). Developments in triaxial testing equipment. *Rock Mech.* **2**, 223–228.

Gerogiannopoulos, N. G. A. (1979). *A critical state approach to rock mechanics.* PhD thesis, University of London.

Goodman, R. E. (1970). The deformability of joints. In *Determination of the in-situ modulus of deformation of rock.* ASTM Special Technical Publication No. 477, pp. 174–196. Philadelphia: American Society for Testing and Materials.

Griffith, A. A. (1921). The phenomena of rupture and flow in solids. *Phil. Trans. R. Soc.* A, **221**, 163–198.

Griffith, A. A. (1925). Theory of rupture. *Proc. 1st Congr. Appl. Mech. Delft*, 1924, pp. 55–63. Delft: Technische Bockhandel en Drukkerij.

Handin, J., Hager, R. V., Friedman, M. & Feather, J. N. (1963). Experimental deformation of sedimentary rocks under confining pressure; pore pressure tests. *Bull. Am. Ass. Petrol. Geol.* **47**, 717–755.

Heard, H. C., Abey, A. E., Bonner, B. P. & Schock, R. N. (1974). *Mechanical behaviour of dry Westerley granite at high confining pressure* UCRL Report 51642. California: Lawrence Livermore Laboratory.

Hencher, S. R. & Richards, L. R. (1982). The basic frictional resistance of sheeting joints in Hong Kong granite. *Hong Kong Engr* Feb., 21–25.

Hobbs, D. W. (1970). The behaviour of broken rock under triaxial compression. *Int. J. Rock Mech. Min. Sci.* **7**, 125–148.

Hoek, E. (1965). *Rock fracture under static stress conditions*. PhD thesis, University of Capetown.
Hoek, E. (1968). Brittle failure of rock. In *Rock mechanics in engineering practice* (eds K. G. Stagg & O. C. Zienkiewicz), pp. 99–124. London: Wiley.
Hoek, E. & Bieniawski, Z. T. (1965). Brittle fracture propagation in rock under compression. *Int. J. Frac. Mech.* **1**, No. 3, 137–155.
Hoek, E. & Bray, J. W. (1981). *Rock slope engineering* (3rd edn). London: Institution of Mining and Metallurgy.
Hoek, E. & Brown, E. T. (1980a). *Underground excavations in rock*. London: Institution of Mining and Metallurgy.
Hoek, E. & Brown, E. T. (1980b). Empirical strength criterion for rock masses. *J. Geotech. Engng Div. Am. Soc. Civ. Engrs* **106**, GT9, 1013–1035.
Horino, F. G. & Ellikson, M. L. (1970). *A method of estimating the strength of rock containing planes of weakness*. US Bureau Mines Report Investigation 7449. US: Bureau of Mines.
Horn, H. M. & Hendron, D. M. (1968). Discussion on Stability analysis for a sloping core embankment. *J. Soil Mech. Fdns Div. Am. Soc. Civ. Engrs* **94**, SM3, 777–779.
Jaeger, J. C. (1970). The behaviour of closely jointed rock. *Proc. 11th Symp. Rock Mech. Berkeley, Calif.* 57–68.
Jaeger, J. C. (1971). Friction of rocks and stability of rock slopes. *Géotechnique* **21**, No. 2, 97–134.
Jaeger, J. C. & Cook, N. G. W. (1969). *Fundamentals of rock mechanics*. London: Chapman and Hall.
John, K. W. (1962). An approach to rock mechanics. *J. Soil Mech. Fdns Div. Am. Soc. Civ. Engrs* **88**, SM4, 1–30.
Krsmanovic, D. (1967). Initial and residual shear strength of hard rock. *Géotechnique* **17**, No. 2. 145–160.
Ladanyi, B. & Archambault, G. (1970). Simulation of shear behaviour of a jointed rock mass. *Proc. 11th Symp. Rock Mech.* pp. 105–125. New York: American Institute of Mining, Metallurgical and Petroleum Engineers.
Ladanyi, B. & Archambault, G. (1972). Evaluation de la resistance au cisaillement d'un massif rocheux fragmente. *Proc. 24th Int. Geol. Congr. Montreal.* Sec. 130, 249–260.
Lajtai, E. Z. (1967). The influence of interlocking rock discontinuities on compressive strength (model experiments). *Rock Mech. Engng Geol.* **5**, 217–228.
Lama, R. D. & Vutukuri, V. S. (1978). *Handbook on mechanical properties of rocks. Vol. IV—Testing techniques and results.* Switzerland: Trans Tech Publications.
Marachi, N. D., Chan, C. K. & Seed, H. B. (1972). Evaluation of properties of rockfill materials. *J. Soil Mech. Fdns Div. Am. Soc. Civ. Engrs* **98**, SM4, 95–114.
Marsal, R. J. (1967). Large scale testing of rockfill materials. *J. Soil Mech. Fdns Div. Am. Soc. Civ. Engrs* **93**, SM2, 27–44.
Marsal, R. J. (1973). Mechanical properties of rockfill. In *Embankment dam engineering, Casagrande Vol.* pp. 109–200. New York: Wiley.
Martin, G. R. & Miller (1974). Joint strength characteristics of a weathered rock. *Proc. 3rd Int. Congr. Soc. Rock Mech. Denver*, **2**, Part A, 263–270.
McClintock, F. A. & Walsh, J. B. (1962). Friction on Griffith cracks under pressure. *Proc. 4th US Congr. Appl. Math., Berkeley.* 1015–1021.
McLamore, R. & Gray, K. E. (1967). The mechanical behaviour of anisotropic sedimentary rocks. *Trans. Am. Soc. Mech. Engrs* Series B, 62–76.
Misra, B. (1972). *Correlation of rock properties with machine performance.* PhD thesis. University of Leeds.
Mogi, K. (1966). Pressure dependence of rock strength and transition from brittle fracture to ductile flow. *Bull. Earthq. Res. Inst., Tokyo Univ.* **44**, 215–232.
Mogi, K. (1967). Effect of the intermediate principal stress on rock failure. *J. Geophys. Res.* **72**, No. 20, 5117–5131.
Muller, L. & Pacher, F. (1965). Modelvensuch Zur Klarung der Bruchgefahr geklufteter Medien. *Rock Mech. Engng Geol.* Suppl. No. 2, 7–24.
Murrell, S. A. F. (1958). The strength of coal under triaxial compression. In *Mechanical properties of non-metallic brittle materials* (ed. W. H. Walton), pp. 123–145. London: Butterworths.
Murrell, S. A. F. (1965). The effect of triaxial stress systems on the strength of rocks at atmospheric temperatures. *Geophys. J.* **10**, 231–281.
Patton, F. D. (1966). Multiple modes of shear failure in rock. *Proc. 1st Int. Congr. Rock Mech. Lisbon*, **1**, 509–513.
Raphael, J. M. & Goodman, R. E. (1979). Strength and deformability of highly fractured rock. *J. Geotech. Engng Div. Am. Soc. Civ. Engrs* **105**, GT11, 1285–1300.
Richards, L. R. & Cowland, J. W. (1982). The effect of surface roughness on field shear strength of sheeting joints in Hong Kong granite. *Hong Kong Engr* Oct., 39–43.
Rosengren, K. J. & Jaeger, J. C. (1968). The mechanical properties of a low-porosity interlocked aggregate. *Géotechnique* **18**, No. 3. 317–326.
Sarma, S. K. (1979). Stability analysis of embankments and slopes. *J. Geotech. Engng Div. Am. Soc. Civ. Engrs* **105**, GT12, 1511–1524.
Schwartz, A. E. (1964). Failure of rock in the triaxial shear test. *Proc. 6th Symp. Rock Mech. Rolla, Missouri*, 109–135.
Seed, H. B. & Sultan, H. A. (1967). Stability analysis for a sloping core embankment. *J. Geotech. Engng Div. Am. Soc. Civ. Engrs* **93**, SM4, 69–83.
Sultan, H. A. & Seed, H. B. (1969). Discussion closure. *J. Geotech. Engng Div. Am. Soc. Civ. Engrs* **95**, SM1, 334–335.
Walker, P. E. (1971). *The shearing behaviour of a block jointed rock model.* PhD thesis, Queens University, Belfast.
Wawersik, W. R. (1968). *Detailed analysis of rock failure in laboratory compression tests.* PhD thesis, University of Minnesota.
Wawersik, W. R. & Brace, W. F. (1971). Post-failure behaviour of a granite and a diabase. *Rock Mech.* **3**, No. 2, 61–85.

VOTE OF THANKS

In proposing a vote of thanks to Dr Hoek, Professor R. E. Gibson said: 'We have listened to a discourse aimed, in the lecturer's own words,

". . . at providing a better understanding of the mechanics of jointed rock mass behaviour: a problem of major significance in geotechnical engineering". To those academics among us who have given attention to this problem, it is recognized as one of great difficulty and fascination. To those practising engineers who are obliged to arrive at decisions based on whatever data and understanding they possess, it can be a daunting responsibility. Evert Hoek's wide-ranging career has given him an unusual understanding and appreciation of both these viewpoints so that he perceives what the engineer needs from research and also the extent to which the uncertainties inherent in nature allow this need to be met.

'Dr Hoek has spoken with authority on a subject of great importance to all geotechnical engineers and has succeeded brilliantly in his aim of providing a better understanding of the mechanics of jointed rock. I am sure that in the future this Lecture will be referred to many times.

'I should like on behalf of us all to congratulate Dr Hoek most warmly on his splendid lecture and to propose now a hearty vote of thanks to him'.

The vote of thanks was accorded with acclamation.

The Rankine Lecture

The twenty-fourth Rankine Lecture of the British Geotechnical Society was given by Professor C. P. Wroth at Imperial College of Science and Technology, London, on 13 March 1984. The following introduction was given by Professor R. E. Gibson, Golder Associates.

It gives me very great pleasure to introduce this evening Professor Wroth, our twenty-fourth Rankine lecturer.

Peter Wroth was born in 1929; he was educated at Marlborough and, after two years' commissioned service in the Royal Artillery, entered Emmanuel College, Cambridge, in 1949 as a Scholar. The intellectual discipline of part II of the mathematics tripos was followed, imaginatively, characteristically and very unusually by part I of the mechanical sciences tripos. After graduating and a short spell as master at Felsted, he returned to Cambridge in 1954 as a research student under the late Professor Roscoe and at a most propitious time—just as it was becoming established as a major centre for soil mechanics research.

Those were the early and exciting days of the 'critical state' ideas and for Wroth they culminated not only in his PhD but, as co-author with Roscoe and Schofield, of the paper *On the yielding of soils*, with the award of the inaugural prize of our society.

Like most young engineers at this stage in their careers, Wroth wisely recognized the need to obtain practical and professional experience. He therefore joined Maunsell & Partners and during the next three years was engaged on the design of pre-stressed concrete bridges and, in particular, the Hammersmith Flyover. Far from regarding this as an undemanding interlude he threw himself with energy and enthusiasm into this work which at that time presented many new problems and uncertainties.

However, by 1961 he was back at Cambridge once more, this time as a Lecturer and Fellow of Churchill College, and again wholly committed to soil mechanics. A glance through a list of his publications since then—and these number more than 70—reveals the remarkable extent of his research interests; they cover analysis, laboratory model testing, observation on full-scale structures, design, both mechanical and civil, and many more besides. They are fundamental in aspect, are without exception relevant to real problems which face civil engineers and are all, as you know, characterized by outstanding clarity of exposition.

Following the death of Professor Roscoe in 1970 Wroth took over the running of the soils group for a while and a small but significant change in the pattern of research occurred. Although laboratory and model studies continued, greater emphasis began to be placed on field measurements and above all on the development of sophisticated means for testing soils in situ. No doubt, this shift, which was to prove so fruitful, reflected Wroth's own committment to civil engineering and came about partly as a result of his own increasing activity as a consultant.

In 1975 he became Reader in Soil Mechanics and most of us thought that he was set to stay at Cambridge until retirement. Not long after this, and quite unexpectedly, I received a letter from him in which he wrote 'I am suffering from a severe attack of *folie de grandeur*... and propose to apply for the Engineering Chair at Oxford'. It came, I might add, as much less of a surprise to his friends than to himself when he was appointed to this prestigious chair. It is, in fact, a chair of Engineering Science and this is singularly appropriate for these two words, engineering and science, aptly embrace the essence of Wroth's ability, the rigour of science and the practical concerns of engineering. We look forward eagerly this evening therefore to a discourse which will reflect these two facets of our discipline.

I have now, on behalf of the British Geotechnical Society, the honour of asking Professor Wroth to give the Rankine Lecture for 1984.

WROTH, C. P. (1984). *Géotechnique* **34**, No. 4, 449–489

The interpretation of in situ soil tests

C. P. WROTH*

The purposes of in situ testing are set out, and the difficulties of the interpretation of the observations are emphasized. These difficulties are due to the complex behaviour of soils together with the lack of control and of choice of the boundary conditions in any field test. One notable exception is the pressuremeter test, from which soil properties can be derived directly without recourse to empirical correlations. The discussion is concentrated on the measurement of undrained shear strength. The results obtained from different tests (triaxial, plane strain, direct simple shear, pressuremeter and vane) are compared by expressing them in terms of the undrained strength ratio s_u/σ_{v0}' as a function of the friction angle ϕ. Special attention is paid to tests in which the principal axes of stress and of strain increment are free to rotate. In such tests, uncertainty exists regarding the definition of failure and the planes of maximum stress obliquity. To derive these functions Matsuoka's failure criterion is used. As a consequence a theoretical hierarchy of strengths is established which agrees qualitatively with experimental evidence. The importance to a designer of this variety of strengths is emphasized. A study is made of the piezocone and the interpretation of the pore pressures in terms of the overconsolidation ratio of the clay tested. A plea is made for the standardization of the equipment, the operation and the interpretation of in situ tests to obtain maximum benefit from them.

L'article décrit les buts des essais in situ et souligne les difficultés de l'interprétation des mesures. Ces difficultés sont dues au comportement complexe des sols, combiné avec le manque de contrôle et de choix des conditions limites qu'on a dans n'importe quel essai in situ. Une exception importante est l'essai pressiométrique—à partir duquel on peut evaluer directement des propriétés du sol sans être obligé de recourir à des correlations empiriques. La discussion se concentre sur la mesure de la résistance au cisaillement non-drainé. On compare les résultats obtenus à partir des différents essais (triaxial, déformation plane, cisaillement simple direct, pressiomètre et scissomètre) en les exprimant en fonction du rapport s_u/σ_{v0}' de la résistance dans l'état non-drainé et en fonction de l'angle de friction ϕ. On étudie plus particulièrement les essais dans lesquels les axes principaux de l'augmentation de la contrainte et de la déformation peuvent tourner librement. Dans de tels cas il existe une incertitude concernant la définition de la rupture et des plans de l'obliquité maximale des contraintes. Afin de trouver ces fonctions on emploie le critère de rupture de Matsuoka. On établit par conséquent une hiérarchie théorique des résistances qui s'accorde qualitativement avec les résultats expérimentaux. On souligne l'importance pour le projeteur de ces diverses résistances. On étudie aussi le piézocône et l'interprétation des pressions interstitielles en fonction du rapport de surconsolidation de l'argile testée. On recommande que l'appareillage, l'exécution et l'interprétation des essais sur place soient uniformisés afin d'en tirer l'avantage maximal.

* Department of Engineering Science, University of Oxford.

INTRODUCTION

In situ testing in geotechnical engineering serves four main purposes:

(*a*) site investigation
(*b*) measurement of a specific property of the ground
(*c*) control of construction
(*d*) monitoring of performance and back analysis.

The first of these, site investigation, is essentially a process of diagnosis, of discovering what the ground consists of at a particular site. This process may consist of direct identification of soil or rock by drilling a borehole, sampling and subsequent inspection on site, or in the laboratory, backed up by index tests and other laboratory tests. Alternatively, in situ tests may be used for the indirect identification of soil type, or more usually of stratification and geologic variation, e.g. by obtaining a continuous profile of the point resistance of a cone penetrometer.

The second purpose, the measurement of a particular soil or rock property, may be adopted either for economic or practical reasons, or more importantly because it is considered essential to measure the property in situ and not in the laboratory.

The third role, that of control of construction, may be an essential part of the satisfactory completion of the works. For example, it might be necessary at a particular site to improve the strength and stiffness of the ground by a means such as dynamic compaction, and the efficiency of the process could be directly monitored by carrying out continuous profiles of piezocone tests (Campanella, Gillespie & Robertson, 1982). Another example is the staged construction of an embankment built on soft ground, with the increase in strength of the underlying clay being directly measured in situ, or monitored indirectly by the decay of excess porewater pressures.

The fourth purpose, the monitoring of performance of geotechnical works, may be a standard procedure such as the continuous observation of movement, of porewater pressures and of quantities of seepage in an earth dam, or in circumstances where there are special problems or uncertainties. A striking example of the latter was the monitoring and subsequent back analysis of the New Palace underground car park at the Houses of Parliament, London, reported by Burland & Hancock (1977).

Most calculations carried out in the past by practising civil engineers for the design of foundations and earthworks have been restricted either to limit analysis for stability calculations or to predictions of settlement. Classical limit analysis is independent both of the deformation characteristics of the ground and of the level of the in situ lateral stress of the undisturbed ground; it depends solely on the groundwater conditions and on the properties of strength and unit weight of the soil or rock. In contrast classical methods of settlement prediction depend only on deformation properties.

In most instances the relevant properties have been evaluated from laboratory tests on supposedly undisturbed samples, and then been used in a simple analysis to lead to designs which have proved to be entirely satisfactory. Why, then, is it necessary to make *in situ* measurements of soil and rock properties?

The main reason for this need is that, as our knowledge of the behaviour of real soils increases, so our appreciation of the inadequacy of conventional laboratory testing grows. The marked consequences of the inevitable disturbance that is caused in any soil specimen, however carefully it has been sampled, transported and reconsolidated in the laboratory, are all too evident. The work at the Building Research Station has shown, for example, that the actual deformation moduli of the ground may be several times greater than those measured in good quality tests in the laboratory on good quality samples, as shown e.g. by Marsland (1973). Consequently predictions of the deformation of the ground around a foundation or excavation based on laboratory data may be grossly overestimated, and the resulting design may be unnecessarily conservative and expensive.

A separate but important advantage of in situ testing is that the soil in question will be tested at the appropriate level of *effective* stress, presuming that disturbance of the ground due to insertion of the instrument has been kept to a minimum.

Apart from good technical reasons for conducting in situ tests, there may be situations where the total cost of site investigation and testing makes them economically attractive, or where they must form the major part of the investigation such as in the exploration of offshore sites for oil production platforms.

In parallel with the major developments that have occurred in the last 25 years or more in experimental techniques, in instrumentation and in the understanding of soil behaviour have been the profound changes in analytical methods made possible by the electronic computer. New methods of numerical analysis not only allow complete solutions to be obtained to complex boundary value problems but also allow the use of non-linear, non-homogeneous, anisotropic—and hence more realistic—models of soil or rock behaviour.

In the past few years there has been a marked growth in the use of in situ tests and in the variety of instruments that have reached a sufficiently developed stage that they can be used with confidence. It is not possible within the limits of this Paper to attempt a comprehensive review of these instruments or of the current state of in situ testing. The purpose of the Paper is to discuss the interpretation and use of the results of in situ tests, and to highlight some of the considerable difficulties and uncertainties associated with them. Most of the discussion is concentrated on

(*a*) in situ tests in clay (most of the principles involved will apply to other soils and rocks to a greater or lesser degree)
(*b*) results of self-boring pressuremeter tests and piezocone tests.

The reasons for this choice are given later.

RELATIONSHIPS BETWEEN SOIL PROPERTIES

The interpretation of data obtained from in situ tests is difficult, and for most tests it is both incomplete and imprecise. A number of separate factors contributes to this unsatisfactory situation. The factors fall into two distinct categories: those due to the behaviour of the soil and those due to the type of test being performed.

Soil behaviour is complex and depends on the complete geological history of the deposit as represented by the size, shape, mineral composition and packing of the particles, the stress history that has been experienced, the pore fluid and other factors. The response of the soil to a particular test will depend on the changes in effective stress that it undergoes, and, further, this response will be inadequately represented by a few simplistic properties such as undrained shear strength, shear modulus, coefficient of consolidation etc. The properties themselves may vary locally to a significant degree both laterally and vertically within the ground, owing to the microfabric of the material and the quirks of its history.

Any in situ test, when considered as a boundary value problem, is beset with difficulties. The boundaries of the problem are unknown and uncontrolled, so that there are insufficient data for a complete solution and for an unequivocal interpretation of the results. The fields of stress increment and strain induced around the instrument by the operation of the test vary significantly with distance from the instrument; this variation is not unique for the type of test but is itself dependent on the stress–strain properties of the soil being tested.

In all but fully drained situations, the non-homogeneous fields of stress cause locally high hydraulic gradients so that some degree of partial consolidation will occur in what is supposed to be—or is interpreted as—an undrained test. This partial consolidation introduces an important rate effect, in addition to that attributable to the viscous nature of soil behaviour.

A further complication arises in that in all in situ tests (except the pressuremeter test) the principal axes of stress rotate within the soil, whereas they do not in the triaxial test, which is used as a standard form of comparison.

In addition, in all experimental work there are limits to the accuracy and reliability of the instrument, a situation which is worse in the field than in the laboratory.

Consequently any interpretation of an in situ test is open to question. To make the most of the interpreted results it is vital to correlate them with the results of all other data, whether from the field or the laboratory, and to draw on all available experience.

The choice of properties that should be used in any attempted correlation is crucial. Any successful relationship that can be used with confidence outside the immediate context in which it was established should ideally be

(*a*) based on a physical appreciation of why the properties can be expected to be related
(*b*) set against a background of theory, however idealized this may be
(*c*) expressed in terms of dimensionless variables so that advantage can be taken of the scaling laws of continuum mechanics.

An illustration of these points is provided by considering correlations of the undrained shear strength s_u of a clay. All soils are basically frictional materials with the strength being provided by the frictional resistance between soil particles governed by the effective stress to which they are subjected. Starting *ab initio*, the first relationship to be explored would be

$$\frac{s_u}{p_f'} = f(\phi) \tag{1}$$

where p_f' is the mean principal effective stress at failure and ϕ is the angle of shearing resistance. In any real situation the value of p_f' will not be known, and it has to be replaced by some other stress variable. If the initial conditions are selected, and the mean principal effective stress p_0' is used, then its relationship with p_f' depends on the excess pore pressures generated during shearing to failure, which in turn depends on the overconsolidation ratio OCR of the clay. Hence a second approach would be to consider the relationship

$$\frac{s_u}{p_0'} = f(\phi, \text{OCR}) \tag{2}$$

However, this relationship will in practice be subject to much uncertainty because the in situ mean principal effective stress p_0' is unlikely to be known or to have been estimated with any accuracy. The single stress variable that can be estimated with most reliability is the in situ vertical effective stress σ_{v0}'. Since this is related to p_0' as a function of OCR, its use in lieu of p_0' will not increase the number of variables in the relationship. Consequently a good engineering compromise is to adopt the expression

$$\frac{s_u}{\sigma_{v0}'} = f(\phi, \text{OCR}) \tag{3}$$

and to define s_u/σ_{v0}' as the *undrained strength ratio.*

Historically in soil mechanics, much use has been made for normally consolidated clays of the relationship suggested by Skempton (1957)

$$\frac{s_u}{\sigma_{v0}'} = 0{\cdot}11 + 0{\cdot}0037\text{PI} \tag{4}$$

where PI is the plasticity index of the clay. Within the context of these arguments, is this a sound relationship, and does it suggest a new variable PI that should be taken into account? At first sight, it is not evident that the undrained strength ratio should be related directly to the plasticity index. However, the value of ϕ can be expected to depend on the shape, size, packing and mineral composition of the clay particles, as will the plasticity index, so the two properties are related in some complex manner. Thus physical reasoning supports Skempton's relationship but suggests that it would be a weaker one than that of equation (3) in which ϕ is preferred to PI.

CONDITIONS AT FAILURE

The majority of in situ tests induce local failure in the soil, and the most commonly deduced property is the undrained shear strength. Much of this Paper is therefore taken up with a detailed and critical look at undrained shear strength.

The symbol used in this Paper for undrained shear strength is s_u, in accordance with common practice in the USA, rather than the symbol c_u as recommended by the British Standards Institution (1975). This is a deliberate choice, because the former relates to 'strength' whereas the latter relates to 'cohesion'; it is argued in this Paper that strength must be interpreted in terms of effective stresses and friction angle, and not total stresses and cohesion.

The basic definition of undrained shear strength is

$$s_u = \tfrac{1}{2}(\sigma_1 - \sigma_3) \tag{5}$$

i.e. half the difference between the major and minor principal stresses, or the radius of the largest Mohr circle. This is an unsatisfactory definition as it neither takes account of the intermediate principal stresses σ_2 nor distinguishes between the different types of test which are well known to give different results for identical soil specimens.

It is essential to distinguish between different test results by an inelegant plethora of suffices as follows:

s_{utc}	triaxial compression test	laboratory
s_{ute}	triaxial extension test	laboratory
s_{upsa}	plane strain active test	laboratory
s_{upsp}	plane strain passive test	laboratory
s_{udss}	direct simple shear test	laboratory
s_{ufv}	field vane test	field
s_{upm}	pressuremeter test	field
s_{ucp}	cone penetrometer test	field

The major question to be faced is how these different measurements of strength are connected. An attempt is made to link them by means of the friction angle ϕ.

The basic concept of an angle of shearing resistance or of internal friction comes from the classical experimental work of Coulomb in which (plane strain) tests were conducted in a shear box. The resulting definition of the friction angle in terms of principal effective stresses is

$$\sin\phi = \frac{\sigma_1' - \sigma_3'}{\sigma_1' + \sigma_3'} \tag{6}$$

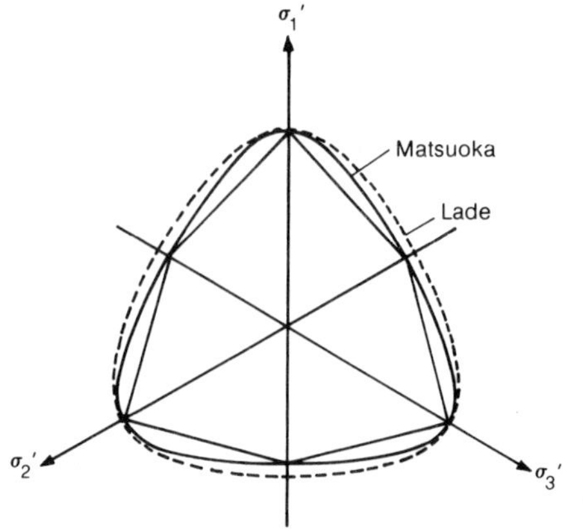

Fig. 1. Sections of failure surfaces in principal stress space

Unfortunately, as for undrained shear strength, this neither allows for the influence of σ_2' nor does it distinguish between different types of test. Because it is absolutely essential to understand soil behaviour in terms of *effective* stresses, and not total stresses, the use of the symbol ϕ_u should be abandoned and the use of the prime in the symbol ϕ' can be dropped, both for convenience and for emphasis.

Furthermore, to make a proper comparison between failure conditions in different tests, the link will be established initially between results of tests on *normally consolidated clays*, so that the friction angle relates not to peak strength but to conditions at the end of a test when there is no further change in volume or effective stresses, i.e. at the *critical state*. Hence ϕ replaces ϕ_{cv}' or ϕ_{cs}'.

Suffices are required to distinguish results, as follows:

ϕ_{tc}	triaxial compression test
ϕ_{te}	triaxial extension test
ϕ_{ps}	plane strain tests.

On the basis of limited experimental evidence it is assumed that the critical state friction angle is the same for all plane strain tests including direct simple shear tests.

To take proper account of the effect of the intermediate principal effective stress σ_2' it is necessary to adopt a generalized failure criterion expressed in terms of all three principal effective stresses. Fig. 1 is a section of principal stress space made by a π plane or octahedral plane, which is perpendicular to the space diagonal. The figure shows the sections of three possible failure surfaces, each of which emanates from the origin in the form of a cone. The inner irregular hexagon is the classical extended Mohr–Coulomb failure envelope given simply by the requirement that

$$\phi = \text{constant} \tag{7}$$

The outer, broken curve is a section of the failure surface proposed by Lade (1972). Since this criterion is expressed in terms of all three principal stresses it is best to make use of the stress invariants

$$\left.\begin{aligned} I_1 &= \sigma_1' + \sigma_2' + \sigma_3' \\ I_2 &= \sigma_2'\sigma_3' + \sigma_3'\sigma_1' + \sigma_1'\sigma_2' \\ I_3 &= \sigma_1'\sigma_2'\sigma_3' \end{aligned}\right\} \tag{8}$$

and to write Lade's criterion as

$$I_1^3/I_3 = \text{constant} \tag{9}$$

An alternative failure criterion is that proposed by Matsuoka (1974) which has the form

$$I_1 I_2/I_3 = \text{constant} \tag{10}$$

and which is represented by the full, inner curve.

Both sections of the Lade and Matsuoka failure surfaces have been drawn so that they coincide with the Mohr–Coulomb criterion for triaxial compression tests (i.e. they pass through the vertices of the hexagon which lie on the positive stress axes). The Matsuoka curve also passes through the other three vertices of the hexagon, whereas the Lade curve does not. The two criteria are very similar, and the curves have the subtle property that the shape varies with the friction angle; as ϕ decreases the shape becomes more circular, and as ϕ increases the shape becomes more triangular.

Of the three criteria, Matsuoka's is chosen for the current analysis for three reasons.

(*a*) It was initially developed from theory and not from curve fitting of experimental data. The theory, which is complex, is based on the concepts of spatially mobilized planes within a soil specimen on which slip is assumed to occur (Matsuoka & Nakai, 1977).

(*b*) It appears to fit experimental data best.* However, the data from complicated laboratory apparatus in which the principal stresses can be independently controlled are notoriously suspect.

(*c*) It is expressed in terms of all three stress invariants.

* Reference can be made to Matsuoka & Nakai (1982), where data are presented for sand from the results of their tests, of Sutherland & Mesdary (1969) and of Ramamurthy & Rawat (1973), and for clay from Shibata & Karube (1965).

A priori, there is no reason to omit the second invariant I_2, as Lade's criterion does.

Bishop (1966) introduced the parameter

$$b = \frac{\sigma_2' - \sigma_3'}{\sigma_1' - \sigma_3'} \tag{11}$$

as a convenient way of expressing the relative value of the intermediate principal effective stress. The value of b varies between zero for triaxial compression and unity for triaxial extension. Its value for plane strain conditions has been much debated, but it has often been taken as 0·5 on the basis of the theory of perfect plasticity.

For the particular case of triaxial compression for which $\sigma_2' \equiv \sigma_3'$ (and $b = 0$) Matsuoka's criterion can be expressed as

$$\begin{aligned} \frac{I_1 I_2}{I_3} &= \frac{(\sigma_1' + 2\sigma_3')(\sigma_3'^2 + 2\sigma_1'\sigma_3')}{\sigma_1'\sigma_3'^2} \\ &= \frac{(\sigma_1'/\sigma_3' + 2)(1 + 2\sigma_1'/\sigma_3')}{\sigma_1'/\sigma_3'} \end{aligned} \tag{12}$$

which can be rewritten in terms of ϕ_{tc} as follows:

$$\begin{aligned} \frac{I_1 I_2}{I_3} &= \frac{(3 - \sin\phi_{tc})(3 + \sin\phi_{tc})}{(1 - \sin\phi_{tc})(1 + \sin\phi_{tc})} \\ &= 9 + 8\tan^2\phi_{tc} \end{aligned} \tag{13}$$

For triaxial extension, an exactly similar derivation leads to

$$\begin{aligned} \frac{I_1 I_2}{I_3} &= \frac{(3 + \sin\phi_{te})(3 - \sin\phi_{te})}{(1 + \sin\phi_{te})(1 - \sin\phi_{te})} \\ &= 9 + 8\tan^2\phi_{te} \end{aligned} \tag{14}$$

which confirms that for this criterion $\phi_{tc} \equiv \phi_{te}$.

It is required to relate, if possible, ϕ_{ps} with ϕ_{tc} by some simple relationship. Satake (1982) has shown that if an associated flow rule is applied to the Matsuoka failure criterion (treated as a yield surface) then for plane strain conditions ϕ_{ps} is the maximum value that ϕ can have (for all values of b). This assumption of an associated flow rule is not valid for peak conditions but is suggested as acceptable for critical state conditions.

By finding the maximum value of the ratio σ_1'/σ_3' (i.e. the maximum value of ϕ) for a fixed value of I_1 (i.e. for one octahedral plane) it can be shown that ϕ_{ps} ($\equiv \phi_{max}$) is given by

$$\sec^2\phi_{ps} + \sec\phi_{ps} = 2\sec^2\phi_{tc} \tag{15}$$

and further that the value of b for plane strain

$$b_{ps} = \frac{\sin\phi_{ps} + \cos\phi_{ps} - 1}{2\sin\phi_{ps}} \tag{16}$$

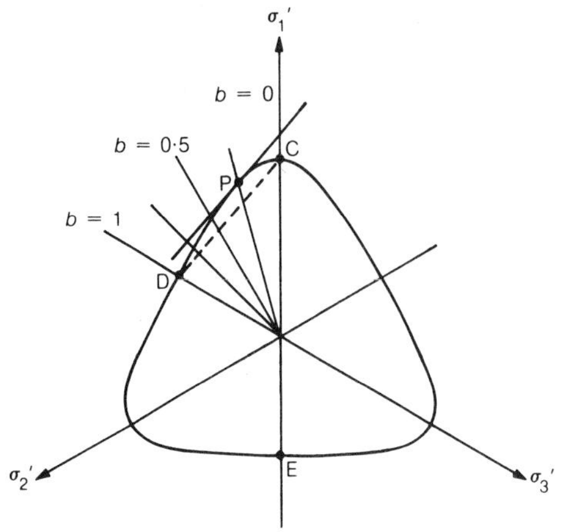

Fig. 2. Plane strain conditions derived from Matsuoka's criterion

or alternatively

$$\left(\frac{\sigma_2'}{\sigma_1'+\sigma_3'}\right)_{ps}=\tfrac{1}{2}\cos\phi_{ps} \quad (17)$$

These findings are illustrated in Fig. 2 where P indicates the point on the failure surface corresponding to plane strain conditions. Note that CD is a line of constant ϕ (the Mohr–Coulomb surface), but that the orientation of such a line changes slightly with the value of ϕ, so that the tangent at P associated with ϕ_{max} will not be parallel to CD. Taking for example a value of $\sin\phi_{ps}=0{\cdot}6$ ($\phi_{ps}=36{\cdot}87°$) then $\phi_{tc}=32{\cdot}5°$ and $b_{ps}=1/3$.

The relationships of equations (15) and (16) are plotted in Fig. 3. The resulting curves can be approximated for engineering purposes by the linear relationships

$$8\phi_{ps}°\approx 9\phi_{tc}° \quad (18)$$

and

$$2b_{ps}\approx 1-\frac{\phi_{ps}°}{100} \quad (19)$$

Equation (18) will be used for relating the results of plane strain tests with triaxial compression tests for a given soil.

UNDRAINED TRIAXIAL COMPRESSION TESTS

The triaxial compression test has become the standard method of obtaining stress–strain and strength properties of soils in the laboratory as part of a conventional site investigation. It is against such a background that the results of in situ tests will be judged in general. Consequently it is valuable to establish a theoretical understanding of the triaxial test, and it is believed that this is best done by means of the framework provided by critical state soil mechanics (CSSM).

The original concepts of this approach to soil behaviour were based on the idea of a critical void ratio conceived by Casagrande (1936), the early triaxial tests on sand by Taylor (1948) and then extended by a detailed analysis of triaxial tests on isotropically consolidated specimens of reconstituted clay by Roscoe, Schofield & Wroth (1958).

In Fig. 4 is presented a comparison between the classical representation of one-dimensional consolidation of Terzaghi and the modern approach for isotropic consolidation of CSSM. The former has voids ratio e plotted against $\log_{10}\sigma_v'$, whereas the latter has specific volume V plotted against $\log_e p'$, with the consequence that the compression and swelling indices are defined differently.

In one-dimensional swelling the principal stresses do not decrease proportionally, so that

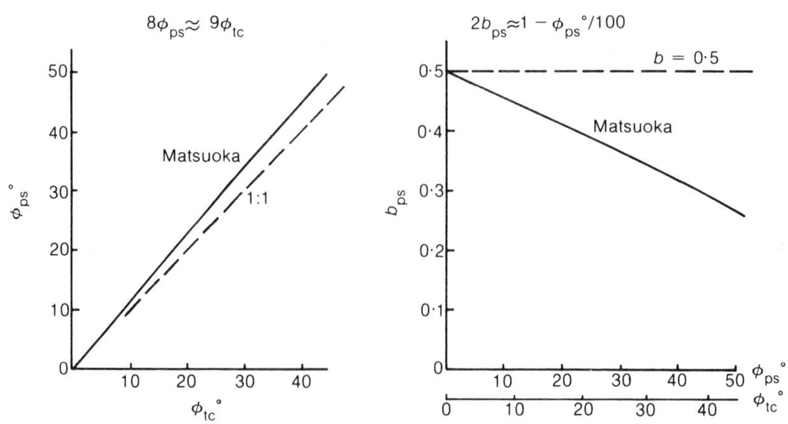

Fig. 3. Relationships for plane strain conditions derived from Matsuoka's failure criterion

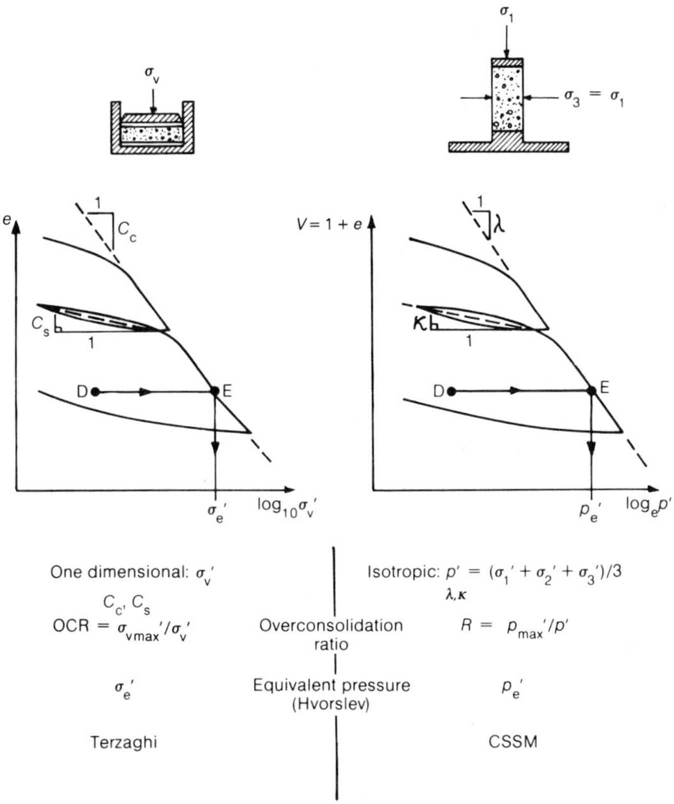

Fig. 4. Definitions of consolidation in one-dimensional and isotropic conditions

care is required with the definition of the overconsolidation ratio which will not be the same in the two plots. For isotropic conditions the overconsolidation ratio is defined as the ratio of the maximum past mean effective stress p_{max}' to the current value p', and it is given the symbol R in accordance with Atkinson & Bransby (1978).

The equivalent pressure of a specimen, introduced by Hvorslev (1937), is defined as the pressure on the normal consolidation line, such as at point E, at the same voids ratio as that of the specimen at state D. This proves to be an elegant and convenient way of converting from the voids ratio (or water content) of a clay specimen into a pressure variable for comparisons in dimensionless form.

The idealized results of undrained triaxial compression tests are represented in Fig. 5 in terms of the state variables p', q and V. A specimen initially normally consolidated at point C undergoes the effective stress path CD whereas an initially overconsolidated specimen at point R experiences the path RS; it is assumed that both specimens reach undrained failure on the critical state line at D and S respectively. The critical state line is assumed to be parallel to the isotropic normal consolidation line ABC in the semilogarithmic plot, and its relative position to be given by the spacing ratio r defined as the ratio of the pressures at C and X which lie on the same swelling line CXR. For the original Cam clay model $r = 2{\cdot}718$ (= e, the base of natural logarithms), whereas for the modified Cam clay model $r = 2$.

In Appendix 1 it is shown that

$$p_s'/p_r' = (R/r)^{\Lambda} \tag{20}$$

where $\Lambda = (\lambda - \kappa)/\lambda$. This parameter was introduced by Schofield & Wroth (1968) because it plays an important role in realistic elasto-plastic models of soil behaviour which incorporate strain-hardening plasticity. It may be termed the *plastic volumetric strain ratio*, being the ratio of the plastic component to the total component of the volumetric strain increment in normal consolidation. This parameter Λ consistently appears as an exponent in the subsequent analyses.

The undrained shear strength in compression is half the deviator stress at failure so that

$$s_{utc} = \tfrac{1}{2}q_s = \tfrac{1}{2}Mp_s' \tag{21}$$

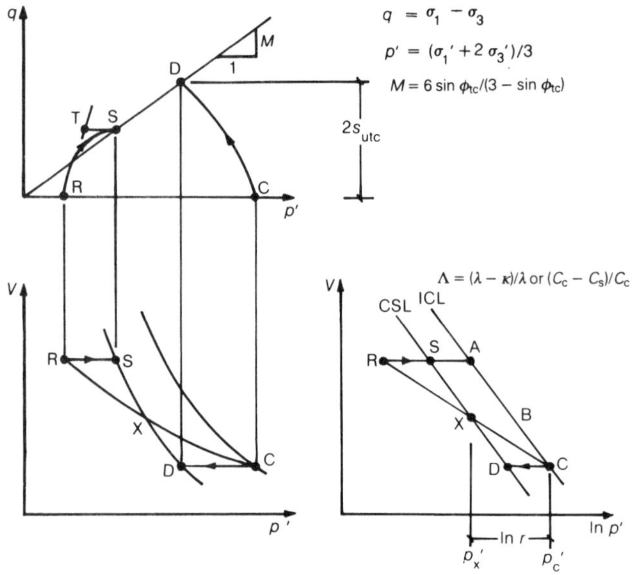

Overconsolidation ratio: $R = p_C'/p_R'$

'Spacing' ratio of ICL and CSL: $r = p_c/p_x' (\approx 2)$

Undrained strength ratio: $s_{utc}/p' = M/2(R/r)^{\Lambda}$

Normalized strength ratio: $(s_{utc}/p')/(s_{utc}/p')_{nc} = R^{\Lambda}$

Fig. 5. Theoretical expressions for undrained strength in triaxial compression tests

Substituting for p_s' from equation (20) gives

$$\frac{s_{utc}}{p'} = \frac{M}{2}\left(\frac{R}{r}\right)^{\Lambda} \tag{22}$$

where for triaxial compression

$$M = \frac{6 \sin \phi_{tc}}{3 - \sin \phi_{tc}} \tag{23}$$

This expression is valid for specimens that have been isotropically consolidated, so that at the start of the compression test $\sigma_{v0}' \equiv p_0'$. Hence equation (22) can be reinterpreted as an expression for the *undrained strength ratio*

$$\frac{s_{utc}}{\sigma_{v0}'} = \frac{M}{2}\left(\frac{R}{r}\right)^{\Lambda} \tag{24}$$

For any given soil M, r and Λ will be constants so that in theory the undrained strength ratio is proportional to the overconsolidation ratio raised to the power Λ. Evidence in support of this finding is presented later.

Experimental evidence supporting these concepts is provided by the results of undrained triaxial compression tests on isotropically consolidated specimens of reconstituted kaolin by Loudon (1967). The effective stress paths for a set of specimens are presented in Fig. 6 where the stresses have been made dimensionless by dividing by the relevant value of the equivalent pressure p_e' in each case. In this plot the critical state line is reduced to a single unique *critical state point* indicated by C.

In the development of the understanding of soil behaviour, the major centres of experimental research in soil mechanics have wisely concentrated their efforts on testing a limited range of soils in a comprehensive manner. Examples that come readily to mind are kaolin, Weald clay, London clay, Boston blue clay and Drammen clay. This philosophy of research has allowed a reliable and detailed picture to be built up of the behaviour of these clays but has the minor disadvantage that there has not been a survey of trends of soil properties for a wide range of clays. In particular there are unfortunately few reliable data on the range of values of the spacing ratio r and the plastic volumetric strain ratio Λ. On the basis of the limited evidence available it seems that neither of these parameters varies significantly for a wide range

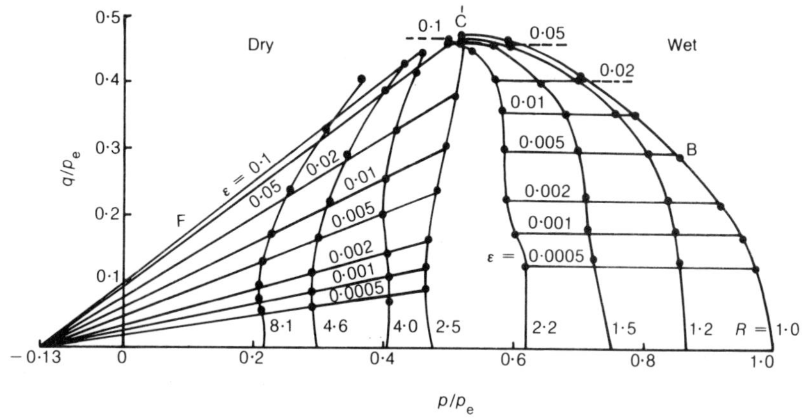

Fig. 6. Effective stress paths for undrained triaxial compression tests on kaolin (after Loudon, 1967)

of clays, and that for present purposes it is reasonable to adopt the approximate values

$$\left.\begin{aligned} r &\approx 2 \\ \Lambda &\approx 0{\cdot}8 \end{aligned}\right\} \quad (25)$$

Use of these values means that the undrained strength ratio for normally consolidated specimens ($R = 1$) can be rewritten by combining equations (23) and (24) to give

$$\left(\frac{s_{utc}}{\sigma_{v0}'}\right)_{nc} \approx \frac{M}{2}\left(\frac{1}{2}\right)^{0{\cdot}8}$$

$$\approx 0{\cdot}5743 \frac{3 \sin \phi_{tc}}{3 - \sin \phi_{tc}} \quad (26)$$

The variation in this ratio with the friction angle ϕ_{tc} is shown in Fig. 7 by the upper curve marked ITC (denoting Isotropically consolidated specimens tested in Triaxial Compression). Note that for a clay that has values of r and Λ that differ from those assumed (equation (25)) the consequence is merely one of scaling the ordinate axis appropriately: the shape of the curve would remain unchanged. For later comparisons the values of the friction angle in plane strain ϕ_{ps} given by equation (18) have been included on the axis.

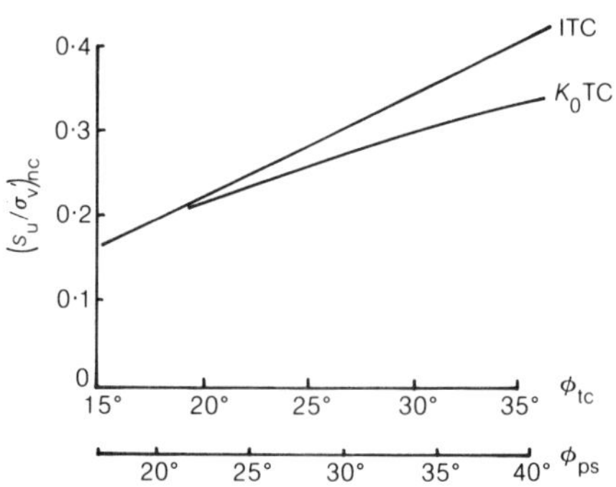

Fig. 7. Variation in the undrained strength ratio with the angle of friction for triaxial tests on normally consolidated clay

Specimens consolidated one dimensionally

In nature, it is usually assumed that soil deposits have become consolidated under one-dimensional conditions. To simulate soil behaviour in the laboratory, it is becoming increasingly the practice to reconsolidate specimens anisotropically to their assumed in situ stresses. The behaviour of such specimens will not be the same as that of isotropically consolidated specimens; this difference must be accounted for when the results are used for predicting field behaviour.

The concepts of CSSM require that the state of an anisotropically normally consolidated specimen of clay lies on the state boundary surface at some point such as B in Fig. 6 (reference can be made to Schofield & Wroth (1968) or Atkinson & Bransby (1978)). It is assumed that failure of this specimen will occur at the critical state given by point C, with the same undrained shear strength as an isotropically normally consolidated specimen at the same water content (i.e. having the same equivalent pressure). However, because the initial stress states are different, the undrained strength ratios will be different; the actual value for the anisotropic specimen will depend on the shape of the state boundary surface, as well as on the value of K_0.

In Appendix 2, it is assumed that the state boundary surface is formed by the elliptical yield surface of modified Cam clay. The analysis gives the following cumbersome expression for the undrained strength ratio:

$$\left.\begin{aligned}\left(\frac{s_{utc}}{\sigma_{v0}'}\right)_{K_0} &\approx \frac{\sin\phi_{tc}}{2a}\left(\frac{a^2+1}{2}\right)^{\Lambda}\\ a &= \frac{3-\sin\phi_{tc}}{2(3-2\sin\phi_{tc})}\end{aligned}\right\} \quad (27)$$

For $\Lambda = 0{\cdot}8$, this expression has been plotted in Fig. 7 as the lower curve marked K_0TC.

There is a substantial difference between the two curves, especially for higher friction angles, so that it is important when undrained strength ratios obtained in triaxial tests are quoted that it is clearly stated whether the specimens were initially isotropically or anisotropically consolidated. This distinction is important for all real soils, even if the idealized model of modified Cam clay is not deemed to be relevant.

Overconsolidated specimens

The undrained strength ratio for overconsolidated specimens is fully specified by CSSM theory and has been expressed by equation (24), i.e.

$$\frac{s_{utc}}{\sigma_{v0}'} = \frac{M}{2}\left(\frac{R}{r}\right)^{\Lambda}$$

The fact that all the effective stress paths in Fig. 6 approach and nearly reach the unique critical state point C means that this relationship is closely followed by the idealized situation of Loudon's tests on isotropically consolidated specimens of reconstituted kaolin.

For testing its application more widely it is better to normalize the undrained strength ratio in the form

$$\frac{s_{utc}/\sigma_{v0}'}{(s_{utc}/\sigma_{v0}')_{nc}} = R^{\Lambda} \quad (28)$$

which has the advantage of being independent of both the frictional coefficient M and the spacing ratio r. This is a powerful relationship well supported by various sets of data, each of which refers to one type of test on one particular clay.

The most comprehensive sets of tests have been carried out at the Massachusetts Institute of Technology where direct simple shear tests have been conducted on a range of clays in the Geonor apparatus (Ladd & Edgers, 1972). During such a test the maximum value of shear stress applied to the specimen, τ_{max}, is observed. The specimens have necessarily been consolidated one dimensionally so that the directly analogous expression to equation (28) is

$$\frac{\tau_{max}/\sigma_{v0}'}{(\tau_{max}/\sigma_{v0}')_{nc}} = (\text{OCR})^{\Lambda} \quad (29)$$

The results of tests carried out on Drammen clay at the Norwegian Geotechnical Institute by Andresen, Berre, Kleven & Lunne (1979) have been replotted in Fig. 8 with both variables plotted with logarithmic scales. These data have been used because the tests are the only ones known to have included specimens with values of OCR as high as 40. Equation (29) suggests that the results should lie on a straight line with gradient Λ, which is indeed the case with $\Lambda \approx 0{\cdot}8$.

Similar data for seven different clays assembled by Ladd & Edgers (1972) were presented in the state of the art report by Ladd, Foott, Ishihara, Schlosser & Poulos (1977) at the International Conference on Soil Mechanics and Foundation Engineering at Tokyo. The bounds to those curves are presented in Fig. 9 in which the

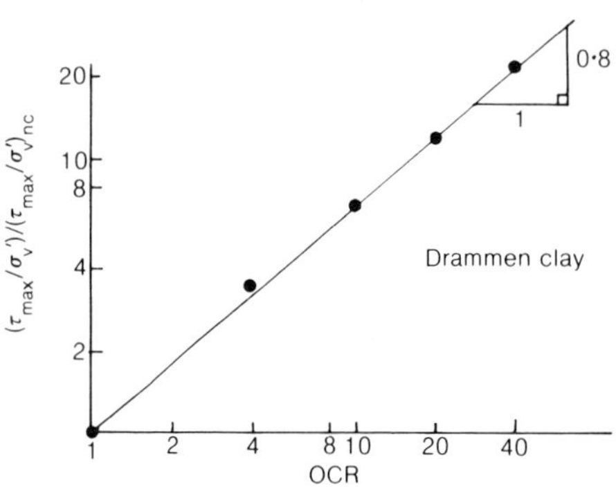

Fig. 8. Variation in the normalized undrained strength ratio with overconsolidation ratio for Drammen clay in simple shear (data from Andresen *et al.*, 1979)

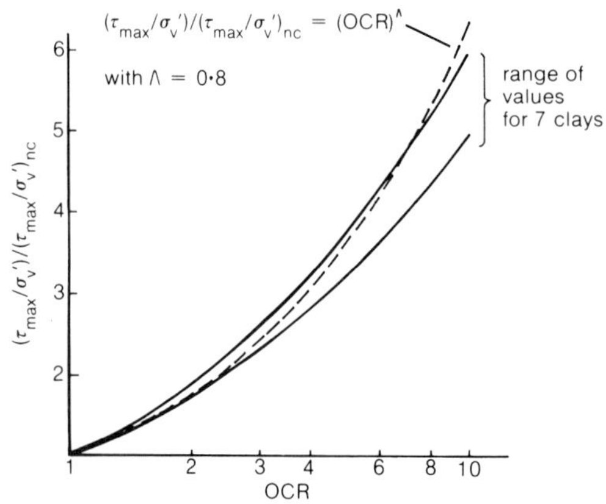

Fig. 9. Variation in the normalized undrained strength ratio with the overconsolidation ratio for various clays in simple shear (data from Ladd & Edgers, 1972)

Table 1. Values of exponent m for equation (30) for five clays (data from Ladd & Edgers, 1972)

	LL:%	PL:%	OCR<7	All OCR
			m_1	m_2
Boston blue clay	41	20	0·802	0·776
Maine organic clay	65	31	0·866	0·802
Bangkok clay	65	24	0·800	0·767
Atchafalaya clay	95	20	0·799	0·795
Connecticut valley varved clay	65	26	0·752	0·685

LL, liquid limit; PL, plastic limit.

normalized undrained strength ratio is plotted with an arithmetic scale, but the overconsolidation ratio has a logarithmic scale.

Ladd *et al.* (1977) suggest that the data can be well defined by the expression (using their symbols)

$$\frac{(c_u/\sigma_{v0}')_{oc}}{(c_u/\sigma_{v0}')_{nc}} = (\text{OCR})^m \qquad (30)$$

'with $m = 0{\cdot}8$, though a better fit is obtained if m is decreased from 0·85 to 0·75 with increasing OCR'. Their finding, which was solely based on empiricism, is confirmed by the CSSM theory. Moreover CSSM theory relates the exponent m with well-recognized physical properties of the clay in question by stipulating that $m \equiv \Lambda$, the plastic volumetric strain ratio.

The original data from the report by Ladd & Edgers (1972) have been replotted in the manner of Fig. 8, and a least-squares regression analysis has been performed to find the best fit of a straight line to each set of data. As Ladd has pointed out, the data do not lie perfectly on a straight line in this plot, but consistently on a shallow convex curve such that a reduced value of m is required for large values of overconsolidation. Accordingly a distinction has been made between excluding all tests with OCR$\geq$7 and obtaining a value for $m = m_1$, say, and including all data and obtaining a lower value of $m = m_2$. The corresponding values of the gradients m_1 and m_2 for the five clays (with sufficient data) are recorded in Table 1; the Boston blue clay was reconstituted, the other four undisturbed.

The importance of the relationship in equation (29) is that if the undrained strength ratio of a clay can be measured or estimated for a normally consolidated specimen then its value can be predicted for other specimens of the clay provided that the degree of overconsolidation is known. This point is returned to later.

RANGE OF IN SITU TESTS

A good indication of the wide variety of in situ tests that have been developed is provided by Table 2. This table is taken from a report by Campanella & Robertson (1983) which forms part of a comprehensive state of the art review of in situ testing. The tests have been listed by the authors in what they consider to be an ascending order of cost and/or complexity.

The list is by no means exhaustive, omitting for example

(*a*) push-in pressuremeter (Henderson, Smith & St John, 1979)
(*b*) push-in spade-shaped pressure cell (Tedd & Charles, 1981)
(*c*) self-boring permeameter (Baguelin, Jézéquel & Le Mehaute, 1974)
(*d*) self-boring plate test (Mori, 1983)
(*e*) electrical conductivity probe (Arulanandan, 1977).

More attention deserves to be paid to various in situ geophysical tests. Pile load tests should also be considered to be a special form of in situ tests; more information could be obtained about the elastic properties of the ground by careful observations during the unloading of the test pile.

The compilers of the table have given their views of the applicability of each test for deducing a variety of soil properties. The two instruments that have the best ratings by far are the self-boring pressuremeter and the piezocone (with friction sleeve). The interpretation of typical results from each instrument is discussed in the following sections.

THE SELF-BORING PRESSUREMETER

The pressuremeter was first conceived by Ménard in 1954, as a tool which was inserted into a preformed borehole and used for the measurement of the strength and stiffness of soils and rocks. Since Ménard's pioneering work there have been many major developments in the pressuremeter, especially in the last 15 years, which can be conveniently divided into four categories:

(*a*) techniques of measurement
(*b*) techniques of insertion
(*c*) new analyses of the expansion test
(*d*) new types of test.

Table 2. Perceived applicability of in situ test methods—update 1982 (after Campanella & Robertson, 1983)*

	Soil type	Profile	Density D_r	Angle of friction ϕ	Undrained shear strength c_u	Pore pressure u	Stress history OCR	Modulus: Young's and shear (E and G)	Compressibility m_v and C_c	Consolidation c_h and c_v	Permeability k	Stress–strain curve	Liquefaction resistance
Dynamic cone	C	A	B	C	C	—	C	—	—	—	—	—	C
Static cone													
Mechanical	B	A	B	C	B	—	C	B	C	—	—	—	B
Electrical friction	B	A	B	C	B	—	C	B	C	—	—	—	B
Electrical piezo	A	A	B	B	B	A	A	B	B	A	B	B	A
Electrical piezo/friction	A	A	A	B	B	A	A	B	B	A	B	B	A
Acoustic probe	C	B	B	C	C	—	C	C	—	—	—	—	C
Dilatometer	B	A	B	C	B	—	B	B	C	—	—	C	B
Vane shear	B	C	—	—	A	—	B	—	—	—	—	—	—
Standard penetration test	B	B	B	C	C	—	—	—	C	—	—	—	A
Seismic cone penetration test downhole	C	C	C	—	—	—	—	A	—	—	—	B	B
K_0 blade	—	—	—	—	—	—	B	—	—	—	—	—	—
Resistivity probe	B	B	A	B	C	—	C	C	C	—	—	C	A
Borehole permeability	C	—	—	—	—	A	—	—	—	B	A	—	—
Hydraulic fracture	—	—	—	—	—	B	B	—	—	C	C	—	—
Screw plate	C	C	B	C	B	—	B	A	B	C	C	B	B
Seismic downhole	C	C	C	—	—	—	—	A	—	—	—	B	B
Impact cone	C	B	C	C	C	—	C	C	C	—	—	—	C
Borehole shear	C	C	—	B	B	—	C	C	—	—	—	C	—
Ménard pressuremeter	B	B	C	B	B	—	C	B	B	—	—	C	C
Self-boring pressuremeter	B	B	A	A	A	A	A	A	A	A	B	A	A
Self-boring devices													
K_0 meter	—	—	—	—	—	—	A	—	—	—	—	—	—
Lateral penetrometer	C	C	B	B	B	—	B	B	C	—	—	—	—
Shear vane	B	C	—	—	A	—	B	—	—	—	—	—	—
Seismic cross-hole	C	C	B	—	—	—	—	A	—	—	—	B	B
Nuclear tests	—	—	A	B	—	—	—	C	—	—	—	—	C
Plate load tests	C	C	B	B	C	—	B	A	B	C	C	B	B

* A, high applicability; B, moderate applicability; C, limited applicability.

In the basic expansion test of the pressuremeter, measurements are made of the pressure and of the increase in volume of the inflatable membrane. In the early equipment used by Ménard these observations were made at the ground surface by monitoring the pressure and volume of the fluid used for inflation. With the development of accurate and reliable electrical transducers placed within the pressuremeter it has been possible to improve the accuracy of these measurements by an order of magnitude or more. For example the feeler gauges incorporated in the Cambridge self-boring pressuremeter are sensitive to movements of the order of 0·005 mm, which is equivalent to less than 0·02% volumetric strain for an instrument with a diameter of 83 mm (as in the latest version).

The self-boring pressuremeter has been developed in parallel in France and Britain, where most of the fundamental work has been done at the University of Cambridge by a succession of research students and fully reported in their

PhD theses: Hughes (1973), Windle (1976), Clarke (1981) and Fahey (1980).

It is presumed that the purpose and details of the self-boring technique are well understood. To minimize the disturbance caused by insertion, various studies have been carried out of the degree of disturbance created by different modes of operation and design of equipment. The geometrical arrangement of the cutting shoe and cutting tool is crucial. Recent work by Clarke (1981) suggests why the push-in pressuremeter may prove to be a successful compromise for offshore use between the complex self-boring pressuremeter and the cruder Ménard instrument.

Initially, the results of pressuremeter tests were interpreted by means of empirical expressions to give parameters for design such as allowable bearing capacity factors and moduli for allowable settlement. The first fundamental interpretation of an expansion test was supplied by Gibson & Anderson (1961) in which the pressuremeter was considered to be infinitely long so that the deformation of the surrounding soil was assumed to be in conditions of axial symmetry and plane strain. One analysis was applicable to undrained expansion tests in clay in which the soil is assumed to undergo zero volume change and to behave as an elastic–perfectly plastic material characterized by a shear modulus G and an undrained shear strength s_u. A separate analysis was achieved for expansion tests in cohesionless soils for which the soil is assumed to behave elastically until failure occurs at a constant effective stress ratio (governed by the Mohr–Coulomb criterion in terms of a friction angle ϕ') and with no volume change.

Both of these analyses have been improved. The undrained expansion test can now be interpreted with no prerequisite assumption of stress–strain properties of the soil being tested, as a result of work independently done by Baguelin, Jézéquel, Le Mée & Le Mehaute (1972), Ladanyi (1972) and Palmer (1972). Unfortunately the interpretation is very sensitive to the disturbance of the soil caused by insertion and to the datum selected for the strain.

The second case, the drained expansion test in a cohesionless material, has been modified by Hughes, Wroth & Windle (1977) to take account of the volume change that occurs after failure has been initiated. The relevance of this new analysis has been confirmed by Fahey (1980) by a special series of laboratory tests under carefully controlled conditions.

To date most pressuremeter tests have consisted of a rapid expansion test either conducted with stress increments applied at regular intervals or under constant rates of strain. With the advent of microprocessors and computer-controlled experiments, as well as additional observations, e.g. pore pressure measurements, it is now possible to carry out new types of test. One such possibility is a 'holding' test for measuring in situ the consolidation characteristics of the soil being tested. Clarke, Carter & Wroth (1979) report the results of such holding tests in which a quick undrained expansion test in a clay to, say, 10% strain is followed by a phase when the membrane is kept automatically at its (enlarged) radius by adjustment of the applied pressure. During this phase the clay undergoes consolidation with a reduction in the total radial stress and excess pore pressure, both of which are continuously recorded. The coefficient of consolidation can be deduced from the results.

Good quality pressuremeter test in clay

In this and the next section examples are given from one site of two good quality pressuremeter tests, one in clay and one in sand. The site is part of the outer harbour at Zeebrugge where it was planned to construct large storage tanks for liquefied natural gas. As part of the overall site investigation, PM Insitu Techniques Ltd were commissioned by Distrigas NV to carry out a profile of self-boring pressuremeter tests; the work was supervised by Tractebel Z acting as engineers.

The soil profile at this site consists of hydraulic-pumped fill to a depth of 15 m, quaternary formations consisting mainly of sands from 15 m to 33 m, Bartoon clay from 33 m to 46 m and below this Pamiselian deposits, predominantly sandy soils.

The self-boring pressuremeter cannot be drilled through soil containing hard inclusions, such as sandstone layers and gravel which occur at this site. It cannot be drilled continuously through sand deposits as it leaves an unsupported borehole which may collapse and make extraction of the instrument difficult if not impossible.

On this site it was therefore necessary to use a drilling rig to provide a borehole which would not collapse and to cope with hard inclusions. The borehole was drilled and cased using a shell and auger rig to a depth of 1 m above the level of a proposed test position. The self-boring pressuremeter was lowered down the borehole and drilled beyond the bottom of the casing to at least 1 m into material presumed to be undisturbed. A pressuremeter test was carried out. In certain instances it was possible to advance the self-boring pressuremeter 1 m further and to carry out a second test. The self-boring pressuremeter was withdrawn from the borehole so that the hole could be drilled further by the shell and auger rig. The one borehole was terminated at a depth of 88 m and included 24 pressuremeter tests.

The results of a pressuremeter test in clay at a depth of 43·4 m are reproduced in Fig. 10. The readings are automatically recorded on site both on magnetic disc and as a hard copy on paper, and subsequently processed and plotted by computer. The pressure ψ applied to the inside of the membrane is plotted against the cavity strain ε which is the observed expansion of the membrane divided by its initial radius a_0. The pressure ψ has been corrected for the (small) strength of the membrane itself, represented by the difference between the ordinates of points O and I.

At the start of the expansion test the membrane fits tightly over the body of the instrument and has the same diameter as the cutting shoe. In theory no expansion of the membrane should be detected until the applied pressure ψ is equal to the in situ total lateral stress in the ground in contact with the pressuremeter. In reality there will be some small compliance of the instrument itself, until at point A (the lift-off pressure), on the expansion curve, the soil starts to deform under increasing lateral stress.

It is usually assumed (both for convenience and for lack of better information) that the lateral stress in the ground is the same in all horizontal directions; this is unlikely to be the case except in very homogeneous deposits subject solely to one-dimensional consolidation during their geological history.

The Cambridge self-boring pressuremeter has three separate feeler arms for measurement of the movement of the membrane, which are symmetrically disposed at 120° around the section of the instrument. By careful scrutiny at suitably large scale of separate plots of ψ against each of the values of ε, three separate estimates of the in situ total lateral stress can be obtained (on the presumption that there has been negligible disturbance of the ground during insertion of the pressuremeter). The results in Fig. 10 are for one of the three feeler arms and indicate a value of σ_{h0} of 646 kN/m^2. The measurement of in situ lateral stress by the self-boring pressuremeter is discussed in detail by Ghionna, Jamiolkowski & Lancellotta (1982) and by Lacasse & Lunne (1982a).

In the original analysis of Gibson & Anderson (1961) for the interpretation of an undrained pressuremeter test in clay, it was assumed that the clay behaves as a perfectly elastic–perfectly plastic material characterized by a shear modulus G, Poisson's ratio $\nu = 0{\cdot}5$ and an undrained shear strength s_u. The analysis leads to the result that, after failure has been initiated in the clay, the applied pressure ψ is linearly related to the logarithm of the *current* volumetric strain $\Delta V/V$. Moreover, if natural logarithms are used the gradient of the line will be equal to the undrained shear strength, denoted here by s_{upm}.

The field results of the pressuremeter test of Fig. 10 have been replotted in Fig. 11 in the

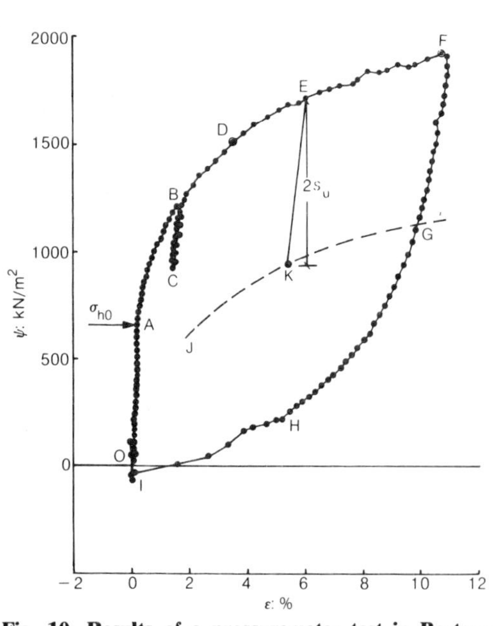

Fig. 10. Results of a pressuremeter test in Bartoon clay, Zeebrugge

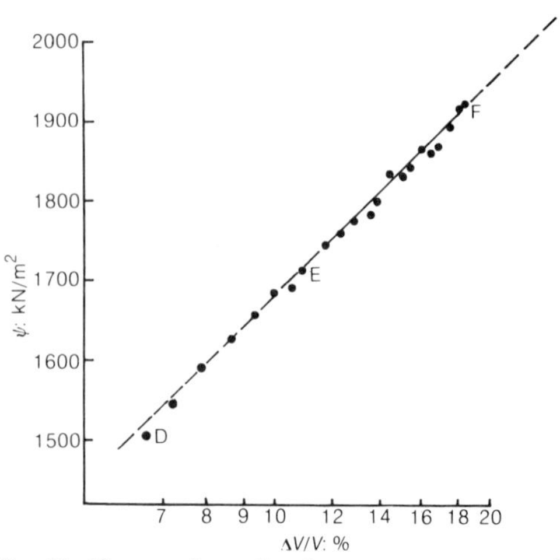

Fig. 11. Processed results of a pressuremeter test in Bartoon clay

above manner as ψ against $\log(\Delta V/V)$. Note that the volumetric strain is directly related to the observed strain by the relationship

$$\Delta V/V = 1-(1+\varepsilon)^{-2} \quad (31)$$

The points D, E and F correspond to those selected in Fig. 10. The results lie remarkably close to a straight line, and that selected has a gradient $s_{upm} = 386\ kN/m^2$. The newer and superior analysis of Baguelin *et al.* (1972), Ladanyi (1972) and Palmer (1972) does not require an assumption to be made about the stress–strain behaviour of the soil. It leads to the result that the local gradient of the curve drawn through the points in Fig. 11 is the shear stress corresponding to the shear strain experienced at that stage of the test by the clay in contact with the pressuremeter. The fact that the curve is so close to a straight line vindicates the assumption implicit in the Gibson–Anderson analysis that the clay behaves as a perfectly elastic–perfectly plastic material.

From experience gained at a large number of sites it is believed that the Gibson–Anderson analysis is satisfactory for design in geotechnical engineering and that the extra sophistication of the 1972 analysis is not warranted, particularly because it is so sensitive to the choice of datum. General experience from a large number of sites indicates that the in situ undrained shear strength obtained in pressuremeter tests is consistently greater than corresponding values from the vane shear test, cone penetrometer test and plate bearing test. This finding is discussed by Mair & Wood (1984) who cite examples quoted by Hughes, Wroth & Pender (1975), Windle & Wroth (1977) and Ghionna, Jamiolkowski, Lacasse, Lancellotta & Lunne (1983).

The shear modulus of the clay can be measured by arranging for an unloading–reloading cycle such as BCB to be included in the expansion test. If the clay behaves as a perfectly elastic material in unloading then BC will be a straight line of gradient $2G$ (using small strain theory). It should be noted that the cycle BCB is markedly linear with very small hysteresis, and gives a value of the shear modulus $G = 47\ MN/m^2$.

In carrying out such an exercise care must be taken not to exceed the elastic limit of the clay during the unloading phase; this restricts the amplitude of the stress cycle that can be applied. For the ideal elastic–plastic material the allowable amplitude is twice the undrained shear strength, as shown by Wroth (1982).

It is therefore possible to construct a curve such as JKG in Fig. 10, which serves as a theoretical boundary to the elastic behaviour in unloading. It should be noted that this boundary fits well with the extent of the linear portion of the final unloading part of the test FGHI.

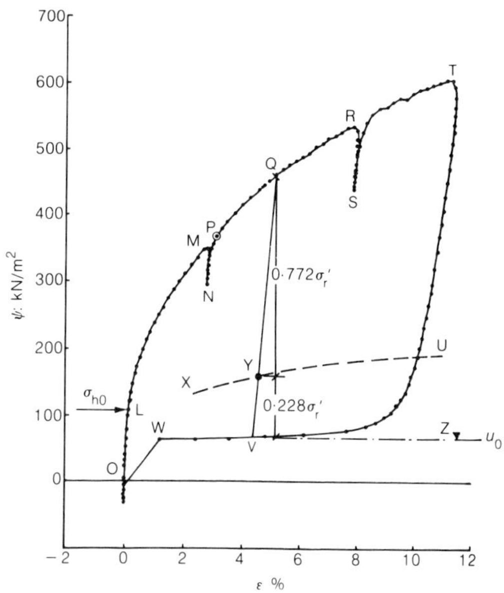

Fig. 12. Results of a pressuremeter test in sand at Zeebrugge

Good quality pressuremeter test in sand

The results of a pressuremeter test in sand (in the same borehole as that for the test in clay) at a depth of 10·6 m are reproduced in Fig. 12. The shape of the curve is different from that of the test in clay of Fig. 10, both in the loading portion MPQRT and in the final unloading portion TUVW. This difference is always observed in good quality tests and can be used as an indicator of the material tested. It can be represented numerically by the factor β defined as (Baguelin, 1982)

$$\beta = \frac{p_{20} - p_5}{p_{20} - p_0} \quad (32)$$

where p_0, p_5 and $_{t\ 20}$ are respectively the applied pressure at 0%, 5% and 20% volumetric strain. Noting that the volumetric strain is approximately twice the plotted strain ε, the value of β for the clay in Fig. 10 is 0·45, whereas that for the sand in Fig. 12 is 0·51.

The same method as that outlined previously can be used for an assessment of the in situ lateral total stress. When the results of Fig. 12 are plotted to a larger scale, the lift-off pressure is given by the point L, i.e. a stress of $107\ kN/m^2$.

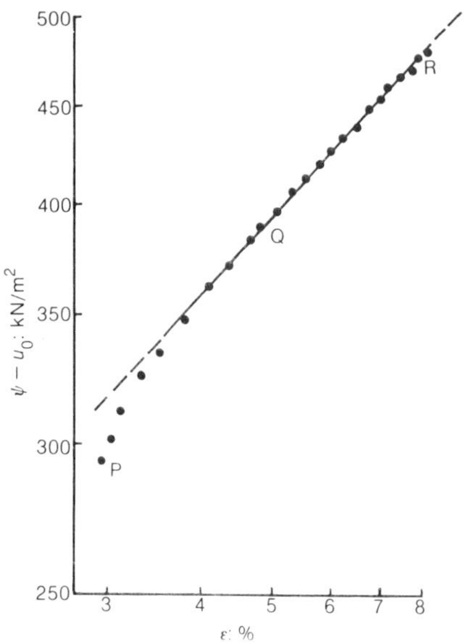

Fig. 13. Processed results of a pressuremeter test in sand at Zeebrugge

In the test on a cohesionless material which is sufficiently permeable, the final part of the unloading curve VW provides valuable information about ground conditions. As the membrane is deflated, ultra-loose sand collapses into the enlarged borehole under very small effective stresses. In the extreme condition at W, the membrane is being pushed back to its initial size by the action of the groundwater, and the pressure recorded inside the membrane is a measure of the ambient pore pressure u_0 given by the horizontal tangent WVZ. (Note that point W is the last one automatically recorded during the test, and the computer when producing the plot of Fig. 12 has inappropriately drawn the chord WO instead of continuing the line VW back to zero strain.)

The value of u_0 is 62 kN/m^2 corresponding to a water-table at a depth of 4·3 m. The water-table in the hydraulic fill was observed to fluctuate between wide limits as the tide level varied. With the estimate that the total vertical stress, at the test depth of 10·6 m, is 160 kN/m^2, this suggests a value of the coefficient of earth pressure at rest, K_0, to be $(107-62)/(160-62)=0{\cdot}46$.

In the analysis of Hughes, Wroth & Windle (1977) for the interpretation of a drained pressuremeter test in sand, it is assumed that the sand behaves elastically before failure and fails at a constant ratio of effective stresses and at a constant rate of dilation. The analysis leads to the result that, after failure has been initiated in the sand, the logarithm of the *effective* radial stress $\psi - u_0$ is linearly related to the logarithm of the strain ε; the gradient of the line s is given by the expression

$$s = \frac{\sin \phi'(1+\sin \nu)}{1+\sin \phi'} \tag{33}$$

where ϕ' is the angle of internal friction and ν is the angle of dilation.

The field results of the pressuremeter test of Fig. 12 for the range PQR have been plotted in this manner in Fig. 13. They approximate to a linear relationship whose gradient $s = 0{\cdot}425$. The adoption of a value of 35° for the critical state angle of friction at constant volume gives a value of $\phi' = 39°$ and $\nu = 9{\cdot}5°$ (see Hughes *et al.*, 1977).

The shear modulus of the sand can be measured in the same way as that for the clay, by carrying out small unloading–reloading cycles such as MN and RS in Fig. 12. The corresponding value of the shear modulus for cycle MN is 31 MN/m^2.

As for the clay there is a limit to the elastic range of behaviour during unloading. For the idealized sand behaviour used in the analysis, the allowable amplitude of the stress cycle QY is

$$\left(\frac{2 \sin \phi'}{1+\sin \phi'}\right)\sigma_r'$$

as illustrated in Fig. 12 (Wroth, 1982).

It is therefore possible to construct a curve, such as XYU in Fig. 12, which serves as a theoretical boundary to the elastic behaviour in unloading. This has been based on the observed value of $\phi' = 39°$ to give ratios of 0·772 and 0·228 used in the plot. It should be noted that σ_r' varies with the stage of the test so that the distance QY would not be the same if Q were to be chosen elsewhere on the virgin loading curve; care must be taken to subtract the ambient pore pressure u_0 from the applied pressure ψ to give the relevant value of σ_r' acting on the sand which is in contact with the membrane. The theoretical boundary fits well with the extent of the linear portion of the final unloading part of the test TUVW.

Correlation of undrained shear strength

The interpretation of undrained shear strength from pressuremeter tests in terms of effective stresses is uncertain with the present state of knowledge. This is because there have been very few attempts to test soil under the relevant combination of consolidation and subsequent shearing.

To study the pressuremeter expansion test in the laboratory by testing single elements of soil, it is necessary to be able to apply to the element the strain path to which the soil elements in the field are subjected. This strain path involves one-dimensional consolidation followed by shearing under conditions of plane strain, at constant volume, in the plane perpendicular to the direction of consolidation. True triaxial devices are the only types of apparatus that can apply this complete strain path in one continuous operation without the need for unloading, trimming and reorientating the sample.

A limited number of such tests has been reported by Wood & Wroth (1977), Wood (1981) and Eden & Law (1980); the results have been compared with those computed from various generalized stress–strain models of soil behaviour, none of which proved to be satisfactory.

For example, one approach is to use the modified Cam clay model, adapted for plane strain conditions, and to derive for the undrained strength ratio for anisotropically normally consolidated specimens (see Appendix 2)

$$\frac{s_{ups}}{\sigma_{v0}'} \approx \frac{\sin \phi_{ps}}{2c} \left(\frac{c^2+1}{2}\right)^{\Lambda} \qquad (34)$$

where $c = 1/(2-\sin \phi_{ps})$. This expression fits data of conventional plane strain active tests satisfactorily, e.g. for Boston blue clay, taking $\phi_{tc} = 30°$ gives $\phi_{ps} = 33 \cdot 75°$, $c = 0 \cdot 6923$ and $s_{ups}/\sigma_{v0}' = 0 \cdot 315$ which compares with an average value of 0·338 from the results reported by Ladd & Edgers (1972). However, if this expression is applied to the simulated pressuremeter tests conducted in the true triaxial apparatus on kaolin by Wood then the predictions are not satisfactory. Wood & Wroth (1977) quote a value of $\phi_{ps} = 26°$ which corresponds to $c =$ 0·640 and $s_{ups}/\sigma_{v0}' = 0 \cdot 259$, whereas the observed value of s_{ups}/σ_{v0}' was as low as 0·179. Clearly the use of the plane strain version of the modified Cam clay model is inappropriate for modelling the data from pressuremeter tests.

Apart from a lack of success in the mathematical modelling of those laboratory tests which simulate pressuremeter tests, there are important differences between the single-element tests conducted under ideal conditions in the laboratory and pressuremeter tests carried out in the field.

The most important difference is that in the ground around the pressuremeter the fields of stress and strain do not remain homogeneous during an expansion test, with the consequence that there are high gradients of excess pore pressure in the radial direction, so that some partial consolidation will occur in a supposedly undrained test. To minimize this effect, most pressuremeter tests are conducted quickly and at strain rates much faster than those normally used in conventional laboratory tests. This, in turn, introduces another undesirable effect in the influence of strain rate on undrained shear strength due solely to increased 'viscosity' which is entirely separate from consolidation effects. Because of the inhomogeneity of stress and strain, the various annuli of soil around a pressuremeter experience different rates of strain, so that it is not possible to make a direct allowance for the higher strain rates used in the field.

The arguments about partial consolidation in a pressuremeter test can be illustrated qualitatively in Fig. 14 by considering what would happen in a test on an idealized soil which is elastic–plastic but obeys the Mohr–Coulomb failure criterion. Fig. 14(a) represents a section of the ground perpendicular to the axis of the pressuremeter showing the radial distribution of excess pore pressure given by the curve FH assuming that no consolidation occurs. Figs 14(b) and 14(c) show the effective and total stress paths, and the stress–strain response of elements of soil D, E, F and G at different radial distances from the pressuremeter. The points with suffix 1 relate to theoretical stress states with no consolidation, whereas those with suffix 2 relate to the more realistic situation in which some consolidation has taken place. The important point is that the partial consolidation means that the effective stress states D_2' and E_2' move up the effective stress failure envelope, and the corresponding soil elements are failing under continually increasing shear stress, i.e. the instantaneous value of the so-called undrained shear strength is increasing monotonically throughout the pressuremeter test.

Although the precise details of Fig. 14 are not correct, the principles it contains apply to a real test on a real soil. In the early days of the development of the self-boring pressuremeter an analysis was attempted by Wroth & Hughes (1972) to interpret a slow expansion test in clay in which full consolidation would occur.

There is no doubt that current practice in the interpretation of rapid pressuremeter tests in clay, which does not allow for the effects either of partial consolidation or of strain rate, leads to overestimates of the undrained shear strength of the undisturbed clay. At present it is not possible to know whether the overestimates are by a margin of 10%, 20% or even 50%, and the margin must be a function of the coefficient of consolidation of the clay in question. There is an urgent

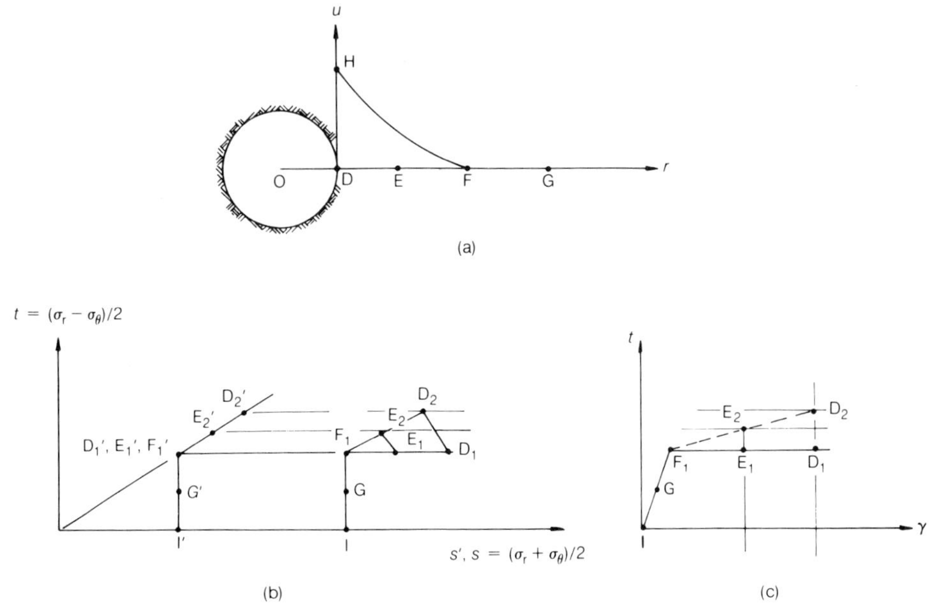

Fig. 14. Changes in the total and effective stress states around a pressuremeter due to partial consolidation during an 'undrained' test

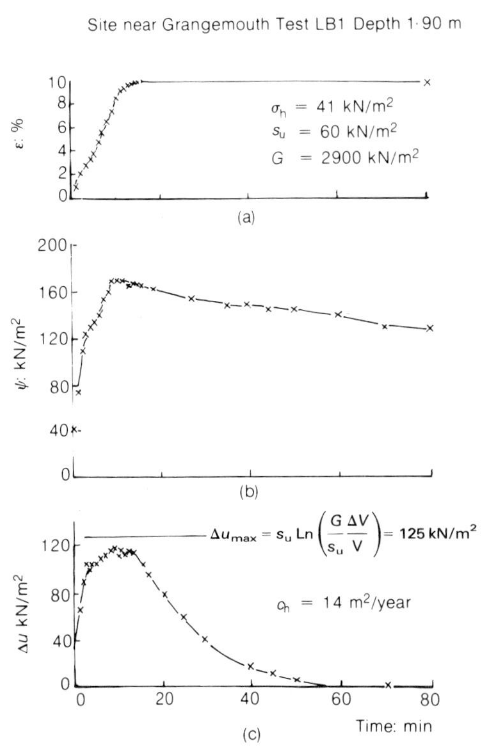

Fig. 15. Typical holding test results (after Clarke *et al.*, 1979)

need for research to be carried out to resolve these uncertainties.

Holding test

A typical set of results from a holding test carried out in soft clay at a site near Grangemouth is shown in Fig. 15 in which the observed quantities ε, ψ and Δu are plotted against time, where Δu is the excess pore pressure measured at the membrane–soil interface.

The initial part of the holding test is carried out in the same manner as an undrained pressuremeter test with a rate of cavity strain of about 1% per minute. By using the method of interpretation already outlined the values of σ_{h0}, s_u and G can be determined. As well as being important in its own right for design, the value of s_u is necessary for the interpretation of the subsequent consolidation data and the rigidity index G/s_u allows a check to be made on the excess pore pressure generated during the expansion phase of the test.

At about 9·5% strain the rate of straining is gradually reduced to zero over about 30 s. The membrane is then held fixed at this inflated radius until completion of the test by automatic reduction of the internal pressure ψ to match the reduction in external total radial stress as consolidation occurs in the surrounding soil. Careful monitoring of the decay of pore pressure allows an estimate to be made of the consolidation characteristics of the soil.

To interpret the data from a holding test a mathematical solution is required for this particular boundary value problem. In addition to the assumptions for an undrained expansion test in clay, it is assumed that during consolidation movements of the soil skeleton and flow of porewater will be entirely radial.

It has been shown (Gibson & Anderson, 1961) that the maximum excess pore pressure after undrained cavity expansion occurs at the membrane–soil interface and has a value

$$\Delta u_{max} = s_u \ln\left(\frac{G}{s_u}\frac{\Delta V}{V}\right) \tag{35}$$

Further, it has been shown that the distribution of excess pore pressure, immediately after expansion, is logarithmic with radius within the zone of yielded soil and is zero in the outer elastic region.

Numerical solutions for the consolidation of an elastic, perfectly plastic soil around an expanded cylindrical cavity have been obtained (Carter, Randolph & Wroth, 1979). These indicate that the total membrane pressure will decrease gradually once membrane expansion stops and as the soil consolidates. This feature is also observed in the field, e.g. see Fig. 15(b).

In principle, it would be possible to use the rate of decay of this total pressure to provide a measure of the horizontal coefficient of consolidation of the soil, c_h. However, the magnitude of this pressure drop is not usually large and a more accurate measure of c_h can be obtained by monitoring the dissipation of the excess pore pressure at the membrane–soil interface.

A closed form solution for the time dependence of the excess pore pressures around driven piles, but also relevant to the pressuremeter problem, has been found by Randolph & Wroth (1979). They assumed that the soil behaves entirely elastically during consolidation. Subsequently, it has been shown that the solution for an elastic soil is sufficiently accurate for an elasto-plastic soil (Carter, Randolph & Wroth, 1979).

In Fig. 16 solutions for the time for 50% consolidation have been plotted against the magnitude of the maximum excess pore pressure Δu_{max} (normalized by s_u). The non-dimensional time factor T_{50} has been used, where

$$T_{50} = c_h t_{50}/r_m^2 \tag{36}$$

and t_{50} is the real-time interval required for the pore pressure at the membrane to reduce to half its maximum value, r_m is the radius of the membrane after expansion and c_h is the horizontal coefficient of consolidation.

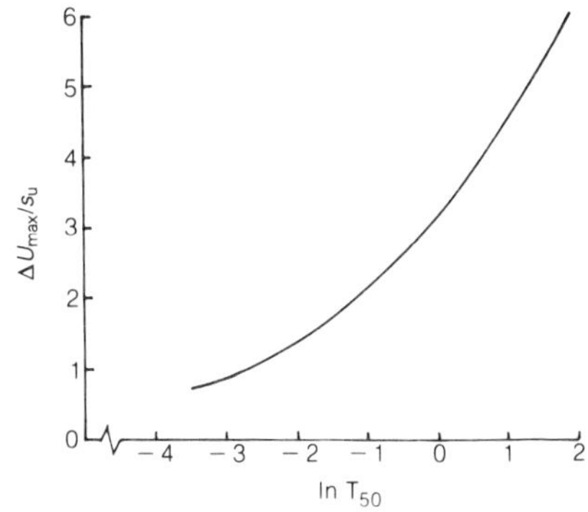

Fig. 16. Time for 50% pore pressure decay at the membrane–soil interface (after Randolph & Wroth, 1979)

The coefficient c_h is related to other soil properties by the equation

$$c_h = \frac{k_h}{\gamma_w} 2G \frac{1-\nu'}{1-2\nu'} \tag{37}$$

where k_h is the horizontal permeability of the soil, γ_w is the unit weight of porewater and ν' is the drained value of Poisson's ratio.

It is possible to fit the theoretical solution for consolidation to the experimental curve of excess pore pressure versus time at the 50% level. This is directly analogous to the standard laboratory procedure of fitting Terzaghi's solution for one-dimensional consolidation to oedometer results. For example, in Fig. 15(c) the maximum excess pore pressure Δu_{max} is approximately 120 kN/m^2 and occurs 10 min after the start of testing. This excess pore pressure has dropped to half its maximum value 25 min after the start. Hence the time t_{50} is given by $t_{50} = 25 - 10 = 15$ min. From the expansion portion of this test a value of the undrained shear strength $s_u = 60$ kN/m^2 is deduced in the usual manner (Gibson & Anderson, 1961). Hence $\Delta u_{max}/s_u = 2$ and a corresponding value of $T_{50} \approx 0{\cdot}33$ can be read from Fig. 16. The final membrane radius in this test was $r_m = 0{\cdot}0352$ m. Thus

$$c_h = \frac{T_{50} r_m^2}{t_{50}}$$

$$\approx \frac{0{\cdot}33 \times 0{\cdot}0352 \times 0{\cdot}0352}{15}$$

$$\approx 2{\cdot}7 \times 10^{-5}\ \mathrm{m^2/min}$$

$$\approx 14\ \mathrm{m^2/year}$$

Clarke *et al.* (1979) quote results of holding tests conducted at Canvey Island and make a comparison between values of the coefficient of

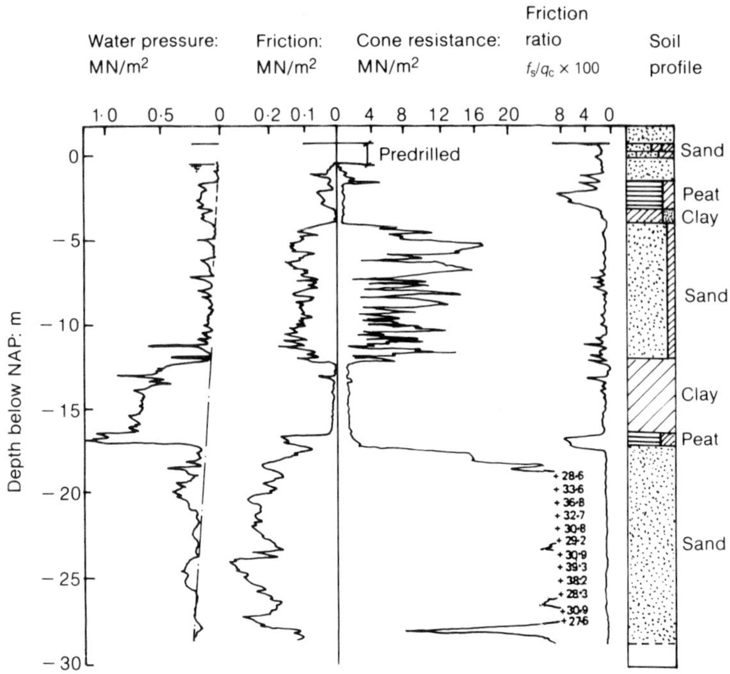

Fig. 17. Results of an onshore piezocone test (after Zuidberg *et al.*, 1982)

consolidation c_h deduced from holding tests with values of c_v obtained from conventional oedometer tests; the former values range from a factor of 30 to 300 greater than the latter values. The field values are much more in line with values back calculated from settlement records of engineering structures.

The coefficient of consolidation, or permeability, is an instance of an important soil property where values measured in the laboratory may be seriously misleading and give rise to gross overestimates for times of settlement. It is suggested that the holding test is an important type of pressuremeter test that should be developed. Apart from allowing realistic estimates of consolidation characteristics to be made, it may prove crucial in the interpretation of undrained shear strength to allow for effects of partial consolidation, as discussed in the previous section.

During the past 15 years the self-boring pressuremeter has been developed to the stage where it has become established as a major tool for in situ measurements of the following soil properties: in situ lateral total stress, shear strength, shear modulus and consolidation characteristics. The range of soils and soft rocks in which it can be used has been extended to include boulder clay, loose, dense and weakly cemented sands, and soft rocks such as mudstone, siltstone and chalk.

THE PIEZOCONE TEST

The basic cone penetrometer test has been used for many years as a standard tool in site investigation especially for establishing quickly and cheaply the soil profile at a particular site. It has grown in importance and acceptance with the recent special needs of offshore site investigation associated with oil exploration.

The first attempts to record porewater pressures during penetration of a probe into the ground were reported separately by Torstensson (1975) and Wissa, Martin & Garlanger (1975). This pioneering work showed the clear promise of incorporating a pore pressure transducer in a cone penetrometer so that continuous profiles of pore pressure could be obtained at the same time as point resistance of the cone and measurement of sleeve friction. The resulting instrument has become known as the piezocone.

The development since then has been rapid, with a variety of instruments on the market with most applications being offshore. Nearly all instruments have been built to the standard dimensions of the Dutch cone (60° apex angle and base area 10 cm^2) but unfortunately the position of the pore pressure transducer has not been standardized. The position has a marked influence on the magnitude of the observations, as discussed later in this section.

Figure 17 shows a typical profile of piezocone

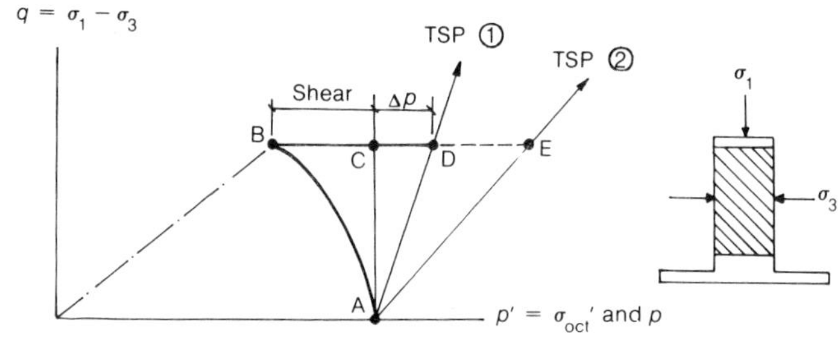

Fig. 18. Interpretation of the excess pore pressures observed in triaxial compression tests

results at an onshore site in the Netherlands reported by Zuidberg, Schaap & Beringen (1982). The soil profile was obtained from an interpretation of the cone resistance, sleeve friction and friction ratio. In the standard Fugro cone the pore pressure transducer is located half-way up the conical tip. The profile of the pore pressure provides dramatic confirmation of the soil profile. The hydrostatic conditions are shown by the chain-dotted line, so that the offset between that line and the continuous recorded profile represents the excess pore pressure. During penetration of the relatively permeable sand the excess pore pressures are small but just positive, whereas the excess pore pressures recorded in the lower stratum of clay are large and positive. The magnitude of the excess pore pressure is an indication of the type of soil and for clays can be correlated in dimensionless form with the overconsolidation ratio. The sharp fluctuations in the pore pressure act as a sensitive indicator of local variations in the soil such as laminations or fine partings of silt. In special circumstances of a dense cohesionless soil that exhibits dilatancy when sheared the excess pore pressures may be negative; an example of a weakly cemented sand is given by Ventura (1983) and of a silt (after dynamic compaction) by Campanella *et al.* (1982).

Confirmation of the type of soil can be obtained by stopping penetration of the cone and observing the dissipation of the pore pressure as consolidation occurs, in exactly the same way as a holding test with a pressuremeter.

It is evident that the piezocone provides important data about soil conditions, but at present this is of a qualitative nature. To advance its potential it is necessary to be able to interpret the data in a reliable and consistent quantitative manner. This requires both a standardization of equipment, test procedures and interpretation as well as a clear understanding of soil behaviour and the excess pore pressures generated during shear. To illustrate some essential points reference is made in the next section to basic soil behaviour as observed in triaxial tests. It is suggested that this pattern of behaviour can be used as a framework for the interpretation of data from the piezocone.

Interpretation of pore pressures in the triaxial test

Consider a saturated specimen of clay that has been isotropically normally consolidated to the stress state represented by point A in Fig. 18. If a conventional undrained triaxial compression test is carried out the effective stress path will be a curve such as AB, with the specimen reaching the critical state at B (see for example the data of kaolin in Fig. 6). The total stress path will be the straight line AD which is necessarily of gradient 3 in the chosen stress space. At failure the excess pore pressure generated within the specimen is represented by the difference between the mean total and effective pressures, i.e. the distance BD. This observed magnitude of excess pore pressure is made up of *two* components: the component BC due to the response of the clay to the shearing process represented by Δq and the component CD due to the change in mean total stress Δp applied to the specimen.

If instead an unconventional undrained compression test had been conducted on the specimen such as one with the total stress path AE (achieved by increasing the cell pressure appropriately as the axial stress is increased) the effective stress path would have remained unchanged as AB. However, the observed magnitude of the excess pore pressure BE would be greater than the value BD measured in the conventional test. This difference cannot reflect any difference in soil behaviour, but merely provides evidence of the different changes in mean total stress Δp in the two tests selected by the operator.

Hence the magnitude of the excess pore pressure measured in a shear test is not a unique property of the soil behaviour but it depends also on the changes in total stress applied externally to the specimen. The first component BC is a unique property of the soil (when tested in triaxial compression) and can be correlated with

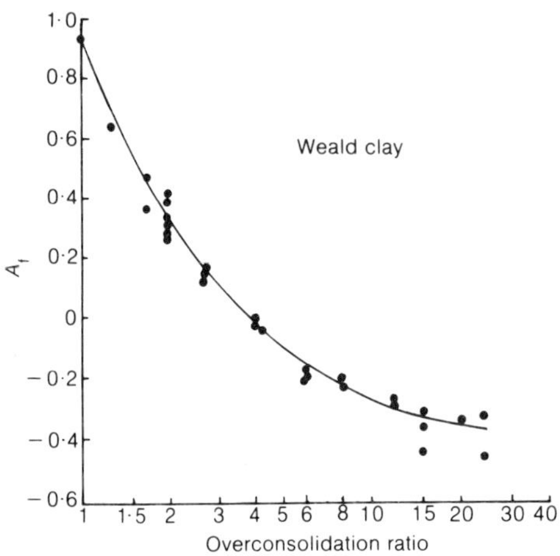

Fig. 19. Variation in the pore pressure parameter A at failure with the overconsolidation ratio for Weald clay (after Bishop & Henkel, 1957)

other properties of the soil specimen such as its overconsolidation ratio. The second component CD or CE gives no information whatsoever about the soil being tested but is just the value of Δp experienced by the specimen.

Consequently it is vital in interpreting observations of pore pressure changes to make due allowance for the change in Δp which accompanies the application of shear stress. This distinction is normally made in the interpretation of pore pressures observed in triaxial tests.

For a saturated clay the pore pressure parameters originally introduced by Skempton (1954) are such that for triaxial compression tests

$$A = \frac{\Delta u - \Delta\sigma_3}{\Delta\sigma_1 - \Delta\sigma_3} \tag{38}$$

and B is taken as unity. Essentially the parameter A is a ratio of the pore pressure change to the change in deviator stress, in which allowance has been made in the numerator for any change in cell pressure $\Delta\sigma_3$, but not unfortunately for any change in mean total stress Δp. This means that the value of A depends on the test conditions as well as soil properties such as the overconsolidation ratio.

To meet this objection Henkel (1960) suggested a revised set of parameters defined so that for a saturated specimen

$$a = \frac{\Delta u - \Delta\sigma_{oct}}{\Delta\tau_{oct}} \tag{39}$$

where $\Delta\sigma_{oct} \equiv \Delta p$ is the change in octahedral normal stress (or mean total stress) and $\Delta\tau_{oct}$ is the change in octahedral shear stress defined as

$$\tau_{oct} = \tfrac{1}{3}[(\sigma_2-\sigma_3)^2+(\sigma_3-\sigma_1)^2+(\sigma_1-\sigma_2)^2]^{1/2} \tag{40}$$

The concepts of CSSM allow a relationship between A and OCR to be derived as follows. In Fig. 5 the isotropically overconsolidated specimen initially at state R reaches the critical state at S. For a conventional triaxial compression test in which the cell pressure is kept constant, $\Delta\sigma_3 \equiv 0$ so that $\Delta p = \frac{1}{3}\Delta q$ and the total stress path is the straight line RT of gradient 3. The excess pore pressure is given by

$$\begin{aligned}\Delta u &= p_t - p_s' \\ &= p_r' + \Delta p - p_s' \\ &= p_r' + \tfrac{1}{3}q_s - p_s'\end{aligned}$$

Hence the pore pressure parameter A is expressed as

$$\begin{aligned}A &= \frac{\Delta u - \Delta\sigma_3}{\Delta\sigma_1 - \Delta\sigma_3} \\ &= \frac{\Delta u}{q_s} \\ &= \frac{p_r' + \frac{1}{3}q_s - p_s'}{Mp_s'}\end{aligned}$$

but using equation (20) this can be written as

$$A = \frac{1}{M}\left[\left(\frac{R}{r}\right)^{-\Lambda} + \frac{M}{3} - 1\right] \tag{41}$$

which is an expression relating A with the overconsolidation ratio R of the specimen and with the soil properties Λ, M and r.

The variation in values of A at failure with OCR for remoulded Weald clay reported by Bishop & Henkel (1957) is reproduced in Fig. 19. The curve drawn by those authors through the experimental points is very closely matched by the prediction of equation (41) with the following values of the soil constants for Weald clay: $\lambda = 0{\cdot}093$ and $\kappa = 0{\cdot}035$ to give $\Lambda = 0{\cdot}6237$, $M = 0{\cdot}95$ and $r = 2$, taken from Table 6.1 of Schofield & Wroth (1968). The quality of fit should be no surprise, as it is another way of confirming how well a remoulded low plasticity clay, such as Weald clay, tested in triaxial compression after isotropic consolidation satisfies the critical state concept.

The foregoing detailed analysis underlines the important relationship between the two dimensionless quantities, the pore pressure parameter A and the overconsolidation ratio. It suggests

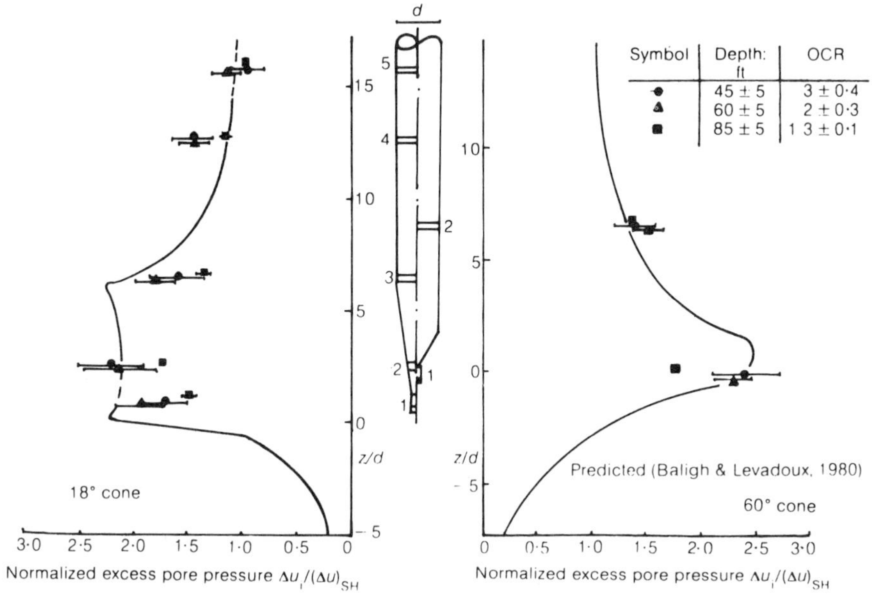

Fig. 20. Normalized excess pore pressures along the face and shaft of 18° and 60° cones during steady penetration in Boston blue clay (after Baligh & Levadoux, 1980)

that A is linearly related to OCR raised to the power of $-\Lambda$, and further that this can be used as a basis for the interpretation of pore pressure data from tests other than triaxial. This is attempted in the next section.

Interpretation of pore pressures in the piezocone test

The quantitative and detailed interpretation of the results of cone penetrometer tests has not yet been achieved because of the complex nature of the strain and stress changes induced in the soil around a penetrating cone. The problem has been studied by a number of workers, notably Baligh & Levadoux (1980), and Fig. 20 is taken from their major report.

In this diagram experimental data are compared with computed results of the distribution of excess pore pressure around a piezocone. The results refer to two different geometries of cone shown in section in the centre of the diagram. The left-hand figure refers to a piezocone with a cone of 18° apex angle and five separate transducers at the positions indicated. The right-hand figure refers to a piezocone with a standard 60° cone and two transducers.

The excess pore pressures Δu occurring on the boundary of the instrument have been normalized by dividing by the 'steady state' value Δu_{sh} that exists a distance up the shaft of the instrument away from the tip; they are plotted against the axial distance from the tip divided by the shaft diameter.

The value of the excess pore pressure varies markedly with the position of the sensor in relation to the cone. It is important that the position of the sensor should be standardized for all piezocones to optimize the interpretation of data, particularly in view of the uncertainties involved and the unavoidable reliance on empiricism to some extent.

It is tempting to suggest that the sensor should be sited at the tip of the cone where the maximum values of the excess pore pressure occur, but this is not the best position for a number of reasons. Campanella *et al.* (1982) list several sound practical reasons why the best location is on the shaft (at position such as No. 2 for the 60° cone on the right-hand side of Fig. 20) rather than on the cone itself. There are three additional and major advantages from a theoretical standpoint in favour of the location on the shaft.

(*a*) A greater *proportion* of the excess pore pressure is due to the component induced by shear compared with the component due to the change in the mean total stress.
(*b*) The gradients of excess pore pressure with axial distance are smaller, so there is greater accuracy.
(*c*) The values are less affected by any change in

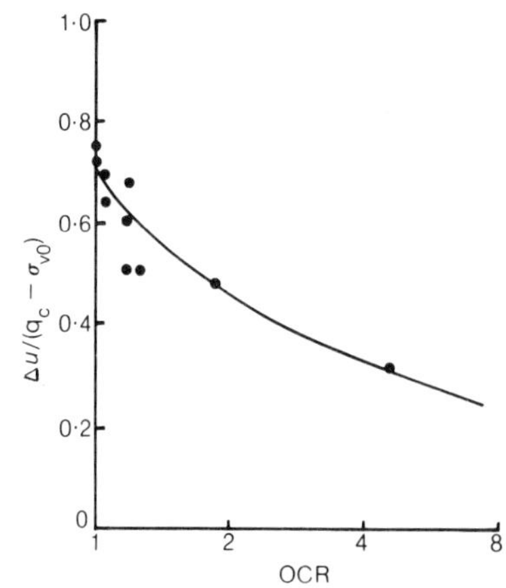

Fig. 21. Variation in the piezocone pore pressure ratio with the overconsolidation ratio at Onsøy

the axial load experienced by the piezocone when the penetration is stopped either for the next drill rod to be added or for a dissipation test to be carried out.

On the basis of the understanding of soil behaviour provided by the long history of experience obtained from high quality triaxial tests, the ideal interpretation of piezocone data would be to derive a pore pressure parameter exactly analogous to Henkel's parameter and to expect this to correlate closely with the overconsolidation ratio, i.e.

$$a = \frac{\Delta u - \Delta\sigma_{oct}}{\Delta\tau_{oct}} = f(\text{OCR}) \qquad (42)$$

Unfortunately there is no means of knowing the changes in octahedral normal and shear stresses; indeed, the changes vary spatially around the piezocone and depend on the unknown properties of the soil which is being penetrated.

In place of a, a dimensionless parameter is required which must be a ratio of excess pore pressure to some measure of shear stress. Robertson & Campanella (1983) discuss the ratios that have been used including

(*a*) u/q_c (Baligh, Azzouz, Wissa, Martin & Morrison, 1981: Tumay, Bogges & Acar, 1981)
(*b*) $\Delta u/q_t$ (Campanella & Robertson, 1981)
(*c*) $\Delta u/(q_c - u_0)$ (Smits, 1982)
(*d*) $\Delta u/(q_c - \sigma_{v0})$ (Senesset, Janbu & Svanφ, 1982; Jones & Rust, 1982; Jefferies & Funegard, 1983)

where q_c is the measured cone end bearing pressure and q_t is the corrected end bearing allowing for the effects of pore pressure acting on the back of the cone. Robertson & Campanella (1983) rightly point out the importance of making this correction, especially as it will vary markedly with the precise mechanical design of the piezocone, and with depth below the water-table.

The first three ratios are all unacceptable, the first because it has the total not the excess pore pressure in the numerator, and all three because the denominators cannot be a proper measure of the shear stress. In an undrained situation the maximum shear stress can only be expressed in terms of principal stresses as a *difference* of two *total* stresses or a *difference* of two *effective* stresses. It is for this reason that the variable $q_c - \sigma_{v0}$ and not q_c is used for deriving values of undrained shear strength from cone penetration tests.

The last ratio, $\Delta u/(q_c - \sigma_{v0})$, denoted B_q by Senneset *et al.* (1982), is the one recommended, therefore, for adoption as the standard parameter for interpretation of piezocone data. It is assumed that proper calibration and corrections have been made so that q_c here replaces q_t.

There are few reported instances of sites where piezocone data can be correlated with reliable data of OCR. One of the best researched sites is that at Onsφy used by the Norwegian Geotechnical Institute. From the information reported by Lacasse & Lunne (1982b) values of B_q have been calculated from the plotted profiles of Δu, q_c and σ_{v0} for depths at intervals of 2 m up to 20 m. These have been plotted in Fig. 21 against the relevant value of OCR taken from the profiles resulting from conventional oedometer tests. The data are given in Table 3.

There is a striking similarity between the results of the piezocone data presented in Fig. 21 and the results of triaxial data in Fig. 19. Indeed, the theoretical background to the triaxial test suggests that B_q might vary linearly with OCR raised to the power $-\Lambda$, in a manner

Table 3. Processed data from piezocone tests from Onsøy

z: m	σ_{v0}: kN/m^2	q_c: kN/m^2	Δu: kN/m^2	B_q	OCR
2	32·5	200	56	0·334	4·58
4	65	240	84	0·48	1·88
6	97·5	280	92	0·506	1·25
8	130	320	122	0·64	1·04
10	162·5	440	141	0·507	1·17
12	195	500	185	0·607	1·18
14	227·5	600	252	0·677	1·19
16	260	680	292	0·696	1·04
18	292·5	750	330	0·721	1·0
20	325	810	365	0·752	1·0

analogous to equation (41). If a value for Λ were known, then such a plot could be made with an expected linear relationship. In practice, field data will not warrant such involved processing, because the variation in OCR in any one stratum is not likely to be great and because of the likely experimental scatter in the data. A plot similar to Fig. 21 should be adequate for correlation.

It has been argued in this section that to derive the maximum benefit from the development of the piezocone

(*a*) the position of the pore pressure transducer should be standardized and should be on the shaft of the instrument and not within the conical tip

(*b*) the interpretation of the data should be standardized by use of the pore pressure parameter $B_q = \Delta u/(q_c - \sigma_{v0})$ in which q_c is the correct total end bearing pressure of the cone.

No attempt is made to consider the derivation of values of undrained shear strength from end bearing data of cone penetrometer tests. At present the interpretation is usually empirical, being based on comparisons with other means of measurement of s_u. Much current research is aimed at a properly formulated method of interpretation, such as that of Senesset *et al.* (1982).

DIRECT SIMPLE SHEAR TESTS

It was pointed out earlier that there will be major differences between tests in which the principal axes of stress and/or strain increment are prevented from rotating and those in which they are free to rotate. The former category of test includes the oedometer, the conventional triaxial tests, most versions of laboratory plane strain apparatus and the triaxial apparatus, and the pressuremeter. The latter category includes the direct simple shear test and most forms of in situ test, i.e. the vane shear test, the plate bearing test and the cone penetration test.

Since most in situ tests, allowing freedom of the principal axes to rotate, are correlated with laboratory tests in which the principal axes are constrained, it is important to assess the differences in behaviour of soil under the two circumstances. Additionally in all real engineering situations individual soil elements experience stress and strain increments such that the principal axes rotate, whereas nearly all mathematical models of soil behaviour are based on data from single-element tests in which no rotation is presumed.

Consequently a detailed study has been made of the results of undrained direct simple shear tests, and it is presented in this section. This study takes further the work of Randolph & Wroth (1981) concerned with the axial capacity of driven piles in clay and is based on the comprehensive series of tests reported by Ladd & Edgers (1972), which have already been referred to.

The tests were carried out in the Geonor apparatus in which cylindrical samples of soil are contained in a rubber membrane reinforced internally by a fine helical wire and subjected to simple shear. 'Undrained' tests are carried out as drained tests in which the total volume of the sample is held constant by suitable adjustment of the vertical (effective) stress.

Undrained tests on normally consolidated specimens

The results of a test on a specimen of normally consolidated resedimented Boston blue clay are shown in the upper half of Fig. 22 in which the observed shear stress τ_h (applied to the horizontal top surface of the specimen) is plotted against the vertical effective stress σ_v'. Both stress variables have been made dimensionless by dividing by the original consolidation pressure σ_{vc}'. The individual points have been plotted directly from the tabulated data in the Massachusetts Institute of Technology report.

The effective stress path ABC traced by these points has two marked features. The first is the very substantial reduction in effective vertical stress that occurs during the test, which would require very large positive excess pore pressures to be generated in an equivalent undrained test. The second is that the maximum value of observed shear stress τ_{max} occurs at point B (after a shear strain of about 5% (see Fig. 25, later)), whereas the maximum stress ratio on the horizontal plane of the specimen occurs at point C (after 30% shear strain).

This raises the question whether 'failure' of the specimen should be identified with state B or state C. It will be shown later that state B does not represent the maximum shear stress on any plane within the specimen; other planes can experience greater values of shear stress at other stages in the test. It is suggested therefore that failure in a frictional material should be associated with the conditions of maximum stress ratio.

The path ABC—and all other such paths on tests of seven normally consolidated clays reported by Ladd & Edgers (1972)—is nearly elliptical in shape. Without attempting a physical explanation of why this might be so, it is worth exploring the consequences of assuming that all such paths can be closely matched by ellipses.

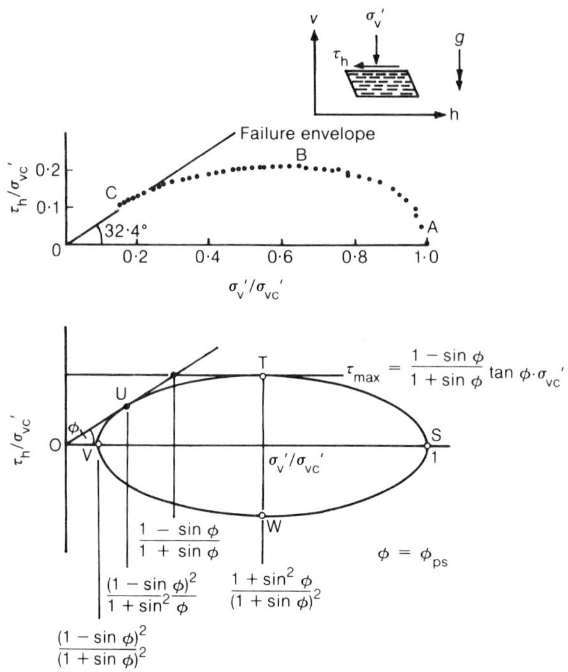

Fig. 22. Undrained simple shear test on normally consolidated Boston blue clay (data from test 204, Ladd & Edgers, 1972)

In the lower half of Fig. 22 a particular ellipse STUVW has been drawn which passes through the initial point S and touches the failure envelope OU at point U. To match the experimental data as closely as possible the ellipse has been drawn to the same scale and the failure envelope has been chosen with the observed value of $\phi = \phi_{ps} = 32{\cdot}4°$. In theory there is an infinite family of ellipses which satisfy these criteria each having a different centre and a different oblateness represented by the relative length of the minor axis TW.

The particular ellipse chosen for simulation is given by the expression

$$4\tau^2 + (\sigma - h)^2 = (\sigma + h)^2 \sin^2\phi_{ps} \qquad (43)$$

where for convenience $\tau = \tau_h/\sigma_{vc}'$, $\sigma = \sigma_v'/\sigma_{vc}'$ and $h = (1 - \sin\phi_{ps})/(1 + \sin\phi_{ps})$. This ellipse has some special properties, which are discussed later, and the abscissae of the main points in explicit form are included in Fig. 22. In particular the maximum value of shear stress, which occurs at point T, is given by the equation

$$\frac{\tau_{max}}{\sigma_{vc}'} = \frac{1 - \sin\phi_{ps}}{1 + \sin\phi_{ps}} \tan\phi_{ps} \qquad (44)$$

The observations made in a direct simple shear test in a Geonor apparatus only provide information about the effective stresses on one plane—the horizontal plane—in a supposedly homogeneous soil specimen. It is not possible to deduce the complete state of stress within the specimen and to draw the relevant Mohr circle. Attempts have been made to meet this deficiency, notably by Soydemir (1976), by using the helical reinforcement encased in the membrane as a transducer for measurement of radial, i.e. horizontal, stress.

As part of a grand strategy of fundamental research into stress–strain behaviour of soils, Roscoe (1953) devised the simple shear apparatus (SSA) in an attempt to be able to subject a soil specimen to uniform conditions of simple shear. The apparatus was developed by a succession of research students who worked predominantly on sand. The two major developments have been

(*a*) the construction of boundary load cells capable of measuring both normal and shear forces acting on the boundaries of the specimen in the SSA
(*b*) the use of the X-ray technique for studying the non-homogeneous strain fields that inevitably develop within the specimen.

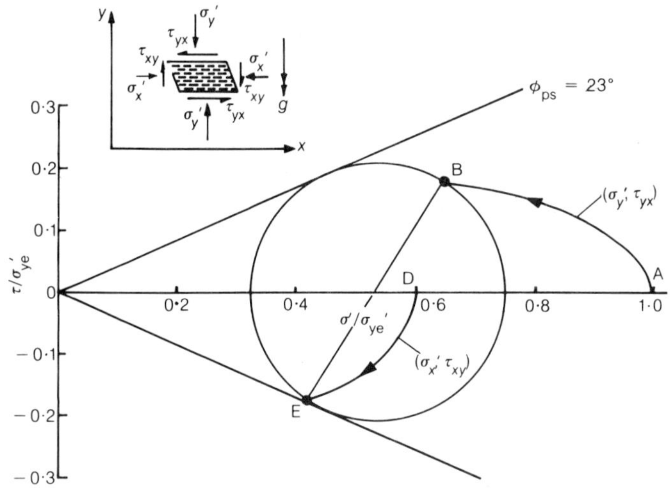

Fig. 23. Effective stress paths and the failure state from an undrained simple shear test on kaolin (data from test 10, Borin, 1973)

The boundary load cells provide enough information for the stress state in the SSA specimen to be estimated.

Little work has been done on clay in the SSA, the only major study to date being that of Borin (1973). The results of one undrained test on normally consolidated reconstituted kaolin are reproduced in Fig. 23. A schematic section through the SSA is shown to define the reference set of co-ordinate axes (x, y). The observed value of stresses on the horizontal and vertical planes respectively (σ_y', τ_{yx}) and (σ_x', τ_{xy}) have been made dimensionless by dividing by the equivalent pressure σ_{ye}', which for an undrained test on a normally consolidated specimen is the same as the consolidation pressure σ_{vc}'.

The effective stress path traced out by (σ_y', τ_{yx}) is the curve AB; the test, unlike those of Ladd and Edgers, was terminated just after the maximum shear stress τ_{yx} was reached at point B. The path AB is essentially the same as that shown in Fig. 22 for a distinctly different type of simple shear test on a different clay, which provides important confirmation of this facet of soil behaviour.

Of more importance is the effective stress path DE traced by the stresses (σ_x', τ_{xy}) on the vertical (or constant x) planes in Fig. 23, in which the usual sign convention has been adopted for Mohr's circle with compressive stresses taken as positive. Because the kaolin specimen was normally consolidated the initial state D corresponds to a value of K_0 of 0·6.

The Mohr circle of stress is drawn for the condition of maximum τ_{yx} with BE as diameter, and it appears to touch the failure envelope given by $\phi_{ps} = 23°$ but at point E (or close to it) and *not* at point B. This suggests that the condition of maximum stress obliquity relates to the *vertical* and not the horizontal planes within the specimen.

Such a supposition is contrary to long-established beliefs in soil mechanics and was first put forward as a possibility by de Josselin de Jong (1972). The essential features of his arguments are presented in Fig. 24 regarding interpretation of failure conditions in direct shear tests. For ease of discussion, the case of a drained test on a soil under constant effective vertical stress is considered. The top part of the figure represents the conventional understanding in which the horizontal planes are ones of relative sliding within the specimen, on which the maximum ratio of shear stress to normal effective stress occurs; the stress state is the point G of maximum stress obliquity.

The bottom part of the figure is related to an alternative mechanism of failure in which the vertical planes are ones of relative sliding within the specimen, on which the maximum stress ratio occurs represented by point K. The sliding on vertical planes is accompanied by a body rotation of magnitude γ in the anticlockwise direction, where γ is the amount of engineering strain applied to the specimen externally. It should be noted that this second mechanism of failure is kinematically compatible with the boundary conditions and requires the application of a *smaller* shear stress.

de Josselin de Jong argues that either mechanism is possible, and that, if the boundary conditions of the test—or the engineering situation—allow the soil element a choice between the two modes, the selected mode is the one which requires the least resistance, i.e. the second mechanism.

The vectors of the principal stress have been sketched in Fig. 24 showing clearly for the two cases the different amounts of rotation of the principal axes of stress and the different magnitudes.

The hypothesis of de Josselin de Jong is confirmed experimentally by the single relevant test carried out by Borin (1973) and plotted in Fig. 23. There is every reason to expect the hypothesis to be valid for the tests on Boston blue clay and the other clays reported by Ladd and Edgers. These ideas have been applied in Fig. 25 to the interpretation of the stress path of Fig. 22.

In Fig. 25, three stages of the test have been examined:

(*a*) the start of the test
(*b*) the condition of maximum shear stress τ_{yx}
(*c*) the ultimate stage of the test.

To match the direction of the effective stress path ABC, the stages and the observed stress–strain curve have been intentionally placed from right to left in the diagram.

For the first stage, the Mohr circle has AD as its diameter and the major principal stress is vertical. For the test in question the initial horizontal effective stress σ_{hc}' is not known, but from many one-dimensional consolidation tests on Boston blue clay an average value of K_0 is 0·5 and this has been assumed to apply to this test also. As soon as shear is applied to the specimen, the Mohr circle moves to the left, reduces in size and rotates. By the time stage (*b*) is reached, the maximum measured shear stress has been attained at point B, the stress state for the vertical planes (σ_x', τ_{xy}) at point E is on the failure envelope (with sliding on these vertical planes) and the direction of the major principal stress has rotated clockwise through an angle of $\pi/4-\phi_{ps}/2$.

Further straining of the specimen causes the Mohr circle to continue to move to the left, and to rotate, but it is now constrained to touch the failure envelope and its point of tangency to move down the envelope towards the origin. During this phase it is believed that the effective horizontal stress σ_x' does not change, with the consequence that the effective stress path EF for the vertical planes is parallel to the τ axis as shown. This particular feature (although not

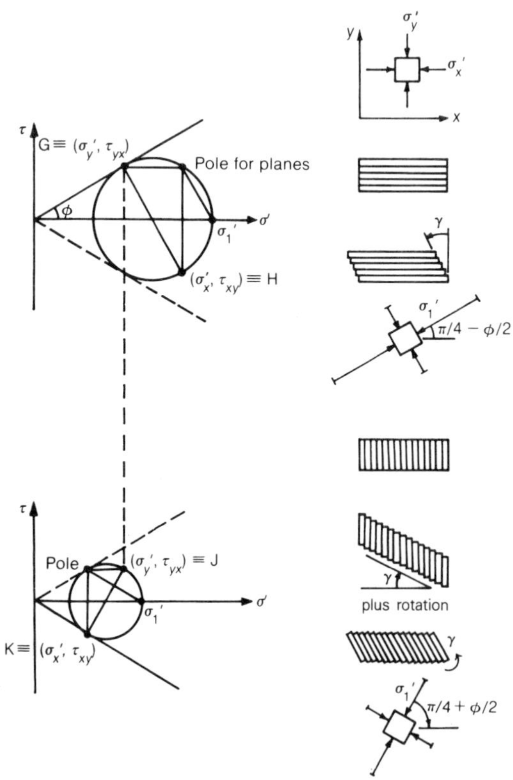

Fig. 24. Possible modes of failure in simple shear (after de Josselin de Jong, 1972)

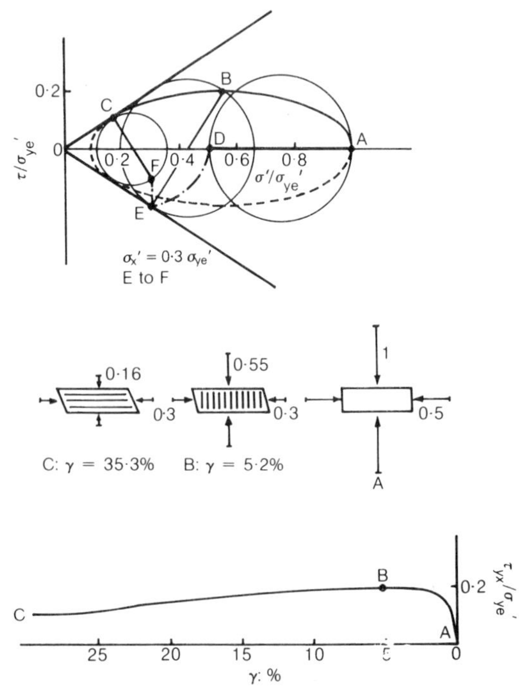

Fig. 25. Behaviour of normally consolidated Boston blue clay in undrained simple shear

supported by physical arguments) is a consequence of the particular choice of ellipse in equation (43).

At the ultimate stage (*c*), the horizontal planes have become the sliding planes experiencing the maximum stress ratio (point C) which satisfies the kinematic requirement of large shear displacements in the horizontal direction. The direction of the major principal stress has rotated clockwise further through an angle ϕ_{ps} to a total rotation of $\pi/4+\phi_{ps}/2$. There is no reason for any change now in the Mohr circle for subsequent shear deformation.

The magnitude of the ultimate shear stress at point C is given by the elliptical representation (see Fig. 22 and equation (43)) as

$$\frac{\tau_{ult}}{\sigma_{vc}'}=\frac{(1-\sin\phi_{ps})^2}{1+\sin^2\phi_{ps}}\tan\phi_{ps} \tag{45}$$

This suggests an apparent sensitivity S for the resedimented Boston blue clay in the direct simple shear test of

$$S=\frac{\tau_{max}}{\tau_{ult}}=\frac{1+\sin^2\phi_{ps}}{1-\sin^2\phi_{ps}} \tag{46}$$

For Boston blue clay with $\phi_{ps}=32{\cdot}4°$, this gives $S=1{\cdot}8$. The same resedimented normally consolidated clay tested in undrained triaxial compression would give a stress–strain curve which monotonically increases to an ultimate (critical state) condition and displays no sensitivity.

This suggestion has the important implication that an apparent sensitivity may occur in tests in which the principal axes are free to rotate, for a soil which would not display any sensitivity in, say, a conventional triaxial test. The consequence is that a greater sensitivity for an undisturbed soil is likely to be observed in a vane shear test than in a triaxial test.

Undrained strength ratios

On reversion to values of the maximum shear stress observed in an undrained direct simple shear test, the assumptions made about the elliptical stress path lead to the expression in equation (44) for the ratio τ_{max}/σ_{vc}'. The relevance of this expression has been tested against all the data reported by Ladd & Edgers (1972) in the plot of Fig. 26. For each test on a normally consolidated specimen, the observed value of this ratio has been plotted against the individual value of ϕ_{ps} observed in that test at the ultimate condition.

A distinction has been made between the four

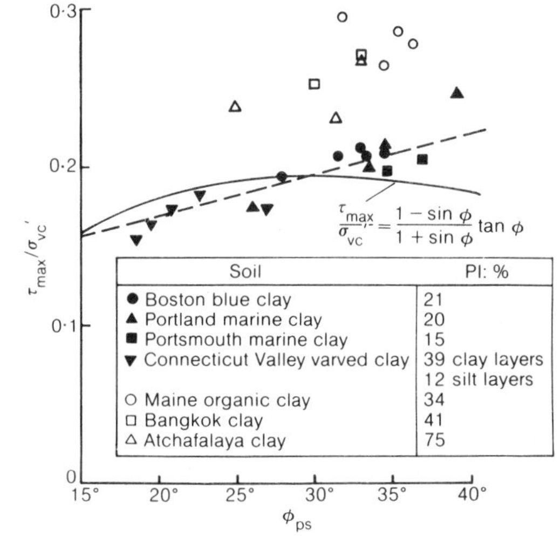

Fig. 26. Undrained strength ratios of seven normally consolidated clays in simple shear tests (data from Ladd & Edgers, 1972)

low plasticity clays displayed by full symbols and the other three higher plasticity clays by open symbols. The former clays show consistently lower strength ratios which are reasonably close to the curve given by equation (44). However, without this speculative theoretical background, the data might have been represented by the broken straight line—which in fact gives a better fit. No suggestion is offered why the three clays of higher plasticity have greater strength ratios.

The discussion so far regarding direct shear tests has been about the maximum value of the observed shear stress τ_{max}. Although this stress may be used appropriately in a limit analysis, it is not the greatest shear stress experienced by the specimen being tested, as pointed out earlier. By definition the greatest shear stress sustained by the specimen is the undrained shear strength s_{udss} given by the radius of the *largest* Mohr circle.

From Fig. 25 it appears that the largest Mohr circle is the *initial* one which has a radius r_{mc} such that

$$\frac{r_{mc}}{\sigma_{vc}'}=\tfrac{1}{2}(1-K_0)\approx\tfrac{1}{2}\sin\phi_{tc} \tag{47}$$

However, the radius of the second Mohr circle (the first occasion that failure has been induced within the specimen) can be obtained by relating it to the stress state at B

$$\frac{r_{mf}}{\sigma_{vc}'}=\frac{\tau_{max}}{\sigma_{vc}'}\sec\phi_{ps}=\frac{\sin\phi_{ps}}{(1+\sin\phi_{ps})^2} \tag{48}$$

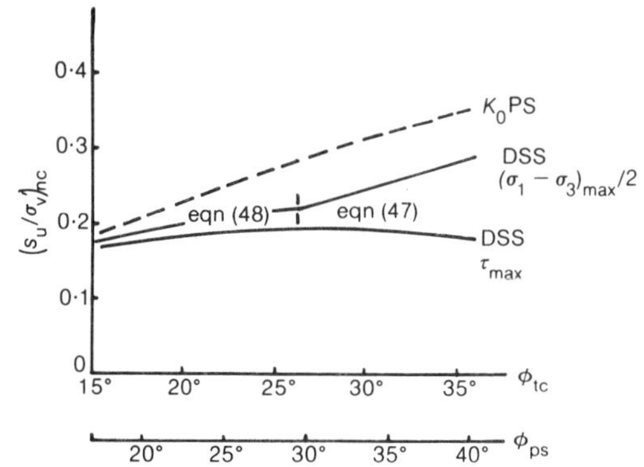

Fig. 27. Variation in the undrained strength ratio with the angle of friction for direct shear tests on normally consolidated clay

Computations show that the first expression is the greater for $\phi_{ps} \approx \frac{9}{8}\phi_{tc} \geqslant 29{\cdot}5°$. Speculation about what might happen between these stages is not warranted, but it is hard to believe that there would be an even larger Mohr circle.

The resulting predictions for the undrained strength ratio for direct shear tests on normally consolidated clay as a function of friction angle are shown in Fig. 27. For comparison the relationship predicted for plane strain tests on anisotropically normally consolidated clay from equation (34) is included. Although these relationships are of questionable accuracy they emphasize the likely differences in undrained shear strength obtained from different tests and the need to take account of these when comparisons are being made.

VANE SHEAR TEST

Basic interpretation

The vane shear test has formed a central role in site investigation, particularly in soft clays, for over 60 years having first been used by Olsson in 1919 according to Flodin & Broms (1981). It is surprising that it is only recently that the test and its interpretation have been studied critically.

The standard method of interpretation (BS 1377) is to assume

(*a*) that the shear stress distribution on both the cylindrical (vertical) surface and the top and bottom (horizontal) surfaces is rectangular, i.e. uniform
(*b*) that the shear strength of the soil being measured is isotropic.

Both of these assumptions need to be examined carefully.

Donald, Jordan, Parker & Toh (1977) have conducted a critical appraisal of the vane test and report results of a three-dimensional finite element analysis of the stress distributions on the putative failure surface created by a vane in an elastic material. Their important finding is reproduced in Fig. 28(a) from which it can be seen that the computed elastic stress distribution, although reasonably rectangular on the sides of the vane, is far from uniform on the top and bottom edges.

Menzies & Merrifield (1980) have carried out some experiments with an ingeniously instrumented vane both in sand and in overconsolidated London clay. Their results naturally show some scatter, but after using regression analysis to give best-fit curves these are compared suitably normalized in Figs 28(b) and 28(c). In each case half the symmetrical distribution is shown, the vane blade having dimensions of 240 mm × 120 mm. The observed shear stress distributions are pleasingly similar to those computed by Donald *et al.* (1977).

In London clay, the assumption of a rectangular shear stress on the vertical cylindrical failure surface is confirmed as reasonable, whereas there is a major difference between the observed distribution on the horizontal failure surfaces and the assumed condition of uniformity. The possible consequences of this departure from the assumptions are now investigated.

Consider a vane of diameter d and height h, and suppose that the shear stress on the top and bottom surfaces is given by the single polynomial expression

$$\frac{\tau}{\tau_m} = \left(\frac{r}{d/2}\right)^n \tag{49}$$

where r is the radial distance from the centre line and τ_m is the maximum value of the shear stress, which is assumed to occur simultaneously along the entire vertical failure surface. The contribution to the torque T_h provided by both top and bottom surfaces is given by the straightforward integration

$$T_h = 2\int_0^{d/2} 2\pi r^2 \tau \, dr = \frac{\pi d^3 \tau_m}{2(n+3)} \tag{50}$$

The conventional assumption of a uniform distribution of shear stress gives a value of $T_h = \pi d^3 \tau_m/6$ which is a special case of equation (50) for $n = 0$.

The contribution to the torque T_v provided by the vertical surface is the same as the conven-

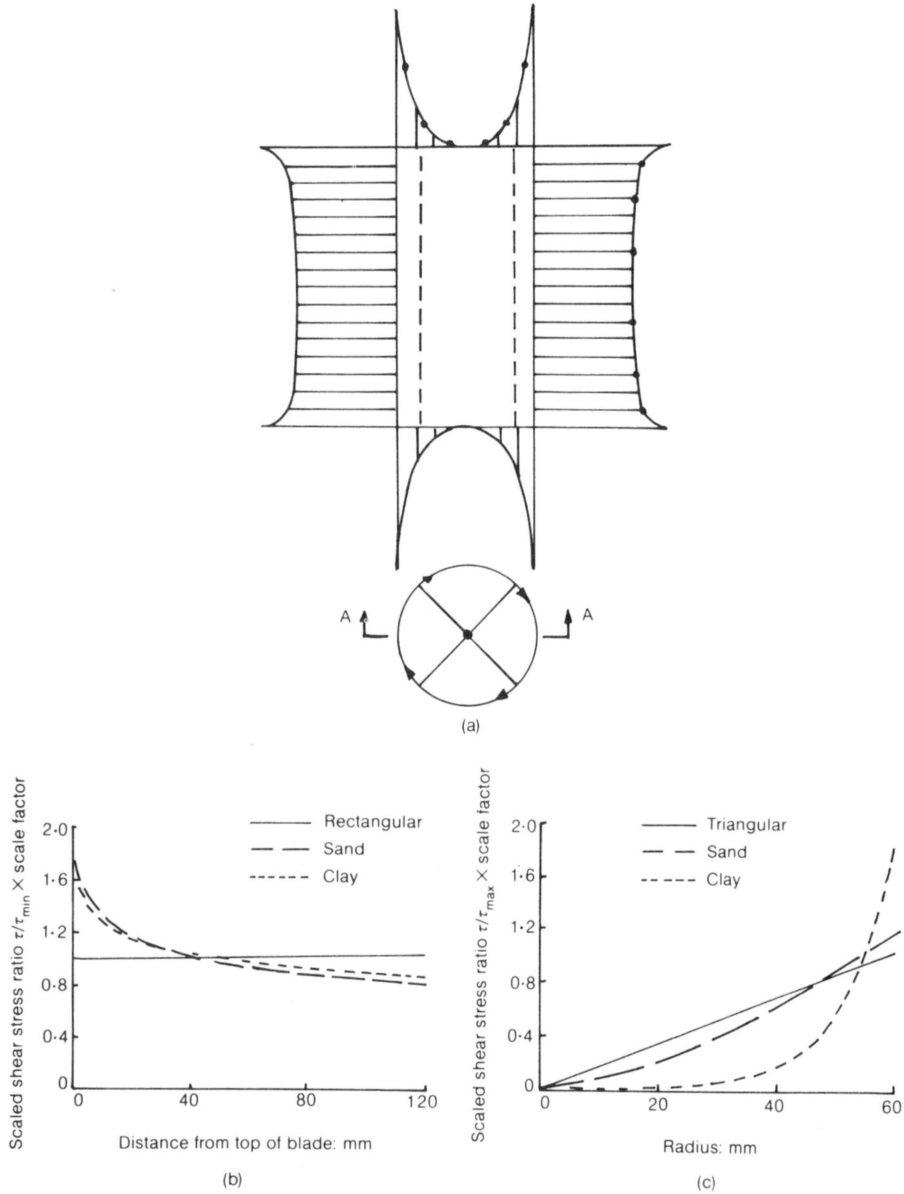

Fig. 28. Theoretical and experimental distributions of shear stress around the blades of a vane: (a) computed elastic shear stress distribution (after Donald *et al.*, 1977); (b) vertical edge; (c) horizontal edge ((b) and (c) are normalized distributions of equivalent shear stress scaled to give equal torque (after Menzies & Merrifield, 1980))

tional, i.e.

$$T_v = \frac{\pi d^2 h \tau_m}{2} \tag{51}$$

Hence the relative proportions of the torques are

$$\frac{T_h}{T_v} = \frac{1}{(n+3)} \frac{d}{h} \tag{52}$$

For the London clay data of Menzies & Merrifield (1980) in Fig. 28(c) an approximate value of n is 5. With the adoption of this value, then for the standard vane with $d/h = 1/2$, the value of T_h/T_v will be 1/16 compared with 1/6 for the conventional interpretation.

This revised interpretation based on equation (49) has two major consequences. The first is

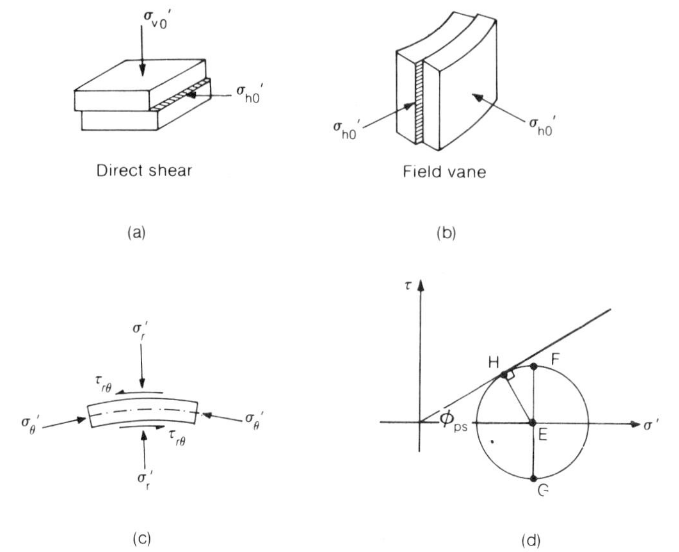

Fig. 29. Possible state of stress around the cylindrical failure surface caused by a vane shear test

that approximately 94% of the resistance to torque is provided by the vertical failure surface compared with 86% in the conventional interpretation. This means that

(*a*) the assumption of isotropy of shear strength is less crucial, and that the dominant shear strength measured is that on vertical planes
(*b*) the method of deducing the degree of anisotropy of strength by using a suite of vanes with different d/h ratios will be strongly affected.

The second major consequence is that the value of the (isotropic) shear strength deduced from a given value of peak torque will be *underestimated* by a factor of $(6/7) \times (17/16)$, i.e. 102/112 or by about 9%. This error is doubly fortunate in that it is conservative and it compensates to some extent for errors caused by rate effects. The standard vane test is carried out at a speed of rotation such that the strain rates applied to the failing soil are considerably higher than in, say, conventional triaxial tests and are thereby associated with higher undrained strengths. It is not possible to calculate the strain rates around the vane without a complete knowledge of the strain field, so the magnitude of the effect cannot be sensibly estimated.

Undrained strength ratio

For a proper comparison with the results of other tests, it is necessary to know the effective stresses that are controlling the undrained shear strength generated at the edges of the vane. The mode of failure is one of direct simple shear. However, it has been shown previously that all but 6% of the resistance is provided by the vertical failure surface, so that it is the mode of failure around the cylindrical surface that dominates events.

Although this failure is one of direct shear, it is not the same as that caused in a conventional direct shear test because of the different orientation of the specimens. This is illustrated in Fig. 29 for normally consolidated clay where the different combinations of effective stress which affect subsequent behaviour are shown. The only attempt at a fundamental study of the vane test under controlled laboratory conditions found in the literature is that by Law (1979); the few data he reported are insufficient for a definite theory to be established.

The following analysis is an attempt to resolve the problem, but it is speculative, not well based on experimental data, and so it must be viewed with caution until further research validates it or not. Fig. 29(c) is a horizontal section through a segment of the vertical cylindrical failure surface around the vane, for which polar co-ordinates are used. Initially the direct effective stresses of concern* are equal to the lateral stress in the ground, $\sigma_{r0}' = \sigma_{\theta 0}' = \sigma_{h0}'$, so that the Mohr circle or interest in Fig. 29(d) is the single point E. As

* Initially the major principal effective stress is the vertical stress, but this is expected to reduce locally to become the intermediate effective stress since plane strain conditions are assumed.

shearing starts it is assumed that these direct stresses do not alter so that, as the shear stress increases, the Mohr circle enlarges (retaining E as its centre) until failure occurs when the circle touches the failure envelope at H.

This means that the maximum shear stress experienced by the vane is given by the simple formula

$$\tau_{max} = \sigma_{h0}' \sin \phi_{ps} \tag{53}$$

which is not at variance with the limited results reported by Law (1979).

If this is so then the maximum value of the observed shear stress τ_{max} is identical with the definition of undrained shear strength, and hence the undrained strength ratio is

$$\frac{s_{ufv}}{\sigma_{v0}'} = \frac{\sigma_{h0}' \sin \phi_{ps}}{\sigma_{v0}'} \approx (1 - \sin \phi_{ps}) \sin \phi_{ps} \tag{54}$$

since for a normally consolidated soil $K_0 \approx (1 - \sin \phi_{ps})$.* The relationship given by equation (54) is plotted in Fig. 30 and compared with the relationship proposed in equation (44) for the ratio τ_{max}/σ_{v0}' obtained in a direct simple shear test. Note that the expressions are such that the ratio

$$\frac{s_{ufv}/\sigma_{v0}'}{\tau_{max}/\sigma_{v0}'} = (1 + \sin \phi_{ps}) \cos \phi_{ps} \tag{55}$$

is nearly constant at a value of $1{\cdot}28 \pm 0{\cdot}02$ over the likely range of $20° < \phi_{ps} < 40°$.

Overconsolidated soils

It has been well established both theoretically and experimentally that the undrained strength

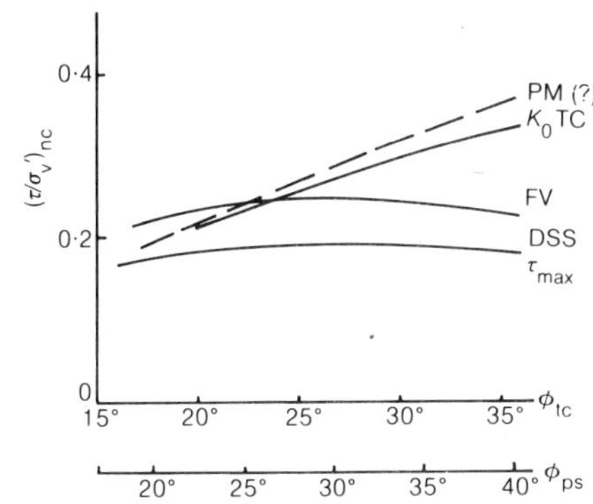

Fig. 30. Variation in the undrained strength ratio with the angle of friction for tests on normally consolidated clay

ratio of a soil, when normalized by the value for the normally consolidated condition, is proportional to OCR to the power Λ (see equation (28)). This finding has been applied to the proposed expression for vane shear tests of equation (54) to generate the family of curves in Fig. 31 for a soil with Λ equal to the typical value of 0·8.

Although the details of this chart are open to question, it is claimed that the general pattern presented is correct. It shows the importance of a small degree of overconsolidation on the undrained strength ratio. Many soft clay sites, supposedly normally consolidated, show small degrees of overconsolidation due to such causes as

(*a*) erosion followed by deposition
(*b*) fluctuating water-table
(*c*) desiccation.

Care must therefore be exercised in correlating the value of the undrained strength ratio for normally or lightly overconsolidated soils with other properties; in particular this applies to Skempton's much used relationship with plasticity index of equation (4).

ENGINEERING APPLICATION

The efforts made in this Paper to correlate measurements of undrained shear strength from different tests have led to a confusing picture of soil behaviour. To put this into perspective an example of an engineering application is presented—that of an embankment (or other

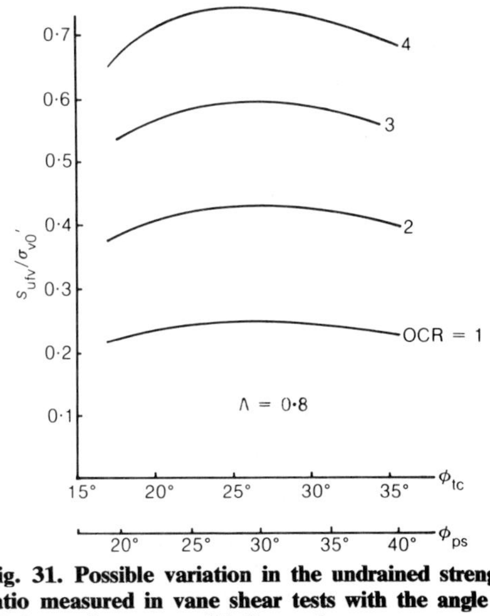

Fig. 31. Possible variation in the undrained strength ratio measured in vane shear tests with the angle of friction and overconsolidation ratio

* Note the preference for $K_0 \approx (1 - \sin \phi_{tc})$ in Appendix 2 and equation (47).

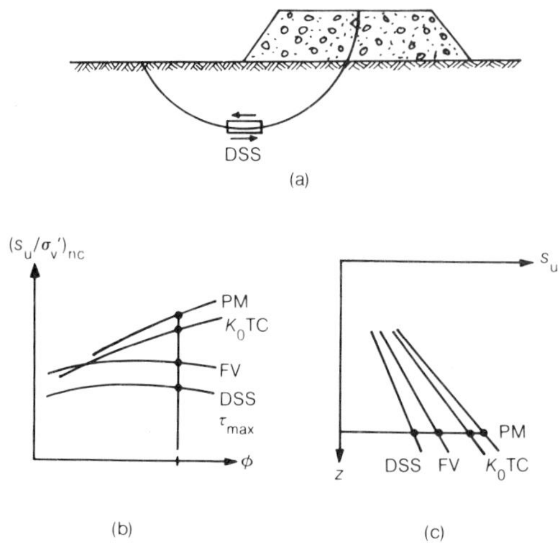

Fig. 32. Possible profiles of the undrained shear strength in a soft clay deposit measured in different tests

surface loading) constructed on a deposit of soft clay, illustrated schematically in Fig. 32(a).

The two major design problems are likely to be the short-term stability of the embankment and foundation, and the long-term settlement and deformations due to consolidation. Amongst other properties, the profile of undrained shear strength will be required, and a decision will face the designer of how this is to be obtained. A choice* of test or tests must be made from among a suite of in situ tests and laboratory tests; for this example two of each are considered—pressuremeter and field vane tests, triaxial tests (on anisotropically reconsolidated samples) and direct simple shear tests.

On the basis of the theoretical expressions developed in this Paper Fig. 32(b) has been prepared. The curve for the pressuremeter test has not been specified, but it is contended that it will be similar to that for plane strain tests on anisotropically reconsolidated samples from equation (34) and Fig. 27 but directly scaled by a factor of say 1·2 owing to strain rate effects and partial consolidation. The curves for the other three tests are those given by equations (27), (44) and (54).

For a given clay, with a given value of ϕ the four tests will give different values of $(s_u/\sigma_{v0}')_{nc}$. If the clay is lightly overconsolidated then each value of the undrained strength ratio should be multiplied by the factor $(OCR)^{\Lambda}$. At a given depth, i.e. for a given value of σ_{v0}', the undrained strengths will form a hierarchy as indicated in Fig. 32(c). The same argument will apply for all depths, resulting in the pattern of profiles that have been sketched. The spacing and even the relative positions of the profiles will depend on the value of ϕ for the deposit investigated.

It is argued that each of these apparently conflicting values of undrained shear strength is correct in its own right, and the designer must choose whichever he considers to be the most appropriate to the problem being analysed and apply to it an appropriate factor of safety.

Two examples of sites which support these ideas are shown in Fig. 33 with profiles of strengths at Porto Tolle (Fig. 33(a)) and Panigaglia (Fig. 33(b)) obtained by Ghionna *et al.* (1983), who were concerned in interpreting and comparing the results of pressuremeter tests with other tests, so that the individual points and scatter bars refer only to pressuremeter data and the straight lines to best-fit profiles of the other test results.

For a classic limit analysis of the short-term stability of the embankment, the direct simple shear test would be the most relevant test, simulating most closely the soil behaviour around the postulated failure surface; due allowance would need to be made for any anisotropy of strength of elements of soil at different positions around the failure surface. The hierarchy of strengths displayed would also rationalize the use of correction factors of less than unity applied to strengths derived from field vane tests as proposed by Bjerrum, Frimann Clausen & Duncan (1972).

In contrast, if a finite element analysis is made for predicting deformations and/or excess pore pressures then the values of soil properties chosen should be the ones relevant to the mathematical model of soil behaviour used in the computer program. Indeed if the model was specified in terms of effective stresses, or was perfectly elastic, a value of undrained shear strength would be irrelevant. If the model is an elastic–perfectly plastic model expressed in total stresses then the undrained shear strength would need to be the radius of Mohr's circle at failure in either axially symmetric or more likely plane strain conditions; the value of τ_{max} on a particular plane from a direct shear test or field vane test would *not* be appropriate.

It should be clear that there is no unique value of the undrained shear strength, and that whatever the circumstances the engineer must

* The cone penetration test has been omitted intentionally because, unlike the other tests, a value of the undrained shear strength cannot be derived directly from the observations; a value for the cone factor N_c has to be used which can only be based on previous comparisons with other tests.

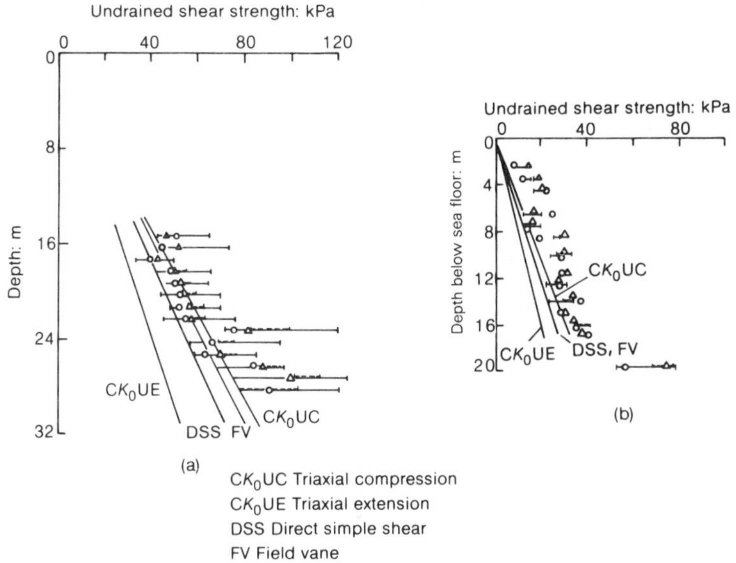

Fig. 33. Profiles of the undrained shear strength measured in different tests at two soft clay sites in Italy (after Ghionna *et al.*, 1983): (a) Porto Tolle; (b) Panigaglia

recognize the differences that exist between data from the various types of test. If the relationships proposed in this Paper are substantiated by further research and experience, then they can be used as a framework for enhancing the interpretation of data from site investigations.

CONCLUSIONS

The present period in the steady development of soil mechanics and foundation engineering is seeing a rapid growth in the range and quality of in situ tests available for site investigation and the measurement of soil properties. The main purpose of this Paper has been to consider the interpretation of the more important and more widely used tests, and to make recommendations for the improvement of their analysis and interpretation. The whole tenor and philosophy of the Paper has been of a speculative nature, attempting to build on established experience and accepted knowledge so that previously unrelated facts can be brought together as part of a broad picture of soil behaviour.

In attempting this ambitious task, the outcome of the work raises more questions than answers, underlines the very complex nature of soil behaviour and exposes the large gaps in current understanding.

The reasons and virtues of carrying out in situ tests have been set out and the need to use dimensionless parameters has been emphasized, particularly when relationships are based on empiricism, without being backed up by theory.

The soil property most often measured in the field is the so-called undrained shear strength of clays, and most of the discussion has centred on this. The definitions of undrained shear strength s_u and the angle of shearing resistance ϕ have been shown to be deficient in that they are defined in terms of the major and minor principal stresses, and they take no account of the effect of intermediate principal stress. Consequently there cannot be a unique undrained shear strength of a soil, and different values will be observed in different tests.

Since the fundamental stress–strain behaviour of soils is of a frictional nature controlled by the effective stresses, it is only possible to link measurements of strength in different tests by means of the friction angle and with a knowledge of the effective stresses. This requires the use of a failure law expressed in terms of all three principal effective stresses.

The failure criterion suggested by Matsuoka has been chosen, from which an expression for ϕ can be found for any stress state on the failure envelope. In particular it allows a simple (approximate) relationship between the observed friction angle in triaxial compression and plane strain tests, namely $\phi_{ps} \approx \frac{9}{8}\phi_{tc}$.

The concepts of CSSM have been used to provide theoretical confirmation

(*a*) of the important relationship between the undrained strength ratio s_u/σ_{v0}' and the overconsolidation ratio

(*b*) that the normalized undrained strength ratio varies as OCR to the power *m* where *m* lies in the narrow range 0·68–0·86 and is typically 0·8. Moreover it has been shown that m is equal to Λ so that it is related physically to the compression indices and is not merely an empirical constant.

The analysis shows that different values of the undrained strength ratio will be observed in triaxial tests according to the type of consolidation, isotropic or one dimensional, so that great care is necessary when results of such tests are used for comparison with field tests.

The major difference between tests in which the principal axes of stress are fixed in direction and those in which they are free to rotate has been emphasized. An attempt has been made to understand the effective stress paths observed in direct shear tests which has exposed two important anomalies, both concerning failure. The first of these is that the maximum value of the applied shear stress observed in the test is not the undrained shear strength as properly defined. The second is that at this stage of the test failure is occurring within the sample but with the planes of maximum stress obliquity perpendicular to those usually assumed, as suggested by de Josselin de Jong. These findings have great importance for the proper interpretation of direct shear tests and the vane shear test. Mathematical expressions to represent the data from both types of test have been suggested, but these are based on slender evidence and urgently require further research.

The two modern in situ tests that hold the most promise are the self-boring pressuremeter and the piezocone. The self-boring technique of insertion—which can be used for other devices also—causes minimal disturbance and allows the possibility of "near perfect" testing of undisturbed soil.

The pressuremeter test is relatively expensive and sophisticated, but it has the great advantage of leading to direct evaluation (without the need for empirical relationships) of the following important soil conditions and properties: in situ lateral total stress σ_{h0}, shear modulus G, undrained shear strength of clays s_u, angle of friction ϕ and angle of dilation ν of sands, and the horizontal coefficient of consolidation of clays c_h.

The piezocone is a tool of great promise for obtaining a rapid and reliable soil profile. The measurement of pore pressures during the penetration of a cone has enhanced its importance considerably. For the civil engineering profession to gain the maximum advantage from this (and other in situ tests) it is essential that the following three aspects are standardized:

(*a*) the geometry of the instrument and in particular the location of the pore pressure transducer
(*b*) the operation of the test, i.e. methods of calibration, rates of testing etc.
(*c*) the interpretation of the results and in particular the choice of the dimensionless parameters to be used.

Theoretical relationships have been developed in this Paper for the undrained strength ratio of normally consolidated clay, each of which can be extended to other values of the overconsolidation ratio by the power law mentioned earlier. Although these relationships can only approximate real soil behaviour (and may be found to be wanting in a quantitative sense) it is claimed that they are relevant qualitatively. They indicate a pattern of behaviour, which leads to a hierarchy of strengths that will be observed in different laboratory and field tests. Consequently it is imperative for a designer to recognize this hierarchy, and to select a strength which is appropriate to the analysis or design procedure being used, and to choose a factor of safety or load factor accordingly.

Finally, it is concluded that there is an important need for further research and development in the conduct and interpretation of both laboratory and in situ testing of soils; this must go hand in hand with practice, so that field experience of full-scale structures can be used to test new theories, new equipment and better interpretation.

ACKNOWLEDGEMENTS

In preparing the Rankine Lecture I wish to acknowledge the substantial help and information that I have received from many people in many countries too numerous to mention individually. Constraints of time and space prevented discussion of other topics of importance in the execution and interpretation of in situ testing of soils, and of other new developments in types of test and of equipment. Their exclusion is regretted but indicates the vigorous expansion that is presently occurring in this area of geotechnical engineering.

The results of pressuremeter tests at Zeebrugge included in this Paper are quoted by kind permission of PM Insitu Techniques Ltd and their clients for the job, Distrigas NV and Tractebel Z.

REFERENCES

Andresen, A., Berre, T., Kleven, A. & Lunne, T. (1979). Procedures used to obtain soil parameters for foundation engineering in the North Sea. *Mar. Geotechnol.* **3,** 201–266.

Arulanandan, K. (1977). *Method and apparatus for*

measuring in situ density and fabric of soils. Patent Application, Regents of University of California.
Atkinson, J. H. & Bransby, P. L. (1978). *The mechanics of soils—an introduction to critical state soil mechanics*. London: McGraw-Hill.
Baguelin, F. (1982). Rules of foundation design using self boring pressuremeter test results. *Proc. Symp. Pressuremeter and its Marine Applications*, pp. 347–360.
Baguelin, F., Jézéquel, J. F., Le Mée, E. & Le Mehaute, A. (1972). Expansion of cylindrical probes in cohesive soils. *J. Soil Mech. Fdns Div. Am. Soc. Civ. Engrs* **98,** SM11, 1129–1142.
Baguelin, F., Jézéquel, J. F. & Le Mehaute, A. (1974). Le perméamètre autoforeur. *Can. Geotech. J.* **11,** 624–628.
Baligh, M. M., Azzouz, A. S., Wissa, A. Z. E., Martin, R. T. & Morrison, M. J. (1981). The piezocone penetrometer. *Proc. Am. Soc. Civ. Engrs Symp. Cone Penetration Testing and Experience, St Louis*, pp. 247–263.
Baligh, M. M. & Levadoux, J. N. (1980). *Pore pressure dissipation after cone penetration*. MIT Research Report MITSG 80–11. Massachusetts Institute of Technology.
Bishop, A. W. (1966). The strength of soils as engineering materials. *Géotechnique* **16,** 91–128.
Bishop, A. W. & Henkel, D. J. (1957). *The measurement of soil properties in the triaxial test*. London: Arnold.
Bjerrum, L., Frimann Clausen, C. J. & Duncan, J. M. (1972). Earth pressures on flexible structures—a state of the art report. *Proc. 5th Eur. Conf. Soil Mech.* **2,** 169–196.
Borin, D. L. (1973). *The behaviour of saturated kaolin in the simple shear apparatus*. PhD thesis, University of Cambridge.
British Standards Institution (1975). *Methods of test for soils for civil engineering purposes*. BS 1377. London: British Standards Institution.
Burland, J. B. & Hancock, R. J. R. (1977). Underground car park at the House of Commons, London: geotechnical aspects. *Struct. Engr* **55,** 87–100.
Campanella, R. G., Gillespie, D. & Robertson, P. K. (1982). Pore pressures during cone penetration testing. *Proc. 2nd Eur. Symp. Penetration Testing, Amsterdam* **2,** 507–512.
Campanella, R. G. & Robertson, P. K. (1981). Applied cone research. *Am. Soc. Civ Engrs Symp. Cone Penetration Testing and Experience, St Louis*, pp. 343–362.
Campanella, R. G. & Robertson, P. K. (1983). *Flat plate dilatometer testing: research and development*. Soil Mech. Report 68. University of British Columbia.
Carter, J. P., Randolph, M. F. & Wroth, C. P. (1979). Stress and pore pressure changes in clay during and after the expansion of a cylindrical cavity. *Int. J. Numer. Analyt. Meth. Geomech.* **3,** 305–322.
Casagrande, A. (1936). Characteristics of cohesionless soils affecting the stability of slopes and earth fills. *J. Boston Soc. Civ. Engrs* 257–276.
Clarke, B. G. (1981). *In situ testing of clays using the Cambridge self-boring pressuremeter*. PhD thesis, University of Cambridge.
Clarke, B. G., Carter, J. P. & Wroth, C. P. (1979). In situ determination of the consolidation characteristics of saturated clays. *Proc. 7th Eur. Conf. Soil Mech.* **2,** 207–213.
Donald, I. B., Jordan, D. O., Parker, R. J. & Toh, C. T. (1977). The vane test—a critical appraisal. *Proc. 9th Int. Conf. Soil Mech.* **1,** 81–88.
Eden, W. J. & Law, K. T. (1980). Comparison of undrained shear strength results obtained by different test methods in soft clay. *Can. Geotech. J.* **17,** 369–381.
Fahey, M. (1980). *A study of the pressuremeter test in dense sand*. PhD thesis, University of Cambridge.
Flodin, N. & Broms, B. (1981). History of civil engineering in soft clay. In *Soft clay engineering*, Chap. 1, pp. 27–156 (eds. E. W. Brand & R. P. Brenner). Amsterdam: Elsevier.
Ghionna, V., Jamiolkowski, M., Lacasse, S., Lancellota, R. & Lunne, T. (1983). Evaluation of self-boring pressuremeter. *Proc. Int. Symp. In Situ Testing, Paris*, **2,** 294–301.
Ghionna, V. N., Jamiolkowski, M. & Lancellotta, R. (1982). Characteristics of saturated clays as obtained from SBP tests. *Proc. Symp. Pressuremeter and its Marine Applications, Paris*, pp. 165–185.
Gibson, R. E. & Anderson, W. F. (1961). In situ measurement of soil properties with the pressuremeter. *Civ. Engng Pub. Wks Rev.* **56,** No. 658, 615–618.
Henderson, G., Smith, P. D. K. & St John, H. D. (1979). The development of a pressuremeter for offshore use. *Proc. Int. Conf. Offshore Site Investigation, London*. Paper No. 10.
Henkel, D. J. (1960). The shear strength of saturated remoulded clays. *Proc. Am. Soc. Civ. Engrs Conf. Shear Strength of Cohesive Soils, Boulder*, pp. 533–554.
Hughes, J. M. O. (1973). *An instrument for in-situ measurement in soft clays*. PhD thesis, University of Cambridge.
Hughes, J. M. O., Wroth, C. P. & Pender, M. J. (1975). A comparison of the results of special pressuremeter tests with conventional tests on a deposit of soft clay at Canvey Island. *Proc. 2nd Aust.-N.Z. Conf. Geomech., Brisbane*, pp. 292–296.
Hughes, J. M. O., Wroth, C. P. & Windle, D. (1977). Pressuremeter tests in sands. *Géotechnique* **27,** 455–477.
Hvorslev, M. J. (1937). *Über die Festigkeitseigenschaften Gestörter Bindiger Boden*. Doctoral thesis, Køpenhavn.
Jefferies, M. G. & Funegard, E. (1983). Cone penetration testing in the Beaufort Sea. *Am. Soc. Civ. Engrs Spec. Conf. Geotechnical Practice in Offshore Engineering, Austin*, pp. 220–243.
Jones, G. A. & Rust, E. A. (1982). Piezometer penetration testing CUPT. *Proc. 2nd Eur. Symp. Penetration Testing, Amsterdam* **2,** 607–613.
de Josselin de Jong, G. (1972). *Roscoe Memorial Symp. Stress–Strain Behaviour of Soils*, Discussion to session II, pp. 258–261 (ed. R. H. G. Parry). Cambridge: Foulis.
Lacasse, S. & Lunne, T. (1982a). In situ horizontal stress from pressuremeter tests. *Proc. Symp.*

Pressuremeter and its Marine Applications, Paris, pp. 187–208.

Laccasse, S. & Lunne, T. (1982b). Penetration tests in two Norwegian clays. *Proc. 2nd Eur. Symp. Penetration Testing, Amsterdam* **2,** 661–669.

Ladanyi, B. (1972). In situ determination of undrained stress–strain behaviour of sensitive clays with the pressuremeter. *Can. Geotech. J.* **9,** 313–319.

Ladd, C. C. & Edgers, L. (1972). *Consolidated-undrained direct-simple shear tests on saturated clays.* MIT Research Report R72-82. Massachusetts Institute of Technology.

Ladd, C. C., Foott, R., Ishihara, K., Schlosser, F. & Poulos, H. G. (1977). Stress-deformation and strength characteristics: state of the art report. *Proc. 9th Int. Conf. Soil Mech., Tokyo* **2,** 421–494.

Lade, P. (1972). *The stress–strain and strength characteristics of cohesionless soils.* PhD thesis, University of California, Berkeley.

Law, K. T. (1979). Triaxial-vane tests on a soft marine clay. *Can. Geotech. J.* **16,** 11–18.

Loudon, P. A. (1967). *Some deformation characteristics of kaolin.* PhD thesis, University of Cambridge.

Mair, R. J. & Wood, D. M. (1984). *A review of the use of pressuremeters for in situ testing.* London: CIRIA.

Marsland, A. (1973). Laboratory and in situ measurement of the deformation moduli of London Clay. *Proc. Symp. Interaction of Structure and Foundation,* pp. 7–17. Birmingham: Midland Soil Mechanics and Foundation Engineering Society.

Matsuoka, H. (1974). Stress–strain relationships of sands based on the mobilized plane. *Soils Fdns* **14,** 47–61.

Matsuoka, H. & Nakai, T. (1977). Stress–strain relationship of soil based on the spatially mobilized plane. *Proc. 9th Int. Conf. Soil Mech., Tokyo,* Special Session 9, pp. 153–162.

Matsuoka, H. & Nakai, T. (1982). A new failure criterion for soils in three dimensional stresses. *Proc. IUTAM Symp. Deformation Failure of Granular Materials, Delft.*

Menzies, B. K. & Merrifield, C. M. (1980). Measurements of shear stress distribution on the edges of a shear vane blade. *Géotechnique* **30,** 314–318.

Mori, H. (1983). In situ plate loading test for dense sandy soil using a self boring instrument. *Proc. Int. Symp. In Situ Testing, Paris* **2,** 353–357.

Palmer, A. C. (1972). Undrained plane strain expansion of a cylindrical cavity in clay: a simple interpretation of the pressuremeter test. *Géotechnique* **22,** 451–457.

Ramamurthy, T. & Rawat, P. C. (1973). Shear strength of sand under general stress system. *Proc. 8th Int. Conf. Soil Mech., Moscow,* **1·2,** 339–342.

Randolph, M. F. & Wroth, C. P. (1979). An analytical solution for the consolidation around a driven pile. *Int. J. Numer. Analyt. Meth. Geomech.* **3,** 217–229.

Randolph, M. F. & Wroth, C. P. (1981). Application of the failure state in undrained simple shear to the shaft capacity of driven piles. *Géotechnique* **31,** No. 1, 143–157.

Robertson, P. K. & Campanella, R. G. (1983). *Interpretation of cone penetration tests,* Parts 1 & 2. Soil Mech. Report Series No. 60. University of British Columbia.

Roscoe, K. H. (1953). An apparatus for the application of simple shear to soil samples. *Proc. 3rd Int. Conf. Soil Mech., Zurich* **1,** 186–191.

Roscoe, K. H., Schofield, A. N. & Wroth, C. P. (1958). On the yielding of soils. *Géotechnique* **8,** 22–53.

Satake, M. (1982). *On equivalent Mohr's circle for granular materials.* Report No. R-205. University of New South Wales.

Schofield, A. N. & Wroth, C. P. (1968). *Critical state soil mechanics.* London: McGraw-Hill.

Sennesset, K., Janbu, N. & Svanø, G. (1982). Strength and deformation parameters from cone penetration tests. *Proc. 2nd Eur. Symp. Penetration Testing, Amsterdam* **2,** 863–870.

Shibata, T. & Karube, D. (1965). Influence of the variation of the intermediate principal stress on mechanical properties of normally consolidated clay. *Proc. 6th Int. Conf. Soil Mech., Montreal* **1,** 359–363.

Skempton, A. W. (1954). The pore-pressure coefficients *A* and *B*. *Géotechnique* **4,** 143–147.

Skempton, A. W. (1957). Discussion on the planning and design of the new Hong Kong airport. *Proc. Instn Civ. Engrs* **7,** 306.

Smits, F. P. (1982). Penetration pore pressure measured with piezometer cones. *Proc. 2nd Eur. Symp. Penetration Testing, Amsterdam* **2,** 877–881.

Soydemir, C. (1976). Strength anisotropy observed through simple shear tests. In *Laurits Bjerrum memorial volume,* pp. 99–113. Norwegian Geotechnical Institute.

Sutherland, H. B. & Mesdary, M. S. (1969). The influence of the intermediate principal stress on the strength of sand. *Proc. 7th Int. Conf. Soil Mech., Mexico* **1,** 391–399.

Taylor, D. W. (1948). *Fundamentals of soil mechanics.* London: Wiley, 358–359.

Tedd, P. & Charles, J. A. (1981). In situ measurement of horizontal stress in overconsolidated clay using push-in spade-shaped pressure cells. *Géotechnique* **31,** No. 4, 554–558.

Torstensson, B.-A. (1975). Pore pressure sounding instrument. *Proc. Am. Soc. Civ. Engrs Spec. Conf. In Situ Measurement of Soil Properties* **2,** 48–54.

Tumay, M. T., Bogges, R. L. & Acar, Y. (1981). Subsurface investigations with piezocone penetrometer. *Proc. Am. Soc. Civ. Engrs Symp. Cone Penetration Testing and Experience, St Louis,* pp. 325–342.

Ventura, P. (1983). Le pénétromètre pièzométrique dans les profils des sous-sols marins. *Proc. Symp. Soil and Rock Investigations by In Situ Testing, Paris* **2,** 425–430.

Windle, D. (1976). *In situ testing of soils with a self-boring pressuremeter.* PhD thesis, University of Cambridge.

Windle, D. & Wroth, C. P. (1977). The use of a self-boring pressuremeter to determine the undrained properties of clays. *Ground Engng* **10,** No. 6, 37–46.

Wissa, A. E. Z., Martin, R. T. & Garlanger, J. E.

(1975). The piezometer probe. *Proc. Am. Soc. Civ. Engrs Spec. Conf. In Situ Measurement of Soil Properties* **1,** 536–545.

Wood, D. M. (1981). True triaxial tests on Boston Blue Clay. *Proc. 10th Int. Conf. Soil Mech., Stockholm* **1,** 825–830.

Wood, D. M. & Wroth, C. P. (1977). Some laboratory experiments related to the results of pressuremeter tests. *Géotechnique* **27,** 181–201.

Wroth, C. P. (1982). British experience with the self boring pressuremeter. *Proc. Symp. Pressuremeter and its Marine Applications, Paris*, pp. 143–164.

Wroth, C. P. & Hughes, J. M. O. (1972). *An instrument for the in situ measurement of the properties of soft clays.* Technical Report Soils TR13. University of Cambridge.

Zuidberg, H. M., Schaap, L. H. J. & Beringen, F. L. (1982). A penetrometer for simultaneously measuring cone resistance, sleeve friction and dynamic pore pressure. *Proc. 2nd Eur. Symp. Penetration Testing, Amsterdam*, pp. 963–970.

APPENDIX 1

Relationship between pressures in triaxial tests

In Fig. 5 the points A and C lie on the normal consolidation line

$$V_a - V_c = \lambda \ln (p_c'/p_a') \qquad (56)$$

Points R and C lie on a swelling line

$$V_r - V_c = \kappa \ln (p_c'/p_r') = \kappa \ln R \qquad (57)$$

But $V_r \equiv V_a$, therefore

$$\lambda \ln (p_c'/p_a') = \kappa \ln R \qquad (58)$$

Points S and X lie on the critical state line, and by similar triangles

$$\lambda \ln (p_x'/p_s') = \kappa \ln (p_x'/p_r') \qquad (59)$$

Subtraction of both sides from $\lambda \ln (p_x'/p_r')$ gives

$$\lambda \ln (p_s'/p_r') = (\lambda - \kappa) \ln (p_x'/p_r') \qquad (60)$$

i.e.

$$\lambda \ln (p_s'/p_r') = (\lambda - \kappa) \ln (R/r) \qquad (61)$$

and

$$\frac{p_s'}{p_r'} = \left(\frac{R}{r}\right)^{\Lambda} \qquad (62)$$

where

$$\Lambda = \frac{\lambda - \kappa}{\lambda} \qquad (63)$$

APPENDIX 2

Undrained strength ratio of a one-dimensionally normally consolidated clay tested in triaxial compression

The concepts of critical state soil mechanics allow an estimate to be made of the ratio of the undrained strength s_{utc} to the consolidation pressure σ_{v0}' of a specimen which has been one-dimensionally normally consolidated.

Figure 34 shows the consolidation and stress space plots for interpretation of triaxial tests used in critical state soil mechanics together with the elliptical yield envelope of modified Cam clay. The equation of this envelope is

$$q^2 + M^2 p'^2 = 2M^2 p_k' p' \qquad (64)$$

where q is the deviator stress and p' the mean principal effective stress. It is assumed that when the specimen is normally consolidated at state J the coefficient of earth pressure at rest is given by the widely adopted expression $K_0 \approx (1 - \sin \phi_{tc})$ so that the ratio of stresses at J is given by

$$\eta_j = \frac{q_j}{p_j'} = \frac{3(1-K_0)}{1+2K_0} = \frac{3 \sin \phi_{tc}}{3 - 2 \sin \phi_{tc}} \qquad (65)$$

but point J must lie on the yield envelope and satisfy equation (64) so that

$$(\eta_j^2 + M^2) p_j'^2 = 2M^2 p_k' p_j' \qquad (66)$$

An undrained compression test on the specimen will bring the specimen to failure at point L on the critical state line at the same specific volume (or water content). Hence the undrained strength, by definition, is

$$s_{utc} = \tfrac{1}{2} q_l = \frac{M}{2} p_l' \qquad (67)$$

but this needs to be related to the initial consolidation pressure σ_{v0}', i.e. σ_{1j}' which can be done by expressing p_l' in turn as a function of (p_j', p_k') then η_j and then σ_{1j}'. The relationship between the pressures at points R, S, X of Fig. 5 established in Appendix 1, equation (60), can be invoked to relate the pressures at points J, K and L as

$$\frac{p_l'}{p_j'} = \left(\frac{p_k'}{p_j'}\right)^{\Lambda} \qquad (68)$$

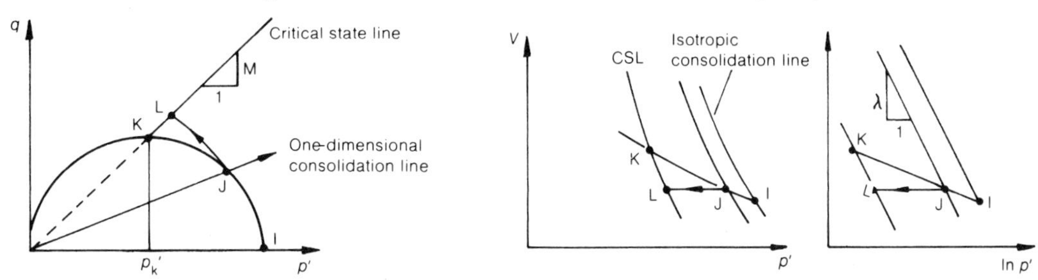

Fig. 34. Yield envelope and critical state line for modified Cam clay

but from equation (66)

$$\frac{p_k'}{p_j'} = \frac{\eta_j^2 + M^2}{2M^2} \tag{69}$$

and from equation (65)

$$\frac{\eta_j}{M} = \frac{3 \sin \phi_{tc}}{3 - 2 \sin \phi_{tc}} \frac{3 - \sin \phi_{tc}}{6 \sin \phi_{tc}}$$

$$= \frac{3 - \sin \phi_{tc}}{2(3 - 2 \sin \phi_{tc})} = a \text{ (say)} \tag{70}$$

and by definition

$$\frac{p_j'}{\sigma_{1j}'} = \frac{1 + 2K_0}{3}$$

$$= \frac{3 - 2 \sin \phi_{tc}}{3} \tag{71}$$

Combining these results

$$\frac{s_{utc}}{\sigma_{1j}'} = \frac{s_{utc}}{p_1'} \frac{p_1'}{p_j'} \frac{r_j'}{\sigma_{1j}'}$$

$$= \frac{M}{2} \left(\frac{p_k'}{p_j'}\right)^{\Lambda} \frac{3 - 2 \sin \phi_{tc}}{3}$$

$$= \frac{3 \sin \phi_{tc}}{3 - \sin \phi_{tc}} \left(\frac{a^2 + 1}{2}\right)^{\Lambda} \frac{3 - 2 \sin \phi_{tc}}{3}$$

$$= \frac{\sin \phi_{tc}}{2a} \left(\frac{a^2 + 1}{2}\right)^{\Lambda} \tag{72}$$

Undrained strength ratio of a one-dimensionally normally consolidated clay tested in plane strain active conditions

An exactly similar analysis to that for the triaxial test can be carried out for plane strain active tests. The same basic assumptions are made except that the stress variables p' and q are replaced by $s' = \frac{1}{2}(\sigma_1' + \sigma_3')$ and $t = \frac{1}{2}(\sigma_1' - \sigma_3')$, and ϕ_{tc} by ϕ_{ps} for failure conditions. However, for one-dimensional normal consolidation, for which the conditions are axially symmetric, it is considered both more appropriate and more convenient to relate the value of K_0 with ϕ_{tc} (not ϕ_{ps}) and use the approximation $K_0 \approx 1 - \sin \phi_{tc}$. Hence

$$\theta_j = \frac{t_j}{s_j'} = \frac{1 - K_0}{1 + K_0}$$

$$= \frac{\sin \phi_{tc}}{2 - \sin \phi_{tc}} \tag{73}$$

and since J must lie on the yield envelope

$$(\theta_j^2 + m^2) s_j'^2 = 2m^2 s_k' s_j' \tag{74}$$

where $m = \sin \phi_{ps}$ and

$$s_{ups} = t_l = m s_l' \tag{75}$$

As above,

$$\frac{s_l'}{s_j'} = \left(\frac{s_k'}{s_j'}\right)^{\Lambda} \tag{76}$$

and

$$\frac{s_k'}{s_j'} = \frac{\theta_j^2 + m^2}{2m^2} \tag{77}$$

$$\frac{\theta_j}{m} = \frac{1}{2 - \sin \phi_{ps}} = c \text{ (say)} \tag{78}$$

and

$$\frac{s_j'}{\sigma_{1j}'} = \frac{1 + K_0}{2}$$

$$= \frac{2 - \sin \phi_{tc}}{2} \tag{79}$$

Combining these results

$$\frac{s_{ups}}{\sigma_{1j}'} = \frac{s_{ups}}{s_1'} \frac{s_1'}{s_j'} \frac{s_j'}{\sigma_{1j}'}$$

$$= m \left(\frac{c^2 + 1}{2}\right)^{\Lambda} \frac{2 - \sin \phi_{tc}}{2}$$

$$= \frac{\sin \phi_{ps}}{2c} \frac{2 - \sin \phi_{tc}}{2 - \sin \phi_{ps}} \left(\frac{c^2 + 1}{2}\right)^{\Lambda} \tag{80}$$

In reality the middle factor is very near to unity; using the relationship of equation (15) or equation (18) it varies from 1·025 for $\phi_{tc} = 20°$ to 1·045 for $\phi_{tc} = 35°$. Within the likely accuracy of all the assumptions it can be taken as unity and the undrained strength ratio in plane strain active conditions reduces to

$$\frac{s_{ups}}{\sigma_{v0}'} = \frac{\sin \phi_{ps}}{2c} \left(\frac{c^2 + 1}{2}\right)^{\Lambda} \tag{81}$$

where $c = 1/(2 - \sin \phi_{ps})$.

Similarly it can be shown that for isotropically normally consolidated clay tested in plane strain active conditions the undrained strength ratio would be

$$\frac{s_{ups}}{\sigma_{v0}'} = \sin \phi_{ps} (\tfrac{1}{2})^{\Lambda} \tag{82}$$

VOTE OF THANKS

In proposing a vote of thanks to Professor Wroth, Professor J. B. Burland made the following remarks.

'There are two characteristics which distinguish Professor Wroth's work and his presentations. The first is clarity and the second is elegance. The lecture we have just heard displays both of these characteristics in full measure. When I was a research student at "The other place" I well remember Peter Wroth's insistence that diagrams should always be kept simple and that in presenting them the axes should first be clearly indicated. This evening's lecture has been an object lesson in its clarity of presentation.

'Few would question the importance of determining directly and with a minimum of disturbance the in situ properties of the ground. There is inevitably an element of competition, even tension, between in situ and laboratory testing and I believe that this is a thoroughly healthy situation. Laboratory testing has the advantage of flexibility in the control of the stress path but suffers the disadvantage of sampling and preparation. Good in situ testing minimizes the disturbance but lacks the flexibility of stress path control. Moreover a careful interpretation of the results is required and this has been the important theme of this evening's lecture.

'In the first part of his lecture Professor Wroth has concentrated on two types of in situ tests—the self-boring pressuremeter and the piezocone. He was one of the first to recognize the potential of the self-boring pressuremeter and the development of the Camkometer, and its interpretation has taken place largely under his direction. I would like to take this opportunity of paying tribute to him for this important contribution. If I may express a personal view I believe that the capability of measuring the in situ horizontal effective stress is still one of the most important features of the Camkometer. The piezocone is less familiar in the UK and Professor Wroth has given us a clear insight into the advantages it offers over the traditional cone test.

'In the last part of his lecture Professor Wroth has tackled one of the most fascinating and fundamental problems of soil mechanics, the undrained strength of clays—a topic which has been touched on by many previous Rankine lecturers. Not only do in situ measurements of undrained strength require careful interpretation but so also do laboratory measurements. His analysis of the simple shear test has been most elegantly presented and I have no doubt that it will be the subject of very careful study and debate.

'Earlier I referred to the healthy tension that exists between laboratory and in situ testing. Professor Wroth has demonstrated that both have a vital role to play and that developments in the one can have important implications for the other. He has also disposed of once and for all any notion that there is such a thing as *the* undrained strength of a clay. Ladies and gentlemen I know that you will agree with me that the lecture has been stimulating, thought provoking and of great value. It is with the greatest pleasure that I propose a hearty vote of thanks to Professor Wroth for delivering the twenty-fourth Rankine Lecture.'

The vote of thanks was accorded with acclamation.

The Rankine Lecture

The twenty-fifth Rankine Lecture of the British Geotechnical Society was given by Professor N. Janbu at Imperial College of Science and Technology, London, on 5 March 1985. The following introduction was given by Professor J. N. Hutchinson, Imperial College of Science and Technology.

'It gives me great pleasure to introduce Professor Janbu as the twenty-fifth Rankine Lecturer.

'Nilmar Janbu was born on 23 August 1921 in Fraena, a remote community on the rugged west coast of Norway, about 200 kilometres from Trondheim. He graduated as a civil engineer from the Technical University of Norway in Trondheim in 1947, his education having been delayed for a couple of years by the war. Immediately after graduation, he left Norway to join that select band of Europeans who have studied under the late Professor Arthur Casagrande at Harvard. Following his Master of Science degree in 1949, Janbu was invited by Casagrande to stay on at Harvard to pursue research into the stability analysis of slopes in terms of dimensionless parameters. His thesis on this subject, published in the Harvard Soil Mechanics Series, led to the award of the degree of Doctor of Science in 1954.

'In 1951 Dr Janbu returned to Norway, initially to work for a year with the consulting firm of Bonde & Company in Oslo. In 1952 he was appointed to two key positions: as lecturer in soil mechanics at the Technical University of Norway, the first such post in the country, and, in parallel, as the leader of the Trondheim division of the newly formed Norwegian Geotechnical Institute. Shortly afterwards Dr Janbu produced his 'Generalised procedure of slices', the first approach to deal satisfactorily with the stability analysis of slip surfaces of general shape. Its basic principles were outlined at the Stockholm conference in 1954: subsequently the method was extended to earth pressure and foundation problems and published fully in 1956 and 1957.

'During the early years of the Norwegian Geotechnical Institute, Dr Janbu's almost legendary computational ability, combined with his fundamental and wide ranging understanding of soil mechanics and his infectious enthusiasm, made him a key figure. He took a leading part in producing NGI Publication No. 16: 'Soil mechanics applied to some engineering problems'. This was published, in Norwegian, in 1956 and contributed, more than any other work, to the adoption, in Norwegian practice, of modern effective stress soil mechanics.

'In 1957, Dr Janbu again took steps to deepen and broaden his experience by returning to the USA. There he worked in Chicago for two years, firstly with M. J. McDermott & Company, Contractors, and secondly with Al. Benesch and Associates, Consulting Engineers. Dr Janbu returned to Norway in 1959 on his appointment to his present position as Professor of Soil Mechanics and Foundation Engineering in the Technical University of Norway, in Trondheim. I first met him in that year and recall vividly an episode which, I now realize, was characteristic. While driving me north from Trondheim to start work on the investigation of the Furre landslide, he made a rapid mental calculation of the mobilized effective friction angle to be expected there and produced a value within a degree of that which we painstakingly obtained six months and many thousands of kroner later!

'In addition to his distinguished work on slope stability. Professor Janbu has made fundamental contributions to the understanding and treatment of soil compressibility and to the design of both spread and piled foundations. He has thus, very naturally, been in much demand as a consultant on major projects both in Norway and abroad, on many technical committees and as a chairman or panel member at numerous conferences.

'While, internationally, Professor Janbu is known primarily for his contributions to theory and practice, his countrymen esteem him at least as much for his excellent and sustained teaching. His ability to attract and inspire the brightest students has, over the past three decades, contributed very significantly to the present strength of geotechnics in Norway. His all-round qualities have been recognized in many ways, for instance by his election as a Member of the Royal Norwegian Academy of Science in 1962 and by the award of the Laurits Bjerrum Memorial Prize in 1983.

'For the past decade or so, Professor Janbu has been deeply and creatively involved in the considerable geotechnical problems associated with the offshore oil industry. On behalf of the British Geotechnical Society, may I say how pleased we are that he has agreed to give the twenty-fifth Rankine Lecture and has chosen this exciting and challenging new field for his theme. We look forward eagerly to hear Professor Janbu's lecture, which I now invite him to deliver.'

Professor N. Janbu

JANBU, N. (1985). *Géotechnique* **35**, No. 3, 241–281

Soil models in offshore engineering

N. JANBU*

The offshore activities in the North Sea have been a very demanding but fruitful challenge to geotechnical engineering for more than a decade now. It soon became evident that there was an urgent need for a better understanding of the behaviour of different types of soil under static and cyclic loading, for drained as well as undrained conditions. The experience from geotechnical engineering onshore was not sufficient, nor were the available models for soil behaviour readily adaptable to practical applications to this wider range of problems. Since about 1960, the Author has addressed himself to the problem of modelling soil behaviour within one consistent framework, namely the classical resistance concept. In particular, over the last 10–15 years this concept has been used extensively for typical offshore problems. During this time the entire staff of the Geotechnical Division at the Norwegian Institute of Technology has taken part in the development of elements of the model. Consequently data input comes from a large number of theoretical and experimental theses, many internal reports and several published papers. The practical result of this comprehensive work is the main content of this lecture.

Depuis plus de dix années les opérations au large dans la Mer du Nord posent des problèmes d'ingénierie géotechnique d'une façon exigeante mais productive. Il est devenu rapidement évident qu'il y avait un besoin urgent d'une meilleure compréhension du comportement des sols différents sous des charges statiques et cycliques dans des conditions drainées et non-drainées. Les connaissances acquises à partir des constructions géotechniques à terre n'étaient pas suffisantes et les modèles disponibles pour l'étude du comportement des sols ne pouvaient pas s'adapter facilement dans la pratique à ces problèmes de plus grande envergure. Depuis environ 1960 l'auteur a étudié la modélisation du comportement des sols dans le cadre exclusif du concept classique de la résistance. Pendant les 10–15 dernières années ce concept a été souvent employé en particulier pour des problèmes se posant typiquement au large. Au cours de cette période tout le personnel de la Division technique de l'institut norwégien de la technologie a participé au développement du modèle. Par conséquent les données proviennent d'une grande quantité de thèses théoriques, de beaucoup de rapports internes et de plusieurs articles publiés. Cette conférence présente principalement des résultats pratiques de ce travail d'ensemble.

* Norwegian Institute of Technology.

INTRODUCTION

In geotechnical design it has long been recognized that the assessment of soil properties is the most important single task. By comparison the details of numerical analyses are of much lesser consequence. This observation is the major reason for the choice of the Lecture topic: soil modelling for offshore engineering problems.

The various types of geotechnical problems which may be involved in the foundation design of a gravity platform are in principle illustrated in Fig. 1 by means of a stress path plot.

The initial in situ stress condition before installation (point I) corresponds most often to low shear stress levels (large safety factors). The installation itself may be carried out in a few days, and the stress path of undrained loading is assumed to end up at point U. After stage 1 at constant load (during summer) some drainage may take place, and the stress path moves to point D, with a larger factor of safety. Environmental loads (storms) may then expose the structure to cyclic loads, leading to a cumulative increase in the pore pressure level, with a temporary reduction in safety. Practical design requires that the critical position of the stress path (point C) for the toe region T lies below a specified design level, which may be given as a reduced level of strength.

From this brief survey it is evident that most of the knowledge required for design is related to soil behaviour within the working stress range. More specifically, the design requires the stress–strain–time behaviour under drained and undrained conditions, for static and repeated loading at different degrees of shear mobilizations, both within and outside the preconsolidation stress regime. In this perspective, the unique assessment of an average failure envelope is by far the simplest, least expensive and most reliable task.

Model requirements

To make a versatile and useful model of behaviour in practice, several requirements should be fulfilled.

(*a*) Each element of the model must be simple and based on classical concepts to the largest possible extent.

(*b*) The model must be capable of describing continuously the soil response to increasing stress levels, from zero up to failure, including the state of failure itself.
(*c*) The basic concepts should be independent of soil type and boundary conditions.
(*d*) If the model is composed of several elements, each element should be usable alone, in simple forecasts or in a combination with others, e.g. in comprehensive computer programs.
(*e*) The model should be easily adaptable to any safety specifications, or any changes in code (say from lumped safety to partial safety coefficients).
(*f*) It should be possible at any time to benefit from collected data and past experience of case histories.

The 25 years of applied research at the Geotechnical Division at the Norwegian Institute of Technology has had, as its long-term aim, the development of engineering models of behaviour that satisfy most of the above requirements. The bulk of this Paper is therefore a summary of this model, with examples of applications to soil testing and design analyses.

The basic elements of the model were completed around 1970 for static stress conditions. The requirements in offshore engineering from the early 1970s have led to the expansion of the model to include cyclic loading. However, this is the first attempt to summarize the elements of the model in a single paper.

Resistance as a unifying concept

The classical resistance concept is used consistently during the build-up of each element of the model. The resistance is a unifying concept, being widely applied in all fields of engineering, where action–reaction systems require analysis.

All media possess resistance against a forced change of existing equilibrium conditions. The resistance of a medium, or of an isolated part of it, can therefore be determined by measuring the incremental response to a given incremental action, Fig. 2. Therefore, by definition

$$\text{Resistance} = \frac{\text{Incremental action}}{\text{Incremental response}} \tag{1}$$

For a non-linear response the resistance is in general defined as the tangent to the action–response curve. For a linear action–response curve the resistance is a constant of proportionality, without a change in definition, e.g. electrical resistance (R or ρ), elastic resistance (E), dynamic resistance (mass), hydraulic resistance (k^{-1}) and heat resistance (C).

In geotechnical engineering, an enormous amount of information is available on failure criteria and strength parameters. By comparison, little general information about the static and cyclic soil behaviour within the design stress range exists. Hopefully, a more consistent use of the resistance concept may, in time, remedy this unacceptable situation. It is readily admitted that this first attempt to summarize the wide applicability of the concept is in many ways inadequate. Nevertheless, it is hoped to spur future research and discussion of the concept.

MODEL ELEMENTS: DEFINITIONS

The purpose of this section is to define the basic elements of the soil model. Each element will be limited to two-dimensional conditions, either plane strain or axisymmetric, where the oedometer condition (no lateral yield) is a special case.

Shear strength

Free water and/or gas bubbles in the voids of mineral soils cannot transfer shear. Hence, the

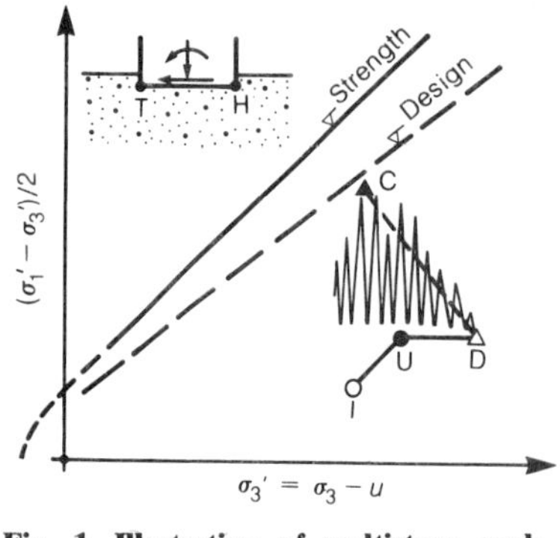

Fig. 1. Illustration of multistage problems in offshore design

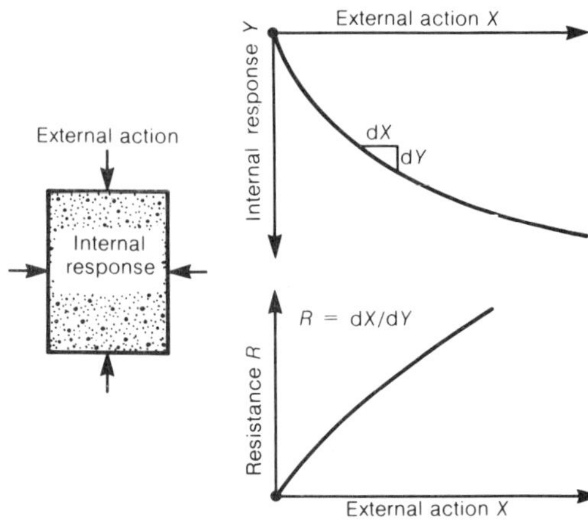

Fig. 2. Definition of the resistance of a material

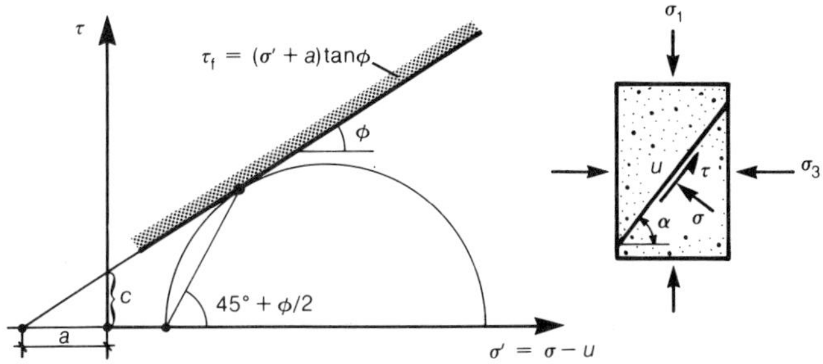

Fig. 3. Definition of linear strength in an ideal material

soil skeleton alone transfers shear stresses. Thus the effective normal stresses govern the internal shearing resistance of granular soils, irrespective of the shear stress level. Therefore the entire stress dependence of the model is based on the effective stress. Any departures from this general principle must be considered as special cases of very limited validity; see e.g. Bishop (1966) and Bishop & Wesley (1975).

The simplest expression for the shear strength τ_f in terms of the effective stress is represented by a straight line of the Coulomb form (Fig. 3)

$$\tau_f = (\sigma' + a)\tan\phi \tag{2}$$

where $\sigma' = \sigma - u$ is the effective normal stress, $\tan\phi$ is the friction of the grain skeleton and a is the attraction (cohesion $c = a\tan\phi$). In practical applications a two-dimensional model and linear strength are all that can be handled adequately.

The use of attraction, instead of cohesion, simplifies greatly all the engineering formulae, since all the coefficients related to the c terms are redundant (Janbu, 1973a). Theoretically, attraction acts as an isotropic prestress, similar to suction; see Sokolovski (1965) and Caquot & Kérisel (1967).

Classical (Hvorslev, 1937) and more recent research (Schmertmann, 1976) indicate that it is logical to expect that cohesion may properly be considered as a product of friction and stress. Since $c = a\tan\phi$ and $c = \kappa p_e$ in the Hvorslev equation, $a = \kappa p_e \cot\phi$. It appears therefore that the product $\kappa\cot\phi$ is a better constant than either κ or $\cot\phi$ alone.

It should be emphasized, however, that for practical engineering purposes it is advantageous to consider a and $\tan\phi$ not as fundamental soil properties but curve fitting coefficients. In this Paper a and $\tan\phi$ are used as effective stress parameters without any subscripts, bars or primes, as it is felt that these basic symbols should be reserved unaltered for the effective stress analyses.

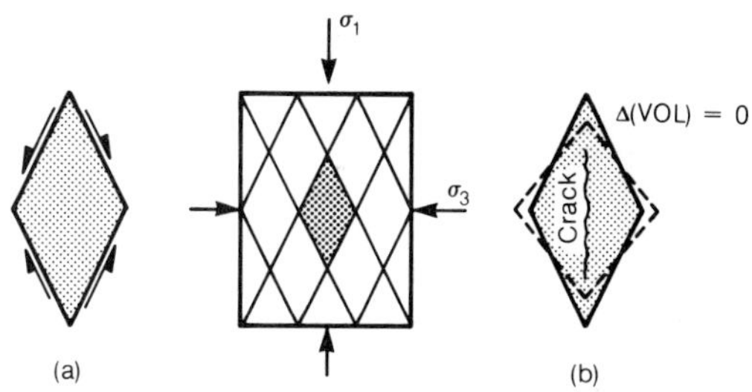

Fig. 4. Major modes of failure observed in soils, rock, concrete and ice: (a) shear failure on conjugate planes; (b) tensile strain cracks in the σ_1 direction

The theoretical case of constant shear strength ($\tau_f = s = c$) is obtained from equation (2) by means of the following limit consideration

$$s = \lim_{\tan\phi \to 0} (a\tan\phi) \tag{3}$$

In the model this case is therefore covered by the condition $\phi = 0$. Hence, the solution of the general case includes the special case.

Modes of failure

In all mineral soils, rocks and concrete two major modes of failure occur in test specimens, namely

(*a*) shear-stress-induced failure, leading to conjugate shear planes, forming rhombic failure elements with their long axis in the σ_1 direction (Fig. 4(a))
(*b*) tensile-strain-induced cracks, in the σ_1 direction, even when the whole stress field is compressive, as seen from a no-volume-change consideration (Fig. 4(b)).

The shear failures are likely to dominate in plastic (ductile) materials, while the strain-induced cracks dominate in brittle materials. Both types of failure may be visible simultaneously in a test specimen.

In dilatant materials (low sensitivity clays) failure takes time to develop. For instance, in a cut slope the soil at the crest may start to crack (giving early warning) and the soil mass may gradually slump for a limited distance and reach a new equilibrium at the displaced position.

In contrast, contractant materials often fail abruptly, without warning and often at small strains. The failed soil may lose most of its strength and become a liquid (quick clay slides, flow slides) and cause damage far outside the original failure zones by flowing downhill at very small gradients.

Degree of mobilization

From a theoretical point of view the simplest definition of the state of equilibrium τ_e in a τ–σ' diagram is of course a straight line with a slope $\tan \rho$ passing through the same origin as the strength line (Fig. 5)

$$\tau_e = (\sigma' + a) \tan \rho \tag{4}$$

where $\tan \rho$ is the mobilized friction.

Any Mohr's circle which is tangent to the equilibrium line will then have the same maximum mobilized friction, irrespective of the stress level, i.e. point C will have to move along the equilibrium line τ_e when constant ρ is assumed.

The (maximum) degree of mobilization f for such a condition becomes

$$f = \frac{\tan \rho}{\tan \phi} \tag{5}$$

As an example, the movement of point C along the equilibrium line may correspond to a drained oedometer condition, where $f_0 \approx 0{\cdot}5$–$0{\cdot}6$.

For a variable ρ point C will move vectorially, as indicated by the broken arrows in Fig. 5. The major part of the experimental results reported herein is related to the development of these vectors, for undrained and drained, static and cyclic loading conditions.

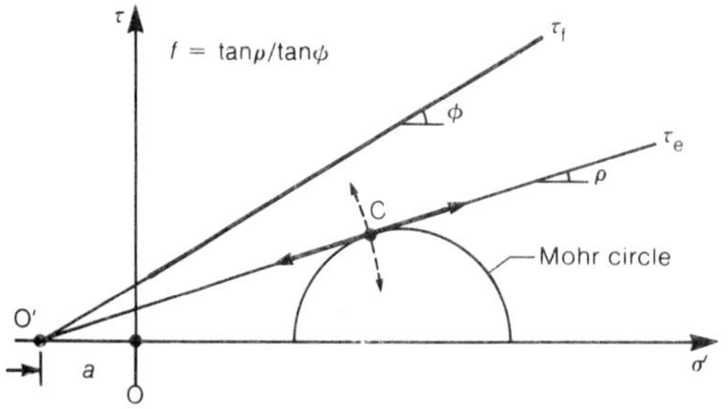

Fig. 5. Definition of the degree of mobilization f

Excess pore pressure

In consolidated undrained tests (CU tests) the total principal stress changes are known and the excess pore pressure is measured. The interpretation of such tests, in terms of pore pressure parameters (A and B) was first published in the mid-1950s (Skempton, 1954; Bishop, 1954). To include the possible effect of σ_2 it was later suggested that octahedral stresses be used in the interpretation (Henkel, 1960). However, because τ_{oct} and $\sigma_d = \sigma_1 - \sigma_3$ are nearly proportional, a simple expression can be used to include σ_2 with sufficient accuracy (Janbu, 1976)

$$\Delta u = \Delta\sigma_m - D\,\Delta\sigma_d \tag{6}$$

where D is the dilatancy parameter

$$\Delta\sigma_d = \Delta\sigma_1 - \Delta\sigma_3$$

$$\Delta\sigma_m = \tfrac{1}{3}(\Delta\sigma_1 + \Delta\sigma_2 + \Delta\sigma_3)$$

For triaxial tests on saturated soils

$$D = \tfrac{1}{3} - A \tag{7}$$

where A is Skempton's pore pressure parameter. Rewriting equation (7) D is defined as

$$D = \frac{\Delta\sigma_m'}{\Delta\sigma_d} \tag{8}$$

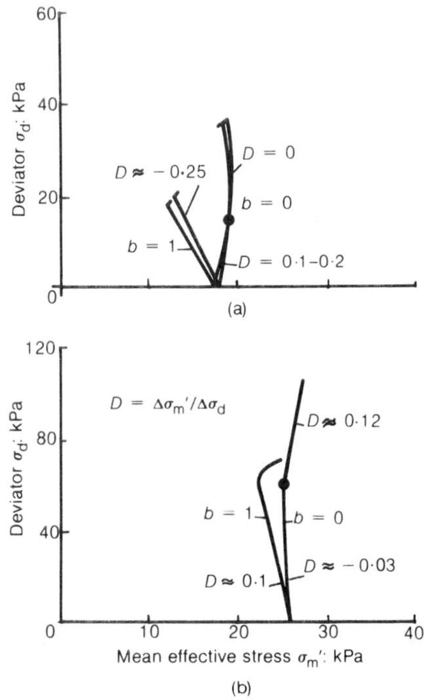

Fig. 6. Pore pressure parameter D **obtained from compression and extension tests: (a) data from Bishop & Wesley (1975); (b) data from Law & Holtz (1978)**

Hence, D expresses the tendency of a soil to change the mean effective stress when subjected to deviatoric stresses under undrained conditions. In general

$D > 0$ for dilatant behaviour

$D = 0$ for elastic behaviour

$D < 0$ for contractant behaviour

It has long been known that the effective stress behaviour of saturated soils is independent of the manner in which the total stress changes are applied to a sample. The important research data of Bishop & Wesley (1975) have been used to analyse the D parameter as shown in Fig. 6(a), while the results of Law & Holtz (1978) are included in Fig. 6(b). Note that D is calculated from the slope of the working stress path.

Normal stress–strain behaviour

For drained tests along the K_0' line (as in an oedometer) the observed stress–strain behaviour is completely described by one curve, namely the $\sigma_1' = \sigma'$ versus $\varepsilon_1 = \varepsilon$ curve. To be able to carry out fundamental studies of the nature of this curve it must be presented on an arithmetic scale. The tangent modulus to the curve is the resistance against deformation, also called the constrained modulus (Janbu, 1963)

$$M = \frac{d\sigma'}{d\varepsilon} \tag{9}$$

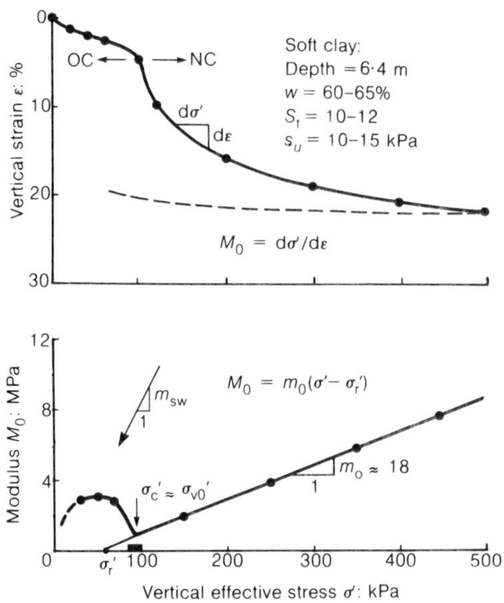

Fig. 7. Definition of oedometer modulus, e.g. for Eberg Clay

The diagrams in Fig. 7 contain the σ'–ε curves for both loading and unloading (swelling) branches from an oedometer test on a normally consolidated clay. The tangent moduli are shown in the same figure as a function of the vertical effective stress σ'.

For a low stress level on the loading branch the resistance M against deformation is large. While the stress increases this high resistance eventually decreases appreciably owing to partial collapse of the grain skeleton. This breakdown of the resistance occurs around the preconsolidation stress level σ_c'. When the effective stress is increased beyond σ_c' the resistance increases with increasing effective stress. The behaviour in the normal consolidation stress range can be approximated by a linear oedometer modulus M_0. Hence, for $\sigma' > \sigma_c'$

$$M_0 = m_0(\sigma' - \sigma_r') \tag{10}$$

where m_0 is the modulus number (say from 10 to 30) and σ_r' is the intercept on the σ' axis and is the reference stress.

Along the swelling branch the modulus decreases almost linearly with decreasing effective stress, and the average slope is equal to the swelling modulus number m_{sw}, which often is 5–10 times m_0.

The definition $M_0 = d\sigma'/d\varepsilon = m_0(\sigma' - \sigma_r')$

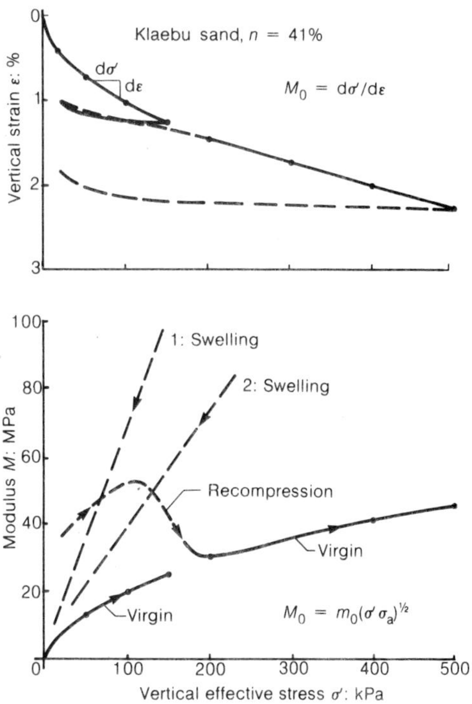

Fig. 8. Oedometer moduli in sand

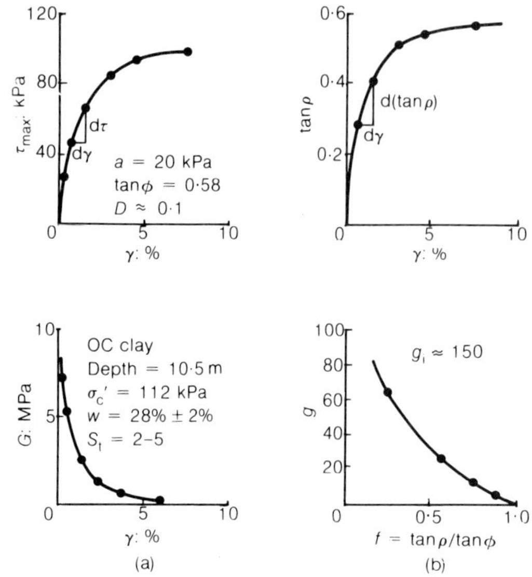

Fig. 9. Definition of shear modulus *G* and shear modulus number g, e.g. for Barnehagen Clay

leads to

$$d\varepsilon = \frac{d\sigma'}{m_0(\sigma' - \sigma_r')}$$

When integrated from σ_c' to σ'

$$\varepsilon_0 = \frac{1}{m_0} \ln\left(\frac{\sigma' - \sigma_r'}{\sigma_c' - \sigma_r'}\right) \tag{11}$$

is obtained. For a normally consolidated clay, $\sigma_c' = \sigma_{v0}'$, equation (11) implies that the simple explanation of the past experiences of linear *e* versus log σ' plots is that the tangent modulus is linearly dependent on the effective stress $M = m\sigma'$, with $\sigma_r' = 0$.

Historically, it is interesting to observe that Terzaghi (1925, pp. 94–95) found this observation (based on tests on dry powder) to be of fundamental importance (von Grundlegender Bedeutung). It is therefore surprising that Terzaghi did not come back to this discovery in later years, except for brief comment.

Figure 8 shows typical curves for preliminary loading, preliminary unloading, reloading and a second unloading of a medium dense sand tested in an oedometer. The tangent moduli variations with stress for the entire test sequence are included.

The virgin modulus curve for sand may frequently be approximated by a parabola

$$M_0 = m_0(\sigma'\sigma_a)^{\frac{1}{2}} \tag{12}$$

where $\sigma_a = 100$ kPa ≈ 1 atm is the reference stress for a dimensionless modulus number m_0. For sands m_0 will often vary between 100 and 500, from a loose to a dense state.

Introducing equation (12) into equation (9) and integrating from σ_c' to σ'

$$\varepsilon_0 = \frac{2}{m_0}\left[\left(\frac{\sigma'}{\sigma_a}\right)^{\frac{1}{2}} - \left(\frac{\sigma_c'}{\sigma_a}\right)^{\frac{1}{2}}\right] \tag{13}$$

For normally consolidated sand $\sigma_c' = \sigma_{v0}'$ is the present effective overburden.

The recompression modulus in Fig. 8 shows that the sand had been statically preloaded, and the preconsolidation modulus gradually decreases when the static preloading is exceeded.

A linear swelling modulus of sand is observed as for clays

$$M_{sw} = m_{sw}\sigma' \tag{14}$$

At times the swelling modulus may intercept the σ' axis at the reference stress σ_r'. The swelling modulus number m_{sw} is often 5–10 times m_0.

Shear stress–shear strain behaviour

In triaxial testing the shear stress is gradually increased from the starting level until failure. The stress–strain behaviour during such a test may be plotted in several ways, as shown in Fig. 9.

From an ordinary τ–γ plot the tangent shear modulus

$$G = \frac{d\tau}{d\gamma} \tag{15}$$

can be obtained where τ is the maximum shear and $\gamma = \varepsilon_1 - \varepsilon_3$ is the maximum deviator strain (often denoted ε_d). The *G* modulus will depend on γ as shown in Fig. 9(a).

The τ–γ curve can be transformed into a $\tan\rho$–γ curve as shown in Fig. 9(b), where $\tan\rho$ is the mobilized friction. The tangent to this curve is dimensionless and is denoted g, where

$$g = \frac{d(\tan\rho)}{d\gamma} \tag{16}$$

The shear modulus number g decreases with increasing mobilization *f* as shown in Fig. 9(b). Typical values of g_i for $\gamma = 0$ are in the range 50–500. Once g has been obtained, the shear modulus *G* can be expressed in terms of σ_1' as follows

$$G = g_1(\sigma_1' + a) \tag{17}$$

where $g_1 = g/N^{\frac{1}{2}}$, and *N* is defined later in equation (31).

It is simple to express *G* in terms of other stresses, like $\sigma_3' + a$ or $\sigma_m' + a$, from the general relationship $G = g(\sigma_n' + a)$ where σ_n' is the effective normal stress on critical shear planes.

Time dependence for static loads

For stepwise loading in oedometers each load step is applied instantaneously and then left constant for some time during which the sample gradually compresses. In total stress terms this is a rheological phenomenon or creep. Considering time t as action and strain as response, the time resistance is automatically defined as the tangent to the ε–t curve (Janbu, 1969)

$$R = \frac{dt}{d\varepsilon} \quad (18)$$

Each part of Fig. 10 has an arithmetic scale. The time resistance R will generally increase with time as shown in the figure.

For saturated clays it is often found that the R–t curve near the origin is parabolic, in accordance with the classical theory of primary consolidation. During this phase the initial excess porewater pressure gradually reduces. After some time the R–t curve approaches a straight line so that for $t \geqslant t_c$

$$R_s = r_s(t - t_r) \quad (19)$$

Introducing equation (19) into equation (18) and integrating between t_c and t

$$\varepsilon_s = \frac{1}{r_s} \ln\left(\frac{t - t_r}{t_c - t_r}\right) \quad (20)$$

is obtained where ε_s is the creep strain, or the so-called secondary consolidation.

The empirical observation of a straight ε_s–log t curve is in mechanical terms equivalent to a linear time resistance $R_s = r_s t$ with $t_r = 0$.

The ε–t behaviour, Fig. 10, is truly a rheological process (creep) both in terms of total and effective stress for all times $t \geqslant t_p$ where t_p is the time for complete dissipation of excess pore pressure $\Delta u = 0$. In most cases, however, t_c is only a fraction of the theoretical value of t_p.

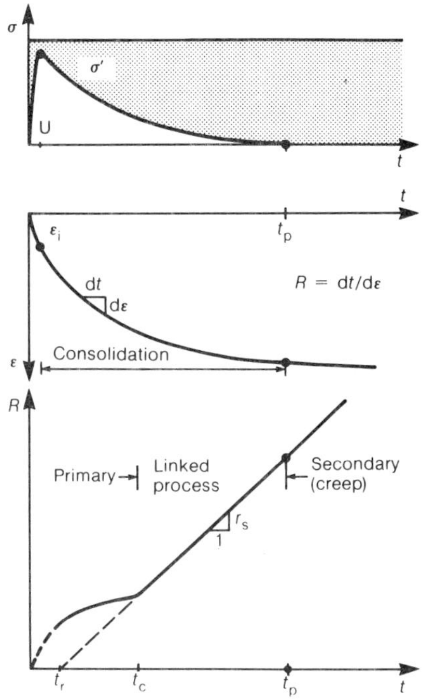

Fig. 10. Definition of time resistance R and creep number r_s for a constant load step in an oedometer

Time dependence during repeated loading

The principle applied in triaxial tests with repeated loads using saturated soils and undrained conditions is shown in Fig. 11. The sample is first consolidated to a known state of effective stress, σ_{10}' and σ_{30}'. For σ_{30} = constant the sample is exposed to repeated changes in σ_1 equal to $\Delta\sigma_1 = \Delta\sigma_d$ = constant.

For undrained conditions the soil responds to load repetitions by changes in the porewater pressure and changes in the vertical strain. Both effects are recorded continuously as a function of the number of load repetitions N. The corresponding curves are shown in Fig. 11 for a test on a soft clay from Eberg. Both curves show an average cumulative increase and an 'elastic' fluctuation of ε and u around the average cumulative trend. The width of the bands of fluctuation is almost constant ($\Delta\varepsilon_N$ and Δu_N).

Considering the number of repetitions N as action and ε as response the cumulative strain resistance R_ε against repeated loading is defined as the slope of the mean curve

$$R_\varepsilon = \frac{dN}{d\varepsilon} \quad (21)$$

In most saturated soils R_ε increases regularly with increasing repetitions. For saturated clays the increase with N is nearly linear, as shown in Fig. 11, where

$$R_\varepsilon = r_\varepsilon(N - N_r) \quad (22)$$

The intercept N_r is difficult to determine precisely and in many cases it is advisable to use $N_r \approx 0$, if possible.

If, for simplicity, $N_r = 0$ and $R = r_\varepsilon N = dN/d\varepsilon$, $d\varepsilon = dN/Nr_\varepsilon$ is obtained. Hence, integration

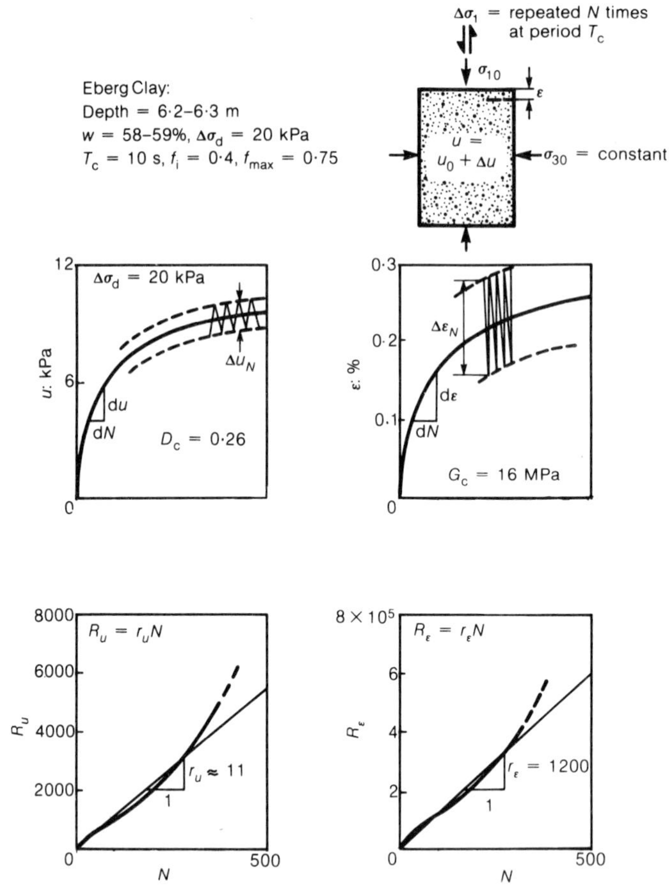

Fig. 11. Definition of cumulative strain and pore pressure resistances R_u and R_ε exemplified by cyclic load tests on a soft clay from Eberg

from $N=1$ to N leads to the cumulative strain

$$\varepsilon_{cu}=\frac{1}{r_\varepsilon}\ln N \qquad (23)$$

From the constant width of fluctuation $\Delta\varepsilon_N$ the cyclic shear modulus

$$G_c=\frac{\Delta\sigma_d}{3\,\Delta\varepsilon_N} \qquad (24)$$

is obtained for $\Delta\sigma_d=$ constant. This value is equivalent to the G_{max} used in vibration analyses for small strains.

The cumulative pore pressure resistance R_u against repeated loading is defined as

$$R_u=\Delta\sigma_d\frac{dN}{du} \qquad (25)$$

where N is the action and u is the response. The introduction of $\Delta\sigma_d$ is to obtain a dimensionless resistance R_u.

In most soils R_u increases gradually with increasing N. In saturated clays the increase is usually linear, in which case

$$R_u=r_u(N-N_r) \qquad (26)$$

The dimensionless pore pressure resistance r_u for clays is almost constant for a wide range of mobilizations, from 0 to 0·8, as will be shown later.

If the R_u–N curve is fitted for small N values so that $N_r\approx 0$, $R_u=r_uN=\Delta\sigma_d\,dN/du$, or $du=\Delta\sigma_d\,dN/N$ is obtained. Integration leads to

$$u_{cu}=\frac{\Delta\sigma_d}{r_u}\ln N \qquad (27)$$

From the constant width of fluctuation Δu_N the

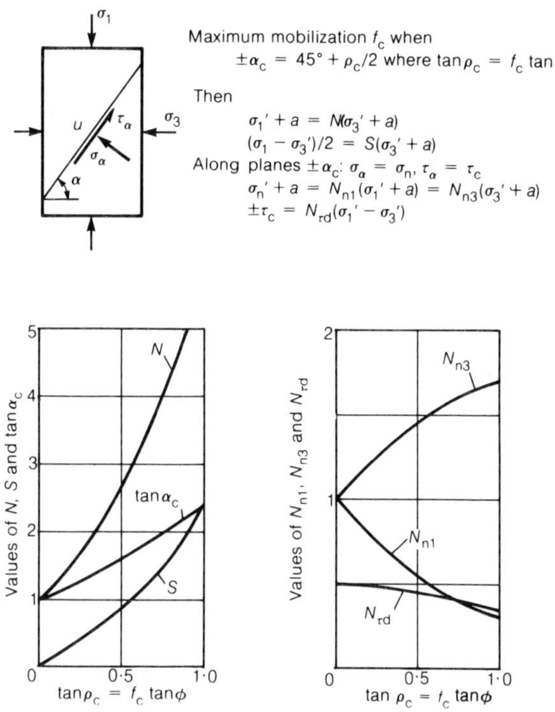

Fig. 12. Stress ratios and inclination of conjugate shear planes in a triaxial test

cyclic D parameter

$$D_c = \frac{1}{3}\frac{\Delta u_N}{\Delta\sigma_d} \tag{28}$$

is obtained where $\Delta\sigma_d = \text{constant}$.

Comments on soil resistances

For normally consolidated clays the stress resistance and the time-dependent resistance are often linearly dependent on stress and time respectively. For example, $M_0 = m_0\sigma'$ and $R = r_s t$ are primary and secondary resistances respectively.

From cyclic tests on clay linear resistances are also obtained for the cumulative strain and the cumulative excess pore pressure. For example, $R_\varepsilon = r_\varepsilon N$ and $R_u = r_u N$ where N is the number of repetitions of a constant change in deviator stress $\Delta\sigma_d$.

In Appendix 1 it is shown that a linear resistance will always lead to a semilogarithmic relationship between response and action, e.g.

$$\varepsilon_0 = \frac{1}{m_0}\ln\left(\frac{\sigma'}{\sigma_0'}\right) \qquad \varepsilon_s = \frac{1}{r_s}\ln\left(\frac{t}{t_0}\right)$$

$$u_{cu} = \frac{\Delta\sigma_d}{r_u}\ln\left(\frac{N}{N_0}\right) \qquad \varepsilon_{cu} = \frac{1}{r_s}\ln\left(\frac{N}{N_0}\right)$$

For simplicity these formulae are based on reference values of zero ($\sigma_r' = 0$, $t_r = 0$, $N_r = 0$) in interpreting the test results. This simplification may not always be possible in practice, and the formulae are then altered accordingly, Appendix 1.

For other conditions the test results may be better approximated by non-linear resistances, in which case one could use a generalized relationship as in Appendix 1: as special examples, $M = m(\sigma'\sigma_a)^{\frac{1}{2}}$ for sand and $M = m\sigma_a = \text{constant}$ for overconsolidated clays.

The dimensionless resistance numbers m and r depend primarily on shear mobilization, as well as on the stress regime (overconsolidated versus normally consolidated). Stresses in soil will therefore be given some consideration before results of multistage testing are presented.

ELEMENTARY STRESS FIELD THEORY

A prerequisite for being able to interpret triaxial test results within the design stress range as well as along the failure envelope is to have simple expressions for the state of stress at any equilibrium condition. For this a consistent tool for stress vectors (or stress paths) which is equally applicable for either the design stress or failure stress conditions is required.

Similarly, to judge the range of subsoil stresses (for which the tests should be run) simple tools are required to express the state of equilibrium stress in the subsoil. Simple stress field theories have been developed for this.

Stresses in triaxial samples

Let the total principal stresses in a triaxial specimen be given as $\sigma_1 > \sigma_2 = \sigma_3$ while the pore pressure is u, Fig. 12. Along a plane inclined at an arbitrary angle α the shear stress τ_α and the normal stress σ_α are well-known expressions from classical mechanics. Hence, an expression for the mobilized friction $\tan \rho$ and/or the degree of mobilization f (Janbu, 1973a) can easily be obtained. This mobilization is a function of α. The maximum value of the mobilized $\tan \rho = f_c \tan \phi$ is obtained for $\alpha = \alpha_c$, where

$$\pm\alpha_c = 45° + \tfrac{1}{2}\rho \qquad (29)$$

or

$$\pm\tan \alpha_c = \tan \rho + (1 + \tan^2 \rho)^{\frac{1}{2}}$$

This means that the critical equilibrium condition (for maximum mobilization) occurs along conjugate shear planes as shown in Fig. 4. The geometry of the rhombic elements bounded by these planes changes with changing mobilization, from squares for $f_c = 0$ to the well-known failure elements with angles $90° \pm \phi$ for $f_c = 1$.

The maximum mobilized friction $\tan \rho$ can now be determined for any given state of effective principal stress σ_1' and $\sigma_3' = \sigma_2'$ from the following *normal stress criterion*

$$\sigma_1' + a = N(\sigma_3' + a) \qquad (30)$$

where

$$N = \tan^2 \alpha_c = \frac{1 + \sin \rho}{1 - \sin \rho} \qquad (31)$$

This state of equilibrium can also be expressed in terms of the maximum *shear stress criterion*

$$\tfrac{1}{2}(\sigma_1' - \sigma_3') = S(\sigma_3' + a) \qquad (32)$$

in which

$$S = \tfrac{1}{2}(N - 1) = \tan \rho \tan \alpha_c \qquad (33)$$

The values of N and S as functions of the mobilized friction are plotted in Fig. 12.

The critical normal stress σ_α' along the conjugate shear planes can be expressed as follows

$$\sigma_\alpha' + a = N_{n1}(\sigma_1' + a) = N_{n3}(\sigma_3' + a) \qquad (34)$$

where

$$N_{n1} = \frac{2}{N + 1} = 1 - \sin \rho \qquad (35a)$$

$$N_{n3} = \frac{2N}{N + 1} = 1 + \sin \rho \qquad (35b)$$

The shear stresses along the conjugate shear planes are

$$\pm\tau_\alpha = (\sigma_\alpha' + a)\tan \rho = N_{\tau d}(\sigma_1 - \sigma_3) \qquad (36)$$

where

$$N_{\tau d} = \frac{N^{\frac{1}{2}}}{N + 1} = \frac{1}{2}\cos \rho$$

These formulae are the basis for stress vector interpretations of triaxial tests as well as the basis for stress fields in weightless soils.

Stress vectors for triaxial test interpretation

The normal stress criterion, equation (30), is a simple straight line in a Cartesian co-ordinate system with axes σ_1' and σ_3'. Since $\sigma_1' > \sigma_3'$ the line of hydrostatic stress $\sigma_1' = \sigma_3'$ represents the lower boundary in Fig. 13, along which $\tan \rho = 0$, i.e. $N = 1$. The upper boundary (at failure) is defined by the slope N_f where $\rho = \phi$.

From the slope N_f of the failure line the friction $\tan \phi$ can be obtained from the formula

$$\tan \phi = \frac{N_f - 1}{2N_f^{\frac{1}{2}}} \qquad (37)$$

For an ideal material with constant attraction a over the entire compressive stress range, the failure line intersects the hydrostatic line at point O′ with co-ordinates $(-a, -a)$. This means that the attraction is in theory the hydrostatic tensile strength of the material, Sokolovski (1965). In stress expressions the attraction acts as an isotropic prestress. The failure line intersects the σ_1 axis at point C, where the ordinate OC is the unconfined compressive strength σ_{uc}. From equation (30)

$$\sigma_{uc} = (N_f - 1)a \qquad (38)$$

Similarly, the unconfined tensile strength σ_{ut} is defined as the intercept on the σ_3' axis, hence

$$\sigma_{ut} = -\frac{N_f - 1}{N_f}a \qquad (39)$$

In the region TOC the principal stresses are of opposite signs, while in region TOO′ both stresses are negative (tensile). It should be emphasized that the combined region COO′T cannot be fully utilized in practice because most materials show non-Coulomb behaviour in

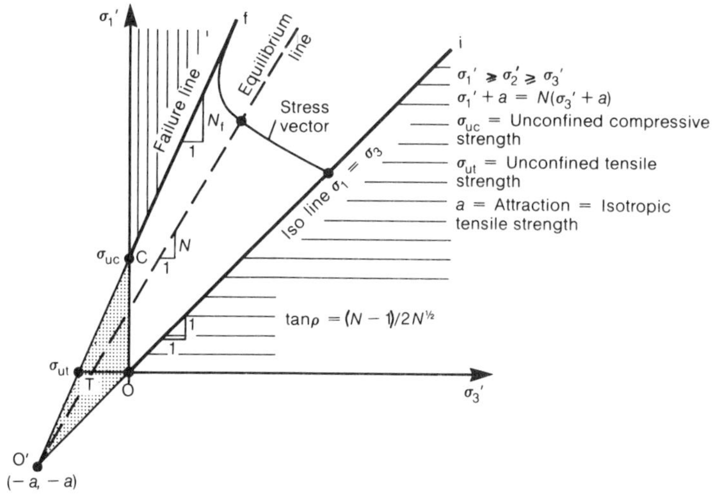

Fig. 13. Normal stress criterion for two-dimensional stress conditions in ideal Coulomb materials

tension (anisotropy, brittleness, low tensile strength).

The purely compressive region is defined by the open zone fCOi in Fig. 13. Within the zone a possible stress vector is indicated, together with a line of equilibrium for constant $\rho<\phi$.

The principal stress diagram in Fig. 13 is impractical for use in triaxial test interpretations, while the shear (or deviator) stress criterion is ideal for that purpose, Fig. 14. In the Cartesian co-ordinate system, $(\sigma_1'-\sigma_3')/2$ versus σ_3', the failure line has an inclination S_f from which the friction

$$\tan\phi=\frac{S_f}{(1+2S_f)^{\frac{1}{2}}} \tag{40}$$

is obtained. The attraction is equal to the negative intercept on the σ_3' axis. Note that the ordinate $(\sigma_1'-\sigma_3')/2$ is always positive since $\sigma_1'\geqslant\sigma_3'$ by definition. This means that both compression and extension test results are located within the triangle formed by the τ_f line and the σ_3' axis.

An arbitrary equilibrium line is also shown in Fig. 14. In particular, for the at-rest condition $(\varepsilon_2=\varepsilon_3=0)$ where the slope of the line is S_0

$$K_0'=\frac{1}{1+2S_0} \tag{41}$$

is obtained. The definition of K_0' has to include attraction as follows

$$K_0'=\frac{\sigma_{30}'+a}{\sigma_{10}'+a} \tag{42}$$

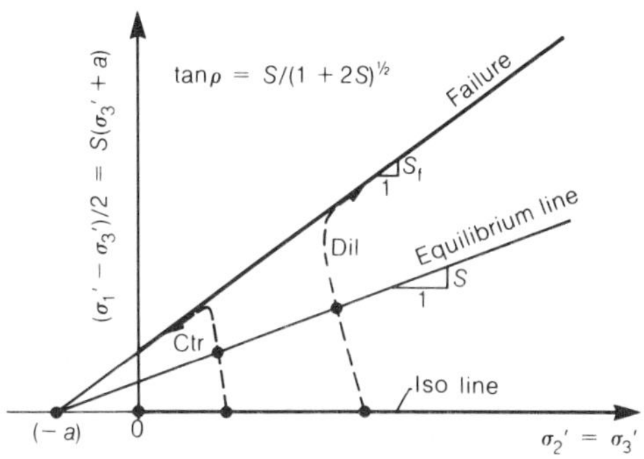

Fig. 14. Deviator stress criterion for two-dimensional stress conditions in ideal Coulomb materials

It should be noted that this definition differs from the conventional $K_0=\sigma_3'/\sigma_1'$.

A combination of a drained K_0' test with a subsequent undrained shear test in a triaxial cell led to an oedo-triaxial test, from which drained settlement parameters and undrained equilibrium shear behaviour parameters are obtained as well as shear strength parameters from the same sample.

Total stress field

The basic elements in stress field theories for weightless soils are briefly summarized in Appendix 2. The case of inclined loads on a horizontal base is used as an example of an application.

Plane strain conditions are assumed. The inclined load on the base is given by the vertical

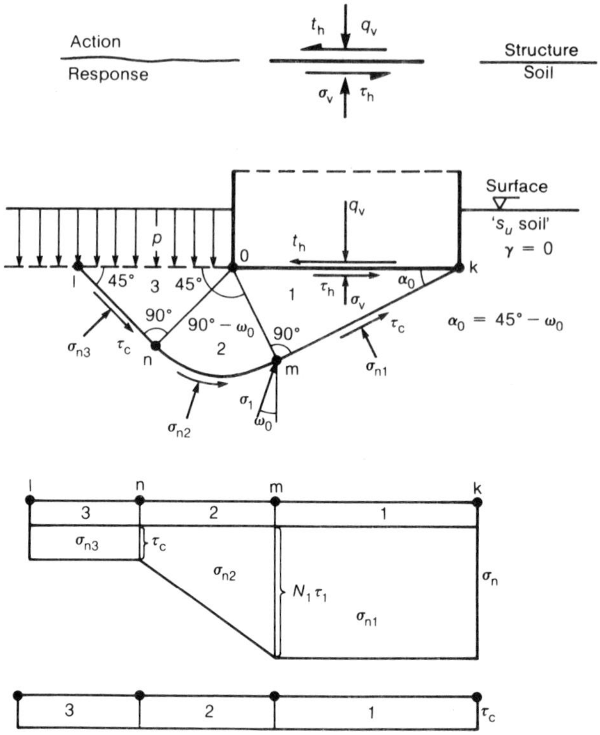

Fig. 15. Total stress fields for inclined loads on weightless soils

component q_v and the horizontal base shear t_h, both uniformly distributed over the width of the foundation, Fig. 15.

The two components q_v and t_h represent the *action* on the subsoil. The soil responds by setting up shear and normal stresses. For a weightless soil with a defined shear stress level there is just one geometry of the stress field, which will be statically and kinematically correct and in complete equilibrium with the action. This field may be called the critical stress field. In Fig. 15 it is assumed that the shear stress at any level of mobilization is independent of normal stress. Moreover, the degree of mobilization is assumed to be constant within the critical stress field. This means that the critical shear stress on all conjugate shear planes within the stress field is assumed to be constant and equal to τ_c. For these assumptions the solution for the equilibrium condition (action $q_v t_h$ = reaction $\sigma_v \tau_h$) is simple.

The stress responses from the subsoil are obtained from the horizontal and vertical equilibria, leading to

$$\tau_h = r\tau_c \qquad (43)$$

$$\sigma_v = p + N_c\tau_c \qquad (44)$$

where

$$r = \sin(2\omega_0) \qquad (45)$$

$$N_c = \pi + 1 - 2\omega_0 + \cos(2\omega_0) \qquad (46)$$

Here ω_0 is the angle of rotation of the principal axis in zone 1.

The equilibrium conditions $\tau_h = t_h$ and $\sigma_v = q_v$ can be satisfied by means of equations (43) and (44) by trial and error through r. First, a value of r is assumed, leading to $\tau_c = t_h/r$, and then equation (44) gives one value of σ_v. By trying several values of r and by plotting σ_v versus r, $r = r_c$ can be determined for $\sigma_v = q_v$. The angle α_0 in zone 1 is then obtained by

$$\tan\alpha_0 = \left(\frac{1 - r_c}{1 + r_c}\right)^{\frac{1}{2}} \qquad (47)$$

The entire geometry of the critical stress field is known when α_0 is calculated; see Fig. 15. The normal stresses along this boundary are given by

$$\sigma_{n3} = p + \tau_c \quad \text{(zone 3)} \qquad (48)$$

$$\sigma_{n1} = p + N_1\tau_c \quad \text{(zone 1)} \qquad (49)$$

where

$$N_1 = \pi + 1 - 2\omega_0 \qquad (50)$$

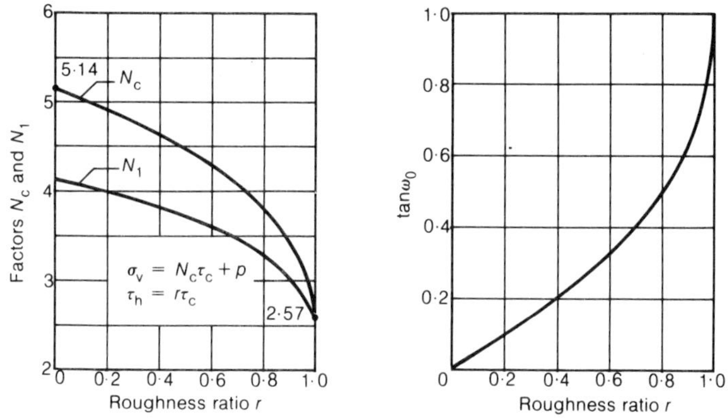

Fig. 16. Stress factors N_c and N_1 and angle of principal stress rotation ω_0 as functions of roughness ratio r

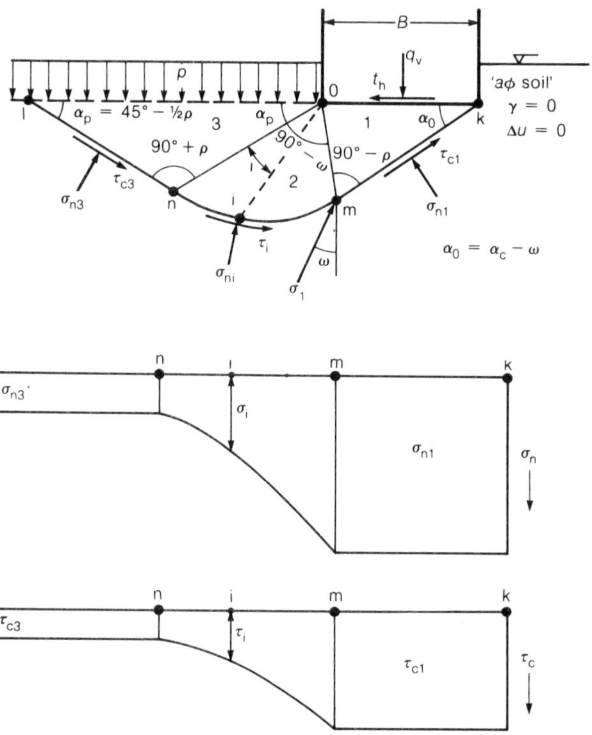

Fig. 17. Effective stress fields for inclined loads on weightless soils

In zone 2 the normal stress varies linearly along the arch, when folded out. The values of N_c, N_1 and $\tan\alpha_0$ are shown as functions of r in Fig. 16. All the formulae are also valid for negative ω_0 between 0° and −45°, corresponding to a negative t_h.

Effective stress fields

Here, the shear stress on conjugate shear planes is assumed to be linearly dependent on the effective normal stress on these planes, i.e.

$$\tau_c = (\sigma' + a)\tan\rho \quad (51)$$

The degree of mobilization is assumed to be constant, i.e. $\tan\rho = f \tan\phi$, where f = constant within the entire stress field.

For a weightless soil, the exact solution for the stress field geometry, and the magnitudes of stresses within this stress field, is easily obtained for a given inclined loading condition, Fig. 17. The horizontal and vertical equilibrium requirements lead to

$$\tau_h = r(\sigma_v' + a)\tan\rho \quad (52)$$

$$\sigma_v' + a = N_q(p' + a) \quad (53)$$

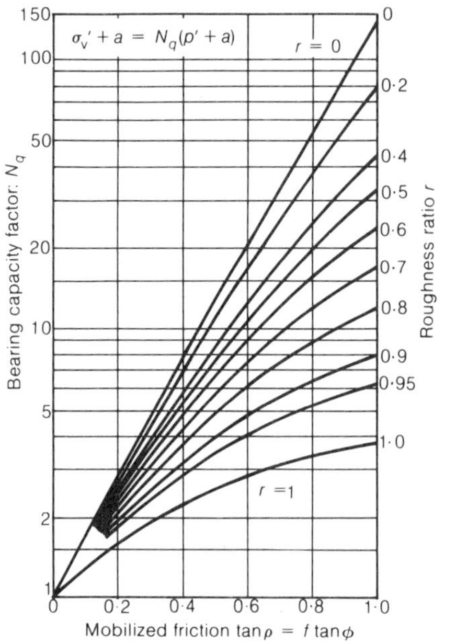

Fig. 18. Bearing capacity factor N_q

corresponding to no excess pore pressure (known stationary, hydrostatic pore pressure condition or $u = 0$). The derivation of the formulae is briefly shown in Appendix 2. The value of N_q is plotted in Fig. 18 as functions of $\tan\rho$ and r.

The geometry of the critical stress field is uniquely defined by rotation ω of the principal stresses in zone 1, when

$$\tan\omega = \left[\frac{1-(1-r^2)^{\frac{1}{2}}}{r}\right]\tan\alpha_c = \tan\omega_0 \tan\alpha_c \quad (54)$$

where

$$\alpha_c = 45° + \tfrac{1}{2}\rho$$

The normal stress along the boundary of the stress field is given by the formulae

$$\sigma_{n1} - p = \lambda_{n1}(q_v - p) \quad \text{(zone 1)} \quad (55a)$$

$$\sigma_{n3} - p = \lambda_{n3}(q_v - p) \quad \text{(zone 3)} \quad (55b)$$

The coefficients λ_{n1} and λ_{n3} are plotted in Fig. 19(a) as functions of $\tan\rho$ and r. Along the arch nm (zone 2) in Fig. 17 σ_{n1}' at point i is calculated as follows

$$\sigma_{ni}' + a = (\sigma_{n3}' + a)\exp(2i\tan\rho) \quad (56)$$

Once the normal stress σ_n' has been determined along the critical shear surface (CSS) the critical shear stress τ_c is given by $(\sigma_n' + a)\tan\rho$.

For the analyses of excess pore pressure under undrained conditions the stress changes $\Delta\sigma_m$ and $\Delta\sigma_d$ are required. Stress field theory gives

$$\Delta\sigma_m = \lambda_m \Delta q \quad (57)$$

$$\tfrac{1}{2}\Delta\sigma_d = \lambda_\tau \Delta q \quad (58)$$

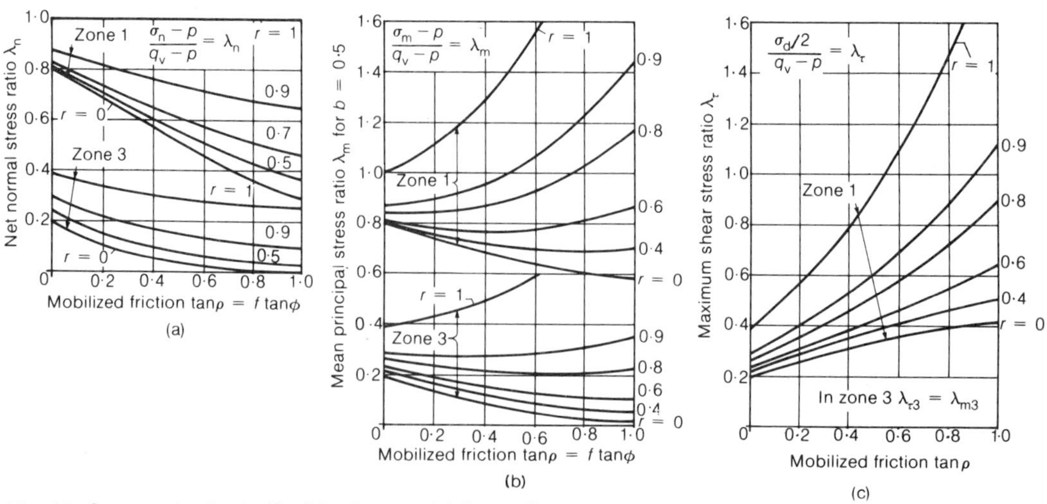

Fig. 19. Stress ratios for inclined loads on weightless soils

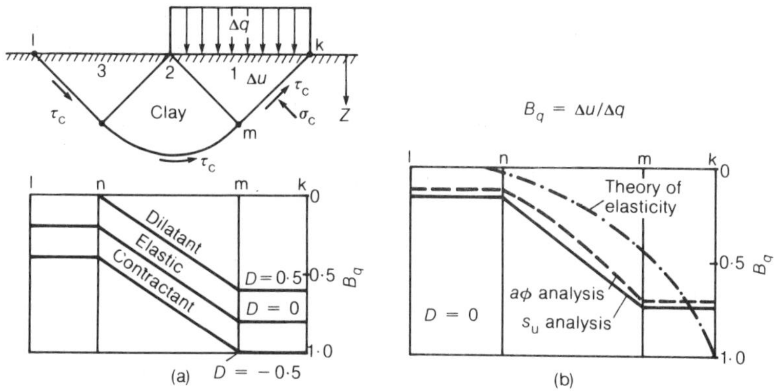

Fig. 20. Excess pore pressure due to a vertical strip load on a weightless soil

In these equations $\Delta q = q_v - p$ and $\Delta\sigma_m = \sigma_m - p$. The values of λ_m and λ_τ in zones 1 and 3 are plotted in Figs 19(b) and 19(c), where λ_m is valid for $\sigma_2 = (\sigma_1 + \sigma_3)/2$ only.

Example

Equations (57) and (58) are used to obtain the pore pressure ratio $\Delta u/\Delta q$ for a strip load on clay, leading to

$$\frac{\Delta u}{\Delta q} = \lambda_m - 2\lambda_\tau D$$

With the aid of Figs 19(b) and 19(c) the values of λ_m and λ_τ can be obtained for known values of r and $\tan\rho$. For vertical loading, the result is plotted in Fig. 20. Note that the excess pore pressure is usually less than the excess load, say 80–60% in medium–stiff clays.

MULTISTAGE LABORATORY TEST RESULTS

Today's advanced laboratory equipment and data systems have made it possible to determine a large number of important soil properties that 10 years ago were out of reach. This applies particularly to behaviour parameters within the working stress range, for both static and cyclic loading.

The principle of the potential of today's laboratory testing and interpretation procedure is illustrated in Fig. 21. For illustration idealized stress paths are shown for

(*a*) an isotropically consolidated undrained triaxial test (CIU) starting at point I
(*b*) an anisotropically consolidated undrained triaxial shear test (CAU) starting at point A
(*c*) one oedo-triaxial test containing both a drained oedo-path AD and the undrained shear path CAU starting at point D

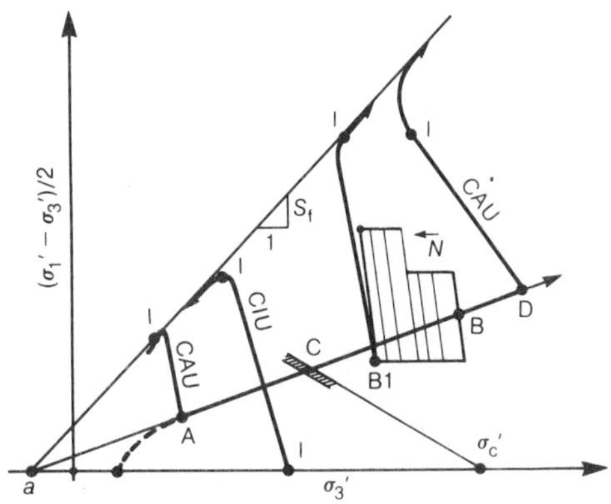

Fig. 21. Potential of multistage testing

(*d*) an undrained cyclic test, starting at point B and containing two stress blocks, after which an undrained static shear test is performed from point B1.

Multistage triaxial testing has been extensively used for more than 10 years, and examples of the results illustrate its potential for future research.

Static triaxial test interpretations

The results of four ordinary CIU tests on compacted sand are shown in Fig. 22(a). The porosity of the samples varied slightly between 39·7% and 40·5%. Stress paths of the four tests are plotted and the ultimate shear strength parameters obtained are

$$\tan\phi = 0{\cdot}75$$
$$a = 20\text{–}25\ \text{kPa}$$

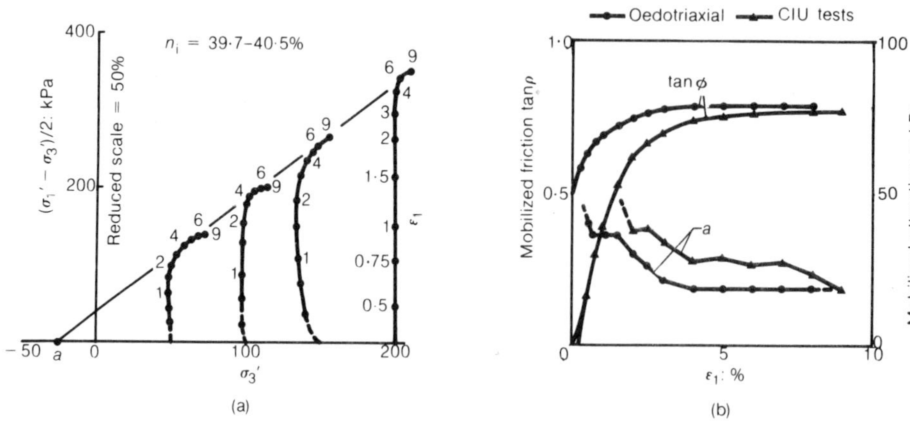

Fig. 22. Mobilization of friction and attraction in dense sand

Moreover, the D parameter is constant in each test and varied between 0·27 and 0·31 in the test series for strains up to 2%.

Strain levels ε_1 along each stress path are illustrated and used to study the mobilization of friction and attraction during loading. By drawing a line through approximately equal strains an a value and a mobilized friction tan ρ can be determined for the strain level. Their variations with strain level are shown in Fig. 22(b).

For the CIU tests the mobilized friction increases linearly up to about 1·5% strain. Then it curves towards an ideal plastic flow limit, where tan ρ = tan ϕ for strains beyond 4%. The attraction is high at small strains, and then it decreases with strain and reaches a constant value for large strains (a = 20–30 kPa). These findings are supported by comparing data in Fig. 22(b) from an oedo-triaxial series reported by Bakken & Westerlund (1974) and conducted on sand at similar porosities.

Figure 23(a) illustrates stress path results of CIU tests conducted on undisturbed Risvollan Clay. For stress path A the shear modulus number g is calculated for several strain increments using equation (16). The friction mobilization f is also calculated and the variation in g with f is shown in Fig. 23(b). The interpretation of the results along path A is as follows.

The D parameter is constant at 0·05 for the failure envelope. Plastic failure occurs for strains between 2% and 10%, with an attraction a = 8 kPa and a friction tan ϕ = 0·56. The sample contracts throughout shearing. The shear modulus number g decreases with increasing mobilization from 200 at f = 0, through 25 at the K_0' conditions to zero at f = 1.

Figure 23 includes, for the same clay, a composite stress path B, containing a compression

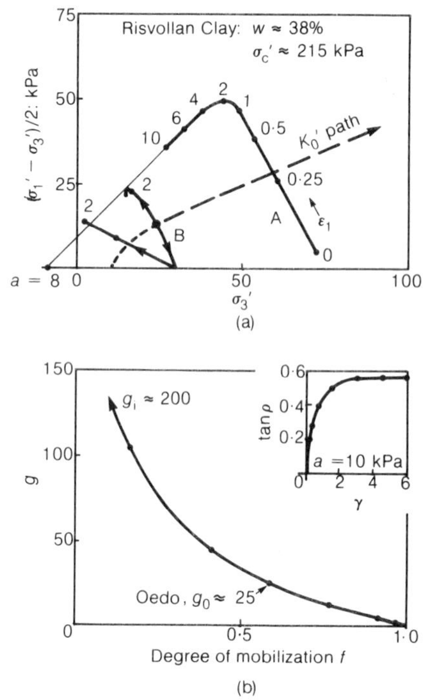

Fig. 23. Undrained shear development in a medium clay

path from the K_0' line, and an extension path from the same point. Note that $\sigma_d = \sigma_1' - \sigma_3'$ is always positive, since $\sigma_1' \geqslant \sigma_3'$. Both branches of path B terminate at approximately the same failure envelope as path A. Moreover, the D parameter is nearly the same on average, although it varies along path B.

On the left-hand side of Fig. 24 are shown stress paths and mobilization curves for CIU tests on Barnehagen Clay ($w \approx 28$–30%). The four tests give a value of tan ϕ of $0{\cdot}58 \pm 0{\cdot}02$ for an average attraction $a \approx 15$ kPa. The lower diagrams show g values versus mobilization. The curves start from 300 at $f=0$ and decrease to 25–35 at K_0' and zero at $f=1$.

The right-hand side of Fig. 24 shows the results of seven special multistage triaxial tests

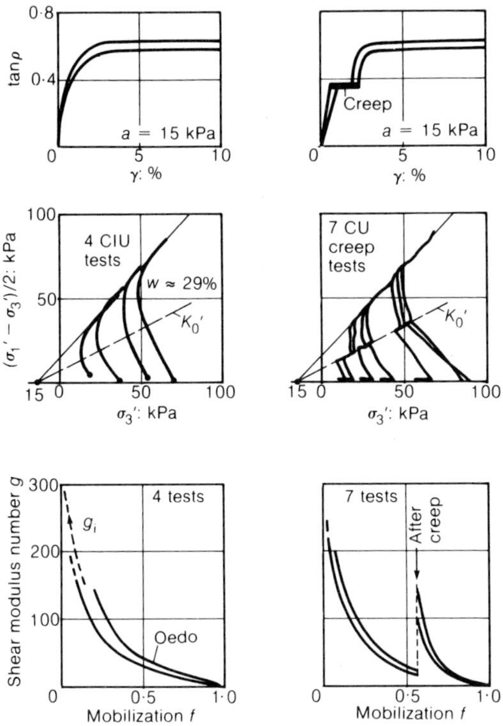

Fig. 24. Influence of undrained creep on shear modulus

on the same clay. First the undisturbed samples were isotropically consolidated to a small stress level (below σ_c'). The samples were left to creep for 2 hours under undrained isotropic stress conditions. Since the pore pressure decreased (the soil dilates) the path moves to the right along the σ_3' axis. Then the samples were sheared undrained to the K_0' line. Here, each sample was left for 16 hours to creep under undrained conditions and then sheared to failure under undrained conditions. The seven multistage stress paths and the corresponding mobilization curves are shown in Fig. 24, together with the g–f curves, represented by a range of variation covering all seven tests. From the plots the following observations can be made. The ultimate shear strength parameters from the seven multistage paths are the same as those obtained from the four ordinary CIU tests. The g–f curves also correspond from $f=0$ to the K_0' condition. However, the dramatic increase in the shear modulus that occurs due to the 16 hours of undrained creep at K_0' is a very interesting and fundamental observation.

Shear strength and shear moduli: statistics

For saturated clays obtained from a given geological region friction is almost uniquely dependent on the natural water content. The attraction, however, for a given water content may depend on several factors, such as preconsolidation, cementation, rate of testing, loading versus unloading and short-term versus long-term conditions. The consistency of the effective shear strength parameters from one geological region is illustrated in Fig. 25. Most Scandinavian clays, except perhaps the high plastic organic clays are included in the evaluation. The variation in friction is particularly moderate.

In natural deposits of dense sand and gravel,

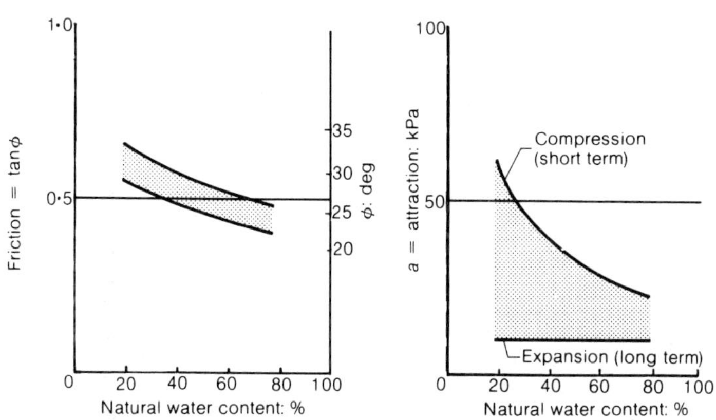

Fig. 25. Typical variations in attraction and friction for Scandinavian clays

and in compacted granular soils, attraction is virtually an expression for the dilatant behaviour of these soils. In fact, the effect of compaction is to increase attraction along with a slight increase in friction.

Triaxial test results for dense sand and compacted crushed rock are often reported in diagrams showing secant friction angle versus effective stress level. Such a presentation is impractical since it assumes *a priori* that $a = 0$ (or $c = 0$). If the same results are plotted in a conventional τ–σ' diagram it is often found that the strength envelope is slightly curved within the applicable stress range and in practice can adequately be defined by one value for attraction and friction.

Typical laboratory results obtained for effective shear strength parameters of clay, silt and sand are shown in Table 1. These results are meant as a guide-line for average values and do not account for the extreme variations. For instance, just a small percentage of clay particles in a silty sand may reduce friction substantially. Organic content acts in a similar way. Moreover, the values of attraction are primarily related to compression only.

It is advisable to present values of friction ($\tan \phi$ instead of ϕ) because friction is the mechanical property that is measured, whereas the angle is not measured, and it is not a mechanical property.

Figure 26(a) contains typical values of the shear modulus number g given as a range between a soft and a dense state for granular soils. The effect of increasing g during undrained creep is also illustrated. It should be noted that further research is needed on this behaviour. For instance, it may be possible that the G modulus for sand may be more adequately represented by a parabolic dependence on effective stress.

Figure 26(b) shows in situ values of G. For comparison the limits of $g = 150$–200 are drawn. The range of some pressuremeter (PM) tests are also indicated for comparison.

Cyclic triaxial tests

In the Author's opinion the behaviour of soil exposed to cyclic loading cannot be rationally explained unless it is interpreted in terms of effective stress.

As an example the results of a multistage,

Table 1. Typical effective strength parameters

Soil	Attraction (a): kPa			Friction ($\tan \phi$)		
Clay	5–10	15–25	30–60	0·40	0·50	0·60
Silt	0	0–10	10–20	0·50	0·60	0·70
Sand (moraine)	0	0–15	15–40	0·60	0·70	0·80
State	Soft Loose	Medium	Stiff Dense	Soft Loose	Medium	Stiff Dense

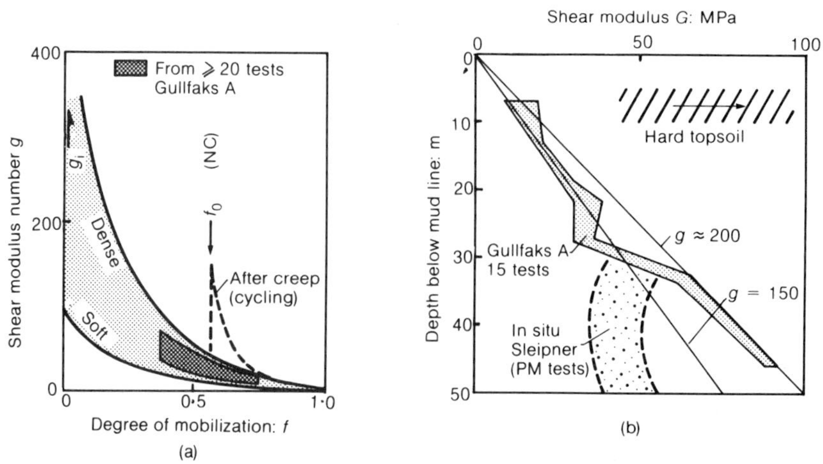

Fig. 26. Shear modulus variations with mobilization and depth: (a) obtained by static CIU tests; (b) CU tests after cycling

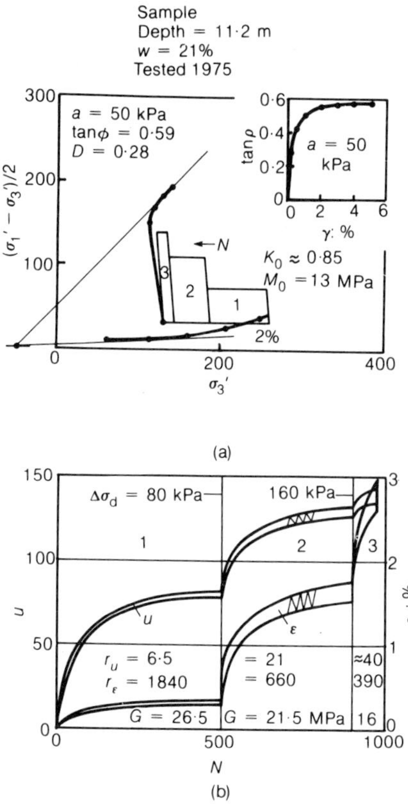

Fig. 27. Multistage static–cyclic tests on clay from the Statfjord field

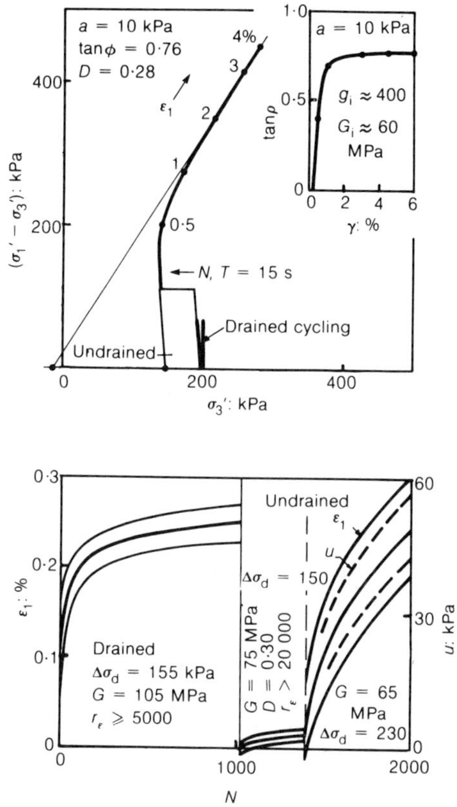

Fig. 28. Multistage static–cyclic tests on dense sand from the Gullfaks field

static–cyclic triaxial test on an overconsolidated clay from the Statfjord A field in the North Sea are shown in Fig. 27. The oedo-path ($\varepsilon_2 = \varepsilon_3 = 0$) leads to an almost constant modulus $M_0 = 13$ MPa, and a K_0' value near 1·0 to begin with, resulting in roughly 0·85 on average. The oedo-path stops at an effective stress level less than one-third of the in situ preconsolidation pressure ($\sigma_c' \geqslant 800$ kPa). The sample was then exposed to repeated loading under undrained conditions at three different deviator stress increments: $\Delta\sigma_d = 80$ kPa, $\Delta\sigma_d = 160$ kPa and $\Delta\sigma_d = 216$ kPa. The cyclic period is 10 s within each of the three constant stress blocks. The total value of σ_3 is kept constant. After the cyclic loading, in three stages, an undrained shear test was performed to failure. Fig. 27(a) contains the oedo-path, the multistage cyclic test paths, the stress path for the final shear test and its mobilization curve.

Effective shear strength parameters of $a = 50$ kPa and $\tan\phi = 0{\cdot}56$ are obtained from the triaxial test after cycling. This result is nearly the same as the average value found for the same clay from several ordinary CIU tests without cycling. Hence, the effective shear strength parameters are not changed appreciably by the cyclic test for this clay. Moreover, the static D parameter after cycling is almost identical with the dynamic D parameter during cycling.

Figure 27(b) shows how the pore pressure and the strain accumulate during the cyclic tests for each stress level. The width of the shaded bands represent the fluctuations of u and ε about the average cumulative curves. In the figure, the data have been smoothed without altering appreciably the engineering results.

The dynamic shear modulus drops slightly with increasing mobilization (from 26·5 MPa via 21·5 MPa to 16 MPa). The cumulative strain resistance r_ε drops substantially with increasing mobilization (from 1840 via 660 to 390), while the cumulative pore pressure resistance increases (from 6·5 via 21 to roughly 40) because the sample dilates more and more as failure is approached. The failure itself is dilatant.

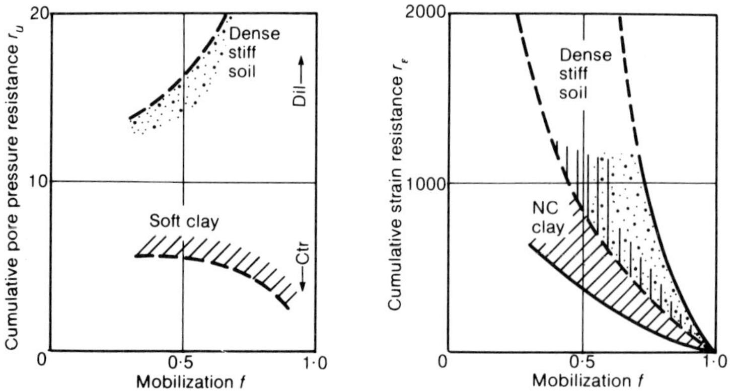

Fig. 29. Variations in cumulative strain and pore pressure resistances

The results of a multistage static–cyclic triaxial test series on a very dense, well-graded sand (moraine) from the Gullfaks A field are shown in Fig. 28. After isotropic consolidation to 200 kPa, the sample was exposed to a drained cyclic test series, containing 1000 repetitions of $\Delta\sigma_d = 155$ kPa with a period of 15 s. Approximately the same deviator stress (150 kPa) was then repeated 380 times undrained, after which 680 repetitions were carried out with $\Delta\sigma_d = 230$ kPa, also undrained. Finally, an undrained shear test to failure was carried out, and its mobilization curve is included in the figure.

The mobilization curve shows a stiff behaviour, where ideal plastic yield is reached at approximately 1% shear strain. Accordingly, the initial static shear modulus is high at 60 MPa, corresponding to $g_i \approx 400$, since $\sigma_i' + a \approx$ 150 kPa. The static D parameter is constant at 0·28 along the entire design stress path, and the effective shear strength parameters are $a =$ 10 kPa and $\tan\phi = 0{\cdot}76$.

The dynamic shear modulus drops from 105 MPa, during drained cycling, via 75 MPa to 65 MPa for the two undrained stress blocks. The cumulative strain resistance r_ε is high (greater than 3000) for all three blocks. The cumulative pore pressure resistance for the last stress block is roughly 40. It is quite possible, however, that more detailed and comprehensive studies of the cyclic behaviour of sand may show that the resistances can be more adequately approximated by parabolas rather than by straight lines.

Examples of typical variations in cumulative strain and pore pressure resistances are summarized in Fig. 29, as obtained from the limited data available so far.

Oedometer test results

Oedometer testing underwent a silent revolution after about 1970 when continuous loading

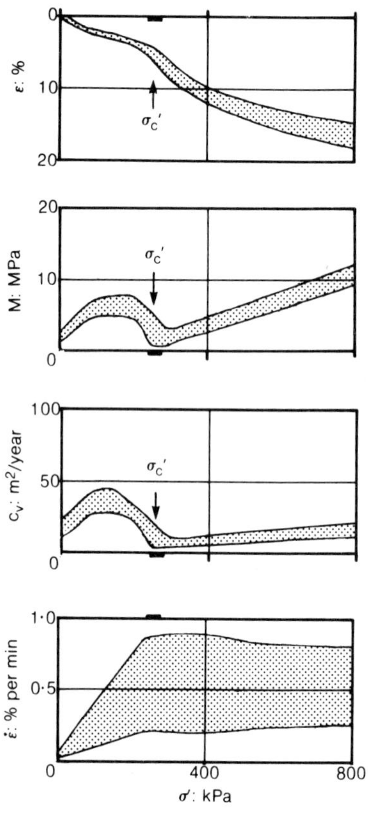

Fig. 30. Results of 12 different CL tests on Risvollan Clay

(CL) tests were introduced (e.g. by using constant rate of strain (CRS), constant gradient or a constant ratio between pore pressure and total stress). The time required for a complete oedometer test was eventually reduced from 14 days to a few hours, in most cases. The CL tests of today are computer controlled and computer interpreted. The results are plotted automatically with arithmetic scales.

An example is shown in Fig. 30 where 12 σ'–ε, M–σ' and c_v–σ' curves are superimposed from 12 different CL tests on Risvollan Clay (reproduced from Janbu, Tokheim & Senneset (1981)). The following main trend is quite clear. The stress–strain curve, as well as the two resistance curves (for M and c_v), identify the preconsolidation pressure σ_c' to nearly the same range of 220–250 kPa, while $\sigma_{v0}' \approx 150$ kPa. The bottom diagram in Fig. 30 shows how the strain rate $\dot{\varepsilon}$ had to be changed to maintain a specified constant ratio $\lambda = du/dp$. This required strain rate increases almost linearly up to σ_c', after which it remains almost constant. Hence, preconsolidation is also evidenced by the distinct change in rate behaviour. All resistances are larger within the preconsolidation stress range than beyond. As the preconsolidation stress is approached, the grain skeleton starts to collapse gradually. This means a reduction in all resistances, whether related to stress or time behaviour. The more brittle the grain skeleton is, the more dramatic the reduction in resistances around σ_c' is.

The Scandinavian quick clays are among the most brittle soils in our part of the world. However, the brittleness of the Canadian quick clays is much more pronounced, as is clearly shown by the two examples in Fig. 31. It is well known that an absolute value of σ_c' is difficult to obtain since σ_c' is somewhat rate dependent, as illustrated in Figs 30 and 31. The general trend is that σ_c' increases with increasing rate, but at the expense of a decreasing modulus when $\sigma' > \sigma_c'$. Altogether, these two effects seem partly to cancel each other in practical applications, when the stress is increased beyond σ_c'.

A large amount of modulus data is now available both from oedometer tests and as back-calculated moduli from settlement case records.

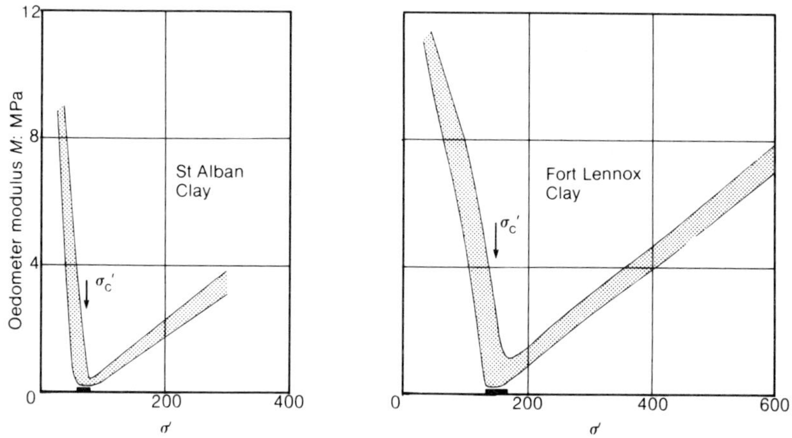

Fig. 31. Moduli variations for two Canadian clays (data from Leahy (1980))

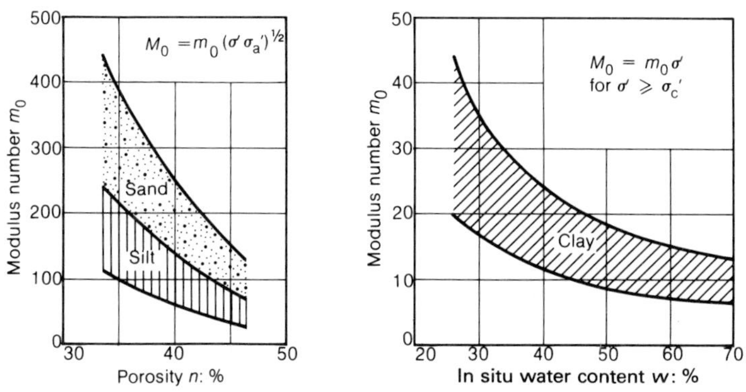

Fig. 32. Typical modulus numbers for normally consolidated sand, silt and clay

Fig. 32 represents a statistical summary of typical stress dependence and modulus numbers for normally consolidated clay, silt and sand, given for the most common ranges of values.

Statistical data for c_v are more scarce. For Scandinavian clays the general trend of variation is illustrated in Fig. 33. The values are obtained by conventional interpretation (see later discussion) for stresses slightly above σ_c'. Therefore the diagram indicates the least values of c_v, since it increases with increasing $\sigma' > \sigma_c'$. The values of c_v for $\sigma' < \sigma_c'$ are much larger.

The creep behaviour of clays (secondary consolidation) is still often determined in oedometers by step loading (SL) tests. When time resistance $R = 1/\dot{\varepsilon}$ is plotted against time (on an arithmetic scale) nearly straight lines are obtained after $t \geqslant t_c$. The slope of such a line is the creep resistance r_s. It has long been known that r_s depends on the stress level, as shown in Fig. 34(a) for a medium clay. Generally, the creep resistance is very large for $\sigma' < \sigma_c'$ and drops radically as σ' approaches σ_c', while r_s has a minimum at σ_c' and thereafter increases slightly for increasing σ'. Therefore it appears that σ_c' is clearly identified from the r_s–σ' curve for natural: see Fig. 34(a).

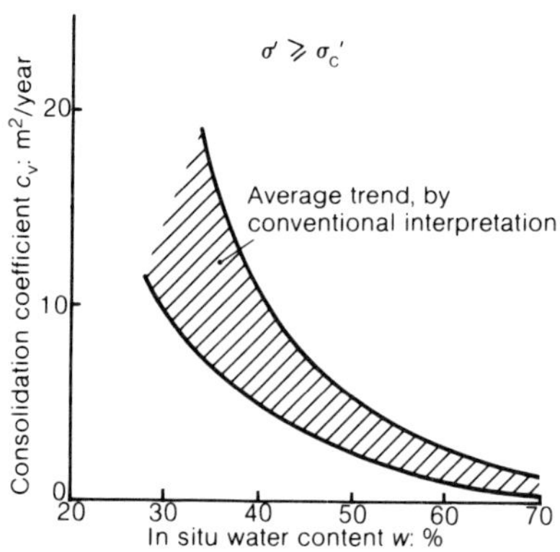

Fig. 33. Typical minimum values of c_v for Norwegian clays

Typical variations in r_s near $\sigma' > \sigma_c'$ for clays of various water contents are illustrated in Fig. 34(b).

ASPECTS OF CLAY BEHAVIOUR

The prediction of the in situ behaviour of clays is still among the most uncertain tasks in geotechnical engineering, despite decades of research and studies of case records. In particular it is the strain rate behaviour that represents the greatest challenge.

Undrained shear in clays

The result of a multistage CIU creep test on a soft clay from Eberg is shown in Fig. 35. From an isotropic effective stress level of $\sigma' = 70$ kPa ($\leqslant \sigma_c'$) the shear stress was increased at constant σ_3' under undrained conditions. At constant shear stress the sample was left to creep undrained for several hours along path A. Then another increase was added to the undrained shear stress at constant σ_3' and again left to creep along path B for several hours, after which an undrained shear test was carried out to failure. The corresponding mobilization curve ($\tan \rho$ versus γ) is also shown in the figure. The curves for strain and pore pressure versus time for paths A and B, and the corresponding resistances R_s and R_u, are also included in the same figure.

It is of particular interest to study the behaviour in a separate undrained creep test along path C–C. From the strain versus time plot for

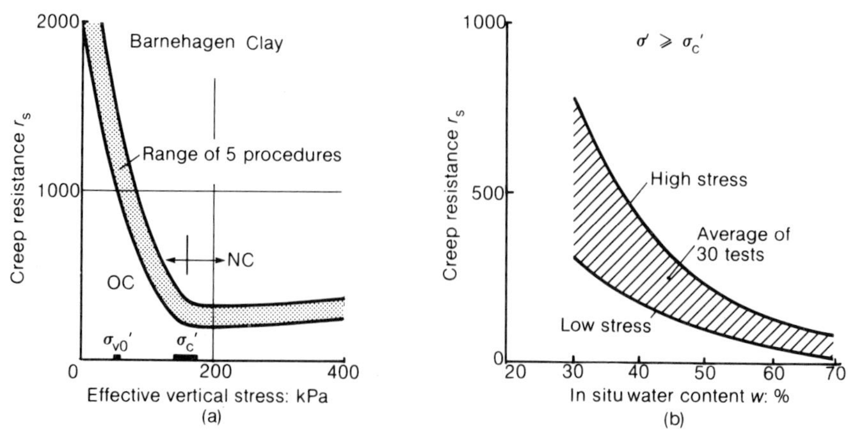

Fig. 34. Examples of variations in creep resistances in clays

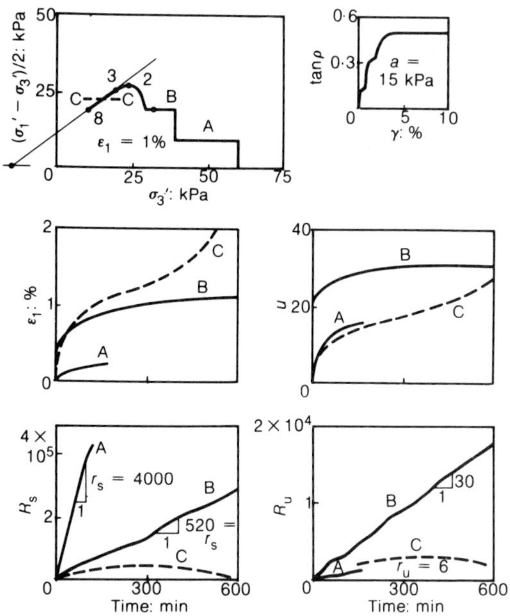

Fig. 35. Example from a multistage test of undrained shear and creep in a soft clay

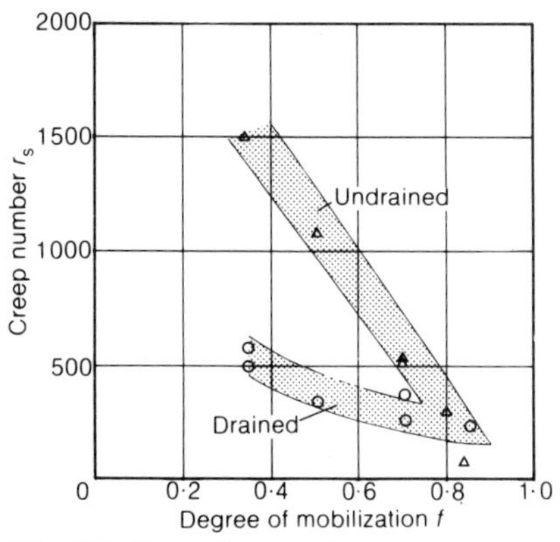

Fig. 36. Examples of drained and undrained creep resistance in soft clay

C–C it is easily seen that the strain rate increases rapidly when the effective stress path passes the failure envelope, i.e. when $\varepsilon_1 \geqslant 1\%$. Consequently, the strain resistance ($1/\dot{\varepsilon}$) starts to decrease as soon as the effective strength envelope is reached. The pore pressure development shows a similar time dependence. These observations are very important for the understanding of soil behaviour near failure.

In a research programme on the same clay, a series of drained and undrained triaxial creep tests was carried out for various mobilizations (Fredriksen, 1983). A summary of the results obtained is given in Fig. 36. The most important finding is that for large mobilizations drained and undrained creep have roughly equal resistances, and the undrained resistance approaches zero for $f > 0{\cdot}8$. However, near the oedo-condition ($f \approx 0{\cdot}5$) the undrained creep resistance is 2–3 times larger than the drained creep resistance.

When testing clays over wide stress ranges it is found that the pore pressure parameter D is constant in the normally consolidated and overconsolidated ranges. However, when σ_c' is passed a gradual change in D is observed as shown in Fig. 37(a). Similar changes in behaviour are also observed in load tests in models and in situ, as illustrated in Fig. 37(b). If the slope of the u–q curve is B_q, D can be estimated from the approximate equation $B_q = 0{\cdot}8 – 0{\cdot}4D$. As a rule $B_q \approx 1$ for $\sigma' > \sigma_c'$, while $B_q \ll 1$ for $\sigma' < \sigma_c'$.

Undrained shear strength

Undrained triaxial tests on soft, saturated clays often show that a maximum level of shear stress can be reached for low strain ($\varepsilon < 1\%$). This maximum shear stress does not change until it reaches a linear effective stress-dependent strength at larger strains (Fig. 38). This constant maximum shear stress level can be called the undrained shear strength s_u of the clay; for instance $s_u = 18–19$ kPa in Fig. 38.

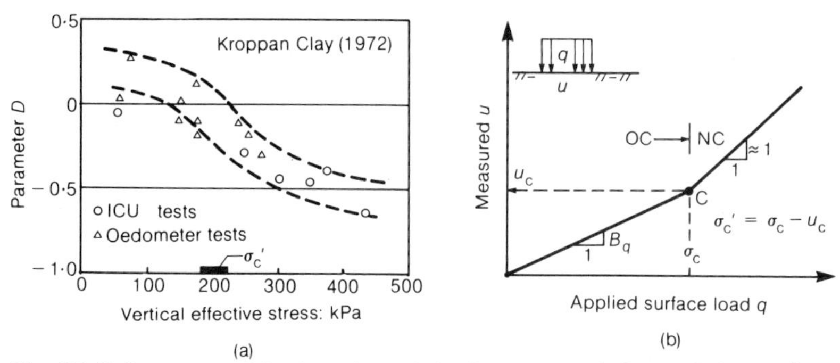

Fig. 37. Laboratory and in situ values of the D parameter below and above σ_c': (a) laboratory test results; (b) principle of the field load test

Theoretically, undrained shear strength can be defined as the level of maximum shear stress that remains constant during large strain changes. The one case where this is appropriate, according to laboratory tests and field experience, is short-term loading on normally consolidated soft clays.

In such cases the maximum shear stress level is theoretically governed by the existing in situ effective stress before soil sampling, σ_{v0}'. The maximum in situ shear stress is therefore given by the equation

$$\tau_{max} = \tfrac{1}{2}(1 - K_0')(\sigma_v' + a)$$

When testing an undisturbed normally consolidated sample, it is theoretically expected that s_u equals the in situ τ_{max}. Using Jaky's formula $K_0' = 1 - \sin\phi$ the approximation

$$s_u = \tfrac{1}{2}(\sigma_{v0}' + a)\sin\phi \qquad (59)$$

is obtained. Laboratory and in situ tests (vanes, cone penetration tests (CPTs)) will in ideal cases correspond fairly well to this equation, as illustrated in principle in Fig. 39(a). The exception is the dry crust, where s_u values are higher than predicted by equation (59). Part of the explanation is the effective stress created by capillary suction during the formation of the dry crust. For instance, for lasting suctions of the order of magnitude of 100–400 kPa the dry crust strength would become 25–100 kPa, for $\phi \approx 30°$ in equation (59).

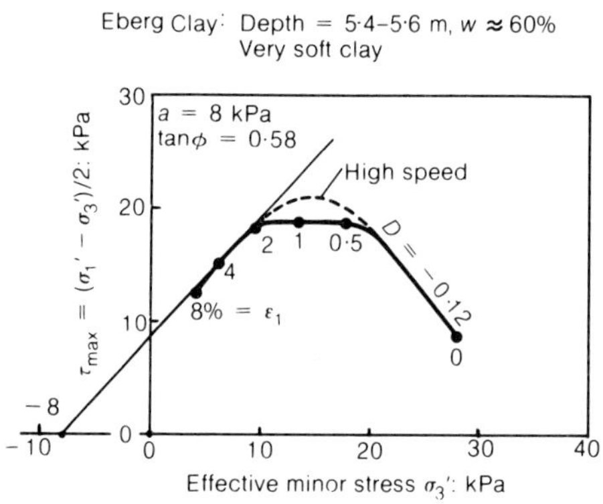

Fig. 38. Consolidated, undrained triaxial test on a soft contractant clay

For classification and identification the undrained shear strength is often determined even in overconsolidated clays. If σ_c' is the maximum past preconsolidation pressure, the maximum in situ shear stress at that time would have been equal to $(1 - K_0')(\sigma_c' + a)/2$. Hence, the approximation

$$s_u \approx \tfrac{1}{2}(\sigma_c' + a)\sin\phi \qquad (60)$$

is used. For instance, if $\sigma_c' + a = 600$–1200 kPa $s_u = 150$–300 kPa would be expected. This range of values covers many of the overconsolidated clay layers found in North Sea oilfields.

Measured values of s_u vary considerably even within a fairly homogeneous layer and particularly for heavily overconsolidated clays. Coefficients of variation of 0·2–0·3 are not uncommon. By comparison numerous triaxial tests on the same clay layer will most often lead to little uncertainty in the shear strength envelope expressed in terms of effective stress. For a selected average attraction, the coefficient of variation for friction is usually $\leq 0{\cdot}1$.

If s_u is an expression for past maximum effective shear stress, s_u and σ_c' must be correlated. However, there is obviously no correlation between σ_c' and σ_{v0}'. Hence, there is no logic in using the overconsolidation ratio (OCR =

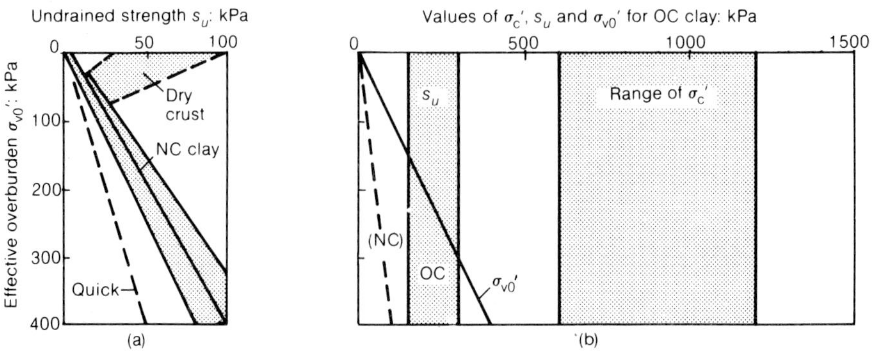

Fig. 39. Undrained strength of clays, with examples of typical overall data for normally consolidated Scandinavian clays onshore and typical ranges for strongly overconsolidated North Sea clays: idealized comparisons

σ_c'/σ_{v0}') in connection with expressions for the shear strength of overconsolidated clays. In practice, the stability of such clays should always be analysed in terms of effective stress, whether long term or short term, or drained or undrained conditions are encountered.

Primary versus secondary consolidation

The time dependences of soft clays have always been of great concern in practice, because the complicated processes are not yet fully understood. The one-dimensional behaviour will be discussed at some length here.

For one load step in an oedometer primary consolidation is by definition due to changes in effective stress σ' during pore pressure dissipation. After pore pressure dissipation, when $\sigma'=$ constant, continued deformation, i.e. creep or so-called secondary consolidation, is now time dependent only.

In reality, the two processes overlap, and in practice it may be of some interest to study whether a linkage time t_c can be found to separate the two processes approximately.

From one-dimensional classical theory of consolidation it is well known that the degree of primary consolidation $U_p \approx 2(T/\pi)^{\frac{1}{2}}$ for $U_p<0{\cdot}5$, or $T_p<0{\cdot}2$. Since $U_p=\varepsilon/\varepsilon_p$ and $T_p=tc_v/d^2$, where d is the drainage path length, the following formula for primary strain during the first phase of consolidation is obtained

$$\varepsilon = 2\varepsilon_p\left(\frac{c_v t}{\pi d^2}\right)^{\frac{1}{2}} \qquad (61)$$

Hence, the primary strain rate $\dot{\varepsilon}_p = d\varepsilon_p/dt$ becomes

$$\dot{\varepsilon}_p = \frac{\varepsilon_p}{(\pi t_0 t)^{\frac{1}{2}}} = \frac{1}{R_p}. \qquad (62)$$

when $t_0 = d^2/c_v$.

The primary strain rate written in terms of the primary time resistance R_p leads to

$$R_p = r_p(t_0 t)^{\frac{1}{2}} \qquad (63)$$

where

$$r_p = \frac{\pi^{\frac{1}{2}}}{\varepsilon_p} = \pi^{\frac{1}{2}}\frac{M}{q}$$

Here, M is the average tangent modulus over the load step q. Equation (63) shows that the primary time resistance is a parabola, in theory.

By comparison, the secondary time rate (equation (19)) is

$$\dot{\varepsilon}_s = \frac{1}{R_s} = \frac{1}{r_s(t-t_r)} \qquad (64)$$

Clay data:
For $\sigma' \leqslant \sigma_c' = 150$ kPa, $M = 5$–10 MPa, $r_s = 1200$ (600), $c_v = 10$ m²/year, $q = 50$ kPa above $\sigma_{v0}' = 50$ kPa
For $\sigma' \geqslant \sigma_c' = 150$ kPa, $M = 15(\sigma' - \sigma_r')(\sigma_r' = 0$–$75$ kPa), $c_v = 2$ m²/year, $r_s = (150)$ 250, $q = 50$ kPa above $\sigma_c' = 150$

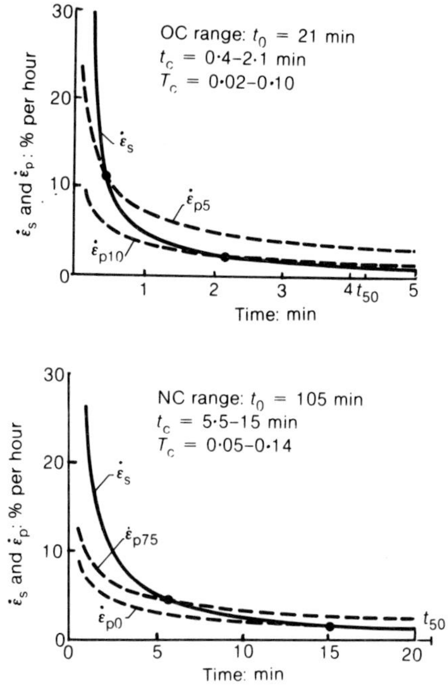

Fig. 40. Comparison of strain rates in primary and secondary consolidation

The two strain rates $\dot{\varepsilon}_p$ and $\dot{\varepsilon}_s$ are equal when their resistances are equal, i.e. $R_p = R_s$, at a linkage time $t = t_c = T_c t_0$. From the formulae (valid only for $T_c \leqslant 0{\cdot}2$)

$$T_c = \frac{t_c}{t_0} = \left[\frac{r_p}{(1-T_r)r_s}\right]^2 \qquad (65)$$

when $T_r = t_r/t_c$. From numerous tests it has been found that $r_p/r_s \approx 0{\cdot}15$–$0{\cdot}20$ on average, when $T_r = 0$–$0{\cdot}5$, leading to $T_c = 0{\cdot}02$–$0{\cdot}15$ for lightly overconsolidated to normally consolidated clays. This finding is emphasized by the two numerical examples in Fig. 40, related to medium clays. The practical conclusion is that the hydrodynamic process is present only for a short time after loading, corresponding to a degree of consolidation of $U_p = 15$–40%. Thus creep (secondary consolidation) dominates even the major part of the primary consolidation in medium clays. The early dominance of creep is more pronounced in soft clay.

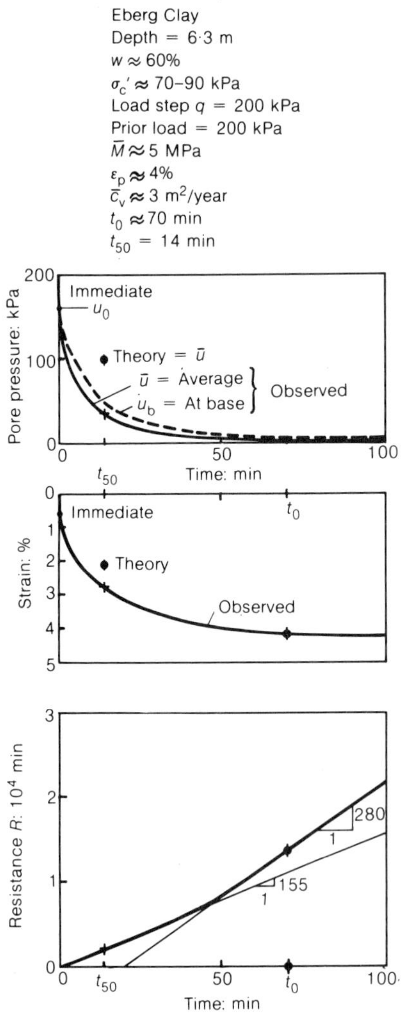

Fig. 41. Observed strain and pore pressure dissipation compared with theory for a soft clay

Strain rates and pore pressure dissipation

The possibility of creep dominance in soft clays deserves further study. From the expression $\varepsilon = f(\sigma', t)$ the following total differential

$$d\varepsilon = \frac{\partial\varepsilon}{\partial\sigma'}d\sigma' + \frac{\partial\varepsilon}{\partial t}dt \tag{66}$$

is obtained, which can be written

$$d\varepsilon = \frac{d\sigma'}{M} + \frac{dt}{R} \tag{67}$$

where $M = \partial\sigma'/\partial\varepsilon$ is the tangent modulus and $R = \partial t/\partial\varepsilon$ is the time resistance.

By studying M at constant time, and R at constant σ', the partial derivatives can be omitted; hence $M = d\sigma'/d\varepsilon$ and $R_s = dt/d\varepsilon$.

For a constant load step q in an oedometer the effective stress becomes $\sigma' = \sigma_0' + q - u$ at any arbitrary time. This means that $d\sigma'/dt = -du/dt = -\dot{u}$. Hence, equation (67) can be written as

$$\dot{\varepsilon} = -\frac{\dot{u}}{M} + \frac{1}{R_s} = \frac{1}{R_p} + \frac{1}{R_s} \tag{68}$$

in which $R_p = -M/\dot{u}$ is the time resistance during the primary consolidation and $R_s = dt/d\varepsilon_s$ is the time resistance (creep) during secondary consolidation.

In the expression for R_p, the rate of pore pressure dissipation $\dot{u}$ is negative, so that R_p is positive. If u_0 is the excess pore pressure at time $t = 0$ the rate of pore pressure dissipation can be used to formulate a pore pressure resistance R_u in the following manner

$$R_u = u_0\frac{dt}{du} = \frac{u_0}{\dot{u}} \tag{69}$$

For a soft clay from Eberg the measured pore pressure dissipation in Fig. 41 led to the resistance

$$R_u \approx r_u(t - t_r) \tag{70}$$

where $r_u \approx -5$ and $t_r \approx 0$ for $t < 50$ min.

Introducing the definition of R_u into the R_p formula, $-M/\dot{u}$

$$R_p = -\frac{M}{u_0}R_u \tag{71}$$

is obtained. In Fig. 41, for the load step of 200–400 kPa, $M \approx 5$ MPa, $u_0 \approx 160$ kPa and $R_u \approx -5t$ which leads to $R_p = 155t$, i.e. $r_p = 155$. The expected parabolic shape of the R_p–t curve is almost erased in this soft clay ($w \approx 60\%$).

The absence of primary consolidation means that the pore pressure dissipation and the strain rate are faster than predicted by the classical theory, as shown in Fig. 41.

Since both R_p and R_s have a similar time dependence for soft clays (after some small initial time) equation (68) can be rewritten to read

$$\dot{\varepsilon} = 1/R \tag{72}$$

where

$$\frac{1}{R} = \frac{1}{R_p} + \frac{1}{R_s} \tag{73}$$

This combined (total) time resistance is shown in the bottom diagram in Fig. 41. Along the time axis are marked the theoretical values of t_0 and $t_{50} = 0{\cdot}2t_0$. The R–t curve shows that there is no distinct parabola, except perhaps near $t \approx 0$. For $t \approx 0$–40 min $R = 155t$, after which $R = 280(t - t_r)$ where $t_r \approx 20$ min.

This shows that the behaviour is predominantly creep of a very simple nature ($R = rt$) during most of what is termed primary consolidation. The classical consolidation process is limited to a very small time after loading, say $t_c < 0{\cdot}1 t_0$. Moreover, Fig. 41 shows that the linear creep resistance increases with increasing time after loading. This effect has previously been observed in research, particularly in soft clays and for load steps of long duration.

From the classical one-dimensional theory of consolidation it is found that the maximum rate of pore pressure dissipation $\dot{u}_b$ at the impervious face occurs for $T_m = 0{\cdot}165$, i.e. for $t_m = 0{\cdot}165 t_0$. This finding can be used to calculate c_v from the observed rate $\dot{u}_b$ by the theoretical formula

$$c_v = 0{\cdot}54 d^2 \left(\frac{\dot{u}_b}{u_0}\right)_{max} \tag{74}$$

The derivation of equation (74) has not yet been published in English. This determination leads to much larger c_v values than those from conventional procedures based on a selected degree of consolidation, say at t_{50}, t_{90} or a constructed t_{100}. For the Eberg clay in Fig. 41 the rate-determined c_v is roughly 30–45 m^2/year compared with about 2–4 m^2/year from conventional interpretations for stresses $\sigma' > \sigma_c'$.

The rate-determined c_v often results in $c_v \approx 25$–50 m^2/year almost independently of the stress level for several types of clay. It is of particular interest to note that this range is nearly equal to the kinematic viscosity of water, which is about 45–30 m^2/year at temperatures of 5–25 °C.

Response analogies

In most normally consolidated clays the creep resistance (when $\dot{u} = 0$) is almost linear (say $R_s = r_s t$) over large time intervals (when $t \geqslant t_c$). The corresponding formula for creep (or secondary consolidation) is

$$\varepsilon_s = \frac{1}{r_s} \ln \left(\frac{t}{t_c}\right) \tag{75}$$

Similarly, the cumulative strain resistance for repeated loading in clay is most often linear (say $R_\varepsilon = r_\varepsilon N$ for $N \geqslant N_0$). The corresponding formula for cumulative strain is

$$\varepsilon_{cu} = \frac{1}{r_\varepsilon} \ln \left(\frac{N}{N_0}\right) \tag{76}$$

Since the time and the number of repetitions are proportional ($t = NT_n$, when T_n is the period of repetition) it is seen that the two formulae are very similar. What is even more astonishing is that the values of r_ε and r_s are of the same order of magnitude for comparable degrees of mobilization. In other words, the cumulative internal soil response is primarily a creep process, or it is governed by the internal creep potential within the soil grain structure.

As a further illustration, the simplest formula for undrained pore pressure build-up during cyclic loading is

$$u_{cu} = \frac{\Delta\sigma_d}{r_u} \ln \left(\frac{N}{N_0}\right) \tag{77}$$

while the build-up of pore pressure during undrained creep in clays exposed to a sudden static load change of $\Delta\sigma_d$ is governed by

$$u_{st} = \frac{\Delta\sigma_d}{r_u} \ln \left(\frac{t}{t_0}\right) \tag{78}$$

corresponding to a linear pore pressure resistance $R_u = R_u t$, for $t \geqslant t_0$. The similarity between the two formulae is again striking, since $t = NT_n$. The dimensionless resistances are also of the same order of magnitude. Hence, the pore pressure generation must be creep dominated. Likewise, the pore pressure dissipation in drained tests is closely approximated by a linear resistance, $R_u = r_u t$, indicating that it is also governed by creep rates.

Natural versus artificial clays

Artificial sediments in research, and very recent sediments in nature, do not exhibit such sharp distinctions in creep behaviour around σ_c'. This difference is very clearly demonstrated in Fig. 42(a). For the natural clays tested by Bishop & Lovenbury (1969) the drained creep resistances r_s are in full agreement with our findings. For instance $r_s = 1200$–1500 for the overconsolidated clay and 200–500 for the normally consolidated clay, when $f = f_0 = 0{\cdot}5$–$0{\cdot}6$. Moreover, the resistance decreases with increasing degree of mobilization.

For comparison, Shibata & Karube (1969) tested artificially sedimented clays, where one test series was performed after stress-induced preconsolidation. Fig. 42(b) shows that there is no clear difference in the r_s values for the two series. The reason must be that it is impossible in a few weeks to duplicate the structural rigidity of an undisturbed natural clay that is 10 000 years old.

This information should indicate the necessity for great care in correlating test results from

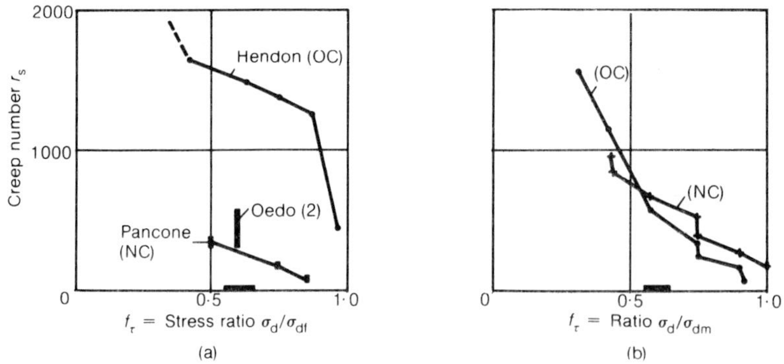

Fig. 42. Creep resistances in natural clays compared with artificial clays: (a) data from Bishop & Lovenbury (1969); (b) data from Shibata & Karube (1969)

remoulded, reconsolidated clays to the behaviour of real natural undisturbed clays, particularly for heavily overconsolidated clays.

GRAVITY PLATFORM ANALYSES: EXAMPLE

A numerical example concerning a gravity platform in deep water will be used to illustrate the application of the various elements of the soil model proposed in this Paper.

Principle of stability analysis

Most offshore stability regulations today call for an ultimate limit state (ULS) design. This means that the environmental loads E are multiplied by a load coefficient (say $\gamma_f = 1{\cdot}3$). For these ultimate loads $E\gamma_f$ a minimum value of the material coefficient γ_m is specified, say

$$\gamma_m \geqslant 1{\cdot}2$$

for an effective stress analysis and

$$\gamma_m \geqslant 1{\cdot}3$$

for a total stress analysis. For comparison with past experiences onshore the lumped safety factor F may also need to be analysed for the serviceability limit state (SLS) in which $\gamma_f = 1$. This comparison may be most appropriate for the ordinary total stress analysis in clays (s_u analysis).

In an effective stress stability analysis the magnitude of the excess pore pressure below the platform is as a rule the most important single variable. This is illustrated in principle in Fig. 43.

For a given loading condition (say in the ULS) the safety level γ_m can be calculated for each assumed state of pore pressure u. Thus a curve showing γ_m versus u is obtained. The safety requirement (say $\gamma_m \geqslant 1{\cdot}2$) intersects the ULS curve for a maximum allowable state of pore pressure u_{max}. The safety requirement is fulfilled if

$$u_b \leqslant u_{max} \tag{79}$$

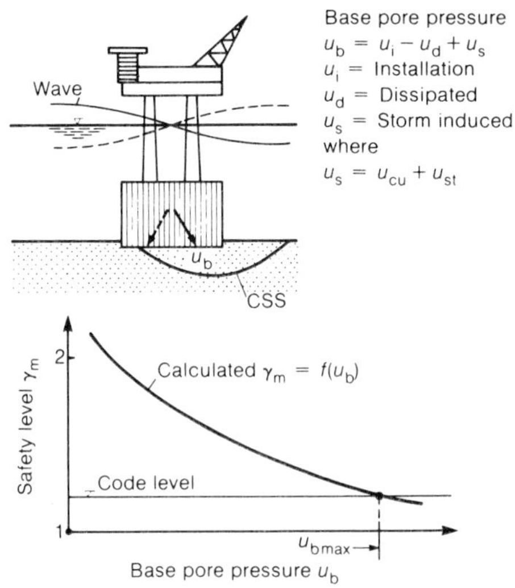

Fig. 43. Stability analyses of gravity platforms: effective stress principle

Here, the average resulting excess pore pressure at the foundation base (u_b) comprises three components, as follows

$$u_b = u_i - u_d + u_s \tag{80}$$

where u_i is the excess pore pressure built up during installation, u_d is the dissipated pore pressure due to drainage between installation and the first storm and u_s is the storm-induced pore pressure during the first storm.

It is often assumed that the first 100 year storm may occur during the very first winter season after installation. This assumption means

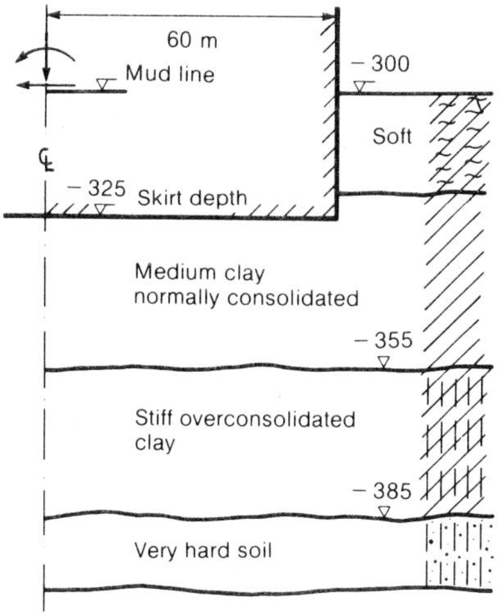

s_u: kPa	a: kPa	tanϕ	D	M:MPa	r_s	r_ε	r_u	g_i
5 45	5	0·50	−0·1	12σ'	≈150			
70 150	15	0·51	0	15σ'	200 300	400–1000 600–1500	7 9	200 250
>250	50	0·60	0·3	25 40	1000 3000			>400
>400								

Fig. 44. Soil and data profile: simplified example

that u_d may be a minimum and u_s a maximum; hence it leads to a maximum possible u_b.

The storm-induced pore pressure contains two components, one cumulative, u_{cu}, and one static, u_{st}. Hence

$$u_s = u_{cu} + u_{st} \tag{81}$$

The static component is usually estimated from total stress field theory for the largest wave in the storm considered.

It has been debated whether a partial coefficient should also be used for the pore pressure. So far this has not been instituted in Norway, partly because the present procedures for obtaining the resultant pore pressure contain a number of conservative components, most of which lead to an overestimation of u. However, there is still room for considerable research in this area.

Design data in soil profiles

In geotechnical engineering offshore a proper assessment of the subsoil conditions is a time-consuming, expensive and all-important undertaking. The subsoil investigations are usually carried out in several stages and at several possible locations, leading to comprehensive reports both from the field investigations and from the laboratory testing programmes. Out of this large amount of information the geotechnical engineer has to extract design values for the various subsoil layers, preferably accompanied with levels of uncertainty.

It is strongly advised to use concentrated data profiles for the important parameters to be used in the different types of analyses required. An idealized example is shown in principle in Fig. 44, as a basis for the numerical examples to follow.

Loads and load transfer

The loads on a gravity platform lead to normal and shear stresses acting along the effective contact area between the platform and the subsoil. These contact forces are preferably considered as an action–reaction system. The action system (q_v, t_h) is obtained from the loads acting on the platform, while the matching reaction system (σ_v, τ_h) must be generated from stresses mobilized in the subsoil.

In this subsection a procedure is given for the calculation of the action stress system, while procedures for estimating the soil reaction (the bearing capacity) are given in the next subsection.

The theoretical foundation level for a gravity platform may either be taken at mud line (no skirts) or at a defined baseline (e.g. underside of skirts); see Fig. 45.

At the theoretical foundation level the characteristic values of the loads are as follows

Q_v = vertical load (permanent and live)

Q_h = horizontal load
M = overturning moment } environmental loads

These characteristic values are obtained as the most unfavourable combination of several types of loading conditions (gravity, wave, wind, earthquake).

For gravity platforms with skirts the loads given at mud line are denoted Q_{vm}, Q_{hm} and M_m. Buoyancy is accounted for in Q_{vm}. Hence,

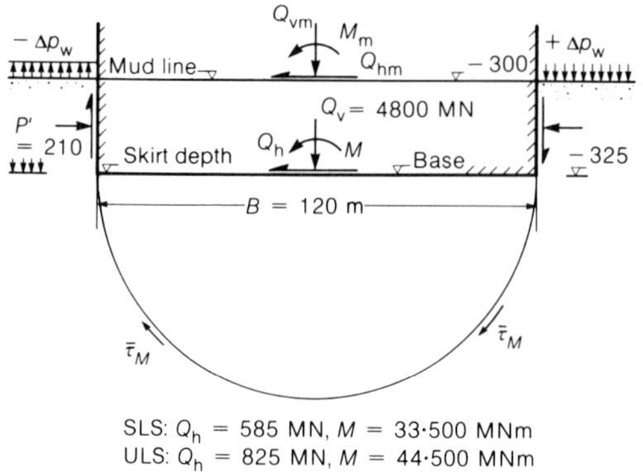

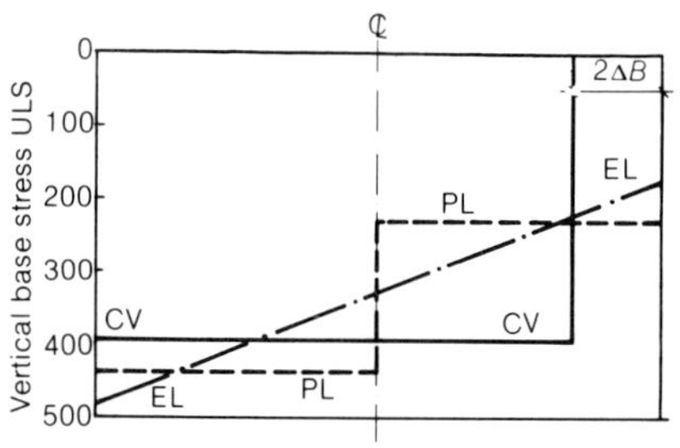

Fig. 45. Mud line forces and load transfer to the foundation baseline

for horizontal sea level the reference water pressure at mud line is zero, while for waves the excess wave pressures are $\pm p_w$ outside the heel (+) and the toe (−) respectively; see Fig. 45.

The given loads at mud line are transferred to the theoretical foundation level (underside of skirts) by means of the following formulae

$$Q_v = Q_{vm} + \gamma' D_s A \tag{82}$$

$$Q_h = Q_{hm} - \Delta Q_h \tag{83}$$

$$M = M_m + Q_{hm} D_s - \Delta M_s \tag{84}$$

in which D_s is the depth of the skirts, γ' is the buoyant unit weight of the soil, A is the total foundation area, ΔQ_h is the resultant horizontal soil reaction along the skirts and ΔM_s is the resultant stabilizing moment due to soil reaction along the skirts.

Initially, the pore pressure is assumed to be hydrostatic, and the reference value of the pore pressure is herein taken as zero at the theoretical foundation level. Single base areas, shaped as a polygon, could either be idealized by an equivalent rectangle, with the dimensions $BL = A$, or a square with sides $B = A^{\frac{1}{2}}$, as is used below.

The overturning moment leads to an eccentricity $\Delta B = M/Q_v$, which means that the effective width B_0 of the idealized area is equal to $B_0 = B - 2\,\Delta B$.

When using relative eccentricity

$$e = \frac{\Delta B}{B} = \frac{M}{BQ_v} \tag{85}$$

the effective width $B_0 = (1-2e)B$, and the effective area $A_0 = (1-2e)A$. The point of application of the resultant forces Q_v and Q_h is the centre of the effective area, about which the moment is zero. The average action stress system at the foundation base then becomes

$$q_v = \frac{Q_v}{A_0} \tag{86a}$$

$$t_h = \frac{Q_v}{A_0} \tag{86b}$$

(short or no skirts). In total stress analyses, and for long skirts, $t_h = Q_h/A$.

For the installation conditions

$$q_{v0} = \frac{Q_{v0}}{A} \tag{87a}$$

$$t_{h0} = 0 \tag{87b}$$

Using linear elastic theory and ideal plastic theory for stress distribution along the foundation base, the edge stresses σ_e can be calculated as follows

$$\sigma_{e\,EL} = \frac{Q_v}{A}(1 \pm 6e) \tag{88}$$

$$\sigma_{e\,PL} = \frac{Q_v}{A}(1 \pm 4e) \tag{89}$$

The possibility of local yield at the edges can now be studied. Fig. 45 contains a direct comparison for the numerical example herein.

The rotational stability, due to the baseline moment (in the ULS) can be estimated approximately by

$$\tau_M = \frac{4e}{\pi}\frac{Q_v}{A} < \frac{1}{\gamma_m}\tau_f \tag{90}$$

where τ_M is the average shear stress required along a semicircle to keep moment equilibrium, while τ_f is the average shear strength along the same circle. In the example in Fig. 45 τ_M = 33 kPa, which is very low compared with the available strength. Hence, the rotational stability is very satisfactory.

For multibase areas it may be necessary to distinguish between rigid and flexible structures. For flexible multibase structures the stability of each individual support could be analysed separately. For rigid multibase structures, the overall stability of the whole composite area should also be investigated. In both cases an (idealized)

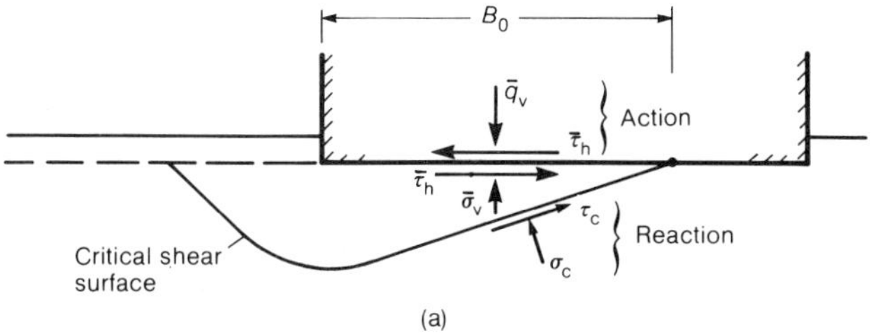

State	q_{vn}: kPa	t_h (s_u): kPa	t_h ($a\phi$): kPa	e
SLS	167	41	48	0·058
ULS	184	57	67·5	0·077

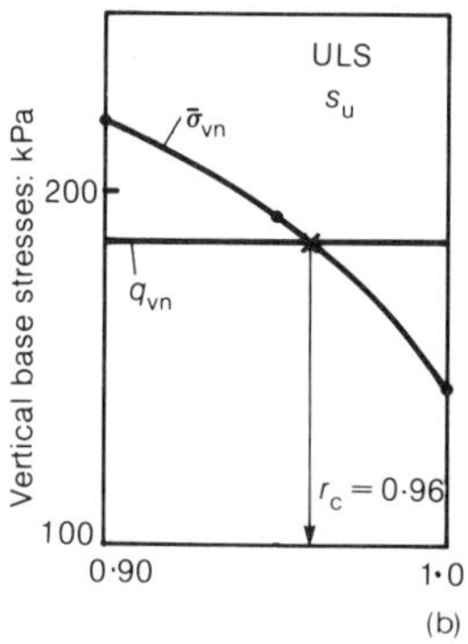

$r_c = 0{\cdot}96$
$\tau_c = 57/0{\cdot}96$ kPa $= 59{\cdot}4$ kPa
$\tan\alpha_0 = (0{\cdot}04/1{\cdot}96)^{1/2} = 0{\cdot}143$
$z_m \approx 0{\cdot}143 \times 101{\cdot}5 = 14{\cdot}5$ m
$\bar{s}_u \approx 75 + 20 = 95$ kPa
$\gamma_m \approx 95/59{\cdot}3 \approx 1{\cdot}6$
$\gamma_m = 75/57 \approx 1{\cdot}3$ along base

Fig. 46. Baseline stresses and an example of the numerical solution of γ_m in an s_u analysis with ULS forces

interaction analysis may be required to obtain proper values of Q_v and Q_h for each separate area.

Examples of stability analyses

The purpose of a stability analysis is to obtain the critical average degree of shear mobilization, defined as the maximum ratio between the average shear stress $\bar{\tau}_c$ required for equilibrium and the average characteristic shear strength $\bar{\tau}_f$

$$f_c = \frac{\bar{\tau}_c}{\bar{\tau}_f} \tag{91}$$

Theoretically, the value of f_c is obtained for a very definite shape and location of the shear surface, i.e. the CSS, the determination of which is also part of the analysis.

In important cases the distribution of the normal stresses σ_n and σ_n', and the shear stress τ_c along the CSS should be determined.

In general, the material coefficient γ_m for ULS forces is given by

$$\gamma_m = 1/f_c \tag{92}$$

while for the lumped safety factor for SLS forces

$$F = 1/f_c \tag{93}$$

The analyses are based on satisfying all three overall equilibrium conditions. Moment equilibrium has already been satisfied by introducing the effective area concept, implying that Q_v acts at the centre of B_0. Horizontal and vertical equilibrium require that the action balances the reaction, Fig. 46.

$$\left.\begin{array}{l}\text{Action due}\\ \text{to loads}\end{array}\right\} \begin{array}{c} t_h = \tau_h \\ \bar{q}_v = \bar{\sigma}_v \end{array} \left\{\begin{array}{l}\text{Reaction}\\ \text{from soil}\end{array}\right.$$

The geotechnical part of the solution is to establish the formulae for the soil reactions τ_h and σ_v. For a weightless soil and plane strain, closed form solutions for σ_v and τ_h have been given in

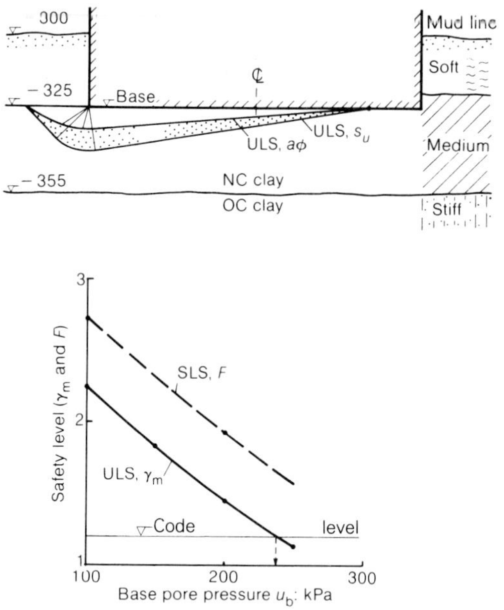

Fig. 47. CSSs, and safety level versus excess pore pressure

a previous section, together with the geometry of the CSS.

For gravity platforms on clay an initial estimate of the stability can be made by a simple total stress (STS) analysis based on the undrained shear strength values s_u obtained without knowledge of the pore pressure, say by vane tests or CPT tests. In such cases the critical equilibrium analysis is very simple indeed. Step by step an STS analysis is carried out as follows, once q_v and t_h have been calculated. Assume a value of r and obtain the corresponding N_c value from Fig. 16 and calculate

$$\tau_c = t_h/r \tag{94}$$

$$\sigma_v = p + N_c\tau_c \tag{95}$$

By repeating this procedure for other values of r a curve of σ_v versus r can be obtained (Fig. 46) and, where $\sigma_v = q_v$, $r = r_c$. This value of r_c determines both the critical value of $\tau = \bar{\tau}_c$, as an average

$$\bar{\tau}_c = \frac{t_h}{r_c} \tag{96}$$

and the geometry of the CSS as previously described. Along the CSS the average $\bar{s}_u$ is obtained from which

$$f_c = \frac{\bar{\tau}_c}{\bar{s}_u} \tag{97}$$

In an effective stress analysis the closed stress field solution is applied in the following manner, once q_v, t_h and u_b are known. Select a value of r and calculate $\tan\rho$ from the equation

$$t_h = r(q_v + a - u_b)\tan\rho \tag{98}$$

observing that $r\tan\rho = f_b$, the base ratio, which is a constant. From this r–$\tan\rho$ combination N_q is obtained from Fig. 18, from which

$$\sigma_v - p' \approx (N_q - 1)(p' + a + d_0\gamma' B_0 - \lambda_u u_b) \tag{99}$$

where $d_0 \leqslant 0{\cdot}5$ and $\lambda_u \leqslant 1$. In this equation the effects of unit weight and pore pressure are included in the stress field solution in an approximate conservative way, compared with the more comprehensive solutions available (Janbu, Grande & Eggereide, 1976).

By repeating this procedure for other r values and plotting a diagram such as Fig. 46 r_c is obtained where $\sigma_v = q_v$. This r_c value leads to $\tan\rho_c = f_b/r_c$ and hence

$$f_c = \frac{\tan\rho_c}{\tan\phi} \tag{100}$$

From f_c, $\gamma_m = 1/f_c$ is obtained in the ULS and $F = 1/f_c$ in the SLS.

The geometry of the CSS is determined by r_c and ρ_c, corresponding to weightless soils. The results of the numerical example are shown in Fig. 47, both for the s_u analysis and for the $a\phi$ analysis. In particular for the effective stress analyses it is seen that

$$u_{max} \leqslant 230 \text{ kPa}$$

to satisfy the code level $\gamma_m \geqslant 1{\cdot}2$. The question is now: can the base pore pressure u_b reach this level, or will it stay below?

Excess pore pressure

The net load after installation $q_{ni} = 120$ kPa, and the installation pore pressure can hence be estimated from

$$u_i \approx (0{\cdot}8 - 0{\cdot}4D)q_{ni} = 96 \text{ kPa}$$

assuming $D = 0$. The net mean total stress and the net deviator stress after installation are also obtained from the idealized stress field theory as follows

$$\sigma_{mi} = 0{\cdot}8q_{ni} = 96 \text{ kPa}$$

$$\sigma_{di} \approx 0{\cdot}4q_{ni} = 48 \text{ kPa}$$

For the ULS loads $q_n = 184$ kPa and $r_c = 0{\cdot}95$. From the stress field theory

$$\sigma_m = 0{\cdot}92q_n = 170 \text{ kPa}$$

$$\sigma_d = 0{\cdot}70q_n = 129 \text{ kPa}$$

Hence,

$$\Delta\sigma_m = 74 \text{ kPa}$$
$$\Delta\sigma_d = 81 \text{ kPa}$$
$$u_{st} = \Delta\sigma_m - D\,\Delta\sigma_d = 66 \text{ kPa}$$

assuming $D = 0{\cdot}1$, while $u_{st} = 74$ kPa if D remains unchanged by cyclic effects.

Since the pore pressure resistance is almost independent of variations in stress level, the cumulative pore pressure can be estimated from a one-block idealization

$$u_{cu} = \frac{\Delta\bar{\sigma}_d}{\bar{r}_u} \ln N \qquad (101)$$

with $\Delta\sigma_d$ = constant, and corresponding to the loading condition for the relevant wave height. In this example $\Delta\sigma_d = 45$–50 kPa is obtained. For a 24 hour storm $N \approx 7200$, and since $\bar{r}_u = 8$

$$u_{cu} \approx 50\text{–}55 \text{ kPa}$$

Hence, the storm-induced pore pressure u_s is approximately 120 kPa. Therefore the total base pore pressure becomes

$$u_b \approx 215 - u_d \text{ kPa}$$

For the conservative assessment of $u_d = 0$ $u_b \approx$ 215 kPa is obtained, corresponding to $\gamma_m \approx 1{\cdot}3 > 1{\cdot}2$.

Storm-induced strains and displacements

Two idealized storm conditions are illustrated in Fig. 48, namely one 6 hour and one 24 hour storm period, each including at least one 100 year wave. Each diagram shows the number of waves N of different degree of wave mobilization f_w relative to the 100 year wave.

The calculation of the cumulative normal strain is herein based on the resistance concept ($R_\varepsilon = r_\varepsilon N$) applicable for a given stress block of constant intensity ($\Delta\sigma_d$) repeated N times. In practical applications to a stochastic variation only a limited number of equivalent stress blocks can be idealized, or the variation must be described continuously. Here, two idealizations are used, namely a one-block average of $\Delta\sigma_d$ and r_ε and a continuous linear idealization.

The selection of the proper r_ε values for different idealizations is done with the aid of an experimentally determined diagram showing r_ε versus the maximum soil mobilization f_{max}, similar to Fig. 29.

Once $\varepsilon_1 = \varepsilon_{cu}$ has been calculated for a layer of thickness H it is easy to show that the vertical and the horizontal cumulative displacements become

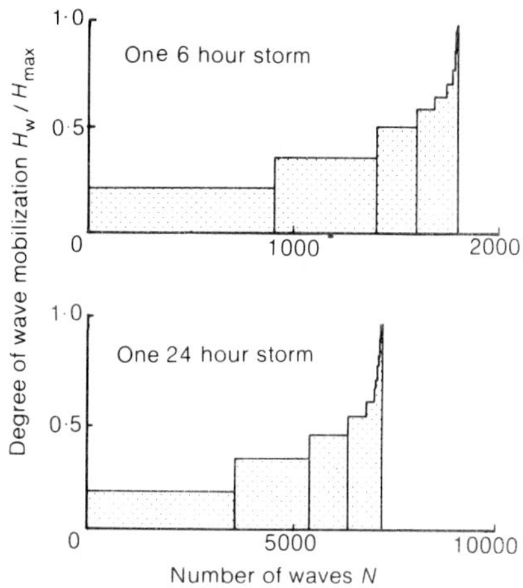

Fig. 48. Two examples of storm idealizations

$$\delta_v = \varepsilon_{cu} H \qquad (102)$$
$$\delta_h = \pm\, \varepsilon_{cu} H \tan\omega \qquad (103)$$

where ω is the rotation of the principal axis from the vertical, due to the transformed base stress system t_h and q_v, relative to the initial condition $t_h = 0$ and $q_v = q_{v0}$.

When idealizing a storm into one average stress block, the estimate of cumulative strain is simply

$$\varepsilon_{cu} = \frac{1}{r_\varepsilon} \ln N \qquad (104)$$

Assuming a linear variation of r_ε with time (and N) the defining equation can be integrated over N for the whole storm from a starting point r_0 to an end point r_1, leading to

$$\varepsilon_{cu} = \frac{1}{r_0} \ln\left[\frac{r_0}{r_1}(N-1) + 1\right] \qquad (105)$$

For a 24 hour storm ($N = 7200$) the one-block average system with $\bar{r}_\varepsilon = 750$ leads to $\bar{\varepsilon}_{cu} = 1{\cdot}2\%$. For a linear idealization with $r_0 = 1200$ and $r_1 = 400$ (escalating storm), $\varepsilon_{cu} = 0{\cdot}9\%$. For the reverse situation (starting with a 100 year wave and decay) corresponding to $r_0 = 400$ and $r_1 = 1200$, $\varepsilon_{cu} = 1{\cdot}9\%$ is obtained. Since a combination of these two calculations is more likely, the average $\bar{\varepsilon}_{cu} = 1{\cdot}4\%$ is considered more representative and is slightly larger than the one-block result. Since the average depth of the critical effective stress fields for the significant wave height is roughly 10 m, the cumulative vertical settlement for the 24 hour storm may be

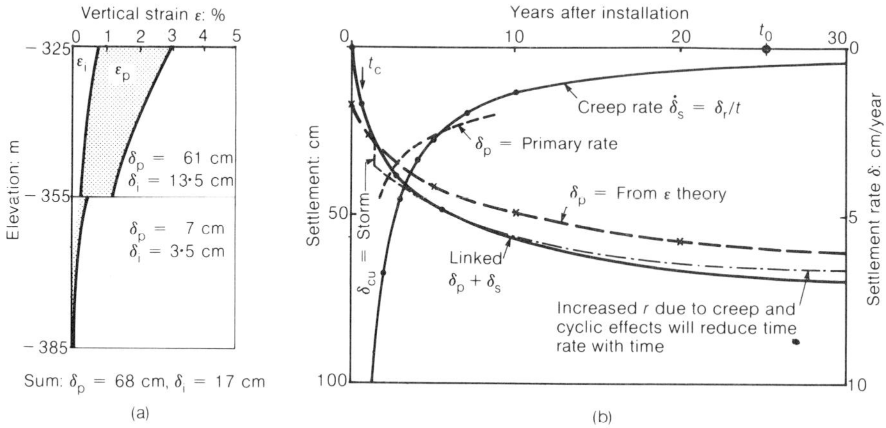

Fig. 49. Estimated settlement and settlement rates

about 10–15 cm. The cumulative effects of later storms are reduced with time because of creep-induced hardening of the grain skeleton.

Settlement analyses

In the following example the compression of the soil within the skirt compartments will not be analysed, because a lengthy discussion would be required for a complete coverage of the important practical problems involved. Fig. 49(a) shows the calculated vertical strain versus depth, from the theoretical foundation level −325 m down to −385 m. For simplicity two soil layers of approximately 30 m thickness each are encountered.

For the normally consolidated upper layer the vertical strain due to primary consolidation is estimated from the formula

$$\varepsilon_p = \frac{1}{m} \ln \left(\frac{\sigma'}{\sigma_{v0}'} \right) \tag{106}$$

For example at −325 m $\sigma_{v0}' = 210$ kPa and since $q_n = 120$ kPa $\sigma' = 210 + 120 = 330$ kPa. For $m = 15$ equation (106) leads to $\varepsilon_p = 3{\cdot}0\%$. The immediate effective stress increase at −325 m is $0{\cdot}2q_n = 24$ kPa; hence $\sigma_i' = 210 + 24$ kPa = 234 kPa. This means that the initial strain ($t = 0$) at −325 m is given by

$$\varepsilon_i = \frac{1}{15} \ln \left(\frac{234}{210} \right) = 0{\cdot}72\%$$

Calculations of ε_p and ε_i have been carried out at several depths throughout both layers for an estimated distribution of the net load increase. The strain distribution with depth is plotted in Fig. 49. The vertical settlement is by definition

$$\delta_v = \int_0^H \varepsilon \, dz \tag{107}$$

and is equal to the area of the ε–z diagrams, leading to $\delta_p = 68$ cm from which $\delta_i = 17$ cm; hence $\delta_c = 51$ cm is due to consolidation. The rate of primary consolidation can be obtained on the basis of the strain–depth distribution procedure published by Janbu (1965). The consolidation strain $\varepsilon_c = \varepsilon_p - \varepsilon_i$ is represented by a trapezoid with $\varepsilon_c = 22\%$ at −325 m dropping to 1·15% at −355 m, with an area of 51 cm. The result of this procedure is shown in Fig. 49 by a broken curve, where creep, or secondary consolidation, is not included. For the time rate estimate the upper 30 m of clay is responsible for δ_c and the layer is assumed to be double drained, so that

$$t_0 = \frac{d^2}{c_v} = 25 \text{ years}$$

using $c_v = 9$ m^2/year.

If the time rate is considered as creep only with a time resistance $R = 1/\dot{\varepsilon}$ the settlement rate $\dot{\delta} = \dot{\varepsilon} H$ of a layer of thickness H would become

$$\dot{\delta} = H/R \tag{108}$$

In layered systems with different resistances $R_n = r_n t$ and thicknesses H_n

$$\dot{\delta} = \frac{1}{t} \sum \frac{H_n}{r_n} = \frac{\delta_r}{t} \tag{109}$$

would be obtained where

$$\delta_r = \sum \frac{H_n}{r_n} \tag{110}$$

Hence, the creep-induced settlement δ_{cr} over a time interval from t_c to t becomes

$$\delta = \delta_r \ln\left(\frac{t}{t_c}\right) \quad (111)$$

because $\delta = \int \dot{\delta}\, dt$.

To obtain a time rate procedure for a linked primary–secondary consolidation process, it is necessary to obtain an estimate of the linkage time t_c. Previous studies indicate that t_c is only a small fraction of t_0 when

$$t_0 = \frac{d^2}{c_v} \quad (112)$$

Since more than 95% of δ_c is completed for $t = t_0$, the approximate relationship

$$t_c = t_0 \exp\left(-\frac{\delta_c}{\delta_r}\right) \quad (113)$$

is obtained from equation (111) when $\delta = \delta_c$ is inserted for $t = t_0$. In this example $\delta_c = 51$ cm and $\delta_r = 3000/250 + 3000/2000$ cm $= 13{\cdot}5$ cm. Hence, $t_c = 0{\cdot}022 t_0 = 0{\cdot}55$ years, because $t_0 = 25$ years for $\bar{c}_v = 9\ \text{m}^2$/year and $d = 15$ m assuming double drainage for the upper layer. Since $t_c = 0{\cdot}55$ years and $\delta_r = 13{\cdot}5$ cm are known, equation (111) is used to obtain the fully drawn δ–t curve in Fig. 49(b) where it is assumed that the 'initial' settlement $\delta_i = 17$ cm is completed in the course of $t_c = 0{\cdot}55$ years. Fig. 49(b) also contains the rate of settlement $\dot{\delta} = \delta_r/t$, which clearly illustrates that the most uncertain predictions are those immediately after platform installation. The main reason is that the basic cause of how the time-dependent process starts (at $t = 0$) is unknown. Hence, the continued use of semi-logarithmic plots (containing no origin) should be discouraged.

Vibrations

During a storm periodic changes take place in the vertical force $\pm\Delta Q_v$, in the horizontal force $\pm\Delta Q_h$ and in the overturning moment $\pm\Delta M$. These force variations will set the platform in motion. The combined effect of the various modes of vibration requires very extensive and complex analyses.

However, if each mode of vibration (vertical and horizontal translation and rocking) is isolated simple analogy models can be used to obtain approximate information of considerable value.

The static displacement at mud line in the vertical (y) and the horizontal (x) directions will, according to elastic half-space theory, become

$$x = \frac{\Delta Q_h}{5GR} \quad (114)$$

$$y = \frac{\Delta Q_v}{6GR} \quad (115)$$

while the corresponding rocking angle ψ about the foundation centre is

$$\psi = \frac{\Delta M_m}{4GR^3} \quad (116)$$

Here, G is the shear modulus of soil and R is the foundation radius (equivalent circle). The analogous spring values correspond nearly to $\nu = \frac{1}{3}$ for all three cases.

For each analogous model the corresponding natural period of vibration can be estimated from

$$T_{nx} = 2\pi\left(\frac{m}{5GR}\right)^{\frac{1}{2}} \quad (117)$$

$$T_{ny} = 2\pi\left(\frac{m}{6GR}\right)^{\frac{1}{2}} \quad (118)$$

$$T_{n\psi} = 2\pi\frac{h}{R}\left(\frac{m}{4GR}\right)^{\frac{1}{2}} \quad (119)$$

Here, h is the height of the equivalent centre of a lumped mass above the mud line.

In the example the value of $R = 67{\cdot}5$ m and mg is assumed to be about 10 000 MN (including the added mass). Since $\sigma' + a = 250$–300 kPa at 25 m depth below the base and $g_i = 250$, $G = 60$–75 MPa for small strains. In the example $G = 70$ MPa is used. For the environmental loads in the SLS static amplitudes of

$$\pm x = 2{\cdot}5\ \text{cm} \qquad \pm\psi = 4\times10^{-4}$$

are obtained, while the vertical displacements $\pm y$ are small because of insignificant variations in Q_v. At 400 m above the foundation base the calculated ψ corresponds to a lateral sway of ± 16 cm for a rigid platform.

The corresponding periods are found to be

$$T_{nx} = 1{\cdot}3\ \text{s} \quad T_{ny} = 1{\cdot}2\ \text{s} \quad T_{n\psi} = 4{\cdot}4\ \text{s}$$

when $h \approx 3R$. If the G modulus is reduced temporarily to 30 MPa, the periods for vertical and horizontal vibration increase to roughly 2 s and the rocking period to nearly 7 s. These results indicate the importance of a more detailed study of the vibration characteristics, including the effect of damping and adequate interactive modelling.

Use of computer programs

The examples given are estimates based on simplified profiles, average data and elementary models to illustrate the applicability of the basic concepts. For more comprehensive and sophisticated numerical handling a large variety of computer programs are available. However, to a large extent our programs use the same basic soil modelling described in this Paper.

CONCLUDING REMARKS

The general philosophy on which this Paper is based may briefly be summarized as follows.

The design of offshore foundations requires knowledge about the behaviour of the subsoil layers when subjected to static, cyclic and dynamic loads. In a safe design, even the most critically stressed regions in the subsoil must have a certain margin of safety. Consequently, it is the soil behaviour within the working stress ranges (serviceability limit states) that are most important to examine.

Soil behaviour is determined by tests in which a soil sample is exposed to a system of external actions (e.g. changing external stresses). The soil sample responds to these external actions and the response is measured (such as the strain and/or the pore pressure). Interpretation of the test results requires that the action–response system be systematically modelled.

The tangent to an action–response curve for a material is the resistance of the material. The resistance concept is a widely used principle in classical mechanics, whether explicitly stated or not. It was therefore decided about 25 years ago to try to model soil behaviour with the aid of the resistance concept.

Definitions have been given for a number of soil resistances related to stress, strain, time and pore pressure behaviour for static and cyclic loading and for drained and undrained conditions. The definitions are independent of the soil type and independent of external action.

All soil resistances are found to depend on the mean normal stress level and the degree of shear mobilization. All soil resistances are larger within the preconsolidation stress region than in the normally consolidated stress region. This means that the preconsolidation pressure σ_c' can be identified by plotting any one of the measured resistances against effective stress (arithmetic plots). Around σ_c' all the resistances are reduced owing to structural breakdown and usually reach a minimum value around σ_c', whereafter the resistances increase or remain constant. The resistance concept has therefore led to several simple diagrams for determining σ_c'. All resistance diagrams lead to similar values of σ_c' for a given soil.

Because of non-linear behaviour it is essential that soil samples be tested in the in situ working stress range. To enable a rapid forecast of the average in situ stresses for inclined loads on gravity platforms, simple closed form solutions from stress field theories for plane strain and weightless soil have been presented.

The theoretical bases for effective stress path plots have been given. Attention is drawn to the importance of indicating the strain along the stress path, or even better to plot separate mobilization curves. This aids judgement of sample quality and proper design strength level.

A consistent use of effective stress path and mobilization plots leads to simple procedures for determining the change in soil resistances with shear mobilizations f. Generally, the resistances are large for isotropic stress ($f = 0$) and decrease rapidly with mobilization towards the oedo-condition ($f_0 = 0{\cdot}5$–$0{\cdot}65$), and continue to decrease to zero as the mobilization approaches failure ($f = 1$).

The shear strength is expressed in terms of the effective normal stress and the corresponding parameters friction and attraction. The term attraction (a, the intercept on the σ' axis) is preferred instead of cohesion (c') mainly because all engineering formulae are thereby simplified, and because its ideal physical meaning is clear (a is the isotropic prestress). Theoretically, the undrained shear strength concept, in its classical sense, is already included in the effective stress expressions, by a limit consideration. Very little has therefore been said about the concept, also because the theoretical and practical aspects have recently been covered thoroughly by Wroth (1984).

As examples of the application of the soil models, simplified analyses have been carried out for a large gravity platform with long skirts in deep water through the softest topsoil layers. Because the subsoil properties and the analytical methods are idealized, this numerical example should be considered as an illustration, in principle, of the type of information required and the type of problems which may have to be analysed. In an actual case, such preliminary analyses are used in predesign, while more comprehensive and detailed analyses, using computer programs, are employed later in the project design.

The settlement problem should be particularly emphasized in connection with this example. It brings out the 'missing link' of knowledge in the time rate behaviour of soft–medium clays. Measurements of time resistance and pore pressure dissipation in oedometers indicate that the hydrodynamic process often disappears shortly after load application, and pure creep behaviour dominates the settlement process. This means that the rate of settlement and the pore pressure dissipation in soft clays may be faster than theoretically predicted immediately after load application, perhaps at the expense of a slower development later. For gravity platforms on soft–medium clay this is of vital importance.

From a practical point of view it is advantageous to have a rapid initial settlement so that a larger portion of the settlement is completed before operation. Consequently, it is important to be able to improve the reliability of *a priori* predictions. At present various empirical procedures for linking primary and secondary settlements are available. Real improvements in the state of the art depend on the results of fundamental research on how the settlement process develops immediately on load application.

For some time, settlement records of embankments on clay have been studied by back calculating the in situ resistances from the observations. Interesting trends of behaviour have already been obtained, but it is too soon to draw definite conclusions yet.

ACKNOWLEDGEMENT

It would be impossible to write a Paper of this scope and content unless research results obtained over a long period by a number of past and present staff members of the Geotechnical Division, Norwegian Institute of Technology, could be drawn from. Particular credit is due to the senior colleagues, Lars Grande, Kåre Senneset and Erik Hjeldnes, for general contributions of a theoretical and experimental nature, and for sharing administrative duties.

Regarding soil modelling important advances were made during the doctoral studies by (chronologically) Fritz Nowacki, Karel Karal, Mete Öner, Lars Grande, Oddvin Tokheim, Geir Westerlund, Arne Skotheim, Torgeir Døssland, Geir Svanø and Steinar Nordal. About 100 diploma theses have also been involved in our long-range research programme on soil modelling.

During the final preparation of the manuscript helpful suggestions were received from Professor David Sego.

The yearly financial support from the Norwegian Council for Industrial and Scientific Research has been particularly helpful because the conditions for the grants were sufficiently flexible to enable research to be directed into the most promising areas at the time.

Finally, permission from Mobil, Statoil and Norsk Hydro to use experimental data from North Sea soils is gratefully appreciated.

BIBLIOGRAPHY

Bakken, Å. & Westerlund, G. J. (1974). *Undersøkelse av attraksjonens og friksjonens variasjon med aksialdeformasjonen i sand*. Internal Report RI 74. Geotechnical Division, Norwegian Institute of Technology.

Bishop, A. W. (1954). The use of pore-pressure coefficients in practice. *Géotechnique* **4,** No. 4, 148–152.

Bishop, A. W. (1966). The strength of soils as engineering materials. *Géotechnique* **16,** No. 2, 91–128.

Bishop, A. W. & Lovenbury, H. T. (1969). Creep characteristics of two undisturbed clays. *Proc. 7th Int. Conf. Soil Mech. Fdn Engng, Mexico City* **1,** 29–37.

Bishop, A. W. & Wesley, L. D. (1975). A hydraulic triaxial apparatus for controlled stress path testing. *Géotechnique* **25,** No. 4, 657–670.

Bjerrum, L. (1967). Engineering geology of normally consolidated marine clays as related to settlements of buildings, *Géotechnique* **17,** No. 2, 83–118.

Caquot, A. & Kérisel, J. (1967). *Grundlagen der Bodenmekanik*. (Translated by G. Scheuch.) Berlin: Springer.

Christensen, S. (1985). *Behaviour of undrained creep and its influence on the shear moduli for a medium clay*. Internal Report 08201-08. Geotechnical Division, Norwegian Institute of Technology.

Crawford, C. B. (1964). Interpretation of the consolidation test. *J. Soil Mech. Fdns Div. Am. Soc. Civ. Engrs* **90,** SM5, 87–102.

Eide, O. & Andersen, K. H. (1984). *Foundation engineering for gravity structures in the Northern North Sea*. Publ. No. 154, 1–148. Oslo: Norwegian Geotechnical Institute.

Fredriksen, F. (1983). *Undersøkelse av en leires krypegenskaper under drenerte og udrenerte forhold*. Diploma thesis, 1–154, Geotechnical Division, Norwegian Institute of Technology.

Grande, L. O. (1976). *Samvirke mellom pel og jord*. Dr ing thesis, Geotechnical Division, Norwegian Institute of Technology.

Grande, L. O. & Eggereide, K. (1976). Effective stress stability analysis for gravity structures. *Proc. Behaviour of Off-Shore Structures Conf., Trondheim* **2,** 452–461.

Henkel, D. J. (1960). The shear strength of saturated remoulded clays. *Proc. Am. Soc. Civ. Engrs Conf. Shear Strength Cohesive Soils, Boulder*, pp. 533–554.

Hvorslev, M. J. (1937). *Über die Festigkeitseigenschaften Gestörter Bindiger Boden*, Doctoral thesis, Copenhagen.

Janbu, N. (1957). Earth pressure and bearing capacity calculations by the generalized procedures of slices. *Proc. 4th Int. Conf. Soil Mech., London* **2,** 207–212.

Janbu, N. (1963). Soil compressibility as determined by oedometer and triaxial tests. *Proc. 3rd Eur. Conf. Soil Mech., Wiesbaden* **1,** 19–25.

Janbu, N. (1965). Consolidation of clay layers based

on nonlinear stress strain. *Proc. 5th Int. Conf. Soil Mech. Fdn Engng, Montreal* **2,** 83–87.

Janbu, N. (1967). Settlement calculations based on the tangent modulus concept. Three guest lectures at Moscow State University. *Bull. No. 2, Soil Mech. Norw. Inst. Technol.*, 1–57.

Janbu, N. (1969). The resistance concept applied to deformations of soils. *Proc. 7th Int. Conf. Soil Mech. Fdn Engng, Mexico City* **1,** 191–196.

Janbu, N. (1970). *Grunnlag i geoteknikk*, pp. 1–426. Trondheim: Tapir Forlag.

Janbu, N. (1973a). Shear strength and stability of soils, the applicability of the Coulombian material 200 years after the ESSAI. In *Norsk geoteknisk forening*, pp. 1–47. Oslo: Norwegian Geotechnical Institute.

Janbu, N. (1973b). Slope stability computations. *Embankment dam engineering, Casagrande volume*, pp. 47–86. London: Wiley.

Janbu, N. (1975). Shear strength of granular soils. *Proc. Nordic Geotechnical Meetings* **1,** 37–50. Copenhagen: Polyteknisk Forlag.

Janbu, N. (1976). Soils under cyclic loading. *Proc. BOSS Conf., Trondheim* **2,** 373–385.

Janbu, N. (1979a). Mechanism of failure in natural and artificial soil structures. *Proc. Int. Symp. Soil Mech. Oaxaca* **1,** 95–124.

Janbu, N. (1979b). Design analyses for gravity platforms. *Proc. Behaviour of Off-Shore Structures Conf., London* **1,** 407–426.

Janbu, N., Grande, L. O. & Eggereide, K. (1976). Effective stress stability analysis for gravity structures. *Proc. Behaviour of Off-Shore Structures Conf., Trondheim* **1,** 449–466.

Janbu, N. & Senneset, K. (1981). Settlements due to drained cyclic loads. *Proc. 10th Int. Conf. Soil Mech. Fdn Engng, Stockholm* **1,** 165–170.

Janbu, N.,Tokheim, O. & Senneset, K. (1981). Consolidation tests with continuous loading. *Proc. 10th Int. Conf. Soil Mech. Fdn Engng, Stockholm* **4,** 645–654.

Karal, K. (1973). *En energimetode for geotekniske stabiliseringsanalyser.* Dr ing thesis, Geotechnical Division, Norwegian Institute of Technology.

Law, K. T. & Holtz, R. D. (1978). A note on Skempton's *A* parameter with rotation of principal stress. *Géotechnique* **28,** No. 1, 57–64.

Leahy, D. (1980). *Contribution a l'étude du comportement oédométrique des argiles.* Master's thesis, University of Laval.

Mesri, G. & Godlewski, P. M. (1977). Time and stress compressibility interrelationship. *J. Geotech. Engng Div. Am. Soc. Civ. Engrs* **103,** GT5, 417–430.

Meyerhof, G. G. (1983). Safety factor and limit states analysis in geotechnical engineering. *Can. Geotech. J.* **11,** 1–7.

Mitchell, J. K. (1976). *Fundamentals of soil behaviour.* London: Wiley.

Motzfeldt, E. (1976). *Spenningsendringer ved belastningsendring.* Internal Report O.7602-1. Geotechnical Division, Norwegian Institute of Technology.

Nordal, S. (1983). *Elasto-plastic behaviour of soils analyzed by the finite element method.* Dr ing thesis, Geotechnical Division, Norwegian Institute of Technology.

Norwegian Petroleum Directorate (1985). *Regulation for structural design of loadbearing structures intended for exploitation of petroleum resources.* (Unofficial translation.) Stavanger: Norwegian Petroleum Directorate.

Nowacki, E. H. F. (1973). *Endimensjonal konsolidering med spennings- og tidsavhengige materialegenskaper.* Dr ing thesis, Geotechnical Division, Norwegian Institute of Technology.

Oner, M. & Janbu, N. (1975). Dynamic soil structure interaction in offshore storage tanks. *Proc. Int. Symp. Soil Mech., Istanbul*, Bull. 9, Geotechnical Division, Norwegian Institute of Technology.

Roscoe, K. H. (1970). The influence of strain in soil mechanics. *Géotechnique* **20,** No. 2, 129–170.

Sällfors, G. (1975). *Preconsolidation pressure of soft, high-plastic clays.* Doctoral thesis, Chalmers, Gothenburg.

Schmertmann, J. H. (1976). The shear behaviour of soil with constant structure. In *Lauritz Bjerrums memorial volume*, pp. 65–98. Trondheim: Norwegian Geotechnical Institute.

Shibata, T. & Karube, D. (1969). Creep rate and creep strength of clays. *Proc. Int. Conf. Soil Mech. Fdn Engng, Mexico City* **1,** 361–367.

Singh, A. & Mitchell, J. K. (1969). Creep potential and creep rupture of soils. *Proc. 7th Int. Conf. Soil Mech. Fdn Engng, Mexico City* **1,** 379–384.

Skempton, A. W. (1954). The pore-pressure coefficients *A* and *B*. *Géotechnique* **4,** No. 4, 143–147.

Skempton, A. W. & Hutchinson, J. (1969). Stability of natural slopes and embankment foundations. *Proc. 7th Int. Conf. Soil Mech. Fdn Engng, Mexico City*, State of the art volume, pp. 291–340.

Skotheim, A. Å. (1979). *Reliability of some models of clay behaviour.* Dr ing thesis, Geotechnical Division, Norwegian Institute of Technology.

Sokolovski, V. V. (1965). *Statics of granular media*, pp. 1–270. Oxford: Pergamon.

Suklje, L. (1970). *Rheological aspects of soil mechanics.* London: Wiley.

Svanø, G. (1981). *Undrained effective stress analysis.* Dr ing thesis, Geotechnical Division, Norwegian Institute of Technology.

Taylor, D. W. (1948). *Fundamentals of soil mechanics.* New York: Wiley.

Terzaghi, K. (1925). *Erdbaumechanik auf bodenphysikalischer Grundlage.* Leipzig: Deuticke.

Tokheim, O. (1976). *A model for soil behaviour.* Dr techn thesis, Geotechnical Division, Norwegian Institute of Technology.

Wroth, C. P. (1984). The interpretation of in situ soil tests. *Géotechnique* **34,** No. 4, 449–489.

APPENDIX 1. DERIVATION OF RESISTANCE FORMULAE

Let y be the response to the action x on a test specimen, Fig. 2. By definition, the resistance of the test material is

$$R = \frac{dx}{dy} \tag{120}$$

The test leads to a resistance that is linearly dependent on the action

$$R = r(x - x_r) \tag{121}$$

Introducing equation (120) into equation (121)

$$dy = \frac{1}{r}\frac{dx}{x - x_r} \tag{122}$$

Integration between $x_0 > x_r$ and x leads to

$$y = \frac{1}{r}\ln\left(\frac{x - x_r}{x_0 - x_r}\right) \tag{123}$$

For clays a number of linear resistances are often obtained, exemplified by equations (10), (14), (19), (22), (26) and (70).

For instance, for normally consolidated clays $M_0 = m_0\sigma'$ and $R = r_s t$ for primary and secondary resistance. Hence, from equation (123)

$$\varepsilon_p = \frac{1}{m_0}\ln\left(\frac{\sigma'}{\sigma_0'}\right) \quad (\sigma' \geqslant \sigma_0')$$

$$\varepsilon_s = \frac{1}{r_s}\ln\left(\frac{t}{t_c}\right) \quad (t \geqslant t_c)$$

When compared with conventional semilogarithmic plots

$$\varepsilon_p = \frac{C_c}{1 + e_0}\log_{10}\left(\frac{\sigma'}{\sigma_0'}\right) \quad (\sigma' \geqslant \sigma_0') \tag{124}$$

$$\varepsilon_s = \frac{C_\alpha}{1 + e_0}\log_{10}\left(\frac{t}{t_c}\right) \quad (t \geqslant t_c) \tag{125}$$

it is seen that

$$m_0 = \frac{1 + e_0}{C_c}\ln 10 \tag{126}$$

$$r_s = \frac{1 + e_0}{C_\alpha}\ln 10 \tag{127}$$

where $\ln 10 = 2{\cdot}3$. Equations (126) and (127) show that both C_c and C_α are incomplete compressibility parameters. Hence, statistical data for C_c and C_α are useless in practice unless e_0 data are also available.

The ratio

$$\frac{m_0}{r_s} = \frac{C_\alpha}{C_c} \tag{128}$$

is often nearly constant (say 0·04–0·06) as also observed by Mesri & Godlewski (1979).

When interpreting several types of soil, a generalized resistance may be needed to handle non-linearity, e.g.

$$R = r x_a\left(\frac{x}{x_a}\right)^{1-b} \tag{129}$$

Introducing equation (120) into equation (129) and integrating between x_0 and x

$$y = \frac{1}{rb}\left[\left(\frac{x}{x_a}\right)^b - \left(\frac{x_0}{x_a}\right)^b\right] \tag{130}$$

As an example let the resistance R be the modulus M and the action $(x = \sigma')$ equal the effective stress; then the resistance number r equals the modulus number m. Hence, from equation (129)

$$M = m\sigma_a\left(\frac{\sigma'}{\sigma_a}\right)^{1-b} \tag{131}$$

For three values of b

$b = 0$	$M = m\sigma'$	e.g. normally consolidated clay
$b = 0{\cdot}5$	$M = m(\sigma'\sigma_a)^{\frac{1}{2}}$	e.g. sand
$b = 1$	$M = m\sigma_a = \text{constant}$	e.g. overconsolidated clay

The case $b = 0{\cdot}5$ corresponds closely to the one-dimensional compression of normally consolidated sands, in which case equation (130) leads to

$$\varepsilon = \frac{2}{m}\left[\left(\frac{\sigma'}{\sigma_a}\right)^{\frac{1}{2}} - \left(\frac{\sigma_0'}{\sigma_a}\right)^{\frac{1}{2}}\right]. \tag{132}$$

where σ_a is the reference stress of 100 kPa (approximately 1 atm).

REFERENCE

Mesri, G. & Godlewski, P. M. (1977). Time and stress compressibility interrelationship. *J. Geotech. Engng Div. Am. Soc. Civ. Engrs* **103,** GT5, 417–430.

APPENDIX 2. STRESS FIELD THEORY

The plane strain stress field theory for a weightless soil leads to simple closed form solutions for inclined strip loads on a semi-infinite body, assuming an ideal Coulomb material with a constant degree of mobilization f. The geometry of the critical stress field (corresponding to maximum f) is given in Fig. 17 for an $a\phi$ material. The stresses within each of the three zones 1–3 are as follows.

Zone 3

In zone 3 the surface load $p = \sigma_3$ while σ_1 is horizontal. Hence the normal stress (equations (34) and (35))

$$\sigma_{n3} + a = N_{n3}(p + a) \tag{133}$$

is constant along the zone boundaries 1–n, and n–0.

Zone 2

The arch nm in zone 2 is a logarithmic spiral of the form

$$R_i = R_n \exp(-i \tan\rho) \tag{134}$$

Since the resultant of $\sigma_{ni} + a$ and τ_i goes through the pole 0, moment equilibrium about the pole leads to $(\theta = \pi/2 - \omega)$

$$\sigma_m + a = N_\theta(\sigma_n + a) \tag{135}$$

where

$$N_\theta = \exp[(\pi - 2\omega)\tan\rho] \tag{136}$$

and $\sigma_n = \sigma_{n3}$, and ω is the rotation of the principal stress in zone 3. At point i on the arch the normal stress becomes

$$\sigma_{ni} + a = N_i(\sigma_{n3} + a) \tag{137}$$

when

$$N_i = \exp(2i \tan\rho) \tag{138}$$

The coefficient N_i defines how σ_n increases from σ_{n3} when $i = 0$ at n to σ_m when $i = \theta = \pi/2 - \omega$ at m.

Zone 1

The derivation of the required stress formulae for zone 1 will be made with the aid of Fig. 50. In zone 1

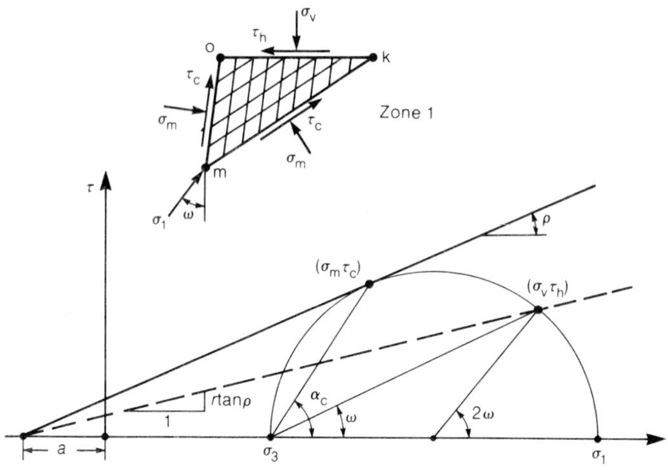

Fig. 50. Mohr circle for the stress conditions in zone 1

the major principal stress must be rotated by an angle ω to obtain vertical ($q_v = \sigma_v$) and horizontal ($t_h = \tau_h$) equilibrium with the normal (σ_m) and shear (τ_c) stresses acting on the conjugate shear planes mo and mk, Fig. 17.

From Fig. 50 by geometric inspection

$$\sigma_v + a = N_\omega(\sigma_m + a) \tag{139}$$

is obtained where

$$N_\omega = \frac{1 + \sin \rho \cos (2\omega)}{\cos^2 \rho} \tag{140}$$

From the slope $r \tan \rho$ defined in Fig. 50

$$t_h = r(\sigma_v + a) \tan \rho \tag{141}$$

where

$$r = \frac{\sin (2\omega)}{N_\omega \cos \rho} \tag{142}$$

In the defining equation (141) for the roughness ratio r it is seen that $r = 0$ means vertical loading, while $r = 1$ means sliding along the base ko. The corresponding ω values are 0 and α_c respectively.

Combining equations (140) and (142)

$$r = \frac{2 \tan \alpha_c \tan \omega}{\tan^2 \alpha_c + \tan^2 \omega} \tag{143}$$

which when solved for $\tan \omega$ yields

$$\tan \omega = \frac{1 - (1 - r^2)^{\frac{1}{2}}}{r} \tan \alpha_c \tag{144}$$

This means that σ_v and τ_h are unique functions of r, $\tan \rho$ and a.

Solution for inclined loads

Once the stresses have been obtained for all three zones, it is a simple matter to obtain the bearing capacity formula, by introducing equation (133) into equation (135) and again into equation (139), leading to

$$\sigma_v + a = N_q(p + a) \tag{145}$$

where

$$N_q = N_{n3} N_\theta N_\omega \tag{146}$$

$$N_q = \tfrac{1}{2}[(N + 1) + (N - 1) \cos (2\omega)] \exp [(\pi - 2\omega) \tan \rho] \tag{147}$$

Hence $N_q = f(\rho, r)$ because ω is a function of $\tan \rho$ and r, equation (144). For $r = 0$ (vertical load) $\omega = 0$, and

$$N_q = N \exp (\pi \tan \rho) \tag{148a}$$

For $r = 1$ (sliding) $\omega = \alpha_c$ and

$$N_q = (1 + \sin \rho) \exp [(\pi/2 - \rho) \tan \rho] \tag{148b}$$

For N_q values as functions of r and $\tan \rho$, see Fig. 18.

Special case $\rho = 0$

Theoretically, when $\rho \to 0$, $a \tan \rho \to \tau_c$. By limit considerations, or simple direct solutions (Fig. 15)

$$\left.\begin{aligned} \sigma_h &= p + \tau_c && \text{(zone 3)} \\ \sigma_m &= \sigma_n + (\pi - 2\omega)\tau_c && \text{(zone 2)} \\ \sigma_v &= \sigma_m + \cos (2\omega)\tau_c && \text{(zone 1)} \end{aligned}\right\} \tag{149}$$

are obtained. Hence, the three equations lead to

$$\sigma_v = p + N_c \tau_c \tag{150}$$

where

$$N_c = 1 + \pi + \cos (2\omega) - 2\omega \tag{151}$$

and since $r = \sin (2\omega)$, when $\tau_h = r\tau_c$

$$N_c = \pi + 1 + (1 - r^2)^{\frac{1}{2}} - \arcsin r \tag{152}$$

is obtained. The diagram for N_c is shown in Fig. 16 as a function of r. For $r = 0$ (vertical loading), $N_c = \pi + 2 = 5{\cdot}14$ and for $r = 1$ (sliding) $N_c = (\pi + 2)/2 = 2{\cdot}57$.

VOTE OF THANKS

In proposing a vote of thanks to Professor Janbu, Professor N. E. Simons made the following remarks.

'Having known and worked with Professor Janbu for many years, it is a pleasure to have the opportunity of proposing the vote of thanks to him on the occasion of the twenty-fifth Rankine Lecture. The first Rankine Lecture dealing with seepage problems under dam foundations was given by Professor Arthur Casagrande. In his introduction to that first lecture, Professor Skempton remarked that he "could feel great satisfaction that what might be called the foundation stone of the Rankine Lectures was being laid by such a distinguished hand".

'In the intervening years, we have seen great practitioners of the art of geotechnical engineering building on that foundation stone with a series of outstanding lectures covering a wide variety of important topics and the published Lectures in *Géotechnique* provide a unique record of the scope of our subject and its development since 1961.

'Professor Janbu this evening has now made his contribution to that record in an exceptional way. His subject certainly is topical. The current issue of the *New Civil Engineer* contains a news item discussing the large settlements which have been observed under the North Sea oil platforms in the Ekofisk field. Oil was first extracted from the North Sea in 1973 and in the same year Bjerrum published his paper "Geotechnical problems involved in foundations of structures in the North Sea". It is therefore timely that the twenty-fifth Rankine Lecture should describe the contribution that Professor Janbu and his Norwegian colleagues have made to solving the immense geotechnical problems involved in North Sea oil exploitation.

'In the first Rankine Lecture, Professor Casagrande stated that he would "make use of theory to supplement empirical knowledge and to enhance sound judgement". That is very much Professor Janbu's approach to engineering practice. He is not only an outstanding exponent of the analytical method in soil mechanics, but he also combines this with sound judgement based on practical experience.

'Most of us here this evening know of and use the Janbu pile driving formula. What is perhaps not so well known is that Professor Janbu has worn out more than one pair of gloves working on a piling rig installing driven piles when obtaining practical experience.

'Professor Janbu has not only made an exceptional contribution to the practice of geotechnical engineering, but he is also a dedicated teacher. Most of the younger geotechnical engineers working in Norway are Professor Janbu's former students and his book "Grunnlag i geoteknikk" is a classic teaching text. The parallel between the first and the twenty-fifth Rankine Lecturers is close and that is perhaps not surprising since Professor Janbu worked with Professor Casagrande at Harvard and obtained his doctorate there.

'Mr Chairman, when a member of our society is invited to lecture abroad it is usually expected that the lecture will be delivered in English and that the proceedings also will be conducted in English. It is at least consistent that when a lecturer from overseas speaks here both the lecture and the proceedings are again conducted in English. That seems a one-sided arrangement so I shall attempt to redress the balance.

'Nilmar, det er nesten tredve år siden vi først traff hverandre i Trondheim. Jeg reiste dit for å hjelpe deg med NGI publikasjon No. 16 som ble utgitt i 1956. Du inviterte meg hjem til middag og vi hadde det rigtig hyggelig sammen. Siden dengang har vi vaert sammen mange ganger og ved mange anledninger, sist i Stockholm i 1981, og det har vaert like hyggelig hver gang. Det er derfor en stor glede for meg å ha denne anledning til å gratulere deg med din enstående fem og tyvende Rankine forelesning og å takke deg på vegne av den Britiske Geotekniske Forening.

'Mr Chairman, it gives me great pleasure to propose a warm vote of thanks to Professor Janbu for his outstanding twenty-fifth Rankine Lecture.'

The vote of thanks was accorded with acclamation.

The Rankine Lecture

The twenty-sixth Rankine Lecture of the British Geotechnical Society was given by Dr A. D. M. Penman at Imperial College of Science and Technology, London, on 4 March 1986. The following introduction was given by Professor A. W. Skempton, Imperial College of Science and Technology.

'Arthur Penman, born in 1922, was educated at Penrith Grammar School and read civil engineering, under Professor Fisher Cassie, at King's College Newcastle. In 1943 Cassie spent part of the summer vacation in the soils laboratory at the Building Research Station, and before leaving he asked one of the station's staff to give a short course of lectures and tutorials on soil mechanics to the students and practising engineers at Newcastle.

'This course was given in May 1944, and Penman seems to have been inspired by what he heard. Thus he leapt at the opportunity, soon afterwards, to join the Building Research Station as an Assistant Grade 3, and caused some consternation at his selection board by insisting on coming into the soil mechanics section, rather than being pushed into research into concrete. He duly arrived at Garston in August 1944 and, a feeling I can well understand, liked it so much that he spent his whole career there.

'Among his first tasks was the development of an automatic apparatus to measure pore pressures without volume change in the triaxial test. He became involved with earth dams in testing London Clay foundations for Walton Reservoir and, with Alan Bishop, investigating leakage of King George V reservoir near Chingford.

'It was Penman who installed the first hydraulic piezometers in this country, at the Usk Dam, in 1951. This job soon hit the geotechnical headlines, for his piezometers in the clay fill showed, after the first construction season, a pore pressure ratio of 0·8. That is to say, the piezometric level was well above the top of the fill, a finding which the engineers were reluctant to believe. I vividly remember installing a stand-pipe, with Stanley Serota's help, to demonstrate the effect. We had to use an exceptionally tall step-ladder to measure the water level, which stood in the pipe at a height of 15 ft above the bank top and confirmed the piezometer readings.

'The list of dams with which Penman has subsequently been involved is a long one. No doubt we shall be hearing about several of them in the Lecture.

'Embankment dams, earthfill and rockfill, are his main interest, but naturally he has worked in other geotechnical fields: oil tank foundations, sheet-pile walls and offshore structures, to mention a few.

'For published papers, which are numerous, he has received Telford premiums and the British Geotechnical Society Prize, on more than one occasion, and was awarded the degree of DSc in 1972.

'One of the endearing aspects of our lecturer is his enthusiasm for soil mechanics, retained throughout his career, and expressed in many ways, not least in his service as a member of Council from 1964 to 1967, chairman of the British Geotechnical Society from 1972 to 1974, and a member of the *Géotechnique* Advisory Panel and of the BNCOLD committee.

'It is most appropriate that Arthur Penman should be invited to give a Rankine Lecture. We have been looking forward to hearing him on a subject to which he has made outstanding contributions.'

Dr A. D. M. Penman

PENMAN, A. D. M. (1986). *Géotechnique* **36,** No. 3, 303–348

On the embankment dam

A. D. M. PENMAN*

'There is a mathematical theory of the combined action of friction and adhesion in earth: but for want of experimental data its practical utility is doubtful'—Rankine

The significance of the embankment dam stems from its importance to mankind in providing one of the cheapest means for storing large volumes of water. Historically this was required for irrigation, but it is also a necessity for hydro-power and supply for industrial and domestic use. The embankment dam was the first type of dam built by man: it is the most numerous type; it is the type most often chosen for a new dam and forms the world's highest dam. It was rivalled by various forms of concrete dam, but developments since the 1930s in geotechnical science, understanding of behaviour through instrumentation and improvement of earth-moving machinery has made it the foremost type of dam throughout the world. Improved methods of stability analysis, utilizing the concept of effective stresses and measured pore pressures, enable safe slopes to be constructed. Weaknesses due to slickensides and/or the effects of failure developing progressively along a potential slip surface clearly must be given due consideration. The simple concept of expressing soil strength, whether peak or residual, in terms of c' and φ' is beginning to be replaced by the concept of a curved failure envelope and analytical methods are available for stability analysis. Despite these advances some slips still occur. Three cases are considered. It may be preferable to design for acceptable movement rather than simply to provide a factor of safety against unacceptable slip failure. Analytical methods utilizing finite element techniques have enabled predictions to be made from deformation parameters. To assess these methods, accurate measurements of movements are required. Movements that occur during temporary cessation of construction can give valuable indications of strains developing within the dam that may cause undesirable reduction of stress or onset of progressive failure. Precise measurements of horizontal movements, even without instruments to measure inside the fill, can reveal a change to unacceptable rates of movement. Arching action, not only across a narrow core, but across a dam from upstream to downstream and between abutments may result in undesirable reduction of total stresses, leading to hydraulic fracture. Failure due to erosion and piping is most dangerous because it can occur while the reservoir is full. Improved design of filters may limit erosion, but it may be better to ensure also that total stresses across any potential plane of fracture through the waterproof system are adequate to prevent hydraulic fracture. Cases of hydraulic fracture and examples of wet seams are discussed. It is probable that many cases of leakage originate in hydraulic fracture. The subsequent degree of erosion under conditions of zero effective stress may be dependent on many factors that require detailed research study.

L'importance du barrage en terre pour l'humanité provient de ce qu'il représente un des moyens les moins coûteux pour amasser de grands volumes d'eau. A une époque reculée de tels barrages servaient pour l'irrigation, mais ils restent encore nécessaires pour la puissance hydro-électrique et pour les besoins industriels et domestiques. Le barrage en terre a été le premier barrage à être construit par l'homme; il constitue toujours le type de barrage le plus nombreux. On le choisit le plus souvent pour un nouveau barrage et ce genre de barrage a été utilisé pour les plus grandes hauteurs. Différentes formes de barrage en béton lui faisait concurrence, mais le barrage en terre reste encore le type le plus important dans tous les pays, grâce à l'évolution des sciences géotechniques depuis les années 1930, à la compréhension approfondie du comportement des barrages réalisée grâce à l'instrumentation et aux perfectionnements effectués dans les machines employées pour les travaux de terrassement. Il est possible de réaliser des pentes stables à l'aide des méthodes améliorées pour analyser la stabilité, basées sur le concept de contraintes effectives et de pressions interstitielles mesurées. Il faut bien entendu tenir compte des faiblesses due aux surfaces de glissement et/ou des effets des ruptures qui peuvent se produire de façon progressive le long d'une surface miroir potentielle. Le concept simple d'exprimer la résistance maximale ou résiduelle du sol en fonction de la cohésion effective c' et de l'angle de frottement effectif φ' commence à être remplacé par celui de l'envelope de rupture non rectiligne, tandis que des méthodes analytiques sont disponibles pour évaluer la stabilité. Malgré tous ces progrès il arrive que des glissements se produisent encore. Trois cas sont décrits en détail. Peut-être est il préférable de calculer le barrage en fonction d'un mouvement admissible plutôt que de fournir seulement un coefficient de sécurité contre la rupture par glissement. On a pu faire des prédictions sur la base des paramètres de déformation à l'aide de méthodes analytiques utilisant des techniques d'éléments finis. Des mesures précises des mouvements sont nécessaires pour évaluer ces méthodes. Des mouvements qui ont lieu à l'occasion des pauses pendant les travaux de construction peuvent donner des indications précieuses concernant les déformations qui se produisent dans le barrage et qui peuvent causer une réduction inadmissible de

* Geotechnical Engineering Consultant, Harpenden.

contrainte ou le commencement de rupture progressive. Des mesures précises des mouvements horizontaux, effectuées même en l'absence d'instruments à l'interieur du remblai, peuvent révéler un changement vers des vitesses inadmissibles de mouvement. Une réduction inadmissible des contraintes totales, entrainant une rupture hydraulique, peut provenir d'un effet de voûte non seulement en travers d'un noyau étroit, mais en travers d'un barrage entre ses côtés en amont et en aval et aussi entre des dispositifs de butée. La rupture due à l'érosion et au renard est très dangereuse, parce qu'elle peut se produire lorsque le bassin de retenue est rempli d'eau. L'érosion peut être limitée par des filtres de construction perfectionnée, mais peut-être vaudra-t-il mieux aussi de prendre des mesures pour que les contraintes totales en travers tout plan de rupture potentiel dans le système imperméable soient suffisantes pour empêcher la rupture hydraulique. Des cas de rupture hydrauliques et des exemples de lignes de séparation humides sont discutés. Il est probable que la rupture hydraulique est à l'origine de beaucoup de cas de fuites. L'érosion qui s'ensuit dans des conditions de contrainte effective nulle peut dépendre de beaucoup de facteurs qui exigent une étude approfondie.

SIGNIFICANCE OF THE EMBANKMENT DAM

The benefit of dams to mankind is undoubted. Their earliest role in providing storage for irrigation water formed a major contribution to the development of our civilization. The oldest dam in the world (Kerisel (1985), quoting Helms), dating from around 4000 BC, was built of earth with a masonry facing, at Jawa in Jordan. In India there was a tradition of dam building that at one time was considered as one of the seven meritorious acts which a man ought to perform during his lifetime (Rao, 1951). During the period of British tenure, many dams were constructed by traditional methods and they were accepted as a means of famine relief giving employment to thousands. The completed schemes, in ensuring crop production, not only paid, but brought happiness and contentment to the people (Buckley, 1898). Today, vast areas of land throughout the world rely for their productivity on irrigation, e.g. southern California fed from the reservoirs at Trinity (142 m), Oroville (230 m) and other large embankment dams.

In Britain, the Industrial Revolution required water for transport, industrial processes and the growing cities. In the 18th century, dams were built to store water for canals; during the 19th century, the majority were for water supply; early in the 20th century, dams specifically for hydropower were constructed in Scotland and Wales to provide electricity for aluminium smelting. Ingots from arc furnaces were produced in June 1896 (Hamilton, 1986) with power from Foyers Dam that had just been completed. It is claimed that Cragside, the home of Lord Armstrong in Northumberland, was the first house to be lit by electricity derived from water power: by arc lamps in 1878 and Swan's incandescent lamps two years later. A small embankment dam that is still operational had been built by estate workers under Armstrong's direction to impound water for a Thomson vortex turbine.

The prospect of almost limitless renewable energy offered by hydro-power excited the world. The International Union of Producers and Distributors of Electrical Energy, realizing the need for development of the specialized knowledge of dam building, conceived the International Commission on Large Dams (ICOLD) in 1928. This new body became a Commission of the World Power Conference in 1930 and has continued to attract new member countries: currently the membership of ICOLD comprises 77 countries. The interchange of experience and dissemination of research findings has made this body of inestimable value: congress transactions and the publications of technical committees form milestones in the development of the subject.

World register

To assess the number, type and size of dams and reservoirs, ICOLD took on the daunting task of compiling a world register: the 1985 edition contains information from 116 countries and has been used to construct Fig. 1. In general, to be eligible for entry, a dam's height must exceed 15 m, but for the four countries (China, Japan, India, USA) with more than 1000 dams, only those exceeding 30 m height have been included. Thus the register excludes large numbers of small dams, many of which are of the embankment type, e.g. Britain is shown to have 411 embankment dams and 125 of other types of dam, whereas it has been estimated that there are more than 2000 subject to the Reservoirs (Safety Provisions) Act of 1930 in the country. An exact number will not be known until implementation of the 1975 Reservoirs Act produces the required national register. The height of dams given by this latest edition of the world register is from lowest foundation rather than from stream bed level.

Accepting these limitations, Fig. 1 shows the increase in the number of large embankment dams during the period 1800–1985. Since 1955, the number has been increasing at an almost constant rate of 200 per year. Clearly the increase is a response to the accelerating increase of world population. The height of embankment dams is of considerable geotechnical interest in view of the stresses and water pressures developed: the world's highest exceeded 100 m in 1926, 200 m in 1968 and 300 m in 1980.

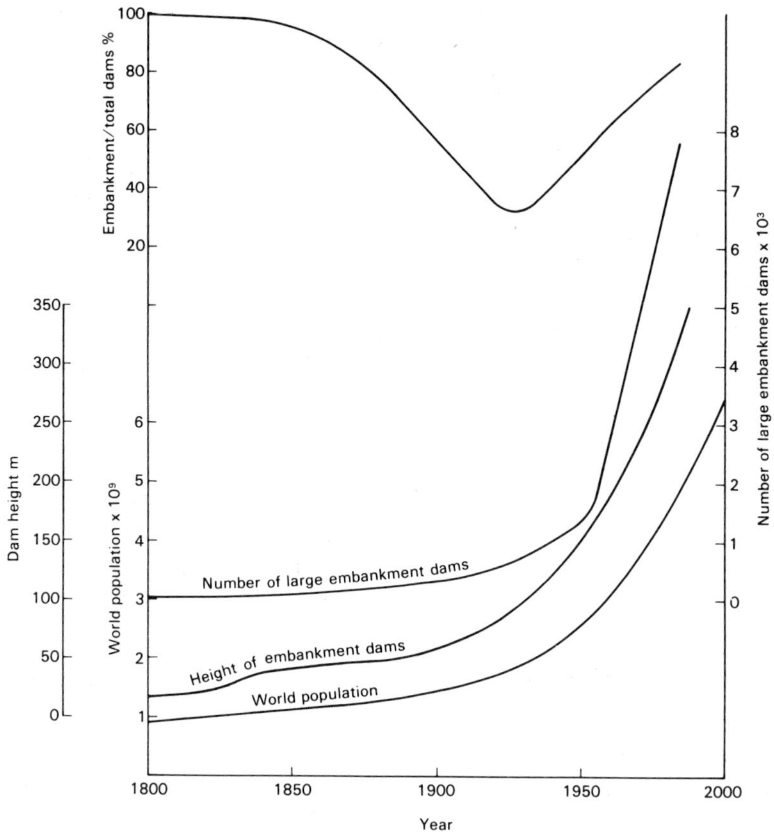

Fig. 1. Embankment dam statistics 1800–1985

Currently the world's highest dam is the Russian embankment dam Nurek, 300 m. For about 18 years, before it reached full height, the world's highest dam was the concrete gravity Grand Dixence (285 m) in Switzerland, completed in 1962. During that period, first Oroville (230 m, USA, 1968) and then Mica (242 m, Canada, 1972) were the highest embankment dams.

Mica was exceeded by Chicoasen, 261 m, completed in Mexico in 1980. Two projected embankment dams in India are Tehri, 261 m, and Kishau, 253 m, both in Uttar Pradesh and expected to be completed during 1985–90. Rogun, the 335 m embankment dam in Russia will be the world's highest dam when it is completed.

Almost all the world's dams were of the embankment type before 1800. During the 19th century, concrete technology and methods of structural analysis evolved. Combined with the fact that there were many sites available with sound rock foundations at shallow depths, this created an increasing interest in concrete dams. The effect was to reduce the ratio of embankment to total number of dams built, from the approximately 100% value that had previously existed. This ratio for dams built in a five-year period (Fig. 1) shows that by the turn of the century there were more large concrete than large embankment dams being built. The ratio reached its lowest value of 33% in the late 1920s, but has now recovered and of all large dams built recently more than 80% are of the embankment type.

Improved standing

The embankment dam represents an almost purely geotechnical problem. Much of the recovery can be attributed to improvements in design methods due to developments in soil mechanics since publication of *Erdbaumechanik* in 1925 and the introduction of instrumentation to reveal dam behaviour. These have enabled a wider range of local soils and rocks to be used, heights to be vastly increased and very difficult foundations to be accepted.

The introduction of the internal combustion engine to power earth-moving machinery has also played a major role. In addition to providing large concentrations of power for excavating strong soils and rock and compacting them effectively, the cost has always been lower than muscle power. Construction in India (Strange, 1898) could involve 20 000 people excavating soil with pickaxes and long hoes and transporting it as headloads of about 0·009 m^3. In a dam of 25 m height, the fill was spread in thin layers, watered and compacted by foot: the layer thickness was allowed to increase with dam height from 0·07 m to a maximum of 0·15 m. In Spain, the 28 m high Ponton de la Oliva dam was built by 1500 convicts, 200 labourers aided by 400 beasts of burden and four steam engines (Smith, 1970). Compaction was by herds of animals driven backwards and forwards over the fill.

In current construction, it is not uncommon to find compaction by vibrating rollers of rockfill in 2 m layers.

The equipment used to construct Grand Maison (140 m) had a total power during the 1983 construction season of 52 500 h.p. The consumption of diesel fuel was 1·88 l/m^3 of placed fill. During 1981, when fill levels were lower, the consumption was 1·60 l/m^3. At Sulby (60 m) rockfill dam on the Isle of Man the consumption was 1·85 l/m^3 and at Carsington (35 m) 1·75 l/m^3. The total volume of Grand Maison is 12·9 hm^3 and its construction used about 22×10^6 l of fuel oil. Compared with the amounts of irreplaceable oil consumed routinely for heating, electricity generation and transport, this is a very small amount and represents a sound investment, contributing to the supply of hydro-power—a replaceable resource.

The efficiency of earth placing equipment has made it attractive for concrete dam construction. The use of rollcrete (Lowe, 1962) and the development of special mixes (Dunstan, 1981) have led to the rolled concrete dam. There are many examples in various parts of the world: those in Japan have been described by Yamauchi, Harada, Okada & Shimada (1985).

Potential danger

The large potential energy of water stored behind a dam makes its uncontrolled release dangerous. Dam failures that have allowed rapid escape of reservoir water are relatively few, despite the ever increasing number of dams. An analysis by Schnitter (1979) of information collected by ICOLD showed that the percentage of embankment dams built in a given year, that subsequently failed allowing the release of water, has fallen at least tenfold during the first half of this century. da Silveira (1984), analysing further material collected by ICOLD on deterioration, has shown that the probability of failure of embankment dams has fallen from 0·028 in the period 1900–20 to 0·0035 during 1960–75. The causes of failure of embankment dams are almost equally divided between

(*a*) erosion by overtopping
(*b*) rotational slips
(*c*) internal erosion.

Improved hydrological studies and methods of predicting flood flows are reducing overtopping risks but there is a geotechnical requirement to improve resistance to accidental overtopping. British Flood Studies Reports since 1933 have provided design methods that have largely overcome overtopping, but recent introduction of the concept of a probable maximum flood (PMF) has indicated that the spillways of many old dams are inadequate. The return period of a PMF cannot be well defined, but is clearly very long. The inadequate spillways of many dams more than 100 years old have successfully prevented overtopping and a solution to the problem may be to improve erosion resistance to permit emergency overtopping.

The 140 year old Toddbrook Dam (20 m) near Whaley Bridge has had a wide concrete auxiliary spillway built over its crest, and Mackey (1985) has described the reinforcement with interlocking concrete cellular blocks of the crest and downstream slope of an old dam. A current research programme by the Construction Industry Research and Information Association is studying the effectiveness of various forms of surface reinforcement that may be used on low embankment dams.

Failure by rotational slip usually occurs during construction, before there is water in the reservoir: various aspects will be discussed in the next section.

Failure by internal erosion is much more dangerous because it can occur suddenly, with a full reservoir. It is the most serious current geotechnical problem relating to embankment dams. Various aspects including hydraulic fracture will be discussed in the Paper.

The following four examples illustrate failure by overtopping and internal erosion.

Estrecho de Rientes (45·7 m) built 1755–89 in Spain was probably the world's highest embankment dam at that time. The reservoir filled for the first time in February 1802 and the dam breached in April releasing a flood that destroyed part of the town of Lorca, drowning 600 people. The exact cause of failure was not determined.

South Fork (21·9 m), Pennsylvania, was built of earthfill with a 1:2 upstream slope and a downstream shoulder of rockfill at 1:1·5. The crest width and freeboard were 3 m. Overtopping occurred during the day on 31 May 1889 and the

Fig. 2. Breach in Dale Dyke after failure in 1864

dam withstood 0·5 m depth of water over the crest for $3\frac{1}{2}$ hours before a breach formed: the resulting flood caused the loss of 2209 lives.

Dale Dyke (29 m), England, failed on first filling in March 1864. Fig. 2 shows the breached dam from upstream. The dam contained a central narrow puddled clay core and had twin cast iron pipes of 0·45 m dia. passing through it to the outlet control valves at the downstream toe. In a reassessment of the Dale Dyke failure, Binnie (1978) uncovered evidence, ignored by the enquiry, of a whirlpool seen during a calm period several days before the failure, indicating a substantial flow entering the upstream slope at about three-quarters of the dam height. The reason that this did not cause visible flow from the downstream slope may be due to the loose nature of the fill and the presence of a large rockfill toe which acted as a drain. Subsequently, Binnie (1981) found evidence that there had been a large issue of water from the foot of the embankment where the breach occurred. He also uncovered evidence that substantial steps had been left in the longitudinal section of the cut-off trench. During construction a 'spring' had been found and excavation was continued to expose the source, leaving an almost vertical face 10 m high and another 3 m high under the central part of the dam. Differential settlement of the puddled clay across the discontinuities could have produced sufficient reduction of total stress to permit hydraulic fracture. Binnie (1877) condemned vertical steps in the floor of cut-off trenches, stating

> 'owing to the unequal depths of puddle that occurs at the steps, the superincumbent weight has caused that on the deeper side to settle more than that on the higher, and so produce a vertical crack or fault in the puddle which has led to serious consequences'.

Teton Dam (93 m), Idaho, is the highest embankment dam to have failed, Fig. 3. It was built mainly from silt across the Teton River canyon in volcanic rocks containing interconnecting open joints and voids. The site was chosen for the situation of the reservoir to store irrigation water for the surrounding silt-covered plains, not because it was suitable for a dam. It failed on first filling when the reservoir level was only 1 m below the spillway gate cill (9·2 m below the crest). No instruments had been placed in the dam and assessment of behaviour depended on visual observations. Although a regular inspection was made for any signs of leakage near the downstream toe as the reservoir approached top water level, the initially very low regional water-table combined with the high overall per-

Fig. 3. Teton Dam

meability of the bedrock to prevent any water appearing at the ground surface until two days before failure.

On 3 June 1976, small, clear springs were found on the right abutment about 450 m downstream of the toe. On the next day some patches of wetness appeared, closer to the toe of the dam, but they gave no cause for concern. On 5 June at 7.00 a.m. there was a steady flow of water coming from the toe, adjacent to the right abutment. Muddy water was soon found to be issuing from the downstream slope of the dam itself and within a few hours backsapping had produced a substantial channel carrying a considerable discharge of earth-laden water and a whirlpool had formed in the reservoir in line with the discharge. Despite the efforts of two bulldozers to check the backsapping, the channel rapidly eroded back to the crest of the dam, which breached at 11.57 a.m., less than five hours after seepage was first seen coming from the dam itself. Six hours later, the 27 km long reservoir was essentially empty and the breach in the dam showed that $2{\cdot}5 \times 10^6$ m^3 of fill adjacent to the right abutment had been lost. Because the developments before failure occurred during daylight and warning had been given to the authorities, most people were able to escape the ensuing flood.

PORE PRESSURE AND STABILITY

Dam engineers are faced with the sometimes conflicting requirements of watertightness and stability. Relatively fat clays were used for many almost homogeneous section dams in India (Strange, 1898) and success may be partly attributed to construction by thousands of workers carrying fill as headloads and compaction in 75–150 mm layers by the feet of men and animals. Homogeneous sections were preferred to avoid stress variations caused by differential settlements in fills of different materials. It was also felt desirable to raise the fill uniformly during construction. Failures were not uncommon and were associated with the downstream fill becoming excessively wet. Slip surfaces in pure black soil were seen to be smooth, of unctuous appearance, striated by the small particles of contained grit.

During the mid 19th century, it was thought that there was a maximum height to which an embankment dam could safely be built. French engineers were reported as placing the limit at 18 m. Rawlinson (1883), the government-appointed inspector of the 1864 failure of Dale Dyke, said that he knew that some engineers had been greatly tormented with leaks from beneath earthen embankments and it had been put on record that no dam, if it had to retain a depth of more than 18 m of water, ought to be constructed of earthwork. He went on to point out, however, that many already existed at greater heights. As late as 1914, Uren (1914) indicated that the limit was 24 m.

> 'Beyond this height, though many have safely exceeded this in England—notwithstanding the theories of French engineers upon the subject—unless carried out with the utmost care and under the strictest supervision, they are troublesome to erect and treacherous when filled, being liable to sudden and unforeseen slips.'

Early piezometers

Slips that occurred during construction could be dug out and replaced with stronger fill, but it was a different matter with slips that occurred when impounding. There was a desire to see whether water was getting through into the downstream fill. At Waghad Dam (32 m) begun in 1881 in the Násik Collectorate of India, constructed of a plastic expansive clay ($w_L = 70$, $w_p = 35$), a major slip occurred during construction. The homogeneous section had reached 29·5 m and the slip carried the downstream toe out about 20 m. It was stabilized by drainage using rock-filled trenches cut through to the rock foundation. Attempts to continue construction caused further movement, so the crest was moved upstream and lowered from the design height by 5·1 m as shown by Fig. 4. To check on the position of the phreatic line when the reservoir was filled, the British engineers installed standpipe piezometers in 1907 (Nagarkar, Kulkarni, Kulkarni & Kulkarni, 1981). This may be one of the earliest installations of piezometers in an embankment dam.

The problem of instability in homogeneous dams caused by the phreatic surface reaching the downstream slope was apparent in the USA. Attempts were made to measure its position in cased holes bored into the fill after construction. A rising water level in these holes before impounding had begun revealed construction pore pressures.

A systematic study of the behaviour of embankment dams was started by the US Bureau of Reclamation (USBR) in 1936. The slow response time of open holes was quickly recognized and the Goldbeck earth pressure cell (Goldbeck & Smith, 1916) was modified by placing a carborundum disc in front of the pressure-sensitive diaphragm to form a remote reading piezometer. These were placed in holes bored on completion of construction and sealed in with concrete backfill. Pore pressures as high as $r_u = 0{\cdot}7$ were measured and found to be slow to dissipate.

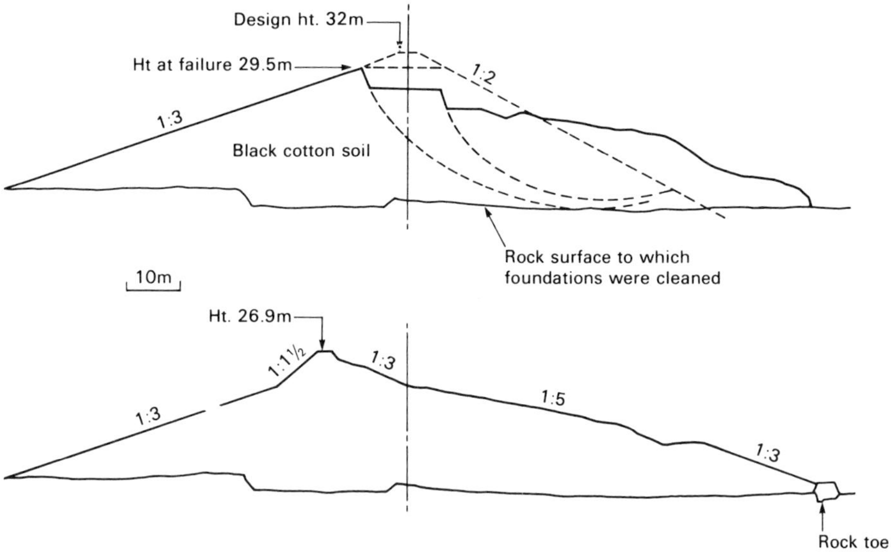

Fig. 4. Waghad Dam: original and redesigned section

Walker (1948) expressed the view that it was therefore not surprising that so many dams had failed during or immediately after construction. Unfortunately, these findings led to a concerted effort to reduce construction pore pressures by the use of low placement water contents, which has had a marked effect on dam design and subsequent behaviour.

Research by the USBR into the behaviour of fill produced methods for predicting construction pore pressures from fill properties and placement air and water contents. Boyle's law gave the pressure of the air, and its solution into the pore-water was governed by Henry's law. If the fill was sufficiently compressible and contained little air, the air soon went into solution as total pressures increased and after that $\delta u = \delta\sigma$.

It was found that, with soils that were compacted too dry, subsequent saturation under load could produce a sudden reduction in volume, i.e. collapse settlement. As the placement water content was increased, a value was reached when collapse settlement ceased and this was regarded as a suitable lower limit. It was thought that an upper limit could be specified, according to the magnitude of pore pressure that could be tolerated, but under the overburden pressures of high dams even placement at the lower limit did not prevent pore pressures from developing. A rule therefore evolved that the placement water content should be 1–3% dry of optimum.

When referring to optimum water content, it is necessary to specify the compaction energy and, in relation to variations from optimum, knowledge of I_p is valuable. The standard Proctor test uses 596 kJ/m^3 whereas the USBR compaction test uses 1135 kJ/m^3 and the modified AASHO uses 2630 kJ/m^3. Clearly the USBR optimum water content will be lower than that for the standard Proctor test, but the effect of reducing the water content by 3% will be much more marked in a soil of low plasticity such as a silt than it would be for a fat clay.

Charles (1979) showed that the change in moisture content required to produce a required change in c_u of a clay fill was a linear function of I_p

$$\delta c_u = -\frac{2{\cdot}3Bc_u\,\delta w}{I_p}$$

where B is a constant which for most clays is about 2.

It is clearly more straightforward to specify a required value of c_u for core fill than to specify placement w. Strength specification has been used in Britain during the last two decades (Kennard, Lovenbury, Chartres & Hoskins, 1979).

It is of interest to note that the twin tube hydraulic piezometer stemmed from damage to a modified Goldbeck unit. Pore pressure measurements were made by increasing the air pressure inside the cell until movement of the diaphragm broke an electrical circuit. The air pressure was then thought to equal the pore pressure. A 6 mm copper tube containing an electrical wire connected the cell to the instrument house, where a Bourdon gauge measured the air pressure. Condensation in the cell could provide continuation

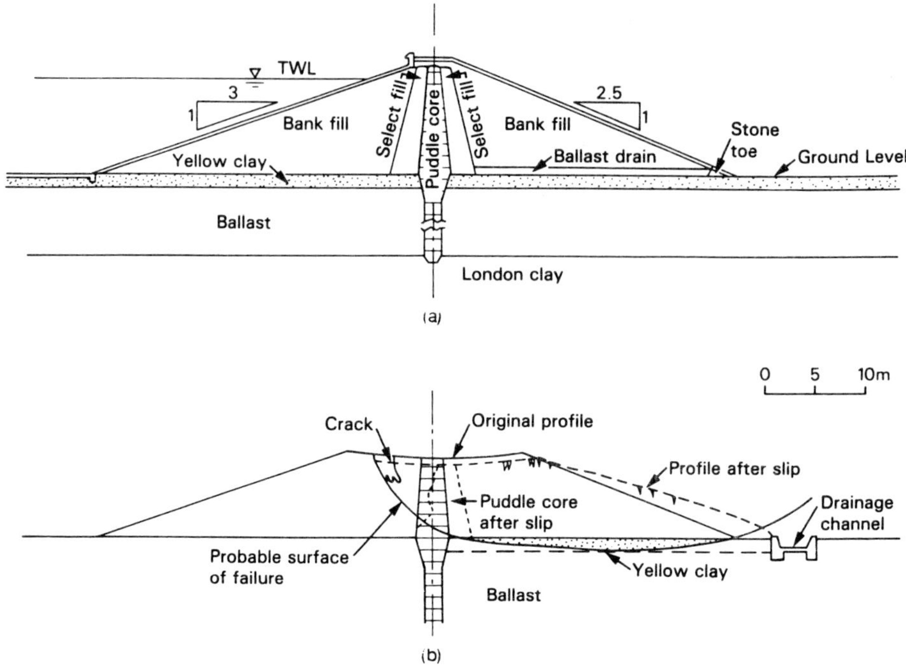

Fig. 5. Chingford Dam, showing the position of the slip surface: (a) section of the bank as designed; (b) section through the bank after slip

of the electrical circuit until excessive diaphragm deflexions had been caused; however, cells with ruptured diaphragms allowed the Bourdon gauges to respond to pore pressure all the time. A second connecting tube was provided so that the tubes could be filled with water to improve the response time.

Speed of construction

Whatever the concept by Victorian engineers of water in fill, advice given by Strange (1898) would allow for pore pressure dissipation and improve stability. He recommended that the fill should not be raised more than 9 m during one season and that a completed dam should be left for at least one season to consolidate before filling the reservoir.

The effect of rapid construction was demonstrated by the failure of Chingford Dam during construction in 1937. The bank, to be 10·4 m high, was built with Caterpillar D8 tractors pulling tracked Athey tipping waggons. The fill level was raised 4 m during the month before the rotational slip. Investigation by the Building Research Station (BRS) (Cooling & Golder, 1942) showed that the slip surface passed through the puddled clay core (Fig. 5) and a layer of soft yellow clay. This had been left in the foundation under the gravelly downstream shoulder at the insistence of the owner/designer, against the recommendation of the contractor. The failure involved about 100 m length of bank that moved out bodily 4·3 m, causing the fill surface to drop about 0·6 m.

Values of c_u were measured with undrained tests on 'undisturbed' samples on site in the portable autographic compression apparatus and at the laboratory in ring shear under zero normal load. (Jenkin had developed his ring shear apparatus expressly to obtain c distinct from φ—Cooling (1936).) For the yellow clay, average $c_u = 14$ kN/m^2 and, for the puddled clay, $c_u = 10$ kN/m^2. When used in a two-circle modification of the Swedish stability analysis, these values gave a factor of safety of unity.

Eight months after the failure, when rebuilding began, tests on the yellow clay under the removed fill showed $c_u = 36$ kN/m^2: an increase of 22 kN/m^2 since failure. The vertical load on the clay layer was about 144 kN/m^2 and Skempton calculated values of pore pressures at the centre of the layer for various times after the application of load based on consolidation theory and laboratory values of c_v. At 37 days and 277 days, corresponding to the times of failure and rebuilding, the calculations showed $u = 115$ kN/m^2 and $u = 5$ kN/m^2 respectively, i.e. an increase in effective stress of 110 kN/m^2. Drained shear box tests

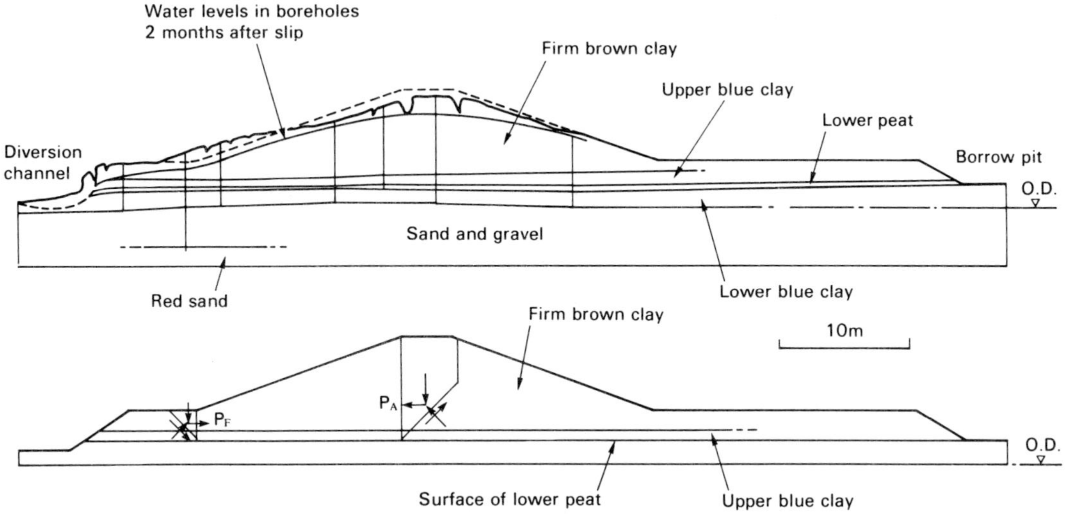

Fig. 6. Flood bank for the River Don at Thorpe Marsh

showed this to cause an increase of $c_u = 30$ kN/m^2, which compared favourably with the increase measured in the field.

Cooling and Golder, acknowledgeing Skempton's contribution, remarked that with a more permeable foundation layer, such as silt, the method of calculation could be used to control the rate of construction so that the strength of the foundation material should at no time be exceeded.

The events surrounding this failure were classic in the annals of British geotechnical engineering because of the interest that they aroused among dam engineers and the attention they drew to the research of the BRS. To protect the contractor's interests, Mowlem's Agent, Wynne-Edwards (who became President of the Institution of Civil Engineers for 1964–65 and was knighted as First Chairman of the Council of Engineering Institutions) brought Terzaghi to the site: he unreservedly approved the BRS report. The evident need for a commercial laboratory to carry out site investigation and to test samples led to the formation of Soil Mechanics Ltd.

Rapid construction also led to the failure of Muirhead (27 m) in 1941. Wartime conditions demanded early completion of the dam and the introduction of track laying machinery accelerated placement from 2500 m^3/week to 12 230 m^3/week and enabled the upper fill to be placed in a few months. A rotational slip occurred when dam height reached 21·9 m and an investigation by the BRS showed that it passed through the lower wetter fill. It was not a brittle failure and after initial movement of about 1·3 m further placement of 0·46 m of fill caused 0·15 m horizontal movement. The upstream slope was stabilized by rockfill toe weighting and, as at Waghad, the crest was moved upstream and lowered 5 m below design level. Although pore pressures were not measured, the failure was clearly due to high values that had developed under the rapid height increase. As a result, standpipe piezometers were placed in the fill of the nearby Knockendon Dam that was currently under construction. These piezometers, placed in 1944, were probably the first to be used in an embankment dam in Britain. This work was described by Banks (1948, 1952). They showed $r_u > 0{\cdot}5$ but stability had been ensured: after Muirhead, design was modified to include a key of granular fill placed in excavation through the existing downstream fill and a substantial toe weighting berm upstream.

An early example of analysis using effective stresses was that of the failure of a small flood bank in 1948. The 5·5 m bank (Fig. 6) built of brown clay ($c_u = 29$ kN/m^2), on a softer blue clay containing a layer of peat ($c_u = 13$ kN/m^2) failed shortly after construction, when a new river channel was excavated, thereby removing toe support. A total stress analysis gave $F = 1{\cdot}6$ and it was suggested that the shearing resistance of the peat under and beyond the toe had been reduced by the redistribution of pore pressure along the layer from the higher construction values under the bank. This concept was checked by placing 12 hydraulic piezometers in the peat across the section of new bank to be built on the opposite side of the channel. Measured values and details of the analysis were given by Ward, Penman & Gibson (1955). Like Chingford, failure was along a soft layer in the foundation, but the

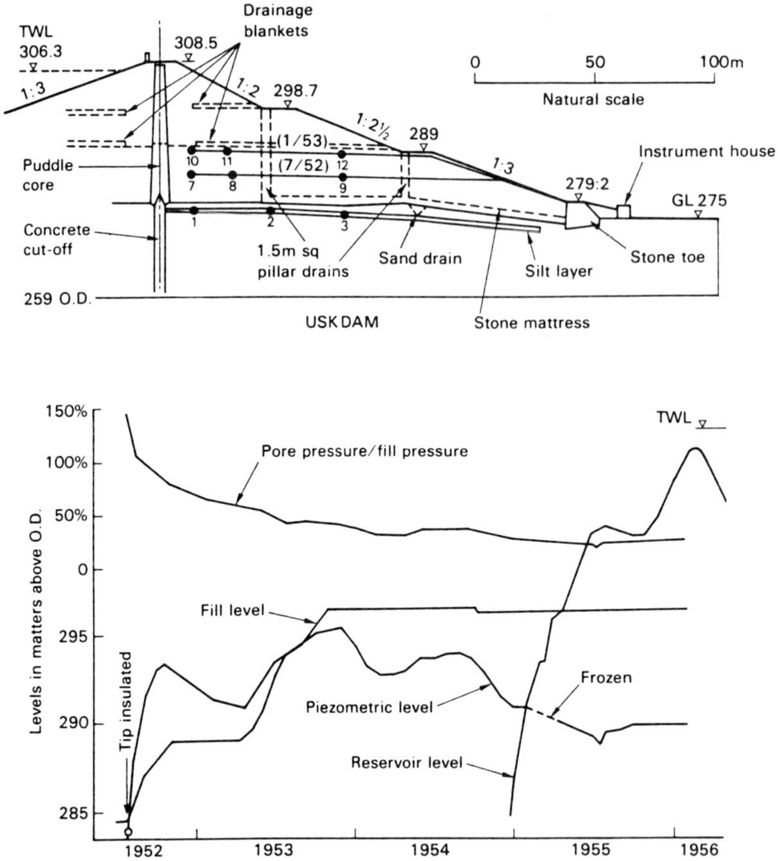

Fig. 7. Usk Dam: section and measured pore pressure at tip 9

failure surface was analysed by wedge shapes rather than by circular arcs.

Usk Dam

The BRS became involved in the Usk Dam (33 m) when a silt layer was discovered in the foundation during excavating for a stilling basin. After designing sand drains for the silt, two tube hydraulic piezometers developed from the USBR type were installed to check pore pressures in the silt layer. It was then agreed with the consultants that the apparatus would be extended so that tips could be placed in the fill to observe construction pore pressures for research interest. The findings were rather remarkable.

Three tips were placed at mid-height of the first season's fill during July 1952, at the positions shown by Fig. 7, and measured values gave $r_u > 1{\cdot}5$. Even though the initial pressures quickly fell, pore pressures towards the end of the winter shut-down period were too high for stability of the completed dam. Under the direction of Skempton and Bishop, standpipe piezometers were driven into the fill on an adjacent section and confirmed the high pressures measured by the two tube piezometers—decorators' step-ladders had to be used to measure water levels in the standpipes! Details of this experience have been given by Penman (1979).

Stability was ensured by changes in the design and construction method. Horizontal drainage layers were placed in the fill to reduce drainage path lengths and placement water content was reduced by winning the fill with face shovels instead of scrapers. By the end of construction, r_u values had fallen sufficiently to satisfy effective stress stability analysis based on the Swedish method of slices.

This experience led to the use of horizontal drainage layers in the shoulders of numerous dams constructed of clayey fill, e.g. Selset, Derwent, Staunton Harold, Altnahinch, Diddington, West Water, Chelmarsh, Backwater and Carsington. Horizontal drainage layers at a vertical spacing of 3 m had been provided in the moraine fill forming part of the downstream shoulder of

Harspranget Dam (50 m) built in Sweden 1946–51 (Westerberg, Pira & Hagrup, 1951).

The problem of $r_u > 1{\cdot}5$ in the Usk fill was particularly intriguing because it is known that pore pressures in clayey fills must be below atmospheric (i.e. there must be pore suctions) to enable the fill to support construction machinery. To avoid unacceptably deep ruts $c_u > 40$ kN/m^2 (Dennehy, 1979) and this implies pore pressures that are less than -50 kN/m^2 near the fill surface. It is evident that a piezometer tip must be strong enough not to be affected by total pressures and that its intake filter must be fine enough to exclude soil particles to separate porewater pressure from total pressure. The filters at Usk were 51 mm dia. carborundum discs similar to but of much larger area than those used by the USBR. Cooling, from his experience with building stones and the use of the suction plate apparatus, suggested that, if porewater suctions were to be measured, the pores of the intake filter should be small enough not only to exclude soil particles but also to exclude air. Owing to surface tension and curvature of the meniscus in a partly saturated fine-grained soil, the pressure of air in the pores is greater than that of the water. In a similar way, a saturated, fine-pored intake filter can prevent the ingress of air because of the differential pressure set up by the curved menisci at the entrance to each pore. For the filter to be successful, it must have uniform-sized pores: special materials are required to obtain good permeability with small enough pore sizes. Rogers (1935) designed a tensiometer to measure porewater suction in agricultural soil: its unglazed earthenware pot could support suctions of 0·8 bar. Black, Croney & Jacobs (1958) used a tensiometer with a sintered glass filter with a pore size of less than 1·5 μm that could measure a suction of 1 bar. Work at the BRS and Imperial College led to the fine-pored piezometer tip (Fig. 8) described by Bishop, Kennard & Penman (1960) that is now in general use for hydraulic piezometers in fill.

A comparison was made of coarse and fine filter units at Chelmarsh Dam by installing an electrical piezometer with a coarse filter adjacent to a Bishop tip. The measured pressures, shown by Fig. 9, indicate that the coarse filter instrument responded to pore air pressure, while the fine filter enabled the much lower porewater pressures to be measured. Increasing total pressures, as the fill was raised, reduced the difference between the two pressures: had the dam been higher, saturation would have occurred, making the pressures the same.

The higher pore air pressures measured by coarse filters leading to $r_u > 1$ usually only occurred under small overburden pressures. Penman (1956) reported two other cases, in addition to Usk, where initially $r_u > 1$. In both cases r_u fell below unity after 1–2 m of fill had been placed. At Usk the average was 3 m with a maximum of 4·3 m of fill.

Fig. 8. Bishop tip during installation

High pore pressures were measured in the moraine core of Hyttejuvet Dam during construction in 1964. From the Usk experience it was thought that this might in part be due to the use of coarse intake filters. As a check, two pairs of piezometer filters, one coarse and the other fine pored to give a high air entry pressure, were placed side by side in this part-saturated fill. Also, to check on the high pressures deeper down in the core, various types of high air entry pressure piezometers were installed in boreholes. The results of this comparative study of filters with different thicknesses and fineness showed somewhat surprisingly that there were no significant differences in the measured pore pressures. Details of this work were given by Dibiagio & Kjaernsli (1985).

Beavan, Colback & Hodgson (1977) have given examples of pore pressures measured in the cores of six dams and have shown that it is not uncom-

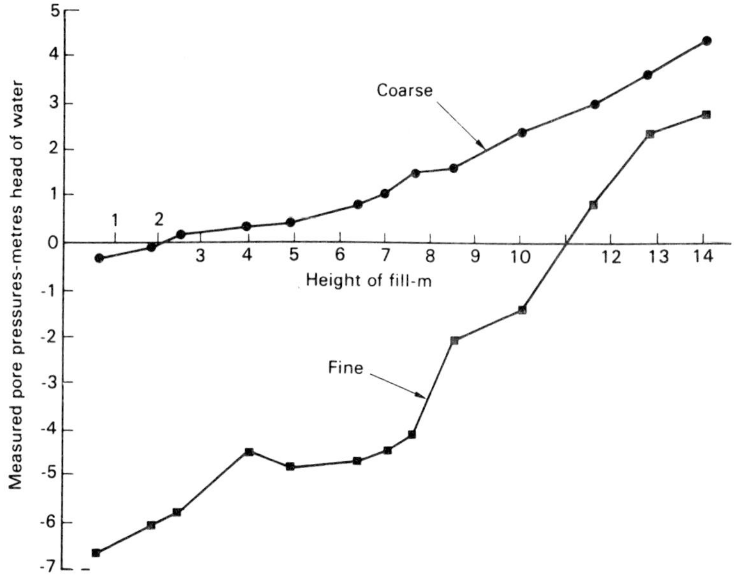

Fig. 9. Pore pressures at Chelmarsh Dam

mon for $\delta u = \delta\sigma_0$ during the earlier stages of construction. At Kielder (52 m) a piezometer in the upstream clay blanket, just upstream of the core, gave $\delta u = 1{\cdot}78\sigma_0$ as the fill approached full height. This was probably due to arching of the type observed at John Martin Dam, discussed in the section on total pressures and arching later.

Slip analysis

Slope stability analysis in terms of effective stresses by the Swedish slices method was made much more rigorous by Bishop (1955) and programmed for computer by Little & Price (1958). Non-circular analysis by computer was provided by Morgenstern & Price (1965) and methods of analysis have been developed, for example, by Janbu (1957), Sarma (1973) and Celestino & Duncan (1981). Several computer programs are now commercially available: with an input of correct values for γ, c', φ' and r_u and the ability to ensure that a realistic shape of slip surface is being analysed, it should be possible to design dams with stable slopes. Skempton (1985) has pointed out that in practice the factor of safety cannot be obtained with an accuracy greater than about $\pm 10\%$ and due allowance must be made.

The fabric of soil (Rowe, 1972) is broken up by excavation from borrow and spreading and compacting as fill. Thus some fills may approach an 'ideal' uniform soil with fairly constant values assigned to c' and φ'. It is recognized (de Mello, 1977) that the failure envelope is frequently curved, so that φ' is dependent on confining pressure and c' may be less than is often assumed. Charles & Soares (1984a, b) have given methods of analysis that cater for these variations.

If the failure surface is likely to go through the foundation, conditions may be very different. Skempton (1964) drew attention to the effect of fissures and joints and the presence of any existing slip surface in reducing φ' towards φ_r', introducing the 'residual factor' R, which may be considered as the proportion of the total slip surface in the clay along which its strength has fallen to its residual value. In the case of reactivation of an old slip, $R = 1$, whereas in clays which are not fissured or jointed (new puddled clay and possibly some compacted clayey fills), the value of R approaches zero.

A variation in the stress–strain properties of different types of material through which a potential slip surface may pass, i.e. clay core, compacted shoulder fill and foundation, can lead to progressive failure. Strains within a soft clay core may be sufficient to carry a stiffer shoulder fill or foundation clay over the peak to various reduced residual values. Such reductions, occurring at first locally, can throw extra stress on to adjoining sections of the forming slip surface, leading to progressive failure.

In the light of this knowledge, it is interesting to consider three recent failures involving the upstream shoulders of dams.

Three examples of slips

At Açu Dam (40 m) a slip occurred on 15 December 1981 during construction when it was

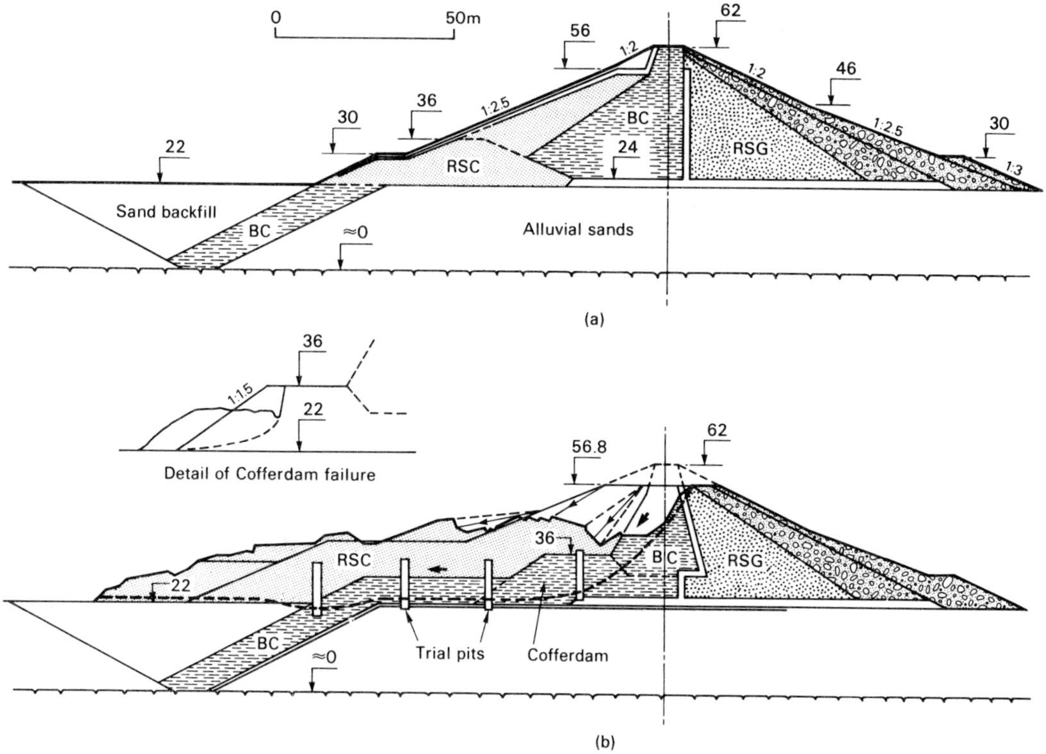

Fig. 10. Açu Dam: (a) original design; (b) failed section

still 5·2 m below crest level. The foundation was 22 m of sand overlying bedrock with a water-table close to ground level. A dark grey–black flood plain silty clay (BC) and an unsaturated red terrace clayey sand and gravel (RSC) were available as fill. A fill with greater gravel content (RSG) was obtained by selection in the borrow area. An original design for the dam is shown by Fig. 10(a). An international specialist consultant was called in to advise on instrumentation and rapid drawdown conditions. He suggested that material RSC would not be sufficiently impervious to provide the link between below ground cut-off and the core. He therefore proposed a modified design, to which the dam was built, with a horizontal layer 7 m thick of material BC over the sand to connect the cut-off to the core.

At an early stage of construction, a coffer-dam 14 m high with an upstream slope of 1:1·5 was built from material BC just upstream of the core position (Fig. 10(b)) in little more than a month. Just after construction, two slips occurred, as shown by the coffer-dam detail, each about 150 m long and separated by some distance, involving about 16×10^4 m^3 of the clay fill. In the absence of pore pressure measurements, a back analysis by total stresses was used and showed $c_u = 49$ kN/m^2.

The coffer-dam was repaired with a slope of 1:2.5 and the clay layer was placed over the sand, initially in accordance with good practice at a placement water content that was wet of optimum. It had apparently been agreed that for material BC $\varphi' = 20°$, but discussion on the values to be assigned to c', probable pore pressures and the use of circular arcs in stability analysis were still continuing as construction reached a height of 34·8 m.

The failure occurred in about 30 minutes over a length of about 600 m, causing the construction surface to fall 15 m and the upstream toe to move 25 m horizontally. First signs had been tension cracks in the level construction surface along the downstream edge of the core. Trial pits revealed slip surfaces in the lower part of the clay blanket, as shown by Fig. 10. A back analysis using total stresses gave $c_u = 48$ kN/m^2, showing remarkable agreement with the value obtained from back analysis of the coffer-dam failures. Conventional unconfined compression tests gave values of $c_u =$ 80–90 kN/m^2. Intensely laminated layers were found in the trial pits, thought to have been caused by the earth-moving and compacting machinery remoulding, shearing and laminating the soft silty clay. The clay has been found to have a high salt content and strength parameters $c' = 10$ kN/m^2 and $\varphi' = 18°$. A back analysis using effective stresses with these parameters has given an average value for core and blanket of $r_u = 0{\cdot}4$ (Costa Filho, 1984).

Details of this failure were published by de Carvalho (1982), de Mello (1982) and Pessoa (1982).

The construction of San Luis Dam (116 m) in California during 1963–67 incorporated several low hills in its $3\frac{1}{2}$ mile length. It forms the largest off-stream reservoir in the USA—its $2{\cdot}5 \times 10^9$ m^3 capacity is pumped from the Californian Aqueduct. It was first filled in 1968.

The slip in the upstream slope, described by Kramer (1982) was first discovered on 14 September 1981 and found to be moving at 150 mm per day after an extended drawdown. It occurred where the dam was 61 m high and had been constructed over a hillside, as shown by Fig. 11. By 10 October 1981, it was moving at up to 300 mm per day and had formed a steep scarp 7·6–9·1 m high, 12–15 m upstream of the crest blacktop road which contained cracks 25 mm wide. The reservoir level was already below the toe of this part of the dam and a toe weighting berm was constructed to stabilize the slip.

An investigation found that a layer up to 6 m thick of highly plastic clayey head deposit had been left on the lower part of the 1:4 foundation slope and had softened under full reservoir. For the material with $I_p = 45$–50, $\varphi' = 16$–$17°$. Consultant T. M. Leps, who took part in the investigation, has provided the added information that in tests with modest deformation values fell to $\varphi_r' = 12°$. Further analysis has been made by Chugh (1986).

Four tunnels pass through the bedrock to the intake structure and any leakage from them could pass through some of the sandstone seams to affect the clayey head deposit. Piezometers have been installed and have indicated a slight rise in pore pressure when water was being pumped into the reservoir.

The difficulty of predicting an event of this nature is emphasized by the facts that on 25–28 February 1981, only seven months beforehand, the Department of Water Resources (DWR) and the USBR had reviewed the dam as part of the 'Safety evaluation of existing dams programme'. In addition the DWR–USBR had completed the five-yearly inspection on 4 September 1981, only ten days beforehand.

The failure of the upstream slope of Carsington Dam (35 m), almost at the end of construction, has been described by Skempton & Coats (1985). Fill placement had been stopped for three days by heavy rain and a longitudinal crack at the downstream edge of the clay was found at 7.30 a.m. on Monday 4 June 1984 when the surface was being inspected to see whether it was suitable for placing the remaining 1–2 m height of fill to bring the dam to full height. As the crack continued to open, attempts were made to seal it and other parallel cracks which formed, with hand-placed fill compacted by a light, self-propelled roller. Movement accelerated during the next two days, culminating in a large movement during the night of 5–6 June, which exposed a section of the slip surface 10 m high (Fig. 12) that had passed through the core, causing a horizontal movement at the upstream toe of 13 m. A section where the failure was initiated is shown by Fig. 13. As at San Luis, a small hill had been incorporated in the dam, providing a foundation that sloped upstream. The appearance of the dam was similar to that of Açuafter failure.

The bedrock was a grey carboniferous mudstone, weathered near the surface and covered by a layer of yellow clay. This, with weathered mudstone, was used for the core and less weathered mudstone was used for the shoulders. A grout curtain cut-off was placed some distance upstream of the centre line and connected to the core by a thick wedge of clay that formed an upstream extension of the base of the core. This connected to the layer of yellow clay that was left in place under the shoulder fill on the sloping formation. The clay core, base extension and clay layer formed a convenient path for the slip surface.

Almost a year before the slip, the contractor had constructed a toe weighting berm (in July–August 1983) to support the highest part of the dam. This had followed a reassessment of stability, the details of which were contained in a report by Kennard (1983). Failure started at one end of the berm, then spread along the dam length to include the major sections, leaving a back scarp almost 500 m long. Work begun during the afternoon of 4 June 1984 to extend and raise the berm was unable to halt the slip.

The dam was well instrumented with twin tube hydraulic piezometers in the core and its base extension, cross-arm settlement gauges etc. Several piezometers had been placed in the weathered mudstone below the yellow clay layer, but there were none at the mid-thickness of the layer to indicate actual pore pressures on that part of the slip surface.

The core had been built to a strength specification and tests made with a hand vane pushed into exposed slip surfaces three days after the

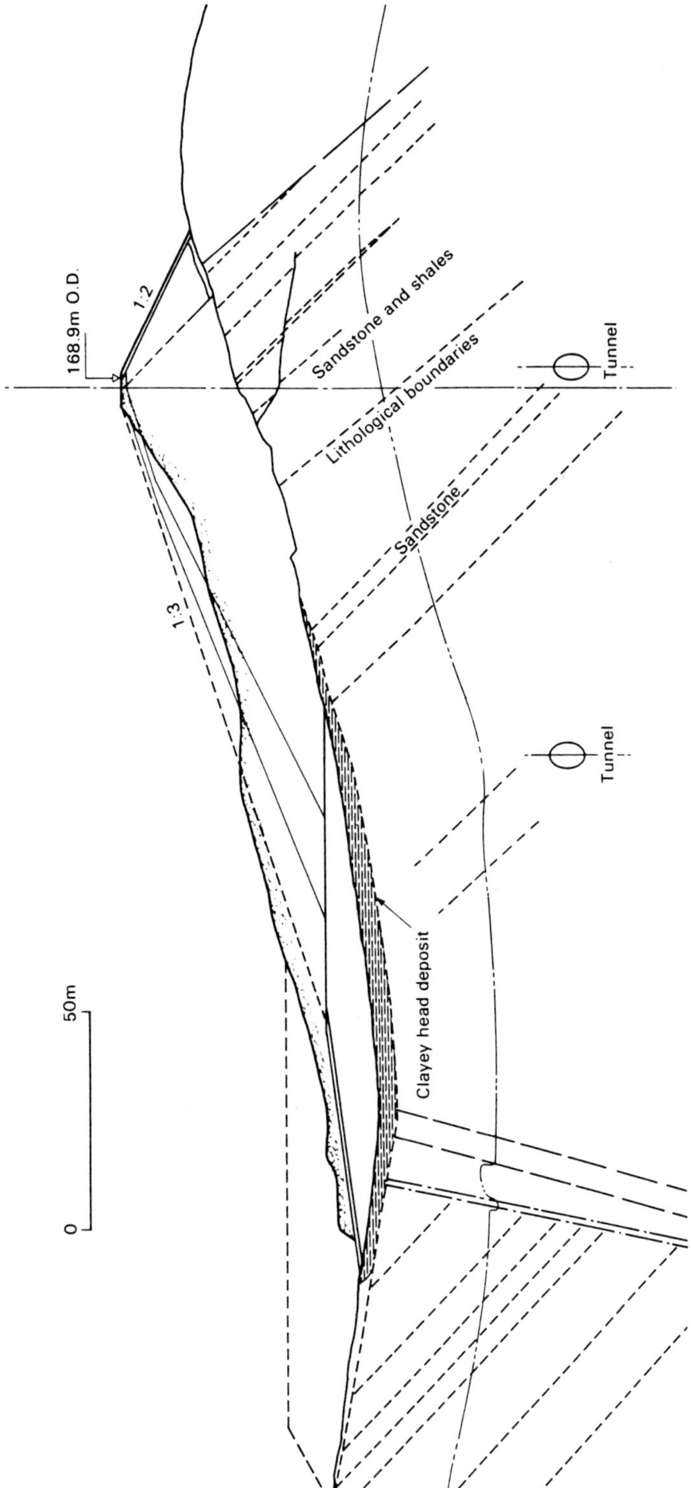

Fig. 11. San Luis Dam: failed section

Fig. 12. Back scarp of slip at Carsington

failure confirmed $c_u \approx 70$ kN/m^2. A total stress analysis based on the shape of the actual failure surface required $c_u \approx 50$ kN/m^2.

The traditional circular arc type of failure surface of the computer programs used in design checks were not applicable to the failure surface that developed and Skempton & Coats (1985) report a multiple wedge back analysis that used shapes closely approximating to the surface observed during the failure investigation. This analysis, using known values of pore pressure in the core and its base extension, but assuming $r_u = 0$ for the layer of yellow clay and using peak values for strength parameters, indicated a factor of safety of 1·4. This value only fell to unity when consideration was given to prefailure strains in the clay (reducing c' to zero) and the existence of pre-existing shear surfaces in the yellow clay layer. Geological evidence showed that the upper part of the yellow clay containing some pebbles was part of head deposits formed by periglacial solifluction: a downslope movement accompanied by shearing. A detailed examination of the clay layer amounting to a total length of 62 m disclosed pre-existing shears occupying 36 m length.

At the major section over some areas where the river had removed the yellow clay, the mudstone shoulder fill is believed to be founded on the weathered mudstone. Failure of this section was probably induced by failure of adjacent sections as the slip developed along the length of the dam, but potential progressive failure may have caused this section to have a relatively low factor of safety. These aspects have aroused new interest in the concept of progressive failure along a potential slip surface and have led to the development of more detailed analytical methods. Sideways propagation of a slip from an initial failure position has also been given more detailed consideration and publication of this work is expected after final reports on Carsington have been completed.

These three cases show the common features of clay layers under the upstream shoulder and similar overall size of structure. The shapes of the failure zones at Açu and Carsington were very similar and an adversely sloping foundation existed at San Luis and Carsington. It can be argued that adequate methods of analysis existed that could have provided satisfactory designs, had all the relevant information been obtained from the sites. The failures occurred in countries that have particularly high standards of geotechnical engineering and considerable experience of advanced methods for embankment dam design.

At Açu it must be presumed that allowance had not been made for the effects of progressive failure and that predicted pore pressures had been underestimated. The absence of piezometers prevented a better reassessment of failure conditions. It is agreed that, to form a watertight flexible core, the clay fill should be placed wet of optimum, but if an upstream blanket must be used the most careful consideration should be given to maintaining a practical balance between the desired non-cracking, flexible behaviour, variations in stress–strain properties and the requirements of overall stability.

At San Luis and Carsington the slip surface passed through foundation clays that had previously suffered downslope creep or shear movements. Due allowance had not been made for this at San Luis because the presence of these slope wash materials was not detected by the predesign site investigation. At Carsington, the design had failed to recognize the weaknesses of the existing, relatively thin layer of yellow clay (close enough to the surface to have been removed), nor had it made adequate allowance for the effects of progressive failure, which were accentuated by the upstream extension of the core base.

DEFORMATIONS AND ACCEPTABLE MOVEMENTS

While there is no doubt that shear failure produces unacceptable movements, it is not easy to define the magnitude of movements that are

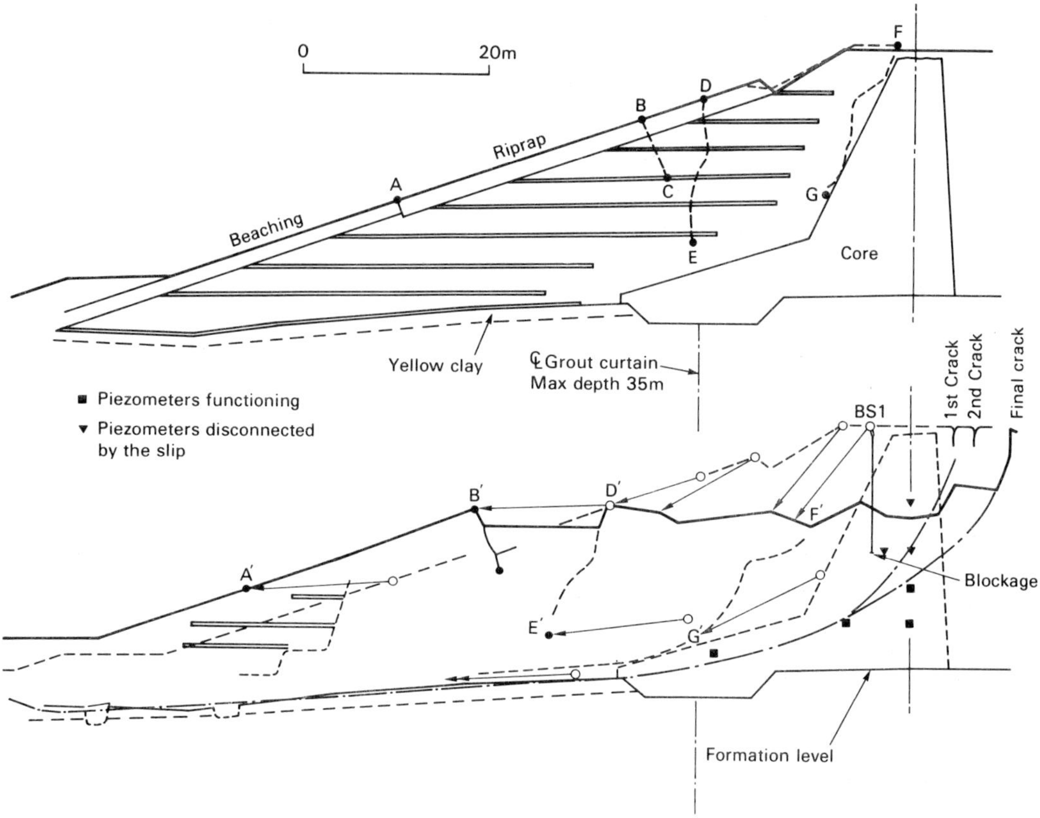

Fig. 13. Carsington Dam: section near origin of failure

acceptable nor, at the design stage, to predict movements. Observations of settlements were among the first to be made on embankment dams because of concern for adequate freeboard and because the development of surveying instruments enabled measurements to be made easily and fairly accurately. Consolidation theory enabled predictions to be made and it became usual to build fill to above the desired crest level, making an allowance such as $0{\cdot}015H$ for post-construction settlement. Often settlements were less than expected and in several Swedish dams with moraine cores the amount of measured crest settlement could be accounted for by consolidation under self-weight of only the top quarter or third height of core. This led to concern that continuing consolidation of the lower part could induce horizontal cracks through the core.

The effect of differential movements between different fill zones and between fill and abutments, while not apparent during continuous construction, can cause visible cracking during halts in construction. At Mattmark Dam (100 m), Switzerland, an avalanche damaged the work-camp, causing construction to be stopped when the dam was at about half-height, in August 1965. After two months, transverse cracks across the core and down the upstream slope appeared near the right abutment. After observation for nearly two years, the cracks were repaired by trenching and backfilling. No further cracking was seen during remaining construction and no abnormal leakage developed when the reservoir was filled.

There was a similar experience at Duncan Dam (36 m) built over a buried canyon 380 m deep in the valley of the Duncan River, British Columbia. Settlements exceeding 4 m were expected. To minimize damage, fill near the abutments was kept low while construction of the central length continued, with the intention of completing the abutment sections after settlement. Several weeks after construction had stopped in the abutment areas, transverse cracks 25–80 mm wide were found right across the core. These were repaired by trenching to depths of 12 m. About 6% bentonite was added to the core material to increase I_p from 4 to 20. Details have been given by Gordon & Duguid (1970).

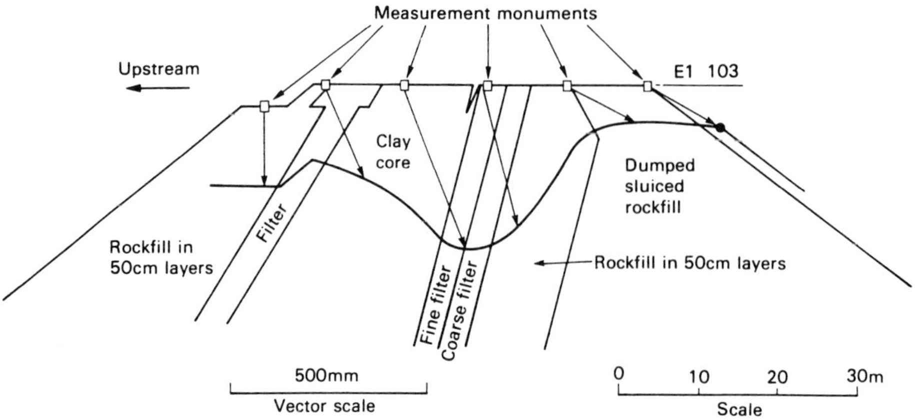

Fig. 14. Djatiluhur Dam: surface movements 10 January–13 April 1965

This experience illustrates construction deformations that, without halts in fill placement, would not have been observed. It is also evident that useful measurements of vertical and horizontal movements may be made on the surface of the fill during enforced rest periods such as winter shut-down.

Such an approach was used at Djatiluhur Dam (100 m) when longitudinal cracking was seen at the downstream edge of the core, in the construction surface, only 12 m below design crest level. Impounding had begun and the reservoir had recently exceeded the height of the coffer-dam, introducing water into the upstream rockfill.

Construction was stopped and six measuring stations placed across the surface as indicated by Fig. 14 (Sherard, 1973). Measurements of both vertical and horizontal movements made during the period 10 January–13 April 1965 confirmed that the upstream rockfill was settling more than the downstream shoulder and that the width was increasing. The measurements also showed that the clay core was settling more than the shoulders.

At Hyttejuvet (93 m) measurements made during the winter shut-down of 1964–65, when the dam was 44 m high, showed 8 cm settlement of the wet moraine core and no settlement of the supporting gravel fill.

These aspects will be discussed further in the section on total pressures and arching.

To obtain more information about the behaviour of puddled clay cores, the BRS developed two settlement gauges that were installed during 1957 in Selset Dam (37 m). A vertical plate gauge, consisting of vertical, telescopic plastic tubing with steel plates threaded over it, was installed at two positions on the core centre line. Because the puddled clay was placed by hand and compacted by foot, the tubing could be kept above fill level, supported by temporary frames. Plate levels were found by lowering an induction coil through the tubing.

To overcome problems with vertical tubes in machine-placed shoulder fill, a sealed water overflow settlement gauge was used, based on that of Spangler (1933). The buried unit was connected to an instrument house by three plastic tubes: one connecting the overflow to a transparent standpipe, one to balance the air pressure in the buried unit and one as a drain for overflowed water. These tubes were placed in trench through the fill. The mistake was made of using small, 2·5 mm bore tubing (piezometer tubing) for the water level. The air and drain tubes were 13 mm bore. The equilibrium time was so long that the level of the overflow was found by plotting the rate of fall in the standpipe and extending the straight line to zero rate. Equilibrium time is proportional to d^4, so by increasing the tube bore to 6 mm, almost the largest size that can be cleared of air by flushing, the time was reduced to about one minute (depending on distance), making it practical to repeat readings to check behaviour and to improve accuracy. The same 6 mm bore tubing was used for the air balance: it had been found extremely difficult to remove water that had entered the 13 mm bore air tube. A description of the current apparatus was given by Penman, Charles, Nash & Humphreys (1975).

Consolidation of the puddled clay was small during the time of construction as evident from the small settlements that occurred during the two winter shut-down periods. Measured settlements during active construction, however, were much greater than those of the adjoining fill, indicating a good deal of plastic flow and widening of the lower part of the core. This must have caused

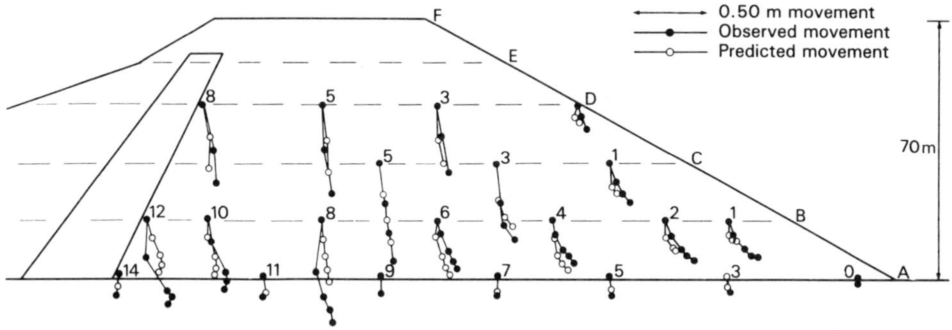

Fig. 15. Predicted and observed movements at Scammonden Dam

compression of the shoulder fill in the horizontal direction and horizontal movements may have extended to the toes.

Horizontal movements

To measure horizontal movements within the fill, the BRS developed the horizontal plate gauge and at the same time developed methods for predicting these movements. The gauge, which used 70 mm dia. telescopic plastic tube, laid to a slight outward fall, passing through metal plates, was similar to that used in Gepatch Dam in Austria. The BRS gauge used a General Post Office duct motor to pull an induction coil and a water overflow unit into the tube so that both the horizontal and the vertical position of each plate could be measured. The tube was laid from instrument chambers built into the downstream slope and by measuring their position by precise surveying from stable reference pillars, built into the valley sides outside the zone of influence of the dam, the position of each plate could be measured in relation to these independent stable positions. An overall accuracy of ± 3 mm was obtained with this system.

First installation of these gauges was in Scammonden Dam (70 m) during 1967. The tube diameter was considered to be too large for it to be taken through the clay core so measurements were confined to the downstream shoulder which in this dam, because of a motorway on the crest and a sloping core, was particularly wide. Plates were placed at positions that could be used as nodes of a finite element grid, and the tubes terminated about 5 m inside the clay core, so that plates could be placed at the interfaces between core and filter, and filter and rockfill shoulder.

Predictions of movement were made by an analysis using finite elements and properties of the rockfill obtained with a large oedometer. The work was described by Penman, Burland & Charles (1971) and results showing predicted and observed movements are given by Fig. 15.

This work was repeated when Llyn Brianne (90 m) was constructed. At both dams, the downstream face of the cores moved downstream during construction, indicating that horizontal pressure from the clay exceeded that from the rockfill shoulders causing compression of the downstream shoulder, movement of the downstream slope and expansion of the core width.

Movements caused by impounding

The thrust in a downstream direction from water impounded in the reservoir (4000 t/m length of dam over the central part of Llyn Brianne) was expected to cause some downstream movement. The fact that numerous dams have been designed with clay cores curved upstream in plan illustrates that the expectation was general. It was with some surprise, therefore, that it was found that there was no downstream movement of the measuring points on the downstream faces of the clay cores of both dams as the reservoirs rose to three-quarters of full height. The accuracy of the measurements was sufficient to show that after construction was complete the face of the core began to move slightly upstream owing to a reduction in the total pressure associated with dissipation of construction pore pressures. These small horizontal movements of the downstream face of the core continued during initial impounding.

To make a joint that will be watertight, as with a gasket between steel flanges, the total pressure normal to the joint must exceed the water pressure. At the interface between core and foundation, the normal total pressure must exceed the maximum water pressure that will be imposed by the reservoir: a principle which must apply to any potential surface passing through a clay core. The observed lack of downstream core movement on impounding was consistent with total pressures from the core, supported by the shoulders, that exceeded reservoir water pressure: a feature that could be regarded as essential for successful water

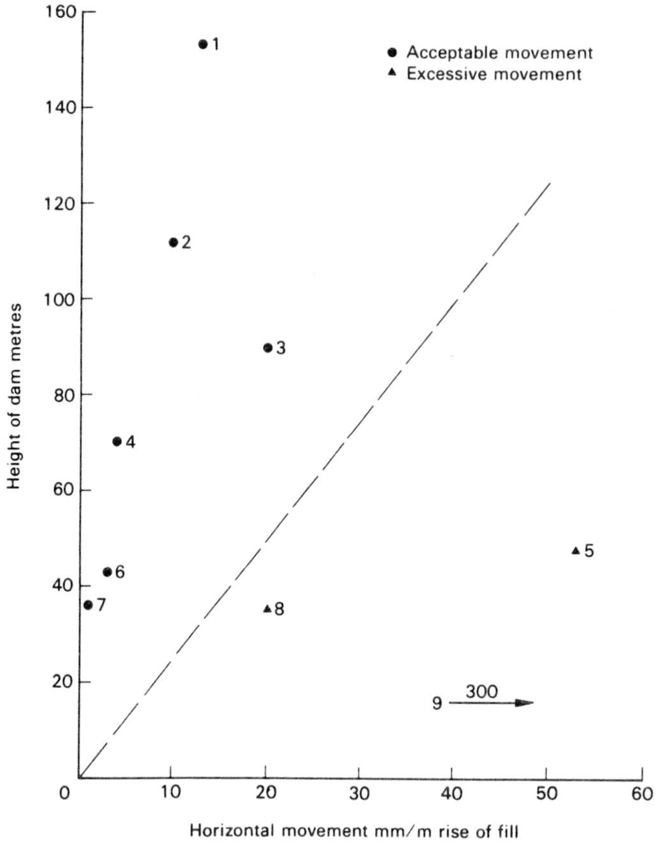

Fig. 16. Rates of horizontal movements observed during construction (dam details are given in Table 1)

retention. This prestressing by the core of the shoulders to pressures greater than those imposed by the reservoir water can also be regarded as a valuable feature in ensuring the adequacy of shoulders and foundations before impounding.

Warning of failure

The use of movement measurements to control construction and to warn of impending failure is not straightforward and depends on potential modes of failure, stiffnesses and brittleness of fill and foundation. Penman & Charles (1974) published values of maximum rates of horizontal movements measured as millimetres per metre rise of fill that had been observed during construction of eight large dams. Two of these had suffered non-brittle shear failure and showed relatively large (unacceptable) horizontal movements that occurred slowly and stopped after fill placement had ceased. These rates have been plotted against the height of dam in Fig. 16. Details of the dams are given in Table 1.

Measurements of the horizontal movements of the upstream slope of Carsington were not begun until half-way through the second placing season (1983) after the toe berm had been built, and results are shown by Fig. 17. The maximum rate of movement observed during the remainder of that season was 20 mm/m rise of fill and is shown in Fig. 16. Settlement of the fill surface was measured on five pegs across the major section during the last two-thirds of the winter shut-down period (Fig. 18). There was negligible apparent settlement of the downstream shoulder fill during the first eleven weeks of measurements, but 75 mm relative to the downstream shoulder in the middle of the clay core and a tilting settlement of 30–45 mm in the upstream fill. Six weeks later, the central settlement had increased to 100 mm, giving a clear indication of relative movement between core and downstream shoulder. A constriction, then a blockage formed in the cross-arm gauge (see Fig. 18) at a position that was consistent with a potential slip surface through the core and core extension. At the same time, the piezometer CU1 ceased to operate. Its connecting tubes

Table 1

Dam	Type	Height: m	Average slope*	Maximum rate of horizontal movement measured: mm/m height of fill
1 Gepatsch	Rockfill with central clay core	153	1:1·5, d	13
2 Blowering	Rockfill with central clay core	112	1:1·9, d	10
3 Llyn Brianne	Rockfill with central clay core	90	1:1·75, d	20
4 Scammonden	Wide rockfill with upstream sloping clay core	70	1:1·8, d	4
5 Galisteo	Earthfill	48	1:3·2, u	53
6 Backwater	Earthfill	43	1:3, d	3
7 Derwent	Earthfill with clay core	36	1·2–1·15, d	1
8 Carsington	Earthfill with clay core	35	1:3, u	20
9 Muirhead	Earthfill with puddled clay core	22	1:3, u	330

* d, downstream; u, upstream.

pass along a common trench with those from CC1 which continued to function, indicating damage to the tube in the stretch between the two piezometers.

During the winter shut-down, the horizontal movement of the toe was only about 25 mm. This imbalance of observed movements is indicative of potential progressive failure.

No further measurements could be made on the five pegs when fill placement recommenced. The added weight of fill caused further measured horizontal movements. For comparison with other dams, these toe movements have been divided by the half-width of the dam at the level of the observations and these values are shown plotted against dam height in Fig. 19. This plot shows the acceleration of horizontal movement associated with impending slip failure. Examples are also given of observed movements of two rockfill dams.

The shape of stress–strain curves depends on the material and the type of test. In general, as failure is approached, small stress increases cause large strain increments (Fig. 20). Under a controlled rate of strain, a brittle clay can exhibit considerable post-failure stress reduction, but under stress control, where the failure stress is maintained, post-failure strains are large and occur rapidly. The horizontal movements (strains) at Carsington followed this pattern.

With non-brittle materials, such as remoulded silty clay (boulder clay), post-peak strain rates may not be large and dam failure, e.g. Muirhead, may involve deformations that only continue until the dam shape has changed sufficiently to restore equilibrium: this may be assisted by pore pressure reduction caused by shearing dilation. Movement measurements, however, as shown in Fig. 19, can be used to indicate the approach of instability, even though subsequent movements may not be as alarming as with a brittle clay.

Upstream membrane

The use of an upstream membrane, by avoiding thrust from a clay core, should enable a much smaller section to be used for the dam. Unsatisfactory behaviour of dumped rockfill dams with upstream membranes had put this design out of favour, but the realization that stable rockfill required sufficient fines content to support each large piece of rock and to fill the spaces between the large pieces completely, to prevent rotation, has produced, in combination with compaction by vibrating rollers, some very satisfactory upstream membrane rockfill dams built since 1970, e.g. Foz do Areia (160 m), Salvajina (148 m), Alto Anchicaya (140 m), Khao Laem (130 m), Cethana (110 m). Penman (1971) and Penman & Charles (1976) have discussed rockfill for dams and suggested that the amount of fines should be limited so that in situ $k > 1 \times 10^{-3}$ cm/s to avoid construction pore pressures. Such a value allows plenty of water to be used to improve workability in the knowledge that surplus can safely drain. If the permeability is lower, the fill should be regarded as earth rather than rockfill.

Winscar Dam (50 m) was the first in England to use an upstream asphaltic membrane (Collins & Humphreys, 1974). The BRS fitted horizontal plate gauges passing right through to contact the

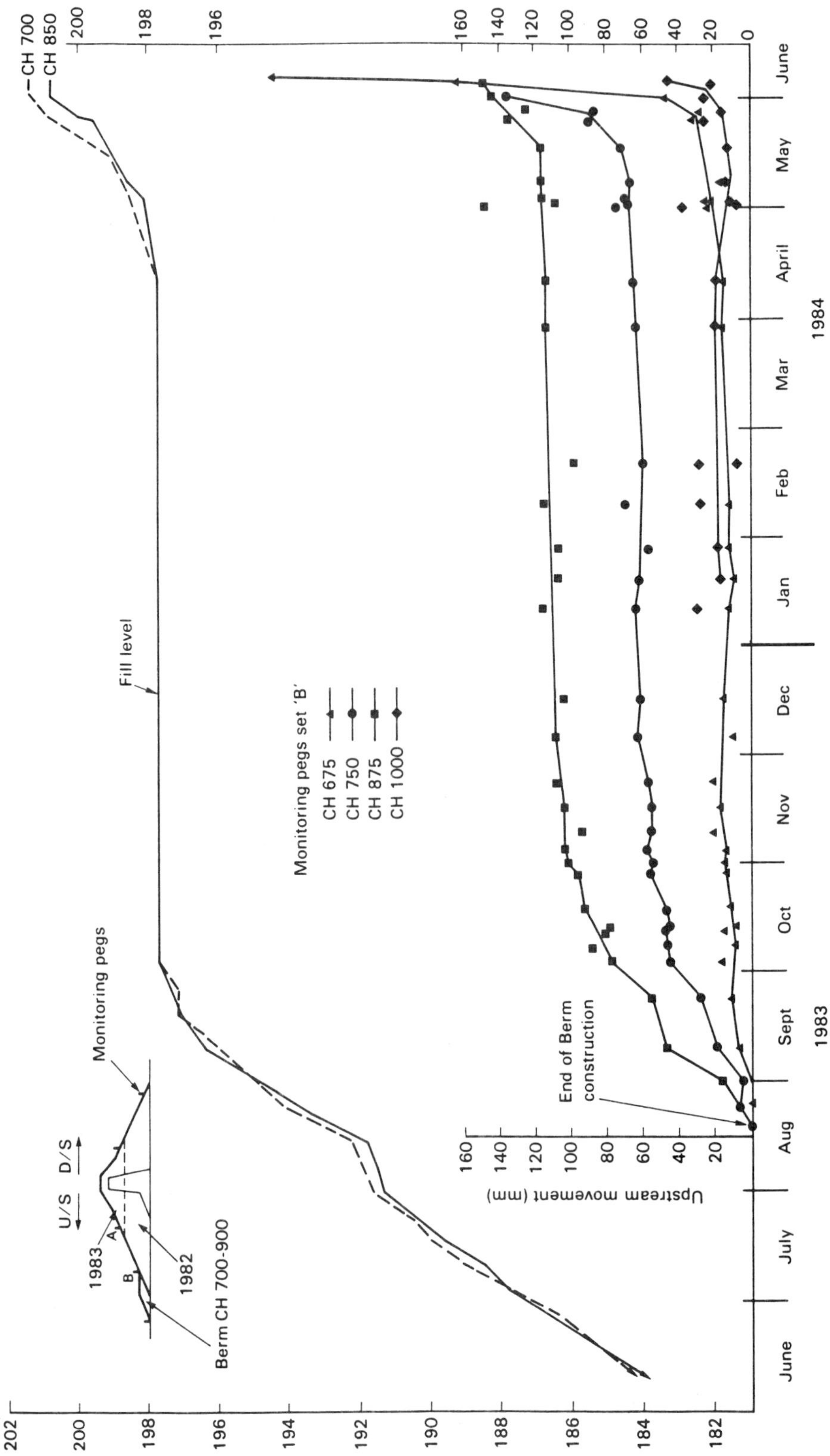

Fig. 17. Measured horizontal movements at Carsington Dam

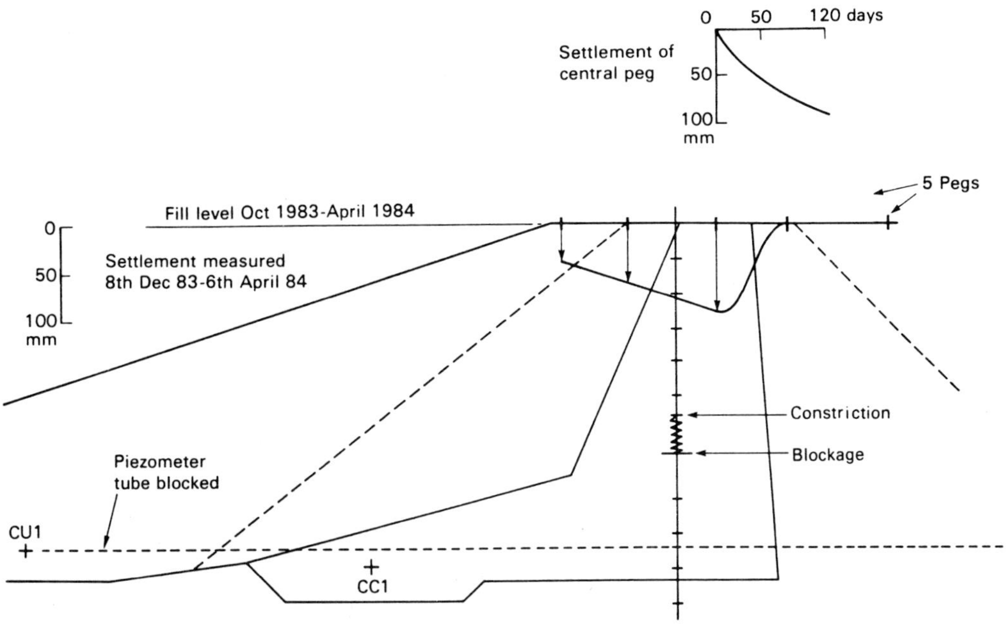

Fig. 18. Conditions at the end of the 1983–84 winter shut-down

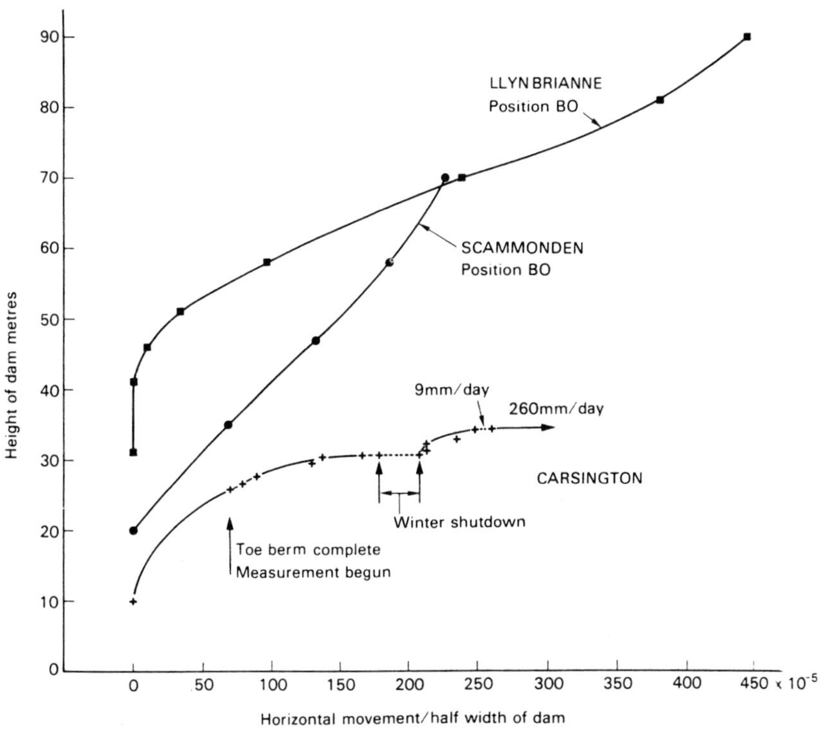

Fig. 19. Horizontal movements related to dam height

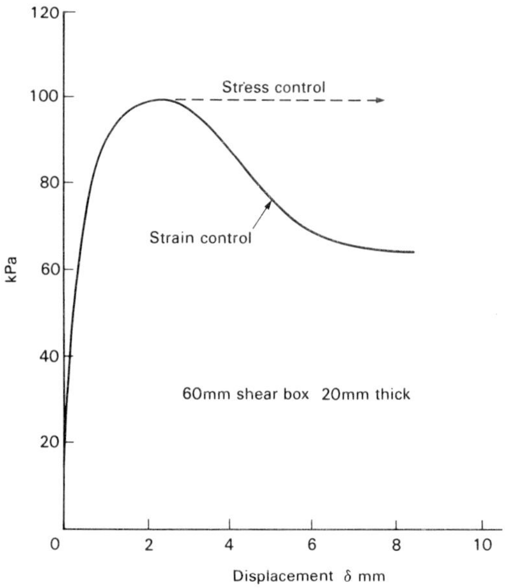

Fig. 20. Stress–strain curve for clay

underside of the membrane (Fig. 21) to observe not only movement of the sandstone rockfill but also to measure impounding deflexions. These deflexions were found to be less than given by a class A prediction made by Penman & Charles (1975) as shown by Fig. 22. The reservoir did not reach full height on first filling and there were several reversals before top water level (TWL) was reached for the first time. It has been suggested by Penman & Charles (1985a) that the smaller movements were due to a reduction in shear stress in the rockfill during impounding. During construction, the value of σ_1'/σ_3' in large zones of the fill could be expected to remain sensibly constant, while the mean effective stress $(\sigma_1' + 2\sigma_3')/3$ increased, whereas the reservoir water pressure on the membrane would reduce $\sigma_1' - \sigma_3'$. Results from a triaxial test made to simulate these conditions are shown in the lower part of Fig. 22. The shear strain $\varepsilon_1 - \varepsilon_3$ increased during the 'construction phase', but during 'impounding' there was negligible shear strain indicating a stiff condition. This approach was used to revise the predictions: the revised values, also shown by Fig. 22, are in much closer agreement with the observed deformations.

These membrane deflexion measurements were only made at the three levels of the horizontal plate gauges and at the crest. They were not able to measure local distortion at the connection between the membrane and the toe structure: a joint that is under maximum reservoir pressure.

To overcome this difficulty, deflexions of the

Fig. 21. Upstream end of a horizontal plate gauge

asphaltic membrane of Marchlyn Dam (72 m) were measured by the BRS with an inclinometer mounted in a trolley which could travel underwater from crest to toe on rails secured to the membrane (Penman & Hussain, 1984). The concrete toe structure (Baines, Newman, Hannah, Douglas & Carlyle, 1983) was cast into a shallow excavation to cut through slightly weathered bedrock and formed a cap to the grout curtain. It contained an inspection/drainage gallery and was shaped to accept the edge of the membrane and to provide a cushion for the inevitable differential movements. The structure was designed so that there was only about 2 m depth of rockfill immediately behind the wall, but in view of the 70 m head of water and the relative compressibilities of concrete (13.5×10^6 kN/m^2) and the rockfill (25×10^3 kN/m^2) several millimetres differential movement was expected. The amounts measured under full reservoir are given by Fig. 23.

There is a danger that unforeseen foundation conditions can lead to the use of undesirably high toe walls (Penman, 1982). There have been cases where the toe position was decided from the results of site investigation and the grout curtain constructed. Site stripping then revealed deeper than expected weathering, necessitating the use of

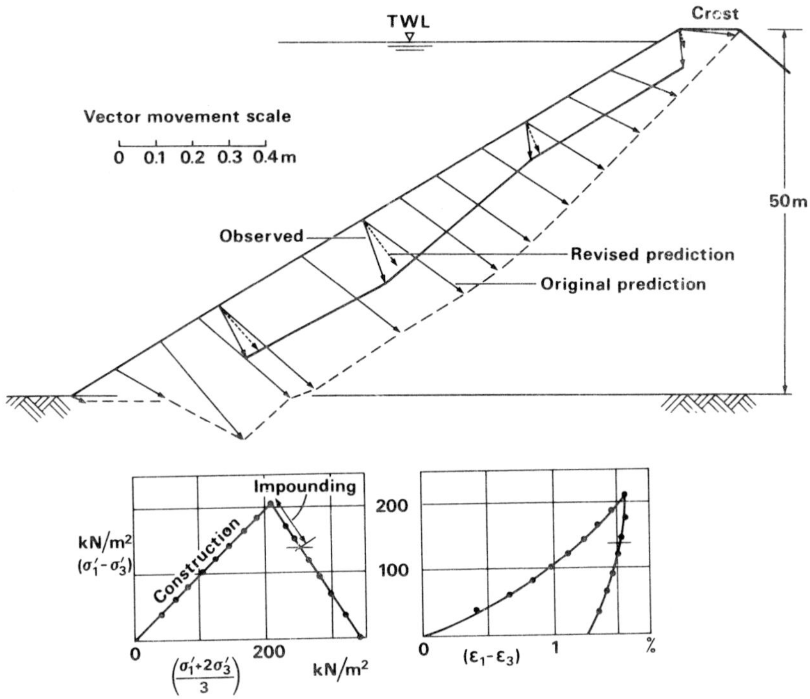

Fig. 22. Winscar Dam: movements during impounding and results from rockfill tests

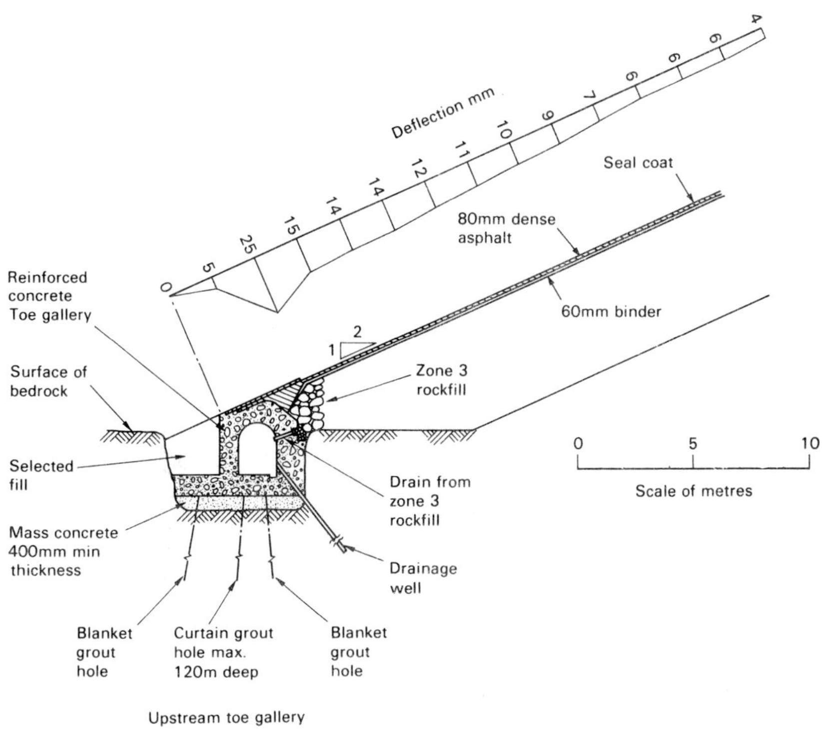

Fig. 23. Marchlyn Dam: membrane deflexion near the toe structure

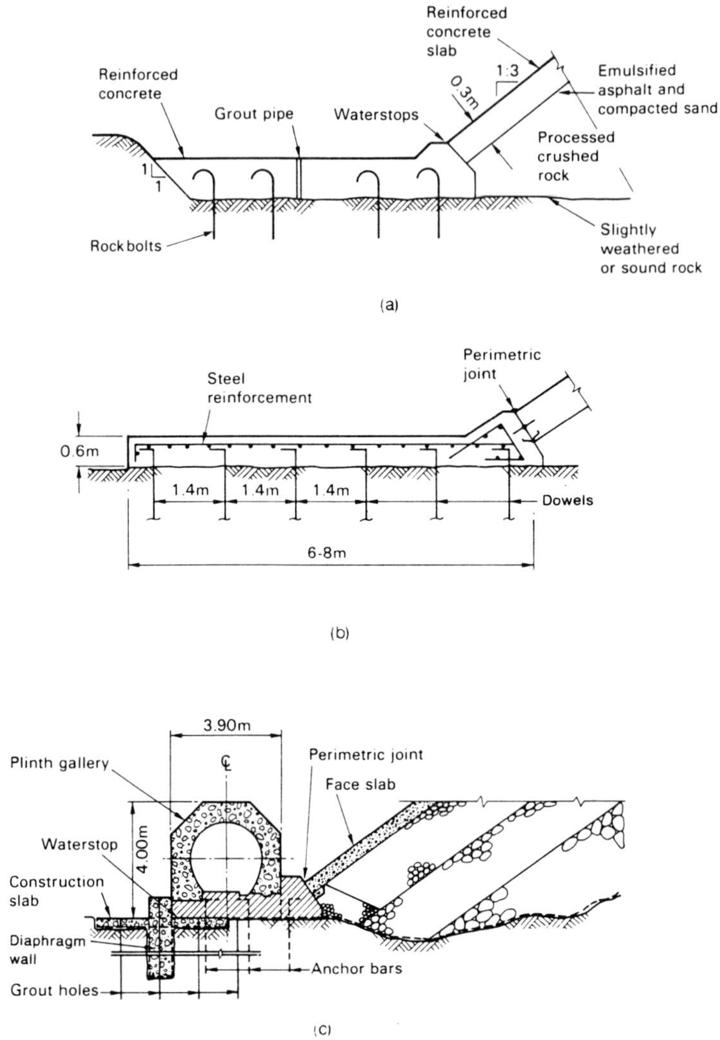

Fig. 24. Toe slabs of three concrete-faced rockfill dams: (a) Deer Creek; (b) Salvajina; (c) Khao Laem

a wall over the curtain. Difficulty in obtaining good compaction behind the wall created an even more compressible zone and resulting deformations were enough to tear the membrane. Sherard (1985a) has also drawn attention to the danger from a high toe wall causing damage to the membrane.

There has been a move with reinforced concrete membrane dams to reduce the height of the toe structure to a minimum by use of a toe slab, as shown by Fig. 24. The design for Deer Creek (85 m) (Hollingsworth, Conner & Anderson, 1985) and Salvajina (148 m), the second highest concrete-faced rockfill dam currently in service in the world (Sierra, Ramirez & Hacelas, 1985) provided flat slabs tied down to the bedrock. At Khao Laem (130 m), founded on karst, the plinth slab was attached to a diaphragm wall and fitted with an inspection gallery placed over the slab so as not to introduce undesirable height (Watakeekul, Roberts & Coles, 1985).

Central asphaltic core

At Megget (55 m), the first British dam to be built with a central vertical asphaltic core, it was hoped to see more clearly the effect of impounding on downstream movement. The shoulder fill was a well-graded gravel that readily developed a high density under vibrating rollers (site control density tests often indicated negative air voids!)

Fig. 25. End of a horizontal plate gauge to be connected to the asphaltic core with concrete

and as a result was very stiff. The horizontal plate gauges for this dam were modified by providing an induction coil sensing unit at each steel plate, connected by steel pipes to form a continuous string that remained inside the telescopic plastic tubes (Penman & Charles, 1982). This removed the slight error caused by elastic stretch of the steel measuring tape used with the original version and gave an improved accuracy that proved necessary with the stiff fill.

As with the clay cores, the plate gauges were not taken through to the upstream side, but the last plate was encased in concrete abutted to the downstream face of the asphaltic core so that any movement could be detected (Fig. 25). Predictions obtained with an analysis using finite elements and oedometer results (Penman & Charles, 1985b) were of negligible horizontal movements during construction. Observations agreed with this prediction (Fig. 26, plates A6, B5 and C4) and indicated that the asphaltic core did not exert a large lateral thrust on the shoulder fill.

During impounding, the core acted as a thin membrane and transmitted reservoir water pressure to the downstream fill, although the total pressure increase in the downstream direction was limited by the reduction in upstream fill pressure due to submergence. The resulting downstream movements were less than predicted possibly because of the reduction in shear stresses in the downstream fill under the action of increasing σ_3.

Accuracy of measurement

The improved horizontal plate gauge has enabled the horizontal position of plates in relation to a reference plate at the instrument chamber to be measured with an accuracy to $\pm 0{\cdot}5$ mm. Use of this apparatus at Kotmale Dam (90 m) has been described by Gosschalk & Kulasinghe (1985).

Movements of the instrument chambers and other positions on a dam have been related to stable reference stations built outside the stress influence of the dam itself by precise surveying. Three-dimensional triangulation with a half-second theodolite and/or trilateration with electronic distance measuring equipment with an accuracy to 2 ppm $\pm 0{\cdot}5$ mm have given measurements accurate to ± 2 mm when instruments and targets could be always mounted exactly. Reference stations must be stable structures embodying mounting plates to suit the instruments so that they can be replaced precisely. The use of portable tripods over reference marks at ground level does not lead to high degrees of accuracy.

Detailed layout planning is needed to ensure maximum possible angular and/or distance changes from the expected movements and to be sure that sight-lines will not become obstructed during construction.

In general, bending of theodolite sight-lines can be reduced by keeping them well above the ground surface. Distance measurements are less affected by low level sight-lines. Theodolites with a built-in horizon produced by reflecting light from a liquid surface in practice give more accurate vertical angle readings than those with a separate bubble. Early construction of the reference stations can be of help in setting-out for the dam.

Repeat zero readings should be taken to establish the shape of the survey layout and the position of every new monument built on the dam. A suitable computer program is required to reduce instrument readings to co-ordinates of the observed positions.

TOTAL PRESSURES AND ARCHING

Although there had been concern that the use of different types of fill in the various zones of a dam would cause differential settlements and undesirable internal stresses, an observation of actual conditions could not be made until satis-

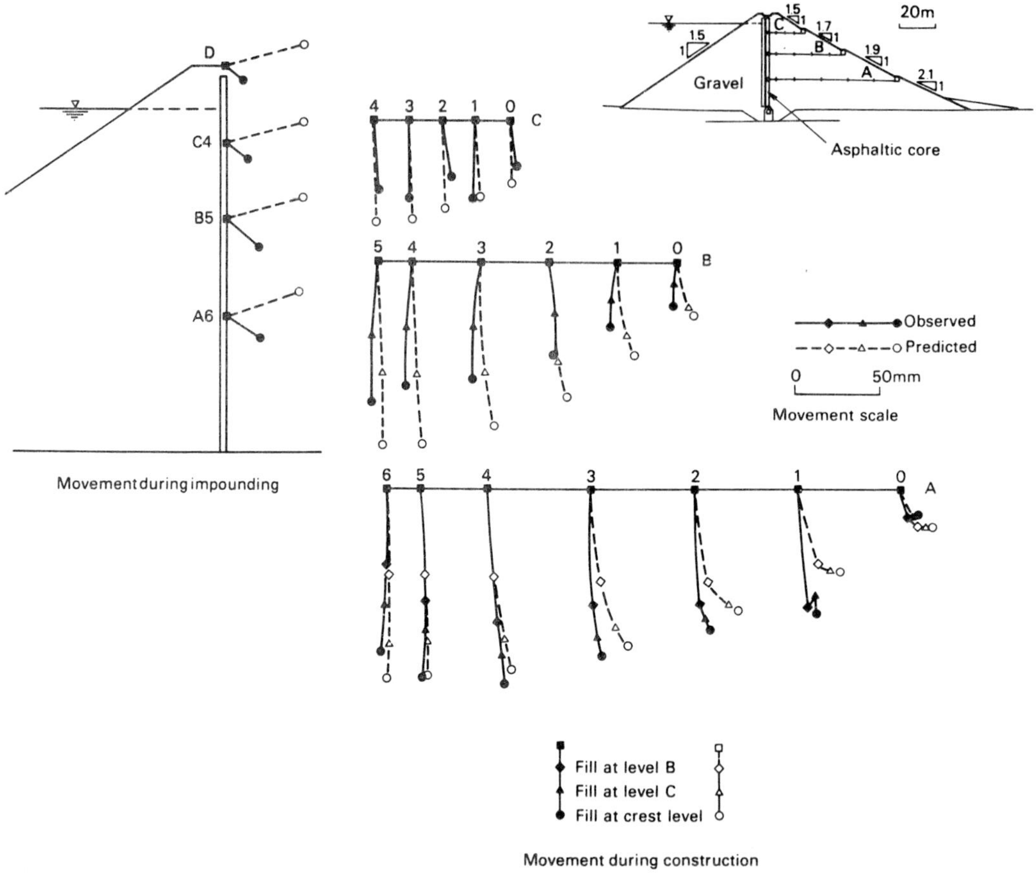

Fig. 26. Megget Dam: movements during construction and impounding

factory earth pressure cells had evolved. To be ideal, a buried cell should have the same deformation characteristics as the element of fill it replaces. In practice a thin, stiff cell may give satisfactory measurements: installation details are more likely to cause irregularities.

A research programme on distribution of total pressures was begun in 1940 by the US Corps of Engineers. The Waterways Experimental Station developed a relatively thin cell that was less compressible than most fills. Load was transmitted through oil to a central diaphragm fitted with stuck-on resistance strain gauges. It was 610 mm dia. × 25 mm thick, described in Waterways Experimental Station report (1942).

A line of 20 of these cells was placed to measure vertical pressures from upstream to downstream of the John Martin Dam (36 m) during the winter 1941–42, as shown by Fig. 27. The results showed a pronounced arching action across the dam section. σ_v averaged $0{\cdot}36\sigma_0$ in the wide core, whereas in the adjacent shoulders $\sigma_v = 1{\cdot}37\sigma_0$.

Taylor (1947) in his review of these results considered that the observed pressures, particularly those in the core, were too low to be reasonable. There had been a blizzard during installation and, although the greatest care was taken to exclude frozen backfill, it was felt that it might be more compressible than adjoining fill.

The summation of all the measured pressures was less than the overburden and Taylor suggested that the ratio of observed to true vertical pressure was 0·5 in the core and 0·7 in the shoulders. Even with these corrections (shown in Fig. 27), the arching action across the section remains clear and illustrates the marked reduction in vertical total pressure that can occur even in a wide core.

Taylor was not critical of the compressibility of the cells: he felt that the low measured pressures were due to pocketing action and emphasized the

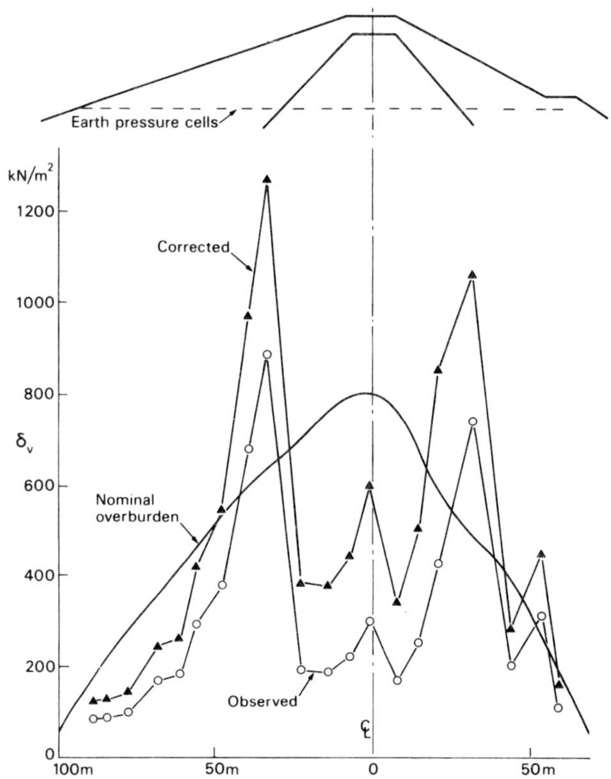

Fig. 27. Total pressures: John Martin Dam

importance of placing procedures. The working life of the cells, however, was less than two years.

Narrow cores

At about the same time, the Swedish State Power Board developed an oil-filled cell, connected by a small brass tube to a pressure gauge in an instrument house. It was criticized for acting as a thermometer, but in practice, once buried in fill, it remained at almost constant temperature. It was 270 mm dia. × 20 mm thick, described by Magnusson (1948) and included in the instrumentation of Medskogsforsen Dam (28 m) constructed during 1942–44.

The settlement of some narrow cores had been found to be less than expected from consolidation under σ_0 and it was feared that this could indicate horizontal cracks through the cores. To ensure watertightness, 'during the first years preceding consolidation of the fill' (Löfquist, 1951a), it became the practice to provide a thin reinforced concrete wall on the downstream side of the moraine cores.

Groups of the hydraulic cells, orientated to measure σ_v, σ_a and σ_h, were placed in the 3–4 m wide clay core of Harspranget (50 m): measured pressures at the end of the second placing season are shown in Fig. 28. The 0·4 m thick concrete wall had been 'lubricated' with a 1 mm coating of bitumen and the clay, placed in 0·1–0·2 m layers, was compacted by tractors, producing $\gamma = 21{\cdot}9$ kN/m³. Placement $w = 12$–14% and $k = 3$–30×10^{-8} cm/s. In general the lowest stress was σ_a (in the direction along the dam axis), not σ_h, and σ_v was less than $\sigma_0/2$ over much of the height.

In discussing these results, Löfquist (1951b) commented that arching might give rise to a risk of horizontal cracks in the core. When discussing Trollope (1957) he went further by stating that arching may take place in the longitudinal as well as the transverse direction of a dam and that, if σ_v decreased to a value lower than the pressure of water from the reservoir at the same level, the situation might be considered as detrimental to the watertightness of an earth core (Löfquist, 1957). It was almost ten years before hydraulic fracture was recognized as the cause of leakage at Hyttjuvet Dam.

Silo action

The narrow puddled clay core of Selset Dam was typical of British practice and an estimate of

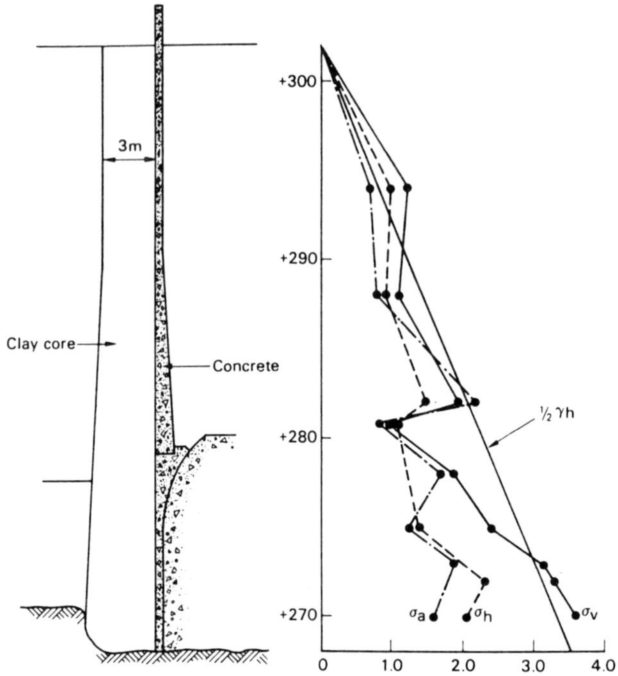

Fig. 28. Total pressures in the core of Harspranget Dam

the vertical total stress could be obtained by considering the core to be supported on either side by vertical walls to which it adhered perfectly. The settlement gauges had shown differential movements of more than 0·4 m between the core and shoulders, sufficient to mobilize the maximum c_u of the puddled clay.

$$\sigma_v = h\left(\gamma - \frac{c_u}{a}\right)$$

where h represents the height of core above the considered position, $2a$ is the core width and γ is the bulk density (more detailed analyses have been given by Bishop (1952), Anagnosti (1965) and Blight (1973)).

Total pressure cells were not placed in the core, but Bishop & Vaughan (1962) compared calculated values of σ_v with measured u. If in the wet, soft clay $u = \sigma_v$, then good agreement was reached when $c_u = 15{\cdot}8$ kN/m^2, an average value for the upper part of the core. Overburden and pore pressures are shown by Fig. 29.

To avoid hydraulic fracture, the total pressure on any plane passing through the core must exceed the reservoir water pressure. The total pressure in the direction of the dam axis is given by

$$\sigma_a = K_0(\sigma_v - u) + u$$

In the case where $u = \sigma_v$ then $\sigma_a = \sigma_v$ and if $u > \gamma_w h$, where h is the height of reservoir water over the position considered, then the condition is satisfied that $\sigma_a > \gamma_w h$.

Thus, if end of construction pore pressures, expressed in terms of a head of water, remain above the TWL during impounding, the risk of hydraulic fracture should be considerably reduced, a conclusion reached by Penman & Charles (1979).

An earth pressure cell 280 mm dia. × 38 mm thick, developed at the BRS (Thomas & Ward, 1969) used vibrating wire sensing units to measure the deflexion of stiff diaphragms on both faces. A group of five was placed at about quarter height on the centre line of the rolled clay core—one of the first in Britain—of Balderhead Dam (48 m) (Fig. 30). To avoid pocketing, they were very carefully installed on prepared surfaces in a wide, shallow excavation and orientated to measure σ_v, σ_a and σ_h as well as in two directions at 45° to vertical in an upstream–downstream direction. The cells were unique in that they could measure pressure independently on their two faces. It was found that the face (generally upper) against which fill was compacted gave a higher reading than that placed on the prepared flat surface. This was attributed to slight bedding errors. Results at end of construction and after

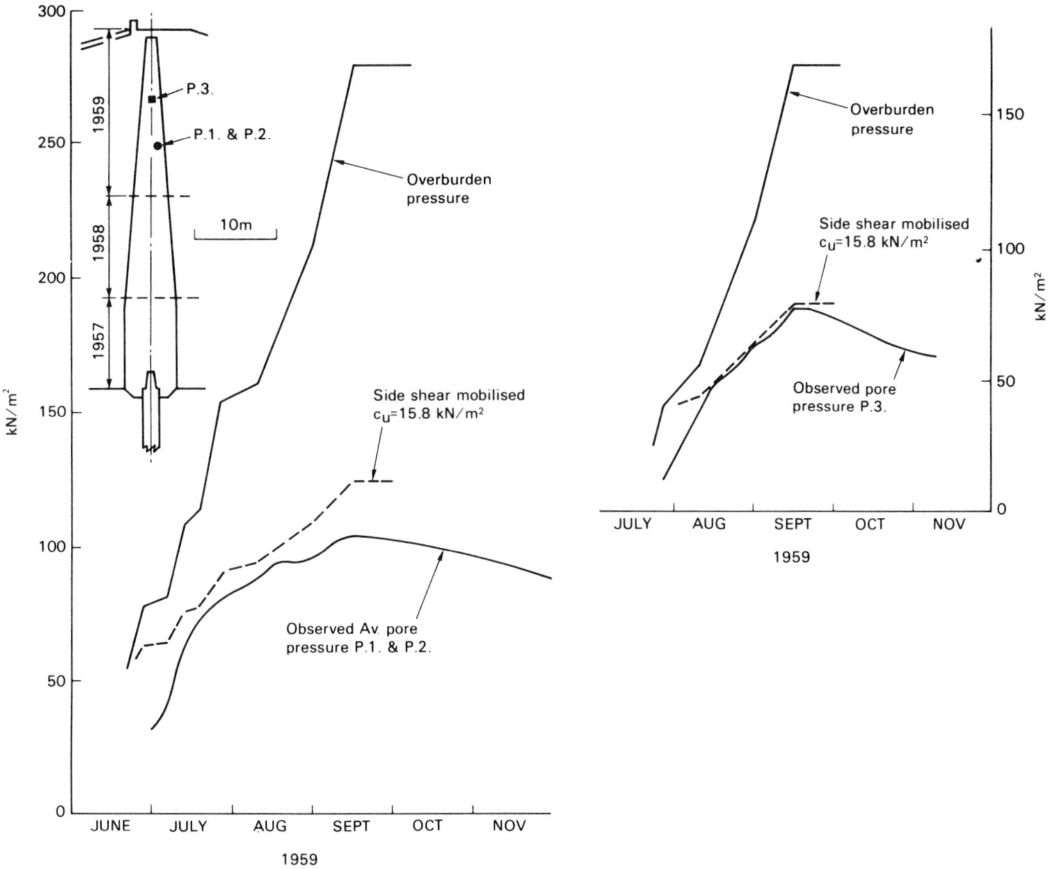

Fig. 29. Arching action and pore pressures in Selset core

impounding and partial drawdown are given in Table 2.

The boulder clay used for the core had $w < w_{opt}$ and was easy to place by machine. During the first season, when the earth pressure (EP) cells were placed, a combination of watering and rainfall increased w to about the specified range of $w_{opt} + 1$–3%. Because of variations in w_{opt}, the specification was changed to $w_p - 3\%$ to give a condition equivalent to $\omega_{opt} - 1\%$, related to a property that was easier to measure. Strength specification had not been put into practice at that time.

The rate of increase of construction pore pressure $\delta u = 0{\cdot}5$–$0{\cdot}7\delta\sigma_0$ in this first season's fill. The second season's fill was somewhat drier and in the upper part $\delta u = 0{\cdot}15$–$0{\cdot}2\delta\sigma_0$. Measured values of u and reservoir pressure in the upper part of the core are given in Table 3.

At the end of construction, the lowest measured total pressure σ_h exceeded the water pressure from a full reservoir, and pore pressures measured by a piezometer associated with the pressure cells (C4) equalled the reservoir pressure, a situation which would avoid hydraulic fracture. Unfortunately, pore pressures in the upper part of the core at the end of construction were much lower.

Hydraulic fracture

Just before the reservoir became full for the first time in 1966 (Vaughan, Kluth, Leonard & Pradoura, 1970), there was a marked increase in seepage and early in 1967 small depressions were found in the crest. A few months later, a sink hole 3 m wide and 2·5 m deep formed over the upstream edge of the core and the underdrain flow was found to be cloudy. Lowering the reservoir had a disproportionate effect of rapidly reducing seepage and returning it to a clear condition.

The detailed study that ensued established this as the first recognized British example of hydraulic fracture by reservoir water pressure to cause leakage and erosion of a clay core. Flight auger holes (0·4 m dia.) drilled dry through the crest

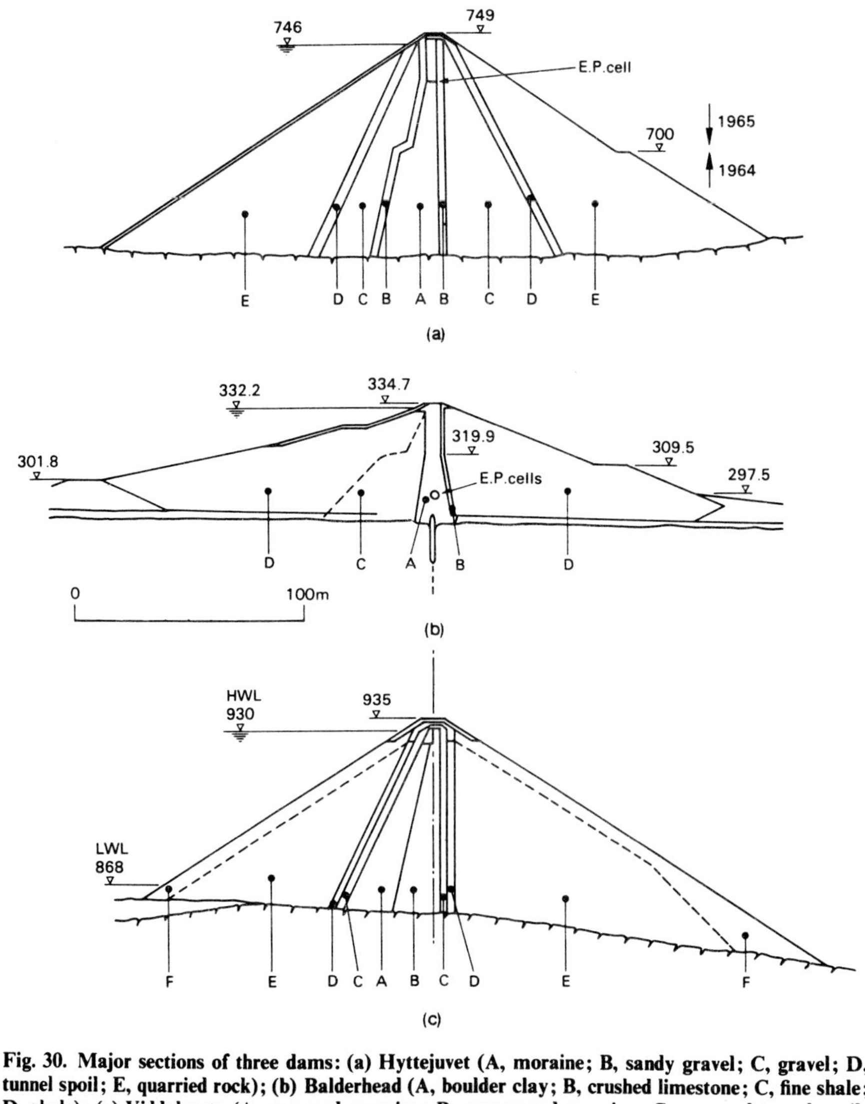

Fig. 30. Major sections of three dams: (a) Hyttejuvet (A, moraine; B, sandy gravel; C, gravel; D, tunnel spoil; E, quarried rock); (b) Balderhead (A, boulder clay; B, crushed limestone; C, fine shale; D, shale); (c) Viddalsvatn (A, screened moraine; B, unscreened moraine; C, screened tunnel spoil; D, unscreened tunnel spoil; E, tunnel spoil and quarry run; F, quarry run)

encountered soft clay and water that rapidly rose to reservoir level, at a depth of 14 m. It was thought at the time that this position of the hydraulic fracture might relate to core shape. Core width was designed to be one-third height with a minimum of 6 m at the crest. Because of a shortage of suitable clay, instead of the usual uniform decrease in width from base to crest, the top 18 m of core was made 6 m wide with vertical sides (Fig. 30).

In Norway there had been a similar experience with Hyttjuvet Dam (Kjaernsli & Torblaa, 1968) at about the same time as Balderhead, and with Viddalsvatn (Vestad, 1976) a few years later. Moraine was used for the rolled, central cores of both dams.

At Hyttjuvet (93 m) high construction pore pressures measured in the core after the end of the first placing season, when the dam was at about half-height, caused a design change to make the core much narrower in the upper half, to accelerate drainage. The resulting core shape (Fig. 30) had vertical sides for the upper part, similar to Balderhead. An earth pressure cell was placed 21 m below crest to measure the vertical pressure. Measured values are given in Table 4.

Table 2. Balderhead core: pressures at cell cluster, 37·9 m below crest*

Maximum after completion, Dec. 1964				3 years later
Direction	Cell	Total pressures		
		Reading: kN/m²	Average: kN/m²	Average: kN/m²
σ_h	1	414	420	363
	2	425		
45°	3	532	529	481
	4	525		
σ_v	5	701	663	637
	6	625		
45°	7	493	489	451
	8	485		
σ_a	9	489	480	392
	10	470		
σ_0			807	807
Reservoir water pressure	When at TWL		357	263
u	C4	357	345	207
	C5	333		

*
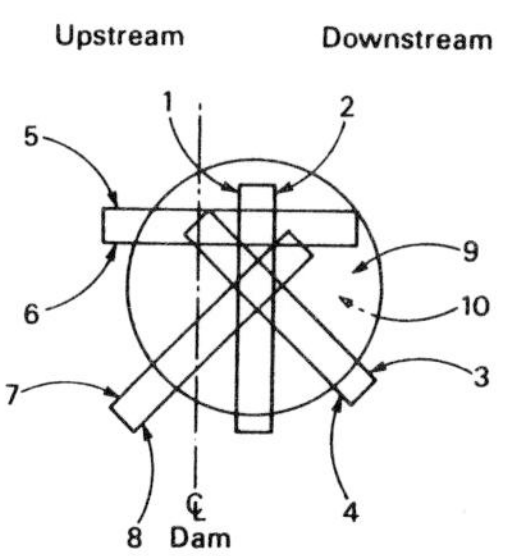

Table 3. Balderhead core: pore pressures in the upper part at the end of construction December 1964

Tip	Depth below crest: m	u: kN/m²	Reservoir water pressure when at TWL: kN/m²	Difference: kN/m²
C13	6·2	−13	41	−54
C12	9·9	29	78	−49
C9	21·2	32	191	−159
C7	27·4	73	253	−180
C8	27·6	66	255	−189

As with Balderhead, instrument measurements indicated arching in three ways

(*a*) $\sigma_v < \sigma_0$
(*b*) $\delta u < \delta\sigma_0$
(*c*) core settlements were less than expected from laboratory consolidation tests.

Also, post-construction crest settlements were remarkably uniform, not reflecting valley shape. This suggested that settlement was occurring only in the upper few metres of core.

Measured leakage increased from 1–2 l/s to 60 l/s as the reservoir level rose from 7 m to 5 m of TWL on first filling. Before lowering could

Table 4. Hyttejuvet core: pressures at a cell 21 m below the crest

	σ_v: kN/m^2	Reservoir pressure: kN/m^2	σ_0: kN/m^2	u: kN/m^2
End of construction, Nov. 1965	174	0	460	165
Reservoir 60% full height, June 1966	128	0	460	145
Reservoir full, Oct. 1966	226	170	460	225

begin, the rate decreased to about 40 l/s by a self-healing process. It did not return to the 1–2 l/s values until the reservoir had been lowered 29 m below TWL. 420 m^3 of grout were injected into the upper 30 m of core. After seven yearly cycles of filling and emptying over a range of about 70 m, a crater of 13 m^3 volume, about 3 m wide, appeared in the crest over the upstream edge of the core. Probing and boreholes showed the fill to be very loose below the crater and it was regarded simply as a surface manifestation of loss of material in the seepage water over a long period. No change was observed in the small, clear seepage at the time when the crater appeared.

Leakage at Viddalsvatn (75 m) also occurred towards the end of first filling, as the reservoir rose over the last 5 m almost to TWL in 1972. Concentrated leaks of dirty water emerged abruptly several times at the downstream toe, with short duration flows of 60–140 l/s. During second filling in 1973, as the reservoir reached TWL, leakage increased from less than 10 l/s to 98 l/s in four days, decreased to 63 l/s over the next two days then suddenly reached a peak of 210 l/s. This decreased to 35 l/s as the reservoir was lowered 1 m during the following week.

Leak positions were determined by releasing a tracer at a depth of 4–5 m from a boat moving slowly along while seepage flow was at its maximum. Two positions found were confirmed later when craters appeared in the crest.

The rapid variations in flow rate indicate that after the moraine had fractured under reservoir pressure the sides or roof of the fissures collapsed, causing temporary blockage, a phenomenon not observed in a clay core.

Established leakage positions along the lengths of the three dams (Fig. 31) lay over some feature of the valley shape that could have induced tensile strains by the 'beam bending' concept of Lowe (1970). This indicates that, although the cross-sectional shape and stiffness of the cores could account for arching, leading to hydraulic fracture, the positions where failure occurred probably suffered further reduction of total stresses due to longitudinal strain. This was not verified by instrumentation.

The failure of Teton (93 m) in 1976 on first filling of the reservoir (Fig. 3) may have been initiated by hydraulic fracture. The failure and subsequent investigation have been described in three detailed reports (Independent Panel, 1976; Interior Review Group, 1977, 1980) as well as by several papers, e.g. Penman (1977), Seed & Duncan (1981) and Leonards & Davidson (1984). It was discussed during the International Workshop on Dam Failures held at Purdue University, 1985.

The valley shape is given by Fig. 32. Preliminary grouting trials had shown that the joints in the volcanic rocks forming the valley sides were too open for a grout curtain to be formed until the depth exceeded 21 m. It was therefore decided to replace the grout curtain by a cut-off trench over this depth on both abutments, as indicated by Fig. 32. The base of the trench was made 9 m wide, to accommodate the drilling rigs for the grout curtain, and the sides made as steep as possible to minimize excavation. The right trench was excavated by blasting, which tended to loosen the rock. On the left side, pre-split blasting techniques were used, which caused less dis-

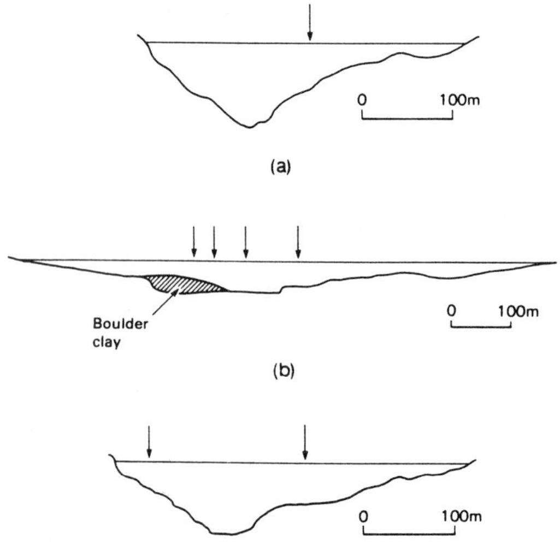

Fig. 31. Positions of core damage: (a) Hyttejuvet; (b) Balderhead; (c) Viddalsvatn

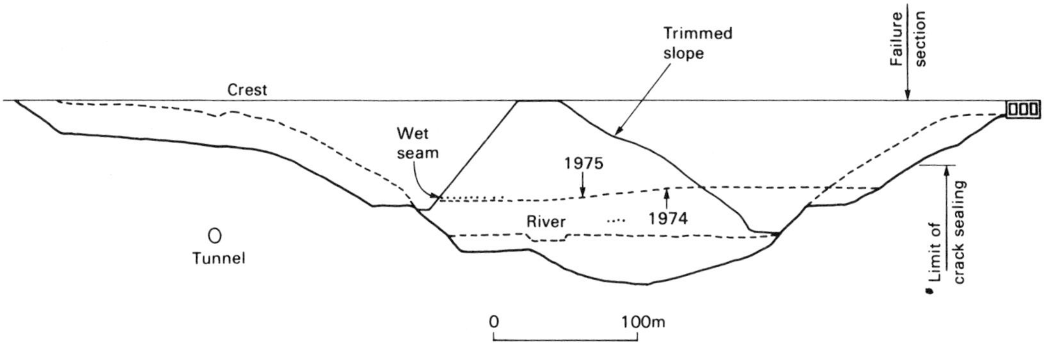

Fig. 32. Teton Dam: longitudinal section

turbance to the surrounding rock and formed smoother surfaces.

The mistake was made of placing the silt fill of the very wide core in these trenches without adequate preparation of the rock surfaces. The specification did not require special treatment to seal the numerous fissures and joints. Site personnel, concerned at the condition of the rock surface, poured grout or concrete into the wider cracks, but even this *ad hoc* treatment was discontinued above a height of about 50 m above river level, a position in the right trench close to the line of failure (from whirlpool to erosion channel in the downstream slope).

The grout cap, cast into a shallow slot in the trench floor, was only 0·9 m wide and finished more or less flush with the floor surface (instead of being brought up as a wall to key into the silt fill). At the time of failure, the reservoir head over the grout cap on the failure section was 30 m, causing a hydraulic gradient of 33 over the cap, a value high enough to cause trouble in the silt if there were the slightest imperfection. The silt could easily be washed into cracks with widths of 0·2 mm or larger.

An analysis of arching action in the trench, using finite element techniques (Seed, Duncan & Bieber, 1976), showed that the normal total stress on the transverse section where failure occurred was less than the reservoir pressure as shown by Fig. 33. It was therefore possible for hydraulic fracture to have initiated the piping erosion through the core.

Leonards & Davidson (1984), from a detailed study of the placement control density tests, found that a layer of slightly drier silt had been placed across the right trench close to the failure position. A volume reduction on wetting this layer (collapse settlement) would further reduce the total stresses in the fill and ensure hydraulic fracture.

Knowledge of pre-failure behaviour of Teton was limited by the absence of instrumentation. Failure occurred rapidly, but it is not known for how long water had been passing through the core into the very permeable bedrock downstream, where the regional water-table was about 54 m below river level. The dangers of placing an erodible core material in direct contact with fissured bedrock had been well demonstrated: (Fetzer, 1977) at East Branch Dam (56 m) in 1957 and at Fontenelle (39 m) (Fig. 34) in 1965 (Bellport, 1967). Both dams were saved from complete failure only because their reservoirs could be lowered fast enough to prevent the erosion from forming a breach.

It was not unknown in British practice a century ago to protect core fill in cut-off trenches with a thick concrete coating over fissured rock. The following extract is from the specification for Winterbourne Reservoir, dated 1885.

> '. . . If it should be considered necessary by the Engineer, the Contractor shall protect the sides of the puddle trench, or portions of the same, by means of concrete walls on either side of the puddle. . . . The concrete walls shall be not less than 18 inches thick (0·46 m), built plumb and parallel with the centre line of the trench, smoothly faced against the puddle, and shall quite fill up all fissures in the rock and all irregularities in the sides of the trench.'

The Teton trenches could have been lined with a thick coating of reinforced concrete, tied to the grout cap and bedrock, to provide a non-erodible, reasonably watertight cut-off above the grout curtain. If the concrete coating had been extended to cover all the rock over the core contact surface, the potential for erosion of the core could have been substantially reduced.

The effectiveness of a smooth concrete contact surface to provide a leak-proof joint with a core is demonstrated by the numerous successful

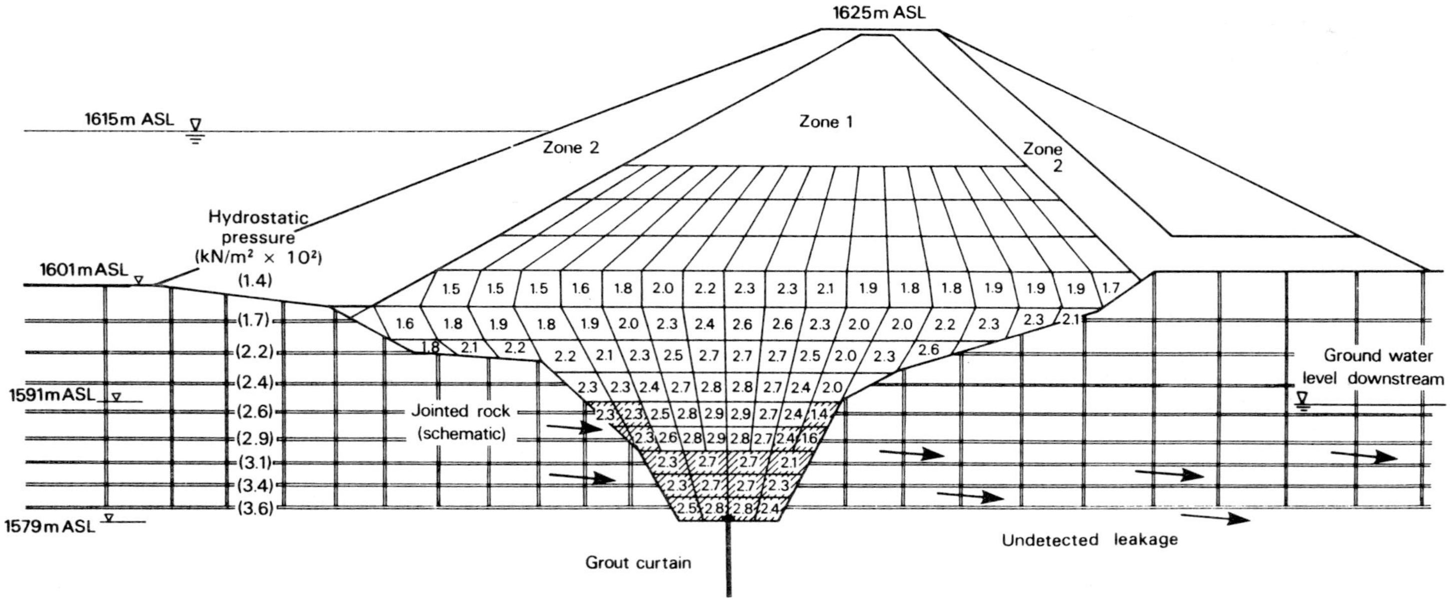

Fig. 33. Failure section: computed values (kN/m²) of normal stress on the transverse section and hydrostatic pressure of reservoir water (shaded zones indicate where hydraulic fracturing could occur)

Fig. 34. Erosion damage at Fontanelle Dam

examples of almost vertical connections between concrete spillway structures and embankment. Measurements of total pressures on such a contact were made with BRS vibrating wire EP cells mounted flush in the concrete surface at Cow Green Dam. A core placement specification $c_u = 50\text{–}100\ \text{kN/m}^2$ resulted in the measured total and pore pressures shown by Fig. 35. Although dissipation of construction pore pressures reduced the total pressures, they were still comfortably above the reservoir pressure nine years after completion.

Wet seams

Although little new geotechnical information derived from the Teton failure, the subsequent detailed investigation was of considerable value. Because of the similarity between the cut-off trenches on right and left, it was hoped that the failure trigger mechanism might be revealed by exposing the left abutment and trench. An excavation of $0{\cdot}68 \times 10^6\ \text{m}^3$ was made (the failure had removed $2{\cdot}5 \times 10^6\ \text{m}^3$ adjacent to the right abutment) and the cut-off trench carefully exposed. Although the fill had become soaked on the upstream side and had slumped away from local overhangs, no clear signs of piping were found. It may be concluded that failure occurred on the right because there were more fissures than in the left trench and the contact surface was rougher. River erosion had exposed rock cliffs on the right side just upstream of the dam and there is little doubt that reservoir water could readily reach the upstream side of the trench. Collapse settlement of the silt fill may also have played a part in explaining why the right side failed first.

Of more interest was a horizontal wet seam discovered near the bottom of the excavation on its right side, passing almost through the wide core from upstream to downstream. When it was exposed on 5 October 1977, 16 months after

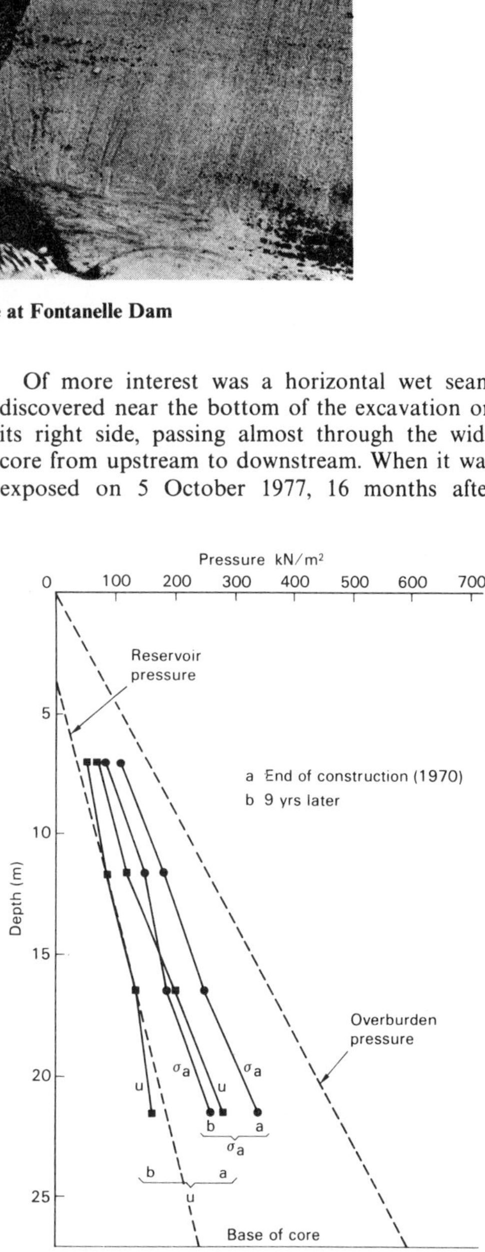

Fig. 35. Total pressures measured at the concrete–clay interface, Cow Green Dam

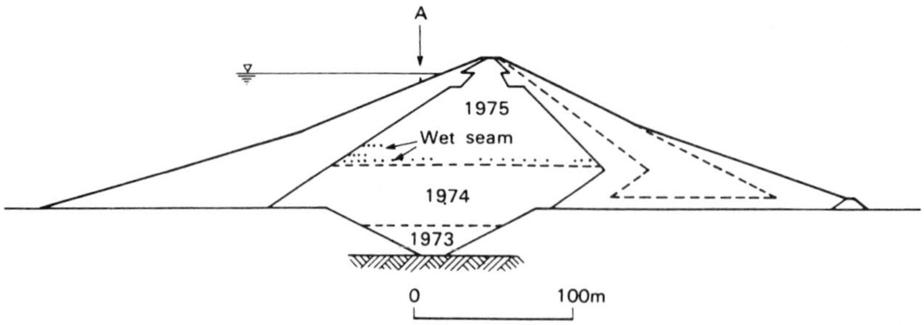

Fig. 36. Teton Dam: major section

failure and 30 months after placement, water was seeping from it and continued for several months. Investigation by adit and boreholes showed that there were thin wet seams (about 0·2 m thick) extending over about five acres in the remainder of the dam. The majority of seams were found to be slightly above the fill surface during the 1974–75 winter shut-down (Fig. 36). Had the wet seams been below the winter surface, it was thought that they might have been caused by melting ice lenses, although it was difficult to see how a sufficient volume of water could have been drawn out of the dry silt fill to form ice lenses.

After the 1974–75 winter shut-down, fill placement began on the lower, left side during periods of rain and snowfall. During the first few weeks there was a shortage of inspectors and it has been suggested that the wet seams originated in layers of wet fill and the effects of ponded surface water. This seems improbable because it would be impossible to operate the heavy placing machines on very wet silt. The ability of the silt fill, placed on average 1·3% dry of optimum, to absorb water was demonstrated when a borehole, put down into the left trench for hydraulic fracture tests, lost 13 m^3 as it was drilled from 30–45 m depth. During the subsequent excavation, no evidence of the water was found. Sherard (1985b) has suggested that the wet seams may have been caused by hydraulic fracture due to reduction of σ_v by arching action both upstream to downstream and between abutments. This action could develop if the fill below the shut-down surface settled, possibly by collapse settlement on wetting by river water. That this hydraulic fracture did not lead to failure may be due to the large width of the core and collapse of the roof of the fracture to form a looser saturated seam of silt with a low enough permeability to prevent erosive flow velocities.

Water in this extensive, loose, saturated seam could become trapped when the breach cut the arch between abutments, allowing the weight of the dam to seal the edges. Some upstream–downstream arching may have remained. One of the survey stations (Fig. 36, A) that had settled 49 mm during the seven months of impounding settled a further 58 mm on failure: this may have been influenced by unmeasured horizontal movement towards the breach. Standpipe piezometers installed in some of the exploratory boreholes about four months after first exposure of the seam(s) showed a maximum pressure of only 1·5 m head of water, which fell to 0·5 m in six months.

While this concept has not been accepted by some other engineers concerned with Teton, Sherard has observed wet seams in other cores. At Yard's Creek (24 m) (USA) they were exposed in a trial pit made through the core for another purpose and at Manicouagan 3 (108 m) (Canada) and Guapo (60 m) (Venezuela) they were found by using a very special exploratory boring technique that required no drilling fluid. The wet seams would not have been discovered at Teton had the exploratory excavation not been taken to the exceptional depth of 70 m. There may be many undetected cases of wet seams in the cores of existing dams.

Wet cores

In Britain, after the experience at Balderhead, there was a return to wetter cores and those of Scammonden (70 m) and Llyn Brianne (90 m) were placed wet of optimum. Groups of the BRS vibrating wire EP cells (Figs 37 and 38) were installed in the middle of the core at three levels during construction of Scammonden (Penman & Mitchell, 1970). Measured values showed that the total pressures at the end of construction (a in Fig. 39) comfortably exceeded the expected water pressures from a full reservoir, with lowest values (σ_h) in the direction across the core. Effective stresses gradually increased with pore pressure dissipation and total pressures decreased towards steady state conditions. After ten years, some of the EP cells failed: total pressures derived from the remaining cells after 14 years are shown at b, in Fig. 39. Pore pressures, measured by five hydraulic piezometers placed across the core at

Fig. 37. Preparing a 45° surface for an EP cell

each level, show a relatively uniform fall from reservoir pressure upstream to the filter drain at the downstream face. It can therefore be expected that total pressures in the core will increase from mid-width, where the measurements were made, towards the upstream face of the core.

At Llyn Brianne, oil-filled EP cells 250 mm dia. and only 10 mm thick were placed in both core and rockfill shoulders. Measured values, described by Carlyle (1973) showed that there was no risk of hydraulic fracture. The cells (Glotzl manufacture) have a simple diaphragm transducer that uses the principle of hydraulic fracture in its operation. A small plastic diaphragm, subjected to the pressure of the oil enclosed in the cell, is pressed against a flat surface containing two holes. Flexible tubes connected to these holes form an external oil circuit with a pump and pressure gauge in an instrument house. When the pressure of oil applied to one of the holes exceeds that in the cell, the diaphragm is pushed away from the flat surface, allowing oil to return through the second connecting tube. The volume change in the cell is small enough not to affect the earth pressure. When pumping is stopped, leakage continues under the diaphragm until pressures balance. Microscopic inspection of some of the flat surfaces has shown that they contain machining marks which form fine ridges between the holes. These will cause stress concentrations at their contact with the diaphragm and assist in making a seal with very little excess pressure from the oil enclosed in the cell.

A thin oil is normally used in the circulating system and must be kept very clean. Any fine particles (soil in the tubes during installation etc.) may damage the sealing surface or lodge under the diaphragm. The oil must also be chemically compatible with the diaphragm: damage has been caused by use of the wrong oil.

This type of EP cell has the advantages of robust simplicity and small thickness over an area

Fig. 38. Three of a group of five EP cells

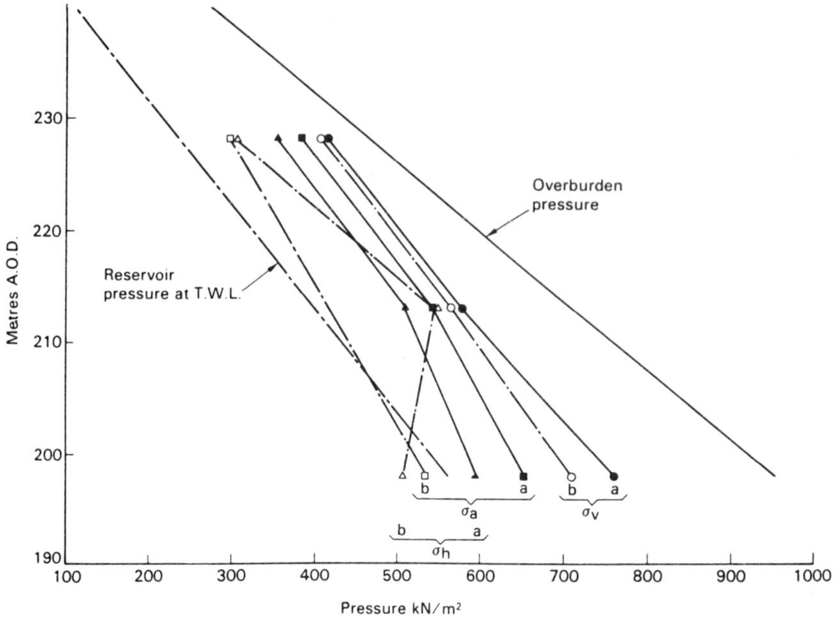

Fig. 39. Total pressures in the clay core of Scammonden Dam on completion in 1969 (a) and in March 1984 (b)

large enough to minimize cell action. It has become one of the most widely used EP cells for total stress measurements in dams.

Old dams

A large number of the world's dams contain no instruments and inspecting engineers may have to resort to various forms of in situ testing. In Britain, many dams built before 1950 had cores of puddled clay and the BRS has begun studies of their condition. Various methods have been used to measure total and effective stresses in these cores, including hydraulic fracture tests from piezometers, measurements with spade-shaped, oil-filled EP cells and the self-boring pressuremeter. This work has been described by Penman & Charles (1981) and Charles & Watts (1986). The spade-shaped cells have been pushed into the clay to measure σ_h and σ_a, but there has been no means for measuring σ_v. Currently an apparatus is being developed at the BRS which will insert EP cells to measure σ_v from a 150 mm dia. vertical borehole. The cells can be pushed up to 0·6 m away from the borehole.

CONCLUDING DISCUSSION AND REMARKS

The major international organizations, ICOLD and the International Society for Soil Mechanics and Foundation Engineering, had their origins in the 1930s and through their regular conferences have provided a unique exchange of experience and development through interaction of ideas and methods. The contribution from soil mechanics has played a vital role in improving design methods for embankment dams, helping to bring them to the leading position that they hold today.

One of the first problems to be tackled was that of slope stability: it was sometimes assumed that the factor of safety of a dam was synonymous with the factor of safety against a slip. Design improvements and analysis using effective stress required testing by back analysis of failures and knowledge of pore pressures in the field. In general, the verification of design assumptions relies on detailed observation of the behaviour of the dam. Knowledge of soil behaviour and the sophistication of stability analysis for complex slip surfaces has reached such a stage that there can be considerable confidence that safe slopes can be designed.

The fact that some slips still occur in recently designed dams reflects a misapplication of design methods rather than weaknesses in the methods. Misunderstanding of site conditions through what may later be condemned as an inadequate site investigation may lie at the root of some failures, but when consideration is given to methods of procurement for site investigation it is perhaps not surprising: lowest tenders based on lengths of holes drilled and numbers of samples taken is not an approach that is likely to reveal the hidden secrets of the site.

Part of the trouble is due to the apparent complexity and procrastination in the promotion of a new scheme. Reeve (1986) in his Presidential address has drawn attention to the long lead time that is common to the construction of a dam and gave an example where this time was more than 13 years. The site investigation was made after four years and was sufficient to show that the scheme was feasible, about nine years before construction started. During that time the design was modified, not in the light of further site investigation, but to satisfy some further twist in the restrictions imposed by planning requirements. There have been cases where rockfill for the shoulders has had to be replaced by weathered mudstone from the reservoir area because of an objection to expansion of an existing quarry. Unfortunately, once approval has finally been obtained, it is expected that the dam will be built in minimum time, at breakneck speed, to provide return on capital with minimum delay. For a potentially dangerous structure with an unspecified life span, this may seem less than ideal.

However intense the site investigations may have been, the exposures made once construction starts form a valuable supplement. There is much to be said for a 'design-as-you-go' approach and it is perhaps unfortunate that so many specifications and contractual arrangements restrict its use. Speed of construction also operates against it, leaving little time for reassessment of design when unexpected features are revealed.

Speed helps to produce maximum pore pressures at the end of construction, making this the most dangerous time for slips. This has the advantage that the failure does not usually involve the release of reservoir water.

The execution of complex analyses by computer can give a false sense of security. In the knowledge that large numbers of slip surfaces have been subjected to refined analysis, the fact that none of the surfaces passed along some obvious feature such as a soft clay layer and that unrealistic values were chosen for c' and φ' may be overlooked.

It can be argued that, as a protection against progressive failure, design using peak values for φ' should accept $c' = 0$. It is now recognized that the use of c' with a constant φ' was only a convenient way of representing the strength characteristics of a soil. There are many more accurate ways: one of the more practical is the use of a curved failure envelope, as discussed by Charles & Soares (1984a, b) who have proposed new methods for stability analysis of slopes.

It may be more rational to design for acceptable movements rather than to guard against the completely unacceptable situation of slip failure by the use of small factors of safety. Provided that realistic parameters for the deformation properties of fill and foundation can be obtained, there are two- and three-dimensional finite element programs that will enable predictions to be made. Their accuracy must be assessed by field measurements and in this connection the horizontal plate gauge has proved invaluable. It can only be used in conjunction with a system for precise surveying that will show movements of the dam in relation to remote, fixed positions. Such observations, even without the horizontal plate gauge, will determine whether movements are becoming unacceptable during construction and, in some circumstances, could show the development of undesirable trends in time for remedial works to be carried out.

A difficulty with this approach is to determine the magnitude of acceptable movements. Progressive failure in brittle clays may develop with remarkably small external movements, whereas, provided that outlet works and other structures have been positioned to allow for it, other fills may undergo large deformations without damage. This may not be acceptable, however, when the watertight element of the dam takes the form of an upstream membrane. There is always a problem of differential movement at the plinth where the membrane bridges from compressible fill to rigid structure. Minimum height and therefore minimum depth of fill behind the structure will help to limit movements, but special flexibility to allow some movement is always needed.

Perhaps the most dangerous form of failure is that developing through erosive seepage under conditions of an almost full reservoir. It can be argued that all cases originate in some form of hydraulic fracture, i.e. total stresses in some part of the water retaining structure are insufficient to withstand the reservoir water pressure and the resulting leaks are able to erode material. The apparently simple check on this situation of observing the volume and quality of leaking water is often most difficult to make. Permeable ground downstream of the waterproof element (and particularly when the groundwater table is low as at Teton) makes it impossible to collect all seeping water. Indications may be given by the pressure distribution measured by foundation piezometers and some water may be collected for examination from relief wells, underdrains etc., but there is a need for the development of satisfactory (preferably automatic) systems that will detect increasing leakage and turbidity.

Water may be lost from reservoirs in many ways not connected with the dam. There are many well-known examples in karstic rock where undetected caverns have opened up sufficiently to prevent the reservoir from filling, but without causing any danger to the dam. To a lesser extent, seepage may occur through the foundation, particularly in fissured rock, without erosion or development of destabilizing pressures. At the contact, however, between clay core and foundation, it is desirable to ensure that total pressures always exceed the reservoir pressure. To this end, not only should core material be sufficiently flexible, but the surface should be relatively smooth. On a rock foundation, it may be necessary to provide a concrete coating.

It is difficult to avoid discontinuities in rolled clay fill. The natural fabric was almost completely destroyed in puddled clay by raising its water content (to reduce c_u) and thoroughly mixing it in a powerful pug mill and there is little evidence of discontinuities being introduced by the method of placing. Inspection of some old cores has shown that some fabric develops with ageing and there was evidence of slickensides, presumably caused by differential movements. There are also many cases of cores built of clays that had been wetted and mixed with shovels on the core itself, not passed through a pug mill.

Rolled clay cores, placed by large machines, may not only retain some of the original fabric, preserved in lumps of fill, but also contain shear surfaces and discontinuities left by the placing machines. It is therefore of even greater importance to ensure that total pressures across these features are sufficient to avoid hydraulic separation by reservoir water pressure.

Towards the end of the puddled clay era, unsuccessful attempts were made to mechanize the traditional compaction by foot. Rolled clay cores were made stronger to accommodate the heavy machines, but it is encouraging to see recent examples showing a return to much softer flexible cores. The core for Kielder Dam (Millmore & McNicol, 1983) was designed to have construction pore pressures that were higher than the reservoir pressure to avoid the possibility of hydraulic fracture, following the recommendations of Penman & Charles (1979). Laboratory tests had shown that this could be achieved by limiting the core strength to about $c_u = 100$ kN/m^2.

A very wet core was placed in the Monasavu Dam (85 m) in Fiji (Knight, Worner & McClung, 1982). The residual soil ($I_p = 52$) was at about 20% wet of Proctor optimum in the borrow area and the monthly rainfall of 0·2–0·4 m prevented natural drying. It was placed by light bulldozers fitted with wide tracks and had $c_u = 14$–26 kN/m^2, close to the values of the old puddled clay. Currently, the core of Wadaslintang (120 m) in Indonesia is being constructed from a heavy, reddish-brown clay, soft enough to require marsh tractors.

Leakage seldom corresponds to a laboratory-determined value of k for the core material. Seeps through imperfections may do little harm if the velocity is low enough and the material does not readily erode. If a discontinuity is forced open by water pressure, soil particles forming the walls of the fracture find themselves under zero effective stress and their resistance to being dragged away by water flow depends on their attraction to or interlocking with their neighbouring particles. This ability may be a function of the tensile strength of the soil, measured in terms of effective stresses. Empirical tests such as Sherard's pinhole test (Sherard, Dunnigan, Decker & Steele, 1976) give a practical guide to the erosion resistance of soils, but there is room for fundamental research into this problem.

No designer can remain satisfied with his design without knowledge of how well it is working in practice. Design improvements rely on feedback from the structure on its behaviour under normal and abnormal conditions. Instrumentation in dams is essential to obtain this information and proper provision should be made in all new schemes to ensure collection of accurate, pertinent records of behaviour.

Research into the behaviour of embankment dams goes hand in hand with geotechnical research and there is a need for this work to be carried out and guided by an independent body representing the interests of owners, consultants and contractors. Costs must be shared and this may be best effected by government funding. Such research and development will continue to support an industry responsible for dams not only in this country, but it is also essential to maintain our involvement in dam projects worldwide.

There is no doubt that progress achieved in the geotechnical field has been fundamental in the satisfactory development of the embankment dam as the major type of dam in use today.

ACKNOWLEDGEMENTS

The Author wishes to thank Mr B. Boden, and through him, both the British Geotechnical Society Committee (of which he is Chairman) for the invitation to give the Rankine Lecture and the Geotechnics Division of the BRS (of which he is Head) for encouragement and advice.

Dr J. A. Charles, Head of the Dams Section, made numerous valuable contributions during the preparation of the Paper. Much of the material has been drawn from work carried out by the Author while with the BRS: permission to use it has been kindly given by the Director, who also made available drawing office and photographic facilities. The Author owes a debt of gratitude to the owners, consultants and contractors with whom he has worked and to his colleagues at the BRS, including Mr D. Burford and Mr K. S. Watts. Mrs Penman assisted with references and in both reading and typing the text.

REFERENCES

Anagnosti, P. (1965). An analysis of stresses and deformations in the wide clay core of a rockfill dam. *Proc. 6th Int. Conf. Soil Mech. Fdn Engng, Montreal* **2,** 447–450.

Baines, J. A., Newman, V. G., Hannah, I. W., Douglas, T. H. & Carlyle, W. J. (1983). Dinorwig pumped storage system. *Proc. Instn Civ. Engrs, Part 1* **74,** 635–680.

Banks, J. A. (1948). Construction of Muirhead reservoir, Scotland. *Proc. 2nd Int. Conf. Soil Mech. Fdn Engng, Rotterdam* **2,** 23–31.

Banks, J. A. (1952). Problems in the design and construction of Knockendon Dam. *Proc. Instn Civ. Engrs, Part 1* **1,** No. 4, 423–443.

Beavan, G. C. G., Colback, P. S. B. & Hodgson, R. L. P. (1977). Construction pore pressures in clay cores of dams. *Proc. 9th Int. Conf. Soil Mech. Fdn Engng, Tokyo* **1,** 391–394.

Bellport, B. P. (1967). Bureau of Reclamation experience in stabilising embankment of Fontenelle earth dam. *Trans. 9th Int. Congr. Large Dams, Istanbul* **1,** 67–79.

Binnie, A. R. (1877). Water supply, rainfall, reservoirs, conduits and distribution. *Lectures at School of Military Engineering, Chatham.* Chatham: Royal Engineers Institute.

Binnie, G. M. (1978). The collapse of Dale Dyke dam in retrospect. *Q. J. Engng Geol.* **11,** 305–324.

Binnie, G. M. (1981). *Early Victorian water engineers.* London: Telford.

Bishop, A. W. (1952). *The stability of earth dams.* PhD thesis, University of London.

Bishop, A. W. (1955). The use of the slip circle in the stability analysis of slopes. *Géotechnique* **5,** No. 1, 7–17.

Bishop, A. W., Kennard, M. F. & Penman, A. D. M. (1960). Pore pressure observations at Selset dam. In *Pore pressure and suction in soils*, pp. 91–102. London: Butterworths.

Bishop, A. W. & Vaughan, P. R. (1962). Selset Reservoir: design and performance of the embankment. *Proc. Instn Civ. Engrs* **21,** 305–346.

Black, W. P. M., Croney, D. & Jacobs, J. C. (1958). *Field studies of the movement of soil moisture.* Road Research Technical Paper No. 41, HMSO, London.

Blight, G. E. (1973). Stresses in narrow cores and core trenches in dams. *Trans. 11th Int. Congr. Large Dams, Madrid* **3,** 63–79.

Buckley, R. B. (1898). Discussion on Reservoirs in India. *Minut. Proc. Instn Civ. Engrs* **132,** 213–217.

Carlyle, W. J. (1973). The design and performance of the core of Brianne dam. *Trans. 11th Int. Congr. Large Dams, Madrid* **3,** 431–455.

de Carvalho, L. H. (1982). Discussion on Q55. *Trans. 14th Int. Congr. Large Dams, Rio de Janeiro* **5,** 551–554.

Celestino, T. B. & Duncan, J. M. (1981). Simplified search for noncircular slip surface. *Proc. 10th Int. Conf. Soil Mech. Fdn Engng, Stockholm* **3,** 391–394.

Charles, J. A. (1979). General report: methods of treatment of clay fills. In *Clay fills*, pp. 315–321. London: Institution of Civil Engineers.

Charles, J. A. & Soares, M. M. (1984a). Stability of compacted rockfill slopes. *Géotechnique* **34,** No. 1, 61–70.

Charles, J. A. & Soares, M. M. (1984b). The stability of slopes in soils with nonlinear failure envelopes. *Can. Geotech. J.* **21,** No. 3, 397–406.

Charles, J. A. & Watts, K. S. (1986). The measurement and significance of horizontal earth pressures in the puddle clay cores of old earth dams. Submitted to *Proc. Instn Civ. Engrs.*

Chugh, A. K. (1986). Variable factor of safety in slope stability analysis. *Géotechnique* **36,** No. 1, 57–64.

Collins, P. M. G. & Humphreys, J. D. (1974). Winscar reservoir. *J. Instn Wat. Engrs* **28,** No. 1, 17–46.

Cooling, L. F. (1936). Discussion on Shearing resistance of soil. *Proc. 1st Int. Conf. Soil Mech. Fdn Engng, Harvard* **3,** 39–40.

Cooling, L. F. & Golder, H. Q. (1942). The analysis of the failure of an earth dam during construction. *J. Instn Civ. Engrs* **19,** No. 1, 38–55.

Costa Filho, L. de M. (1984). Private communication.

Dennehy, J. P. (1979). The remoulded undrained shear strength of cohesive soils and its influence on the suitability of embankment fill. In *Clay fills*, pp. 87–94. London: Institution of Civil Engineers.

Dibiagio, E. & Kjaernsli, B. (1985). Instrumentation of Norwegian embankment dams. *Trans. 15th Int. Congr. Large Dams, Lausanne* **1,** 1071–1101.

Dunstan, M. R. H. (1981). *Rolled concrete for dams: a laboratory study of the properties of high flyash content concrete.* CIRIA Technical Note 105, Construction Industry Research and Information Association, London.

Fetzer, C. A. (1977). Seepage problems of earth and rockfill dams on rock foundations. *Int. Wat. Pwr Dam Constr.* **29,** No. 8, 38–44.

Goldbeck, A. T. & Smith, E. B. (1916). An apparatus for determining soil pressure. *Proc. Am. Soc. Test. Mater.* **16,** No. 2, 309–319.

Gordon, J. L. & Duguid, D. R. (1970). Experiences with cracking at Duncan dam. *Trans. 10th Int. Congr. Large Dams, Montreal* **1,** 469–485.

Gosschalk, E. M. & Kulasinghe, A. N. S. (1985). Kotmale dam and observations on CFRD. *Proc. Am. Soc. Civ. Engrs Symp. Concrete Face Rockfill Dams, Detroit*, pp. 379–395.

Hamilton, D. M. (1986). Private communication.

Hollingsworth, H., Conner, T. R. & Anderson, V. E. (1985). Design of Deer Creek dam. *Proc. Am. Soc. Civ. Engrs Symp. Concrete Face Rockfill Dams, Detroit*, pp. 541–558.

Independent Panel (1976). *Failure of Teton dam.* Report to US Department of Interior and State of Idaho, US Government Printing Office, Washington.

Interior Review Group (1977). *Failure of Teton dam.*

Report on findings by Department of Interior Review Group, US Government Printing Office, Washington.

Interior Review Group (1980). *Failure of Teton dam.* Final report by Department of Interior Review Group, US Government Printing Office, Washington.

Janbu, N. (1957). Earth pressures and bearing capacity calculations by generalised procedure of slices. *Proc. 4th Int. Conf. Soil Mech. Fdn Engng, London* **2,** 207–212.

Kennard, M. F. (1983). *Carsington Reservoir—report on consideration of embankment stability.* Confidential Report No. 1362 to Shephard Hill. Rofe, Kennard & Lapworth, Sutton.

Kennard, M. F., Lovenbury, H. T., Chartres, F. R. D. & Hoskins, C. G. (1979). Shear strength specification for clay fills. In *Clay fills,* pp. 143–147. London: Institution of Civil Engineers.

Kerisel, J. (1985). The history of geotechnical engineering up until 1700. *Proc. 11th Int. Conf. Soil Mech. Fdn Engng, San Francisco, Golden Jubilee volume,* pp. 3–93.

Kjaernsli, B. & Torblaa, I. (1968). Leakage through horizontal cracks in the core of Hyttejuvet dam. *Norw. Geotech. Inst. Publ.* No. 80, 39–47.

Knight, D. J., Worner, N. M. & McClung, J. E. (1982). Materials and construction methods for a very wet clay core rockfill dam at Monasavu Falls, Fiji. *Trans. 14th Int. Congr Large Dams, Rio de Janeiro* **4,** 294–303.

Kramer, R. W. (1982). Discussion on Safety of dams in operation. *Trans. 14th Int. Congr. Large Dams, Rio de Janeiro* **5,** 166–169.

Leonards, G. A. & Davidson, L. W. (1984). Reconsideration of failure initiating mechanisms for Teton dam. *Proc. Int. Conf. Case Histories in Geotechnical Engineering, St Louis* **3,** 1103–1113.

Little, A. L. & Price, V. E. (1958). The use of an electronic computer for slope stability analysis. *Géotechnique* **8,** No. 3, 113–120.

Löfquist, B. (1951a). Calculating a concrete core wall. *Trans. 4th Int. Congr. Large Dams, New Delhi* **1,** 111–133.

Löfquist, B. (1951b). Earth pressure in a thin impervious core. *Trans. 4th Int. Congr. Large Dams, New Delhi,* **1,** 99–109.

Löfquist, B. (1957). Discussion on Cracking in earth dams. *Proc. 4th Int. Conf. Soil Mech. Fdn Engng, London* **3,** 261–262.

Lowe, J. (1962). Discussion on Use of rollcrete in earth dams (unpublished). *1st Am. Soc. Civ. Engrs Water Resources Engineering Conf., Omaha.* (Reproduced in part in *Proc. Int. Conf. Rolled Concrete for Dams, 1981,* pp. W-1–W-5. London: Construction Industry Research and Information Association.)

Lowe, J. (1970). Recent development in the design and construction of earth and rockfill dams. *Trans. 10th Int. Congr. Large Dams, Montreal* **5,** 1–28.

Mackey, P. G. (1985). Rehabilitation to meet reservoir safety and flood criteria. *Trans. 15th Int. Congr. Large Dams, Lausanne* **4,** 899–919.

Magnusson, G. (1948). Research methods and instruments for the measurement of stresses and deformations in earth dams. *Trans. 3rd Int. Congr. Large Dams, Stockholm* **2,** Q9, R32.

de Mello, V. F. B. (1977). Reflections on design decisions of practical significance to embankment dams. 17th Rankine Lecture. *Géotechnique* **27,** No. 3, 281–354.

de Mello, V. F. B. (1982). *A case history of a major construction period dam failure,* pp. 63–78. Brussels: Comité d'Hommage au Professeur E. de Beer.

Millmore, J. P. & McNicol, R. (1983). Geotechnical aspects of the Kielder Dam. *Proc. Instn. Civ. Engrs, Part 1* **74,** 805–836.

Morgenstern, N. R. & Price, V. E. (1965). The analysis of the stability of general slip surfaces. *Géotechnique* **15,** No. 1, 79–93.

Nagarkar, P. K., Kulkarni, R. P., Kulkarni, M. V. & Kulkarni, D. G. (1981). Failure of a monozone earth dam of expansive clay. *Proc. 10th Int. Conf. Soil Mech. Fdn Engng, Stockholm* **3,** 491–494.

Penman, A. D. M. (1956). A field piezometer apparatus. *Géotechnique* **6,** No. 2, 57–65.

Penman, A. D. M. (1971). Rockfill. *J. Instn Highw. Engrs* **18,** No. 12.

Penman, A. D. M. (1977). The failure of Teton dam. *Ground Engng* **10,** No. 6, 18–27.

Penman, A. D. M. (1979). Construction pore pressures in two earth dams. In *Clay fills,* pp. 177–187. London: Institution of Civil Engineers.

Penman, A. D. M. (1982). Materials and construction methods for embankment dams and cofferdams. General Report. *Trans. 14th Int. Congr. Large Dams, Rio de Janeiro* **4,** 1105–1228.

Penman, A. D. M., Burland, J. B. & Charles, J. A. (1971). Observed and predicted deformations in a large embankment dam during construction. *Proc. Instn Civ. Engrs* **49,** 1–21.

Penman, A. D. M. & Charles, J. A. (1974). Measuring movements of embankment dams. *Proc. Symp. Field Instrumentation in Geotechnical Engineering,* pp. 341–358. London: Butterworths.

Penman, A. D. M. & Charles, J. A. (1975). Predicted deformation of the upstream membrane of a rockfill dam. *Ground Engng* **8,** No. 6, 47–48.

Penman, A. D. M. & Charles, J. A. (1976). The quality and suitability of rockfill used in dam construction. *Trans. 12th Int. Congr. Large Dams, Mexico City* **1,** 533–556.

Penman, A. D. M. & Charles, J. A. (1979). The influence of their interfaces on the behaviour of clay cores in embankment dams. *Trans. 13th Int. Congr. Large Dams, New Delhi* **1,** 695–714.

Penman, A. D. M. & Charles, J. A. (1981). Assessing the risk of hydraulic fracture in dam cores. *Proc. 10th Int. Conf. Soil Mech. Fdn Engng, Stockholm* **1,** 457–462.

Penman, A. D. M. & Charles, J. A. (1982). An improved horizontal plate gauge. *Géotechnique* **32,** No. 3, 278–282.

Penman, A. D. M. & Charles, J. A. (1985a). Behaviour of rockfill dam with asphaltic membrane. *Proc. 11th Int. Conf. Soil Mech. Fdn Engng, San Francisco* **4,** 2011–2014.

Penman, A. D. M. & Charles, J. A. (1985b). A comparison between observed and predicted deformations of an embankment dam with a central asphaltic core. *Trans. 15th Int. Congr. Large Dams, Lausanne* **1,** 1373–1389.

Penman, A. D. M., Charles, J. A., Nash, J. K. T. L. & Humphreys, J. D. (1975). Performance of culvert under Winscar dam. *Géotechnique* **25,** No. 4, 713–730.

Penman, A. D. M. & Hussain, A. (1984). Deflection measurements of the upstream asphaltic membrane of Marchlyn dam. *Int. Wat. Pwr Dam Constr.* **36,** No. 9, 33–37.

Penman, A. D. M. & Mitchell, P. B. (1970). Initial behaviour of Scammonden dam. *Trans. 10th Int. Congr. Large Dams, Montreal* **1,** 723–747.

Pessoa, J. C. (1982). Discussion on Q55. *Trans. 14th Int. Congr. Large Dams, Rio de Janeiro* **5,** 680–682.

Rao, K. L. (1951). Earth dams ancient and modern in Madras State. *Trans. 4th Int. Congr. Large Dams, New Delhi* **1,** 285–301.

Rawlinson, R. (1883). Discussion on Waterworks. *Minut. Proc. Instn Civ. Engrs* **74,** 163–167.

Reeve, D. A. D. (1986). Presidential address 1985. *Proc. Instn Civ. Engrs, Part 1* **80,** 1–12.

Rogers, W. S. (1935). A soil moisture meter. *J. Agric. Soc., Part 3* **25,** 326–343.

Rowe, P. W. (1972). The relevance of soil fabric to site investigation practice. 12th Rankine Lecture. *Géotechnique* **22,** No. 2, 195–300.

Sarma, S. K. (1973). Stability analysis of embankments and slopes. *Géotechnique* **23,** No. 3, 423–433.

Schnitter, N. J. (1979). Discussion on Deterioration or failure of dams. *Trans. 13th Int. Congr. Large Dams, New Delhi* **5,** 488–493.

Seed, H. B. & Duncan, J. M. (1981). The Teton dam failure—a retrospective review. *Proc. 10th Int. Conf. Soil Mech. Fdn Engng, Stockholm* **4,** 219–238.

Seed, H. B., Leps, T. M., Duncan, J. M. & Bieber, R. E. (1976). Hydraulic fracturing and its possible role in the Teton dam failure. In *Failure of Teton dam.* Report to US Department of Interior and State of Idaho, Appendix D, US Government Printing Office, Washington.

Sherard, J. L. (1973). Embankment dam cracking. In *Embankment dam engineering,* Casagrande volume, pp. 271–353. New York: Wiley.

Sherard, J. L. (1985a). The upstream zone in concrete-face rockfill dams. *Proc. Am. Soc. Civ. Engrs Symp. Concrete Face Rockfill Dams, Detroit,* pp. 618–641.

Sherard, J. L. (1985b). Lessons learned from the Teton dam failure. *Proc. Int. Workshop Dam Failures, Purdue University, 6–8 Aug.* to be published. Amsterdam: Elsevier.

Sherard, J. L., Dunnigan, L. P., Decker, R. S. & Steele, E. F. (1976). Pinhole test for identifying dispersive soils. *J. Geotech. Engng Div. Am. Soc. Civ. Engrs* **102,** GT1, 69–85.

Sierra, J. M., Ramirez, C. A. & Hacelas, J. E. (1985). Design features of Salvajina dam. *Proc. Am. Soc. Civ. Engrs Symp. Concrete Face Rockfill Dams, Detroit,* pp. 266–285.

da Silveira, A. F. (1984). Statistical analysis of deteriorations and failures of dams. In *Safety of dams,* pp. 55–60 (ed. Serafim). Rotterdam: Balkema.

Skempton, A. W. (1964). Long-term stability of clay slopes. 4th Rankine Lecture. *Géotechnique* **14,** No. 2, 77–101.

Skempton, A. W. (1985). Residual strength of clays in landslides, folded strata and the laboratory. *Géotechnique,* **35,** No. 1, 3–18.

Skempton, A. W. & Coats, D. J. (1985). Carsington Dam failure. *Failures in earthworks,* pp. 203–220. London: Telford.

Smith, N. A. F. (1970). *The heritage of Spanish dams.* Madrid: Spanish National Committee on Large Dams.

Spangler, M. G. (1933). *The supporting strength of rigid pipe culverts.* Bulletin 112, Iowa State College of Agriculture and Mechanical Art.

Strange, W. L. (1898). Reservoirs with high earthen dams in western India. *Minut. Proc. Instn Civ. Engrs* **132,** 130–199: Discussion, 200–272.

Taylor, D. W. (1947). *Review of pressure distribution theories, earth pressure cell investigations and pressure distribution data.* Soil Mechanics Fact Finding Survey—Progress Report, Waterways Experimental Station, Vicksburg.

Thomas, H. S. H. & Ward, W. H. (1969). The design, construction and performance of a vibrating-wire earth pressure cell. *Géotechnique* **19,** No. 1, 39–51.

Trollope, D. H. (1957). The systematic arching theory applied to the stability analysis of embankments. *Proc. 4th Int. Conf. Soil Mech. Fdn Engng, London* **2,** 382–388.

Uren, F. C. (1914). *Waterworks engineering.* Bristol: Castle Litho.

Vaughan, P. R., Kluth, D. J., Leonard, M. W. & Pradoura, H. H. M. (1970). Cracking and erosion of the rolled clay core of Balderhead dam and the remedial works adopted for its repair. *Trans. 10th Int. Congr. Large Dams, Montreal* **1,** 73–93.

Vestad, H. (1976). Viddalsvatn dam: a history of leakage and investigations. *Trans. 12th Int. Congr. Large Dams, Mexico* **2,** 369–390.

Walker, F. C. (1948). Experience in the measurement of consolidation and pore pressures in rolled earth dams. *Trans. 3rd Int. Congr. Large Dams, Stockholm* **2,** Q9, R58.

Ward, W. H., Penman, A. & Gibson, R. E. (1955). Stability of a bank on a thin peat layer. *Géotechnique* **5,** No. 2, 154–163.

Watakeekul, S., Roberts, G. J. & Coles, A. J. (1985). Khao Laem—a concrete face rockfill dam on karst. *Proc. Am. Soc. Civ. Engrs Symp. Concrete Face Rockfill Dams, Detroit,* pp. 336–361.

Waterways Experimental Station (1942). *Interim report—pressure cell installation—Arkabutla dam.* Waterways Experimental Station, Vicksburg.

Westerburg, G., Pira, G. & Hagrup, J. (1951). Description of some Swedish earth and rockfill dams with concrete core walls and measurement of the movements and pressure in the filling material and the core walls. *Trans. 4th Int. Congr. Large Dams, New Delhi,* Q13, R11.

Yamauchi, T., Harada, J., Okada, T. & Shimada, S. (1985). Construction of Tamagawa dam by the RCD method. *Trans. 15th Int. Congr. Large Dams, Lausanne* **2,** 89–115.

VOTE OF THANKS

In proposing a vote of thanks to Dr Penman, Professor J. B. Burland made the following remarks.

'I had the privilege and stimulation of working closely with Dr Penman at the Building Research Station for 14 years. The lecture just presented captures much that typifies his unique style and approach as well as exemplifying his life-long love for embankment dams, first kindled by Professor Skempton.

'Arthur Penman is a resourceful researcher and a thoroughly practical engineer. He is also an enthusiastic sailor. An expedition to a dam site with Arthur, as with a sailing expedition, is an adventure not to be undertaken lightly by the faint hearted. When in harbour with a force seven gale blowing outside Arthur will suggest that we should go out and "just give it a try". Those who have sailed with him will know that once under sail it will take a hurricane to turn him back. Similarly obstacles to instrumenting a dam are there to be overcome and overcome they always are. His enthusiasm is infectious and sweeps all before it and we have experienced that enthusiasm this evening.

'The resourcefulness with which Dr Penman pursues his research is displayed by the immense range of instrumentation he has developed, some of which he has described here. It is in the conception and design of this instrumentation that his skill as a mechanical engineer is displayed and his ideas are widely used all over the world.

'Arthur Penman's devotion to embankment dams is reflected in the many case histories that he has referred to. He has been responsible for the instrumentation in many of them and has developed a deep understanding, almost an intuitive feel, for embankment dam behaviour. The lecture will form a most valuable source of reference for practitioners and students alike.

'I referred to Dr Penman's unique style and I am reminded of a course on research management which he and I attended some years ago. One of the lectures was about communication. The lecturer brought with him an object of complex geometric form. The intention was to choose some unsuspecting soul to describe the object for the class to draw on the basis of the verbal description alone. If the lecturer's intention had been to demonstrate the consequences of poor communication he failed miserably as he happened, quite by chance, to choose Arthur Penman to describe the object. There followed a step by step description of such clarity that everyone on the course drew the object perfectly!

'We have had the benefit of listening to an enthralling, enjoyable and thoroughly stimulating lecture presented with the clarity that we have come to expect from Dr Penman. It is with the very greatest pleasure that I propose a hearty vote of thanks to Dr Penman for a most memorable twenty-sixth Rankine Lecture.'

The Rankine Lecture

The twenty-seventh Rankine Lecture of the British Geotechnical Society was given by Professor R. F. Scott at Imperial College of Science and Technology, London, on 24 March 1987. The following introduction was given by Professor A. N. Schofield, Cambridge University.

'The British Geotechnical Society and the Institution of Civil Engineers invite to London a succession of Rankine Lecturers whose essential qualification is that we all know of them and want to hear them lecture. Professor Scott is eminently qualified in this respect, but his choice as the twenty-seventh Rankine Lecturer is doubly appropriate, for he also is a Scot and a graduate of Glasgow University, where the Scottish engineer and physicist who is commemorated in the lecture held the Chair of Engineering from 1855 until his death in 1872.

'Ronald Fraser Scott was born in London on 9 April 1929; he went to school in Perth. After graduating in civil engineering as a bachelor of science of Glasgow University in 1951, he studied in the USA, obtaining his doctorate of science under Professor D. W. Taylor at the Massachusetts Institute of Technology in 1955. He worked from 1955 to 1957 for the US Army Corps of Engineers in Boston and from 1957 to 1958 for Racey, MacCallum and Associates of Toronto, but his academic inclinations led him in 1958 to become a professor of civil engineering in the California Institute of Technology.

'The California Institute is unique: Caltech's faculty includes many most distinguished professors and it has about an equal number of very able students. Professor Scott has attracted successive Caltech students to soil mechanics for nearly 30 years, and although he has never been able to have a large group there is excellence and continuity in their work. Other schools such as Harvard have seen soil mechanics come and go, but there has been soil mechanics at Caltech for 67 years.

'The American Society of Civil Engineers awarded Professor Scott the Huber research prize in 1968, the Norman medal in 1972, the Middlebrooks prize in 1981 and the Terzaghi lectureship in 1983. He is the author of four books: in 1963, *Principles of soil mechanics*; in 1968 (with Schoustra), *Soil: mechanics and engineering*; in 1975 (with Bolt, Horn and MacDonald), *Geological hazards*; in 1981, *Foundation analysis.* He was eminent 20 years ago in lunar soil mechanics and is eminent now in centrifuge dynamic modelling, to name only two of the fields of his activity.

'On behalf of the British Geotechnical Society I welcome Professor Scott and invite him to deliver his Rankine Lecture on the topic of Failure in geotechnical engineering.'

Professor R. F. Scott

Scott, R. F. (1987). *Géotechnique* **37**, No. 4, 423–466

Failure

R. F. SCOTT*

'Failure: an ill-coined and late word'—Skeat,
An etymological dictionary of the English language, 1893

The Pennsylvania Disaster (1889 Johnstown Flood)
by William McGonagall, Scottish poet and tragedian

''Twas in the year of 1889, and in the month of June,
Ten thousand people met with a fearful doom,
By the busting of a dam in Pennsylvania State,
And were burned and drowned by the flood—Oh pity their fate!

The embankment of the dam was considered rather weak,
And by the swelled body of water the embankment did break,
And burst o'er the valley like a leaping river,
Which caused the spectators with fear to shiver . . .'

An attempt is made to classify geotechnical failures, and these are illustrated by several examples. The experiences involved in trying to obtain remotely the properties of the granular material on the moon and Mars are summarized. Descriptions are given of several landslides in southern California but which have aspects in common with other landslide events in other parts of the world. The failure of two dams is discussed briefly. The problem of fault rupture as a hazard in its own right, apart from the associated generation of strong ground motion, is treated from the points of view both of faulting mechanics and also the effect of the displacements on structures. Methods of analysis including the finite element and finite difference techniques are considered with respect to the examination of failures. Discussion of the constitutive relations used in such methods is given, especially regarding stable and unstable material behaviour. Since the propagation of slip or rupture surfaces is important in many of the examples, emphasis is given to the implementation of numerical approaches in which unstable material behaviour can be employed, and which can give rise to the generation of slip surfaces or zones. Examples are given of the results of a discrete element method and the use of dynamic relaxation with the finite difference technique applied to some typical problems such as embankments, slopes and punch indentation. Attention is finally given to the questions raised in the design of structures, and the analysis and use of failures in this regard.

KEYWORDS: analysis; case history; failure; finite elements; landslides; site investigation; soil properties; stability; stress analysis.

Plusieurs examples de ruptures géotechniques sont cités en vue de les classifier. Des expériences sont décrites dont le but était d'obtenir à distance les propriétés des matériaux pulvérulents sur la lune et sur Mars. Des descriptions sont données de plusieurs glissements de terrain qui ont eu lieu en Californie du Sud mais qui avaient des ressemblances avec d'autres survenues ailleurs dans le monde. La rupture de deux barrages est discutée brièvement. Le problème de la rupture par faille analysé comme risque proprement dit, indépendamment du resultat de fort mouvement du terrain, est traité au point de vue de la mécanique des failles et aussi de l'effet des déplacements sur les constructions. Les méthodes analytiques, y compris les techniques à éléments finis et à différences finies sont traitées en fonction de leur application à l'étude des ruptures. Les relations des constituants employées dans de telles méthodes sont discutées, surtout en ce qui concerne le comportement stable et instable des matériaux. Comme la propagation des surfaces de glissement ou de rupture est importante dans beaucoup des exemples décrits, on accentue l'emploi des méthodes numériques dans lesquelles on peut utiliser le comportement instable des matériaux menant à la génération de surfaces ou de zones de glissement. Des exemples sont présentés des résultats d'une méthode à éléments discrets et aussi de l'emploi de la relaxation dynamique à l'aide de la technique des differénces finies avec application à des problèmes typiques tels les remblais, les pentes et l'indentation au poinçon. Finalement des considérations sont données concernant l'analyse et l'emploi des ruptures en relation avec les constructions.

* California Institute of Technology.

INTRODUCTION

When the invitation to deliver the 27th Rankine Lecture was made, the subject of failure was very much in the public eye, so that it became almost an obvious topic for the lecture. A few months before, the Mexican earthquake of 19 September 1985 had caused much damage and many deaths and injuries, principally in Mexico City. That earthquake, as is usual with seismic disturbances, had at least two surprises: the low levels of acceleration in the near-source area along the Mexican coast and the high levels of acceleration in the valley of Mexico. On 13 November 1985 the eruption of the Nevado del Ruiz volcano in Colombia, South America, led to mudflows (lahars) which took many lives. Lastly, as far as immediate events were concerned at that time, in 1986 the failure of the Challenger booster engine and the loss of the spacecraft and crew left an entire space programme in disarray.

These events led me to consider the instances of failure in my professional life, since the incidents closely paralleled my interests at various times in my career. A few examples of failures are given which I used to learn something about deformational and failure processes in geotechnical engineering, and then I go on to ratiocinate about the role that failure and its analysis plays in the geotechnical field.

FAILURE CLASSIFICATION

Failures of geological and soil materials can be divided into a number of classes. The following tentative arrangement based on personal observation is suggested. There are three components of a failure; for it to be understood in the most general sense, information is needed on all three. They are *mechanism*, *properties* and *analysis*. We must have a clear picture of the *mechanics* involved in a failure before any subsequent steps can be taken. The initial emphasis of failure investigations is usually directed to the elucidation of the mechanism. Examples are the failure of dams, as described briefly later, or the collapse of a structure in, say, an earthquake. The sequence of events is important. Secondly lies the determination of both the qualitative and the quantitative nature of the material *properties* which rendered the failure possible. Was the soil easily erodible? Did it lend itself to hydraulic fracturing? Was it susceptible to creep? Was the stress–strain relation unstable? Then, subsequently, what were the peak and residual shearing strengths, modulus values etc.? When these two components have been classified, *analyses* can be attempted, incorporating mechanism and properties. Sometimes the examination of a failure only leads to the conclusion that it was caused by lack of application of already known principles. On other occasions a new mechanism or a consequence of material property variation becomes apparent, and then new precautions and analytical procedures are added to our lexicon.

Each failure can be classified in a broad sense by the extent to which the contribution of each of these components is understood. Here such a classification is illustrated by several examples, which refer directly to the brief case histories described subsequently. Other combinations of the three components may be readily inferred.

(*a*) An analysis can be performed based on measured material properties in standard tests, and which offers an acceptable explanation of the failure. When this condition pertains, a failure can be deliberately induced to evaluate the material properties. Here, mechanism, properties and analysis are all established.
(*b*) A mechanism is apparent, and an analysis can be performed, but its employment with material properties measured in standard tests does not give a result corresponding to observation.
(*c*) The cause or causes are minor defects in the material, or unclear mechanisms, which may be destroyed in the failure or which cannot be detected by investigation, are not amenable to analysis and can only be speculated about in post-failure investigations.
(*d*) The mechanism is clear, and a dependence on material properties is also relatively clear, but the material properties cannot be obtained for analysis, because of the nature of the material, economics, the dimension involved or the statistical variation of the material characteristics.
(*e*) For design purposes, a failure analysis is required but it is not obvious how it should be done since possible failure mechanisms are not clear. Prototype or model tests are called for.

The safety of existing or proposed structures is evaluated according to assumed mechanisms of failure and known properties, and therefore lies in class (*a*). If a possible failure mechanism requiring a lower load or stress than needed by the assumed mechanisms exists, it will operate and the structure is in trouble. It may eventuate that a failure occurs, which is entirely explicable, but only after more field or laboratory tests have been made. On subsequent study, the failure which occurs can fall into any of the other four categories. Category (*a*) constitutes the state of the art of failure analysis; new results come from failures which fall into the other groups. For those failures which lie in category (*b*), other processes, or mechanisms, have to be adduced to provide an explanation. On occasion, this has led to new results (Skempton, 1964) and altered analyses.

In many circumstances it is not possible to determine the detailed events leading to a failure, but design or construction methods can be proposed for future structures which will lessen the chances of a similar event. Peck (1981) has described such cases, which I place in category (*c*). Many failures involving difficult-to-sample dry or crumbling soils and fractured rocks may be placed in the realm of category (*d*) because the in-place behaviour cannot be related to the broken or partial samples obtained. Landslides involving cubic kilometres of material have occurred (Bolt, Horn, MacDonald, & Scott, 1975) and any reasonable assessment of material property, or perhaps even of original geometric configuration, is impossible to arrive at. There are conditions where the analysis or failure configuration is not obvious and small-scale or full-size tests are required to establish a failure mode to lead to an analytical model: these are referred to category (*e*). In some cases, legal ramifications are such that no involved party wishes to investigate the detailed nature of the failure.

The failures described in the following will be identified with these categories, wherever possible. Benjamin Baker's (1881) famous remark, 'if an engineer has not had some failures [with retaining walls], it is merely evidence that his practice has not been sufficiently extensive', is still applicable today.

FAILURES ON OTHER PLANETS

> 'It's all very pretty but I don't see that it proves anything'—comment by Senior Wrangler after reading 'Paradise Lost' (Crowe, 1967)

The moon

The first question of a serious nature concerning failure, and which affected my life, as it turned out, for many years, was directed to me from an engineer at the California Institute of Technology (CIT) Jet Propulsion Laboratory (JPL) in Pasadena. He said, 'How far would a sphere about 3 ft in diameter, weighing about 100 lb on earth and travelling about 100 ft/s penetrate into soil on the moon, if there were soil on the moon?'. The JPL began, and continues, as a laboratory of the CIT, which manages it for the US National Aeronautics and Space Administration (NASA). After the shock of the Soviet Union's first Sputnik, JPL had been given responsibility for planning the unmanned exploration of the solar system. At the time of the question (1959) planning was under way for the generally ill-fated series of spacecraft called 'Rangers'. That programme proceeded fitfully and frenetically, punctuated by regular failures, none of which, except for the final millisecond or two, involved any geomechanics.

At one stage it was intended to include a lunar impact capsule (Fig. 1) on the spacecraft; the capsule would be released and decelerated to fall, relatively softly, on to the lunar surface. The capsule included a seismometer to record tremors—moonquakes—on the moon. The development of the Ranger sphere gave the seismological and earthquake engineering communities the 'Ranger seismometer', which never recorded a moonquake, but which still provides excellent service as a sturdy but sensitive sensor in structural vibration studies. The JPL engineers were becoming knowledgeable about spacecraft design and performance by following Baker's precept but were lacking information on the mechanical properties of granular materials. Some studies of impact and penetration were made (Roddy, Rittenhouse & Scott, 1963), which required an estimation of a failure mechanism, and the properties of a dry sand at lunar gravity were estimated (Scott, 1964) to give some small assistance to the capsule design process. The penetration studies were of assistance later in the instrumentation of ocean floor coring and penetration devices (Scott, 1967a, 1970).

However, an intact sphere never reached the moon, since the first six spacecraft failed (Hall, 1977), causing a considerable diminution in the scientific squabble as to which scientific instruments should first be placed on the lunar surface. For experimental equipment, the last three spacecraft carried only cameras, with which to photograph the surface as the unretarded spacecraft approached it rapidly. A succession of pictures resulted, revealing the lunar surface at diminishing distances. The most remarkable feature was the similarity of the lunar surface at all scales, as each frame showed a lunar surface of lateral dimension decreasing from tens of kilometres to 30 m, containing only a random array of craters of assorted sizes. I do not know whether an intersecting mass of circles, similar at all viewing scales, can be described in terms of fractals (Mandelbrot, 1983), but I learned that, in much of science, increasing resolution does not bring with it increasing understanding; most funding agencies have yet to learn this.

The failures and delays in a programme whose planned 1962 termination was stretched out thereby to 1965 meant that the function of the missions had changed from one of scientific inquiry to one of support of the Apollo programme, announced by President Kennedy in 1961, long after initiation of the Ranger series. In the meantime my own extraterrestrial curiosity had been aroused by the interaction with the JPL

Fig. 1. Ranger spacecraft: the spherical hard landing capsule appears on top of the spacecraft (photograph by courtesy of JPL/NASA)

and, besides working on the studies already mentioned, I naturally wondered about the composition of the lunar surface. At the time there were two schools of thought

(*a*) the surface was mostly volcanic, and therefore likely to be relatively hard and rock like with volcanically generated craters

(*b*) the observed features were due to meteorite impacts, which would have generated a fair amount of granular debris of all sizes on the surface.

My natural inclination lay towards (*b*), and therefore in 1963 I proposed to NASA that a soil mechanics experiment be designed and flown on the next series of spacecraft being planned: Surveyor. The proposal was accepted, but it was a long time before the equipment went to the moon.

Since it had taken six failures to achieve the first Ranger success, plans were made for nine Surveyors; clearly the task of landing a survivable vehicle softly on the lunar surface was immensely more complex than simply sending a man-made meteorite to it. Of these, the first six were termed engineering vehicles, since their task was to test the feasibility of the concept. Science (which to NASA involves anything not pertinent to spacecraft functions and which includes what I would call engineering) was to be left to follow-on missions beginning with Surveyor 7; my surface experiment was identified with this definition. To indicate the difficulty of the task, I should point out that in the period 1963–65, there were five failures of Russian lunar soft landing spacecraft before their first success.

The Surveyor spacecraft (Fig. 2) had three legs equipped with non-linear gas pressurized shock absorbers and crushable pads; the final stages of the flight to the lunar surface were controlled by three so-called vernier rocket engines whose thrust and direction were under the command of a complex Doppler radar and inertia sensing system involving the spacecraft's attitude, range to the surface and velocity. These engines were to be shut down when the vehicle was a few feet above the lunar surface so that it would fall vertically under the gentle gravity of the moon, to strike the surface at about 3 m/s. Since no active tactile experiments were planned for the first six spacecraft, one of my students and I had calculated what would happen when the spacecraft hit a soil surface on a $g/6$ planet. With the infinite attendant combinations of three-dimensional spacecraft velocity components and surface slope and roughness, the computations were simplified to just the case of simultaneous contact of all three footpads at a range of velocities on a variety

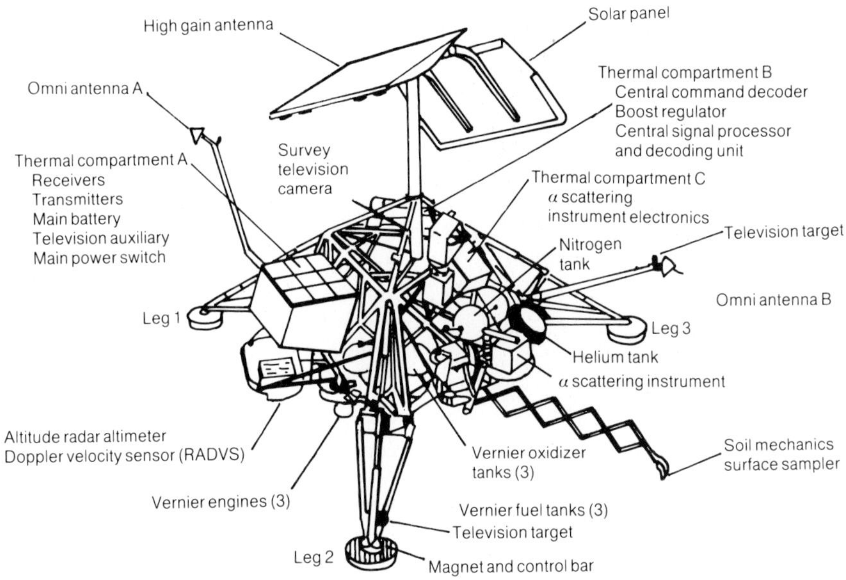

Fig. 2. Surveyor spacecraft (source: NASA)

of soils of varied mechanical properties ('soft', 'medium', 'hard' etc.) under lunar gravity and vacuum conditions. For engineering reasons involved with failure analysis of the spacecraft, the velocity at contact, the contact times on the three footpads and the force history in the shock absorbers were all to be measured and sent telemetrically back to earth. These data would only be of value to failure analyses if the impact occurred under something approaching the design assumptions—the associated ranges of the various parameters were known as the 'nominal' values.

Eventually, after many delays, the Atlas Centaur rocket combination containing Surveyor folded in its cocoon was launched on 30 May 1966. In view of the previous history of the lunar programme, no one was especially optimistic about the chances of success. The launch, however, and subsequent mid-course correction, went flawlessly. In the final stages of flight, to the small group in the Space Flight Operations Facility, the telemetry indicated an incredible degree of coincidence with the nominal mission. This has become a familiar (until Challenger) event, via the newspapers and evening television news broadcasts, but to that group at that time it was completely unexpected. The trajectory was excellent, the main retro-rocket fired and was discarded appropriately, the vernier engines started and operated as planned, and eventually the final, absolutely nominal stages of the descent were counted through to a perfect touchdown. The dazed euphoric group could only expect that the telemetry or camera would fail. They did not. A few minutes after touchdown, the velocity, contact times and shock absorber force histories were available. All three footpads had made contact with the lunar surface within 10 ms of each other, at a vertical impact velocity of 3·5 m/s. The computations for this had been prepared in graphical form on the first sheet of paper in my stack (case (*a*)), and a comparison established rather close bounds on the mechanical (soil) properties of the lunar surface material, within minutes of the first contact. Since the first picture had not yet been obtained, it was of considerable interest that, since the contact measurements and calculations demonstrated that the surface was neither rigid (rock) nor extremely soft, it probably consisted of granular material, although perhaps, to be broad minded, a porous, crunchy lava could not be excluded. The first picture, some time later (Fig. 3), of the nearest footpad confirmed the granular nature of the surface. The soil model obtained at the time has not changed in essence since. This and all spacecraft tests of planetary surfaces may be termed a category (*a*) failure.

In the early stages of Surveyor spacecraft planning, the design included a device for sampling the lunar surface material, so that the material could be supplied to chemical testing equipment. A preliminary model of this sampler was made by Hughes Aircraft Company in 1962; I tested it on soil and decided that, with some modifications and some instrumentation, it could serve as a useful device for soil property determinations. It formed the basis of my lunar surface experiment proposal. During the years up to the first flight, the modifications were incorporated in the sampler (Scott, 1967b), and extensive tests were conducted on a variety of possible lunar surface

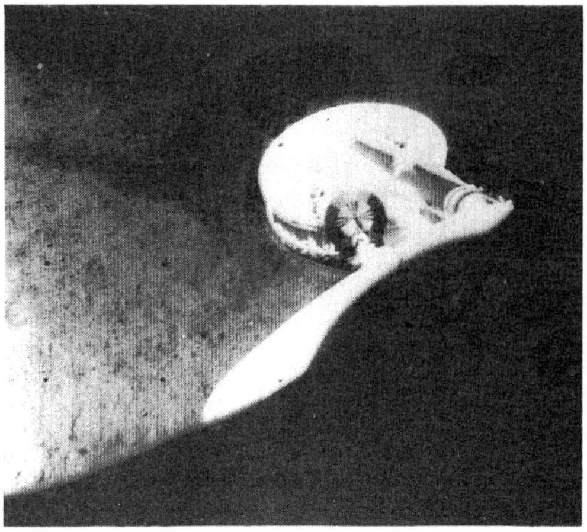

Fig. 3. Surveyor 1 footpad picture (photograph by courtesy of JPL/NASA)

materials from soils to rocks. The device could be extended, retracted, elevated, lowered and moved in azimuth by electric motors. An additional motor opened and closed the bucket door (see Fig. 2), enabling the sampler to pick up and drop lunar soil and rocks. It was possible to take pictures of the sampler at all points in its operational range from the Surveyor television camera.

The components of follow-on Surveyor spacecraft had already been manufactured and were in various stages of assembly, since subsequent launches were to be made every few months, at times dictated by the relative positions of the moon and earth. Surveyor 2 was complete and ready to go later in 1966 and was therefore not available for any modification at all; the schedule meant that the first vehicle capable of alteration was the third Surveyor, scheduled for an April 1967 launch. Efforts were directed to having the surface sampler accepted for flight on that spacecraft.

Because of the spacecraft's design, it was not possible to equip the surface sampler with force or displacement measuring instrumentation, but it was possible to estimate the force quite closely by using measurements of motor current, which was monitored. After some practice, displacements and positions could be estimated and subsequently measured from successive television pictures of the sampler.

In the latter part of 1966, the decision was made to fly the sampler on Surveyor 3 which, after final checks of the spacecraft, including the surface sampler, was launched on 17 April 1967. This flight proceeded without incident until the spacecraft touched down on the moon on 20 April. On command, the surface sampler deployed correctly.

My main objective with the surface sampler was to determine the mechanical properties of the lunar surface by failing it. To this end the tip of the surface sampler had been redesigned to present a flat area of about 1 in × 2 in to the lunar surface when the bucket was closed, and an area of about 0·1 in × 2 in when the door was wide open. The principal experiments available were bearing capacity/penetration tests caused by driving the bucket down into the surface at various extension distances. The maximum force available and the angle of contact with the surface varied with the distance. Trenching tests were also possible by pushing the open bucket down into the lunar surface and then retracting the sampler towards the spacecraft. A bearing test could be done in the bottom of such a trench, which could be deepened by successive passes of the sampler. Impact tests could be performed, and it was possible to pick up soil and small rocks to make an approximate estimate of weight and to relocate them.

From the tests performed, all of which involved failing the lunar surface material (Fig. 4), it was possible to calculate values of cohesion and friction for the lunar material and to estimate its density approximately. The soil was surprisingly, and to some extent, considering the effort involved, disappointingly normal in its behaviour and properties. The only somewhat unexpected feature was its slight cohesion, which did not appear to result from cementation but was persistent. In a press conference, trying to explain this feature to reporters, I described the soil as having the behaviour of a 'damp sand', hastening to make sure that they did not understand me to mean that it *was* damp, since there is no free water on the moon. A few years later, when soil was retrieved by the Apollo astronauts for study in terrestrial laboratories, it appeared that the cohesion was due to a random mosaic of positive and negative charges on the particles' surfaces, caused by solar wind bombardment. Adjacent particles therefore experienced attractive forces. From all points of view, the Surveyor 3 mission was a considerable success.

Surveyor 4 failed, but Surveyor 5 and Surveyor 6, without the surface sampler, reached the moon and performed well. The sampler was again installed on Surveyor 7, which, because of the delays in the programme, had been determined to be the last mission. That spacecraft, which also carried a deployable surface chemistry experiment, landed successfully near the large crater Tycho Brahe in January 1968. Apart from surface testing operations similar to those performed previously, the sampler was used to move the chemistry box about the lunar surface to test both undisturbed and reworked soil (Fig. 5). The

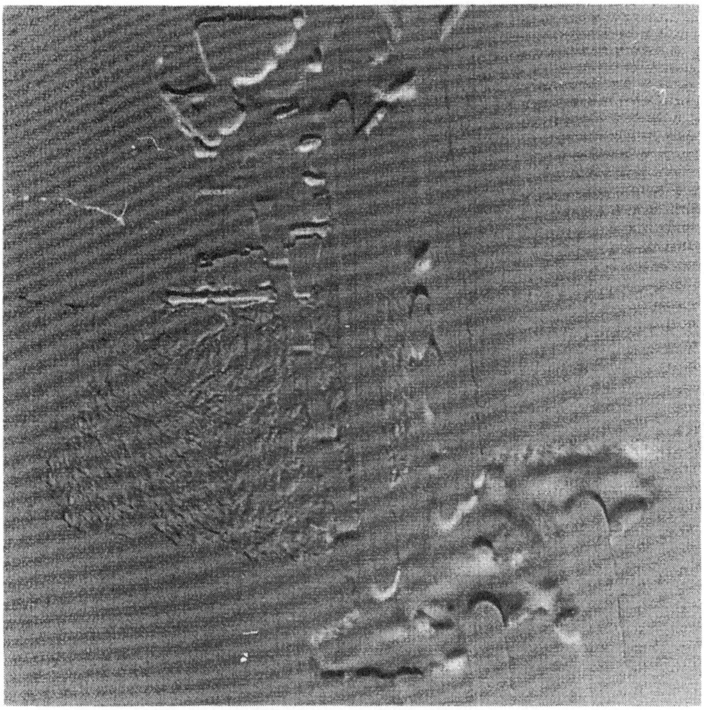

Fig. 4. Lunar surface test with sampler: two successive pictures in a penetration experiment were digitized and subtracted from each other to give this difference picture; this shows more clearly than either original picture the extent of surface disturbance, which can be used for property calculations (photograph by courtesy of JPL/NASA)

mechanical soil properties were similar to those determined before (Scott, 1967c, 1968; Scott & Roberson, 1968, 1969). With this mission, the Surveyor period was over.

By this time, I was becoming more heavily engaged in the Apollo programme as a consultant on lunar soil mechanics. However, I had one more remote sampling function to perform that year (1968). Plans were being initiated for an unmanned spacecraft to land on Mars in 1976, with a full complement of experiments. NASA's confidence had obviously increased. Requests for

Fig. 5. Panorama of sampler test area around Surveyor 7: the α scattering experiment box is on the left-hand side (photograph by courtesy of JPL/NASA)

spacecraft design proposals had gone out to several US aerospace companies, including the Martin-Marietta Company (as it was then) in Denver, Colorado. Martin-Marietta hired me as a consultant to prepare preliminary designs of a spacecraft surface sampling tool for Mars, to perform essentially the same functions as on Surveyor, but with greater emphasis on obtaining samples of the Martian surface and delivering them to chemistry and biology experiments on board the spacecraft. Among other considerations, I suggested the use of a furlable tube. In the ensuing spacecraft design competition, Martin-Marietta was the successful bidder.

NASA records indicate that the design of the Apollo lunar module including landing gear had been finalized in 1964, to include, among other engineering details, footpads about 1 m in diameter. The engineering information provided by the Surveyor mission had not been used in the design. The lunar surface properties indicated that the lunar module's footpads only needed to have a diameter of about 1 ft—there would be no question of substantial penetration into any of the lunar surfaces that had been encountered.

Another area of interest was the tools to be employed by the astronauts on the lunar surface. They included sampling tubes, to be driven into the lunar soil with a hammer. These tools had been designed by the geological experiment team, none of whom had probably ever taken a *soil* sample. The astronauts had to be provided with a hammer, because geologists carry a hammer, in spite of the hazards of using it on the lunar surface, while the wielder was encased in a spacesuit, with a highly vulnerable visor. No other option was considered. A sampling tube had also been designed geologically, with a thick wall and a cutting edge flared on the *inside*, so that the tube's inner diameter was substantially smaller than that of the cutting edge itself. Since, in all the Surveyor experiments, the lunar soil had turned out to be dense, even remarkably so (Scott, 1968; Jaffe, 1973; Carrier, 1973), it was apparent that such a design, requiring strong shearing and deformation of the sampled material, with resulting soil volume expansion, would have little success on the moon. I prepared an alternative design of a thin-walled tube, with the usual soil mechanics feature of a constricted cutting edge, so that the soil could expand into a larger diameter after it had passed the opening, and some other features to facilitate sample handling. This design was not accepted. It was with some personal satisfaction, and a great deal of professional distress, that I subsequently watched, at Mission Control in Houston, as Armstrong and Aldrich on Apollo 11, and Conrad and Bean on Apollo 12, flailed away at their recalcitrant sampling tubes with little success. A new sampling tube (Fig. 6) appeared on the Apollo 15 mission, virtually identical in design with that of my proposal.

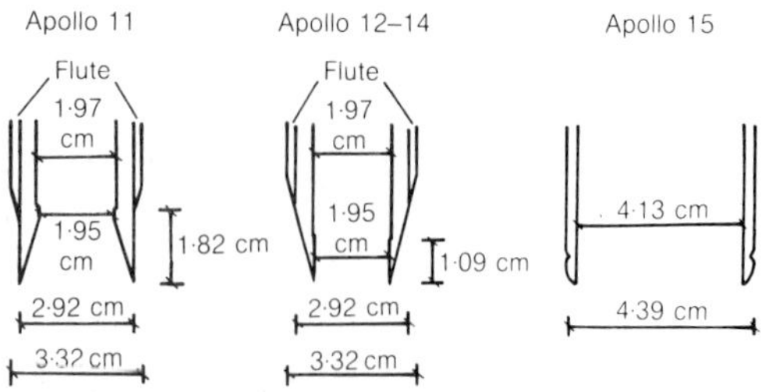

Fig. 6. Apollo sampling tubes: comparison of core tube bits (source: NASA)

However, before the flight of Apollo 11, my consulting arrangement with NASA had been terminated and I joined a soil mechanics experiment team, whose other members were W. D. Carrier III, N. C. Costes and J. K. Mitchell, which functioned in support of the Apollo group, participating in often acrimonious pre-flight meetings where scientists fought for their share of the astronauts' time on the lunar surface, as the schedule for the extravehicular activities of the crew was being assembled. Our tasks included analysis of the final stages of the lunar module's descent, the landing, sampling, coring and trenching activities, astronaut mobility on the lunar surface, soil mechanics aspects of other equipment placed on the surface and the performance of the wheeled lunar vehicle that was eventually used (Fig. 7). These activities have been well documented (Costes, Carrier, Mitchell & Scott, 1970; Scott, Carrier, Costes & Mitchell, 1971; Mitchell, Bromwell, Carrier, Costes & Scott, 1972) in reports and papers that will go unread until lunar bases are actively planned, if, indeed, they are employed even at that future time.

The astonishing success of Apollo 11 raised the immediate question of what to do next. At that time there was still a question of navigation on the moon; how close could the lunar module be brought down to selected co-ordinates? It was decided that an Apollo 12 landing near the location of Surveyor 3 would be an interesting test, while, at the same time, a visit to the 30 month old (April 1967 to Apollo 12 flight of November 1969) spacecraft might provide other useful information. That is where Apollo 12 landed in November 1969. Conrad and Bean, the astronauts, took some cable cutters with them as they pranced over to Surveyor, where they cut off the television camera and a portion of the surface sampler including the bucket (Fig. 8). When they returned, I was given charge of the returned

Fig. 7. Astronaut's footprints on the edge of a small crater: the penetration depth increases at the edge of the crater slope as the soil approaches failure (photograph by courtesy of NASA)

surface sampler for preliminary examination and photography, soil removal and cleaning. These operations were performed in a 'clean room' at Hughes Aircraft Company (the original makers of the surface sampler). I was able to do some simple experiments on the adhesion of the lunar soil to the painted surface of the bucket, which contained about 100 g of lunar soil that had been left in it at the end of the operation in 1967 (Scott & Zuckerman, 1971).

I had proposed some soil mechanics tests on the returned lunar soil to determine its properties in a relatively precise way, but standard soil experiments, such as the triaxial test, require 100 g or more of soil, and the Space Agency did not want to devote that much material to soil mechanics, even in a non-destructive test. However, after a time, the NASA Lunar Receiving Laboratory agreed to release some soil for soil mechanics studies—1·103 g. The vial containing the sample arrived while I was in England at Cambridge University in 1972 on sabbatical leave, and a miniature triaxial apparatus was designed since it is possible to place 0·8 g of soil into a triaxial cylinder 0·25 in (6·23 mm) in diameter and 0·5 in (12·7 mm) high. It was built in the Cambridge University Engineering Department machine shops (Fig. 9). One of the Department's undergraduates became interested in the project and performed some tests with me as a final year project (Boddam-Whetham, 1973). Since it was required to test the material with minimal disturbance to its natural state, the confining pressure was imposed by subjecting the specimen to a vacuum. Calibration tests were carried out on fine Leighton Buzzard sand, which was also tested both in standard 1·5 in dia. and larger 4 in dia. triaxial tests (Fig. 10). The apparatus and tests have been described (Scott, 1973) but detailed results have never been published. Some of the triaxial test data are shown in Figs 11 and 12. It

Fig. 8. Returned portion of the surface sampler

can be seen from these sand tests that there was little evidence of a scale effect on the *failure* behaviour of the material. It seems possible, from a shearing strength point of view, for sufficiently fine soil, to test all soil samples at 6 mm diameter taken, say, from 10 mm boreholes!

One of the interesting points to appear was the generation of slip planes in the dense samples at all test sizes. It has sometimes been postulated that a certain amount of shearing displacement rather than strain must take place before slip planes develop (Palmer & Rice, 1973). These tests in a miniature apparatus do not confirm this supposition, but indicate rather the usual appearances of slip surfaces, relatively independent of size. Centrifuge tests on model piles confirm this behaviour, since the axial load versus displacement behaviour scales well with prototype pile tests in terms of both peak load *and* displacement. Similar effects can also be observed with at least some slope tests. It can be seen in Fig 11 that the initial loading behaviour of the large specimens is much stiffer than that of the very small samples, at similar densities. The lunar soil has a higher shearing strength than the terrestrial sand, at the same void ratio.

Fig. 9. Miniature triaxial test equipment: the micrometer is driven by friction from the electric motor at the bottom left-hand side

Mars

Since the earth, with its $1g$ of gravity acceleration, can be considered to be an ideal centrifuge in which to perform model tests of full-scale experiments on the $\frac{1}{6}g$ moon or $\frac{3}{8}g$ Mars, I had performed lunar and Martian model experiments in the laboratory at $\frac{1}{6}$ and $\frac{3}{8}$ linear scales respectively to study the geometry of soil failed by the sampling equipment. Those experiences and the Cambridge visit in 1972–73 stimulated my interest in centrifuge work. By the time of the 'sunset' of the lunar manned programme, plans were well under way for the Mars Viking missions. Properties of Mars soil were to be obtained, as on the moon, by failing the soil in bearing and trenching experiments performed by the surface sampler.

Before the flights to Mars, NASA had established a 'physical properties experiment' and team, consisting of R. W. Shorthill, R. E. Hutton, H. J. Moore, C. R. Spitzer and myself. The proposed experiments were described by Shorthill, Hutton, Moore & Scott (1972).

The first (of two) Viking spacecraft (Fig. 13) arrived at Mars on 19 June 1976 and landed successfully on the surface in the northern hemisphere at 22·3° N, 48° W, on 20 July 1976, after rejection of the pre-launch landing site, about 1000 km away, when high resolution Orbiter

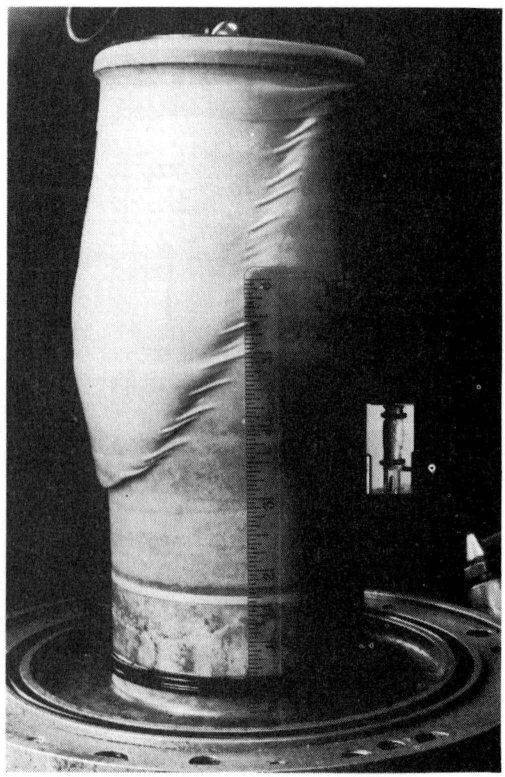

Fig. 10. Small and large samples after failure

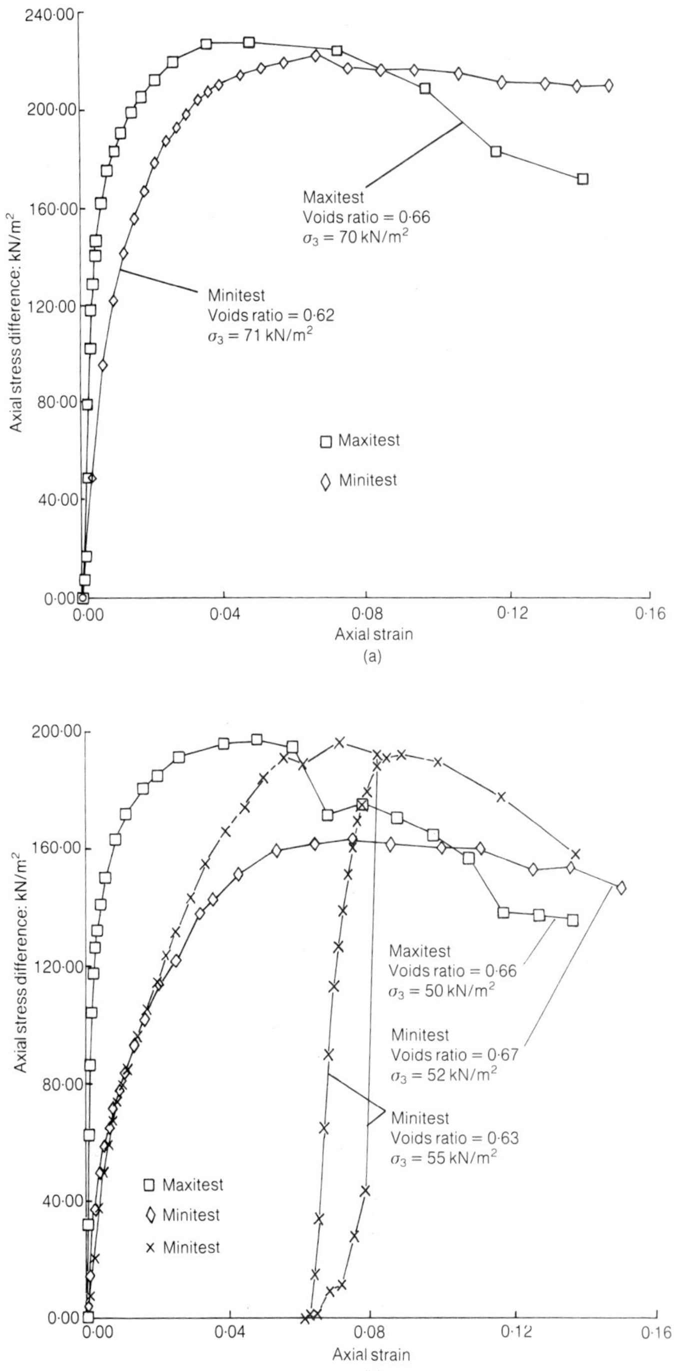

Fig. 11. Triaxial test results on Leighton Buzzard sand, for small samples compared with large samples: (a) confining pressure 70 kN/m^2; (b) confining pressure 50 kN/m^2

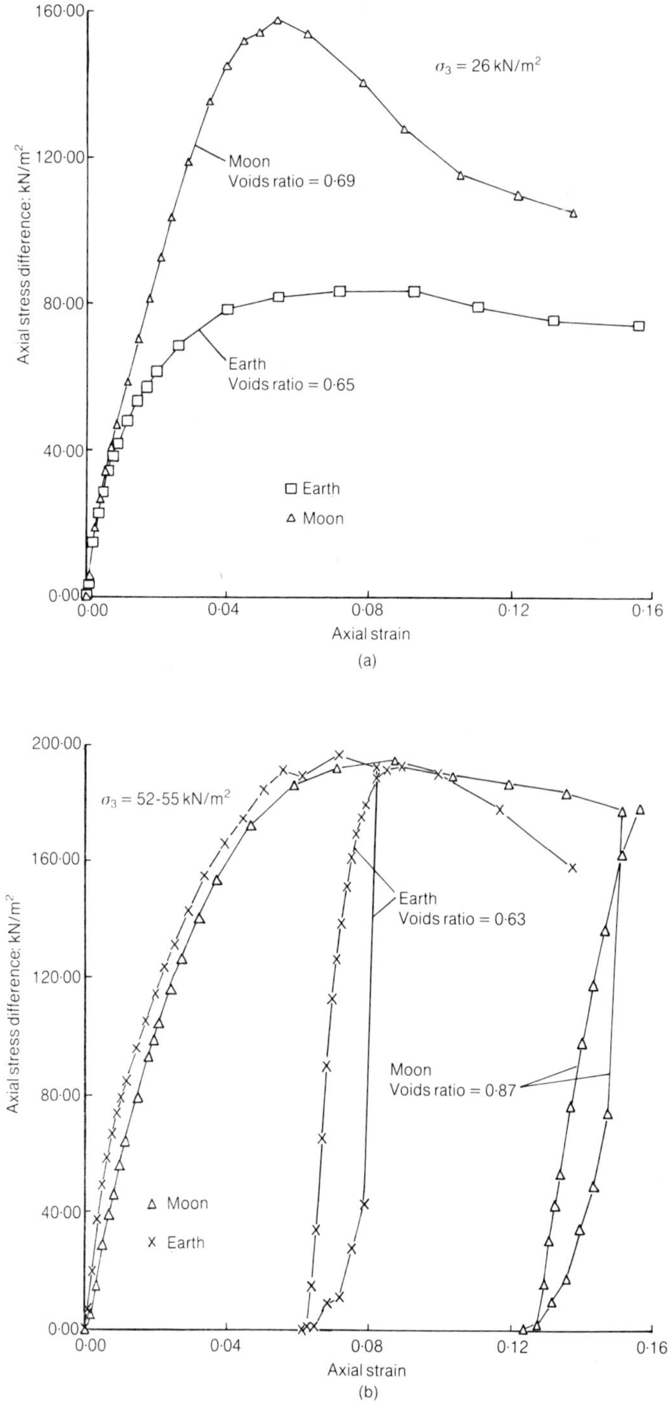

Fig. 12. Small test results on Leighton Buzzard sand and lunar soil: (a) confining pressure 26 kN/m², same void ratio; (b) confining pressure 52–55 kN/m², different void ratios

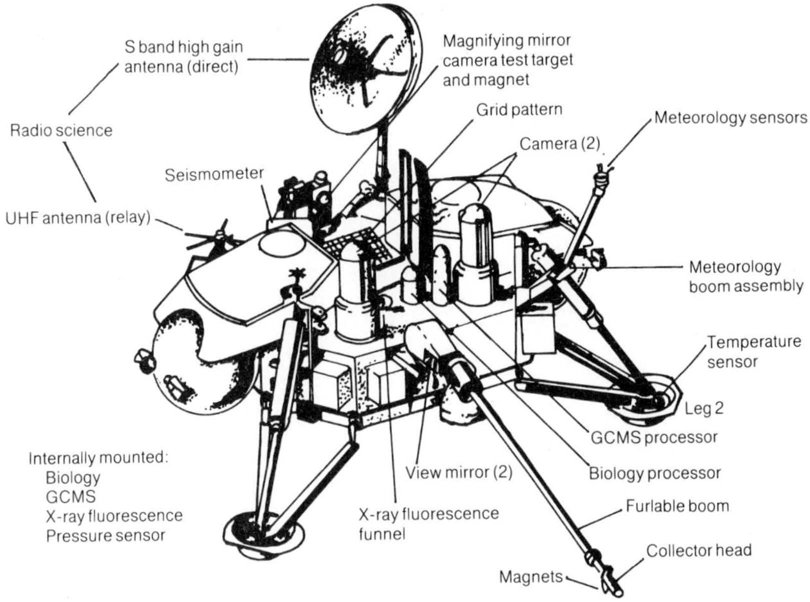

Fig. 13. Mars Viking spacecraft (source: JPL/NASA)

pictures of the surface were examined (Shorthill, Hutton, Moore, Scott & Spitzer, 1976). The biologists, seeking to maximize the possibility of finding life, had requested that the spacecraft be put down in a 'warm, low, wet' spot, all these terms being relative to Mars. The nature of the Martian surface had been less well known in advance than the moon's, but periodic Martian dust storms had disclosed the probable existence of granular material. The spacecraft possessed two imaging systems, operating on a different principle from the Surveyor's television camera, but providing an image (which could be in colour and stereo) by scanning the Martian terrain.

Early pictures showed a surface not, in general features, much different from that of the moon, but differing in a few atmospheric-related details. Little fillets of granular material were apparent behind individual rocks, and the tracks of small rock fragments moved by the descent engines' exhaust across the soil could be seen. Eventually the sampler went to work, dug trenches, performed bearing tests, moved rocks, sieved the soil into conical piles and obtained samples and supplied them to the other experiments (Fig. 14). The measurements of force and displacements enabled the Martian material's mechanical properties to be determined (Shorthill, Moore, Scott, Hutton, Liebes & Spitzer, 1976). Viking 2 landed six weeks later on the surface of Mars and performed a similar sequence of experiments (Moore, Hutton, Scott, Spitzer & Shorthill, 1977). Landers 1 and 2 transmitted data until November 1982 and April 1980 respectively. Although the experimental results revealed some anomalies, no unequivocal determination of the existence of life was made. The soil had the properties of loose to medium–dense fine sand.

LANDSLIDES AND DAMS

'All observation must be for or against some view, if it is to be of any service'—C. H. Darwin

At the same time as the Mars landers were working on the surface, cameras on the Orbiters were recording an astonishing variety of Martian landscape, including multiple landslides of sublime dimensions. If present conditions prevailed at the time of the slips, then they occurred in dry material with a very low gas pressure, yet many of the slides exhibit the character of flows. Is the gas pressure, low as it is, still sufficiently high to generate liquefaction during the sliding process, or was water present, or was gas released or generated during the shearing process by crushing of porous grains or by diffusion from physically adsorbed gas at particle contacts or in cracks in rocks? Similar, enormous slide/flow features have been observed on the moon (Fig. 15) in an almost complete absence of atmosphere. They are, of course, well known on earth (Fig. 16) where the mechanism of flow is still the subject of controversy. Shreve (1966) has suggested an air cushion below the flowing debris enabling it to travel great distances on flat slopes with minimal disturbance, although it is difficult to substantiate this mechanism with calculation. The relative run-out distance of such a slide seems to depend

Fig. 14. Viking sampler test (photograph by courtesy of JPL/NASA)

on the mass of material involved, being greater for larger volumes.

Portuguese Bend

A somewhat different circumstance attends the Portuguese Bend landslide in the Palos Verdes peninsula, California (Jahns & Vonder Linden, 1973; Ehlig, 1982; Vonder Linden & Lindvall, 1982). It occurred in an area with a general slope of about 6·5° composed of the marine sedimentary rocks of the Monterey Formation of Miocene age, associated with volcanic intrusions. The Monterey Formation consists of siliceous mudstones, siltstones and shales, and includes bentonitic tuff layers and basalt intrusives of volcanic origin. It is underlain by Tertiary basalt and Mesozoic schist. Aerial photographs taken and geological studies before and after the Second World War, at a time when there had been little or no development, showed clearly that the region was a complex of large prehistoric landslides, of which the date of one was ascertained to be about 16 000 years Before Present (Vonder Linden & Lindvall, 1982). Ancient slide masses at higher elevations near the crest of the peninsula may have moved as long ago as 200 000 years. The slides may have been related to sea level changes and to tectonic uplift of the peninsula during glacial and interglacial episodes. In the early 1950s the area was subdivided and houses were built over a substantial portion of one of these ancient landslides. To provide access to them an extension of an existing road was begun by the County of Los Angeles down into the landslide area. In 1956 cracking of the road fill and breaking of water pipes was observed. A portion of the prehistoric landslide had been reactivated (Fig. 17). Investigation revealed a slide plane in a highly tuffaceous section of shale, in bentonitic clays at a depth of about 100 ft below the surface. Flow of the slide material destroyed or damaged most of the homes over an area of approximately 300 acres; of the original 156 dwellings, only 22 remained by 1982, all of which were supported by underpinnings, and, in some cases, jacks, which are periodically readjusted.

In spite of attempts at stabilization, the landslide, with a volume of about 40 million yard^3, has continued to move up to the present day (30 years), at rates of up to 20 ft/year, after the initially more rapid motion. In a period of unusually low rainfall in the 1970s, movement slowed down considerably, leading to hopes that the slide would stop, but a few wet years occurred at the end of that decade, and the movement rate increased again. The slide at present is displacing several feet a year.

Since 1956, movements of other portions of the ancient landslides have occurred, and some other areas are also displacing at the present time. There are also sections of the prehistoric landslides which are not currently developed, and the question arises at intervals about the possibility of construction being carried out safely in such areas. In the Portuguese Bend area, it seems clear that the clay of the bentonitic tuff has a residual friction angle equal to the average slope of the slide plane (Skempton, 1964), of about 6–7°. Since the last prehistoric slide, assuming that it occurred in the same material on about the same slope angle, the clay accumulated enough cohesion, probably by desiccation, with associated chemical changes and particle bonding (Bjerrum, 1967), to maintain stability of the slope under the gradual natural processes of erosion, and high and low rainfall periods, for, by geological inspection, at least a few hundred years (estimated on the basis of the appearance of some head scarps), and in some cases tens of thousands of years.

Rainfall has several effects on the slope, by

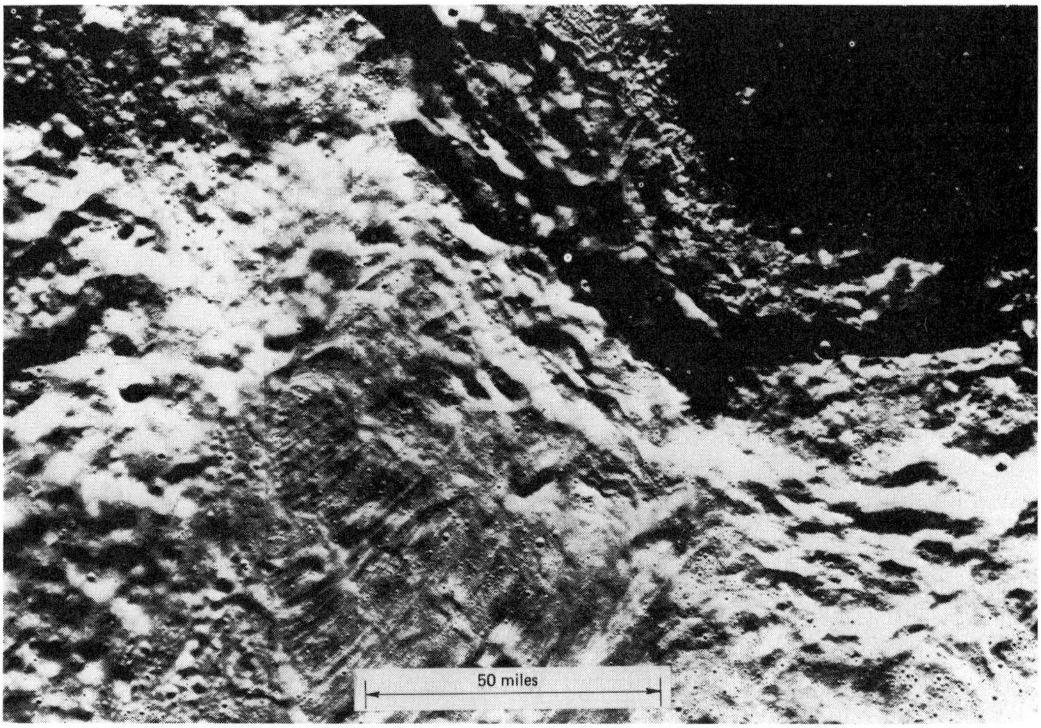

Fig. 15. Flow slide near Tsiolkovsky on the far side of the moon (photograph by courtesy of NASA)

simply increasing the weight of the overburden, by exerting lateral pressures where it fills cracks, by raising the water-table, and thus decreasing effective stresses, and by infiltrating the clay to alter its properties. Measurements of the rate of slope movement during periods of rainfall (Ehlig & Bean, 1982) have clearly indicated an increase in the rate of movement within a few hours of rainfall, showing the sensitivity of the stability to the influx of water into fissures, and to the increase in weight of the sliding mass, before any diffusion of the rain-water into the underlying clay could have occurred. Indeed, in intact areas outside landslide boundaries, the rise in water-table lags rainfall by several months (Ehlig & Bean, 1982).

What was the mechanism of the reactivation of the ancient landslide? Was it a change in acting load due to engineering activities and continuous erosion of the toe of the slide which emerges at the coast, or an alteration of the clay properties by infiltration of groundwater both from rainfall, which had occurred throughout geologic time, but now from household water use and with changed drainage patterns? If the process, as described by Bjerrum (1967), of a rupture surface propagating with time from the toe of the slope, took place, then it occurred without observation until the final breakthrough. The deformations accompanying such a rupture propagation are fairly small up to this point and are frequently noticed by residents, as mentioned again later, but are not generally interpreted correctly until the landslide develops fully. During the road construction referred to earlier, approximately 150 000 ton of fill were added to the north-east section of the ancient landslide. It was here that movements were first observed in August 1956. In subsequent litigation, the court ruled that the work on the road had triggered the landslide which spread from the north-east corner of the ancient event to cover a much larger area in subsequent months. Total damages assessed against the County of Los Angeles amounted to US $9·5 million. Although the cause has therefore been established legally, doubts remain about the real cause, particularly about the role that groundwater played in the failure mechanism.

How can the degree of stability of adjacent areas, whether underlain or not by ancient landslides, be established? The conventional soil engineering approach is to bore holes, to take samples, to perform soil mechanics laboratory tests, and to engage in slope stability analyses with and without the modifications to be imposed by the proposed new construction. However,

Fig. 16. Sherman Glacier flow slide, photographed $2\frac{1}{2}$ years after the Alaskan earthquake (photograph by courtesy of Austin Post, US Geological Survey)

there are zones of the bentonitic material at various depths, old slide surfaces can sometimes be recognized, and sometimes may not be, and the bentonitic tuff itself is difficult to sample, although cores can be obtained with care. In its natural state, it is apparently relatively dry, friable, much jointed and fractured, and after sampling it is almost impossible to prepare for a conventional, or even unconventional, shearing strength test. If, with much diligence and patience, a few specimens are prepared, the results are erratic as the samples rupture and shear along arbitrary cracks, at a wide range of strengths. The only feasible course is to remould the samples with water (what chemistry?), to consolidate or overconsolidate them and then to shear them in a ring shear apparatus or other equipment which allows sufficient displacement to give the residual shear strength. The analysis itself can be simple, since geometrically it has the form of an infinite slope failure, with possibly a breakout through a more competent layer to the free surface. Failures such as this may perhaps be placed in category (*d*) cited earlier.

At a potential development area, situated on sloping ground, with lateral dimensions of 1 km or more, by how much does the peak strength of the clay exceed its residual strength, or is the peak strength irrelevant? Is the residual friction angle of 6·5° (laboratory tests of pure sodium montmorillonite clays give residual friction angles of about these values (Olson, 1974)) indicated by the Portuguese Bend landslide uniform over the entire region? How will the strength be affected by future water influx to the region? When an area is developed in naturally semiarid southern California, the input of water into the ground increases markedly, from landscape irrigation and household use. In the Portuguese Bend area, no piped sewage system was included in the development (this is also true in adjacent developed areas which are also on ancient landslides); instead each house was supplied with a septic tank and a seepage pit. Household water was thus delivered directly into the ground and has been estimated at least to double the water quantity entering one of the landslides in the area (Ehlig & Bean, 1982). From the events that have occurred, the only safe approach is to assume that the site is clearly of marginal stability and must be stabilized by

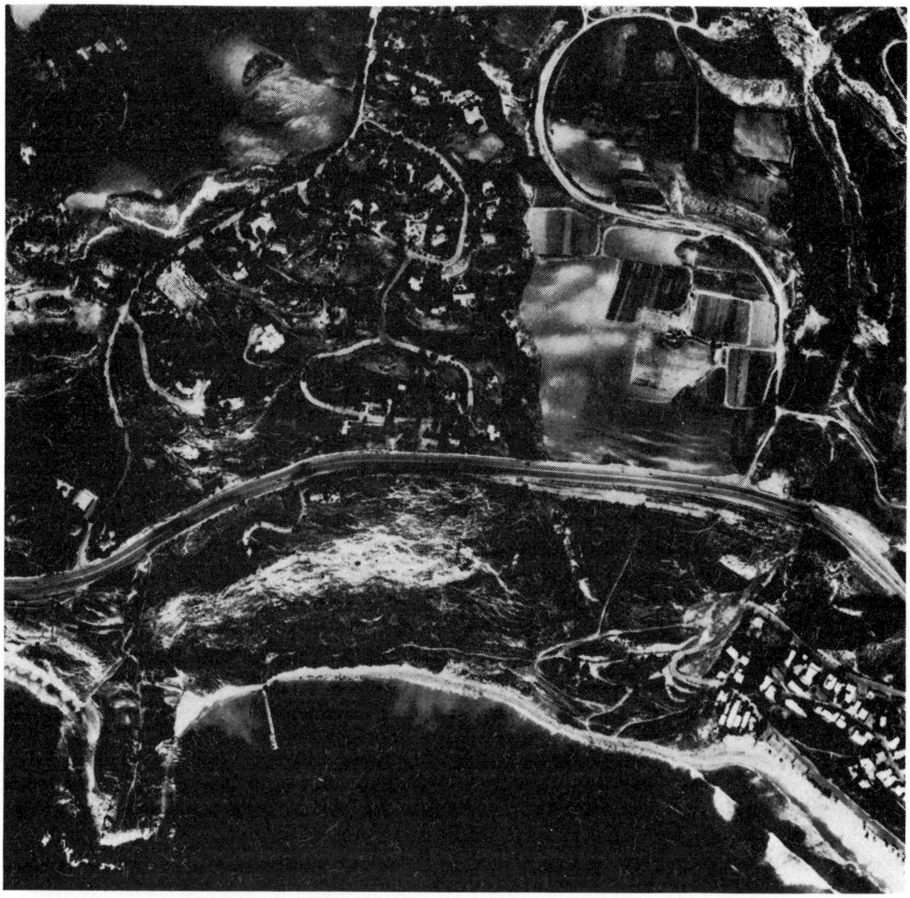

Fig. 17. Portuguese Bend landslide, 1974: the head scarp is visible at the top of the photograph; side shear zones are indicated by road patches; the toe of the slide extends from the peninsula at the bottom left-hand side to where the housing begins at the bottom right-hand side (photograph by courtesy of Metrex Aerial Surveys Inc.)

strong measures (buttress fills in trenches of substantial depth) unless development is to be prohibited, except perhaps as a public or private park. Even in this case, irrigation would have to be eliminated or minimized, and it would be desirable to monitor water levels or pore pressures in the soil.

This is not a solitary opinion. A recent paper (Smelser, 1987) describes a 9 million yard^3 landslide, adjacent to the San Andreas fault south of San Francisco, that has moved in whole, or in part, since at least 1900. A tunnel, railroad and a highway through the area have successively been abandoned because of ground movements. The author of the paper cited concludes, 'The dynamic geology and the history of landsliding in the Mussel Rock area strongly indicate that the land is unsuited for permanent structures'. He suggests that it be used as a park for recreation and 'educational geologic field trips'. Since subtle movements may be occurring in ancient landslide areas, it would also be useful to conduct periodic precise surveys over them, preferably as part of a regional programme, but especially for several years in any area proposed for development.

Similar montmorillonitic clay zones are present in other locations in California and are usually identified with particular geological sequences, such as the Capistrano or Monterey formations. On occasion, the layer of volcanic ash laid down in an ancient volcanic event was perhaps only 1 mm or so thick and was subsequently covered by silt or sand deposits, which now make up weakly cemented siltstones or sandstones. With hydrothermal alteration, montimorillonite clay forms and is then present in these formations in laminae which may be 1 mm or less in thickness. It is accordingly difficult to identify in samples obtained from field boreholes. However, the thin layers have the shearing characteristics of montmorillonite and form weak zones among the siltstones and sandstones. Where tectonic processes have tilted the beds, which have later been eroded, hillside areas are found with the layers sloping at angles of a few degrees. Presumably on much steeper angles landslides removed the material in geological time. In the naturally arid state, and with sufficiently small angles, such slopes are relatively stable. However, if the slope angle is about 7–10°, stability is only present until some change occurs.

The process of destabilization of such a slope occurs by continuous erosion at the toe, by tectonic tilting, and the factor of safety lessens with both the weight and infiltration of rainfall. Human development inevitably introduces water of variable chemistry into the natural material, continually reducing the factor of safety, as it is taken up by montmorillonite layers. Under these conditions, it may not require an exceptionally wet winter to cause the additional erosion or increment of weight which incites the soil mass to slide on the fine clay layers, but this coincidence is often observed. When sliding takes place during or immediately after the wet season, it may be assumed that weight and possibly erosion, if the toe of the slope extends to a natural stream bed, are the direct causes of the slide. Frequently, slope failures occur several months or a year after the particularly wet season; in such circumstances, it must be expected that the slide was the result of additional water absorption by the clay layer after diffusion of the water through the overlying layers, or a delayed rise in the water-table.

Highland Park

One of these landslide events occurred in 1969, during a particularly wet winter, in Highland Park, Los Angeles. The movement occurred quite suddenly, with about 15 m of displacement in a few hours, and several houses were destroyed (Fig. 18). At the suggestion of a colleague, a student and I visited the slide a few hours after the major motion. The slide block had left a canyon at its upper end, and we climbed down into this chasm with a forlorn hope of identifying a slide surface. To this end one or two pits were dug in the debris at the toe of the uphill surface of the slide block. Nothing was particularly apparent, so during the next couple of hours we explored the remainder of the slide, marking a map with the cracks, shear zones and deformation observed. Before leaving, the test pits were visited again, and it was surprising and pleasing to find that a distinct fresh shear plane had developed, as the upper part of one pit was apparently attached to a still sliding block, whereas the lower part constituted unmoving bedrock. The movement was about 1 in/h. Some equipment and a recorder to measure the displacement continuously were put together and brought to the slide. A sample of the records obtained over the next few days is shown in Fig. 19. The slide was still moving, and slowing down, but not smoothly. It was moving in exponential increments, separated by astonishingly similar intervals of time. When these were apparent on the recorder plot, it could be stated within a fraction of a second when the next movement would occur. Disconnecting the recorder, exchanging gauges and using a different measurement system, all of which occasioned no effect on the movements, removed any doubt that the records obtained were due to the motion of the soil mass. The increments occurred on time over the period of equipment removal, substitution and replacement. During the subsequent month the movements became somewhat smaller in amplitude, but the slowing displacement mainly resulted from increasing intervals of time between individual motions. Another test hole, with instrumentation, located further uphill from the first, but still adjacent to the same soil block, gave results indicating that the block moved caterpillar fashion, with a movement propagating uphill at the block–bedrock interface, at a speed of about 0·3 m/s (Scott, 1978). The cumulative motion of the incremental record corresponded to the movements of survey monuments on the slide surface, indicating that the identified slide plane was, indeed, the only slip surface present. Samples of the siltstone rock including the zone on which slip occurred showed that the layer was montmorillonite a few tenths of a millimetre in thickness. Surveys showed that the slope was quite uniform, with an angle of 13°. In this case, the area had been populated with houses for many years.

Bluebird Canyon

In 1979, a landslide occurred in the Capistrano Formation at Bluebird Canyon in Laguna Beach, California. Many of the features were similar to previous landslide observations, but in this case the slide occurred in October, about six months after the last of the heavy winter rains, and again at the location of an ancient landslide. The new slide plane was a metre or two above the identified old plane. Once again, houses had been present at the site for many years, but the delay between rain and slippage implies that the slide may have resulted from changes to the sliding layer by diffusion or rising water-table rather than to the immediate increase in weight of the mass. Here the slope of the sliding surface was about 10°, but the surface was only a few hundred

Fig. 18. Head scarp of the Highland Park landslide, California, in oblique aerial view

feet long. The Bluebird Canyon slide was subsequently stabilized with compacted soil keys (Leighton and Associates, 1979).

In these slides, it would be of considerable interest to detect the early stages of the movement. When residents are questioned after a landslide, it is common to find instances of prior movement evidenced by cracks in driveways or structures, sticking doors and broken pipes in the ground. These may be observed as much as a year or two before the slide event. Generally, however, they are not communicated to engineers, or even among neighbours, and are usually attributed, in the young developed area of Los Angeles suburbs, to settlement of construction fills, and adjustments of the ground following construction excavation and fill operations, which also cause such disturbances. The problem of early detection is similar to that of recording the strong ground motion of earthquakes, which was such a difficult task 50 years ago. The location and time of the event are unpredictable, so that instrumentation must simply be set out and maintained for years, in the hope that it will be functioning when movement occurs. Self-contained acceleration recorders, which incorporate only inertial reference, were possible and were developed for earthquakes, whose economics justified deployment of instrumentation. For landslides, equipment is necessarily more complicated, and in some respects the problem is more difficult, since the movement is relative. For each station, two points are required, one on a potentially sliding mass and one on stable ground, and the distance between them must be continuously monitored. In a possibly unstable area, it is not easy to identify where these two points are to be located, it is difficult to devise a means for making the measurement in a place of public access, where extension wires, for example, will have a short lifetime, and continuous recording requires power, maintenance and record retrieval. Possibly, as for earthquake recorders, equipment could be devised to remain quiescent until relative displacement exceeded a predetermined value, but the problem of continual checking remains. When an earthquake occurs, the fact is immediately widely known, and the recorders can be examined expeditiously. The movement of slopes is more subtle, more localized and therefore requires repeated instrumentation inspection. A movement, in addition, may take place for days to perhaps years before a catastrophic slide occurs, if it does, and the sociological impact of data interpretation and warning, as is involved with earthquake prediction, is a significant question.

Baldwin Hills Reservoir

The main embankment of the Baldwin Hills Reservoir failed in December 1963; the failure has been written about extensively, most recently in a symposium at Purdue University in 1985 (Leonards, 1987). Briefly, the reservoir was constructed in an excavated natural valley in which the underlying material consisted of poorly cemented or uncemented siltstones and sandstones in a loose, friable state. In design (late 1940s) it was recognized that reservoir water

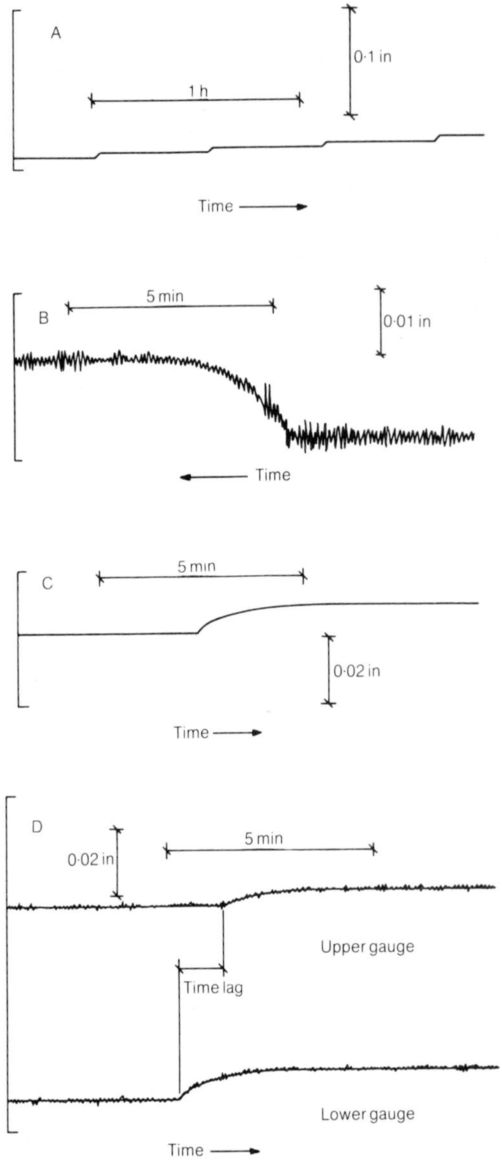

Fig. 19. Highland Park records: curve A, several events; curve B, one event, no filtering; curve C, one event, filtered; curve D one event at two separate stations

should be prevented from reaching those soils, so an impervious asphalt membrane was placed on the ground surface, a pea gravel drain system across the entire area of the reservoir built on top of it and the whole covered by a reservoir lining of 4–10 ft of compacted sandy clayey silt. It was expected that any leakage through the lining would be conducted by the drains to a drainage gallery below the reservoir, where it would be measured. Before and at the time of construction, the geological investigation disclosed that the site was intersected by several geological faults associated with the main Newport–Inglewood fault. In addition, it was recognized that the reservoir lay at the north-east edge of a subsidence zone caused by oil, water and gas withdrawal from the Inglewood oilfield. At the time of construction (1950) the maximum settlement at the reservoir due to subsidence was 0·4 m; at the time of failure it had increased to 0·6 m. In the design phase, engineering memoranda show that the designer and consultant board were concerned about the faults, but mostly from the point of view that the presence of the faults indicated the possibility of an earthquake, rather than with the consideration that one or more of the faults would rupture, displacing the material on each side and causing abrupt differential movements in the reservoir lining and embankments. In fairness, the idea that earthquake vibrations were caused by fault rupture and slippage was not so clearly understood at the time as it is now. Many earthquakes had occurred (and do occur) without fault slippage and differential displacement at the ground surface.

On completion of the structure, a complete programme of vertical and horizontal movement and crack surveys, drain leakage, water levels etc. was initiated. Data were recorded during the life of the reservoir. Early in its existence, the concrete drainage gallery which passed through two of the faults below the reservoir cracked, and cracked again. Brass plugs were established across the cracks and the measurements made on them were added to the data file. The cracks opened at an exponentially increasing rate up to failure. Flows into the drainage galleries varied erratically. Differential displacements and cracking at the embankment crest were observed along the line of the fault. However, no other obvious indications of trouble were observed until 14 December 1963, the day of the failure. That morning, the caretaker heard the sound of rushing water in the spillway which was enclosed in the dam, and whose inlet was well above the current reservoir water level. A short time later a vortex was observed in the reservoir near the main embankment, and water emerged from the downstream face at its contact with the natural ground, and along the trace of one of the faults. Futile attempts were made to plug the openings which had appeared in the upstream face of the embankment, but failure continued until a portion of the dam collapsed a few hours after the first observation of water flow.

The various failure investigations (Hudson & Scott, 1965; Casagrande, Wilson & Schwantes, 1972) concluded that the reservoir floor, including the drain system and the asphalt lining, had broken early in the reservoir's life along the lines of two of the faults, causing seepage through the lining to pass into the natural material along the faults and below the drainage system. For years water flowed into the fault zones and eroded large cavities in the soft material; presumably for much of the time this water, which was passing under the dam, was not in very large quantities, and so flowed undetected into the groundwater or into the unmonitored drainage system in the canyon downstream of the dam. It is surmised that, by the day of the failure, one of the cavities had enlarged sufficiently that the floor of the reservoir broke above it, releasing reservoir water into the chain of cavities along the fault. After this, total failure was inevitable.

The principal question which arises is what was the cause of fault displacement? It can be variously attributed to the following.

(*a*) Either a sudden geological movement within a day or two of the failure (this was ruled out, since an earthquake record study showed no relatable events) or a long-term creep of the fault akin to that observed on other faults occurred. This obviously would have to be occurring before and after the reservoir failure. There is evidence of fault movement near the reservoir before and after failure.

(*b*) There have been oil-related movements of the subsidence region which existed long before reservoir construction and continue to the present day. Oil, water and gas withdrawals and water repressurization have been taking place in the oilfield so that ground movements both down and up have occurred. The stresses and ground movements in the reservoir, which sat on the tension zone around the edge of the subsidence basin, were directly related to both withdrawal and injection procedures in the oilfield, but were the discontinuous fault movements caused by the pressure changes?

(*c*) A third possibility is the loads imposed by the reservoir itself. The load of the embankments is substantial, but the principal burden imposed on the underlying material is that of the 50 ft depth of water in the reservoir. The detailed post-failure investigations by means of trenches across the faults show signs of greater natural soil disturbance on the upper side (hanging wall) of the 70° dipping faults, although the evidence is not entirely convincing. It has been postulated (Casagrande *et al.*, 1972) therefore that the pressure of the reservoir water on the top of the reservoir lining caused greater consolidation and compaction on one side of each fault than on the other, which resulted eventually in rupture of the lining and drainage system as described.

What can be done to analyse such a failure? The measured horizontal ground displacements can be included, for example, in a large linearly elastic finite element model of the region of the reservoir along with estimated, but reasonable, values of the linearly elastic material properties. Two-dimensional models of this type were constructed as part of one of the investigations, including the possibility of tensile cracking (Casagrande *et al.*, 1972). The properties were checked by matching the observed rebound on reservoir unloading after the failure, but the initial stress state in the ground was not known. The calculations indicated that the oilfield-caused subsidence might induce a tensile zone to develop at the reservoir in spite of the reservoir water load. That is subsidiary evidence of a problem area, but not directly related to the failure. Since water undoubtedly, at some stage, broke through from the reservoir into the fault zones and may have accumulated there, an appropriate water pressure, to a guessed-at depth, can be included in such an analysis, and the results indicate lateral displacements and vertical movements due to the relief of shearing stress along the near-vertical fault surface. These movements can be compared also with the observed uplift in the reservoir vicinity, to confirm or refute ideas relating the uplift to repressurization programmes. However, any numerical analysis of this kind cannot contribute substantially to establishing the mechanism of failure of the Baldwin Hills failure. The undoubtedly non-linear material properties are not known in detail, and their time dependence may be important. Seepage, of the kind that was surmised, could not be included, and there was no mechanical model of the piping process. This leaves only speculative arguments about what went on. In such circumstances the analytical engineer is, and is likely to remain, helpless. However, the situation with respect to lessons learned for future design is not hopeless. There is clearly a good recognition of the basic causes of failure at this site. As recommendations, it is possible to say the following.

(*a*) Use extreme caution in a canyon or valley containing faults. Such sites are difficult to avoid in some regions of the world, such as California, where, to a large extent, the valleys are fault controlled. The site must be selected not only for the usual reasons, but to minimize the fault hazard. It is necessary to identify the faults, to date their movements if possible, and the amount and sense of those movements. The conclusions about such movements obtained from geological studies are usually largely qualitative and controversial. Could, on any basis, a fault movement of 8–9 in have been predicted for the Baldwin Hills Reservoir? When all the information is brought together, a defensive design may be possible.

(*b*) Do not build in a region of substantial subsidence. This recommendation is easier to implement, but may not be entirely avoidable. Numerical quantities of the magnitude and extent of subsidence are easier to obtain, however, and are more reliable than for faulting.

(*c*) The care which accompanies the design of drainage systems and controlled seepage zones in any dam must be reinforced to be sure that any displacements, abrupt or gradual, will not result in uncontrollable leakage.

(*d*) Monitoring systems are required that truly give information on what is happening in the structure in adequate time to carry out remedial measures in unforeseen consequences. If the event occurs (fault rupture, for example) on which design concern is centred, the monitoring arrangement must continue to function.

(*e*) The results of periodic surveys must be closely inspected and interpreted by engineering staff.

Teton Dam

The failure of the Teton Dam in Idaho in 1976 (Independent Panel, 1976; Leonards, 1987) was accompanied by similar intangibles. The post-failure investigation indicated conditions which left many engineers uncomfortable, without leading at the same time to definitive results that pin-pointed the exact mechanism of failure. Finite element analyses of the stress state around the key cut-off trench indicated stresses which raised concern about the possibility of hydraulic fracturing in the material as the reservoir level was being raised. However, discussion also centred on a 'wet zone' running entirely through the earth dam structure, and which was attributable to inadequate removal of fill which had been frozen during one winter of construction. As for the Baldwin Hills Reservoir, the Teton Dam failure will concentrate designers' attention on certain suspicious aspects of dam design, although the study of the failure itself has not led to a single conclusive mechanism. More analytical work, in the light of attendant uncertainties, is not warranted, and, indeed, it is not clear what conditions a possible analysis could simulate, nor what useful results would ensue. Both these engineering failures are placed in category (*c*).

GEOLOGICAL FAULTS, EARTHQUAKES AND CONSTRUCTION

'. . . not of practical importance to the engineer'—W. C. Unwin in 1911 Institution of Civil Engineers' Presidential Address concerning the efforts of 19th century elasticians

As discussed earlier, earthquake occurrence and slope stability events have much in common. In an earthquake, the accumulation of shearing stress due to mechanical processes which are only vaguely understood, on slip surfaces which have generally been ruptured before (an earthquake has never yet been observed to originate from a rupture in fresh rock), but whose properties are poorly defined, leads to a dynamic rupture event which radiates waves in a complex pattern. A landslide is caused by the development of shearing stresses, due to largely unknown changes in the causative mass (except in circumstances such as earth embankments, where material and geometry are relatively closely controlled), in soil or rock material which may or may not possess previous rupture zones, joints or cracks, to a point at which slipping occurs. Sliding may be a relatively static or a dynamic event. In both cases the mechanical causes are obscure, although less so in landslides, and the stress–displacement (constitutive) properties of the potential rupture zone, as well as the physical-chemical influences on it, are imperfectly understood. Since it is not possible to predict adequately the sliding hazard or safety of a specified natural slope, even when its geometry is well known and the possible sliding material has been sampled and tested, the frequent claims for funding for 'earthquake prediction', which has the same problems on a larger scale, in unknown materials which cannot be sampled, frequently offshore, at sites which cannot be precisely specified and are subject to unquantifiable stress fields, should be viewed with scepticism.

In site investigations for large structures, which, in the USA, used to include nuclear power plants, and still involves dams, oil and liquefied natural gas storage facilities, offshore oil structures and pipelines, faults are frequently identified during geological mapping. When they are, the question of their activity is foremost. Stringent guide-lines for this determination with regard to nuclear power plants have been issued by the US Nuclear Regulatory Commission. The difficulty of determining when a fault last displaced, or how many times it has moved in the last 10 000 or 10 million years is substantial. Investigations for nuclear power plants in the 1960s and 1970s generated much more quantified activity on the part of geologists, and trenching of alluvial materials became an important part of site investigations.

Once an answer has been obtained to the question of how often, how much and the probability of a movement in the 50 or 100 year life of the construction, the engineering task remains of designing the structure for the shaking to be caused by the causative fault rupture, or, on occasion, for the displacement which might be caused by rupture of a fault passing under the structure.

Propagation

In many parts of the world, including California, earthquake faulting originates in bedrock, overlain in some locations with many thousands of feet (as much as 20 000 ft (6 km) in the Imperial Valley) of alluvial soil materials. During an earthquake, the fault in the rock displaces as rupture propagates along it. It is surprising that this abrupt discontinuity in displacement continues on a slip surface running through kilometres of soil to the ground surface, where it appears as a sharp disruption of surface features. As mentioned earlier, the fault displacements observed to the present have occurred where previous faults are located, but since many faults (the San Andreas fault is an example) are characterized not by one break but by a zone of failure 1–2 km wide, new material must be, on occasion, disrupted. In the investigations in the early 1970s for a nuclear power plant in the San Joaquin valley of southern California, a fault was detected by seismic profiling means in alluvium many thousands of feet deep, near the site. The discontinuity did not extend all the way to the ground surface, but showed a gradually decreasing fault offset up to a level at which it could no longer be detected several hundred feet below the ground surface. One interpretation was that this was an old fault, on which activity had occurred many times as the alluvium was being deposited, so that the older alluvium evidenced all the displacement, but the youngest material recorded only the most recent movement. At some time, faulting activity ceased, but alluvial deposition continued, to give the upper few hundred feet of undisturbed deposits. The conclusion was that the fault was inactive. No historical earthquakes had been associated with the fault but the brevity of that record in California does not give useful information.

Another interpretation is that dislocation events might occur at bedrock of insufficient magnitude of displacement to rupture the soil all the way to the ground surface. A relative displacement across the fault would also be observed which diminished with the vertical distance from bedrock, but the implication of this interpretation is that the fault might be active, and shaking from a subsequent rupture event might be of importance to the power plant design. Depending on the amount, a further bedrock displacement or displacements might extend the rupture in the soil all the way to the ground surface. The interesting question was how much displacement would be required at bedrock to cause rupture just to reach the surface of soil of a specified depth? This problem could be placed in category (*d*).

Although faults have a variety of habits, two configurations contain all the features of particular engineering interest: a normal fault of vertical dip without lateral motion and a strike slip fault with no vertical movement. An analytical study of displacement in a material of fairly real soil properties increasing with depth did not seem to be achievable, so I began with a finite element analysis of the normal fault. At the time, failure could only be represented by a bilinear model incorporating a Mohr–Coulomb failure condition, and this was incorporated in the calculation. The results of several studies were presented (Scott & Schoustra, 1974) in terms of the amount of displacement required to develop 'failure' (by the criteria of the model) all the way to the ground surface or not, but the detailed results were only described in an unpublished report. In the calculations, failure was shown by shading those elements which reach the failure condition as the displacement is increased, as illustrated, for one example, in Fig. 20.

Since the bilinear model is stable, no slip surface appears, only a diffused yielding zone, which spreads upwards through the soil from the bedrock discontinuity. In the real soil, the rupture surface presumably develops because of an unstable stress–strain relation, but this could not be represented in the finite element computation. It was thought, however, that the stable model would give a conservative (too high) estimate of the required bedrock displacement (which it probably did). A disturbing feature of these analyses of the fault problem was that the failure region did not propagate in the correct way. If a basement region is uplifted as shown in the study, a slip line should appear, diverging from the uplift block to the left-hand side with reference to Fig. 20, as demonstrated in both model tests of anchors and slip line analyses. The numerical results demonstrated a yielding region propagating to the right-hand side. Would this be changed by the incorporation of a more complex and descriptive constitutive model for the soil in the analysis? In general, this is not known, because such models are not yet widely available in finite element codes. However, an analysis by a finite difference method, incorporating a non-linear model with a yield surface (Roth, Scott & Austin, 1981) was performed in association with the centrifuge faulting study illustrated in Fig. 21 and

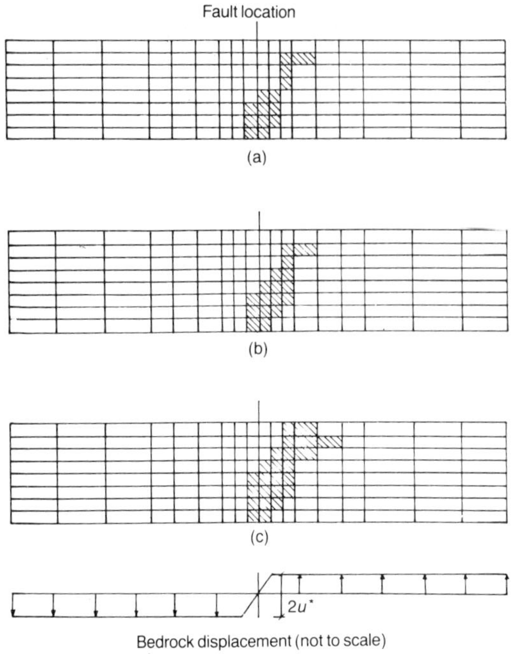

Fig. 20. Bilinear finite element solution for a vertical fault displacement of successively larger amounts (the yielded region is represented by the shaded elements; depth of section shown, 800 m; width shown, 4500 m, half of the finite element model; unit weight, 16·7 kN/m^3; cohesion 49 kN/m^2; friction angle, 25°; *E* varies linearly with depth to a value of 157 MN/m^2 at the base; Poisson's ratio, 0·4): (a) $u^* = 20$ m; (b) $u^* = 22$ m; (c) $u^* = 25$ m

produced the results shown, in which the calculations exhibit a reasonable match to the experimental results. More will be said about finite element and finite difference approaches to yield and failure problems later. The centrifuge study (category (*e*)) was performed to give more information on the question of the effect of an isolating zone of soil between a faulted bedrock and a liquefied natural gas tank to be placed on the soil surface and how an analysis might be performed.

Los Angeles Subway

The same subject matter, faults, has an impact on every kind of engineering venture in seismically active areas. It was decided several years ago that Los Angeles should have a subway to bring it into line with other major cities. Preliminary route studies and soil and geological investigations were initiated. It is not possible in the city to find any line joining suburbs to working areas which does not traverse one or more faults, some of which are considered active. All subway structures in the region must be designed for the

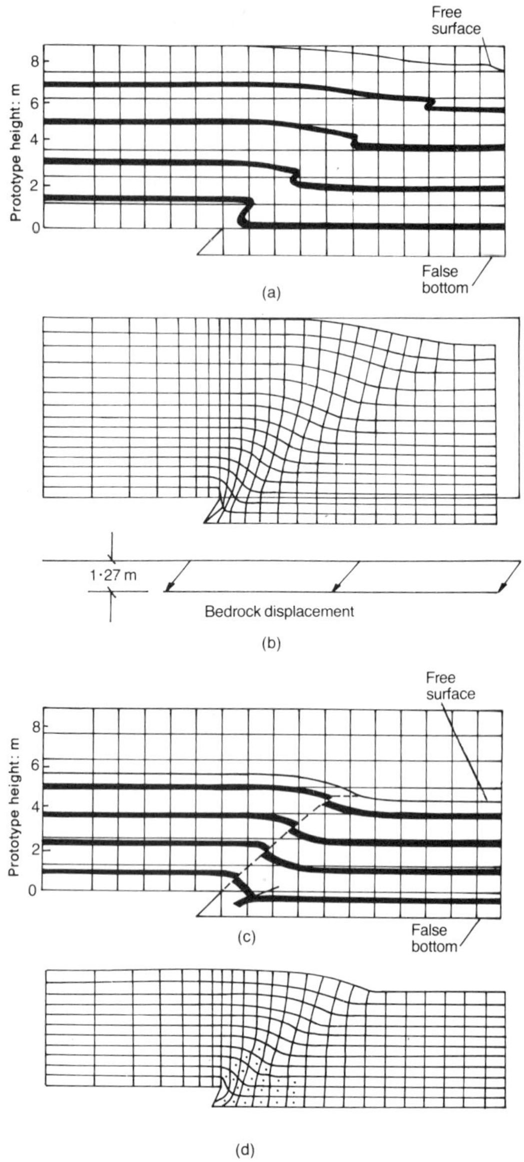

Fig. 21. Dynamic faulting through loose sand in a centrifuge: (a) centrifuge test result, loose sand; (b) finite difference analysis; (c) centrifuge result, remoulded site material; (d) finite difference analysis (the dots show elements indicating tension) (after Roth *et al.*, 1981)

expected level of shaking produced by an earthquake (the two usually employed events are the 'maximum expectable' and the 'maximum credible'; interestingly for potentially the largest earthquake-producing fault, the San Andreas fault in southern California, the two terms coalesce), but the crossing of an active fault by a subway tunnel introduces different design considerations. In consonance with the design philos-

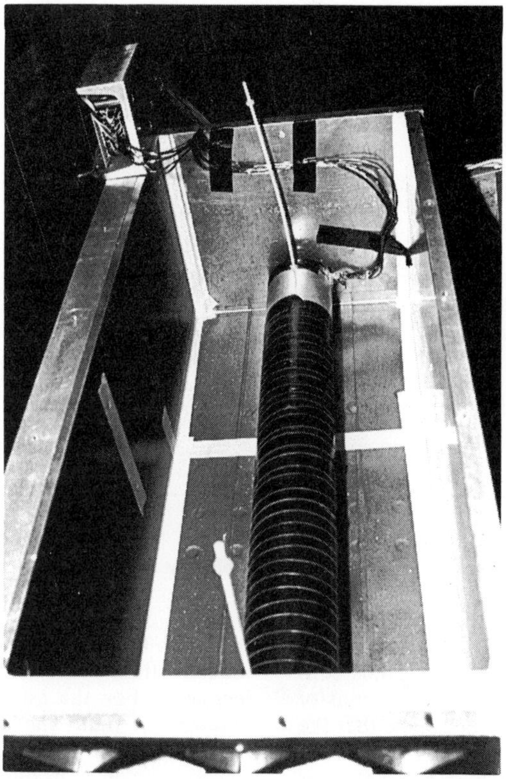

Fig. 22. Los Angeles Rapid Transit District centrifuge tunnel model (after Lindvall, Richter and Associates (1984))

ophy employed for above-ground structures, the subway tunnel should be expected to survive a large displacement without danger to the travelling public, although subsequent repairs and maintenance inevitably would be required. In other words, after the largest possible rupture on a fault through which the subway passes, the tunnel should remain open and accessible, even though vehicular movement may not be possible.

There are several subsidiary questions associated with this engineering problem. Those concerned with the probability of rupture at a fault, the magnitude of relevant relative displacement and the location of this displacement in the fault zone are not of concern here. The particular engineering problem is, given a specific displacement amount, what is the effect on the tunnel section and what special design might be required to maintain the tunnel open after displacement? It is assumed that the soil properties and tunnel section have been identified. The mechanics of this problem are again difficult to address analytically, it therefore falls in category (*e*), and some model experiments in the centrifuge were proposed, with the same test equipment as employed in the faulting study described before. The proposed tunnel was to be composed of precast concrete segments, bolted together, but it was not practical, for a preliminary study, to build a 1:50 scale model closely simulating this anisotropic real structure. Instead an aluminium tube was chosen, whose cross-section reasonably scaled the stiffnesses EI of the tunnel in both longitudinal and ring directions, and which could be easily instrumented with strain gauges and displacement transducers. The most worrying fault is the so-called Hollywood fault, a normal fault of 45° dip; the tunnel intersects it at 90° to its strike. The apparatus (Fig. 22) was already arranged to simulate a 45° dipping fault, so that no substantial alterations were required. The model tunnel was placed on 'bedrock', surrounded with soil compacted in different tests to densities spanning the field values, the centrifuge brought up to scale speed and the fault actuated. Some of the stresses recorded on the aluminium tunnel are shown in Fig. 23 for a fault displacement designed so that the aluminium would not yield.

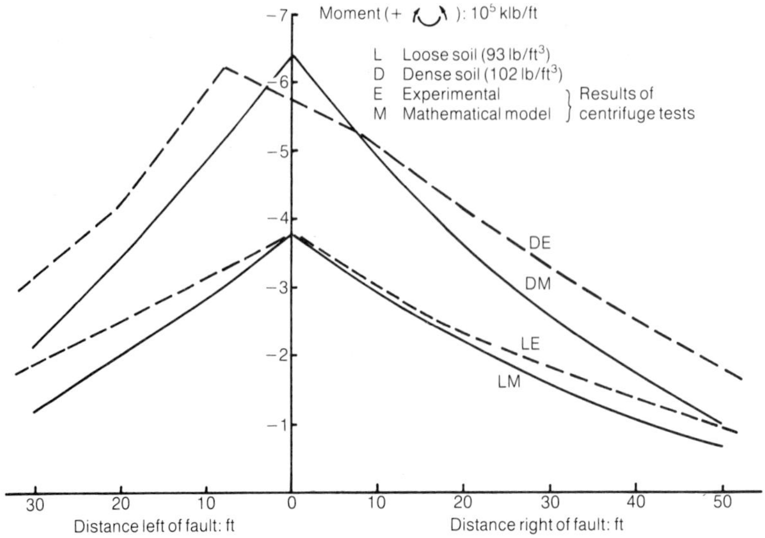

Fig. 23. Tunnel moments developed by fault displacement (after Lindvall, Richter and Associates (1984))

A complete finite element simulation of the model condition including yielding soil was considered briefly and rejected partly on grounds of economics, but mostly because of doubts that the soil behaviour could be properly (or even reasonably) represented. Instead a much simpler Winkler foundation type of representation was devised as shown in Fig. 24. The non-linear displacement elements were calibrated by matching the numerical model to the centrifuge test results with the boundary conditions imposed on the aluminium tube. The tube was finite in the tests whereas the subway tunnel will be effectively infinite in length. The stresses imposed on the tunnel under the condition of infinite length were obtained from the numerical model by maintaining the element constants derived by the test comparison, but increasing the number of elements to distances on each side of the fault at which negligible effects on the tunnel were observed. In additional studies with the numerical model, the fault displacements were also increased. For the estimated fault displacement required for the tunnel design, the stresses determined were greatly in excess of the strength of the tunnel segments. A different section of subway passing through this fault zone is therefore called for, possibly in the form of a steel tube, to maintain the integrity of the opening (Lindvall, Richter and Associates, 1984).

Discontinuous structures

'There is a limit to what we can do with numbers, as there is to what we can do without them'—N. Georgescu-Roegen

The engineering situations discussed so far have all fallen into an area in which continuum mechanics solutions are broadly applicable, although discontinuous results in the form of slip

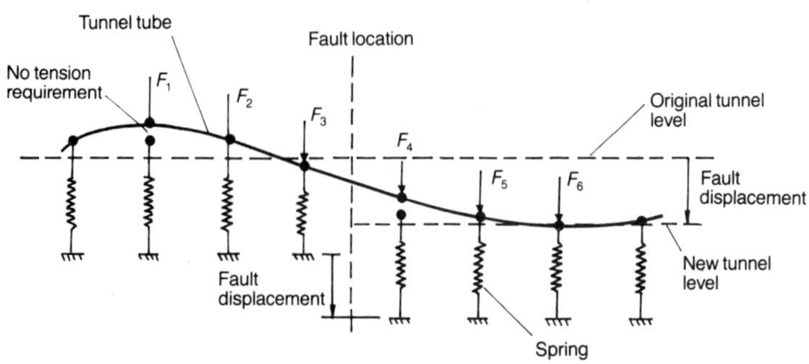

Fig. 24. Modified Winkler model of a tunnel–soil system, for simplified analysis (F_1–F_6 are overburden soil forces) (after Lindvall, Richter and Associates (1984))

surfaces emerge in the material behaviour. In much of rock mechanics, however, the principal characteristic of the material is not its continuous nature but its discontinuities. Individual blocks of intact material make contact with one another along fractures, fissures and cracks. In the response of a mass of such material to applied loads the properties of the intact blocks are of relatively less importance than the properties of contacts along the fracture surfaces. At the boundaries between soil and rock mechanics, problems involving this qualitative difference in behaviour arise where soft rocks are concerned, or possibly heavily overconsolidated and fissured clays. It is unfair, in these circumstances, to expect the application of a continuum mechanics method to give results reasonably representing the response of the jointed real material.

A difficulty of this nature arose a few years ago when it became necessary to analyse the seismic response of a ridge joining two separate concrete dams, Juncal Dam, in a water project in California (Fig. 25). The ridge itself formed a dam for the reservoir, in a roughly earth dam configuration, but it consisted of highly fractured and jointed soft rock. Following the 1971 San Fernando earthquake and the near failure of the Lower San Fernando Dam, questions were raised by the State of California Division of Safety of Dams (DSOD) regarding the stability of all major dams in California. The DSOD required all substantial dams to be inspected and analysed; depending on the consequences a dam might be left alone, rehabilitated, operated under reduced reservoir elevation or taken out of service. An order of priority based on the consequences of failure was established. In due course, the Juncal Dam system came to be examined to assess its seismic safety. The usual geological and soils investigation led to the identification of faults which could cause earthquakes relevant to the stability of the dam system, and the assessment of earthquake magnitudes and strong ground motion which could be developed by these 'design' earthquakes (Lindvall, Richter and Associates, 1982). The safety of the two concrete dams at these levels of shaking could be assessed by methods which had been presented or developed during the integrity

Fig. 25. Juncal Dams and Jameson Reservoir, California (after Lindvall, Richter and Associates (1982))

analysis programme, and which had been accepted by the DSOD, but the fractured rock ridge posed a different problem.

Earlier conventional 'pseudodynamic' slope stability calculations employing horizontal and vertical static accelerations to represent an earthquake's effect on the structure had raised questions about the stability of the ridge, but the assumption involved in these analyses required a relatively continuous slide surface through the jointed and fissured rock mass. No such continuous through-going fracture could be identified (Fig. 26) in the soil and geological investigation, and another approach was sought for an improved, and more realistic, analysis. The finite element method was considered, but it was not clear how the relevant material properties could be obtained; because of the fractured state of the rock mass, intact cores were difficult to obtain, and their properties were not pertinent in any event. Some large plate loading tests were performed in the walls of trenches near the crest of the ridge, to give information on the integrated response of the fractured mass, with a wide scatter of results. It was additionally not clear how the behaviour of the material in the core of the ridge would correspond to that of the near-surface material tested. Other testing expedients were considered. Finally, even if a finite element model were constructed and many analyses run to span a plausible range of continuum material properties, including non-linear behaviour, experience has shown that failure mechanisms are not displayed, but rather plastic deformational patterns emerge which are consistent with the continuum formulation. It is possible to formulate finite element geometry incorporating cracks, but the implementation of such a program for a dynamic input which would cause crack opening, closing and sliding was seen to be a formidable task, involving considerable development effort. A discrete element approach was more logical. Here the individual rock blocks between fractures would be modelled, with an appropriate fracture geometry. The material properties efforts would be directed towards a determination of the conditions at the block interfaces. Perhaps this effort would fall in category (*d*).

On request, Dr P. Cundall (now of the University of Minnesota) made his discrete element program, UDEC2, available, and the code, which was formulated for static problems, was modified to handle dynamic seismic conditions. In its original form the code functions according to a finite difference procedure along dynamic relaxation (Otter, Cassell & Hobbs, 1966) lines, so that, for the static problem, the forces in the system acting on each block at one time are used to calculate pseudoacceleration, velocities and displacements of the block, and from these the forces are recalculated for the next time step. After a time the velocities die down and the static solution emerges. The procedure has been widely used in the study of particulate behaviour (Cundall, 1976; Cundall & Strack, 1979; Walton, 1981) and will be referred to again later. After modification to handle a dynamic situation, the UDEC code was employed in a seismic analysis of the Juncal ridge in the configuration shown in Fig. 27 under the required earthquake acceleration versus time horizontal and vertical base inputs. The reservoir water pressure was included in the calculations. A larger number of blocks was desirable and would now be computationally feasible. As with the finite element computational procedure, a potentially bewildering amount of information on each

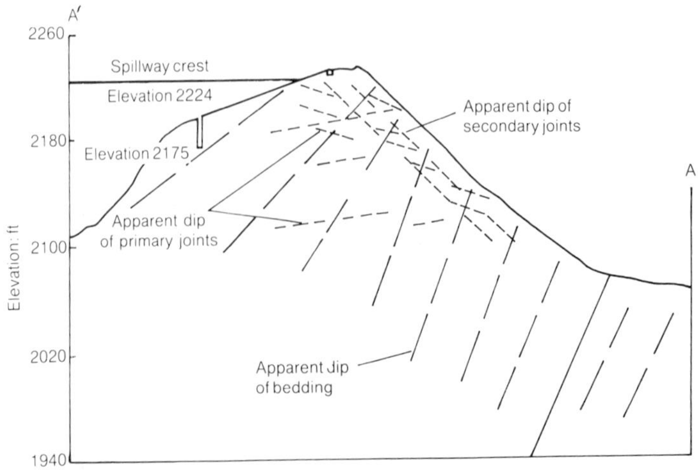

Fig. 26. Observed and inferred fracture pattern in Juncal ridge: cross-section through the auxiliary ridge (after Lindvall, Richter and Associates (1982))

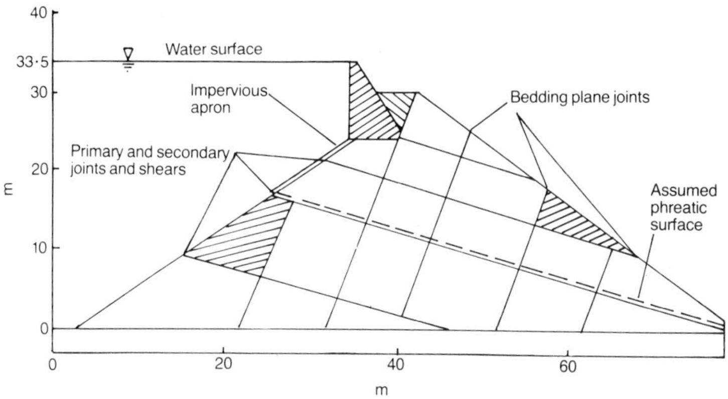

Fig. 27. Discrete element model of Juncal ridge (after Lindvall, Richter and Associates (1982))

block is available from the solution—accelerations, velocities and displacements in two components, as well as the same rotational quantities. A history of the making and breaking of contacts between the blocks can also be accessed. If a failure mechanism wishes to develop, the computation permits it to do so. If a complete and obvious failure does not occur, particular engineering items of interest are the crest displacement history and final displacements and the crest acceleration history.

Engineering decisions on the safety of the structure and operation of the reservoir must be based on these quantities, and these judgements are perhaps the most difficult aspects of the entire process. In the solution for the Juncal ridge various properties for the ridge substance were assumed for various trial analyses; one of the results felt to be representative of real behaviour is shown in Fig. 28. No definite failure mechanism emerged, but a lateral crest displacement of 1·5 m is evident; the vertical movement is almost as large. Since the ridge is 30 m high, this represents a movement of about 5% of the height. What is an acceptable vertical or horizontal displacement of an earth dam or an earth-dam-like structure during a strong earthquake? Santa Felicia Dam (80 m high) experienced only small vertical and lateral displacements during the 1971 San Fernando earthquake at crest accelerations of 0·2*g* with no apparent distress or hazard (Abdel-Ghaffar & Scott, 1979). In Mexico, in September 1985, La Villita Dam (60 m high) displaced 0·3 m vertically (0·5% of its height) with peak crest accelerations measured at 0·69*g*, whereas El Infiernillo Dam (140 m high) in the same earth-

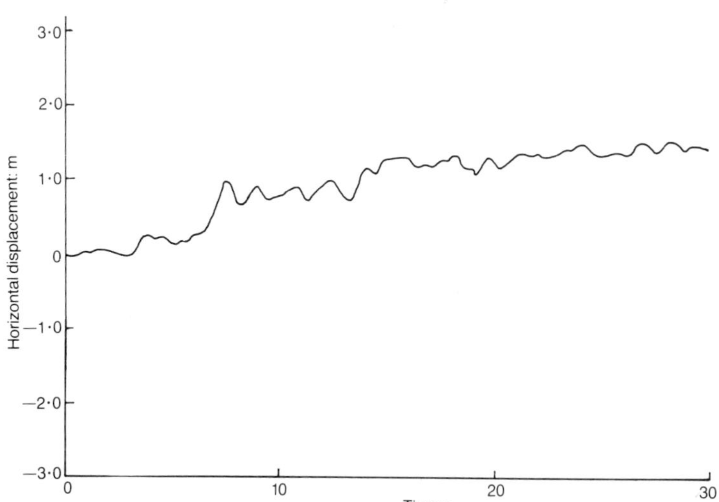

Fig. 28. History of displacement of the crest block of Juncal ridge for a selected base input earthquake acceleration history (after Lindvall, Richter and Associates (1982))

quake had a maximum vertical movement of 0·1 m (0·07% of its height) at accelerations of about 0·3g. Both these dams have experienced several earthquakes and have undergone cumulative horizontal and vertical movements of 0·7 m (1·2%) (La Villita) and 0·45 m (0·3%) (El Infiernillo). For La Villita Dam, the displacements are proportionately larger and have caused concern. The Upper San Fernando Dam (30 m high) displaced about 1 m vertically and horizontally (3% of its height), at peak accelerations of about 0·6g in the San Fernando earthquake, with clear evidence of a failure mechanism developing in the downstream direction, but without going to completion. Much larger movements occurred in the Lower San Fernando Dam, but it did fail in the structural sense, although no water was released. Pleasant Valley Dam in California, a rolled fill dam, approximately 120 ft high, has experienced shaking from several earthquakes, of magnitudes from 5·7 to greater than 6, producing peak accelerations at the dam site estimated to be as high as 0·3g with no measured vertical or horizontal displacements of the crest, although minor cracking along the crest has been observed. The displacement was considered sufficiently large at the Juncal ridge to warrant certain rehabilitative measures.

There are other engineering uses in the realm of failure for discrete element computer codes such as those pioneered by Cundall. A few years ago it was realized that there might be earthquake stability problems for statuary at, in particular, the Getty Museum in Los Angeles, or at other museums in seismic areas. At the Getty Museum there were, indeed, hazards. Many of the busts were located on stone plinths which were not anchored to the terrazzo floor. A quick pushing experiment in the absence of a security guard showed that such an arrangement was very fragile. In this case, experiments on a shaking table would be easy to conduct and conclusive, but in their absence it was decided to model the bust and pedestal discretely. The base of the simulated plinth was subjected to a typical Los Angeles design earthquake acceleration input, but this selection was unnecessary since within a second or two of initiation of the event, at accelerations of only a few per cent of gravity, the unfortunate response was apparent (Fig. 29). It is a great pity that some of the few relatively undamaged specimens of Greek sculpture to survive should be at risk in the 20th century in Los Angeles. It is likely that a substantial proportion of missing noses, forearms and legs of Greek marble statues has resulted from seismic events. Incidentally, these figures, in many museums, which lack one or both legs, are sometimes solely supported on a vertical steel bar cemented into a hole drilled in the remaining portion of the leg, and imbedded in a base block. Structurally this is a poor solution to the problem, since it is quite possible, given the period of the lowest mode of vibration of such an arrangement, that, even without overturning, the marble will simply break above the end of the steel bar in the leg, and the statue will fall. Base isolation might be a feasible method of protecting sculptures in seismic areas.

In a different form discrete element models have been employed in studies of the micromechanics of granular media, usually with the grains represented in the form of circular discs, in a two-dimensional simulation. Both static (soil-like) and dynamic (flow) problems have been examined by this approach (Cundall & Strack, 1979; Werner & Haff, 1986; Walton, 1981; Campbell & Brennen, 1985). Some light has been cast on flow processes by the introduction of a pseudotemperature and by the ability to examine at will the mechanisms of momentum transfer in a moving granular mass under conditions where laboratory experiments are difficult if not impossible to perform. When the slope of the flume is small, the mass of grains flows continuously, without gaps (Fig. 30(a)), but in a numerical experiment (Campbell & Brennen, 1985) conducted at a slope angle of 40°, the mechanism of movement became totally different, as a more-or-less intact layer of grains moved at high velocity some distance above the base supported by the impacts of a few grains bouncing backwards and forwards between the base of the layer and the floor of the flume (Fig. 30(c)). Presumably momentum transfer between these particles and the flow provided an equivalent supporting pressure to keep the flow in suspension. In the experiments, no fluid and thus no fluid pressure was included. This might be a possible explanation for the long runout distances of some large landslides, such as illustrated in Figs 15 and 16, for which the air cushion mechanism (Shreve, 1966) is difficult to validate on mechanical grounds.

Attempts so far in the use of micromechanical modelling as a means of elucidating the static mechanical behaviour of granular masses (Cundall & Strack, 1979; Matsuoka, 1983) have not proved fruitful in providing principles to guide, say, the construction of suitable macroscopic constitutive relations. Such calculations are computationally intensive for any reasonable number of grains, even in two dimensions, and most studies have been confined to only a few hundred grains, which is probably not a sufficiently large number statistically. However, the recent introduction of parallel processing computers offers a natural approach to increasing the number of grains involved, and eventually extending the simulation to three dimensions. In

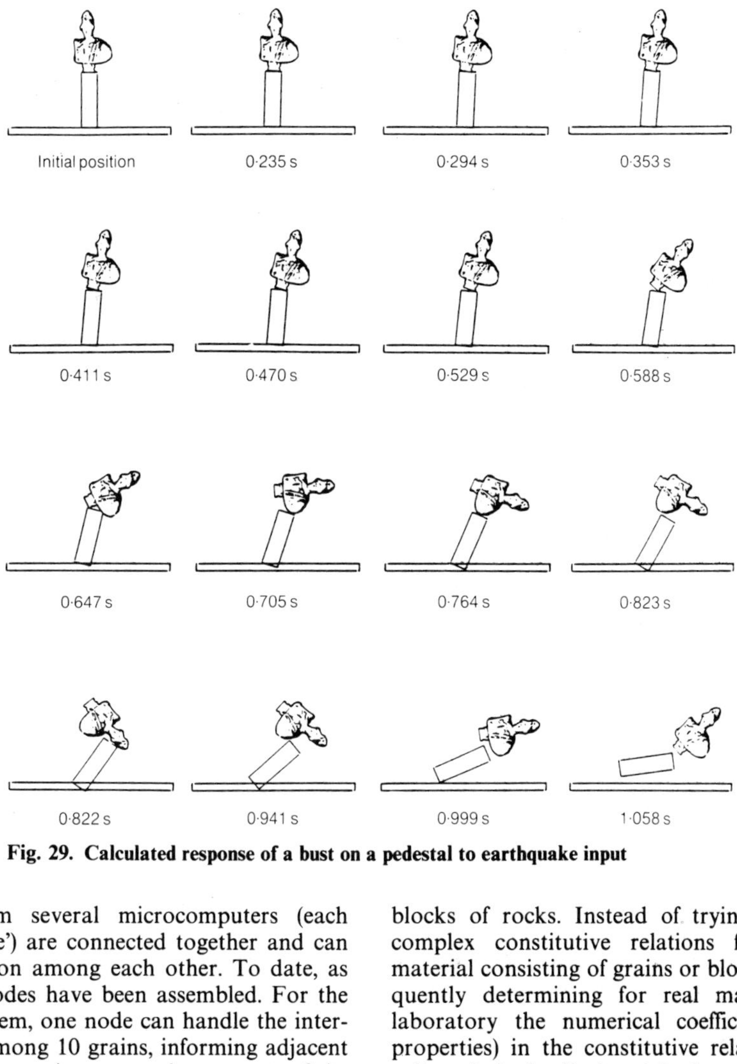

Fig. 29. Calculated response of a bust on a pedestal to earthquake input

such a system several microcomputers (each termed a 'node') are connected together and can pass information among each other. To date, as many as 64 nodes have been assembled. For the granular problem, one node can handle the interactions, say, among 10 grains, informing adjacent nodes when a grain passes in or out of its borders, and the whole system can therefore cover the interactions between 640 grains. Although some housekeeping calculations are required, the computational time is more closely associated with the time required to process 10 grains than that involved in a serial calculation of a mass of 640 grains. Programming at present is difficult, but when such computers become operational and relatively easy to use it may be possible to treat problems of thousands of particles. This obviates questions, such as arose at the Juncal ridge, of the effect of a small number of blocks in the simulation on the final result.

This possibility, when applied to problems such as Juncal Dam, which is an initial attempt at the process to be described, presents a new avenue of approach to the study of the behaviour of granular media at all scales from clay particles to blocks of rocks. Instead of trying to arrive at complex constitutive relations for masses of material consisting of grains or blocks, and subsequently determining for real materials in the laboratory the numerical coefficients (material properties) in the constitutive relations—a difficult task in itself—it might be possible just to simulate the entire granular mass. There are three aspects to this. One is the assumption in such a suggestion that the behaviour at grain contacts and the grain geometry are more important than the properties of the grain itself, and that the properties of granular media in the mass derive substantially from the grain contact forces and movements. Another is that the behaviour of grains at contacts can be adequately modelled. (In this respect clays will present more difficulty than sands or gravels.) The last aspect is the necessity of including a sufficiently large number of grains to be statistically representative of the real material, which will generally consist of many more particles. There should also be sufficient grains in the model that the smallest dimension of any continuous structure (wall, pile, footing), associated with the granular mass, spans many

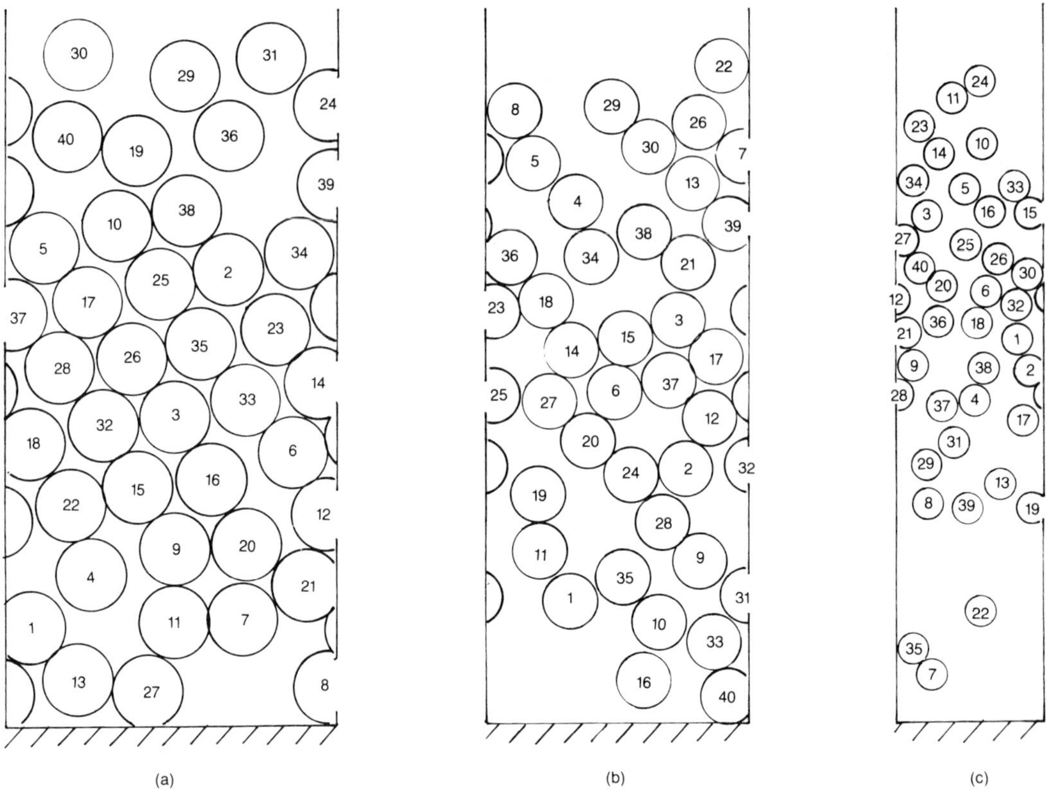

Fig. 30. Computer model of two-dimensional granular medium flow (from left to right) in a sloping chute at various angles (the left- and right-hand boundary conditions are periodic, i.e. a particle emerging from the right-hand boundary is inserted at the same velocity, acceleration etc. through the left-hand boundary) (after Campbell & Brennen (1985)): (a) 20°; (b) 30°; (c) 40°

grains. 'Many' in this case may be 10 or more. If the assumptions involved are valid, many of the problems of constitutive relation formulation and material property determination are avoided, but replaced with questions of grain representation and what happens at the contacts between grains. At least for coarser soil grains, and those minerals which are relatively hard, so no grain crushing occurs, the latter set of difficulties is less imposing than the former. For the less mathematically inclined, the simulation of large displacement processes by the micromechanical approach avoids the difficulties involved with finite deformations in solid mechanics theory, with Lagrangian and Eulerian strains, second Piola–Kirchhoff stress tensors etc., since the equilibrium equations are dealt with at the particle level, and grain displacements follow the kinematic constraints imposed by neighbouring grains.

One example of this approach may be given here. The problem was a two-dimensional one of a single, more massive grain, striking a mass of other similar grains at an impact velocity normal to the surface. Fig. 31 gives a series of snapshots of the penetration process, showing the formation of a crater and ejection of individual grains. Simulations of identical initial particle geometry and contact distribution can be performed, in which, for example, only the mass of individual grains or the intergrain friction or cohesion is altered from test to test. In this way the variation of, say, penetration depth with these properties can be examined. Similar numerical tests have been carried out by Werner & Haff (1986) in their studies of sand grain saltation, an important process in the movement and formation of sand dunes. The effect of the properties cited on the behaviour of a static triaxial test can also be examined (Corkum & Ting, 1986). Eventually three-dimensional models of many grains will bring the simulations closer to real materials. Discrete models at the block level have some utility in the study of, say, mechanisms of deformation and failure of masonry structures, such as arches (Heyman, 1980) and medieval cathedrals (Mark, 1982).

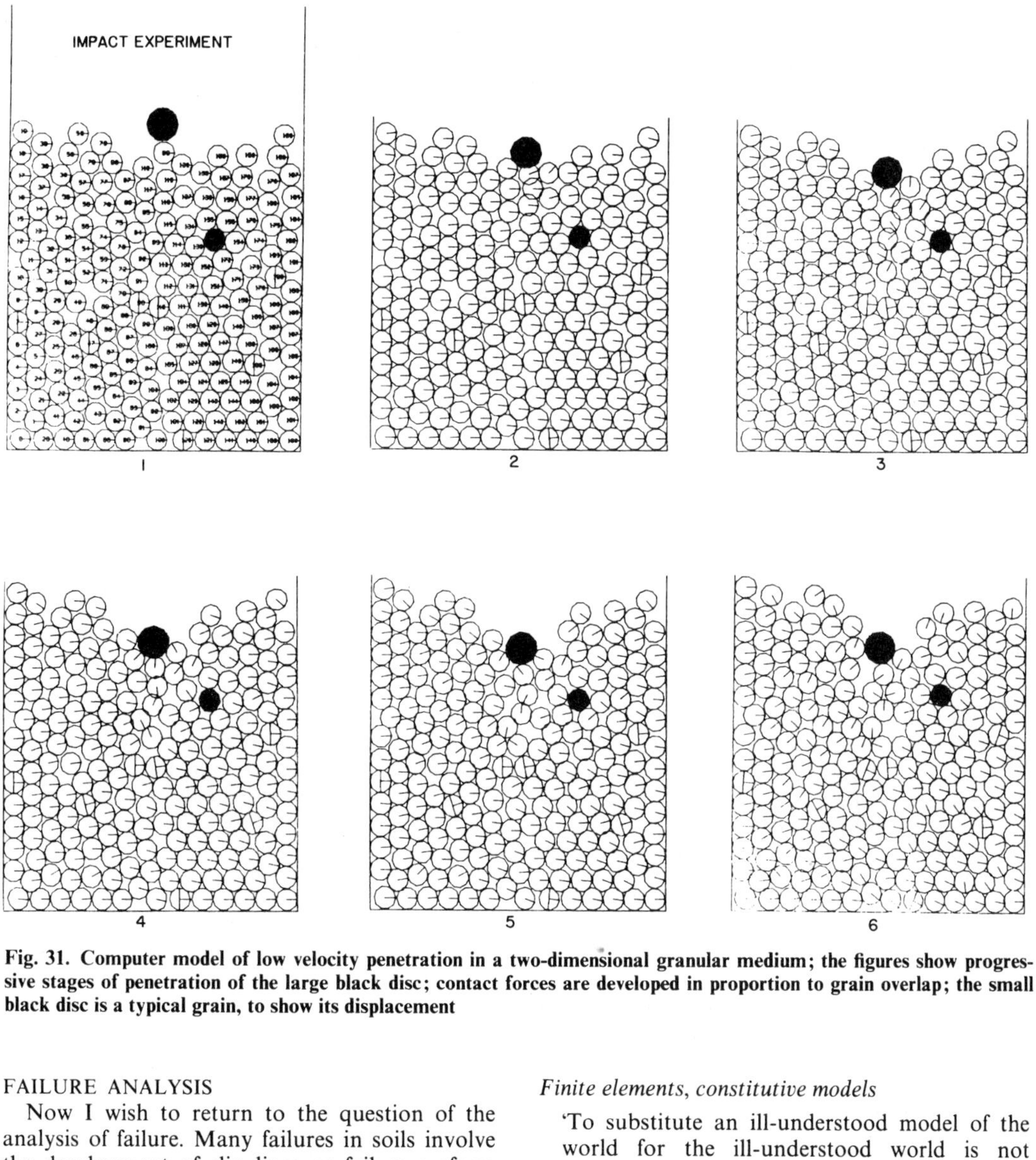

Fig. 31. Computer model of low velocity penetration in a two-dimensional granular medium; the figures show progressive stages of penetration of the large black disc; contact forces are developed in proportion to grain overlap; the small black disc is a typical grain, to show its displacement

FAILURE ANALYSIS

Now I wish to return to the question of the analysis of failure. Many failures in soils involve the development of slip lines or failure surfaces where the material shears, exhibiting displacement discontinuities or shearing zones with widths, in some cases, of only 10 or 20 grain diameters. In other cases, where the material is less dense, failure is associated with extensive zones of shearing as the soil densifies and hardens as it is sheared. These two behaviours were classified by Terzaghi (1943) as 'general' and 'local' failures respectively and are generally thought to be related to unstable and stable material stress–strain responses respectively, although rate dependence is also invoked by metal plasticians (Asaro, 1983; Lemonds, Asaro & Needleman, 1985) as a causative factor.

Finite elements, constitutive models

'To substitute an ill-understood model of the world for the ill-understood world is not progress'—Boyd & Richerson (1986)

A method of analysis, which has evolved over the last 30 years to handle deformation problems in geotechnical engineering, is the finite element methodology. In the linear area of material behaviour, or where the problem conditions permit a reasonable assumption of almost linear response, since no soils are ever linear in their behaviour, the method is clear cut and has been demonstrated (Burland, Simpson & St John, 1979; Casagrande *et al.*, 1972) to give reasonable correspondence with measurements if the relevant material properties can be assessed correctly. More often the finite element analysis indicates the properties which the material must exhibit in the field to give the observed displacements, and how much they differ from test values obtained in the laboratory. When it comes to highly non-linear behaviour of the soil, as it approaches or reaches failure, another situation prevails. Bilinear finite element models were first constructed 20 years ago to represent this condition (Hoeg, Christian & Whitman, 1968) and continue to be used in more sophisticated forms (Smoltczyk, 1982) in the analysis of failure, with various requirements imposed to account for the stress at which the soil yields or fails. Without special treatment, difficulties arise with the matrix operations if the slope of the second part of the curve is made negative, so the bilinear technique has been restricted to stable materials. When this is done solutions are obtained in which the yielded or failure region of the material is usually shown shaded. As successively higher loads are applied to the system, the shaded areas grow and spread. This behaviour may be representative of local failure, consistent with the material property employed. Several examples of this behaviour, taken from a variety of papers, are shown in Fig. 32.

It would be desirable to apply the finite element technique to the problem of slope stability. If a constitutive relation could be used which included a yield or failure condition, then in principle it would appear that, when the proper boundary conditions were applied to a slope which was in a potentially unstable state, a failure region would form in the calculation, in which both kinetics and kinematics were properly accounted for, and the failure mechanism would become evident. At present, with the method, as mentioned before, only stable material behaviour can be included, to obviate numerical difficulties, and the result of an analysis is a picture of a diffuse yielded region, unlike that which develops in the majority of real slope failures, for example. This result appears in finite element analyses of other problems also, such as studies of the bearing capacity of a strip footing. More detailed examination (Burridge, 1987) shows that the development of yielded regions depends on the shape and arrangement of the elements employed. In one study (Prévost & Hughes, 1981) yielding was 'seeded' by including an element in the system that was weaker than adjacent elements, so that it would fail first, and this appeared to give rise to a restricted failure zone, but no application or follow-up to that result appears to have been made. In research on the yielding behaviour of metal specimens in tension at Brown University (Peirce, Asaro & Needleman, 1982; Asaro, 1983) a rate-dependent constitutive relation was proposed and employed in finite element calculations. However, to obtain yielded zones of limited width, another seeding subterfuge was employed by making the boundaries of the problem sinusoidal in shape rather than straight. Two sine functions, of long and short wavelength, were superimposed, the first to give the sample a slight hourglass shape and the second to give rise to local stress concentrations. In a soil mass, much more so than in a metal, there are substantial local variations in soil properties, particularly in terms of strength, and the boundaries of a region are anything but regular. However, the major material and geometrical irregularities control the growth and propagation of a rupture surface in practice, since the principal features of many failures are similar, even in disparate geometries and material properties. Developments in the finite element method will enable the problem encountered in treating unstable material behaviour to be overcome. Advances occur in this area so rapidly that these remarks may soon be outdated.

In the past 10 years there has been a growing interest in the development of constitutive relations to represent the behaviour of soils more realistically than can be obtained by the use of a bilinear elasto-plastic model (Scott, 1985a). Most of these representations involve incremental plasticity, and some of them simulate the overall soil response to loading, or even to a load–unload cycle applied to a soil sample in single-element triaxial or shear tests, quite well. It would therefore have been expected that by this time a variety of finite element programs would have been developed incorporating such material models, and that they would be in wide use in the prediction of soil behaviour in the mass in various two- and even three-dimensional boundary condition problems. Although some codes have been written, they involve incremental plasticity relations of the simpler kind and have been used in only a few studies. In one investigation a cap model was employed in a finite element simulation of the Long Beach, California, subsidence bowl (Kosloff, Scott & Scranton, 1980). Perhaps the earliest complete model is the well-known Camclay model (Schofield & Wroth, 1968), which is a very useful teaching tool in describing the behaviour of normally or slightly overconsolidated clays, when it is applied to the description of a triaxial test. Even with the few constants incorporated in the model, the tracing of a model material's response to a simple loading path applied to a single element is a remarkably complex task, which generally must be handled numerically. The inclusion of this model in a finite element program involves many difficulties (Naylor & Pande, 1981), but it has been implemented (Almeida & Ramalho-Ortigão, 1982)

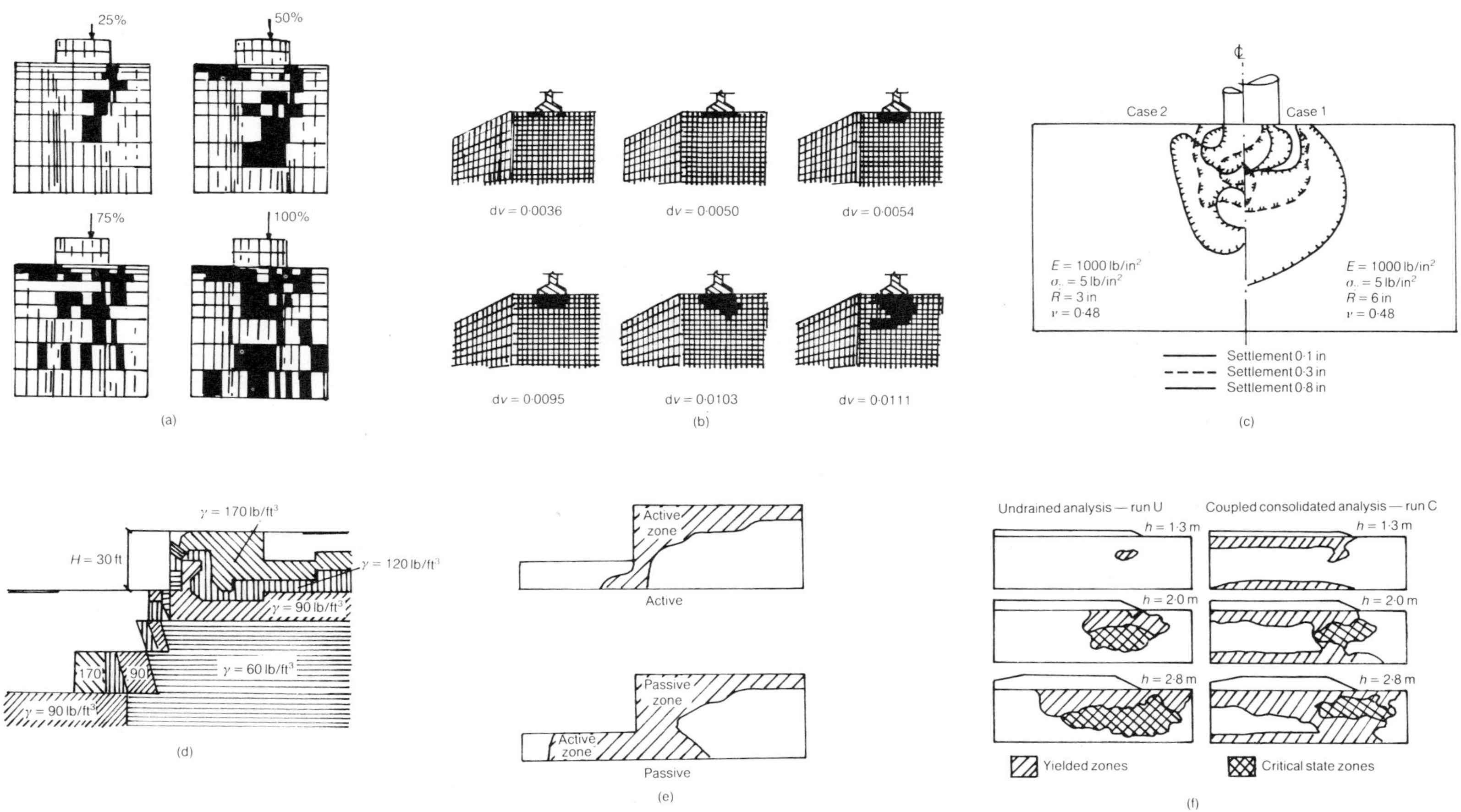

Fig. 32. Finite element solutions for elastic–plastic models (in each case the shaded regions indicate yielding under progressively increasing load or penetration): (a) after Smoltczyk (1982); (b) after Kiousis, Voyiadjis & Tuman (1986); (c) after Yamada & Wifi (1977); (d) after Snitbhan & Chen (1976); (e) after Christian *et al.* (1977); (f) after Almeida & Ramalho-Ortigão (1982) ((c) represents an axially symmetric problem; the other figures represent plane strain problems)

although it is an option in a commercially available program. The bounding surface model has been incorporated in a code (Herrmann, Dafalias & De Natale, 1982) but it has not been widely employed on a variety of soil engineering problems as yet. Many finite element codes using different types of yield surface and flow rules have been described (Christian, Hagmann & Marr, 1977; Davidson & Chen, 1976; Carter, Booker & Davis, 1977; Snitbhan & Chen, 1976). Numerical and analytical solutions to problems associated with pile behaviour have been obtained using the Camclay representation and another model due to Davis and co-workers (Davis, Scott & Mullenger, 1984; Mullenger, Scott & Davis, 1984), but the latter has not yet appeared as a finite element code.

There is no doubt that the coding of these mathematical models is a difficult and trying task; the programmer must have a deep knowledge of the mathematics of the model, as well as of the programming process. In many cases numerical stability problems may be encountered, especially when loading–unloading paths are followed.

When such a code is written, it must be tested, preferably against analytical solutions, but also against field or model test results, before confidence can be gained in its application under all loading conditions. Except for the almost trivial case of linearly elastic response, to which all the models can be reduced, there are *no* analytical solutions against which a numerical solution can be tried. Other difficulties present themselves in the circumstances when experimental data form a basis for comparison. There is the ever-present boundary condition question: do the numerical and field or model test boundary conditions truly correspond? However, the major obstacle has to do with material properties. For either a field or model experiment comparison, the material constants in the constitutive model to be used have to be assessed from conventional laboratory tests. For some models, the number of constants may be as large as 20, and their identification from laboratory tests is a difficult task.

When, as inevitably occurs, the results of the numerical computations differ from the test behaviour, the difference may be due to malfunctioning of the finite element code or may be caused by, say, anisotropic soil behaviour that is not brought out in the single-element laboratory tests. In even the simplest of circumstances, checking the solution provided by such a calculation is a difficult exercise. Schad (1985) presents results showing substantial differences between the results produced by using different algorithms in the same, stable, non-linear finite element code. In addition, none of the constitutive models can currently represent the behaviour of a very dense sand or heavily overconsolidated clay, where failure is accompanied by the development of slip surfaces (Fig. 10). It follows that an associated code lacking these constitutive features, if it existed, could not properly simulate the development of slip surfaces in soils. Some studies (Sture & Ko, 1976; Tarzi, Kalteziotis & Menzies, 1982) have included a strain softening model, but do not discuss resulting displacements or strains. Thus any prototype or model test in which a footing, retaining wall or slope failure (to pick a few examples) occurs with accompanying slip surfaces cannot be represented by a finite element calculation with any of the current models, including the simplest bilinear representation. However, these are frequently the conditions of most interest to a designer. In finite element models, the choice of element and its associated shape function controls the possible deformation modes of the element. For that class of material which exhibits failure surfaces or slip zones during loading, a different calculational approach may be useful.

Dynamic relaxation, finite differences

> "When someone says 'I want a programming language in which I need only say what I wish done', give him a lollipop"—Perlis (1982)

For static problems one possibility is the use of the technique called 'dynamic relaxation', in association with the finite difference method. In this approach the computations are set up as for a dynamic problem, but in which the damping is artificial. When a load or displacement is applied, usually as a ramp function in pseudotime, the system responds dynamically with accelerations, velocities and displacements varying in time, but relatively rapidly settles down to a steady state solution. The method is a modification of the original relaxation approach of Southwell (1940) and Cross (1932), but here the relaxation process is controlled automatically by the damping. It seems to have been suggested first by Newmark (1959) but was developed by Day (1965), Otter *et al.* (1966, 1967) and Rushton (1968, 1972) during the 1960s. Chaplin (1971) referred to the procedure as 'metadynamic relaxation'. The term dynamic relaxation seems generally to imply that a finite difference method will be used to solve the associated differential equations. Presumably, a fictitious damping could be introduced with the finite element method also, and could be employed in the solution of non-linear static problems. The advantage of the finite difference method is that it is an explicit technique, and thus does not include matrix operations. Underwood

(1983) indicates also that the convergence rate is slower if finite elements are used. The significant advantage of the dynamic relaxation technique used with the finite difference operation lies in its application to non-linear problems, and particularly those involving unstable material behaviour. It was at one stage a competitor with the finite element method for linear problems, but has diminished in popularity because of the convenience of several finite element features.

In the calculational process, the forces acting on an element are used to calculate the accelerations, from which the velocities and displacements are obtained. The strains can be derived from the displacements of adjacent elements and substituted in appropriate constitutive relations to give the stresses, from which the forces are computed to begin the cycle again. The mass or density and viscosity employed are fictitious parameters that are necessary for the progress of the calculations, since they do not appear in, and do not affect, the final static solution. The direct relation of the stresses to the strains means that non-linear or even unstable stress–strain relations present no difficulties for the numerical process. No matrices are formulated, and the calculation proceeds explicitly step by step through the mesh configuration, one pass for each time step. In the first computations by this method of the deflexions and stresses in concrete dams, for example (Otter *et al.*, 1966, 1967), the material behaviour was assumed to be linear, but the advantages for non-linear response were recognized early and the approach was extended to buckling.

One of the problems that has been referred to frequently in this discussion is that of the stability of slopes. It is an unsatisfactory condition of this situation that an upper-bound solution still has to be obtained by a trial and error method, involving an assumption of a slip surface, followed by an equilibrium analysis of its degree of stability (Bishop, 1955). With all the variations proposed on the details of this approach (Morgenstern & Price, 1965), it is still generally necessary to perform many trials to obtain the worst case. A real worst case mechanism may not be noticed. Basing an analysis solely on the equilibrium of an assumed surface ignores both the constitutive relations of the material and the kinematics of the attendant deformations, as is well known. The variational technique proposed by Baker & Garber (1978), which has not yet been used in routine stability analyses in practice, does generate a failure surface as a result of solving a variational problem.

The finite difference method has recently been adapted by Silling (1986) with the inclusion of finite deformations in a theoretical study of the conditions around a fracture or crack in a plastically deforming material (Abeyaratne, 1981; Knowles & Sternberg, 1978). The equations describing that problem are elliptic if the material behaviour is linearly elastic and become parabolic or hyperbolic (as in soil mechanics) when yielding is encountered. Silling included in his dynamic relaxation finite difference code (CHIMP) a trilinear material behaviour, Fig. 33, of a markedly unstable nature and examined the transition from elliptic through hyperbolic to elliptic response. It seemed that his program might be applied to the problem of interest here: the development of discrete slip surfaces in an unstable material. Having made the necessary adjustments a bar compression test (an unconfined compression plane strain test) was run with the material property given in Fig. 33. The results showed a propagation of slip zones from the corners of the loading plates as displacements increased, until a complete failure mechanism was evident. No weak element or boundary perturbations were necessary, since the rigid end boundary provided a singularity. With further assistance from Silling, a plane strain punch indentation problem has also been examined, with the results shown in Fig. 34 for successive punch displacements. The results of a plane strain punch or bearing capacity test on an essentially half-space region are presented in Fig. 35, along with a granular model test performed several years ago. The technique has been applied (Silling, 1985) to the slope stability problem (Burridge, 1987) where gravity is the loading condition, with the results shown in Fig. 36.

Although this approach needs substantial further development, and needs to be studied in much more detail, it offers the possibility of establishing a failure mechanism in soil mechanics and other structural problems. When the geometry and soil properties of a particular situation (say a slope stability case) have been established, and the problem arranged in the dynamic relaxation

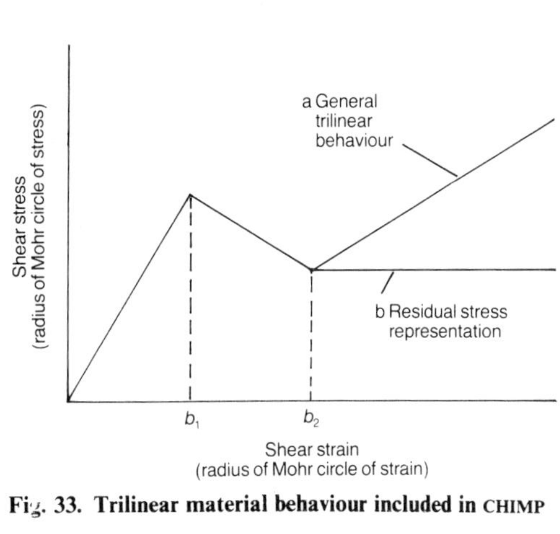

Fig. 33. Trilinear material behaviour included in CHIMP

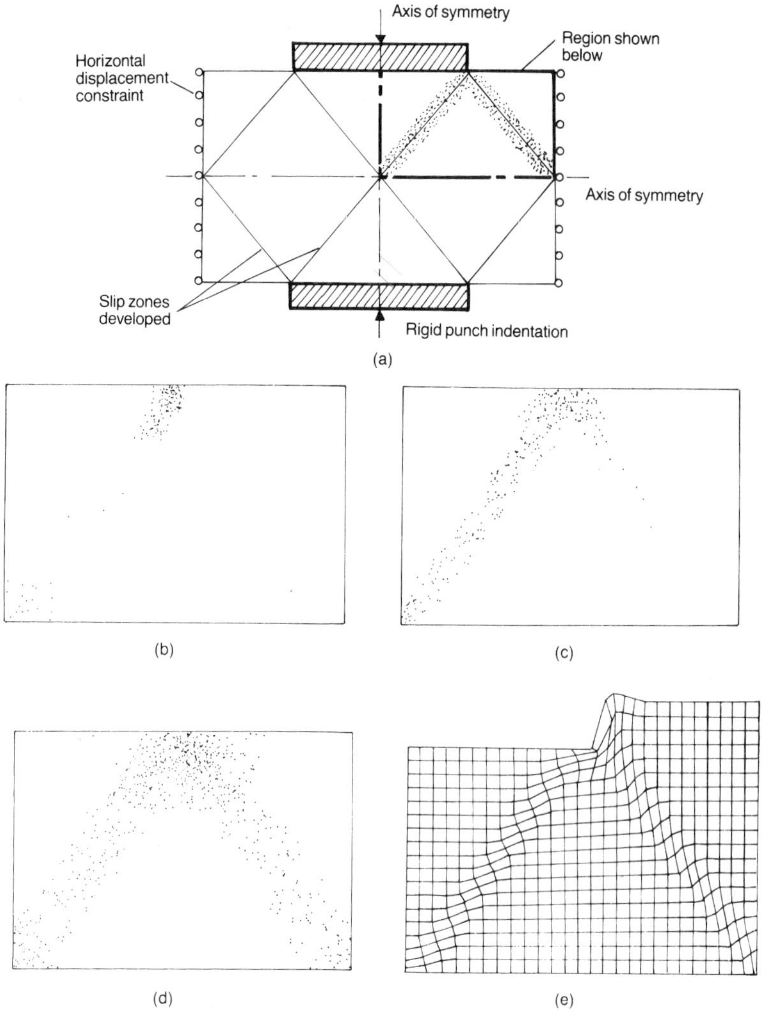

Fig. 34. Plane strain punch indentation test with CHIMP: (a) test illustrated (the stippled region indicates shear strains greater than the strain at the peak shearing test); (b) early stage in indentation; (c) later stage; (d) complete indentation mechanism; (e) exaggerated deformation of the mesh at the same stage as (d)

finite difference format, the resulting calculation traces the development of a slip surface, if the conditions are right for its manifestation, through the material. For a sufficiently high loading, a collapse mechanism will ultimately be generated. At lower loads, a partially penetrating yield zone or no slip surface will develop. For a given load, one computation gives the mechanism, without trial and error, although, of course, another calculation is needed for a different load, to find, say, a failure value. The effect of a different geometry, soil layers or soil properties may be examined. Once a sufficiently high load has revealed a failure mechanism in one analysis, conventional upper-bound analysis can be employed to give a lower upper-bound load. The speed of advance of computer developments is such that to cite a current state is almost meaningless in only a year or two, but the program employed to generate Figs 33–35 was used on an IBM PC AT machine, which required approximately one hour to run 200 time steps of a finite difference network composed of about 600 nodes. The complete development shown in the figures requires 600–700 steps. On a somewhat larger machine (VAX 780) the entire running time is reduced to a few minutes.

CONCLUSION

'It's frightful that people who are so ignorant should have so much influence'—George Orwell (said of V. Gollancz, his publisher)

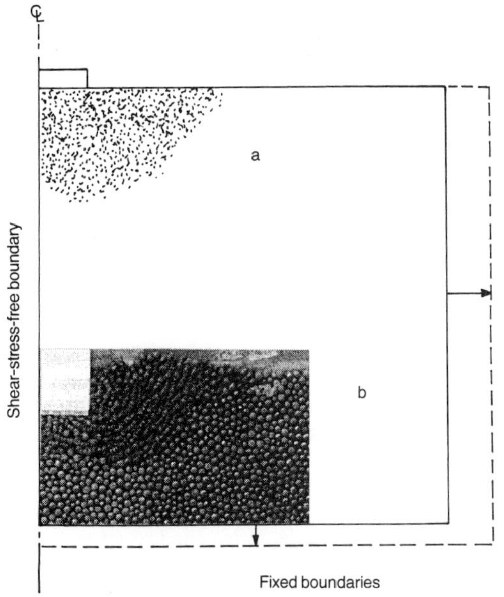

Fig. 35. Plane strain punch test with CHIMP on an infinite region—shear strains in excess of the peak, along with double exposure of a two-dimensional punch experiment (performed in 1969) in a granular medium composed of steel rods: the computational boundaries (broken lines) are more distant than shown

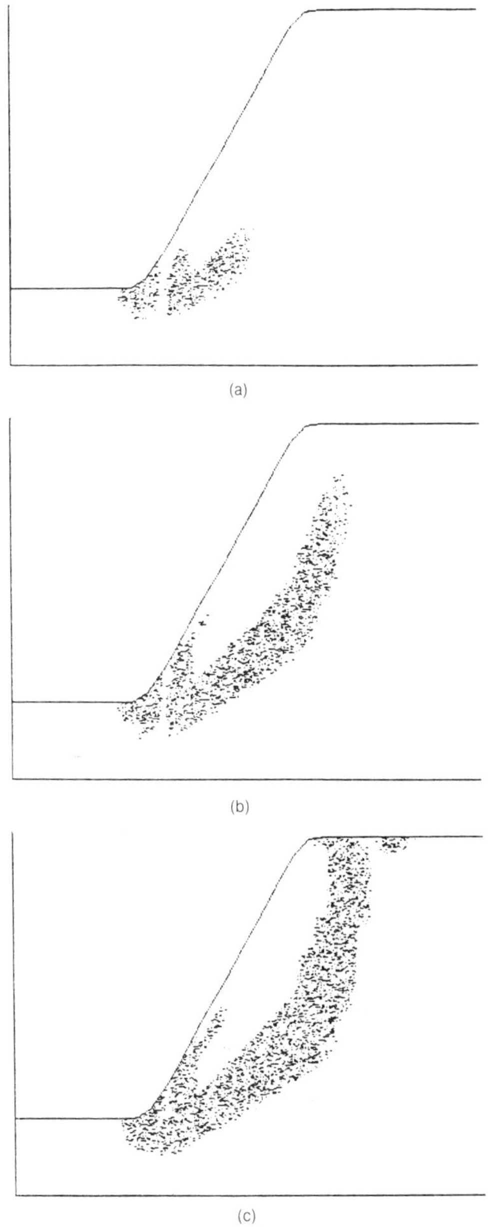

Fig. 36. Slope stability solution with CHIMP with gravitational acceleration exceeding the failure value, applied as a ramp function: (a) excess shear strains at an early stage; (b) at a later stage; (c) final excess shear strain picture (after Burridge (1987))

This has been a highly personal view of the question of failure, taken in a broad context, in geotechnical problems, including a description of some field examples, research and a brief exposition of opinions. At the present time, and for many years, many of the usual practical problems in soil engineering have been solved to a level which might be termed economically satisfactory. In looking at the everyday state of practice in terms of static problems, the most recent paper referred to in many commercial soil engineering reports is the well-known one by Bishop (1955) on slope stability analysis! In the area of soil dynamics, progress is still required for practical solutions to seismic conditions, and there are other practical problems, involving soft soils, unusual soils (volcanic or calcareous sediments) and large structures which still need study.

In many areas of soil behaviour, however, intellectual pursuits have been developed which are almost divorced entirely from engineering practice. One example is the mechanics of granular media, in which (Horne, 1965, 1969) an investigation on the behaviour of assemblages of circular discs or spheres can proceed without reference to the deformations of real sand. Another is the study of constitutive relations, although its eventual application to the solution of practical problems is still a goal. Many of the details of this particular discipline, after the initial solid mechanics foundations were laid a decade or more ago, have reduced to the curve fitting games that are so familiar in traditional soil mechanics and engineering practice. Certain energy conditions give rise to the requirement that the plastic strain increment vector be normal (associative flow rule) to a yield surface, say, but experimental results show that real frictional soils do not know this. Consequently, to fit a particular mathematical model to a real soil response, a non-associative rule is adduced, and the plastic strain increment vector is taken at any angle to the yield surface that fits the experimental results. The angle can be varied as the test proceeds. If such an empirically adjusted model is included in an analytical or finite element solution, the uniqueness of the solution is no longer assured. In various models, both static and dynamic, the shear or bulk modulus variation with applied hydrostatic pressure or shear strain is approximated by some convenient fitting function describing experimental results under some simplified boundary conditions. On occasion a non-linear variation in modulus is approximated by a piecewise linear representation. If the situation is a dynamic one, the calculation is stopped in full flight, all the moduli of different soil elements are switched to new values, and the program is set running again. Clearly the real dissipative plastic nature of the soil response cannot be represented in this way. The use of computers permits almost any variation of properties to be incorporated in an *ad hoc* fashion without regard to energy relations or thermodynamic principles. Is it correct to do this? I am sure that it is not and I feel far from comfortable with the situation.

In any event, the result of a calculation can never been checked. What can it be checked against? It cannot be checked against experience in the field because the necessary measurements of stress, displacement, acceleration etc. are almost never available, or at such few points as to be of little assistance, or, quite properly, the pressure transducers are not to be trusted. Nor can the result be checked against an 'analytical' solution, because they do not exist. I have worried about this problem elsewhere after finding errors in a published slope stability computer program (Scott, 1985b). In examining all computer solutions, what Dawkins (1986) calls 'the principle of personal incredulity' should be continuously employed.

What is left is only judgement, the magic quality of Peck (1981), but in the complex geometries, layering and material properties of the cases that such solutions may be or are applied to, where does judgement enter, and how can it be reliable?

I once had a discussion with Arthur Casagrande, in the general context of the investigation into the failure of the Baldwin Hills Reservoir, on the subject of earth dam design. Talking about design and analysis, he pointed out that I did not grasp the realities of the design situation fully. He said that a mature engineer with experience and judgement (himself), after looking over the site, geology, soil properties etc. very carefully and meticulously, thoroughly absorbing himself in the site conditions, sketched the design on a piece of paper (it might be modified subsequently as more information became available during construction), and it was then drawn up more formally by draughtsmen. The resulting cross-sections, he said 'are then given to young chaps like you to do your analyses on and show that the design is correct'. Calling on the famous remark by Baker (1881) to the effect that experience was based on failure, I asked him about failure of dams that he had designed that way. He said, with some irritation, that none had failed, and I then tried to find out in consequence what the basis of his judgement was that a given dam section, built of a particular soil in a certain valley, subjected to almost unquantifiable, say, seismic risks, was safe. I also tried to enquire why, when, for a large earth dam, every single degree of flattening of the upstream and downstream slopes costs more than US $1 million dollars, earth dams always come out with slope angles of 2·5:1 or 3:1. I was unsuccessful in obtaining answers to both of these questions, and am not much wiser 16 years later. Little is learned from an unfailed dam, but failures cannot be afforded. Are not more failures needed, in general, in field and model tests, to which analyses can be applied, to ensure that too much money is not being spent unnecessarily? Should not more effort be put into contrived and controlled failures of special structures? Has judgement grown stale when applied to the complex situations that face engineers today?

ACKNOWLEDGEMENTS

The investigations and studies performed by the Author and reported here could not have been carried through without the assistance of many colleagues. It is well known that large teams are engaged particularly in space missions, whose success depends on their efforts. It would not have been possible to have tested the lunar surface without the assistance of many NASA, JPL and Hughes Aircraft Company workers. The Author's immediate colleagues at JPL were F. I. Roberson and M. C. Clary, electrical engineers, who ensured the function and performance of the Surveyor sampler; Roberson and the Author commanded the device on the moon. On Mars, and the Apollo programme, the Author's co-workers are listed on the referenced papers. C. E. Lindvall was kind enough to review the description of the Portuguese Bend landslide, which was discussed frequently together. At Highland Park, former students K. Zuckerman and T.-D. Lu assisted with the instrumentation. M. M. Baligh carried out the finite element fault propagation calculations, and P. B. Burridge set up and ran the Metrorail tunnel experiments on the Caltech centrifuge. The analysis was provided by a colleague, J. Hall, with whom the Author has had many discussions on finite element methods. P. Cundall's distinct element code was modified by J.-P. Bardet, who ran all the simulations in the Juncal ridge study and provided the statue computations. The finite difference computation leading to the slope stability result is a part of research performed by P. B. Burridge.

This Paper could not have been completed without the hard work and patience of Sharon Beckenbach of the Division of Engineering and Applied Science at Caltech. Caltech provided the unique opportunity to pursue the space ventures, for which the Author is grateful.

BIBLIOGRAPHY

Abdel-Ghaffar, A. M. & Scott, R. F. (1979). Analysis of earth dam response to earthquakes. *J. Geotech. Engng Div. Am. Soc. Civ. Engrs* **105,** GT12, 1379–1404.

Abeyaratne, R. (1981). Discontinuous deformation gradients away from the tip of a crack in anti-plane shear. *J. Elasticity* **11,** 373–393.

Almeida, M. S. S. & Ramalho-Ortigão, J. A. (1982). Performance and finite element analyses of a trial embankment in soft clay. *Numerical models in geomechanics* (eds R. Dungar, G. N. Pande and J. A. Studer), pp. 548–558. Rotterdam: Balkema.

Asaro, R. J. (1983). Micromechanics of crystals and polymers, *Adv. Appl. Mech.* **23,** 1–112.

Baker, B. (1881). The actual lateral pressure of earthwork. *Minut. Proc. Instn Civ. Engrs* **65,** 140–186.

Baker, R. & Garber, M. (1978). Theoretical analysis of the stability of slopes. *Géotechnique* **28,** No. 4, 395–411.

Banichuk, N. V., Kartvelishvili, V. M. & Chernousko, F. L. (1972). Numerical solution for an axisymmetric problem of the pressing of an indentor into an elastic-plastic medium. *Proc. Acad. Sci. USSR Mech. Solids,* No. 1, 50–57.

Bishop, A. W. (1955). The use of the slip circle in the stability analysis of slopes. *Géotechnique* **5,** No. 1, 7–17.

Bishop, A. W. (1967). Progressive failure–with special reference to the mechanism causing it. *Proc. Geotech. Conf., Oslo* **2,** 142–154.

Bjerrum, L. (1967). Progressive failure in slopes of overconsolidated plastic clay and clay shales. *J. Soil Mech. Fdns Div. Am. Soc. Civ. Engrs* **93,** SM5, 3–49.

Boddam-Whetham, P. N. (1973). *A miniature triaxial apparatus for testing 1 gram samples of lunar soil.* Report on 3rd year undergraduate research project, Cambridge University Engineering Department.

Bolt, B. A., Horn, H., MacDonald, G. A. & Scott, R. F. (1977). *Geological hazards,* 2nd edn. Berlin: Springer.

Boyd, R. & Richerson, P. J. (1986). *Culture and the evolutionary process.* University of Chicago Press.

Burland, J. B., Simpson, B. & St John, H. D. (1979). Movements around excavations in London clay. *Proc. 7th Eur. Conf. Soil Mech. Fdn Engng, Brighton* **1,** 13–30.

Burridge, P. B. (1987). *Failure of slopes.* PhD thesis, Division of Engineering and Applied Science, California Institute of Technology, Pasadena.

Campbell, C. S. & Brennen, C. E. (1985). Chute flows of granular materials; some computer simulations. *J. Appl. Mech.* **52,** 172–178.

Carrier, W. D. (1973). Lunar soil grain size distribution. *Moon* **6,** 250–263.

Carter, J. P., Booker, J. R. & Davis, E. H. (1977). Finite deformation of an elasto-plastic soil. *Int. J. Numer. Analyt. Meth. Geomech.* **1,** No. 1, 25–44.

Casagrande, A., Wilson, S. D. & Schwantes, E. D. (1972). The Baldwin Hills Reservoir failure in retrospect. *Proc. Conf. Performance of Earth and Earth-Supported Structures* **1,** 551–588. New York: American Society of Civil Engineers.

Chaplin, T. K. (1971). Metadynamic relaxation applied to automatic analysis of slabs, plates, and beams on elastic foundations. *Proc. Symp. The Interaction of Structure and Foundation, Birmingham,* pp. 76–83. Birmingham: Midland Soil Mechanics and Foundation Engineering Society.

Christian, J. T., Hagmann, A. J. & Marr, W. A. (1977). Incremental plasticity analysis of frictional soils. *Int. J. Numer. Analyt. Meth. Geomech.* **1,** No. 4, 343–376.

Corkum, B. T. & Ting, J. M. (1986). *The discrete element method in geotechnical engineering.* Publication 86–11, Department of Civil Engineering, University of Toronto.

Costes, N. C., Carrier, W. D., Mitchell, J. K. & Scott, R. F. (1970). Apollo 11 soil mechanics investigation. *J. Soil Mech. Fdns Div. Am. Soc. Civ. Engrs* **96,** SM6, 2045—2080.

Cross, H. (1932). Analysis of continuous frames by distributing fixed-end moments. *Trans. Am Soc. Civ. Engrs* **96,** 1–10.

Crowe, M. J. (1967). *A history of vector analysis.* Notre Dame: University of Notre Dame Press.

Cundall, P. A. (1976). Explicit finite difference methods in geomechanics. *Numerical methods in geomechanics* (ed. C. S. Desai), pp. 132–150. New York: American Society of Civil Engineers.

Cundall, P. A. & Strack, O. D. L. (1979). A discrete numerical model for granular assemblies. *Géotechnique* **29,** No. 1, 47–65.

Davidson, H. L. & Chen, W. F. (1976). Nonlinear analyses in soil and solid mechanics. *Numerical methods in geomechanics* (ed. C. S. Desai), pp. 205–218. New York: American Society of Civil Engineers.

Davis, R. O., Scott, R. F. & Mullenger, G. (1984). Rapid expansion of a cylindrical cavity in a rate-type soil. *Int. J. Numer. Analyt. Meth. Geomech.* **8,** 125–144.

Dawkins, R. (1986). *The blind watchmaker.* London: Longman.

Day, A. S. (1965). An introduction to dynamic relaxation. *Engineer, Lond.,* **219,** 218–221.

Dunlop, P. & Duncan, J. M. (1970). Development of failure around excavated slopes. *J. Soil Mech. Fdns Div. Am. Soc. Civ. Engrs* **96,** SM2, 471–493.

Ehlig, P. H. (1982). Mechanics of the Abalone Cove

Landslide including the role of ground water in landslide stability and a model for development of large landslides in the Palos Verdes Hills. *Landslides and landslide abatement, Palos Verdes Peninsula, southern California.* Field trip 10, pp. 57–66. Anaheim: Geological Society of America.

Ehlig, K. A. & Bean, R. T. (1982). Dewatering of the Abalone Cove Landslide, City of Rancho Palos Verdes, Los Angeles County, California. *Landslides and landslide abatement, Palos Verdes Peninsula, southern California.* Field trip 10, pp. 67–79. Anaheim: Geological Society of America.

Gould, J. P. (1960). A study of shear failure in certain tertiary marine sediments. *Proc. Res. Conf. Shear Strength of Cohesive Soils, Boulder,* pp. 615–641. New York: American Society of Civil Engineers.

Hall, R. C. (1977). *Lunar impact: a history of Project Ranger.* NASA SP-4210, US National Aeronautics and Space Administration, Scientific and Technical Information Office, Washington DC.

Hammond, R. (1957). *Engineering structural failures.* New York: Philosophical Library.

Herrmann, L. R., Dafalias, Y. F. & De Natale, J. S. (1982). Numerical implementation of a bounding surface soil plasticity model. *Proc. Int. Symp. Numerical Models in Geomechanics,* pp. 334–343. Rotterdam: Balkema.

Heyman, J. (1980). The estimation of the strength of masonry arches. *Proc. Instn Civ. Engrs.,* Part 2 **69,** 921–937.

Hill, R. & Hutchinson, J. W. (1975). Bifurcation phenomena in the plane tensile test. *J. Mech. Phys. Solids* **23,** 239–264.

Hoeg, K., Christian, J. T. & Whitman, R. V. (1968). Settlement of strip footing on elasto-plastic soil. *J. Soil Mech. Fdns Div. Am. Soc. Civ. Engrs* **94,** SM2, 431–445.

Horne, M. R. (1965). The behaviour of an assembly of rotund, rigid, cohesionless particles, parts I and II. *Proc. R. Soc. A* **286,** 62–97.

Horne, M. R. (1969). The behaviour of an assembly of rotund, rigid, cohesionless particles, part III. *Proc. R. Soc. A* **310,** 21–34.

Hudson, D. E. & Scott, R. F. (1965). Fault motions at the Baldwin Hills Reservoir site. *Bull. Seism. Soc. Am.* **55,** No. 1, 165–180.

Independent Panel (1976). *Failure of Teton Dam.* Report to US Department of the Interior and the State of Idaho, US Government Printing Office, Washington.

Jaffe, L. D. (1973). Shear strength of lunar soil from Oceanus Procellarum. *Moon* **8,** 58–72.

Jahns, R. H. & Vonder Linden, K. (1973). Space–time relationships of landsliding on the southerly side of the Palos Verdes Hills, California. *Geology, seismicity, and environmental impact.* AEG Special Publication, 123–138. Los Angeles: University Publishers.

Kerr, P. F. & Drew, I. M. (1969). *Clay mobility, Portuguese Bend, California.* Special Report 100, pp. 3–16. California Division of Mines and Geology.

Kiousis, P. D., Voyiadjis, G. Z. & Tuman, M. T. (1986). A large strain theory for the two-dimensional problem in geomechanics. *Int. J. Numer. Analyt. Meth. Geomech.* **10,** 17–39.

Knowles, J. K. & Sternberg, E. (1978). On the failure of ellipticity and the emergence of discontinuous deformation gradients in plane finite elastostatics. *J. Elasticity* **8,** 329–379.

Kosloff, D., Scott, R. F. & Scranton, J. (1980). Finite element simulation of Wilmington Oil Field subsidence: I, linear modeling; II, nonlinear modeling. *Tectonophysics* **65,** 339–368; **70,** 159–183.

Leighton and Associates (1979). A guidebook for visiting selected Southern California landslides. *Japanese–American Field Conf., Irvine.*

Lemonds, J., Asaro, R. J. & Needleman, A. (1985). A numerical study of localized deformation in bicrystals. *Mech. Mater.* **4,** 417–435.

Leonards, G. A. (ed.) (1987). *J. Engng. Geol.,* Special issue on dam failures, to be published.

Lindvall, Richter and Associates (1982). *Investigation of seismic stability of Juncal Dam.* Report to Montecito Water District, California.

Lindvall, Richter and Associates (1984). *Centrifuge and numerical studies to evaluate effect of fault displacement on Metrorail project.* Report to Southern California Rapid Transit District.

Lu, T. D. & Scott, R. F. (1972). The distribution of stresses and development of failure at the toe of a slope and around the tip of a crack. *Proc. Symp. Applications of the Finite Element Method in Geotechnical Engineering, Vicksburg.*

Mandelbrot, B. B. (1983). *The fractal geometry of nature.* New York: Freeman.

Mark, R. (1982). *Experiments in Gothic structure.* Cambridge: Massachussetts Institute of Technology.

Matsuoka, H. (1983). Deformation and strength of granular materials, based on the theory of 'compounded mobilized planes', and spatial mobilized plane. *Advances in the mechanics and the flow of granular materials* (ed. M. Shahinpoor) **II,** 813–836. Houston: Gulf.

Mitchell, J. K., Bromwell, L. G., Carrier, W. D., Costes, N. C. & Scott R. F. (1972). Soil mechanical properties at the Apollo 14 site. *J. Geophys. Res.* **77,** No. 29, 5641–5664.

Moore, H. J. II, Hutton, R. E., Scott, R. F., Spitzer, C. R. & Shorthill, R. W. (1977). Surface materials of the Viking landing sites. *J. Geophys. Res.* **82,** No. 28, 4497–4523.

Morgenstern, N. R. & Price, V. E. (1965). The analysis of the stability of general slip surfaces. *Géotechnique* **15,** No. 1, 79–93.

Mullenger, G., Scott, R. F. & Davis, R. O. (1984). Rapid shearing in a rate-type soil surrounding a cylindrical cavity. *Int. J. Numer. Analyt. Meth. Geomech.* **8,** 141–155.

Naylor, D. J. & Pande, G. N. (1981). *Finite elements in geotechnical engineering.* Swansea: Pineridge.

Newmark, N. M. (1959). A method of computation for structural dynamics. *J. Engng Mech. Div. Am. Soc. Civ. Engrs* **85,** EM3, 67–94.

Olson, R. E. (1974). Shearing strengths of kaolinite, illite, and montmorillonite. *J. Geotech. Engng Div. Am. Soc. Civ. Engrs* **100,** GT11, 1215–1229.

Otter, J. R. H., Cassell, A. C. & Hobbs, R. E. (1966). Dynamic relaxation. *Proc. Instn Civ. Engrs* **35,** 633–656.

Otter, J. R. H., Cassell, A. C. & Hobbs, R. E. (1967).

Discussion on Dynamic relaxation. *Proc. Instn Civ. Engrs* **37,** 723–750.
Palmer, A. C. & Rice, J. R. (1973). The growth of slip surfaces in the progressive failure of over-consolidated clay. *Proc. R. Soc. A.* **332,** 527–548.
Peck, R. B. (1967). Stability of natural slopes. *J. Soil Mech. Fdns Div. Am. Soc. Civ. Engrs* **93,** SM4, 403–417.
Peck, R. B. (1981). *Where has all the judgment gone?* Publication 134, pp. 1–5, Norwegian Geotechnical Institute, Oslo.
Peirce, D., Asaro, R. J. & Needleman, A. (1982). An analysis of non-uniform and localized deformation in ductile single crystals. *Acta Metall.* **30,** 1087–1119.
Perlis, A. (1982). *Epigrams on programming.*
Prévost, J.-H., Cuny, B., Hughes, T. J. R. & Scott, R. F. (1981). Offshore gravity structures: analysis. *J. Geotech. Engng Div. Am. Soc. Civ. Engrs* **107,** GT2, 143–165.
Prévost, J.-H., Cuny, B. & Scott, R. F. (1981). Offshore gravity structures: centrifugal modeling. *J. Geotech. Engng Div. Am. Soc. Civ. Engrs* **107,** GT2, 125–141.
Prévost, J.-H. & Höeg, K. (1975). Soil mechanics and plasticity analysis of strain softening. *Géotechnique* **25,** No. 2, 279–297.
Prévost, J. H. & Hughes, T. J. R. (1981). Finite element solution of elastic–plastic boundary-value problems. *J. Appl. Mech.* **48,** 69–74.
Rice, J. R. (1976). The localization of plastic deformation. *Proc. 14th IUTAM Congr. Theoretical and Applied Mechanics, Delft.* Amsterdam: North Holland.
Roddy, D. J., Rittenhouse, J. B. & Scott, R. F. (1963). Dynamic penetration studies in crushed rock under atmospheric and vacuum conditions. *A.I.A.A. J.* **1,** No. 4, 868–873.
Roth, W. H., Scott, R. F. & Austin, I. (1981). Centrifuge modeling of fault propagation through alluvial soils. *Geophys. Res. Lett.* **8,** No. 6, 561–564.
Roth, W. H., Scott, R. F. & Cundall, P. A. (1986). Non-linear dynamic analysis of a centrifuge model embankment. *Proc. 3rd US Nat. Conf., Charleston.* Earthquake Engineering Research Institute.
Rushton, K. R. (1968). Dynamic-relaxation solutions of elastic-plate problems. *J. Strain Anal.* **3,** No. 1, 23–32.
Rushton, K. R. (1972). Buckling of laterally loaded plates having initial curvature. *Int. J. Mech. Sci.* **14,** 667–680.
Schad, H. (1985). Computing costs for FEM analysis of foundation engineering problems and possible ways of increasing efficiency. *Int. J. Numer. Analyt. Meth. Geomech.* **9,** 261–275.
Schofield, A. N. & Wroth, C. P. (1968). *Critical state soil mechanics.* London: McGraw-Hill.
Scott, R. F. (1964). Report on soil mechanics and foundation engineering aspects of the Alaskan earthquake of March 27, 1964. *Report on analysis of earthquake damage to military construction in Alaska, 27 March 1964,* Appendix II. Washington DC: Engineering Division Office of Chief of Engineers, US Army.
Scott, R. F. (1967a). In-place soil mechanics measurements. *Marine geotechnique.* Champaign: University of Illinois.
Scott, R. F. (1967b). Soil mechanics surface sampler experiment for Surveyor. *J. Geophys. Res.* **72,** No. 2, 827–830.
Scott, R. F. (1967c). The feel of the Moon. *Scient. Am.* **217,** No. 5, 34–43.
Scott, R. F. (1968). The density of the lunar surface soil. *J. Geophys. Res.* **73,** No. 16, 5469–5471.
Scott, R. F. (1970). In-place ocean soil strength by accelerometer. *J. Soil Mech. Fdns Div. Am. Soc. Civ. Engrs* **96,** SM1, 199–271.
Scott, R. F. (1973). Lunar soil mechanics. *Proc. 8th Int. Conf. Soil Mech. Fdn Engng, Moscow* **4.2,** 177–190.
Scott, R. F. (1978). Incremental movement of a rock-slide. *Rockslides and avalanches* (ed. B. Voight) **1,** Ch. 18, 659–668. Amsterdam: Elsevier.
Scott, R. F. (1980). Slope stability studies in the centrifuge. *Int. Symp. Landslides, New Delhi.*
Scott, R. F. (1985a). Plasticity and constitutive relations in soil mechanics. *J. Geotech. Engng* **3,** No. 5, 563–605.
Scott, R. F. (1985b). Software certification: foreseeing (sic) problems. *Civ. Engng* **55,** No. 2, 6.
Scott, R. F., Carrier, W. D. III, Costes, N. C. & Mitchell, J. K. (1971). Apollo 12 soil mechanics investigation. *Géotechnique* **21,** No. 1, 1–14.
Scott, R. F. & Craig, M. J. K. (1980). Computer modeling of clay structure and mechanics. *J. Geotech. Engng Div. Am. Soc. Civ. Engrs* **106,** GT1, 17–33.
Scott, R. F. & Ko, H. Y. (1968). Transient rocket-engine gas flow in soil. *A.I.A.A. J.* **6,** No. 2, 258–264.
Scott, R. F. & Roberson, F. I. (1968). Soil mechanics surface sampler: lunar surface tests, results and analyses. *J. Geophys. Res.* **73,** No. 12, 4045–4080.
Scott, R. F. & Roberson, F. I. (1969). Soil mechanics surface sampler, (Surveyor VII). *J. Geophys. Res.* **74,** No. 25, 69–110.
Scott, R. F. & Schoustra, J. J. (1974). Nuclear power plant siting on deep alluvium. *J. Geotech. Engng Div. Am. Soc. Civ. Engrs* **100,** GT4, 449–459.
Scott, R. F. & Zuckermann, K. A. (1971). Examination of returned Surveyor III surface sampler. *Proc. 2nd Lunar Science Conf.* **3,** 2743–2751. Cambridge: Massachusetts Institute of Technology.
Shorthill, R. W., Hutton, R. E., Moore, H. J. & Scott, R. F. (1972). Martian physical properties experiments: the Viking Mars lander. *Icarus* **16,** No. 1, 217–222.
Shorthill, R., Hutton, R. E., Moore, H. J., Scott, R. F. & Spitzer, C. R. (1976). Physical properties of the Martian surface from the Viking I lander: preliminary results. *Science* **193,** 805–809.
Shorthill, R. W., Moore, H. J., Scott, R. F., Hutton, R. E., Liebes, S., Jr, & Spitzer, C. R. (1976). The soil of Mars (Viking 1). *Science* **194,** 91–97.
Shreve, R. L. (1966). Sherman Landslide, Alaska. *Science* **154,** 1639–1643.
Silling, S. A. (1985). *CHIMP—a computer program for finite elastostatics.* Technical Report 54, Division of Engineering and Applied Science, California Institute of Technology, Pasadena.
Silling, S. A. (1986). *Singularities and phase transitions in elastic solids: numerical studies and stability analysis.* PhD thesis, California Institute of Technology, Pasadena.
Skempton, A. W. (1964). Long-term stability of clay slopes. *Géotechnique* **14,** No. 2, 77–101.

Skempton, A. W. (1970). First-time slides in over-consolidated clays. *Géotechnique* **20,** 320–324.
Skempton, A. W. & Hutchinson, J. (1969). Stability of natural slopes and embankment foundations. *Proc. 7th Int. Conf. Soil Mech. Fdn Engng, Mexico City,* State of the art volume, pp. 291–340.
Slade, M. A., Lyzenga, G. A. & Raefsky, A. (1984). Modeling of the surface static displacements and fault plane slip for the 1979 Imperial Valley earthquake. *Bull. Seism. Soc. Am.* **74,** No. 6, 2413–2433.
Smelser, M. G. (1987). Geology of Mussel Rock landslide. *Calif. Geol.* **40,** No. 3, 59–66.
Smith, I. M. (1970). Incremental numerical solution of a simple deformation problem in soil mechanics. *Géotechnique* **20,** No. 4, 357–372.
Smith, I. M. & Hobbs, R. (1974). Finite element analysis of centrifuged and built-up slopes. *Géotechnique* **24,** No. 4, 531–559.
Smoltczyk, U. (1982). Use of nonlinear models in engineering practice: some personal experiences. *Numerical models in geomechanics* (eds R. Dungar, G. N. Pande and J. A. Studer), pp. 535–547. Rotterdam: Balkema.
Snitbhan, N. & Chen, W. F. (1976). Finite element analysis of large deformation in slopes. *Numerical methods in geomechanics* (ed. C. S. Desai), pp. 744–758. New York: American Society of Civil Engineers.
Southwell, R. V. (1940). *Relaxation methods in engineering science.* Oxford: Clarendon.
Sture, S. & Ko, H.-Y. (1976). Stress analysis of strain-softening materials. *Proc. 2nd Int. Conf. Numerical Methods in Geomechanics, Blacksburg* **1,** 580–590.
Tarzi, A. I., Kalteziotis, N. A. & Menzies, B. K. (1982). Element analysis of strip footings on strain-softening isotropic clay. *Numerical models in geomechanics* (eds R. Dungar, G. N. Pande and J. A. Studer), pp. 740–748. Rotterdam: Balkema.
Terzaghi, K. (1943). *Theoretical soil mechanics.* New York: Wiley.
Underwood, P. (1983). *Computer methods for transient analysis* (eds T. Belytschko and T. J. R. Hughes), pp. 245–265. Amsterdam: Elsevier.
Vardoulakis, I. (1985). Stability and bifurcation of undrained, plane rectilinear deformations on water-saturated granular soils. *Int. J. Numer. Analyt. Meth. Geomech.* **9,** 399–414.
Vonder Linden, K. & Lindvall, C. E. (1982). Mechanics of the Abalone Cove Landslide including the role of ground water in landslide stability and a model for development of large landslides in the Palos Verdes Hills. *Landslides and landslide abatement, Palos Verdes Peninsula, southern California.* Field trip 10, pp. 49–56. Anaheim: Geological Society of America.
Walton, O. (1981). *Particle dynamics modeling of geological materials.* Report UCRL-52915, University of California.
Werner, B. & Haff, P. (1986). *The impact process in aeolian saltation: two-dimensional studies.* Brown Bag Reprint Series BB50, Division of Physics, Mathematics and Astronomy, California Institute of Technology, Pasadena.
Wilkins, M. L. (1969). *Calculation of elastic–plastic flow.* Report UCRL-7322, Lawrence Radiation Laboratory, University of California at Berkeley.
Wood, W. L. (1967). Comparison of dynamic relaxation with three other iterative methods, *Engineer, Lond.* **224,** No. 5835, 683–687.
Yamada, Y. & Wifi, A. S. (1977). Large strain analysis of some geomechanics problems by the finite element method. *Int. J. Numer. Analyt. Meth. Geomech.* **1,** No. 3, 299–318.
Zienkiewicz, O. C. & Cormeau, I. C. (1974). Viscoplasticity, plasticity, and creep in elastic solids: a unified numerical solution approach. *Int. J. Numer. Meth. Engng* **8,** 821–845.
Zienkiewicz, O. C., Humpheson, C. & Lewis, R. W. (1975). Associated and non-associated visco-plasticity and plasticity in soil mechanics. *Géotechnique* **25,** No. 4, 671–689.
Zienkiewicz, O. C. & Pande, G. N. (1977). Time-dependent multilaminate model of rocks—a numerical study of deformation and failure of rock masses. *Int. J. Numer. Analyt. Meth. Geomech.* **1,** No. 3, 219–247.

VOTE OF THANKS

In proposing a vote of thanks to Professor Scott, Professor R. E. Gibson made the following remarks.

'In the *Proceedings of the Institution of Civil Engineers* for January 1954 there is a discussion contribution by Mr R. F. Scott of Cambridge, Massachusetts, on a paper entitled "Numerical solution of some problems in the consolidation of clay", written by two students of the Institution, Lumb and Gibson. On reading that discussion two things will be discovered: first, how very much further Mr Scott had already gone along this road than the two authors and secondly how clearly his objectives were stated, how well his arguments were marshalled and deployed etc.—in short, just how well it was written.

'This was an early intimation of that talent for clarity of exposition which we have witnessed in the context of a Rankine Lecture. Of course, Professor Scott's wide ranging interests and outstanding competence have given far more than that. This twenty-seventh Rankine Lecture has dealt in a fascinating and stimulating way with failure, a subject of central importance and concern to all geotechnical engineers. Much is learned from the intended failures, and from the history of the discipline it is known that even more may be learned from those which are unintended. In the lecture there has been something for everyone and much to think about.

'On behalf of the British Geotechnical Society I have the honour to propose a hearty vote of thanks to Professor Scott for a most memorable lecture.'

The Rankine Lecture

The twenty-eighth Rankine Lecture of the British Geotechnical Society was given by Professor H. B. Sutherland at Imperial College of Science and Technology, London, on 28 March 1988. The following introduction was given by Professor A. McGown.

'When Professor A. W. Skempton introduced the first Rankine Lecturer in 1961, he reported that it was the decision of the British National Committee of the International Society of Soil Mechanics and Foundation Engineering to have an annual lecture given by a person of distinction in the field of soil mechanics. It was further decided to name the lectures after Professor Rankine, the first British engineer to make a published contribution to the subject. Thus, it is most appropriate that the twenty-eighth Rankine Lecturer is Professor Hugh Brown Sutherland, a most distinguished soil mechanics engineer and teacher who has much in common with Rankine and with the first Rankine Lecturer, Professor Arthur Casagrande. In the first case, Hugh Sutherland was, until very recently, Head of the Department of Civil Engineering at the University of Glasgow—a position once held by Rankine. In the second, he was both pupil and research associate to Casagrande when at Harvard University.

Hugh Sutherland is a man whose interests range well beyond soil mechanics. In his younger days he was a keen sportsman, gaining his blue at soccer. He has also had a lifelong interest in athletics, rugby, and that uniquely Scottish sport: playing the bagpipes.

In 1936, Hugh Sutherland became an indentured civil engineer to the City Engineer of Glasgow and undertook his professional training at the Royal Technical College, now the University of Strathclyde. In 1940 he moved to Oscar Faber and Partners in London, but returned to Glasgow in 1942 to lecture to Army Officer Cadets at the University of Glasgow. In 1946 he went to Harvard to undertake his Master's degree under the instruction of Professors Terzaghi and Casagrande. Following this, he worked as a research associate for Professor Casagrande, then went on to Canada to work with Dr Legget at the National Research Council.

In 1948 he once again returned to Glasgow University, as a Lecturer in the Department of Civil Engineering. Over the following years, he attained recognition for his teaching, research and administrative contributions to the Department and in 1966 he was appointed to the Cormack Chair of Civil Engineering, which position he retained until December 1986.

During this period he undertook wide ranging research and consultancy work, from traffic vibrations in Canada to heavy foundations for nuclear power stations in Japan. His most recent research interests have been in the fields of tunnel linings, reinforced earth and anchorages for offshore structures.

Hugh Sutherland has, nevertheless, always managed to participate fully in the activities of the professional engineering societies and institutions in the United Kingdom and overseas. He was the founder Chairman of the Scottish Geotechnical Group. He has served on the Councils of the Fellowship of Engineering, the Institution of Structural Engineers and the Institution of Civil Engineers. From 1982–84 he was a Vice-President of the ICE. His services to overseas communities through UNESCO and other bodies have also been widely recognized, including the award of an Honorary Citizenship of the city of Winnipeg.

With this extensive background of professional involvement, practical experience, research and teaching, I am sure that we shall find Professor Sutherland's lecture on the uplift resistance of soils to be most interesting and informative. It is therefore with much pleasure that I invite Professor Sutherland, on behalf of the British Geotechnical Society, to give the Rankine Lecture for 1988.

Professor H. B. Sutherland

Sutherland, H. B. (1988). *Géotechnique* **38**, No. 4, 493–516

Uplift resistance of soils

H. B. SUTHERLAND*

The uplift resistance of soils is normally associated with the design of foundations for structures such as transmission line towers where an adequate margin of safety is required against failure. This represents the design for safety case. The Author describes a scheme for the water coolant system at Sizewell A nuclear power station, where eight shafts were jacked up from tunnels to the sea-bed through 6 m of cohesionless soil. This represents the design for failure case. Existing theories based on the design for safety case were inadequate and the required jacking pressures, to overcome the uplift resistance of the soil, were successfully determined from laboratory model tests where dimensional similarity was satisfied between model and prototype. The Sizewell and field test results were compared with predictions from centrifuge tests and finite element predictions which were carried out later for the design for safety case. For cohesionless soils the centrifuge tests gave similar results to the model and field tests, whereas the finite element predictions grossly underestimated the ultimate uplift resistance for other than loose sands. A comparison is also made between predictions of the uplift resistance of cohesive soils from model tests and from theoretical studies, including finite element methods. Recent field tests on belled-out piers in cohesionless soils are compared with the results from the original Sizewell model tests. Attention is drawn to the increasing number of specializations within geotechnical engineering, the fragmentation that could develop between geotechnical engineers because of these specializations, and the necessity of keeping an open mind on, and a broad approach to, the solution of problems in geotechnical engineering.

KEYWORDS: anchors; failure; model tests; field tests; centrifuge modelling, finite elements.

La résistance des sols à l'arrachement est généralement associée aux fondations de constructions telles que les mâts des lignes de transport d'énergie, où on a besoin d'une marge suffisante de sécurité vis à vis de la rupture. Ceci correspond à un projet avec coefficient de sécurité. L'article décrit un système de refroidissement à l'eau employé à la centrale nucléaire de Sizewell A, dans lequel huit puits furent soulevés par moyen de vérins à partir de tunnels jusqu'au fond de la mer à travers 6 mètres de sol sans cohésion. Ceci représente un exemple d'une construction à la rupture. Les théories qui existaient déjà et qui étaient basées sur un project avec coefficient de securite se montraient insuffisantes, tandis que la pression des vérins nécessaire pour maîtriser la résistance du sol à l'arrachement furent effectivement déterminée à partir d'essais sur modèle en laboratoire, dans lesquels les dimensions du modèle et du prototype étaient pareilles. Les résultats des essais effectués à Sizewell et sur chantier furent comparés avec des prédictions d'essais en centrifuge aussi bien qu'avec des prédictions d'eléments finis obtenues plus tard à pour des projects avec coefficient de sécurité. Dans le cas des sols sans cohésion les essais en centrifuge produirent des résultats semblables à ceux des essais sur modèle et sur chantier, mais au contraire les prédictions aree'éléments finis sous-évaluèrent énormément la résistance limite à Varrachement pour des sables autres que les sables meubles. Une comparaison est aussi faite entre des prédictions de la résistance à l'arrachement de sols cohévents sur la base d'essais sur modèle et sur chantier et d'études théoriques, y compris des méthodes à éléments finis. Des essais récents effectués sur des pilers évasés dans des sols sans cohésion sont comparés avec les résultats des essais sur modèle originaux effectués pour Sizewell. L'article souligne le risque de fragmentation couru par le génie géotechnique en raison d'une augmentation des spécialisations et attire l'attention sur le fait que les ingénieurs doivent avoir l'esprit ouvert et s'occuper de façon compréhensive des problèmes auxquels cette technologie fait face.

A wide range of topics have been the subject of previous Rankine Lectures and, while I have had diverse interests in geotechnical engineering over the years, I decided to choose a subject which has not been previously dealt with and in which we, in Glasgow, have had a continuing interest over the last 25 years or so.

There are many aspects of the uplift resistance of soils. All cannot be covered in a lecture, no more so than could all the aspects of bearing

* University of Glasgow Trust.

capacity be covered in one paper. I intend to limit my presentation to foundations embedded in cohesionless and cohesive soils. I am excluding grouted or other forms of anchor; topics which have been dealt with at length in the literature.

As far as the uplift resistance of soils is concerned, I propose to consider the experimental work on models which was initially carried out before the development of centrifuge and finite element techniques, and to make comparisons between the results from all these methods. I also wish to introduce the unusual concept of designing to ensure failure as against our normal approach of designing to ensure safety.

Scott (1987) suggested that there were three components of failure on which information is required. These are: the mechanism of failure; the properties of the materials involved; and the subsequent analysis. If the mechanism of failure is not clear, then prototype or model tests are needed for the formulation of design procedures.

Scott quoted Sir Benjamin Baker's remark with respect to retaining walls that 'if an engineer has not had some failures it is merely evidence that his practice has not been sufficiently extensive'. That is the kind of remark to which we would all, no doubt, subscribe provided it happens to somebody else.

However, there are few opportunities of inducing full-scale failures. One can recall Terzaghi recounting how he had deliberately produced a slope failure in the construction of a dam in India. Arthur Casagrande induced failures in the preliminary excavations for the construction of the Winnipeg Floodway and subsequent slope analysis had great difficulties in confirming that the failure had actually occurred at a factor of safety of unity. Following their example, in 1948 I took the opportunity of excavating vertical slopes to failure in an overburden consisting of varved clays (Steep Rock Iron Mine in Canada) in order to find the undrained shear strength of the clay. However, in my youth and enthusiasm, I was at that time unaware of the potential legal risks of causing a full-scale failure, a factor which often precludes this type of site investigation.

Scott (1987) reviewed methods of analysing actual and potential failures and pointed out the empiricism that is often necessary in formulating any analytical or finite element solution. He posed the question as to whether more failures are required through field or model tests to ensure that money is not being spent unnecessarily.

The study of premeditated or induced failures proposed by Scott and others is intended to give an improved understanding of the mechanism of failure, leading to a more satisfactory approach to a design for safety. It is most unusual, however, to have to design for failure—the success of the operation being dependent on that failure occurring. In the case of designing for failure we have to sweep away the conservatism that often shrouds our design activities and behind which we often shelter with hidden factors of safety.

UPLIFT RESISTANCE OF COHESIONLESS SOILS

Engineers would normally associate resistance to uplift with, for example, the design of pylon foundations. This is the design for safety case.

My first encounter with the uplift problem was in connection with the design and construction of the water coolant intake system at the Sizewell A nuclear power station on the Suffolk coast. It was certainly not a design for safety problem. The scheme involved the construction of four tunnels driven from the shore under the North Sea. The tunnels were 3·35 m dia. and were formed from cast iron segments. They were driven through cohesionless soil under compressed air conditions.

Two of the tunnels were for the water intake, and two for the discharge. The position of the intake was approximately 600 m beyond the low-water mark. The depth of water at the intake position was 6 m at mean tide and the sand cover over the tunnel was fixed at approximately 6 m. A greater cover would have increased the compressed air pressure required and less cover would have increased the risk of a blow at the tunnelling face.

One problem in the design of the scheme was the form of the connection between the tunnels and the sea-bed. A conventional solution of sinking shafts or caissons from which the tunnels could be driven was very much at the mercy of the elements. Consideration was therefore given as to whether a large proportion of the work necessary to achieve a break-through to the sea-bed could be carried out from within the confines of the tunnels. The feasibility of jacking a shaft from the tunnel to the sea-bed was investigated and it was shown that a satisfactory water inlet could be achieved.

Figure 1 shows the design concept. It involved raising two shafts each of 2·39 m dia. from each of four tunnels; a total of eight shafts. For a variety of reasons, solid caps to the shafts were selected and the problem became one of assessing the uplift resistance of a circular footing being pushed up to the sea-bed instead of the more conventional problem of a foundation being pulled out of the ground.

This was the design for failure case. The hidden factor of safety that often applies in the design of conventional foundations under compressive or

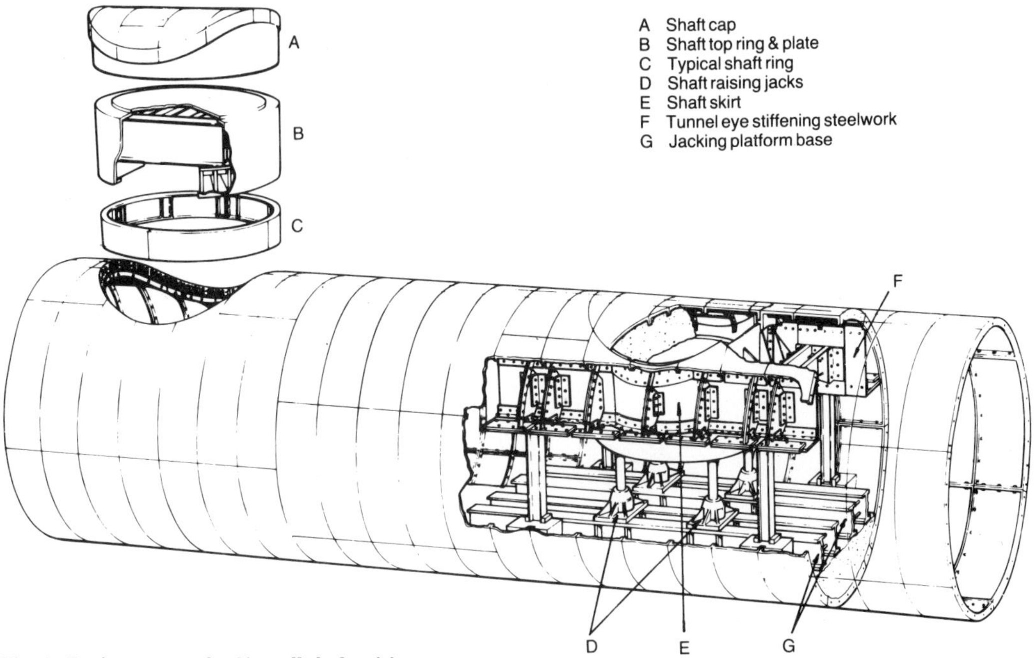

Fig. 1. Design concept for Sizewell shaft-raising

tensile loads would be a positive embarrassment, and a critical factor in the design was to determine and provide adequate jacking capacity to ensure failure. When the design was being considered there was practically no literature on uplift resistance. What little there was concerned the design of footings under uplift loads, i.e. the design for safety case. It is interesting that this topic is still hardly mentioned in modern texts.

The one reference in a text (Teng, 1962) outlined two methods which are shown in Fig. 2. The cone method gave the uplift resistance as the weight of soil contained within an arbitrarily defined surface, any frictional resistance being neglected. An apex angle θ of 60° was stated to be potentially unsafe. In the cylinder method failure was assumed to occur on a cylindrical surface equal in diameter to the foundation, with friction being mobilized along that surface. This was stated to be the lower limit and very conservative.

A method published by Balla (1961) is also shown in Fig. 2. He assumed a failure surface which was circular, starting with a vertical tangent to the base slab and curving outward from the vertical axis to meet the ground surface at an angle of $(45° - \phi/2)$. He based this failure surface on a limited number of model pull-out tests in an air dried sand with ϕ equal to 36° and 38°. The total uplift resistance was taken as the weight of soil contained within the failure surface plus the frictional resistance developed over that surface.

Rocha (1957) in a celebrated paper drew attention to the potential advantages of the use of

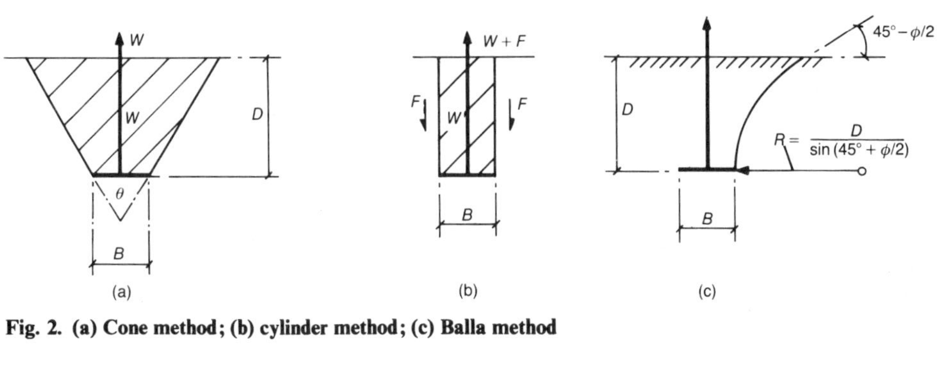

Fig. 2. (a) Cone method; (b) cylinder method; (c) Balla method

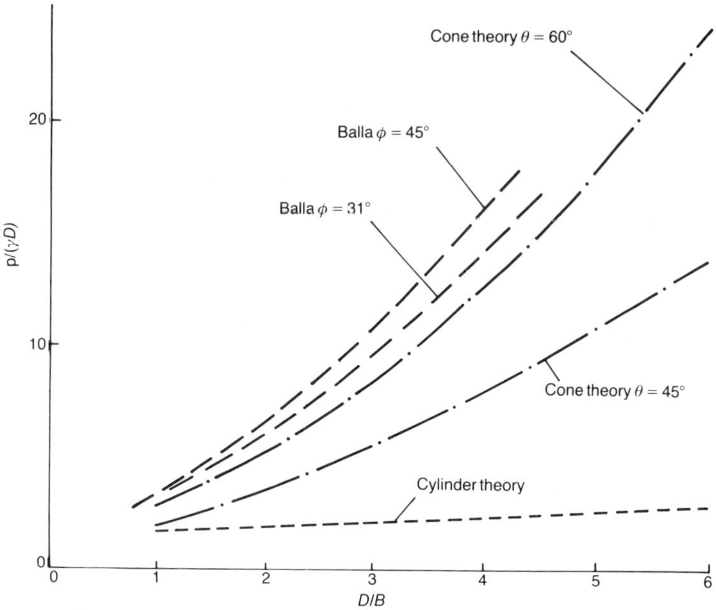

Fig. 3. Dimensionless plot of $p/(\gamma D)$ against D/B for cone, cylinder and Balla (1961) methods

models in soil mechanics and pointed out the limited use made of them compared with the use of models in hydraulic and structural problems. For model tests to be valid, dimensional similarity must be established between the model and

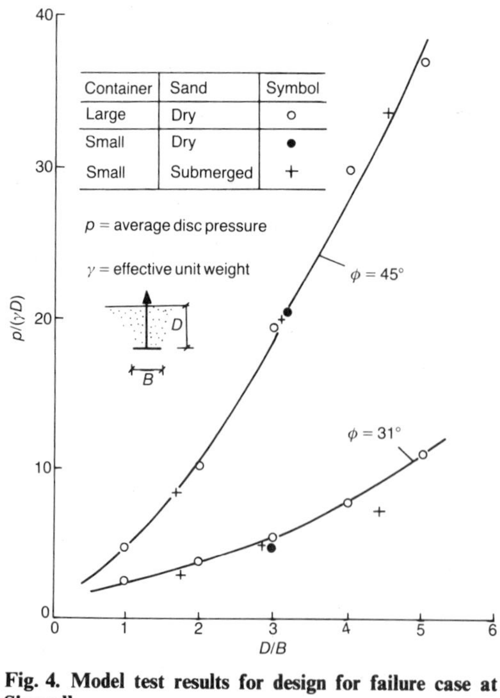

Fig. 4. Model test results for design for failure case at Sizewell

the prototype, and this was investigated, (as described by Sutherland, 1965) for the case of the cohesionless soil at Sizewell. The following simple relationship between dimensionless groups was obtained

$$\frac{p}{\gamma D} = f\left[\frac{D}{B}, \phi\right] \tag{1}$$

where p is the jacking pressure at the closed end of the vertical shaft, γ is the effective unit weight of the cohesionless soil, D is the depth of soil overburden above the closed end of the shaft, ϕ is the angle of shearing resistance of the sand and $P = (p\pi/4)B^2$ is the total jacking force representing the uplift resistance of the soil.

Figure 3 shows the predictions for the cone, cylinder and Balla theories plotted in these dimensionless terms. These methods were concerned with the design for safety case. The variations in predictions demonstrated their inadequacy for the design for failure case at Sizewell.

The design problem had to be solved quickly. There was no time for an extensive theoretical or experimental study. A series of 45 laboratory model tests was carried out whereby discs of different diameter B were pulled out from different depths D, from sand placed in very loose and very dense conditions. The corresponding ϕ values were 31° and 45° respectively. The tests were carried out on dry and saturated sands and covered the range of D/B values and sand conditions anticipated at the site. The results are

Fig. 5. Jacking section being lowered into trial excavation

shown in Fig. 4. The capacity of the jacking system used to raise the shafts was determined from these model tests and provision was made for jacking the shafts through a very dense sand. Having determined the jacking capacity, the design and construction could proceed.

A total of eight shafts had to be raised from the tunnels and before these operations, a trial shaft was raised from an excavation on shore to develop the necessary techniques. Fig. 5 shows the jacking section being lowered into the trial excavation and Fig. 6 shows the jack in position in the tunnel.

Figure 7 gives a view looking vertically up a completed shaft. There was always the possibility of inadequate jacking capacity having been provided and the blank flanges shown would have permitted action to be taken to facilitate jacking (e.g. jetting) if the soil and other conditions had been different from those anticipated. However, as the tunnels were being driven, in situ samples of the sand were tested in order to assess the ground conditions at the shaft raising locations. Consolation was obtained from the thought that there would be less resistance on forcing a 2·39 m diameter shaft upwards than by pushing a 3·35 m diameter tunnel bodily downwards.

The field jacking results are shown on Fig. 8. Shaft 1 was the onshore trial. Shafts 2, 3, 4 and 5 were grouped in the intake area. The results from these four shafts were consistent and indicated a ϕ value in this area of about 42° based on the model test results. Shafts 6, 7, 8 and 9 were

Fig. 6. Shaft-raising jacks in position in tunnel

Fig. 7. Vertical view up a completed shaft

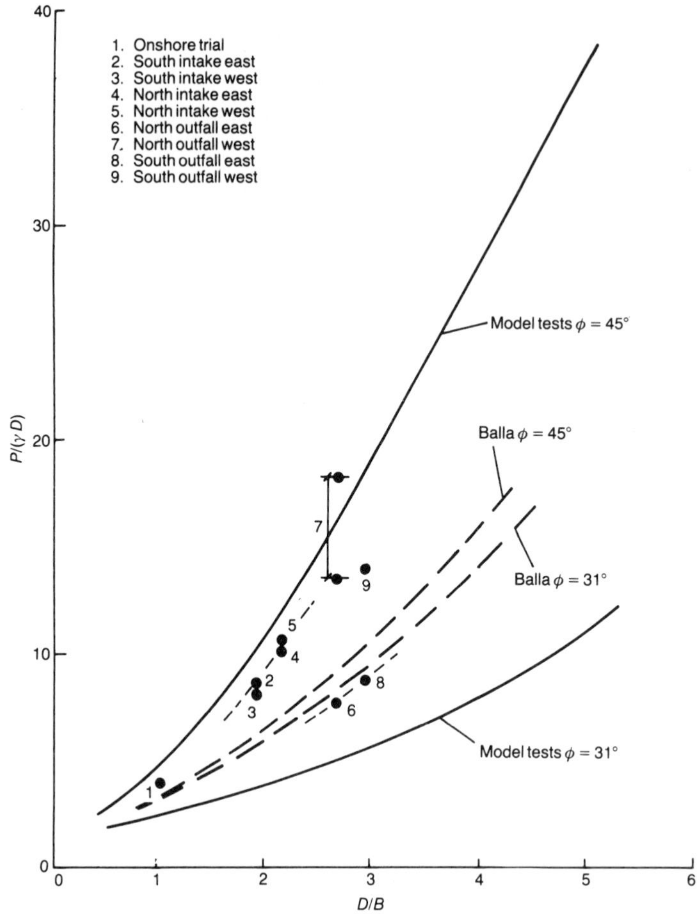

Fig. 8. Comparison of model test results, field results, and Balla (1961) theory

grouped in the outfall area. At this location the in situ tests indicated that the sand was less dense. Shafts 6 and 8 gave reasonably similar results. Shafts 7 and 9 required jacking pressures that were greater than would have been expected. However, during tunnelling at the West Intake a 'blow' occurred and approximately 20 tonnes of cement had to be injected. This grouting obviously affected the uplift resistance in the area of these two shafts, and demonstrated the unexpected hazards that can arise.

Also shown in Fig. 8 are the predictions from the Balla theory for the ϕ values in the model tests. These show that the uplift resistance pressures from Balla for $\phi = 31°$ are about 90% of those obtained for $\phi = 45°$. This small range implies that the frictional properties of the sand have little effect on the uplift resistance, in contrast to the results from the model tests. The Balla theory was the most recent at the time of the investigation, but the model studies and site tests indicated that this theory underestimates the loads mobilized in dense cohesionless soils. This would constitute an error on the safe side for pylon foundations, and an error on the unsafe side for shaft raising. For loose cohesionless soils the uplift resistance would be overestimated giving safe errors for shaft raising and unsafe errors for pylon foundations.

The technique of shaft raising from tunnels was successful at Sizewell and the work was carried out at substantial savings in cost compared with conventional methods. Model tests with the models observing dimensional similarity between model and prototype, formed the basis of assessing the uplift resistance of the sand overlying the tunnels. It is interesting to note that, when discussing the use of model tests in soil mechanics problems, Roscoe (1968) pointed out the significance of the dimensional analysis approach used in the Sizewell tests. Analytical methods of assessing uplift resistance available at the time were based on considerations of design for safety. They were inadequate for the prediction of actual ultimate uplift resistance. The shaft raising operations gave nine full-scale tests to failure against which predictions from any future theoretical or experimental work could be assessed.

This first encounter with the problem of assessing the uplift resistance of cohesionless soils indicated that there was no satisfactory theory due to the difficulty of defining the shape of the failure surface. This difficulty has also bedevilled more recent attempts to develop a satisfactory theory.

Constraints of time meant that the model tests for Sizewell had to be restricted to the cases of loose and dense sands. El-Rayes (1965) extended this work to a range of sand density and ϕ values with a view to ascertaining a failure surface shape that was universally applicable. From his tests he proposed a straight line failure surface but this had limited success.

The work in connection with Sizewell was carried out in the range of shallow anchors. The categories of shallow and deep anchors in cohesionless soils will be defined and dealt with later in the Paper.

UPLIFT RESISTANCE OF COHESIVE SOILS

The success of the shaft raising operations in cohesionless soils at Sizewell led to an investigation of the uplift resistance of cohesive soils. The principal interest was in shaft raising where failure had to be obtained, but there was also potential application to the design for safety case involving pylon foundations and also to the emerging field of anchored foundations for offshore structures.

By the time work commenced in Glasgow, a number of theories had been advanced for cohesive soils. Compared with cohesionless soils there was a clearer definition between shallow and deep anchors. With shallow anchors, any variation in the anchor depth has an appreciable influence on the ultimate uplift resistance. For deep anchors, the anchor depth has little influence on the uplift capacity and the failure is a local failure. The problem is simpler than for cohesionless soils in that the concept of general and local failure is similar to that for foundations imposing compressive loads on cohesive soils. In bearing capacity problems, that capacity is expressed in terms of a factor N_c. In uplift capacity problems, the corresponding factor is F_u, where

$$F_u = \frac{p_u - \gamma D}{c} \tag{2}$$

Model tests for cohesive soils similar to those carried out for cohesionless soils were considered and Davie & Sutherland (1977) examined the similarity conditions required between model and prototype for cohesive soils. A number of soil and size parameters were considered; dimensionless groups of these parameters were established and their relative importance ascertained. The principal dimensionless groups to be satisfied can be expressed as

$$\frac{p}{\gamma D} = f\left[\frac{c}{\gamma D}, \frac{D}{B}\right] \tag{3}$$

or

$$\frac{p}{\gamma B} = f\left[\frac{c}{\gamma B}, \frac{D}{B}\right] \tag{4}$$

In the cohesionless soil case the factor ϕ which influences the shear strength is dimensionless, whereas the undrained shear strength of a cohesive soil c has dimensions and this, amongst other factors, makes a substantial difference to the similarity conditions to be satisfied. If dimensional similarity is to be maintained between model and prototype, then a much smaller value of c than exists in the prototype must be used in the model.

Theories have been developed for both shallow and deep anchors, and the predictions from these theories can be considered. Balla's theory could also be applied to cohesive soils. Other theories have been presented by Mariupol'skii (1965); Vesic (1971 and 1972); Matsuo (1967); and Meyerhof & Adams (1968). Each of these authors made differing assumptions as to the shape of the failure surface or the form of failure. Fig. 9 compares the theoretical predictions for shallow anchors. Three of the theories give identical curves for F_u against D/B. The Balla and Matsuo theories express F_u as a function of $c/(\gamma B)$ as well as D/B. With the exception of Balla the theories are in relatively good agreement.

Unlike the shallow anchor theories, the deep anchor theories are not directly comparable since each includes different parameters in its solution. Mariupol'skii took account of the compressibility of the soil through the term ω which is a function of the reciprocal of the coefficient of volume compressibility. Vesic (1972) also considered the influence of soil compressibility using I_r, the rigidity index, and Δ, a measure of the average volume strain in the plastic zone. Vesic defines I_r as G/c, the ratio of the shear modulus of the soil to its cohesion. The modification by Ali (1968) can be taken into account in computing values of F_u from Vesic's deep anchor theory. Meyerhof & Adams (1968) stated that they use the bearing capacity factors $N_c = F_u$ postulated by Meyerhof (1951), but in fact they use only the value 9·34 pertaining to a rigid incompressible material, and take no account of the compressibility of the clay

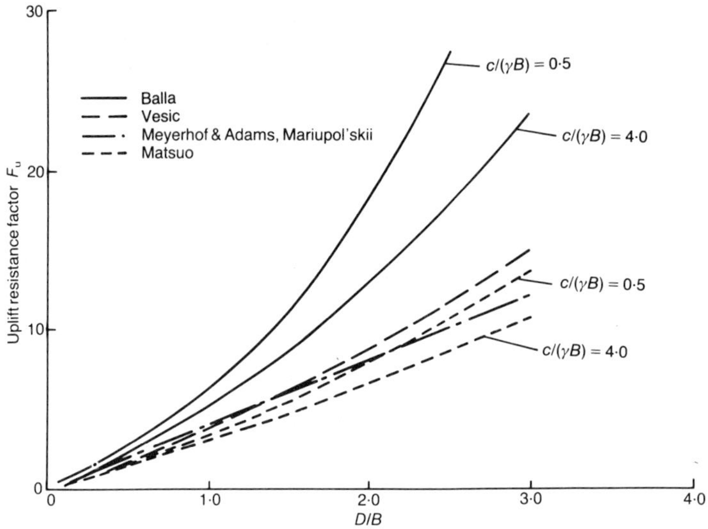

Fig. 9. Comparison of shallow anchor theories for cohesive soils

which Meyerhof (1951) indicated could reduce the bearing capacity factor to 7·0.

As shown in Table 1, a comparison can be made between the F_u values predicted from the deep anchor theories for the cases of rigid incompressible soils and nonrigid compressible soils by choosing appropriate values for the soil constants. The comparison indicates that the degree of rigidity or compressibility of the soil can have a significant effect upon the uplift resistance of deep anchors predicted by Vesic and Mariupol'skii, an effect which Meyerhof & Adams did not take into account in their approximate theory, but which is indicated by Meyerhof's earlier work.

In order to compare the predicted uplift resistance values from the various theories, a series of model tests were carried out by Davie (1973). Clays of low shear strength were required to provide $c/(\gamma B)$ and $c/(\gamma D)$ values in the models corresponding to those existing in typical prototypes. No naturally occurring clays were available to satisfy these conditions, and an artificial clay designated glyben was used for most of the tests. Glyben is a mixture of sodium bentonite and glycerine. Undrained tests showed this material to be almost completely frictionless and its advantages as a laboratory material are

(*a*) different strengths can be readily obtained by varying the proportions of the bentonite and glycerine
(*b*) it is completely insensitive to handling
(*c*) there is negligible evaporation of the glycerine at room temperatures.

The latter two properties permit repeated tests using the same batch of material. Disadvantages of glyben are that pore-water pressure measurements are precluded and that samples cannot be sedimented from a slurry and must be formed by compaction.

Sixty-five model tests were carried out by Davie (1973) using different D and B values giving a range of D/B from 0·4 to 8·0. Both push-out and pull-out tests were used. Pull-out tests simulate pylon foundations—push-out tests apply to

Table 1. Values of F_u from deep anchor theories

Type	Theory	Uplift resistance factor, F_u	Values of parameters included in solution
Rigid, incompressible	Mariupol'skii	7·00	$\omega = 200$
	Vesic	9·94	$I_r = 300$, $\Delta = 0$
	Meyerhof & Adams	9·34	
Nonrigid, compressible	Mariupol'skii	4·60	$\omega = 25$
	Vesic	5·15	$I_r = 10$, $\Delta = 0{\cdot}02$

the shaft-raising problem. Two methods of preparation for the glyben test samples were used. In the first method, the material was kneaded manually in layers. These samples were termed 'layered'. Laboratory vane tests were carried out on the material at various stages of compaction which indicated that a consistent shear strength could be obtained throughout a sample. However, tensile strength tests indicated that the bond in tension between the kneaded layers was considerably weaker than in equivalent samples of unlayered material. To remove this weak tension bonding effect, a second method of compaction was devised in which the samples were prepared outside the container by kneading together small balls of material so that there was no preferred orientation. These samples were termed 'non-layered'. Horizontal and vertical coloured marker layers were placed in some samples before testing and the deformation and cracking patterns observed in the samples at ultimate uplift resistance gave guidance on the failure mechanism. Typical cases of a shallow anchor, a deep anchor, and an anchor falling between these categories can be considered. No suction was allowed to develop in the cases presented.

Figure 10 shows the results for a shallow anchor case with $D/B = 1{\cdot}5$. Failure occurred by a combination of yielding above and below the anchor due to excess shear and tensile stresses, and cracking in the material. Shallow anchor theories predict failure by shear failure alone, and do not take into account tension cracking and anchor movement. This is a general type of failure since the yielded material extends from the anchor to the surface. Fig. 11 shows the failure behaviour for a deep anchor with $D/B = 4{\cdot}5$. A length of white thread positioned around the area of the clay which was visibly affected by the anchor movement, shows the limited extent of the local failure, i.e. a failure restricted to a region in the vicinity of the anchor plate only and not extending to the surface of the sample.

Figure 12 shows the failure behaviour for the intermediate case with $D/B = 3$. Some cracking is apparent, but considerably less than in the shallow case. There is also less displacement of the soil at and near the surface than in the shallow anchor case, but more than in the deep anchor case. Ultimate failure at this intermediate case consists of a combination of general and local failure characteristics.

Figure 13 shows plots of F_u against D/B for representative theories, for the Glasgow model tests and also for the limited number of field tests available at the time, notably from Meyerhof & Adams (1968). The latter theory gives $F_u = 9{\cdot}34$ for a deep anchor, while the Vesic approach (by taking soil compressibility into account by means of the rigidity index) gives F_u values ranging from 5·1 to 9·9 for compressible to incompressible clays. The Glasgow test results are shown for the layered and non-layered tests. The layered material possessed less structural resistance in the vertical plane and also had a weak tension bond between the layers and this led to a lesser uplift

Fig. 10. Deformation and cracking at ultimate uplift resistance at $D/B = 1{\cdot}5$ for a shallow anchor case

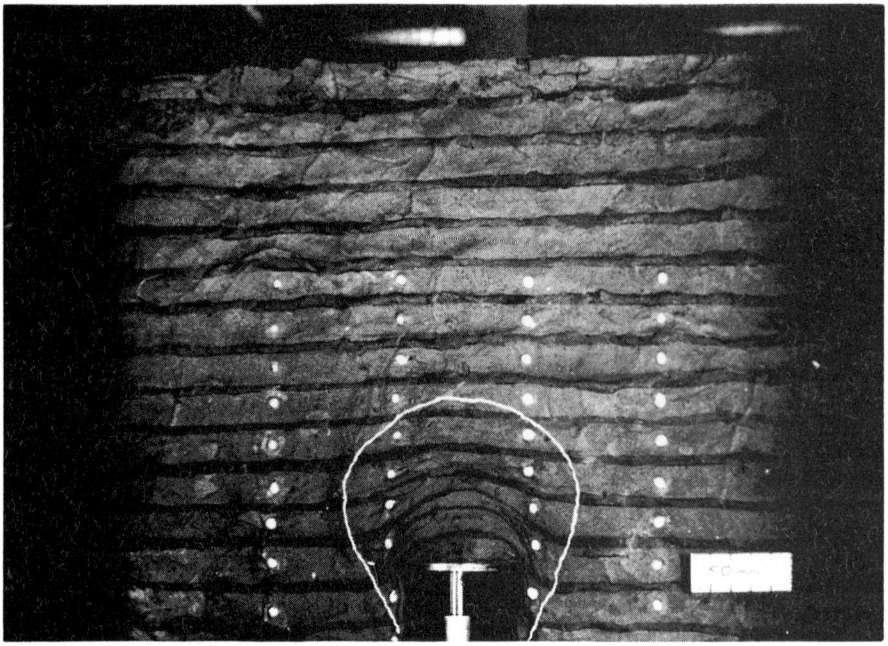

Fig. 11. Deformation and cracking at ultimate uplift resistance at $D/B = 4{\cdot}5$ for a deep anchor case

resistance than for the non-layered material. The influence of these factors was also observed in the displacement measurements (Davie & Sutherland, 1977).

The transition from shallow to deep anchors for both theory and experimental work can be seen. For the shallow anchors the model test results indicate that the theories overestimate the uplift resistance. This is also apparent from the field test results, and has been noted by later investigators. For the Glasgow test material, the Vesic rigidity index was established as about 45

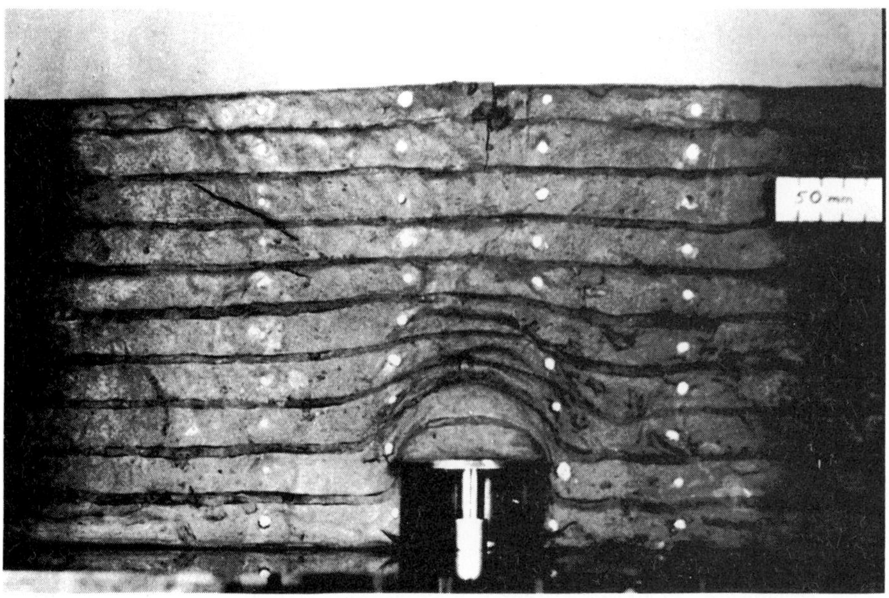

Fig. 12. Deformation and cracking at ultimate uplift resistance at $D/B = 3$ for an intermediate case

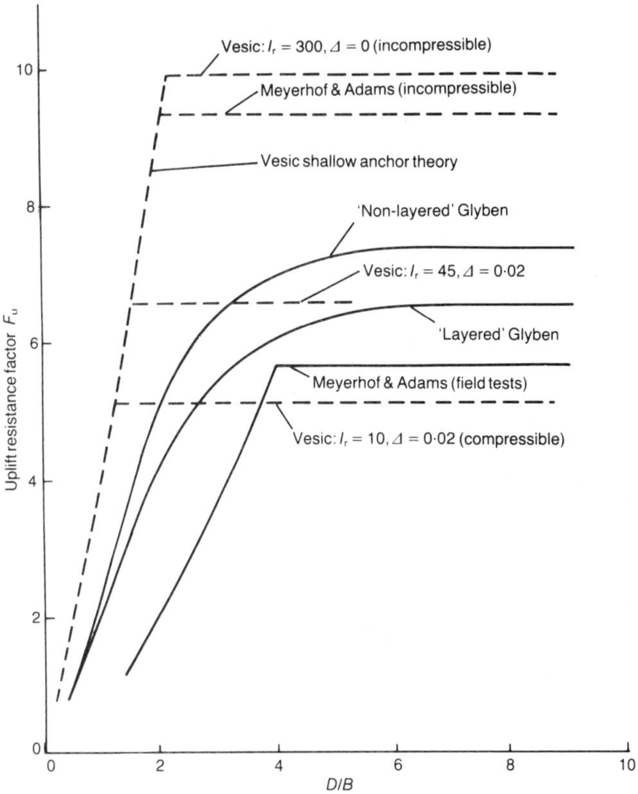

Fig. 13. Comparison of theoretical, model and field test results for cohesive soils

and the Vesic prediction for this value gave reasonable agreement with the Glasgow tests. The model tests therefore appear to support the Vesic theory and confirm that the compressibility of the clay has a significant influence on uplift capacity. Meyerhof & Adams noted that the majority of their field tests were at relatively shallow depths in stiff, brittle clays. They considered that the strengths mobilized in these were considerably less than the laboratory peak values and were possibly as low as the residual values. More field test results are required, however, and as yet there has not been an opportunity for shaft-raising in cohesive soils.

UPLIFT RESISTANCE OF LAYERED SOILS

At this stage in the Glasgow investigations, the primary interest was still in the shaft-raising aspects of the ultimate uplift resistance of soils. The techniques of shaft-raising through cohesionless soils, based on model tests, had been successfully carried out at Sizewell. A model study had also been carried out on clay. The question arose of the uplift resistance of a layered soil where tunnelling might be carried out in a clay overlain by sand, and shafts raised from the tunnel. There was also a developing interest in the uplift resistance of sea-bed anchors for use in the foundations of offshore structures. In this application, consideration could be given to an anchor installed in a surface clay layer subject to the limitation of a maximum value of approximately $9c$ in the uplift capacity of the anchor. However, this uplift capacity could be increased by placing an overlay of cohesionless material.

An experimental programme was carried out by Stewart (1985). Glyben was again used to simulate the purely cohesive soil. A shallow anchor with $D/B = 1{\cdot}5$ was first placed in the glyben and its uplift capacity determined. Sand overlays of equal thickness to the glyben layer, and three and five times that thickness, were investigated. The sand was placed at relative densities of 39% and 88·7%, i.e. at relatively loose and dense states. A plot of load against displacement is given in Fig. 14 and it can be seen that the cohesionless soil overlay significantly increases the ultimate uplift capacity of the anchor compared with its value when embedded

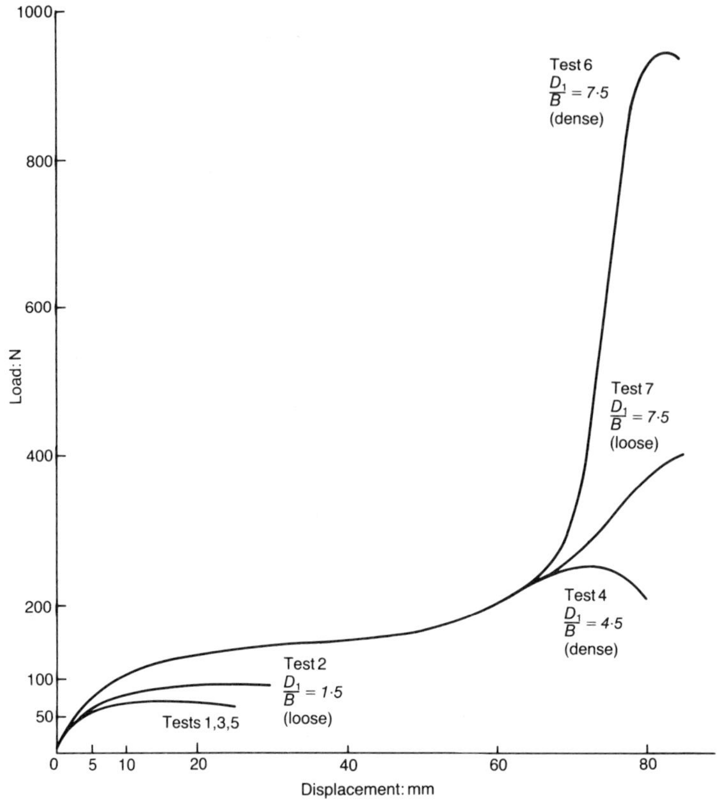

Fig. 14. Load–displacement curves to failure for layered soils

in clay alone. This increase in uplift capacity is due to two main factors. The first is the additional overburden pressure which converts the original shallow anchor with $D/B = 1{\cdot}5$ into a deeper anchor, albeit up to a limiting value of a maximum of approximately $9c$. The second is the mobilization of the frictional resistances of the overlays. The increases can be identified separately for the individual tests and can then be compared with the results from tests on cohesionless soil alone at corresponding D/B values and similar relative densities. Good agreement was obtained. What is also apparent is that a substantial displacement of the sand/clay interface must occur before any additional frictional resistance of the sand can be mobilized.

This work was primarily undertaken in connection with the shaft-raising problem, i.e. the design for failure condition. It yields useful information in principle and indicates that the contribution of the clay to the total uplift resistance would be relatively small for any appreciable thickness of overlying sand. In order to mobilize the frictional resistance of the sand, substantial displacement would be required but as this would probably be preceded by a punching shear failure in the clay the surface displacement would be reduced. However, the design for safety case is another matter, since displacement considerations become a limiting factor.

For the case of an anchor embedded in a clay layer overlain by cohesionless material, Fig. 14 indicates that the shear strength contribution to uplift from the overlying sand is not mobilized until large anchor displacements occur, a situation which would not normally be tolerated in practice. Therefore, there would be little, if any, real benefit to uplift capacity by placing a cohesionless material overlay on an anchor embedded in clay. If a sand overlay was to be used, a more sensible solution would be to place the anchor on the surface of the clay layer and then place the overlay of cohesionless material. By so doing the uplift resistances of the overlay would be mobilized immediately at small acceptable displacements.

MODELLING OF UPLIFT RESISTANCE OF CLAYS

When the Sizewell model tests on cohesionless soils were started, centrifuge testing methods and finite element methods were in their infancy.

Model testing, centrifuge testing, and finite element methodology each has its merits and one or other of them may be the most suitable approach under particular circumstances.

Davie (1973) in a study of cohesive soils, included some preliminary finite element work and also carried out an investigation into the relative roles of model and centrifuge testing in the determination of the uplift resistance of foundations embedded in clays. The cases of shallow and deep anchors can be considered separately (as reported by Davie & Sutherland, 1978).

For shallow anchors in clay, the theories of Matsuo (1967); Vesic (1971); and Meyerhof & Adams (1968) give reasonably similar predictions of the ultimate uplift resistance in terms of the weight of the volume of soil above an anchor, and of the shear strength of the soil developed over the surface of this volume. The contribution of soil weight can be calculated from these theories and is plotted against $c/(\gamma B)$ as the solid curve (a) in Fig. 15.

Dimensional similarity for shallow anchors will generally be possible if $c/(\gamma B)$ can be satisfied for both model and prototype, and solid curve (a) indicates a decreasing soil weight contribution to total uplift resistance as $c/(\gamma B)$ increases. A very stiff prototype clay with a $c/(\gamma B)$ value of 3 can be considered. If the prototype clay was used in the model, and with a model scaling factor of 20, the $c/(\gamma B)$ value for the model would be 60 and dimensional similarity between model and prototype would in no way be satisfied. However, solid curve (a) shows that the contribution of soil weight to total uplift resistance would be less than 10% for the cases of $c/(\gamma B) = 3$ and $c/(\gamma B) = 60$. No significant advantage would therefore be gained from obtaining dimensional similarity by using a very soft clay in the model. However, if the prototype clay is very soft giving a prototype $c/(\gamma B)$ of say 0·2, then if the prototype soil is used in the model the $c/(\gamma B)$ for the model would be 4·0 with a scaling factor of 20. Again, there would not be dimensional similarity between the model and the prototype, but in this case solid curve (a) indicates that the contribution of soil weight to total uplift resistance in the prototype will be 55% compared with approximately 5% in the model, a quite unsatisfactory outcome. In simple models, the low $c/(\gamma B)$ values of a very soft prototype clay can only be modelled by conducting tests with either very large models or with a soil with practically zero shear strength. Both choices are undesirable since very large models are normally not economically feasible and the failure characteristics of extremely soft clays can be quite different from those of stiffer clays because of their viscous nature. In this situation, recourse may have to be made to centrifuge tests or consideration might be given to surcharging the surface of the soil in the model.

Solid curve (b) in Fig. 15 shows the relationship between the ratio of model to prototype ultimate uplift resistance pressure and prototype $c/(\gamma B)$

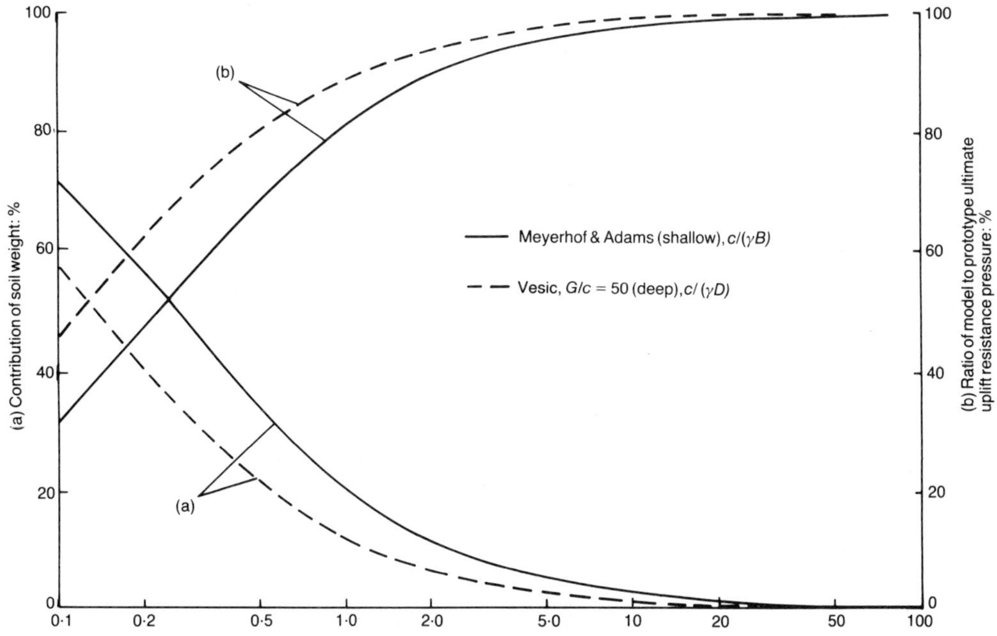

Fig. 15. Contribution of soil weight to total uplift resistance and the ratio of the model to prototype ultimate uplift pressure for the same soil in model and prototype and model scaling factor of 20

and indicates the extent of the discrepancies that arise from dissimilar prototype and model $c/(\gamma B)$ values. The curve indicates that if the prototype $c/(\gamma B)$ value is about 2 or above, the ultimate uplift resistance pressure obtained from the model will be at least 90% of the prototype value. However, for low prototype $c/(\gamma B)$ values, the model ultimate uplift resistance pressure can be as low as 30% of that of the prototype. In this case, the model pressure applied to prototype design would produce a very conservative design for footings, but could lead to appreciable under-provision of jacking capacity in a shaft-raising situation.

For deep anchors the total uplift resistance pressure is independent of the anchor diameter B and a more appropriate dimensionless factor is $c/(\gamma D)$. The importance of dimensional dissimilarity between prototype and model can be examined by considering Vesic's (1972) deep cavity theory as modified by Ali (1968) for deep anchors. The Vesic theory incorporated a rigidity index G/c and, using a typical rigidity index value of 50 and assuming the clay to be incompressible in the plastic zone, the relative contribution of soil weight to total uplift resistance pressure can be plotted against $c/(\gamma D)$ as shown by the dotted curve (a) in Fig. 15. The relationship indicates a decreasing soil weight contribution as $c/(\gamma D)$ increases. The dotted curve (b) in Fig. 15 shows the ratio of the model and prototype values of total uplift resistance plotted against $c/(\gamma D)$. For deep anchor prototype $c/(\gamma D)$ values greater than about 1, no significant advantage is achieved by obtaining a dimensionally similar model. However, for the low prototype $c/(\gamma D)$ values that would generally apply to soft prototype clays, a dimensionally dissimilar model could lead to considerable underestimation of the prototype uplift resistance pressure.

CENTRIFUGE TESTS FOR UPLIFT RESISTANCE

Reference has been made to situations where centrifuge tests would be more suitable than model tests when investigating the uplift resistance of cohesive soils. However, there are very few publications concerning the prediction of uplift resistance of soils using centrifugal testing. Ovesen (1981) investigated the uplift capacity of anchors embedded in cohesionless soils using both model and centrifuge tests. For cohesionless soils, one would not expect much difference between the results from dimensionally similar models and centrifuge tests, as the body forces have little influence according to the simplified dimensionless groups.

Ovesen's tests were made on a dense sand which was stated to have a relative density of 109%, and also on a loose sand at a relative density of 38%. Ovesen stated that he had difficulty in interpreting the triaxial test results on the sand and that the angles of internal friction of the sands were estimated using an empirical formula given in the Danish Code of Practice for Foundation Engineering. This formula gave 37·9° for the dense sand (which appears low in terms of the stated relative density) and 29·6° for the loose sand. The tests were carried out over a small range of D/B values up to $D/B = 3{\cdot}5$.

Ovesen concluded that his model and centrifuge test results compared well for the loose sand but that scale errors existed in the model tests results for dense sand which might lead to an overestimation of the uplift capacity of dense sands compared with the centrifuge results. He devised a formula for the uplift capacity of circular and square anchors on the basis of curve-fitting the centrifuge model test results, and not by assuming a form of failure surface as in previous theoretical solutions. He applied his formula to five field tests. The first of these consisted of pulling out a square plate buried to 0·75 m and inclined at 45° with $D/B = 1{\cdot}68$. The other four field tests were from Matsuo (1968) with very small D/B values ranging from 0·5 to 0·92. Ovesen mentioned the Sizewell paper on shaft-raising by Sutherland (1965) but did not compare the predictions from his empirical formula with the Sizewell field or laboratory results.

In Fig. 16 the Ovesen formula is shown, along with a comparison of its predictions with the Sizewell results. The prediction for the loose sand compares well with the Sizewell model results, as would be anticipated from Ovesen's model and centrifuge tests. For the dense sand condition, however, the Ovesen prediction underestimates the Sizewell values by about 20–25%. The Ovesen empirical formula is simple to apply and would be a good basis for the design for safety case, although it would underestimate the uplift resistance for the shaft raising case in dense soils. The formula applies over a restricted range of D/B up to about 3·5, but its derivation may have been adversely affected by the use of an empirical formula for obtaining the angles of internal friction. Judged in engineering terms, the centrifuge test results were in good agreement with the direct model tests carried out originally for the Sizewell investigation and the further model tests by later investigators.

FINITE ELEMENT METHODS FOR UPLIFT RESISTANCE

Finite element methodology has evolved over the last 25 years or so and Zienkiewicz (1987) has

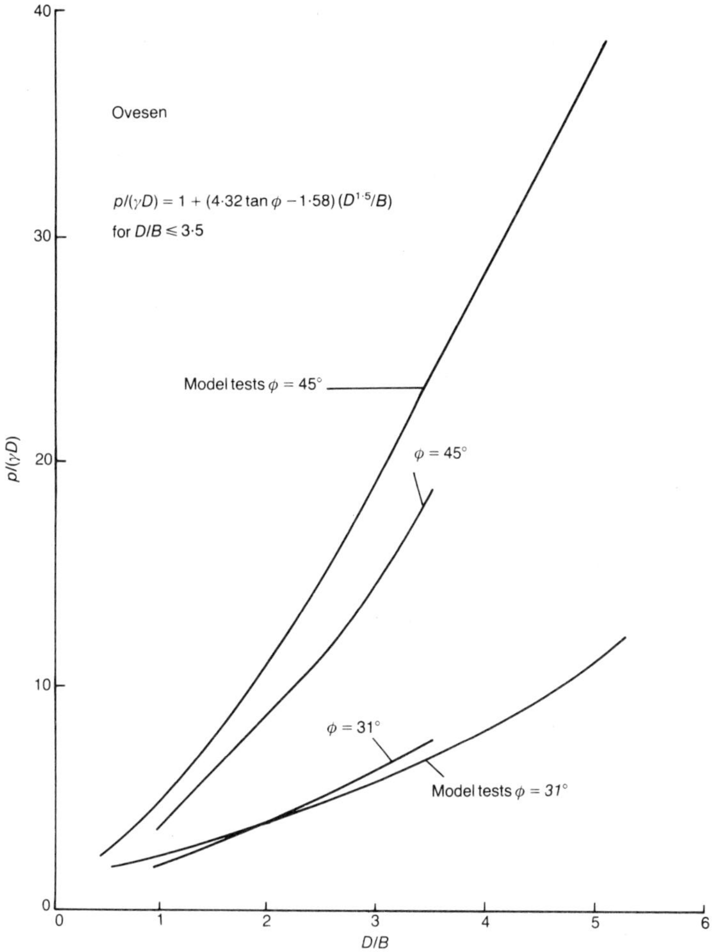

Fig. 16. Comparison of predictions by Ovesen (1981) with Sizewell model tests

given an excellent review of its development. He mentions the importance of comparing computational predictions of collapse with experimental evidence when initiating a new computational procedure. This point was also emphasized by Scott (1987) and was the basis of his plea for more tests to failure in models and prototypes. As far as soils are concerned, the finite element method has been successful in deformation problems where an assumption of linear response is reasonable. Where the soil behaves in a nonlinear fashion, as most soils do as they approach or reach failure, the problem becomes more difficult and in recent years there has been a great deal of work in developing constitutive relationships to more realistically represent the behaviour of soils.

Scott pointed out the empiricism which often has to be resorted to in the definition of mathematical models representing the behaviour of soils, and also the potential dangers of such an approach unless the predictions can be checked. Zienkiewicz was being realistic when he commented on the dangers of being lulled into a false sense of security by assuming that 'if the approach is by the finite element method it must be correct'. Practically every research student in Glasgow in recent years, and Glasgow is no exception, has felt obliged to devote at least one section of his thesis to finite elements. This is excellent from the training point of view, provided a realistic approach is taken to the predictions that result.

The attempts by the Glasgow research students to date have not been particularly successful when the finite element approach has been applied to the uplift resistance problem, and especially so when the predictions were compared with the field tests at Sizewell in the shaft-raising operations. However, other investigators have been active in this field.

Rowe & Davis (1982a) described a theoretical investigation on cohesionless soils in which consideration was given to the effect of anchor embedment, friction angle, dilatancy, initial stress state k_0 and anchor roughness for anchor plates with either a vertical or horizontal axis. Their theoretical solution was based on an elasto-plastic finite element analysis using a soil structure interaction theory presented earlier by their co-workers at the University of Sydney. The soil was assumed to have a Mohr–Coulomb failure criterion and either an associated or a non-associated flow rule. The predictions from the theory were compared with the results from vertical pull-out tests on anchors embedded in dry sand. The sand was essentially in a loose state with relative densities of 20% and 32%. The anchors had length to breadth ratios ranging from 1 to 9·75 and embedment ratios D/B of 1–8. The authors claimed reasonable agreement between their theoretical predictions and the results from their limited range of laboratory tests on the relatively loose sands. The theoretical results were then presented in influence or design charts to be used in hand calculations to obtain an estimate of anchor capacity for what were described as a wide range of geometries and soil types.

The authors illustrated the use of their design charts by a worked example to obtain the uplift resistance of a strip footing embedded in sand. The resulting calculated value is low and is not much greater than that obtained by using the original simple procedure of considering the uplift resistance to be due to the weight of the soil alone contained within an arbitrarily defined failure surface where any friction mobilized along the failure surface is neglected. In view of the low predicted uplift capacity from this worked example it is interesting to examine in principle the design chart arising from the finite element analysis. In order to make a comparison with the model and centrifuge test results, the Rowe & Davis approach can be expressed as

$$p/(\gamma D) = F_\gamma R_\psi R_K R_R \quad (5)$$

where F_γ is a basic anchor capacity factor and R_ψ, R_K and R_R are correction factors for the effect of soil dilatancy, initial stress state and anchor roughness respectively.

Figure 17 reproduces the plot of F_γ against ϕ for D/B values of 1–8. Fig. 17 shows that the predicted increase in F_γ is comparatively small (16–34%) for an increase in ϕ from 30° to 45° over the range of D/B values. These F_γ values are very low compared with the results of model and centrifuge tests. The correction factor R_ψ depends on the angle of dilatancy and varies from 1 to a maximum of about 1·4 for a recognized range of

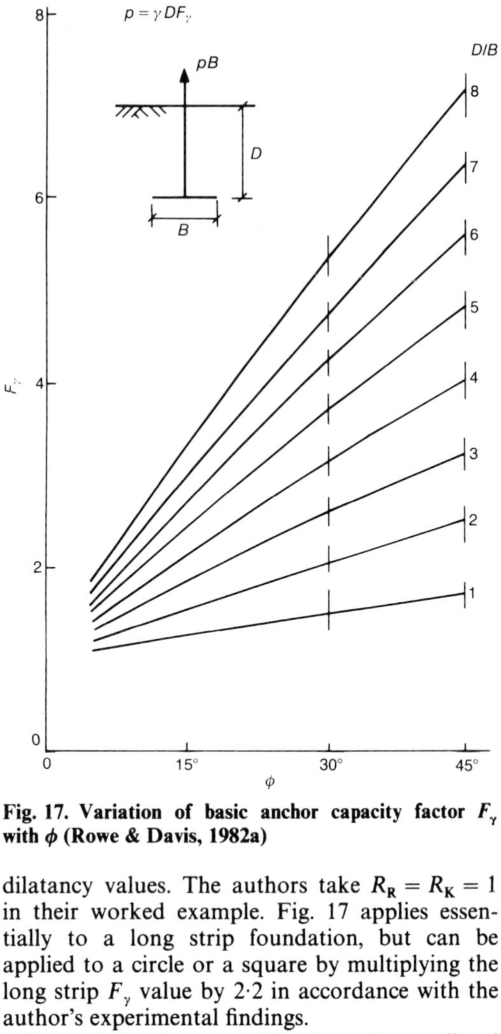

Fig. 17. Variation of basic anchor capacity factor F_γ with ϕ (Rowe & Davis, 1982a)

dilatancy values. The authors take $R_R = R_K = 1$ in their worked example. Fig. 17 applies essentially to a long strip foundation, but can be applied to a circle or a square by multiplying the long strip F_γ value by 2·2 in accordance with the author's experimental findings.

By applying these modifications, the predicted values of uplift resistance can be compared with the Sizewell test values for dense and loose sands using appropriate values for the angles of dilatancy. The comparison is shown in Fig. 18 This finite element approach by Rowe & Davis is intended for the design for safety case and gives very conservative predictions for sands other than those in a very loose state. Its use in predicting the uplift resistance in the design for failure case would have unfortunate consequences as far as the shaft raising operation is concerned.

Vermeer & Sutjiadi (1985) referred to the work of Rowe & Davis (1982a) and they also used a finite element analysis in their theory for the uplift capacity of shallow anchors embedded in sand. They assumed that the failure surface is inclined to the vertical through the edge of the anchor at an angle equal to the angle of dilatancy for the sand and claimed that their finite element

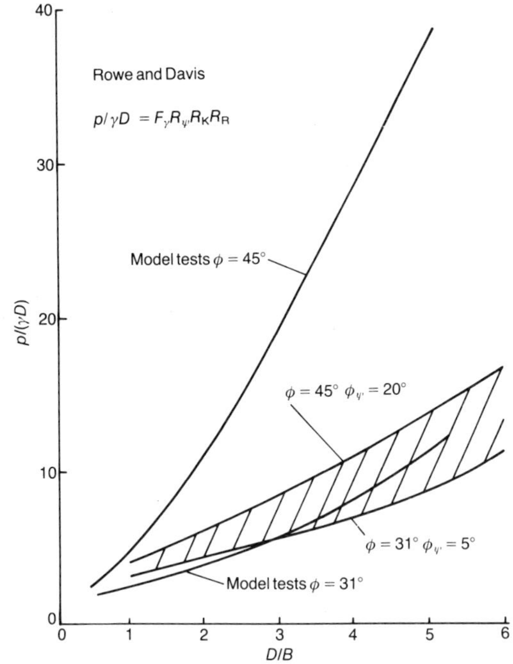

Fig. 18. Comparison of Rowe & Davis (1982a) predictions with Sizewell model tests

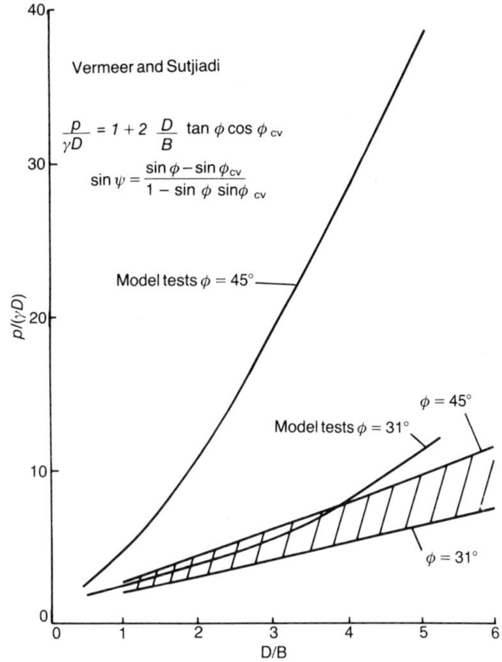

Fig. 19. Comparison of Vermeer & Sutjiadi (1985) predictions with Sizewell model tests

analysis validates this assumption. They derived the simple formula for $p/(\gamma D)$ shown in Fig. 19, involving D/B, the angle of internal friction and the friction angle at the critical state. Fig. 19 shows the predictions of uplift capacity for a circular or square anchor compared with the Sizewell model results. Vermeer & Sutjiadi stated that their formula is conservative and Fig. 19 confirms that their finite element approach is even more conservative than the finite element approach of Rowe & Davis and would have even more embarrassing repercussions if applied to the shaft-raising operation. Both these finite element based methods of predicting the uplift resistance of cohesionless soils made no comparison of their predictions with large scale available field test results. Both methods were aimed at the design for safety case and both gave very conservative predictions for other than very loose sands.

Rowe & Davis (1982b) published a companion to their paper on sands in which they applied their finite element approach to anchors embedded in clays. They referred to the finite element study on clays by Davie (1973) and developed it further. The Rowe & Davis approach placed emphasis on the effect of local yield on the load deflexion response, and in some cases this resulted in a displacement related practical failure load which might be less than or equal to the ultimate load actually measured or calculated. Any method of modifying the collapse load can only be obtained by some arbitrary rule and finite element analysis is no different from any other analytical method as far as making assumptions is concerned.

The assumption Rowe & Davis made was that the practical failure load is reached when the displacement is a selected multiple of the displacement which would have occurred if the conditions had remained entirely elastic. They chose 4 as the multiple and their practical failure load is described as the K_4 failure load. It corresponds to an apparent stiffness of one quarter of the elastic stiffness. A K_{10} failure load would correspond to an apparent stiffness of one tenth of the elastic stiffness. For shallow anchors the K_4 practical failure load is identical with the ultimate collapse load. For a very deep anchor the K_4 practical failure load is about 75% of the ultimate collapse load.

Figure 20 compares the previously presented theoretical and experimental results for clays with the Rowe & Davis K_4 and K_{10} practical failure loads. Both practical loads are within the range of those of the previous investigators although the Rowe & Davis K_4 failure load lies well towards the lower limit, i.e. the value for the Vesic compressible material. Rowe & Davis recommended that a factor of safety of 2·5 to 3 should be applied to their K_4 practical load. They also state

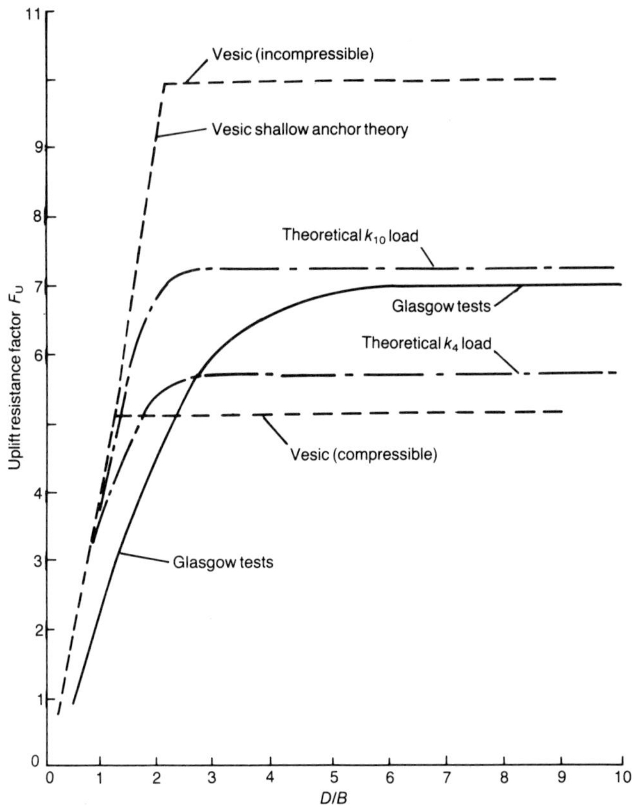

Fig. 20. Comparison of Rowe & Davis (1982b) predictions for clay with other theoretical and model tests

that for anchors with D/B greater than 3, the K_4 definition of failure gives a conservative estimate of the failure load (as is apparent from Fig. 20). Their application of the finite element method to clays gives conservative results which are, however, much less conservative than those from their application of the finite element method to sands.

FURTHER MODEL TESTS ON COHESIONLESS SOILS

The cases of shallow and deep anchors in clays have now been considered, but the work on sands has been concerned essentially with shallow anchors. Fadl (1981) carried out an extensive series of model tests on shallow and deep circular anchors. One hundred and thirty-seven tests were performed, covering a range of D/B values up to 25, on vertical and inclined anchors and with a sand at three relative densities of 25%, 50% and 85%, corresponding to angles of friction of 33·6°, 36·5° and 41·5° respectively. Push-out tests with displacement control were used. Push-out tests permit unrestricted measurement of surface movements, and displacement control tests permit the observation of pre- and post-peak loads and the corresponding displacements. Measurements at surface level were made in order to define the extent of the failure surface. The shape of the failure surface was investigated using sand layers separated by a tracer layer of cement. After an anchor had failed, the sand-bed was moistened to set the cement, and then excavated to expose and measure the shape of the failure surface. The change from the shallow to the deep anchor could be assessed from the surface displacement measurements and the change occurred at what was designated as H/B, the critical depth ratio.

Figure 21 is a plot of Fadl's results for $p/(\gamma D)$ against D/B for vertical circular anchors and for the three relative densities of sand tested. $p/(\gamma D)$ increased at an accelerating rate as D/B increased, until a change to a convex curve occurred at criti-

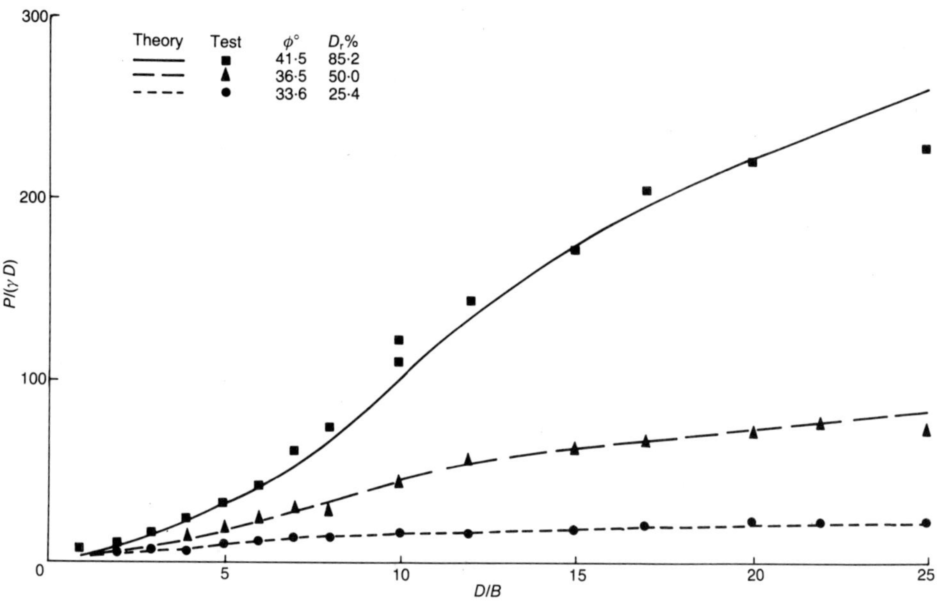

Fig. 21. Comparison of Fadl (1981) predictions with model test results

cal depth ratios (H/B) similar to those observed in the soil surface displacement measurements. Beyond the critical depth ratios the rate of increase in $p/(\gamma D)$ decreased until $p/(\gamma D)$ reached an almost constant value at high D/B values.

The theoretical and experimental work on clays indicated that their compressibility had a substantial influence on ultimate uplift capacity. The relative density of a sand is an indication of its compressibility and tests show that the relative density has an important influence on the ultimate resistance and displacement characteristics of anchors embedded in sand.

From an analysis of his experimental work, Fadl (1981) deduced an empirical relationship between the angle of inclination of the failure surface to the vertical, and the relative density. Relationships between the critical depth ratio, the relative density and the angle of friction were also obtained. From what was essentially a sophisticated curve-cutting exercise applied to the extensive experimental results, a simplified general theory for shallow, deep and inclined anchors in sand was developed as reported by Sutherland, Finlay & Fadl (1982). The predictions from this theory are also shown in Fig. 21 and compare well with the experimental results. Satisfactory checks were also obtained with the Sizewell laboratory and field results, and with the experimental results of other investigators. As will be discussed, a further opportunity of checking the Fadl predictions arose subsequent to his work.

OFFSHORE APPLICATIONS OF EMBEDDED ANCHORS

The investigations on embedded anchors (with the exception of the Sizewell shaft-raising operation) have been concerned with the design for safety case and in recent years, with the expansion of offshore exploration and production, particularly in deeper waters, it has been claimed that anchors embedded in the sea-bed have significant potential based on economic and technical considerations (*see* Andreadis & Harvey, 1981).

Taylor, Jones & Beard (1975) describe a number of types of embedded anchors and their methods of installation ranging from jetting to vibratory, direct driving and explosive embedment. There have also been developments in the United Kingdom on the design and method of embedment of offshore anchors.

The Hydropin anchor (Kerr, 1976) has been developed and patented by the National Engineering Laboratory, Scotland. It buries itself into the sea-bed by fluidizing and pumping. Hydropins of 0·3 m, 0·61 m and 1·22 m dia. have been buried up to 9 m deep. Kerr reports the results from five Hydropins which were embedded in a cohesionless soil sea-bed and tested to failure under a vertical pull. The D/B values ranged from 1 to 22, but no details of the soil properties were given, other than particle size distribution. Nevertheless, the $p/(\gamma D)$ values can be assessed for each test and correspond, with one exception, to the values which would be anticipated for a very loose sand, which is not surprising in view of the method of installation by jetting and fluidization. If, subsequent to installation by jetting, a vibratory technique could be developed, the uplift capacity of devices such as the Hydropin could be substantially increased.

Andreadis & Harvey (1981) describe a fluke-type anchor developed at Queen Mary College, London. The anchor is driven into the sea-bed to the required depth and the flukes are then opened mechanically. They also present a design procedure for embedded anchors in cohesionless soils which deals with vertically embedded plate and fluke anchors. The design method is concerned with static and repeated loading, and the base from which the design method is developed is a plot of $p/(\gamma D)$ against D/B for the ultimate uplift capacity of anchors under static loading. This plot, which was developed from model tests and some prototype tests on small fluked anchors, is reproduced in Fig. 22.

The $p/(\gamma D)$ relationship with D/B is given for three cases of relative density from 65% to 100% over a range of D/B values from 2 to 12. This plot is of particular interest as it allows a direct comparison to be made with the original Sizewell model tests and also with the predicted uplift capacity from the Fadl (1981) simplified theory, a theory which has already been shown to give reasonable agreement with the Sizewell and other model test results.

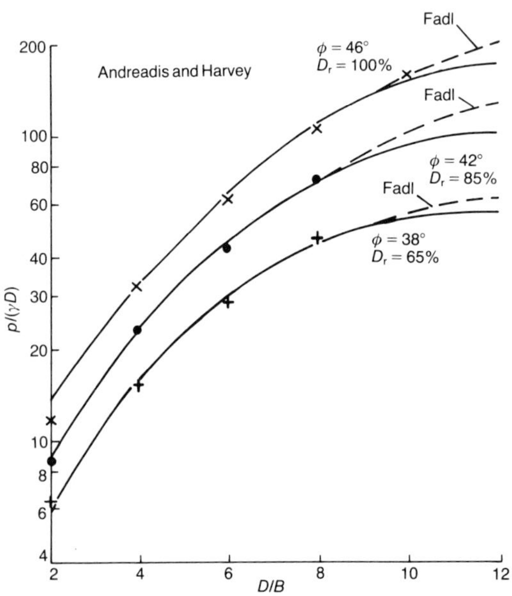

Fig. 22. Comparison of Fadl (1981) predictions with Andreadis & Harvey (1981) basic design curves

In Fig. 22 the theoretical predictions from Fadl are plotted on the Queen Mary College design chart for the three cases of relative density, and these give almost identical agreement over a range of D/B values up to about 10. Variations occur only when the Fadl deep anchor case is reached. Andreadis & Harvey (1981) pointed out that installation procedures can produce disturbance of the adjacent soil which can reduce the life of an anchor when subjected to repeated loading, although the same anchor when subjected only to static loading may not show serious deficiencies. There is, they commented, a need for a comprehensive programme of large-scale repeated loading tests.

FOUNDATIONS FOR TRANSMISSION LINE TOWERS

In North America, there has been a substantial upsurge of interest in recent years in foundations for transmission line towers. This interest follows on from the pioneering work carried out 20 or so years ago by Meyerhof & Adams (1968) under the auspices of the Ontario Hydro Electric Company. There have recently been two important publications: Symposium on the uplift behaviour of anchor foundations in soil (1985); and Geotechnical special publication No. 8 (1987)—Foundations for transmission line towers. Both of these publications were sponsored by the Geotechnical Engineering Division of the American Society of Civil Engineers.

There has been substantial sponsorship of research on uplift resistance in the U.S.A. by bodies such as the Electrical Power Research Institute. In addition, Geotechnical special publication No. 8 contains papers on investigations which are supported by power companies such as the Tennessee Valley Authority, the Niagara Mohawk Power Corporation, the Houston Light and Power Company, the Intermountain Power Agency and the Southern California Edison Company. It will be interesting to see if the proposed privatization of the power industry in the United Kingdom will lead to similar sponsorship of field research.

Up until the publication of Geotechnical special publication No. 8, there had been a dearth of published full-scale field data, apart from the Sizewell shaft-raising results, and the need for this type of data had been commented upon by a number of investigators. However, this publication produced an abundance of field test data for anchor-type foundations embedded in cohesionless soils.

There were over 20 field tests on belled piers from different sites (Konstantinidis, Pacal &

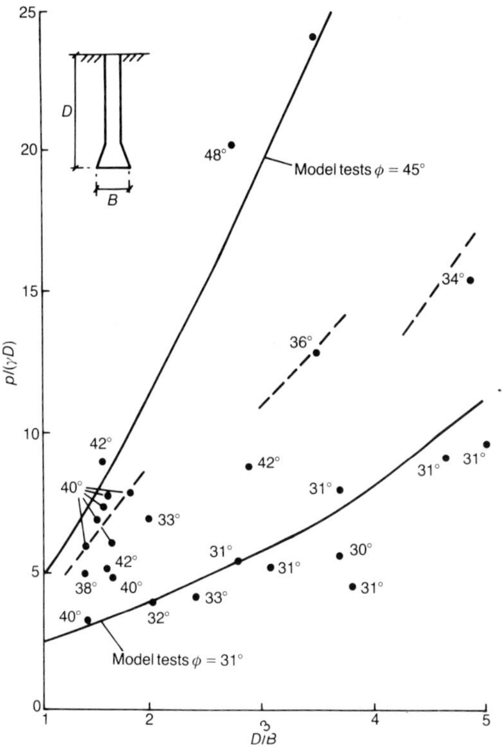

Fig. 23. Comparison of field tests on belled piers with Sizewell model test results

Shively, 1987; Tucker, 1987). The piers were up to 5·2 m deep and 1·14 m wide, and covered a range of D/B values from 1·38 to just over 5. Information was given on unit weight and angle of friction for each test—this enabled each test to be compared with the Sizewell laboratory and field results, as shown in Fig. 23. The results are almost embarrassing in that the comparison of these field tests with the Sizewell model tests (on which the Sizewell shaft-raising operation was based) is extremely good over a range of friction angles from 30° to 48°. There are very few anomalous results and, even including these anomalies, the average of the belled pier results is 91% of the average value from the model tests.

It would appear from Fig. 23 that we are back to where we were 20 years ago, as far as prediction of the ultimate uplift resistance of anchors or foundations embedded in sand is concerned. The model test results would still form the basis of prediction, but the Fadl empirical theory based on more extensive model tests could also serve. The belled piers constitute a design for safety case. The finite element-based theories previously discussed would lead to very conservative estimates of the uplift resistance of the belled piers for cases other than very loose sands. The inadequacy of these theories in the design for failure case has already been commented upon.

SUMMARY AND CONCLUSIONS

This Rankine Lecture was not intended to be a comprehensive review or state of the art lecture on the uplift resistance of soils. I have quoted the work of only a few investigators; work which I consider representative of various approaches to the problem. There are many other investigators, and the fact that I have not dealt with their contributions does not mean that I am unaware of them.

Even though the subject matter of this lecture was quite different from that of Professor Scott's (1987) the two could almost be regarded as complementary in principle. Professor Scott concluded his lecture with a plea for more contrived and controlled failures with a view to their analysis for the design for safety case, to ensure a more economic design of structures. I approached the concept of failure from two points of view with respect to the uplift resistance of soils.

Initially, my involvement was with the design for failure case; a rather unusual concept for geotechnical engineers, as the success of the shaft-raising operation at Sizewell depended on failure actually occurring. What resulted from Sizewell was not one, but nine full-scale tests to failure. These failures could have acted as the yardstick that Scott advocated, against which the predictions from analytical or other solutions could be compared for the design for safety case for the uplift resistance of soils.

The design for failure case at Sizewell was not based on analytical methods, as the few such methods which were available at the time were concerned with the design for safety case, and gave conflicting and unsatisfactory results. The design method at Sizewell was based on model tests where dimensional similarity was established between the model and the prototype. Since this original model work was carried out, centrifuge model testing and finite element methodology have become increasingly popular and have been applied to the design for safety case for the uplift resistance of soils. Surprisingly, the predictions from these methods were never compared with the nine full-scale failure tests on cohesionless soils at Sizewell. When such a comparison is made, the centrifuge model tests on cohesionless soils (Ovesen, 1981) gave reasonable agreement with the Sizewell model tests, over the restricted range of the Ovesen tests. This would be anticipated, as the dimensional analysis used in the model tests, while simplified, was approximately correct.

As far as the finite element analyses were concerned, the predictions for clay were reasonable, but conservative compared with the model tests and other methods of analysis. The finite element predictions for cohesionless soils, however, were most unsatisfactory. They gave highly conservative results for other than loose sands and could not have formed the basis for the successful shaft-raising design at Sizewell as a design for failure case. I make this point as the shaft-raising concept was not a one-off operation. It has been considered for use at a number of locations and has been used on at least one other project, and I understand that it is presently under active consideration at another location. So in this connection, engineers have still to deal with the design for failure case.

It is emphasized that the foregoing comments are not a condemnation of the finite element method of analysis. I admire and respect the pioneering work by Professor Zienkiewicz and others, and their successful applications. However, Scott (1987) has pointed out the difficulty of obtaining a constitutive model for dense sands in particular, and this has been confirmed by the finite element predictions for dense sands, which I have previously considered.

I have discussed the application of model tests, centrifuge tests and finite element methods to the problem of the uplift resistance of soils. These are only a few of the many analytical and experimental techniques which are now available to the geotechnical engineer. When my interest in soil mechanics started 45 years ago, I and most of my colleagues had undergone a basic education and training in civil and structural engineering. The advent of specialisms such as soil mechanics and prestressed concrete created groups of specialists and also led to fragmentation among civil engineers. For example, a certain mysticism seemed to develop about soil mechanics and it worried me that often all the engineers who designed the structure above the ground wanted from the soil mechanics man was a figure for bearing capacity and settlement irrespective of how it was obtained. I know that this situation does not now exist to the same extent but it has taken many years to produce the change.

Just as a fragmentation occurred among civil engineers a generation or so ago, I now detect the beginnings of a fragmentation among geotechnical engineers because of the specialisms that are developing within geotechnical engineering. The demands of mastering a particular specialism are great, and can, if care is not taken, produce a degree of tunnel vision which can lead to intolerance towards other methods. As I have attempted to demonstrate in this lecture, there is no one method or procedure which is a panacea for all our ails and problems, and we must keep as open a mind as possible. This applies equally to the engineers young and old; experienced or naive.

On the question of an open mind, I remember, as a young research associate at Harvard, asking Terzaghi with a certain degree of impertinence and trepidation why he was saying something which was different from what he had been saying previously. His retort was 'Sutherland, once I cannot change my mind I shall have become a fossil'.

We should also guard against the problem of taking ourselves and the results of our research and other work too seriously and personally. I can well appreciate why this should be so, coming from the University of Glasgow (now approaching the 550th anniversary of its founding) where anything most people achieve is only a ripple on the surface, except for men such as Rankine and Kelvin, who only emerge once in a 100 years or so.

I can well remember getting my comeuppance at the seventh European Conference on Soil Mechanics held in Brighton. I had been acting as Chairman at a Session, and also contributing. I was feeling quite pleased with my efforts and afterwards, at the conference dinner in the Brighton Pavilion, we were being entertained by a Scottish group of pipers and dancers. One of my so-called friends persuaded (or bribed) the pipe major to march across the hall and he placed his bagpipes in front of me with a salute. Having dined rather well I took up his challenge. Afterwards I was inundated with requests (mainly from Europe, I may say) inviting me to come and lecture, provided I brought my bagpipes. There was no mention of a soil mechanics subject matter for the lecture. We should all be serious in our endeavours but we should not take ourselves too seriously.

ACKNOWLEDGEMENTS

The Sizewell model studies were carried out at the University of Glasgow on behalf of Taylor Woodrow Construction Ltd and their consulting engineers for the Sizewell offshore works, C. S. Allott & Son (now Allott and Lomax).

Among many participants in the research work on the uplift resistance at Glasgow, the Author is particularly indebted to R. D. McKenzie and M. K. El-Rayes for their assistance in the original model tests, and to J. Davie, W. Stewart, M. O. Fadl, and in particular, T. W. Finlay, for their contributions to subsequent work. Parts of the research work were carried out with grants from the Marine Technology Board of the (then) Science Research Council of the United

Kingdom, and this support is gratefully acknowledged.

REFERENCES

Ali, M. S. (1968). *Pull-out resistance of anchor plates and anchor piles in soft bentonite clay.* MSc thesis, Duke University, Durham, S.C.

Andreadis, A. & Harvey, R. C. (1981). A design procedure for embedded anchors. *Applied Ocean Res.* **3,** No. 4, 177–182.

Balla, A. (1961). The resistance to breaking-out of mushroom foundations for pylons. *Proc. 5th Int. Conf. Soil Mech., Paris* **1,** 569–576.

Davie, J. R. (1973). *Behaviour of cohesive soils under uplift forces.* PhD thesis, University of Glasgow.

Davie, J. R. & Sutherland, H. B. (1977). Uplift resistance of cohesive soils. *J. Geotech. Engng Div. Am. Soc. Civ. Engrs* **103,** GT9, 935–952.

Davie, J. R. & Sutherland, H. B. (1978). Modelling of clay uplift resistance. *J. Geotech. Engng Div. Am. Soc. Civ. Engrs* **104,** GT6, 755–760.

El-Rayes, M. K. (1965). *Behaviour of cohesionless soils under uplift forces.* PhD thesis, University of Glasgow.

Fadl, M. O. (1981). *The behaviour of plate anchors in sand.* PhD thesis, University of Glasgow.

Kerr, N. (1976). A self-burying anchor of considerable holding power. *Proc. 8th Offshore Technol. Conf., Houston,* OTC 2466, 447–455.

Konstantinidis, B., Pacal, A. J. & Shively, A. W. (1987). Uplift capacity of drilled piers in desert soils. *Foundations for transmission line towers.* Geotech. Special Publn. No. 8. New York: ASCE.

Mariupol'skii, L. G. (1965). The bearing capacity of anchor foundations. *Soil Mech. and Fdn Engng* **2,** No. 1, 26–32. (Translated from Osnovaniya Fundamentyi Mekhanika Gruntov.)

Matsuo, M. (1967). Study on the uplift resistance of footing. *Soil & Foundation* **7,** No. 4, 1–37.

Meyerhof, G. G. (1951). The ultimate bearing capacity of foundations. *Géotechnique* **2,** No. 4, 301–332.

Meyerhof, G. F. & Adams, J. I. (1968). The ultimate uplift capacity of foundations. *Can. Geotech. J.* **5,** No. 4, 225–244.

Ovesen, N. K. (1981). Centrifuge tests of the uplift capacity of anchors. *Proc. 10th Int. Conf. Soil Mech., Stockholm* **1,** 717–722.

Rocha, M. (1957). The possibility of solving soil mechanics problems by the use of models. *Proc. 4th Int. Conf. Soil Mech., London* **1,** 183–188.

Roscoe, K. H. (1968). Soils and model tests. *J. Strain Analysis* **3,** No. 1, 57–64.

Rowe, R. K. & Davis, E. H. (1982a). The behaviour of anchor plates in sand. *Géotechnique* **32,** No. 1, 25–41.

Rowe, R. K. & Davis, E. H. (1982b). The behaviour of anchor plates in clay. *Géotechnique* **32,** No. 1, 9–23.

Scott, R. F. (1987). Failure. 27th Rankine Lecture. *Géotechnique* **37,** No. 4, 423–466.

Stewart, W. (1985). Uplift capacity of circular plate anchors in layered soil. *Can. Geotech. J.* **22,** 589–592.

Sutherland, H. B. (1965). Model studies for shaft raising through cohesionless soils. *Proc. 6th Int. Conf. Soil Mech., Montreal* **2,** 410–413.

Sutherland, H. B., Finlay, T. W. & Fadl, M. O. (1982). Uplift capacity of embedded anchors in sand. *Proc. 3rd Int. Conf. Offshore Structures, Cambridge, Massachusetts* **2,** 451–463.

Taylor, R. J., Jones, D. & Beard, R. M. (1975). *Handbook for uplift resisting anchors.* Port Hueneme, California: U. S. Navy.

Teng, W. C. (1962). *Foundation design.* Englewood Cliffs: Prentice-Hall.

Tucker, K. D. (1987). Uplift capacity of drilled shafts and driven piles in granular soils. *Foundations for transmission line towers.* Geotech. Special Publn No. 8. New York: ASCE.

Vermeer, P. A. & Sutjiadi, W. (1985). The uplift resistance of shallow embedded anchors. *Proc. 11th Int. Conf. Soil Mech., San Francisco* **4,** 1635–1638.

Vesic, A. S. (1971). Breakout resistance of objects embedded in ocean bottom. *J. Soil Mech. Fdns Div. Am. Soc. Civ. Engrs* **97,** SM9, 1183–1206.

Vesic, A. S. (1972). Expansion of cavities in infinite soil mass. *J. Soil Mech. Fdns Div. Am. Soc. Civ. Engrs* **98,** SM3, 265–290.

Zienkiewicz, O. C. (1987). Computer prediction in engineering. *Sci. & Tech. Review,* No. 2, 5–19.

VOTE OF THANKS

In proposing a vote of thanks to Professor Sutherland, Professor P. A. Green made the following remarks.

'We have had the privilege of hearing this evening the twenty-eighth Rankine Lecture, presented by Professor Hugh Sutherland, and I am sure you will agree with me that it was an outstanding and most memorable lecture. Professor Sutherland has had a long and distinguished career, during which he has put particular emphasis on teaching. His success is demonstrated by the many Glasgow graduates who now hold high and important positions in civil engineering. Unfortunately, until now Professor Sutherland's research work in soil mechanics has been more appreciated overseas than in Britain—which is our loss. For this reason, I was delighted when I heard that the British Geotechnical Society Committee had at last recognized his unique contribution to soil mechanics by asking him to present the 1988 Rankine Lecture.

'I have always had great admiration for Hugh Sutherland and for his professional achievements. This admiration was particularly enhanced in 1985 when, as Chairman of the BGS and on behalf of the ISSMFE, I asked him to prepare a review of the International Society's finances—one of the many non-engineering tasks which he has carried out in parallel with teaching and research. He supported me at the Council meetings during the San Francisco Conference and it goes without saying that his financial report was explicit, precise and exactly what was required.

'During these meetings he displayed all of the

characteristics I have come to admire in our Scottish colleagues: he penetrates to the heart of issues without being distracted by irrelevant details; he is full of sound common sense; he is always practical; he appreciates and is tolerant of other people's points of view; and he tackles any job with considerable humour. Most of these characteristics have been clearly demonstrated in this evening's lecture. He focused on a clearly defined practical problem, he has used common sense when the theoretical results appear to make nonsense and, not least, he was able to make the lecture both humorous and entertaining.

'I am sure you will agree that we must unanimously express our thanks to Professor Sutherland for an outstanding Rankine Lecture.'

The Rankine Lecture

The twenty-ninth Rankine Lecture of the British Geotechnical Society was given by Professor H. G. Poulos at Imperial College on 22 March 1989. The following introduction was given by Professor C. P. Wroth.

Harry Poulos was born in Australia in April 1940. He went to the University of Sydney to study Civil Engineering, and graduated with First Class Honours in 1961. He stayed on to do research in soil mechanics under the inspired leadership of Professor Ted Davis. I believe that Harry Poulos was very fortunate to join this burgeoning research group, which became established by the 1970s as the premier group in Australia—indeed in the whole of the Southern Hemisphere. I think that it is appropriate that I pay passing tribute to Professor Davis who had such an important influence on Professor Poulos' career. Davis graduated from University College, London in 1940, and after war service, some practical engineering experience and a spell at the Road Research Laboratory (as it then was) left the United Kingdom in 1952 to take up a post as Senior Lecturer in Soil Mechanics at the University of Sydney. Ted Davis, in his quiet and effective manner, set about the task of getting research going as well as establishing the teaching of soil mechanics in one of the most respected schools of Civil Engineering. Davis saw the need to give a sound scientific basis to the practice of geotechnical engineering. His own interests were devoted to the development of a unified theory of foundations, and to extending the theory of plasticity to soil and rock stability problems.

Besides being a scholar of distinction, Davis had a charming personality, and he attracted around him a talented and dedicated group—the most notable members being Professor Poulos and Professor John Booker. It is my lasting regret that I failed to persuade the committee of our society, ten years ago, to invite Professor Davis to give a Rankine Lecture.

After completing his PhD on three-dimensional settlement analysis, Harry Poulos had nine months' experience with a firm of consulting engineers before returning to the University in 1965 as Lecturer. He flourished in the Davis stable, which provided an ideal environment for a young and ambitious academic; as a consequence his progress was rapid. A Senior Lecturership in 1969 was followed by a Readership in 1972, at the age of 32. Doubtless a personal Professorship would have followed but, in 1981, Professor Davis suddenly died and Harry Poulos found himself unexpectedly inheriting the leadership of the soil mechanics group and, a year later, succeeding to Davis' Chair. To take over such responsibilities, in the wake of a great man who has left his own personal and distinctive mark on the affairs of a University is by no means easy—as I know from my own experience. Harry Poulos took this challenge in his stride—indeed with marked success—so much so that only three years later, in 1984, he was appointed Head of the combined School of Civil and Mining Engineering. Having had the good fortune to spend seven weeks on leave in that Department in 1985, I can testify to how well Harry Poulos was and is carrying out his duties as Head of Department and how well he is respected in his local community.

His research has been unusually productive and his CV lists 3 books, 6 chapters in other books, 53 papers of which he is sole author and 80 joint papers. His main interest fall into three categories: behaviour of pile foundations, foundations in marine carbonate sediments and settlement of foundations. It is for the first of these that he is best known: his name is indelibly identified with the elastic analysis of the performance of piles and pile groups—and indeed it is no surprise that he has chosen 'Pile Behaviour—Theory and Application' as the title of tonight's lecture. He has made major contributions to our understanding of the behaviour of piles and piled foundations, and he is a world authority on this topic.

His two books written with Ted Davis have become classics. The first, 'Elastic Solutions for Soil and Rock Mechanics', must be a standard reference book on the shelves of any self-respecting geotechnical engineer, second only in importance as a reference source to the Navdocks Manual. The second book, 'Pile Foundation Analysis and Design', is no doubt effectively updated by the written version of his Rankine Lecture. His latest book, 'Marine Geotechnics', published in 1988, is set to be equally successful.

He has received many prizes and awards, including the Warren Medal of the Institution of Engineers of Australia and, most recently, the John Jaeger Memorial Award of the Australian Geomechanics Society for 1988.

From these comments you might get the impression that Harry Poulos was solely an academic dedicated to his teaching and research. This is far from the case, for he has carried out an impressive list of consulting assignments (too numerous to elaborate), mostly in Australia, but in other countries also. He is in great demand as an invited speaker at international conferences all over the world, renowned for his fluent delivery, entertaining style and well-informed and challenging comments.

Harry Poulos is a distinguished engineer carrying out all his commitments with ease—the four E's—excellence, elegance, economy and efficiency. It is with the greatest pleasure and expectation that, on behalf of the British Geotechnical Society, I invite Professor Harry George Poulos to deliver the twenty-ninth Rankine Lecture.

Professor H. G. Poulos

Poulos, H. G. (1989). *Géotechnique* **39**, No. 3, 365–415

Pile behaviour—theory and application

H. G. POULOS*

This paper discusses the application of theory to the analysis of pile foundation behaviour under axial loading. A classification system is suggested for pile analysis and design procedures, based on the rigour of the underlying theory. It is shown that a number of the analyses in use have a common underlying theoretical basis founded on the boundary element method and that such methods give solutions which are consistent with those from other methods such as the finite element method. The main characteristics of pile behaviour are reviewed for single piles and pile groups subjected to static loading, cyclic loading, and to loading arising from externally-imposed soil movements. Correlations are then summarized for the geotechnical parameters required for the prediction of axial pile behaviour. Using characteristic values of these parameters, design charts are developed for the load capacity and settlement of piles and pile groups. Finally, three case studies are described which demonstrate the sensitivity of pile performance predictions to the method of analysis, the idealization of the soil profile, and the selection of soil parameters. It is shown that the method of analysis is likely to have less effect on the predicted performance than does the geotechnical characterization of the site.

KEYWORDS: **analysis; case history; design; piles; repeated loading; settlement.**

L'article discute l'application de la théorie à l'analyse du comportement des fondations par pieux sous chargement axial. On propose un système de classification basé sur la rigueur de la théorie associée pour l'analyse des pieux et les procédés de construction. On démontre que quelques-unes des analyses employées ont une base théorique commune dérivant de la méthode des éléments limites et que de telles méthodes donnent des solutions qui s'accordent bien avec celles obtenues à partir d'autres méthodes, telles que les éléments finis. Les caractéristiques principales du comportement des pieux sont examinées pour les pieux et les groupes de pieux soumis aux chargements statiques et cycliques aussi bien qu'au chargement résultant des mouvements du sol imposés de l'extérieur. Puis on résume les corrélations pour les paramètres géotechniques nécessaires pour la prédiction du comportement axial des pieux. À l'aide de valeurs caractéristiques de ces paramètres on propose des abaques de construction pour la force portante et le tassement des pieux et des groupes de pieux. Trois cas réels sont présentés. Ils illustrent la façon dont les prédictions du comportement des pieux dépendent de la méthode d'analyse, de la schématisation du profil du sol et de la sélection des parametres du sol. On démontre que la performance prévue est probablement moins influencée par la méthode d'analyse que par la détermination des caractéristiques géotechniques du site.

INTRODUCTION

For many years, the design of pile foundations was based on a combination of empiricism and experience, and the general attitude towards theoretical analysis of pile foundations was exemplified by Terzaghi & Peck (1967), who stated

> '. . . theoretical refinements in dealing with pile problems . . . are completely out of place and can be safely ignored.'

Despite this pessimistic evaluation, the past three decades have seen a gradual change in pile design procedures, from essentially empirical methods, towards methods with a sounder theoretical basis. This change has resulted from a number of stimuli, including the wider use of piling, the recognition that pile foundations do indeed settle and that such settlements must be controlled, and the need to support very large loads on piles, especially for the foundations of offshore structures. In the latter case injudicious extrapolation of previous experience with small onshore piles may be hazardous, particularly as the loads of offshore piles will generally involve a cyclic component, and the soils encountered may exhibit unusual characteristics.

Allied to these stimuli for improved design pro-

* School of Civil and Mining Engineering, University of Sydney.

cedures has been the rapid development of powerful numerical techniques, in particular the finite element method and the boundary element method. These, in conjunction with the now commonplace availability of powerful desktop computers, have made feasible methods of pile analysis which even a decade ago would not have been considered practical for foundation design. Several such analysis methods have been proposed and various claims have been made regarding the superiority of one or other of these methods over alternative approaches. It would therefore appear appropriate at this time to review some of these methods, examine similarities and differences among them, and obtain some insight into the significance of such differences when compared with uncertainties which may arise in the practical application of the methods.

This Paper therefore has the following objectives

(*a*) to present a classification of methods for pile foundation analysis and design
(*b*) to outline a versatile analytical framework which can incorporate several existing methods of analysis
(*c*) to compare alternative analytical procedures
(*d*) to examine the characteristics of behaviour of single piles and pile groups under static axial loading, cyclic axial loading, and when subjected to externally-imposed soil movements
(*e*) to review methods of obtaining the soil parameters required for the prediction
(*f*) to present examples of simplified design charts based on theoretical analysis and characteristic soil parameters
(*g*) to discuss the application of theory to practical cases, and to examine the sensitivity of pile performance predictions to a variety of factors over which the geotechnical analyst has control, including the method of analysis, and the geotechnical characterization of the site.

Attention will be confined to problems involving static or quasi-static axial loading of piles and pile groups in soil. Both load capacity and defor-

Table 1. Categories of analysis/design procedures

Category	Subdivision	Characteristics	Method of parameter determination
1	—	Empirical—not based on soil mechanics principles	Simple in situ or laboratory tests, with correlations
2	2A	Based on simplified theory or charts—uses soil mechanics principles—amenable to hand calculation. Theory is linear elastic (deformation) or rigid plastic (stability)	Routine relevant in situ tests—may require some correlations
	2B	As for 2A, but theory is non-linear (deformation) or elasto-plastic (stability)	
3	3A	Based on theory using site-specific analysis, uses soil mechanics principles. Theory is linear elastic (deformation) or rigid plastic (stability)	Careful laboratory and/or in situ tests which follow the appropriate stress paths
	3B	As for 3A, but non-linearity is allowed for in a relatively simple manner	
	3C	As for 3A, but non-linearity is allowed for by way of proper constitutive models of soil behaviour	

mation are considered, but it is emphasized at the outset that load capacity and deformation are not independent and, although they can be separated with reasonable justification for many problems involving conventional direct loading, their interdependence may be very important for problems involving cyclic loading or external soil movements.

CATEGORIES OF ANALYSIS AND DESIGN PROCEDURES

Analysis and design procedures can be divided into three broad categories, depending on the level of sophistication and rigour. An extended classification system of these procedures has been proposed by Poulos & Hull (1989) and is shown in Table 1. Category 1 procedures probably account for most pile design done throughout the world. Category 2 procedures have a proper theoretical basis, albeit simplified, and are being increasingly used for pile deflexion calculations. Such procedures involve the use of simple computational methods or design charts, and generally do not demand the use of a computer. Category 3 procedures involve the use of a site-specific analysis based on relatively advanced analytical or numerical techniques such as the finite element method or the boundary element method. In most cases, such procedures require the use of a computer. Category 3 procedures are frequently used to carry out the necessary parametric solutions and develop the design charts which can then be used as category 2 solutions.

Typical examples of the various categories of procedures for axially loaded piles are shown in Table 2. In choosing an appropriate category of design for a practical problem, the following factors need to be considered

(*a*) the significance and scale of the problem
(*b*) the available budget for foundation design
(*c*) the geotechnical data available
(*d*) the complexity of both the geotechnical profile and the design loading conditions

Table 2. Examples of categorization of methods for evaluation of axial pile response

Category	Axial pile capacity	Settlement
1	Correlations with CPT (e.g. Schmertmann, 1975; De Ruiter & Beringen, 1979). Correlations with SPT (Thorburn & McVicar, 1971); Meyerhof, 1956) Total stress (α) method (Tomlinson, 1957; Semple & Rigden, 1984)	Approximate correlations with pile diameter (Meyerhof, 1959; Frank, 1985) Column deflexion multiplied by a factor (Focht, 1967)
2A	Effective stress (β) method (Burland, 1973; Meyerhof, 1976; Stas & Kulhawy, 1984)	Elastic solutions (Randolph & Wroth, 1978; Poulos & Davis 1980)
2B	Effective stress method (Fleming *et al.*, 1985)	Elastic solutions modified for slip (Poulos & Davis, 1980)
3A	Plasticity solutions for, end bearing capacity (Giroud *et al.*, 1973; Meyerhof, 1963)	Elastic finite element analysis (e.g. Valliappan *et al.*, 1974)
3B	Non- linear load transfer analysis (e.g. Coyle & Reese, 1966; Kraft *et al.*, 1981)	
	Non-linear boundary element analysis (e.g. Poulos & Davis, 1980) Non-linear finite element analysis (e.g. Desai, 1974; Jardine *et al.*, 1986)	
3C	Finite element analysis, including simulation of pile installation (e.g. Nystrom, 1984; Randolph *et al.*, 1979; Withiem & Kulhawy, 1979)	

(*e*) the stage of the design process (i.e. whether a feasibility, preliminary or final design is being carried out)
(*f*) the experience of the designer with the methods being considered.

Small projects with a limited budget for geotechnical work rarely justify more than a category 1 or 2A approach. However, for final foundation design in a major project for which considerable geotechnical data has been obtained, a category 3 approach would be appropriate. If, for the same project, a preliminary design was required based only on limited in situ or laboratory data, a category 2 approach might be useful for carrying out sensitivity studies to identify those parameters which are most significant and need to be determined with greatest care.

There will often be occasions in practice where a category 3 analysis is required to account for some unusual aspect of the soil profile or the unconventional nature of the design loading (e.g. if it involves a significant cyclic component). It is not unlikely that the parameters which are required for such a category 3 analysis will have to be estimated rather crudely, e.g. from empirical correlations which would be more appropriate to category 1 or 2 procedures. Consequently, it cannot be expected that solutions from a category 3 analysis will always be superior to those from a simpler analysis. As will be illustrated in this Paper, the method of analysis frequently is of much less importance than the geotechnical parameters which are selected and the way in which the geotechnical profile is idealized.

ANALYTICAL FRAMEWORK FOR AXIALLY LOADED PILES

In the past two decades, a number of methods have been developed for analysing axially loaded piles, many of which fall into category 3. Most of these methods involve the use of one (or more) of the following analytical techniques

(*a*) simplified analytical methods involving the consideration of independent horizontal 'slices' of pile and soil, e.g. Randolph & Wroth (1978).
(*b*) boundary element methods, employing either load-transfer functions to represent the interface response (e.g. Coyle & Reese, 1966; Kraft *et al.*, 1981) or elastic continuum theory to represent the soil mass response (e.g. Butterfield & Banerjee, 1971; Banerjee, 1978; Banerjee & Davies, 1978; Poulos & Davis, 1980).
(*c*) Finite element methods (e.g. Desai, 1974; Valliappan *et al.*, 1974; Balaam *et al.*, 1975; Ottaviani, 1975; Jardine *et al.*, 1986), in which a variety of constitutive soil models can be utilized, and such factors as soil non-homogeneity and anisotrophy can be taken into account.

Finite element methods offer the most powerful analytical approach in that, not only can non-linear soil behaviour be modelled, but the complete history of the pile can be simulated, i.e. the processes of installation, reconsolidation of the soil following installation, and subsequent loading of the pile (e.g. Nystrom, 1984; Withiam & Kulhawy, 1979). Such analyses are valuable in leading to a better understanding of the details of pile behaviour, but are unlikely to be readily applicable to practical piling problems because of their complexity and the considerable number of geotechnical parameters required.

A reasonable compromise between excessive complexity and unacceptable simplicity is provided by boundary element methods, in which the pile–soil interface is discretized and the characteristics of the soil response are represented in a lumped form by ascribing the behavioural features of the soil to the interface elements. This method has been developed by a number of research workers and is widely used in practice, and attention will therefore be focused on this method. Although a considerable number of formulations exist, most appear to have a common underlying basis. A convenient means of developing a unified analysis is to employ a substructuring technique in which the pile (or piles) and the surrounding soil are considered separately and then compatibility conditions are imposed.

The behaviour of each element is considered at a node which is located at the centre of that element and along a common vertical plane through the pile axis. Fig. 1 shows the division of a single pile into elements, the distribution of free-field soil movements due to some external cause (e.g. swelling or consolidation of the soil mass due to moisture changes or external loading) and the specified distributions of

(*a*) the limiting pile–soil stresses, f_c for compressing loading, and f_t for tensile loading; for shaft elements, the limiting pile–soil resistance will be termed here the shaft resistance, while for base elements, the term end-bearing resistance will be used
(*b*) the local stiffness k_s of the soil.

At this stage, no assumptions are made regarding the nature of f_c, f_t and k_s; these may vary with depth, stress (or displacement) level and with time.

The responses of the pile and soil elements to an increment of axial load ΔP are analysed in

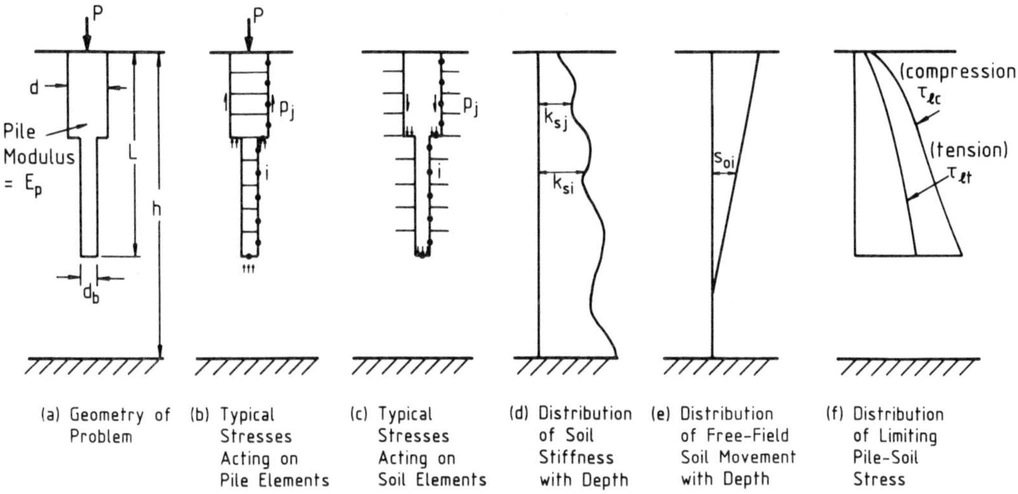

Fig. 1. Basic problem of single axially loaded pile: (a) geometry; (b) stresses acting on pile elements; (c) stresses acting on soil elements; (d) distribution of soil stiffness with depth; (e) distribution of free-field soil movement with depth; (f) distribution of limiting pile–soil stress

turn. Details of the equation that describes incremental displacements of the pile are given by Poulos (1979b). For the soil elements, the incremental displacement of the soil may arise from two sources, the pile–soil interaction stresses $\{\Delta p\}$, and the free-field soil movements $\{\Delta S_e\}$. At any element i, the soil will be in one of three states; non-failure, failure in 'compression', or failure in 'tension'.

In the non-failure state, the incremental displacement of the soil at an element i, $\Delta\rho_{si}$, can then be expressed as

$$\Delta\rho_{si} = \sum_{j=1}^{m} \frac{I_{ij}}{k_{sij}} \cdot \Delta p_j + \Delta_{S_{ei}} \tag{1}$$

where I_{ij} is displacement influence factor for element i due to element j, k_{sij} is soil stiffness for element i due to element j, Δp_j is the increment of pile–soil interaction stress, ΔS_{ei} is the increment of external soil movement at element i and m is the total number of elements into which pile is divided.

In the 'compressive' failure state, the conditions at element i may be expressed as

$$\Delta p_i = f_{ci} - p_i \tag{2a}$$

and for the 'tensile' failure state,

$$\Delta p_i = f_{ti} - p_i \tag{2b}$$

where δp_i is incremental pile–soil stress on element i, p_i is the existing pile–soil interaction stress on element i before the load increment, f_{ti} is the limiting pile–soil interaction stress in tension, at element i, and f_{ci} is the limiting pile–soil interaction stress in compression, at element i.

Equations (1), (2a) and (2b) describe the behaviour of the soil, and may be written for all elements of the system. By equating the expressions for incremental pile and soil displacements, invoking the vertical equilibrium condition, and checking the state of each element during the loading increment, the incremental pile–soil stresses can be determined, and hence the pile displacement at each element can be evaluated. Further details of the numerical procedure are given by Poulos & Davis (1980). By successive application of loading increments, the entire load–displacement relationship for the pile may be determined, including any unloading and reloading sequences.

This analysis becomes less satisfactory as the pile becomes very short or its stiffness relative to the surrounding soil decreases. For pile sockets in rock, the finite element method provides more reliable solutions, e.g. Pells & Turner (1979). The analysis in this section can be used for both the load-transfer analysis (commonly termed the *t–z* analysis), in which the soil resistance is modelled by a discrete spring at each element, and the continuum analysis, in which the soil is modelled as an elastic continuum.

For the load-transfer or *t–z* analysis, the displacement influence factor I_{ij} in equation (1) is zero, except when i = j, in which case $I_{ii} = 1$; in other words, the soil deflexion of a node i depends only on the stress at that node. The soil stiffness in equation (2) can be evaluated either empirically (e.g. Coyle & Reese, 1966) or theoretically (Baguelin & Frank, 1980; Kraft *et al.*, 1981;

Ha & O'Neill, 1983; Chow, 1986; Ooi *et al.*, 1989). Such features as soil non-homogeneity, non-linearity, post-peak softening, and interface dilation or contraction, may be incorporated. The main limitation of the load-transfer analysis is that it can only properly consider a single pile, and cannot be used directly to analyse pile groups. It is possible to modify the single pile t–z curves to allow for group effects (Randolph, 1986) but it is usually necessary to evaluate these group effects using elastic continuum theory (e.g. Chow, 1986).

In the continuum analysis the displacement influence factor I_{ij} (in equation (1)) for all elements is non-zero, and is usually evaluated from elastic continuum theory. The theory of Mindlin (1936) is frequently used for this evaluation (e.g. Poulos & Davis, 1968), although other basic elastic solutions may be employed, for example, the equations of Chan *et al.* (1974) for a layered half-space, as utilized by Chin (1988).

An implicit assumption in the elastic continuum analysis is that, if incremental tensile stresses are developed in the soil mass by the loading of the pile, the soil continues to respond elastically. Such an assumption is reasonable if the overall stress conditions remain compressive (due to the compressive overburden stresses), but may be questionable if the overall stresses become tensile. Unease about this point has led to some criticism of elastic continuum theory (e.g. Leonards & Darrag, 1989) and a preference for the load-transfer approach, as the latter involves no consideration of the stress state in the soil other than at the pile–soil interface. However, load-transfer and continuum solutions for a single pile generally agree well and suggest that any effects of possible tension in the soil are unlikely to influence pile behaviour significantly.

IMPLEMENTATION OF THE THEORY

In this Paper, attention will be concentrated on the continuum analysis. With this approach, it is only possible to derive proper results for quite idealized problems that involve linear soil behaviour and one- or two-layer elastic soil masses. The range of problems for which proper results can be derived can be extended by using the finite element method to evaluate the soil displacement influence factors. However, it is also possible to utilize continuum theory in an approximate manner to obtain solutions for a much wider range of problems of practical interest.

Figure 2 gives an indication of the extensions to the theory which may be accomplished by using the continuum analysis. The extensions are classified into six groups, two of which involve the modelling of the soil, two of which involve the modelling of the piles, and the last two which involve the modelling of the loading on the piles.

Modelling the soil profile

It is convenient to use the theory of elasticity for modelling the soil behaviour. Despite the gross simplification which this model involves when applied to real soil, it provides a useful basis for the prediction of pile behaviour, provided that appropriate equivalent elastic parameters are selected for the soil. A significant advantage of using an elastic model for soil is that it provides a rational means of analysis of pile groups and evaluation of immediate and final movements of a pile. In determining immediate movements, the undrained elastic parameters of the soil are used in the theory, whereas for final movements the drained parameters are used.

Mindlin's equations may be used to obtain approximate solutions for a layer of finite thick-

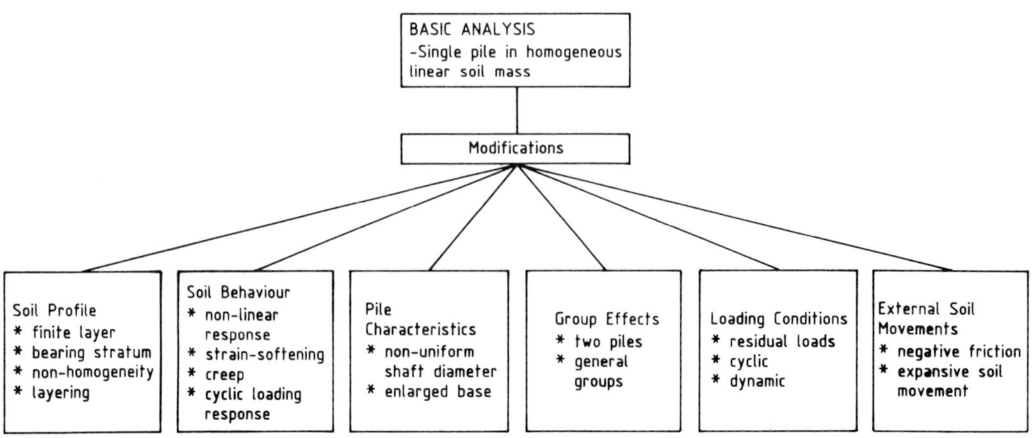

Fig. 2. Basic analysis, and extensions to allow for more realistic modelling

ness by using the Steinbrenner approximation (Steinbrenner, 1934) to allow for the effect of the underlying rigid base in reducing the soil displacements. For piles bearing directly on a stiffer stratum, the reduction in soil displacement may be analysed approximately by using the mirror-image technique described in Poulos & Davis (1980). In both these cases, the soil displacement factors I_{ij} in equation (1) are modified.

In the elastic continuum theory, the soil stiffness k_{ij} in equation (1) is directly related to the Young's modulus of the soil. In a homogeneous soil mass, $k_{sij} = E_s/d_j$ where E_s is the (constant) Young's modulus and d_j is the diameter of element j. If a non-homogeneous soil mass is considered and Mindlin's equation is used to evaluate the displacement influence factors, an approximation is required to determine k_{sij}. Poulos (1979a) has examined alternative assumptions and concluded that the following approximation is generally satisfactory

$$k_{sij} = 0{\cdot}5(E_{si} + E_{sj})/d_j \tag{3}$$

where E_{si} and E_{sj} are the values of soil Young's modulus at elements i and j respectively.

Equation (3) becomes inaccurate if large differences in soil modulus exist between adjacent elements or if a soil layer is overlain by a much stiffer layer. In a subsequent investigation, Yamashita *et al.* (1987) have developed a more general approximation in which E_{si} and E_{sj} in equation (3) are taken as weighted average values of soil modulus, involving the value of soil modulus at all elements along and beneath the pile.

Modelling of soil behaviour

One of the more important features of pile–soil behaviour is the limited stress which can be developed between the soil and the pile. This feature is already incorporated into the analysis in that checks are made to ensure that the pile–soil shear stress does not exceed the specified limiting value. If the interface behaviour at an element is assumed to be perfectly linear until the limiting stress is reached, the interface model will be termed (conveniently, but not strictly correctly*) an 'elastic–plastic' interface model.

It is possible to incorporate non-linear behaviour of the soil, in an approximate manner, by assuming that the soil Young's modulus varies with either stress or strain level. The simplest assumption is of a hyperbolic relationship between soil shear stress and shear strain, in which case the tangent Young's modulus of the soil E_{tan} is given by

$$E_{tan} = E_t\left(1 - \frac{pR_f}{p_f}\right)^2 \tag{4}$$

where E_t is the initial tangent Young's modulus, R_f is the hyperbolic curve-fitting constant, p is the pile–soil stress, p_f is the limiting value of pile–soil stress ($p_f = f_c$ in compression, or f_t in tension).

It is also possible to use alternative models to describe non-linear behaviour, for example, multi-linear models, or the Ramberg-Osgood model. Regardless of which non-linear model is used, it is important to consider unloading and reloading, as well as initial loading. The simplest assumption is to assume that on unloading or reloading, the behaviour is elastic until the previous greatest stress level at that element is reached; thereafter, non-linear initial loading resumes. However, alternative assumptions are possible in which non-linear behaviour recommences at stress levels less than the previous greatest level (Randolph, 1986). Such behaviour can give rise to the accumulation of permanent displacements if repeated loading is applied, but additional parameters are then required in the analysis.

In analysing the behaviour of piles subjected to cyclic loading, at least three aspects of soil response should be considered,

(*a*) the degradation of pile–soil resistances (and possibly soil modulus) under repeated cyclic loading
(*b*) loading rate effects
(*c*) the accumulation of permanent displacements.

One possible approach to incorporating cyclic loading effects has been described elsewhere by the Author (Poulos, 1988b). The effects of cyclic degradation are most significant for shaft resistance, and may be conveniently quantified by means of a degradation factor D_f, where

$$D_f = \frac{\text{shaft resistance at an element after cyclic loading}}{\text{shaft resistance at that element for static loading}} \tag{5}$$

The model of Matlock & Foo (1980) is useful for quantifying the change in D_f as cycling proceeds, and involves only two parameters, a minimum degradation factor D_{lim} and a degradation rate parameter λ.

The effects of loading rate are incorporating by means of a loading rate factor D_R, while permanent displacements developed in the soil during cyclic loading are considered to be equivalent to cyclic loading-dependent external soil

* Such a model does not necessarily imply that the relationship between interface stress and displacement at an element is elastic–plastic; because of the continuum nature of the soil, the slope of this relationship will change as other interface elements reach a failure state.

movements. Other aspects of soil behaviour may also be included, for example, strain-softening interface behaviour (Randolph, 1983; Kraft *et al.*, 1981) and creep (Booker & Poulos, 1976).

Pile characteristics

The analysis can readily accommodate piles of varying diameter or varying stiffness within the formulation of the pile displacement equation. In the case of varying diameter, additional annular elements are introduced at the diameter discontinuities, and values of the limiting pile–soil resistance at these elements, for both compression and tension loading, must be specified.

It is also possible to allow for failure of the pile material itself (either in compression or tension) by limiting the axial stress that can be developed in the pile section. The procedures for making this modification are described in Poulos & Davis (1980).

Pile groups

The continuum-based analysis can readily be extended to analyse a group of axially loaded piles. Each pile is discretized into elements, soil and pile displacement equations are assembled for each element, compatibility of soil and pile displacements is imposed at elements in the non-failure state, and the vertical equilibrium equation for each pile is written. In addition, the pile head conditions must be specified. Usually, a rigid cap connects the piles, so that all piles will undergo an equal head displacement. By specifying this condition, a further set of equations is obtained, which enables the load increment on each pile head to be computed, in addition to the distribution of incremental pile–soil stress and displacement (Hewitt, 1988).

In the evaluation of the displacement influence factors I_{ij}, use may again be made of the elastic solution of Mindlin (1936) to determine the displacement of an element of a pile due to all elements of that pile and the other piles in the group. A convenient approximation for this evaluation has been developed by El-Sharnouby and Novak (1985) which avoids the need for double integration of the Mindlin equation and hence reduces the computation time substantially. A number of methods of implementing the above analysis have been developed, including the following.

(*a*) A complete analysis of the group is performed using a continuum analysis—this has been done by Banerjee & Driscoll (1976) and Poulos & Hewitt (1986).

(*b*) A complete analysis of the group can be performed, using a load-transfer analysis to determine the response of a pile to its own load, and continuum theory to determine the influence of the elements of the other piles—this has been termed the hybrid method and has been used by O'Neill *et al.* (1977) and Chow (1986).

(*c*) Solutions for a two-pile group (obtained from a continuum analysis) can be used to obtain interaction factors that express the increase in head settlement of a pile due to the presence of another pile (Poulos, 1968; Randolph & Wroth, 1979). The interaction factor α is defined as

$$\alpha = \Delta S / S_1 \qquad (6)$$

where ΔS is the increase in settlement of a pile due to the presence of another equally loaded pile and S_1 is the settlement of a single pile under its own load. For a group of piles, the interaction factors may be superposed to develop a set of equations relating the settlement of each pile to the single pile settlement, the load on each pile head, and the interaction factors.

(*d*) A modified form of the interaction factor method can be employed in which different values of soil Young's modulus are used to determine the single pile behaviour and the interaction factors. Higher values are generally used for the latter to reflect the lower level of strain (and hence the greater stiffness) of the soil between the piles as compared with the soil immediately adjacent to each pile. This method is termed the modified interaction factor method, and has been described in detail by Poulos (1988a).

Comparisons between these approaches will be discussed later in the Paper.

Loading conditions

Residual stress effects. Most analyses of pile response assume an initially stress-free pile, although it is well-recognized that residual stresses exist in piles, due to installation effects, particularly in driven piles. Ideally, the driving of a pile should be modelled using a dynamic analysis such as that used by Holloway *et al.* (1978). However, a simpler first approximation can be employed in which the pile (at final penetration) is loaded to failure in compression and then unloaded back to zero load. Poulos (1987) shows that this procedure gives residual stresses in the pile which appear to be reasonably realistic. Leonards & Darrag (1989) point out that assumptions made regarding the soil modulus near the pile tip can significantly influence the computed residual stresses in the lower part of the pile, particularly for piles in sand.

Cyclic loading. At least two methods can be adopted to simulate cyclic loading of a pile

(*a*) a single-step analysis, in which a single iterative analysis is carried out to determine the behaviour of a pile after a specified number of cycles, *N*
(*b*) a cycle-by-cycle analysis, in which the application of each of the *N* cycles is modelled in turn.

Details of these methods are given by Poulos (1983). The cycle-by-cycle analysis is considerably more time-consuming, but much more versatile, in that it can accommodate strain-softening interface behaviour and sequences of irregular cyclic loading.

External soil movements. The effects of external soil movements, such as those arising from soil consolidation due to external loading or dewatering, or from soil heave due to wetting of expansive clay layers, can be analysed directly, provided that the external free-field soil movements at each element (i.e. values of ΔS_{ei} in equation (1)) can be specified. In the case of one-dimensional loading of a clay layer, these movements may be determined from Terzaghi's one-dimensional consolidation theory, and the time-dependency of the pile behaviour can readily be determined (Poulos & Davis, 1980). The free-field movements in an expansive soil layer are less easily computed, although a number of empirical or semi-empirical methods exist (e.g. Blight, 1965; Van Der Merwe, 1964; Cameron & Walsh, 1984).

Computer codes

The preceding analyses almost invariably require computer evaluation, and a number of computer programs have been written for examining various aspects of axial pile behaviour. A selection of these is given in Table 3. This list is by no means exhaustive and there is no doubt a great number of other codes in existence. Also excluded from this list are codes based on finite element analysis.

COMPARISONS BETWEEN ALTERNATIVE THEORIES

The preceding section has discussed alternative approaches to the analysis of pile behaviour, within a single framework of the boundary element method. It is of interest to compare these alternative approaches to determine the sensitivity of the computed behaviour to the approach adopted. At the same time, it is of considerable interest to compare the solutions with an independent analysis method such as the finite element method. Therefore, this section will make such comparisons for three different problems: (*a*) a single axially loaded pile; (*b*) an axially loaded pile group; and (*c*) a pile subjected to downdrag forces by external vertical soil movements.

Single pile

For a single pile having a typical relative compressibility $K = E_p/E_s$ of 1000, (where E_p is the Young's modulus of pile and E_s is Young's modulus of soil) in a deep homogeneous elastic soil mass, Fig. 3 compares two sets of solutions for the settlement as a function of the length-to-diameter ratio L/d, the boundary element approach, and the approximate analytical approach developed by Randolph & Wroth (1978). For $L/d > 15$, the agreement is close, the difference being less than 10%. For smaller values of L/d, the Randolph & Wroth solution gives smaller settlements, possibly due to the effects of the assumption of the soil surrounding the pile being a series of concentric cylinders. A similar measure of agreement is found between the Author's solution and the finite element solutions of Valliappan *et al.* (1974).

Figure 4 shows various solutions for the settlement of a single pile in a Gibson soil layer. There is reasonable agreement among the three solutions compared but as in Fig. 3, the greatest potential for differences appears to be for relatively short piles ($L/d < 15$).

In comparing non-linear solutions for single pile response, the problem considered is that analysed by Jardine *et al.* (1986). They have employed a finite element analysis involving the use of a non-linear soil model, the LPC2 model,

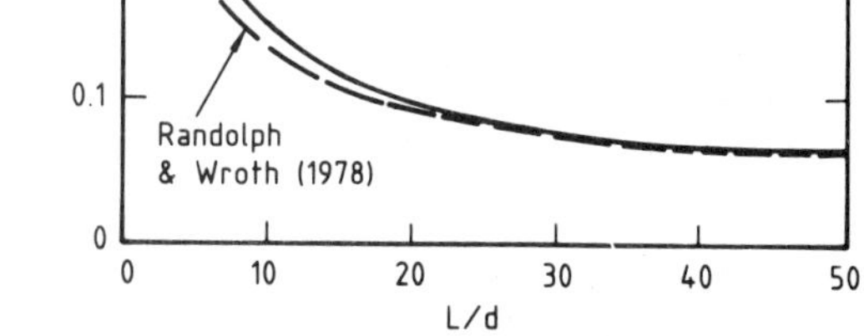

Fig. 3. Comparison between solutions for settlement of single pile in deep uniform elastic soil layer

Table 3. Some computer codes for pile analysis

Problem addressed	Program name	Reference	Remarks
Settlement of single pile	TAPILE	Poulos (1978)	*t–z* analysis continuum analysis
Settlement of pile group (can also be used for single piles)	DEFPIG	Poulos (1980)	Non-linear continuum analysis, using interaction factors
	GAPFIX	Hewitt (1988)	Non-linear continuum analysis, complete solution
	PIGLET	Randolph (1987)	Simplified continuum analysis, using interaction factors
	PGROUP	Banerjee & Driscoll (1976)	complete linear continuum analysis
	PILGPI	O'Neill *et al.* (1977)	Non-linear hybrid analysis
	—	Chow (1986)	Continuum-based non-linear hybrid analysis
Cyclic loading	RATZ	Randolph (1986)	Cycle-by-cycle *t–z* analysis
	AXCYC	Poulos (1988b)	Continuum based cycle-by-cycle analysis
	GAPCYC	Hewitt (1988)	Continuum-based cycle-by-cycle analysis for groups
External soil movements	PNEGA	Kuwabara & Poulos (1989)	Continuum-based analysis for downdrag on end-bearing piles
	PIES	Poulos (1989)	Continuum-based analysis for pile in shrinking or swelling soil

in which the Young's modulus decreases markedly as the axial strain level increases. The pile is 30 m long, 0·75 m in dia. and is located in a homogeneous soil layer 50 m deep. The initial tangent modulus of the soil (for very low strains) is 1056 MN/m^2, Poisson's ratio is taken as 0·49, and a constant shaft resistance of 220 kN/m^2 is assumed. Two values of pile Young's modulus have been considered, 30 000 MN/m^2 and 30 000 GN/m^2 (the latter would be unrealistically stiff in practice).

Analyses have been performed using the following three boundary element analyses

(*a*) an elastic–plastic continuum-based interface model, using a constant soil Young's modulus of 1056 MN/m^2

(*b*) a hyperbolic continuum-based interface model, using an initial tangent soil Young's modulus of 1056 MN/m^2, a constant shaft resistance of 220 kN/m^2, and a hyperbolic curve fitting constant R_f of 0·9 for both the shaft and the pile tip

(*c*) a load transfer analysis in which the interface response at each element is elastic–plastic, the linear portion being derived from the initial tangent soil modulus of 1056 MN/m^2.

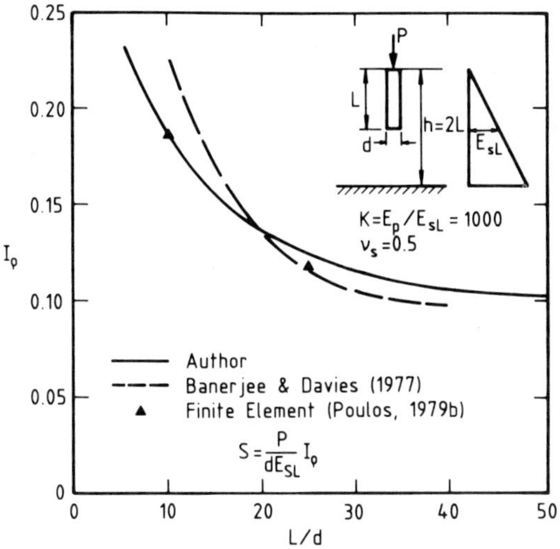

Fig. 4. Comparison between solutions for settlement of single pile in Gibson soil layer of finite depth

Figure 5 shows the comparisons between the four theoretical load–settlement curves. The following observations may be made

(*a*) for the more compressible (and realistic) pile, all four analyses correspond quite closely; indeed, the finite element analysis and the two elasto–plastic analyses agree remarkably well

(*b*) for the stiffer pile, the agreement between the computed load–settlement curves is not as close; at a load of one-half of the ultimate, the predicted settlements from three of the analyses are within ±10%, but as failure is approached, the curves diverge, and the hyperbolic model in particular predicts larger settlements than the other three methods.

It is clear that, for very stiff piles, the details of the pile–soil interface model have a greater influence on the load–settlement behaviour than for more compressible piles. It is, however, encouraging to observe that the simpler category 3B boundary element analyses are capable of predicting a very similar load–settlement response to that from a category 3C non-linear finite element analysis.

Pile groups

For a 3 × 3 group in a homogeneous elastic soil layer, Table 4 shows solutions from four different simplified approaches using computer programs listed in Table 3. There is generally very good agreement among the various solutions, for both very compressible and very rigid piles in a deep layer and in a relatively shallow layer. An exception is the PIGLET analysis for the very compressible pile group, in which case the predicted settlement is significantly greater than the other three methods. This difference appears to be associated with the approximations used in PIGLET for the interaction factors; when these same approximations are used in the DEFPIG analysis for the very compressible piles, similarly large settlements are obtained. However, for stiffer piles, these approximations appear to be quite adequate.

Figure 6 shows solutions for the settlement of groups in a Gibson soil. For a finite soil layer (Fig. 6a), the full boundary element solutions of Banerjee & Davies (1977) are in fair agreement with the DEFPIG solutions, although the latter settlements are larger. For an infinitely deep layer, Fig. 6b shows good agreement between the DEFPIG and PIGLET solutions.

Comparisons between boundary element and finite element methods have been made by Poulos (1976) and Pressley & Poulos (1986). The accuracy of some of the finite element solutions is difficult to assess, but in general they agree reasonably well with the boundary element solutions. Cheung *et al.* (1988) have developed an infinite layer method to obtain two-pile interaction factors for use in a pile group analysis. The

Table 4. Comparisons between solutions for group settlement in homogeneous soil: $L/d = 40$, $\nu_s = 0{\cdot}49$, 3 × 3 group, $s/d = 3{\cdot}0$; $S = (P_G/dE_s)I_G$

Method	Values of I_G			
	$h/L = 1{\cdot}67$		$h/L = \infty$	
	$K = 30$	$K = 30\,000$	$K = 30$	$K = 30\,000$
DEFPIG	0·063	0·021	0·069	0·029
GAPFIX	0·060	0·021	0·069	0·029
PIGLET	—	—	0·105	0·026
Butterfield & Douglas (1981)	0·058	0·020	0·067	0·025

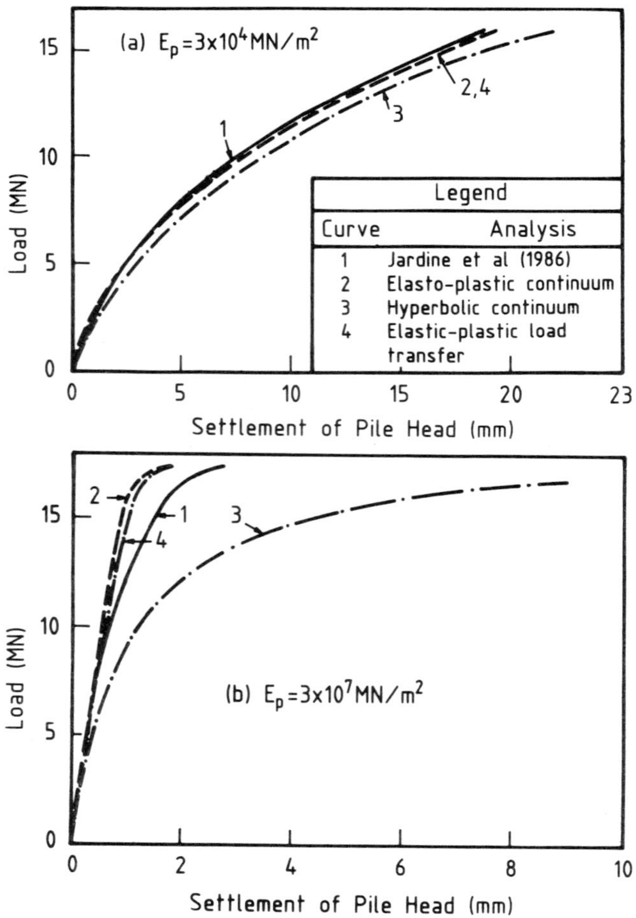

Fig. 5. Comparison between various non-linear analyses of single pile load–settlement behaviour: (a) $E_p = 3 \times 10^4$ MN/m²; (b) $E_p = 3 \times 10^7$ MN/m²

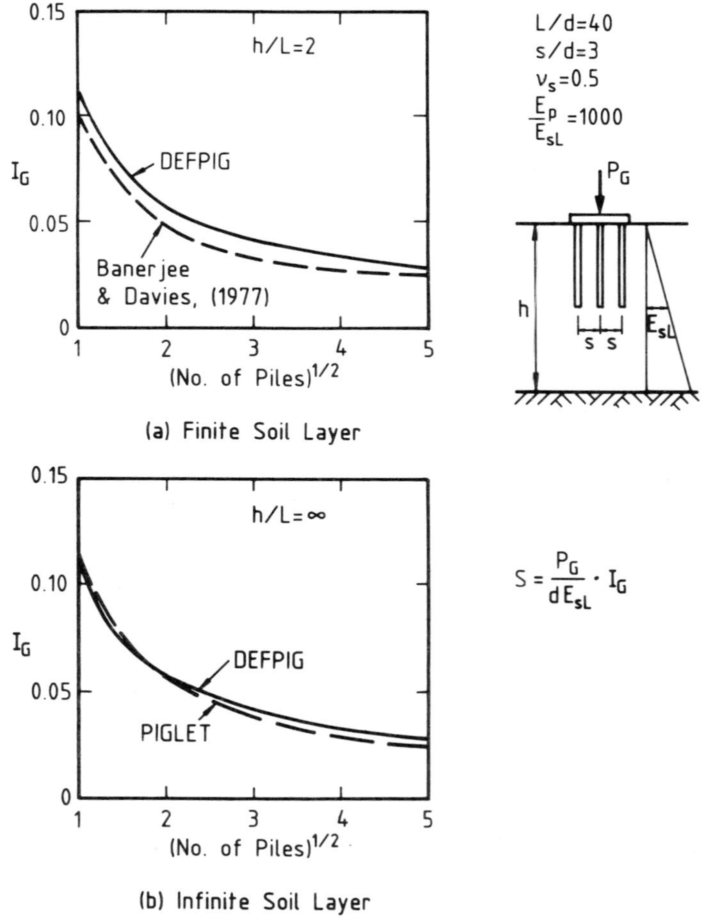

Fig. 6. Comparison between solutions for group settlement in Gibson soil (a) finite soil layer; (b) infinite soil layer

interaction factors thus obtained are in close agreement with the values of Poulos & Davis (1980) determined from the boundary element analysis.

O'Neill & Ha (1982) have compared the behaviour of pile groups predicted by the program DEFPIG with that predicted by a kind of hybrid analysis implemented by means of the program PILGPI. It is found that the two programs give comparable predicted behaviour, although different values of soil Young's modulus are required to give exact numerical agreement. These comparisons reinforce the fact that the value of soil modulus is not unique but must be selected carefully for use with the method of analysis employed.

In summary, while the foregoing comparisons demonstrate some differences among various methods, they also indicate generally satisfactory agreement between the boundary element solutions using interaction factors and solutions from the other approaches. Finite element analyses can be illuminating in that they reveal detailed behavioural characteristics, but it would appear that adequate practical predictions of group settlement can be obtained from simpler approaches based on boundary element analysis.

Single pile subjected to vertical soil movements

Small (1988) has used the finite element method to analyse a single end-bearing pile resting on a rigid base, in a soil layer which is subjected to external loading so that vertical soil movements occur and downdrag (or negative friction) forces are developed in the pile. The problem is illustrated in Fig. 7 which shows the finite element solution obtained for the development of pile head displacement with time as the soil layer consolidates. Also shown is the solution obtained by Poulos & Davis (1980) from a boundary element analysis, and it can be seen that, for a value of drained Poisson's ratio of the soil of 0·3, the agreement between the two solutions is close. There are, however, some differences in the detail of the developed shear stresses along the pile shaft, especially near the soil surface, possibly because of discretization inaccuracies in both analyses. Other independent analyses of the same general problem by means of continuum-based boundary element analyses (e.g. Chin, 1988; Kog *et al.*, 1986) are in general agreement with the Poulos & Davis solutions.

CHARACTERISTICS OF PILE BEHAVIOUR

In this section, some of the more significant characteristics of pile behaviour will be itemized. Most have been derived from theoretical analyses but in many cases are also supported by measurements made from laboratory and field tests. Unless otherwise stated, the solutions described will have been obtained from boundary element analyses based on elastic continuum theory, with an elastic, or an elastic–plastic interface model, using simplified distributions of soil Young's modulus and shaft resistance with depth (either constant or linearly increasing).

Four aspects of behaviour will be considered

(*a*) the load–settlement behaviour of a single pile under static axial loading
(*b*) the settlement of pile groups under static axial loading
(*c*) piles subjected to external soil movements
(*d*) piles subjected to cyclic axial loading.

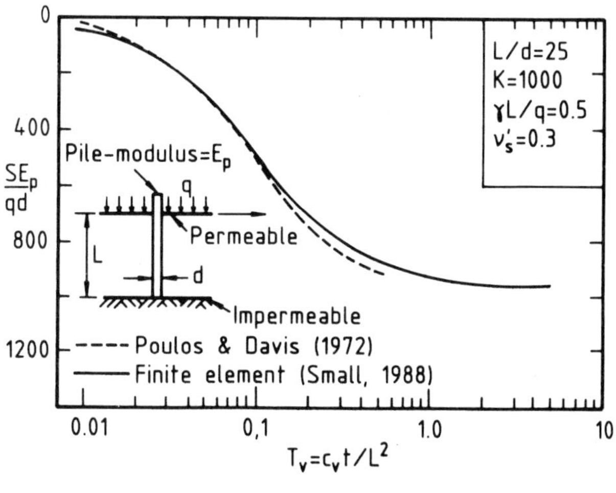

Fig. 7. Vertical deflexion of pile head with time

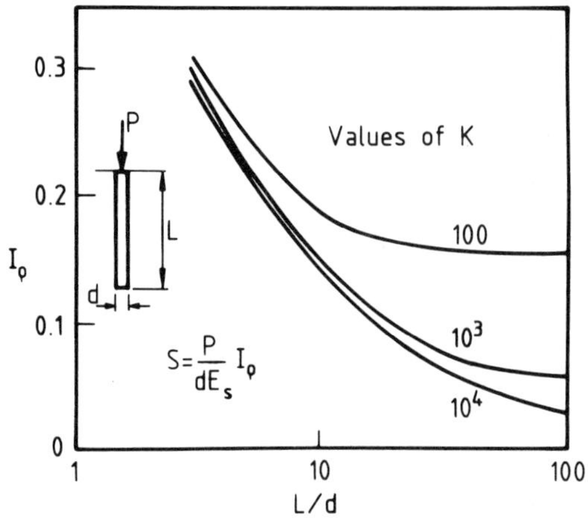

Fig. 8. Influence of dimensionless parameters L/d and K on settlement. Single friction pile in homogeneous soil

Illustrations of the points will be made frequently with reference to hypothetical problems involving realistic soil and pile parameters so that the practical implications of these points may be more readily appreciated.

Single pile under static loading

The settlement of a single pile is governed largely by the following dimensionless parameters

(*a*) the length-to-diameter ratio L/d

(*b*) the pile stiffness factor K, the ratio of the Young's modulus of the equivalent solid pile section E_p to the Y ug's modulus of the soil E_s

(*c*) E_b/E_s, the ratio of the Young's modulus of the bearing stratum at the pile tip to the Young's modulus of the soil.

For the case of a friction (or floating) pile in a homogeneous elastic soil, Fig. 8 shows that the settlement decreases as L/d and K increase. Experimental evidence from model tests in clay presented by Butterfield & Ghosh, 1977) demonstrates that the theory can give a realistic prediction of the effects of L/d.

The settlement of a pile is not significantly influenced by the nature of the bearing stratum if the pile is relatively slender and/or compressible. Fig. 9 compares the settlement of an end-bearing pile relative to a corresponding friction pile, for a typical value of K of 1000. For values of L/d in excess of about 50, the reduction in settlement due to the bearing stratum is less than 40%, even if the bearing stratum is very much stiffer than the overlying soil. Thus, if a reduction in settlement of a long pile is sought, there appears to be little to be gained by founding the pile tip on a stiffer underlying stratum. Increasing the diameter and/or the stiffness of the pile is likely to be more productive. It should also be noted that there is a critical length for a pile, beyond which further increase in length produces no further reduction in settlement. For a friction pile in a homogeneous soil, this critical length is given by the approximate expression (Hull, 1987)

$$L_c/d = \left(\frac{\pi E_p A_p}{E_s d^2}\right)^{0.5} \quad (7)$$

where A_p is the area of pile cross-section.

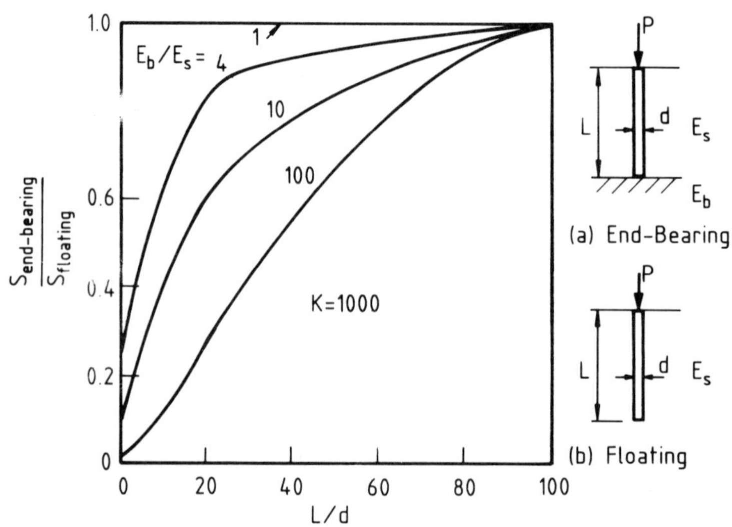

Fig. 9. Relative settlement of end bearing and floating pile. Homogeneous soil

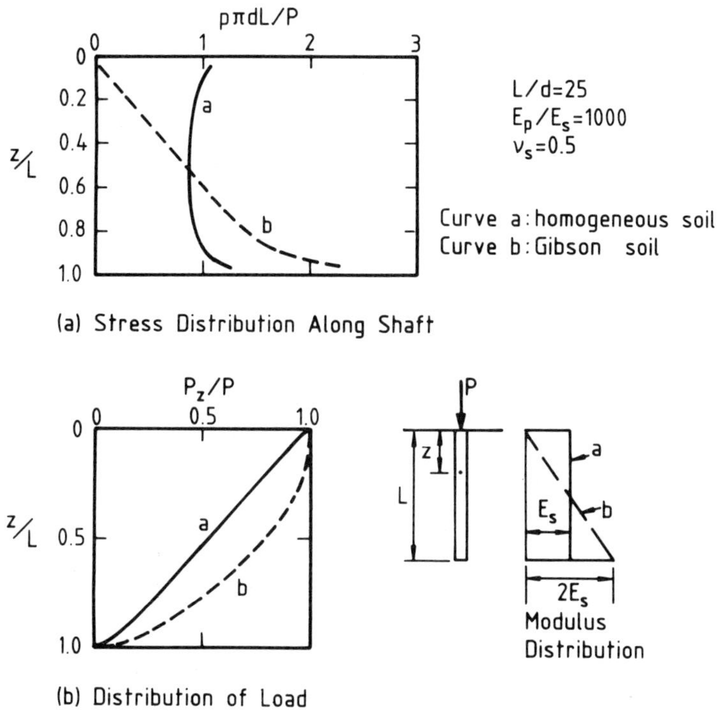

Fig. 10. Influence of distribution of soil Young's modulus on load transfer: (a) stress distribution along shaft; (b) distribution of load

The settlement and load transfer are influenced by the distribution of soil Young's modulus along the pile shaft. An example is given in Fig. 10 for a typical friction pile in two different soils: a homogeneous soil, and a Gibson soil in which the soil modulus increases linearly with depth, from zero at the soil surface. For the homogeneous soil, the distribution of shear stress τ is relatively uniform with depth, whereas for the Gibson soil, τ increases with depth. The similarity between the stress distribution and the distribution of soil Young's modulus may explain why load transfer approaches can give reasonable predictions of pile behaviour. For the same average value of Young's modulus along the pile shaft, the pile head settlements in this case are reasonably similar, and differ by only about 7%. However, the tip settlement for the pile in the Gibson soil is about 18% less than that of the pile in the homogeneous soil.

The major part of the final settlement of a single pile is immediate settlement, and occurs on application of the load because the load is transferred to the soil essentially by shear, with relatively little change in mean stress. For practical values of ν_s' (of the order of 0·3–0·4) and L/d, the theory suggests that the ratio S_i/S_{TF} of immediate to final settlement of a pile in a homogeneous soil is in excess of 0·85, and is almost independent of L/d. This theoretical conclusion is supported by the results of model tests on brass piles in kaolin, reported by Mattes & Poulos (1971) and field maintained loading tests, e.g. Whitaker & Cooke, 1966. A corollary to the above observations is that the *rate* of consolidation settlement of piles in clay is not likely to be an important consideration in design. Time effects stemming from creep at higher load levels are likely to be more important than consolidation time effects (Edil & Mochtar, 1988).

At normal working loads (of the order of 40–50% of the ultimate load), non-linear behaviour of the soil generally does not have a substantial influence on pile settlement. Careful model tests reported by Butterfield & Abdrabbo (1983) support this theoretical conclusion.

For a typical pile in a homogeneous soil, Fig. 11 gives some indication of the potential influence of non-linear soil behaviour on settlement. The ratio of settlement of the pile in a purely elastic soil S_{elas} to the settlement of the pile in a soil with hyperbolic response S, is plotted as a function of the load level P/P_u, where P_u is the ultimate load capacity. The initial tangent Young's modulus of the hyperbolic soil model is assumed equal to the modulus of the purely elastic soil. S_{elas}/S is gener-

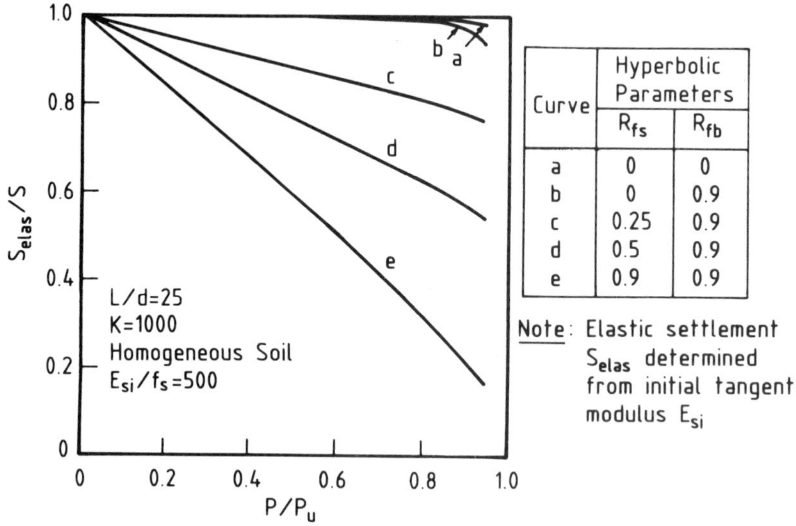

Fig. 11. Example of influence of nonlinearity on computed pile settlement. Hyperbolic interface model

ally less than unity (indicating that non-linearity results in an increase in the settlement), the extent of this increase depending on the hyperbolic parameters adopted. For the parameters considered to be most realistic (curves *c* and *d*), non-linearity causes an increase of between 11 and 25% in the settlement, as compared with the settlement determined from a purely elastic analysis using the initial tangent Young's modulus of the soil. It should be noted that, in this case, the non-linearity arising from pile–soil slip is only significant at load levels close to failure and therefore that the use of elastic theory with an appropriately reduced secant modulus would give a reasonable prediction of settlement. However, for slender compressible piles, pile–soil slip may have a more dominant influence on the non-linearity of the load–settlement response.

Residual stresses that remain in the pile after installation may influence the pile head stiffness and hence the calculated pile head movements. The theoretical analysis of Poulos (1987) has demonstrated that, for cases in which significant residual stresses remain (e.g. for a driven pile in dense sand), the stiffness of the pile head in tension may be smaller than in compression. Fig. 12 illustrates the point for an elastic–plastic interface. If no account is taken of residual stresses, the pile head stiffness in tension and compression is the same, but if residual stresses are allowed for, substantially larger movements can occur in tension than in compression. Under zero net load, the residual stresses are such that tensile pile–soil slip occurs over a significant amount of the pile shaft, and a residual compressive load is developed at the tip. Resistance to applied tensile loading comes primarily from the pile tip, the response of which is less stiff than the pile shaft; therefore, the pile head movements will be greater than if compression is applied.

If the pile–soil interface can strain-soften, both the load–settlement behaviour and the ultimate load capacity will be affected. The ultimate load is no longer statically determinate, but will depend on the relative stiffness of the pile, the ratio of peak to ultimate resistance, and the post-peak behaviour. Randolph (1983) presents solutions for a reduction factor to be applied to pile shaft capacities based on peak values of shaft resistance, in order to allow for the effects of progressive failure along the pile due to strain-softening.

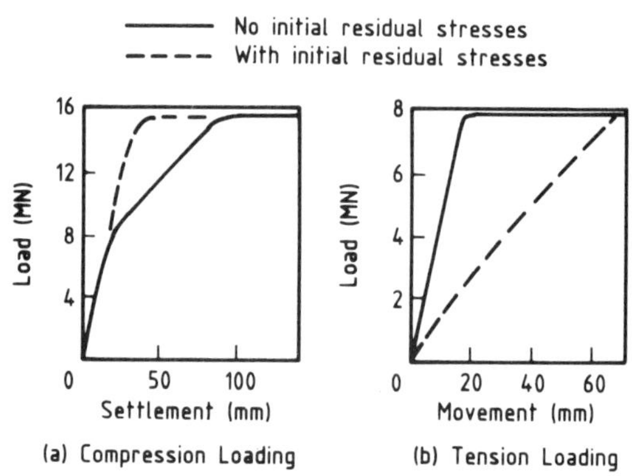

Fig. 12. Effect of residual stresses on load–deflection behaviour pile in sand (Poulos, 1987): (a) compression; (b) tension

These solutions reveal that reductions in the peak load capacity are only likely to be significant for relatively long compressible piles.

Pile group under static loading

Before the development of modern analytical techniques, it was commonly believed that no rational relationship existed between the behaviours of single piles and pile group. In his James Forrest Lecture, Terzaghi (1939) stated

> Both theoretical considerations and experience leave no doubt that there is no relation whatever between the settlement of an individual pile at a given load and that of a large group of piles having the same load per pile.

Such statements quite properly encourage caution in dealing with pile groups that contain very large numbers of piles; these are often better considered as a large block foundation. However, for groups that contain relatively few piles, it is possible to link theoretically the settlements of single piles and pile groups. For such groups, under normal working loads, it is convenient to characterize the influence of interaction between piles on the settlement in terms of two dimensionless quantities

(*a*) for two piles, the interaction factor α, which is defined in equation (6) and expresses the relative increase in settlement due to the presence of another pile
(*b*) for general pile groups, the group settlement ratio R_s, defined as

$$R_s = \frac{\text{settlement of group}}{\text{settlement of single pile at the same average load}} \qquad (8)$$

An alternative quantity to R_s is the group reduction factor R_G, also termed the efficiency factor by Butterfield & Douglas (1981) and Fleming *et al.* (1985). R_G can be defined as

$$R_G = \frac{\text{stiffness of group}}{\text{sum of individual pile stiffness}} \qquad (9)$$

For a group of n piles, R_s and R_G are related as

$$R_s = nR_G \qquad (10)$$

A comprehensive review of the load capacity and settlement of pile groups has been made by O'Neill (1983). Here, attention will be focused primarily on group settlement behaviour, as determined from analyses based on elastic continuum theory. Some of the more significant aspects of behaviour are discussed in the following.

Under working load conditions, pile group interaction depends largely on two sets of dimensionless parameters: those related to the soil and pile characteristics, and those related to the geometry of the piles and the pile group. The important soil and pile characteristics are the pile stiffness factor K, the ratio of E_b/E_s of Young's moduli of bearing stratum to soil, and the distribution of the soil Young's modulus E_s with depth. Fig. 13 illustrates the influence of these factors on the two-pile interaction factor α. It may be seen that α decreases as K decreases, or as E_b/E_s increases, or as the distribution of soil Young's modulus becomes less uniform with depth. Consequently, it should be expected that early published solutions for interaction factors, which are for a rigid pile ($K = \infty$) in a homogeneous mass ($E_b/E_s = 1$ and E_s constant with

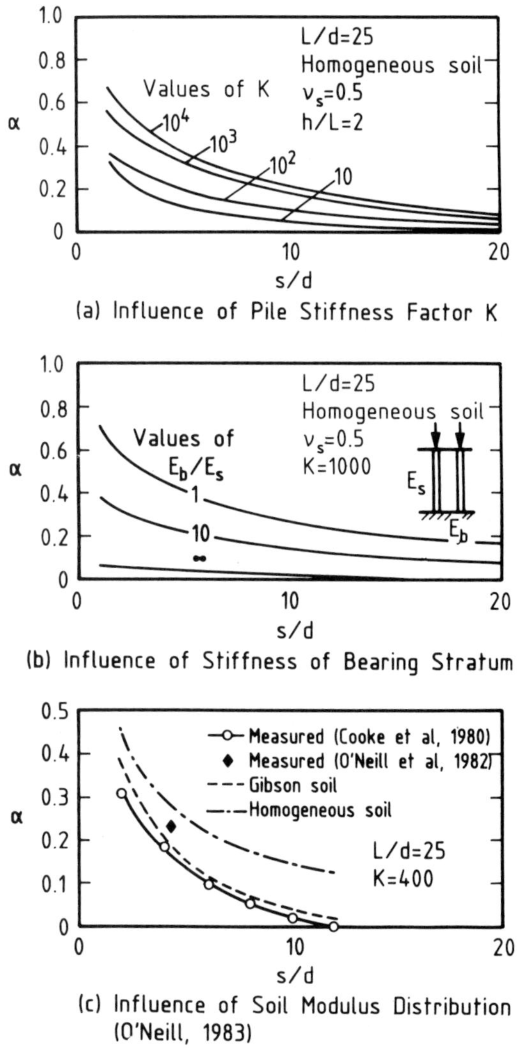

Fig. 13. Effect of soil–pile parameters on interaction factors: (a) pile stiffness factor *K*; (b) stiffness of bearing stratum; (c) soil modulus distribution (O'Neill, 1983)

depth), will overestimate the settlement interaction between piles in more realistic situations. The field measurements shown in Fig. 13(c) (O'Neill, 1983) support this contention; however, the theoretical interaction factors for a Gibson soil agree quite well with the measurements.

The primary geometric factors that influence group settlement interaction are the length-to-diameter ratio L/d, the relative spacing between the piles s/d, and the number of piles in the group. The effects of s/d are apparent from Fig. 13. For a value of s/d of 4 and $K = 1000$, the variation of settlement ratio R_s with L/d and n is shown in Fig. 14 for square groups of friction piles in a Gibson soil of finite thickness $h = 2L$. R_s increases as both L/d and n increase. The influence of L/d is small for L/d values in excess of 25. Fleming *et al.* (1985) have presented results for larger numbers of piles which suggest that R_s can be approximated as follows,

$$R_s \simeq n^{\omega} \tag{11}$$

where n is the number of piles and ω is an exponent which lies between 0·4 and 0·6 for most pile groups. This expression gives results which are reasonably consistent with those in Fig. 14. For a group with a rigid cap, and a given number of piles at a given centre-to-centre spacing, the settlement ratio does not depend to any significant extent on the precise geometrical configuration of the piles, e.g. for a group of 16 piles, R_s for a 4 × 4 configuration is very similar to that for an 8 × 2 configuration.

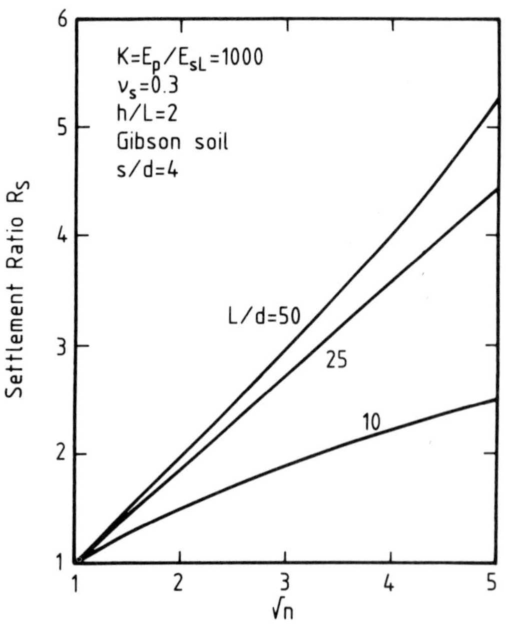

Fig. 14. Influence of geometric parameters on group settlement ratio

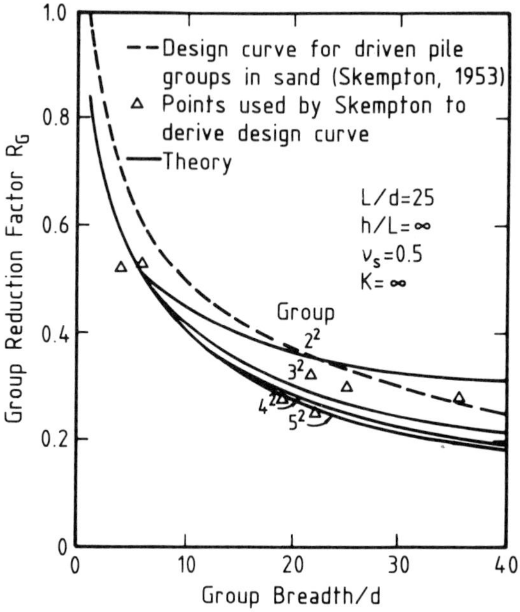

Fig. 15. Settlement against group breadth

For a given set of pile characteristics, the group reduction factor (or settlement efficiency factor) R_G depends largely on the breadth of the group. For groups that contain more than about nine piles, there is an almost unique relationship between R_G and group breadth. Fig. 15 shows a typical plot, together with data used by Skempton (1953) to derive an empirical design curve. Both the trend and magnitude of the theoretical curves agree well with these data.

In a pile group with a rigid cap, the distribution of load among the piles is generally non-uniform. In a square or rectangular group, the corner piles carry the greatest proportion of load, while those near the centre carry least. Poulos & Davis (1980), O'Neill *et al.* (1982) and Chow (1986) show that the theoretical trends are supported by field and model test data.

Interaction among piles in a group may be influenced by the stiffness of the soil between the piles. Most of the published theoretical solutions assume a soil to be horizontally homogeneous, with the soil Young's modulus between the piles the same as the value adjacent to each pile. However, in reality, the soil between the piles undergoes smaller strains and is likely to be stiffer than near the pile–soil interface, and interaction between the piles will be therefore reduced. A simplified analysis of this effect has been made by Poulos (1988a). For groups of piles, the presence of stiffer soil between the piles leads to a smaller settlement ratio R_s and a more uniform distribution among the piles, than is predicted by the conventional analysis.

The relative proportion of immediate settlement S_i to final settlement S_{TF} decreases as the size of the group increases. For groups of piles in either a homogeneous soil, with a typical value of ν_s' of 0·3, the ratio S_i/S_{TF} decreases from 0·93 for a single pile to 0·85 for a 25-pile group. Similar values are found for groups in a Gibson soil. The pile stiffness factor K has little influence on S_i/S_{TF}. For pile groups, the consolidation settlement (and hence the *rate* of consolidation settlement) is more important than for a single pile, but it is still likely to be the minor component of settlement unless soft compressible layers exist beneath the pile tips.

The stiffness of underlying soil layers may have a significant influence on pile group interaction and settlement. Fig. 16 shows an example of the influence of the relative stiffness of the underlying soil on the settlement of a pile group. Clearly, the presence of a softer layer ($E_b/E_s < 1$) may substantially increase the settlement, as compared with the case of a homogeneous soil mass ($E_b/E_s = 1$). Also shown is the settlement computed from the approximate approach described by Poulos & Mattes (1971), in which the group is replaced by an equivalent block. This solution agrees well with the solution computed from the interaction factor method.

The effect on group settlement of the pile cap being in contact with the soil is relatively small unless the pile spacing is large and the group is relatively small. Even for piles at an unusually large centre-to-centre spacing of 10 diameters, the reduction in settlement due to cap contact is only about 5%. Therefore, for most practical purposes, the influence of pile cap contact on settlement at working loads can be ignored.

Piles subjected to external soil movements

There are several circumstances under which loading may be induced in piles by external soil movements, but attention will be confined here to two problems

(*a*) end bearing piles subjected to negative friction by settlement of the surrounding soil; the soil movement is assumed to occur over the entire depth of the soil layer, and to vary linearly with depth

(*b*) floating piles in a relatively stiff expansive clay, subjected to swelling or shrinking move-

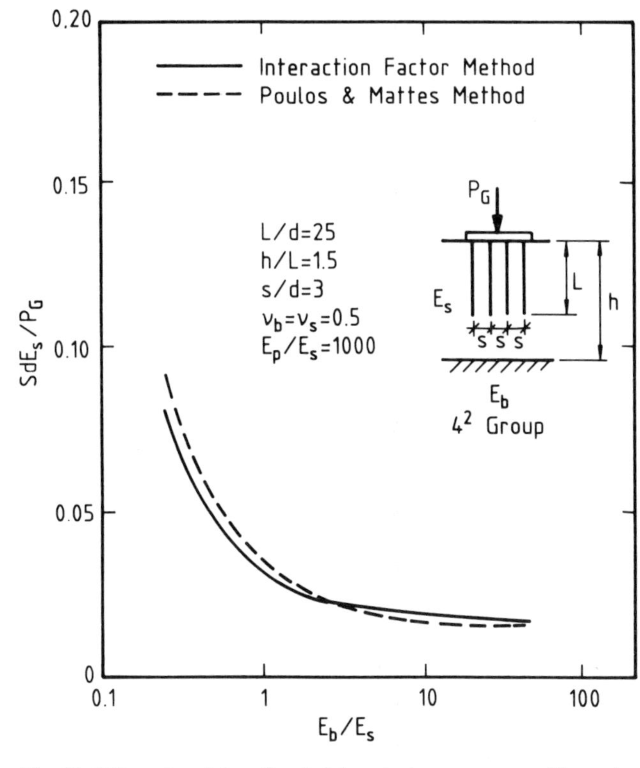

Fig. 16. Effect of modulus of underlying stratum on group settlement

ments; these movements are generally assumed to vary linearly with depth, but to extend to only a limited distance below the surface.

Negative friction on end-bearing piles

Solutions from purely elastic theory show that the movement and downdrag force induced in the pile depend on the dimensionless parameters L/d, K and E_b/E_s (Poulos & Davis, 1980; Chin, 1988). However, slip at the pile–soil interface generally plays a dominant role in pile behaviour. Poulos & Davis (1980) present solutions that indicate the circumstances under which slip occurs along virtually the entire pile shaft. For typical normally consolidated or lightly-overconsolidated soils, if a surface pressure q is applied to the soil, full slip is likely to occur if

$$q > (0{\cdot}3 - 0{\cdot}5)\gamma' L \qquad (12)$$

where γ' is the submerged unit weight of soil and L is the embedded length of pile.

The soil surface movement required to develop full slip depends on the relative stiffness of the pile and the distribution of soil modulus and shaft resistance with depth. Values typically range between about 0·5% of diameter for relatively short stiff piles in soft soils, to 5% of diameter for relatively long piles.

In contrast to the case of conventional axial loading, the development of pile–soil slip is advantageous. It leads to reduced pile movement and downdrag forces as compared with the purely elastic interface condition. A typical example is shown in Fig. 17. Pile–soil slip starts when the soil surface movement is about 10 mm, and is almost complete at about 25 mm. Elastic theory would seriously overestimate the pile force and movement for soil movements in excess of about 20 mm.

Because of the dominant influence of pile–soil slip, the soil model used in the analysis of negative friction generally does not have a major effect on the solution. Fig. 18 shows a typical example in which three models, the elastic–plastic continuum, the elastic–plastic t–z model, and the hyperbolic t–z model, have been used, with the same values of initial tangent Young's modulus and shaft resistance. Before the development of full slip, the continuum model tends to give values of downdrag force and pile movement which are about 30% smaller than the other two methods, but as pile–soil slip develops, the differences become less. It will also be observed that the non-

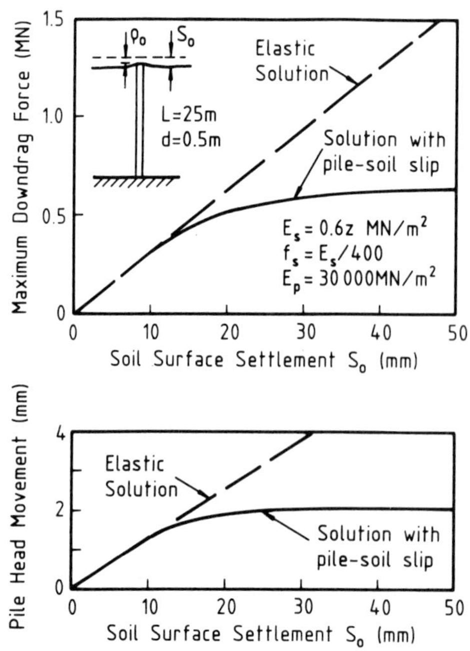

Fig. 17. Influence of pile–soil slip on downdrag force and pile movement

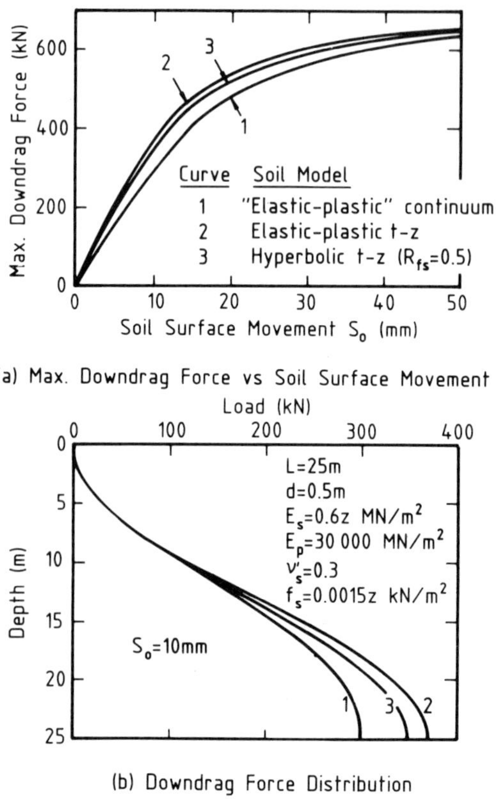

Fig. 18. Influence of soil model on computed downdrag force in end-bearing pile

linear hyperbolic *t–z* model gives results which are quite close to those from the elastic–plastic *t–z* model (i.e. the details of the non-linearity appear to be relatively unimportant).

In contrast to conventional axial loading, group effects stemming from interaction among piles have a beneficial effect in the case of negative friction problems. As compared with a single isolated pile, there is a tendency for reduced downdrag forces and pile movements in the group, especially for inner piles within the group (Chin, 1988; Kuwabara & Poulos, 1989). For inner piles, the downdrag force is reduced to a small fraction of the value for an isolated single pile and even the corner piles have a maximum force only about one-third of the single pile value. Thus, in designing a group of piles to withstand the effects of negative friction, it would be extremely conservative to assume that all piles in the group are subjected to the downdrag force which would be developed in a single pile.

Piles in expansive soils

Because expansive soils are frequently relatively stiff, elastic interface conditions are more likely to be relevant for piles in expansive soils than for piles in soft clay subjected to downdrag. For purely elastic interface conditions, and a relatively short pile in a uniform soil, Fig. 19 shows the effect of the relative depth of soil movement z_s/L and the pile stiffness factor K, on the relative pile head movement ρ_p/S_0 (where S_0 is the soil surface movement). The main factor that influences pile head movement is the depth of soil movement; the influence of pile compressibility is relatively small.

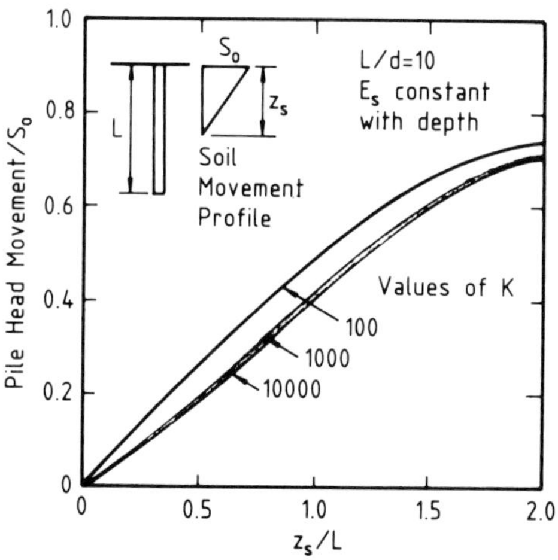

Fig. 19. Elastic solutions for pile movement in expansive soil—uniform pile diameter

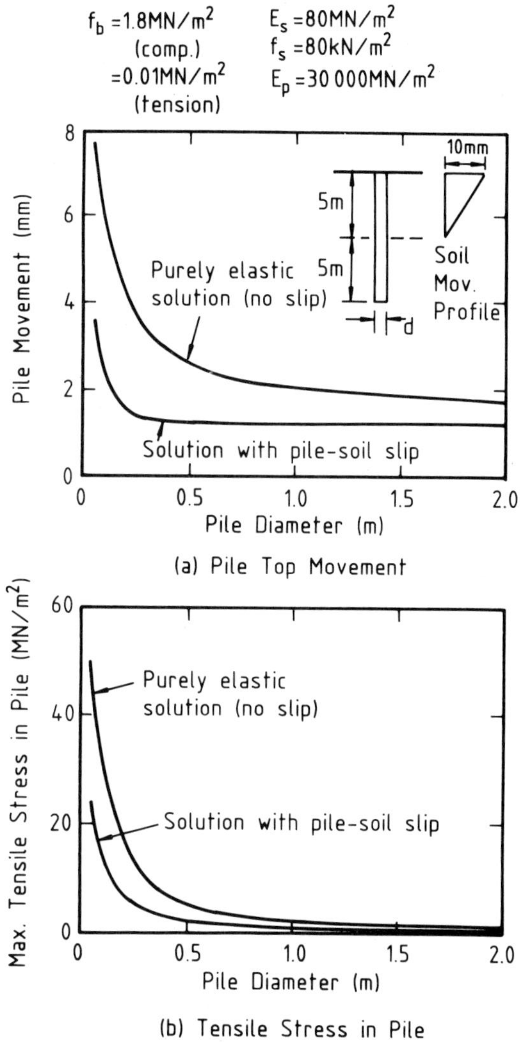

Fig. 20. Effect of pile diameter—pile in expansive soil: (a) pile top movement; (b) tensile stress in pile

For a given length of pile, the diameter has relatively little effect on the movement of the pile. Fig. 20 shows an example of the variation of pile head movement and maximum tensile stress in the pile as a function of diameter. Solutions for both a purely elastic interface, and with consideration of pile–soil slip, are shown. The elastic solutions are conservative and give larger movements and stresses than the solutions incorporating slip. For pile diameters between about 0·2 and 2·0 m, the pile head movement is almost constant when pile–soil slip is accounted for. The maximum tensile stress in the pile increases with decreasing diameter, but does not reach a significant level until the diameter is less than 0·2 m. The important implication of these results is that it is possible to use relatively small-diameter piles to suppress foundation movements in expansive soils. This point was also made by Donaldson (1967) who described the successful use of small-diameter piles to support brick buildings on expansive soils.

The use of an enlarged base tends to reduce the pile movement, but this reduction is only effective for pile length to diameter ratios less than 15. The maximum effect appears to be obtained when the bell is located at or just below the active zone of soil movement.

In a group of piles in expansive soil, there is a tendency for the pile forces and movements to be reduced because of interaction among the piles. This theoretical tendency is supported by measurements reported by Blight (1984) on instrumented piles within a seven-pile group.

Piles subjected to cyclic axial loading

Laboratory test data on model piles suggest that one of the most important effects of cyclic loading is to cause a reduction (or degradation) of pile shaft resistance. This degradation depends on the amplitude of cyclic displacement to which the pile is subjected, and on the number of cycles. There appears to be a threshold cyclic displacement below which no degradation occurs, this threshold being of the same order as the displacement to cause pile–soil slip under static loading. As the cyclic displacement increases beyond this value, there is an increasing loss of skin friction; this loss also increases as the number of cycles increases. There are currently insufficient experimental data to define clearly whether the degradation of shaft resistance depends on the *absolute* or the *relative* cyclic displacement, but some data from model and field tests on grouted piles in calcareous sediments (Lee, 1988) suggests the latter. Fig. 21 summarizes this data and plots the degradation factor for shaft resistance as a function of the normalized cyclic slip displacement (i.e. the total cyclic displacement minus the threshold value) divided by pile diameter *d*. A reasonably consistent relationship is obtained for model pile diameters of between 24 and 77 mm, and field pile diameters between 440 and 589 mm. Fig. 21 shows that major reductions in shaft resistance can occur for this case, with the possibility of the value after severe cyclic loading being less than 10% of the initial value for static loading.

Using data such as that in Fig. 21, it is possible to analyse the behaviour of a pile subjected to cyclic axial loading. Some findings from these analyses are presented here.

A useful means of portraying the behaviour of piles subjected to cyclic loading is by means of a cyclic stability diagram (Poulos, 1988b). This is a normalized plot of mean load against cyclic load (each being divided by the static compressive load capacity Q_c), in which at least three regions can be identified

(*a*) a cyclically stable region in which cyclic loading has no influence on the axial capacity of the pile
(*b*) a cyclically metastable region in which cyclic loading causes some reduction of axial load capacity, but the pile does not fail within a specified number of cycles
(*c*) a cyclically unstable region in which cyclic loading causes sufficient degradation for the pile to fail within a specified number of cycles of load.

In addition, Lee (1988) has identified a fourth zone, a serviceability loss zone, which lies within the metastable zone and in which excessive settlement of the pile develops within a specified number of cycles.

Figure 22 shows an example of a cyclic stability diagram for a 100 m long offshore grouted pile (Lee, 1988) in a soil whose Young's modulus and shaft resistance increase linearly with depth. In this case, stable behaviour will occur only if the cyclic load is less than about 20% of the ultimate compressive static load. Comparisons between theory and model and field test data, presented by Poulos (1988b) and Lee (1988), show quite good agreement and tend to confirm the validity of the concept of the cyclic stability diagram.

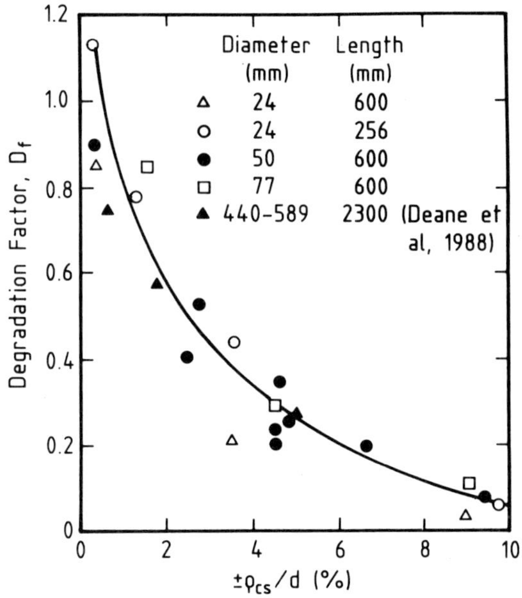

Fig. 21. Effect of normalized cyclic slip displacement on D_f with different pile diametes (after Lee, 1988)

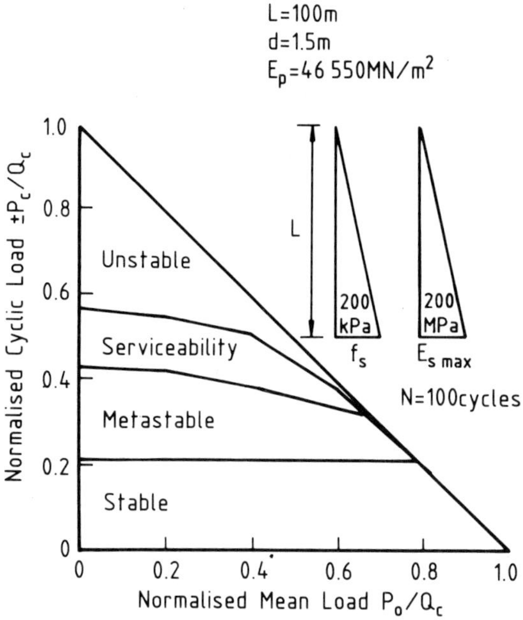

Fig. 22. Typical cyclic stability diagram for offshore pile; $N = 100$ cycles (lee, 1988)

It has been observed that two-way cyclic loading (about zero mean load) has a more severe effect on piles than one-way cyclic loading (for which the minimum load is zero–i.e. the cyclic load equals the mean load). This observation can readily be verified from the cyclic stability diagram. Referring to Fig. 22, for two-way cyclic loading, the unstable zone for $N = 100$ cycles begins at a normalized cyclic load of about 0·57; this is also the maximum load that can be sustained under two-way cyclic loading for 100 cycles. For one-way cycling ($P_c = P_0$), the unstable zone is reached when $P_c/Q_c = 0{\cdot}46$, so that the maximum load that can be sustained under one-way loading conditions is 0·92 Q_c, substantially greater than the value of 0·57 Q_c for one-way loading.

As piles become shorter or stiffer, the positions of the boundaries between the various zones alter; both the stable and unstable zone boundaries increase, but the metastable zone shrinks. Fig. 23 shows the effect of pile length on the zone boundaries for a 1·5 m dia. driven steel tube pile in clay (Poulos, 1988b). For long compressible piles, there is a large metastable zone. Such piles exhibit a gradual decrease in load capacity as the cyclic load level is increased. In contrast, short stiff piles have a very limited metastable zone, and may fail abruptly after small increases in cyclic load above the stable zone. It may thus be considered that long compressible piles exhibit a

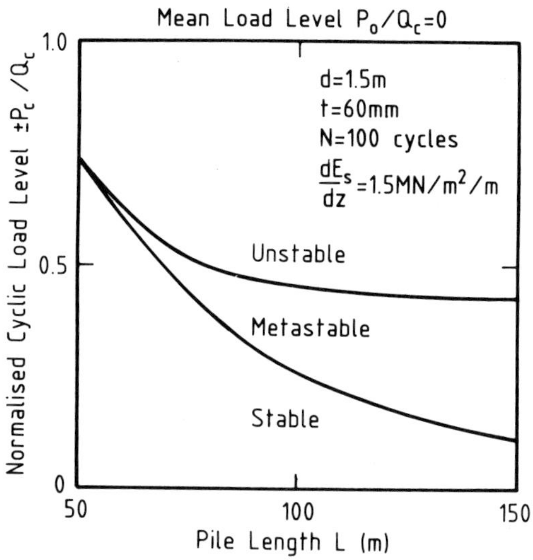

Fig. 23. Effect of pile length on stability diagram zone boundaries. Driven pile in clay (Poulos, 1988b)

ductile cyclic behaviour, while short stiff piles show a brittle cyclic behaviour.

Group effects may influence the cyclic behaviour of the piles, but to a relatively minor extent. Hewitt (1988) has performed analyses of a single pile, and four- and eight-pile groups, using the Matlock & Foo model for degradation of shaft resistance. For two-way loading of relatively stiff piles, these results are shown in Fig. 24 together with laboratory data on model piles in kaolin. The theory shows that the maximum cyclic load

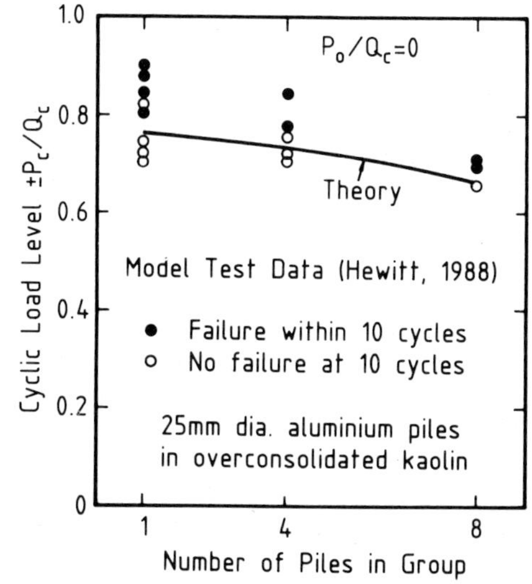

Fig. 24. Influence of number of piles on failure under two-way cyclic loading

that can be sustained decreases as the number of piles increases. The experimental results for 10 cycles are in good agreement with the theoretical curve.

ESTIMATION OF GEOTECHNICAL PARAMETERS

The most significant parameters required for many of the category 2A, 2B, 3A and 3B analyses of pile behaviour under static loading are as follows

(*a*) the shaft resistance f_s
(*b*) the end bearing resistance f_b
(*c*) Young's modulus of the soil E_s
(*d*) Poisson's ratio of the soil ν_s
(*e*) the hyperbolic curve-fitting constant R_f.

For prediction of cyclic axial response, a number of other parameters are required, and some limited data on these is presented by Poulos (1988c). Attention will be concentrated here on the above parameters, and in particular on f_s, f_b and E_s.

For calculation of axial pile load capacity, f_s and f_b must be estimated as accurately as possible. For the calculation of settlement resulting from direct axial loading, the theoretical solutions reveal that the choice of an appropriate value of E_s is generally crucial, unless the piles are long and compressible. For piles in soil subjected to external movement, the pile behaviour is generally much less dependent on E_s and, provided that the soil movement is known,* an approximate estimate of E_s may be adequate, although reasonable estimates of shaft and end-bearing resistance are desirable.

Methods of determining parameters

For evaluating the parameters for static pile response, a number of methods can be contemplated, including

(*a*) laboratory testing
(*b*) appropriate interpretation of field pile load tests
(*c*) empirical correlations with laboratory-determined parameters
(*d*) empirical correlations with the results of in-situ test data.

Conventional laboratory tests, such as triaxial or oedometer tests, are generally not suitable for direct measurement of the soil Young's modulus as they do not follow, even approximately, the stress path which the soil adjacent to the pile follows. Laboratory model pile tests may overcome this deficiency to some extent, but may not accurately reflect the behaviour of prototype piles because of the presence of scale effects, particularly for piles in sand. Some potential exists for more sophisticated tests such as the constant normal stiffness (CNS) direct shear test (Johnston & Lam, 1984; Ooi & Carter, 1987), and this type of test has been used in the design of grouted piles in offshore carbonate sediments (Johnston *et al.*, 1988). However, the direct utilization of laboratory tests for pile design is infrequent in practice, and still requires further research before it can be applied with confidence.

The most reliable means of determining f_s and E_s is by backfiguring from the results of pile load tests. Methods for interpreting the pile load test data have been detailed by Poulos & Davis (1980) and Stewart & Kulhawy (1981), among others. Such methods are particularly effective if the pile is instrumented so that details of the load transfer along the pile shaft are available; it is then possible to determine detailed distributions of soil modulus and limiting pile–soil friction along the pile shaft.

A variant of the pile loading test is the pile section test. This involves the testing of a series of relatively short rigid sections at different depths, in order to determine the distribution of limiting pile–soil friction and soil modulus with depth. In such tests, the sections must be installed in a similar manner to the prototype pile in order to obtain appropriate data. Examples of section tests have been reported by Hyden *et al.* (1988), and Williams & van der Zwaag (1988); in both cases, tests were carried out on grouted pile sections in marine calcareous sediments.

In most practical situations, it is not possible to carry out such testing, at least in the early stages of design. Resort is frequently made to correlations between the pile design parameters and laboratory or field test data.

Shaft resistance f_s

Tables 5 and 6 summarize available methods for determining the shaft resistance f_s from laboratory strength data, for both driven and bored piles. Effective stress approaches can be used for all soil types, whereas a total stress approach is still adopted commonly for piles in clay. The parameters α and β (or K and δ) are usually obtained from empirical correlations, despite the fact that the effective stress β approach is fundamentally sound and falls into category 2.

A summary of some suggested correlations between f_s and the standard penetration resistance N are given in Table 7. Considerable varia-

* The soil movement is treated here as an independent variable, although it will be influenced to some extent by the soil modulus.

Table 5. Shaft resistance f_s for driven piles, determination from laboratory strength data

Soil type	Equation	Remarks	Reference
Clay	$f_s = \alpha c_u$	$\alpha = 1{\cdot}0$ ($c_u \leqslant 25$ kN/m^2) $\alpha = 0{\cdot}5$ ($c_u \geqslant 70$ kN/m^2) Linear variation in between	API (1984)
		$\alpha = 1{\cdot}0$ ($c_u \leqslant 35$ kN/m^2) $\alpha = 0{\cdot}5$ ($c_u \geqslant 80$ kN/m^2) Linear variation in between. Length factor applies for $L/d > 50$	Semple & Rigden (1984)
		$\alpha = \left(\frac{c_u}{\sigma_v'}\right)_{nc}^{0{\cdot}5}\left(\frac{c_u}{\sigma_v'}\right)^{-0{\cdot}5}$ for $\left(\frac{c_u}{\sigma_{vo}'} \leqslant 1\right)$ $\alpha = \left(\frac{c_u}{\sigma_v'}\right)_{nc}^{0{\cdot}5}\left(\frac{c_u}{\sigma_v'}\right)^{-0{\cdot}25}$ for $\left(\frac{c_u}{\sigma_{vo}'} \geqslant 1\right)$	Fleming *et al.* (1985)
	$f_s = \beta\sigma_v'$	$\beta = (1 - \sin\phi')\tan\phi'(\text{OCR})^{0{\cdot}5}$	Burland (1973) Meyerhof (1976)
Silica sand	$f_s = \beta\sigma_v'$ ($f_s \not> f_{slim}$)	$\beta = 0{\cdot}15$–$0{\cdot}35$ (compression) $0{\cdot}10$–$0{\cdot}24$ (tension)	McClelland (1974)
		$\beta = 0{\cdot}44$ for $\phi' = 28°$ $0{\cdot}75$ for $\phi' = 35°$ $1{\cdot}2$ for $\phi' = 37°$	Meyerhof (1976)
		$\beta = (K/K_0) \,.\, K_0 \,.\, \tan(\phi \,.\, \delta/\phi)$ δ/ϕ depends on interface materials (range $0{\cdot}5$–$1{\cdot}0$); K/K_0 depends on installation method (range $0{\cdot}5$–$2{\cdot}0$). K_0 = coefficient of earth pressure at rest, and is a function of OCR	Stas & Kulhawy (1984)
Uncemented calcareous sand	$f_s = \beta\sigma_v'$	$\beta = 0{\cdot}05$–$0{\cdot}1$	Poulos (1988d)

tions occur in these correlations, particularly for bored and cast-in-place piles.

Figures 25 and 26 show values of f_s correlated with static cone resistance q_c. These relationships have been developed by the Author from the correlations suggested by Bustamante & Gianeselli (1982) and cover a wide range of pile types in both clay and silica sand. The classification of these pile types is shown in Table 8. It should be emphasized that several other correlations have been proposed and that wide variations exist between some of these. Fig. 27 shows an example of this variability, for driven piles in silica sand. The potential inaccuracy of shaft capacity prediction using category 1 correlations, especially for loose sands, is clearly demonstrated.

Schmertmann (1975; 1978) proposes a different approach to the utilization of cone data, whereby the pile shaft resistance is related to the measured sleeve resistance of the penetrometer. Corrections are applied, depending on soil type, pile type, relative pile length, and depth below the surface. Robertson *et al.* (1985) have found the method proposed by Schmertmann to provide a more reliable prediction than the direct correlation to q_c, when applied to piles in a clayey silt.

Extensive correlations have been developed in France between shaft resistance and the limit pressure p_l deduced from pressuremeter measurements. Based on the results of over 300 pile load tests at more than 100 sites, Bustamante *et al.* (1987) have proposed correlations, similar in nature to those with the CPT in Figs 25 and 26.

Correlations such as those outlined must always be employed with caution, as a number of other factors may also influence shaft resistance e.g. the presence of overlying layers (Tomlinson, 1977).

Table 6. Shaft resistance f_s for bored piles, determination from laboratory strength data

Soil type	Equation	Remarks	Reference
Clay	$f_s = \alpha c_u$	$\alpha = 0{\cdot}45$ (London clay)	Skempton (1959)
		$\alpha = 0{\cdot}7$ times value for driven displacement pile	Fleming *et al.* (1985)
	$f_s = K \tan \delta\sigma_v'$	K is lesser of K_0 or $0{\cdot}5(1 + K_0)$	Fleming *et al.* (1985)
		$K/K_0 = 2/3$ to 1; K_0 is function of OCR; δ depends on interface materials	Stas and Kulhawy (1984)
Silica sand	$f_s = \beta\sigma_v'$	$\beta = 0{\cdot}1$ for $\phi' = 33°$ $0{\cdot}2$ for $\phi' = 35°$ $0{\cdot}35$ for $\phi' = 37°$	Meyerhof (1976)
		$\beta = F \tan(\phi' - 5°)$ where $F = 0{\cdot}7$ (compression) & $0{\cdot}5$ (tension)	Kraft & Lyons (1974)
Uncemented calcareous sand	$f_s = \beta\sigma_v'$ ($f_s \not> f_{slim}$)	$\beta = 0{\cdot}5$ to $0{\cdot}8$ $f_{slim} = 60$ to 100 kN/m^2	Poulos (1988d)

End bearing resistance f_b

Table 9 summarizes the two usual methods used for assessment of the end bearing resistance of piles using laboratory data. A total stress approach is almost invariably used for piles in clay, whereas an effective stress approach is used for piles in sand. Two main problems arise in the latter case

(*a*) some experimental evidence suggests that a limiting value of f_b may occur when the pile is embedded more than 10 to 20 diameters; no entirely satisfactory method of theoretical analysis has been developed to take this into account and an empirical upper limit to f_b is usually specified

(*b*) the theoretical bearing capacity factor N_q is very sensitive to the angle of internal friction ϕ'; for values of ϕ' in excess of about 35°,

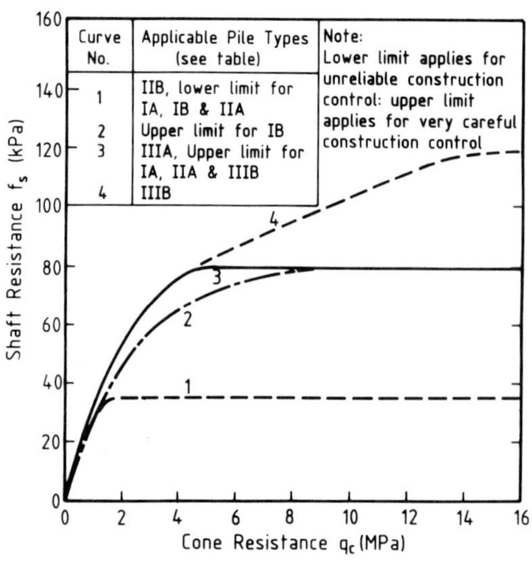

Fig. 25. Design values of shaft resistance for piles in clay (based on Bustamante & Gianeselli, 1982)

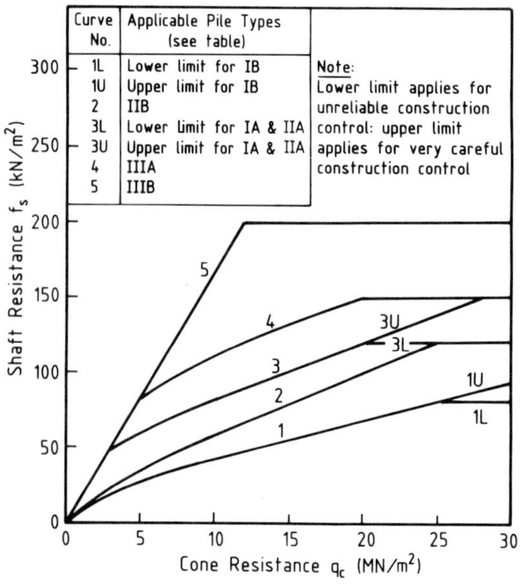

Fig. 26. Design values of shaft resistance for piles in sand (based on Bustamante & Gianeselli, 1982)

Table 7. Correlations between shaft resistance f_s and SPT value, with $f_s = \alpha + \beta N$ kN/m²

Pile type	Soil type	α	β	Remarks	Reference
Driven displacement	Cohesionless	0	2·0	f_s = average value over shaft $\bar{N}$ = average SPT along shaft Halve f_s for small displacement pile	Meyerhof (1956) Shioi & Fukui (1982)
	Cohesionless & cohesive	10	3·3	Pile type not specified $50 \geqslant N \geqslant 3$ $f_s \not> 170$ kN/m²	Decourt (1982)
	Cohesive	0	10		Shioi & Fukui (1982)
Cast in place	Cohesionless	30	2·0	$f_s \not> 200$ kN/m²	Yamashita *et al.* (1987)
		0	5·0		Shioi & Fukui (1982)
	Cohesive	0	5·0	$f_s \not> 150$ kN/m²	Yamashita *et al.* (1982)
		0	10·0		Shioi & Fukui (1982)
Bored	Cohesionless	0	1·0		Findlay (1984) Shioi & Fukui (1982)
		0	3·3		Wright & Reese (1979)
	Cohesive	0	5·0		Shioi & Fukui (1982)
	Cohesive	10	3·3	Piles cast under pentonite $50 \not> N \not> 3$ $f_s \not> 170$ kN/m²	Decourt (1982)
	Chalk	−125	12·5	$30 > N > 15$ $f_s \not> 250$ kN/m²	After Fletcher & Mizon (1984)

small changes in ϕ' can theoretically lead to large changes in N_q, although the effects of soil compressibility are then more important and may reduce the dependence of N_q on ϕ'.

Table 10 shows some empirical correlations between f_b and the standard penetration resistance in the vicinity of the pile tip. These correlations indicate that bored or cast-in-place piles develop a significantly smaller end-bearing resistance than do driven piles.

Bustamante & Gianeselli (1982) suggested correlations between f_b and the average cone penetration resistance value near the pile tip. The correlation factor to the latter value is between 0·3 and 0·55, depending on soil and pile type. These correlations contrast with procedures such as those proposed by Belcotec (1985) and De Ruiter & Beringen (1979), in which a factor of unity is applied to the average value (computed differently than in the Bustamante & Gianeselli approach). However, the latter approaches are confined to driven piles, whereas the Bustamante & Gianeselli approach is more general, simpler to apply, and probably more conservative.

Baguelin *et al.* (1986) and Bustamante *et al.* (1987) have related f_b to the pressuremeter limit pressure p_l by way of a factor k_p which depends on pile and soil types. For non-displacement piles, k_p is 1·2 for clays and silts and 1·1 for sands, whereas for driven piles, k_p is 1·8 for clays and silts, and ranges between 3·2 and 4·2 for sands.

Table 8. Classification of pile types (Bustamante & Gianeselli, 1982)

Pile category	Type of pile
IA	Plain bored piles, mud bored piles, hollow auger bored piles, cast screwed piles Type I micropiles, piers, barrettes
IB	Cased bored piles Driven cast piles
IIA	Driven precast piles Prestressed tubular piles Jacked concrete piles
IIB	Driven steel piles Jacked steel piles
IIIA	Driven grouted piles Driven rammed piles
IIIB	High pressure grouted piles ($d > 0{\cdot}25$ m) Type II micropiles

Soil Young's modulus E_s

Ideally, for piles in clay, a distinction should be made between the undrained Young's modulus, used for calculations of immediate or undrained settlement, and the drained Young's modulus, used for calculations of total settlement of a pile. However, for many clays, the difference between the drained and undrained modulus values is not great and the approximate nature of most correlations makes such a distinction impractical. It is therefore suggested that the correlations presented in this Paper should be considered to apply to

Table 9. End bearing capacity of pile tip, f_b, determination from laboratory data

Soil type	Equation	Remarks	Reference
Clay	$f_b = N_c c_{ub}$	$N_c = 9$ for $L/d \geqslant 3$ c_{ub} = value of c_u in vicinity of pile tip	Skempton (1959)
Silica sand*	$f_b = N_q \sigma_v' \ngtr f_{blim}$†	$N_q = 40$	API (1984)
		N_q plotted against ϕ'	Berezantzev *et al.* (1961)
		N_q related to ϕ', relative density and mean effective stress	Fleming *et al.* (1985)
		N_q from cavity expansion theory, as a function of ϕ' and volume compressibility	Vesic (1972)
Uncemented calcareous sand	$f_b = N_q \sigma_v' \ngtr f_{blim}$	$N_q = 20$	Datta *et al.* (1980)
		Typical range of N_q = 8–20	Poulos (1988d)
		N_q determined for reduced value of ϕ' (e.g. 18°)	Dutt and Ingram (1984)

* For silica and calcareous sands, the above expressions apply for driven piles only.
† Typical limiting values f_{blim} range from 10–15 MN/m^2 for silica sand, and 3–5 MN/m^2 for calcareous sand; the latter value depends on soil compressibility (Nauroy *et al.*, 1986).

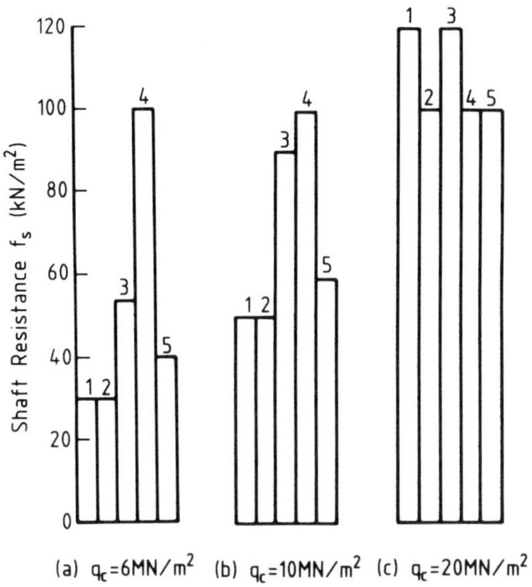

Number	Source of Correlation
1	Bustamante & Gianeselli (1982)
2	Fleming & Thorburn (1984)
3	Verbrugge (1982)
4	Van Impe (1986)
5	This paper

Fig. 27. Example of variations between correlations for shaft resistance against CPT-driven piles in sand: (a) $q_c = 6$ MN/m²; (b) $q_c = 10$ MN/m²; (c) $q_c = 20$ MN/m²

the drained Young's modulus. For all soil types, a further distinction needs to be made, between the tangent value of Young's modulus (if a non-linear interface model is being used) and the secant value of Young's modulus (if a purely linear analysis is being used). Again, it is often difficult to make such distinctions with rough empirical correlations and, unless specified, the Young's modulus referred to here will be a secant value, relevant for normal working load levels of between one-third to one-half of the ultimate load capacity.

For piles in clay, Young's modulus has been correlated often with laboratory-measured undrained shear strength, c_u. Some of these correlations are shown in Fig. 28, and a feature of this figure is the wide spread of correlations. Possible reasons for this spread might include differences in the method of determining c_u, differences in the method of determining the modulus values, differences in the load level at which the modulus was determined, differences in the over-consolidation ratio of the clay between different tests, and differences between the clay types. Callanan & Kulhawy (1985) find that values of E_s/c_u

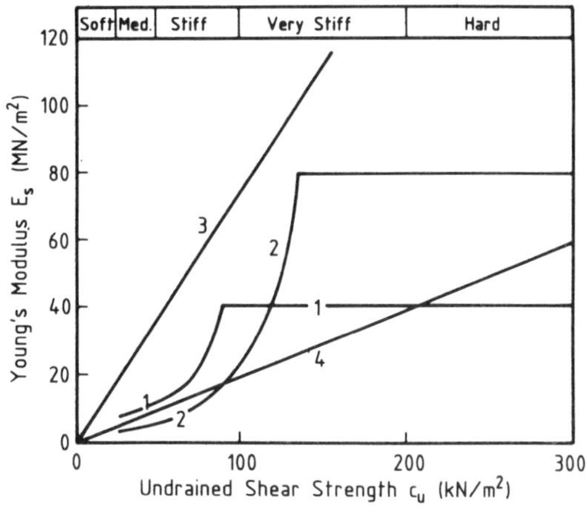

Legend	Remarks	Reference
1	Driven Piles	Poulos (1972)
2	Bored Piles	Poulos (1972)
3	Driven Piles (E_s=750c_u)	Aschenbrenner & Olsen (1984)
4	Bored Piles, lower bound (E_s=200c_u)	Callanan and Kulhawy (1985)

Fig. 28. Correlations for soil modulus for piles in clay (after Callanan & Kulhawy, 1985)

generally range between 200 and 900, with an average value of about 500. These values apply to piles with a length-to-diameter ratio in excess of about 15. For shorter piles, the upper range of E_s/c_u may be greater because of the possible effects of fissuring, desiccation and over-consolidation of the clay near the surface.

Some correlations between Young's modulus and standard penetration test number are summarized in Fig. 29, and show alarming variability. At least some of this variability may be attributed to differences in the determination or definition of the SPT value, but it is clear that the potential for selection of inappropriate values of E_s is great.

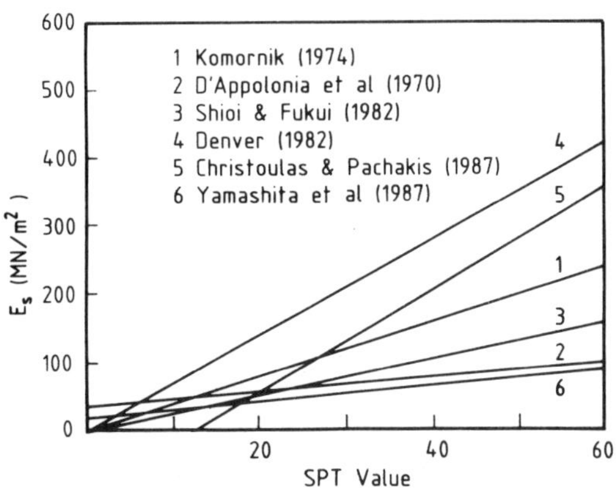

Fig. 29. Comparison between correlations for soil modulus driven piles in sand

Table 10. Correlations between end bearing resistance f_b and SPT value; $f_b = KN$ MN/m²

Pile type	Soil type	K	Remarks	Reference
Driven displacement	Sand	0·45	N = average SPT value in local failure zone	Martin *et al.* (1987)
	Sand	0·40		Decourt (1982)
	Silt, sandy silt	0·35		Martin *et al.* (1987)
	Glacial coarse to fine silt deposits	0·25		Thorburn & MacVicar (1971)
	Residual sandy silts	0·25		Decourt (1982)
	Residual clayey silts	0·20		Decourt (1982)
	Clay	0·20		Martin *et al.* (1987)
	Clay	0·12		Decourt (1982)
	All soils	0·30	For $L/d \geqslant 5$ If $L/d < 5$, $K = 0{\cdot}1 + 0{\cdot}04\ L/d$ (closed-end piles) or $K = 0{\cdot}06\ L/d$ (open-ended piles)	Shioi & Fukui (1982)
Cast in place	Cohesionless		$f_b = 3{\cdot}0$ MN/m²	Shioi & Fukui (1982)
		0·15	$f_b \not> 7{\cdot}5$ MN/m²	Yamashita *et al.* (1987)
	Cohesive	—	$f_b = 0{\cdot}09\,(1 + 0{\cdot}16z)$ where z = tip depth (m)	Yamashita *et al.* (1987)
Bored	Sand	0·1		Shioi & Fukui (1982)
	Clay	0·15		Shioi & Fukui (1982)
	Chalk	0·25 0·20	$N < 30$ $N > 40$	Hobbs (1977)

Table 11 shows some suggested correlations between E_s and cone penetration resistance q_c, and, as with most of the other correlations, the range is large. Two correlations for initial tangent modulus E_{st} are shown, both being derived from dynamic triaxial tests but believed to be relevant for piles.

Correlations between E_s and pressuremeter data have not been extensively developed, although Frank (1985) suggests that the initial tangent modulus can be taken as the initial tangent modulus of the expansion curve from the self-boring pressuremeter. Most of the correlations with pressuremeter data have been made with respect to the axial load transfer curves for a pile. A summary of a number of these correlations, for both pile shaft and pile tip responses, is given by Frank (1985).

Poisson's ratio ν_s

Poisson's ratio of the soil is a necessary input parameter into analyses that involve elastic continuum theory, but its effect is generally quite minor when the solutions are expressed in terms of Young's modulus of the soil. For saturated

Table 11. Correlations between soil Young's modulus E_s and CPT value—driven piles

Soil type	Correlation	Remarks	Reference
Clay and silts	$E_s{}^* = 21{\cdot}0q_c^{1{\cdot}09}$	Various pile types E_s and q_c in MN/m^2	Christoulas (1988)
	$E_s = 15q_c$		Poulos (1988c)
Silica sands	$E_s = \alpha q_c$	$\alpha = 20$–40	Milovic & Stevanovic (1982)
		$\alpha = 5$ (normally-consolidated sands) $\alpha = 7{\cdot}5$ (over-consolidated sands)	Poulos (1988c)
	$E_{st}\dagger = 53q_c^{0{\cdot}61}$	E_{st} and q_c in MN/m^2 Dynamic modulus value	Imai & Tonouchi (1982)
Unspecified	$E_{st} = \alpha q_c$	$\alpha = 24$–30 Dynamic modulus value	Holeyman (1985)
	$E_s = 10{\cdot}8 + 6{\cdot}6q_c$	E_s and q_c in MN/m^2 (for $q_c > 0{\cdot}4$ MN/m^2)	Verbrugge (1982)

* E_s = secant Young's modulus.
† E_{st} = initial tangent Young's modulus.

clays under undrained conditions, ν_s can be taken as 0·5. For clays under drained conditions, ν_s generally lies within the range $0{\cdot}35 \pm 0{\cdot}05$, whereas for silica sands, ν_s is usually within the range $0{\cdot}3 \pm 0{\cdot}1$. Lower values, within the range $0{\cdot}15 \pm 0{\cdot}1$, are applicable for many marine calcareous sediments.

Hyperbolic curve fitting constant R_f

If a hyperbolic interface model is used, R_f defines the degree of non-linearity and can range between 0 (an elastic–perfectly plastic response) and 1·0 (an asymptotic hyperbolic response in which the limiting pile–soil stress is never reached). Limited experience suggests that different values of R_f should be used for shaft and pile tip elements. For the shaft, there is a relatively small amount of non-linearity, and values of R_f in the range 0–0·5 may be appropriate. In contrast, the pile tip response is often highly non-linear, and it is suggested that a value of R_f of about 0·9 may give a reasonable fit with observed behaviour.

In much of the published theoretical work to date, elastic–plastic response has been assumed (i.e. $R_f = 0$ for both shaft and tip). As pointed out by Poulos & Davis (1980) and Frank (1985), displacements at load levels approaching failure are therefore often seriously underestimated. Fortunately however, at normal working loads, the effects of non-linearity are not great and adequate predictions of settlement are often achieved with elastic or elastic–plastic theory (Fig. 11).

DESIGN CHARTS FOR PILES AND PILE GROUPS

Dimensionless category 2 solutions for pile settlement, such as those presented by Butterfield & Banerjee (1971), Randolph & Wroth (1978), Banerjee (1978), Poulos & Davis (1980) and Butterfield & Douglas (1981) are useful for preliminary design purposes and can often form the basis for final design calculations. In applying such solutions, it is necessary for the user to determine the most appropriate simplified soil profile and the values of Young's modulus for the soil in this profile. These determinations may be difficult in the early stages of a project when little quantitative data is available, and it is therefore of value to develop design charts based on hypothetical but realistic soil data. A limited series of such charts is presented in this section for driven and bored piles in both silica sand and clay profiles. The geometrical and soil parameters are defined in Fig. 30. The assumed parameters for the sand and clay profiles are summarized in Table 12 and are based on the Author's judgement and the correlations presented in the preceding section.

For all the charts presented, Young's modulus of the pile, E_p, has been taken as 30 000 MN/m^2. This value is representative of concrete piles and

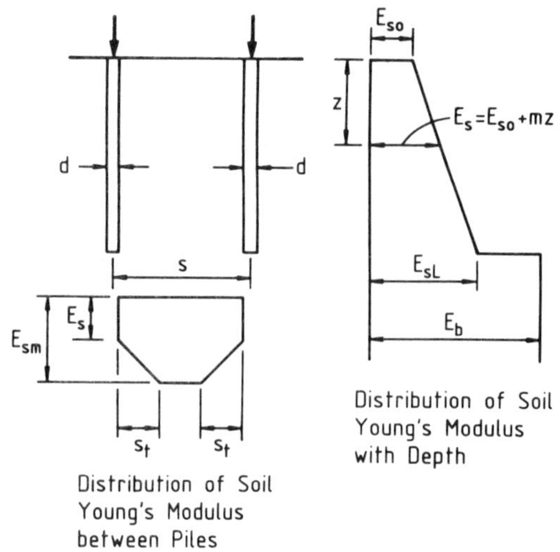

Fig. 30. Definition of soil modulus distributions assumed for design charts

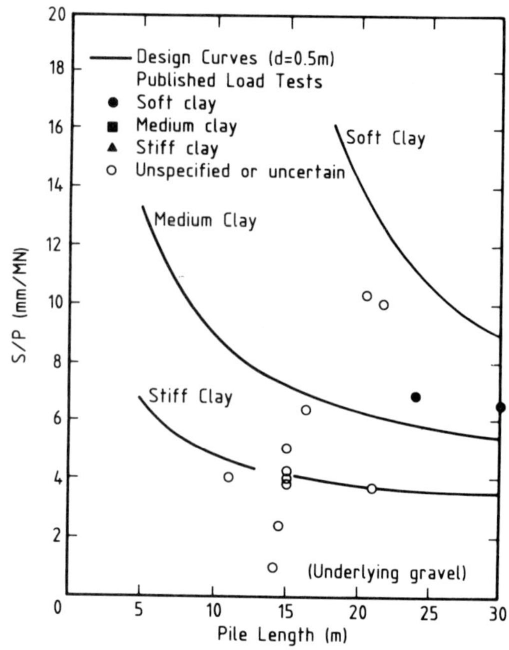

Fig. 31. Design chart for settlement of driven piles in clay (diameter = 0·5 ± 0·1 m)

also of the effective modulus of most steel tube piles, for which the section area lies between 10 and 20% of the gross cross-sectional area.

Settlement of single piles

The charts for single pile settlement have been developed from the closed-form solution developed by Randolph & Wroth (1978), which can be evaluated conveniently by means of a computer spreadsheet program.

Figures 31–34 present theoretical relationships between the pile head settlement per unit load S/P and the pile length L, for driven and bored piles having a diameter in the range 0·4–0·7 m. Similar charts may be developed for other diameters. In each case, a range of soil stiffness or density, representing the likely limits of practical conditions, is considered. Also shown on these charts are observed flexibilities (at average working load levels of 40 ± 10% of ultimate load) from published field load tests. Several have been taken from the useful compilation prepared by Kulhawy *et al.* (1982). The observed values of S/P

Table 12. Parameters used to derive design charts*

Pile type	Soil type	E_{so}: MN/m^2	m: MN/m^3	E_b/E_{SL}	$\mu = E_{sm}/E_s$	ν_s	f_s: kN/m^2	f_b: MN/m^2
Driven	soft clay	0	0·7	1·0	3·0	0·5	25	0·2
	medium clay	20	0	1·0	4·0	0·5	45	0·5
	stiff clay	42	0	1·0	5·0	0·5	—	—
	hard clay	—	—	—	—	—	70	1·8
	loose sand	0	1·5	4·0	5·0	0·3	—	7·0
	medium-dense sand	0	4·0	2·0	3·0	0·3	—	10·0
	very dense sand	0	8·0	1·5	—	—	—	15·0
Bored	medium clay	30	0	0·5	4·0	0·5	—	—
	stiff clay	80	0	0·5	2·0	0·5	—	—
	hard clay	150	0	1·0	—	—	—	—
	loose sand	0	1·0	0·5	—	—	—	—
	medium-dense sand	0	3·0	1·0	3·0	0·3	—	—
	very dense sand	0	6·0	1·0	1·5	0·3	—	—

* Fig. 30 gives definition of parameters.
In all cases, $s_t/d = 3·0$.

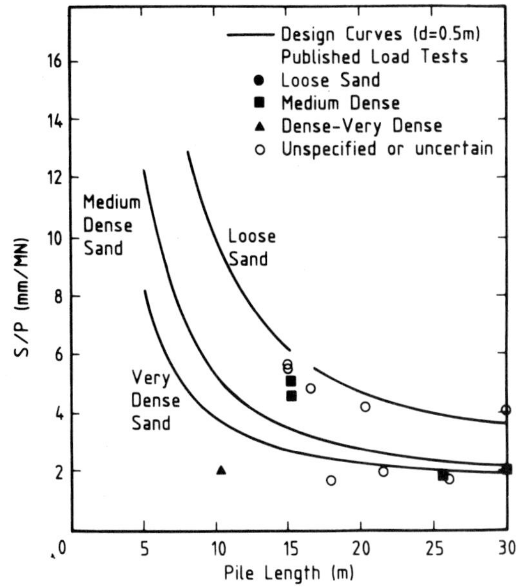

Fig. 32. Design chart for settlement of driven piles in sand (diameter = 0·5 ± 0·1 m)

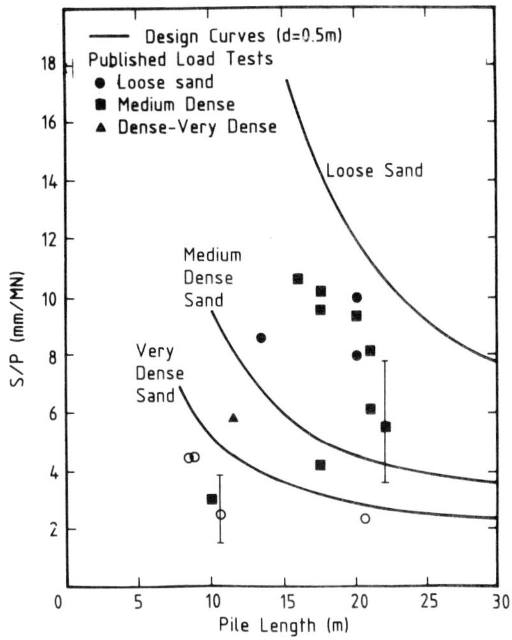

Fig. 34. Design chart for settlement of bored piles in sand (diameter = 0·5 ± 0·1 m)

are generally consistent with the theoretical values, although in some cases, they are less than the computed value for the stiffest or densest soil condition. Thus, it may be expected that the design charts will occasionally give a conservative estimate of settlement.

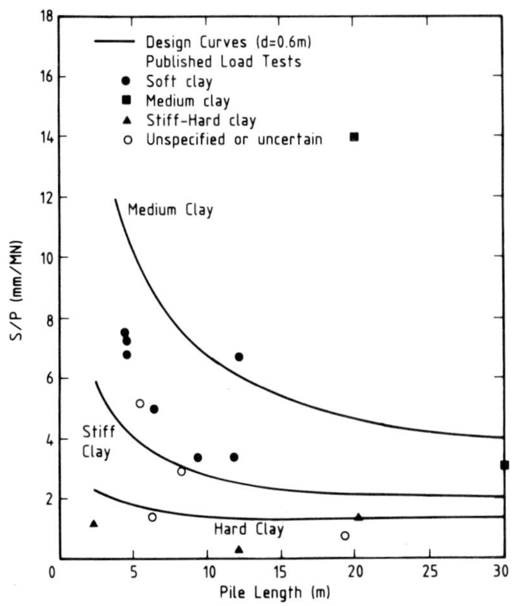

Fig. 33. Design chart for settlement of bored piles in clay (diameter = 0·6 ± 0·1 m)

It is relevant to note that Frank (1985) has summarized practical experience with isolated piles and has suggested that, under the design load, the following settlements are typical

(*a*) for driven piles: 0·9% of diameter (range 0·8–1·2%)
(*b*) for bored piles: 0·6% of diameter (range 0·3–1%).

A recent statistical analysis reported by Briaud & Tucker (1988) suggested that there is a 95% probability that the pile head settlement at half the ultimate load will be 1·25% of the pile diameter or less.

Settlement of pile groups

Following the approach adopted for single piles, charts have been developed to enable a rapid estimate to be made of group settlements for both driven and bored piles in various soil types. The most convenient form of presentation is in terms of the group settlement ratio R_s and, more specifically, the exponent ω of the settlement ratio in equation (11). An elastic boundary element analysis, by means of the computer program DEFPIG, has been used to obtain the solutions, with allowance being made for the higher stiffness of the soil between the piles than at the pile–soil interface (Poulos, 1988a).

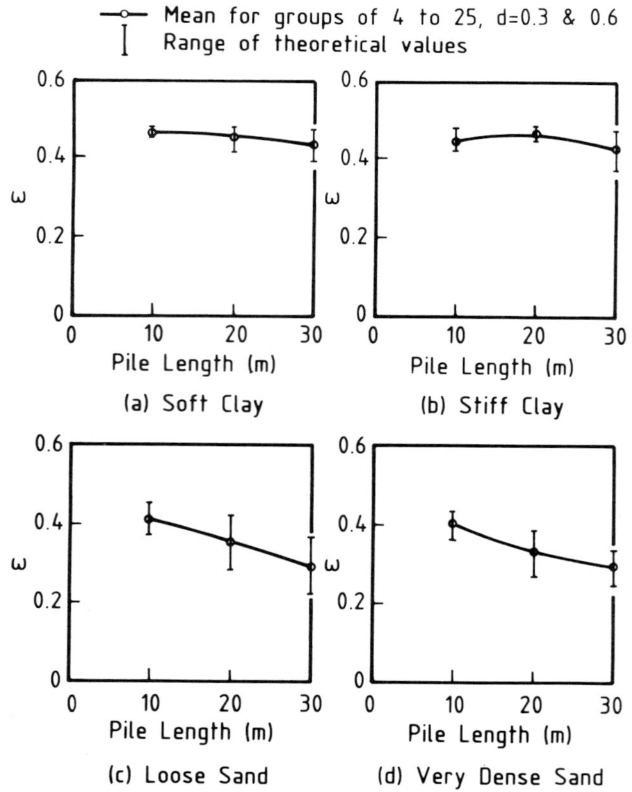

Fig. 35. Settlement ratio exponent ω, driven piles, $s/d = 3$: (a) soft clay; (b) stiff clay; (c) loose sand; (d) very dense sand

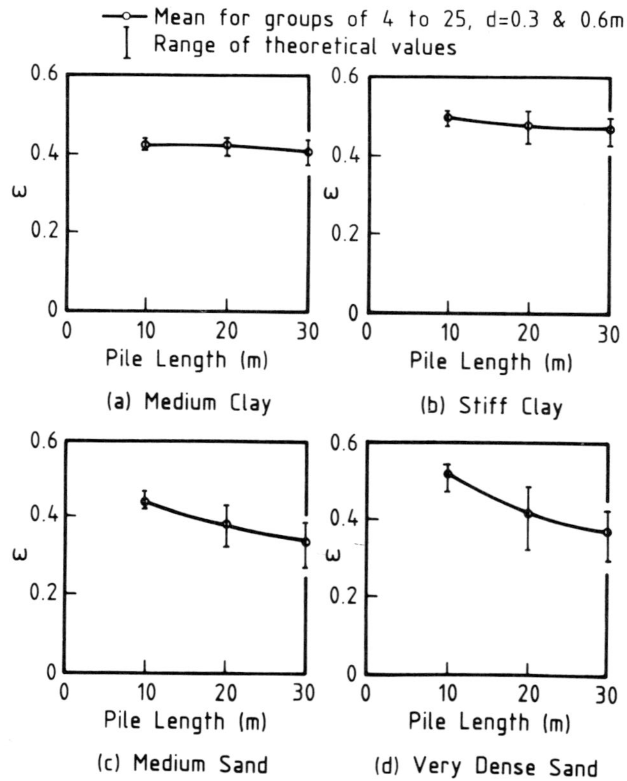

Fig. 36. Settlement ratio exponent ω, bored piles, $s/d = 3$: (a) medium clay; (b) stiff clay; (c) medium sand; (d) very dense sand

The design charts are shown in Figs 35 and 36 as plots of the settlement ratio exponent ω against pile length for four soil types and for driven and bored friction piles of lengths between 10 and 30 m. The centre-to-centre spacing between the piles is assumed to be 3 diameters. The average value of ω for square groups of between 4 and 25 piles, and for diameters of 0·3 m and 0·6 m, is plotted, together with the range of theoretical values of ω for each pile length. For piles in clay, ω is almost independent of pile length, whereas for the piles in silica sand, ω decreases as the pile length increases. The range of values of ω is narrow for shorter pile lengths, but increases as the pile length increases. Over the entire range of pile and soil types, ω varies between 0·41 and 0·51 for 10 m long piles, and between 0·22 and 0·47 for 30 m long piles. There is therefore a clear implication that the use of these design charts is likely to be less accurate for long piles than for short piles.

To illustrate the use of the design charts for both single piles and pile groups, consider the case of a group of eight driven piles (in a 4 × 2 configuration) each 20 m long and 0·6 m dia., in a deep layer of soft clay. The centre-to-centre spacing between the piles is 1·8 m. An estimate is required of the settlement of the group under a total load of 3·2 MN, which is equivalent to a safety factor of about 2·5 on the ultimate capacity.

From Fig. 31, for 20 m piles in soft clay, the value of S/P is about 14·0 mm/MN. Thus, at the average working load of 3·2/8 = 0·4 MN, the settlement of a single pile would be 14·0 × 0·4 = 5·6 mm. From Fig. 35, the average settlement ratio exponent ω for 20 m long driven piles in soft clay is about 0·45. Thus, for an 8-pile group, $R_s = 8^{0{\cdot}45} = 2{\cdot}55$. Therefore, the group settlement is estimated to be 2·55 × 5·6 = 14·3 mm.

Ultimate load capacity of single piles

Although this section focuses on pile settlement, it is relevant to remark that design charts may also be developed for pile load capacity. Such charts are shown for driven piles in Figs 37–39 and are classified as category 1 charts because they are based on the empirical values of shaft and end bearing resistance shown in Table 12. Fig. 37 plots the ultimate shaft resistance against pile length for four pile diameters and three classifications of clay condition, soft, firm and hard. Figs 38 and 39 plot the pile tip cap-

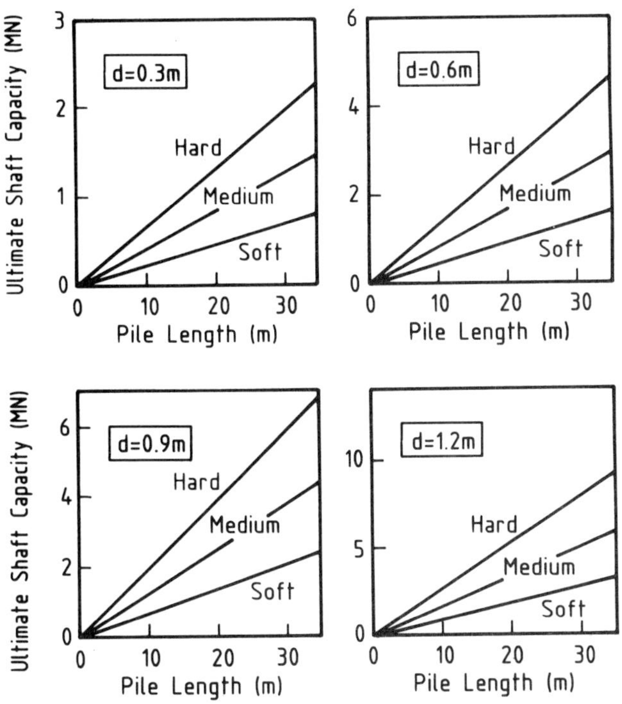

Fig. 37. Design charts for ultimate shaft capacity of driven piles in clay: (a) d = 0·3 m; (b) d = 0·6 m; (c) d = 0·9 m; (d) d = 1·2 m

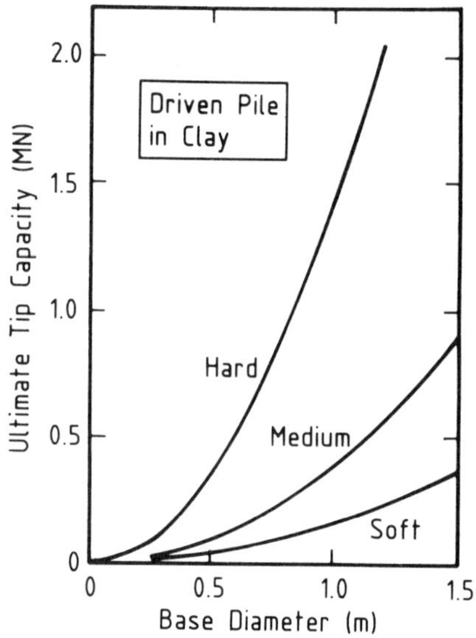

Fig. 38. Design chart for ultimate tip capacity of driven piles in clay

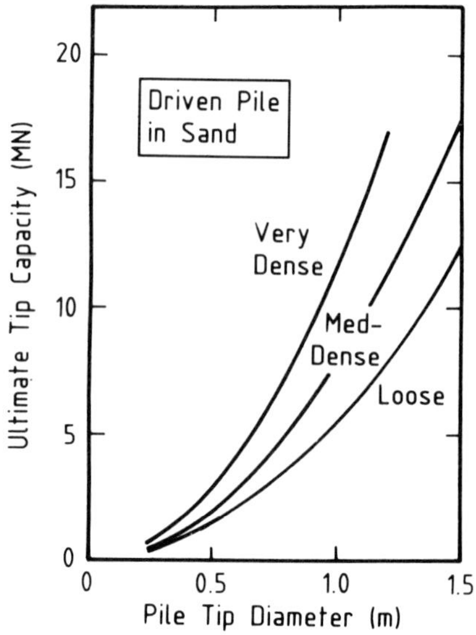

Fig. 39. Design chart for ultimate tip capacity of driven piles in sand

acity, as a function of pile tip diameter, for tips bearing in either clay or sand. Charts such as these are perhaps over-simple and must be used with due caution. Nevertheless, they can give a useful preliminary appreciation of the pile length and diameter requirements to develop a specified ultimate load capacity.

CASE STUDIES

In predicting the behaviour of pile foundations, the geotechnical engineer is faced with a number of decisions, including

(*a*) the method of analysis, and the associated soil model, to be used
(*b*) the way in which the soil profile can be simplified and idealized for the analysis
(*c*) the geotechnical parameters to be used.

The influence of these decisions on the predicted pile performance will be examined with respect to two published case histories, one involving an instrumented single pile, and the second involving a group of instrumented piles. In each case, theoretical calculations will be presented to illustrate the sensitivity of the predicted behaviour to the above factors, and these calculations will be compared with the observed behaviour. As an example of the application of theory to practice without the prior benefit of performance measurements, the results of a recent 'class A' prediction (Lambe, 1973) made by the Author will be presented and compared with observed behaviour.

Case study 1—sensitivity of single pile performance calculations

The case analysed is a pile test described by Gurtowski & Wu (1984). The test pile was located at the site of the West Seattle Freeway in the USA. For the site under consideration (site A) the geotechnical profile is shown in Fig. 40 together with SPT data, which was the only quantitative geotechnical data available. The water table was located about 3 m below the surface.

The test pile was a 0·61 m wide octagonal prestressed concrete hollow pile with a plug at the tip, and was driven to a depth of about 30 m.

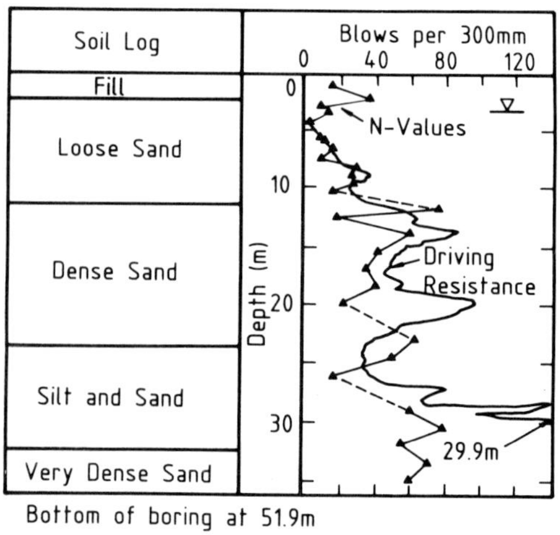

Fig. 40. Geotechnical data for site A (Gurtowski & Wu, 1984)

Table 13. Analysis conditions and parameters for performance sensitivity evaluation

Aspect considered	Standard	Variations
Analysis method and soil model	Boundary element analysis, with elastic–plastic continuum soil model (with secant modulus)	Elastic continuum with secant modulus Elastic–plastic continuum with tangent modulus Hyperbolic continuum Elastic–plastic load transfer model, with secant modulus
Soil profile	Two-layer 0–125 m $\bar{N} = 15$ 12·5 m$^+$ $\bar{N} = 40$	Homogeneous ($\bar{N} = 30$) Gibson soil (N varies from 0 to 60 at pile tip) Detailed profile, as indicated by SPT values
Soil parameters	$E_s = 4N$ MN/m^2 $f_s = 2N$ kN/m^2 $f_b = 0{\cdot}4$ MN/m^2	$E_s = 2{\cdot}5N$ MN/m^2 $E_s = 7N$ $E_s = 36{\cdot}8 + 1{\cdot}04N$ $E_s = 7{\cdot}5N - 94{\cdot}5$
Pile modulus	$E_p = 35\,000$ MN/m^2	$E_p = 15\,000$ MN/m^2 $E_p = 50\,000$

Note: N = SPT value.

A 'standard' set of conditions and parameters was chosen, and the effects of deviations from these standards were examined. The standard values chosen, and the deviations considered, are summarized in Table 13. The soil parameters are derived from the correlations presented in Tables 7, 10 and 12, and Fig. 29.

Figure 41 compares the measured load–settlement behaviour and load distributions with those predicted from the standard analysis. The standard analysis predicts the pile head load–settlement behaviour quite well, and also gives a good prediction of the pile load distribution at three different applied load levels. However, the head load against tip settlement relationship is not well predicted, indicating that perhaps too large a stiffness has been assigned to the soil at and beneath the pile tip.

In order to enable ready appreciation of the effects of deviations from the standard conditions, attention has been concentrated on the pile head settlement, the tip settlement, and the load at mid-depth of the pile, for an applied load of 1·8 MN, which would be a normal working load for this pile.

The influence of the soil model is shown in Fig. 42. In some cases, E_s is taken as the secant modulus, whereas in others, it is taken as the initial tangent modulus. The pile head settlement and pile load at mid-depth are relatively insensitive to the soil model, whereas the tip settlement is influenced more. The two models involving the use of the initial tangent modulus give smaller tip movements than those using the secant modulus; all, however, predict a tip settlement which is much smaller than the measured value.

The effect of the soil profile idealization is shown in Fig. 43. The use of a homogeneous profile leads to a significantly smaller head settlement and mid-depth load than the other three profiles. The use of a detailed profile in which E_s and f_s vary as N varies gives a predicted behaviour which is little different from that for the two-layer profile. The indiscriminate assumption of a homogeneous soil profile can lead to inaccuracy in the predicted pile behaviour, whereas more realistic (but nevertheless simplified) modelling of the soil profile variation gives quite adequate results in this case.

Figure 44 shows the effect of the soil modulus correlation used for the analysis. The larger the soil modulus, the smaller are the settlements and the smaller is the load in the pile at mid-depth. The tip settlement is particularly affected by E_s. However, it is interesting to note that an increase in E_s from 2·5 N to 7 N MN/m^2 reduces the head

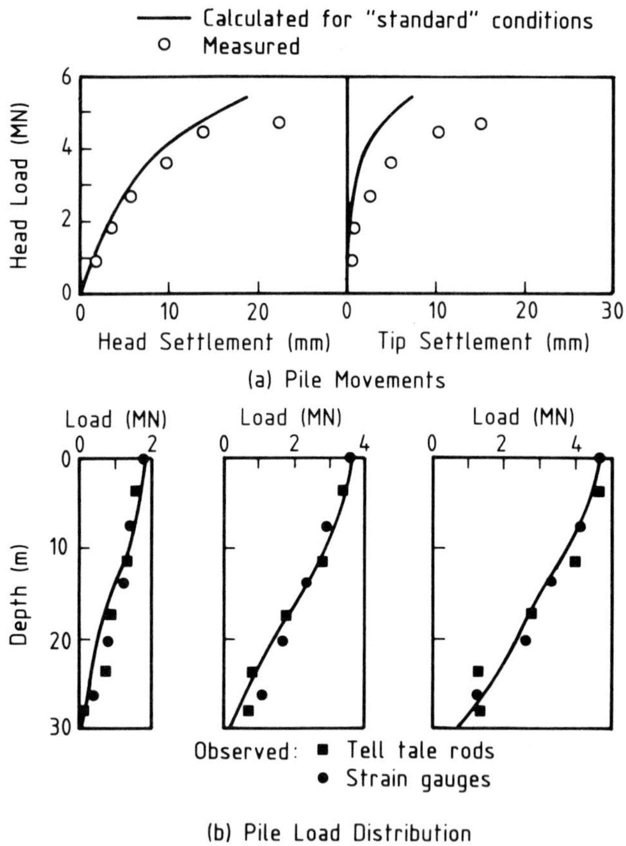

Fig. 41. Comparisons between observed and theoretical behaviour, pile test A (Gurtowski & Wu, 1984): (a) pile movements; (b) pile load distribution

settlement by less than 30%. This occurs because the pile is relatively compressible. In this case, because the piles are concrete, some uncertainty will be present in the choice of the pile modulus E_p. The importance of E_p is demonstrated in Fig. 45. Of all the factors considered, it has the greatest influence on the pile head settlement, although it has little influence on the tip settlement. The load distribution is also affected by the pile modulus, especially in the lower stiff soil layer.

The foregoing study therefore reveals that the head settlement of a relatively long compressible pile may be influenced as much by the pile modulus as by the soil model, the soil profile idealization, or the soil modulus correlation adopted. The soil modulus may have a substantial influence on the pile tip settlement, while the soil profile idealization can influence both load distribution and pile head settlement significantly. The effect of the soil model is not great, except for the tip settlement. Thus, in predicting the settlement of long compressible piles, it would seem desirable to attempt to determine the pile modulus as accurately as possible.

Case study 2—sensitivity of pile group performance calculations

The sensitivity of theoretical predictions of pile group settlement and load distribution will be discussed with respect to the well-documented case study described by O'Neill *et al.* (1981; 1982).

Figure 46 summarizes the geotechnical data at the test site which was located at the University of Houston. The site consists of various layers of stiff to very stiff clay, and geotechnical data is available from standard penetration tests, cone penetration tests, pressuremeter tests, unconsolidated undrained triaxial tests, laboratory consolidation tests, and seismic cross-hole tests.

Vertical load tests were performed on full-scale pile groups and single piles, with measurements being made of settlement and load distribution

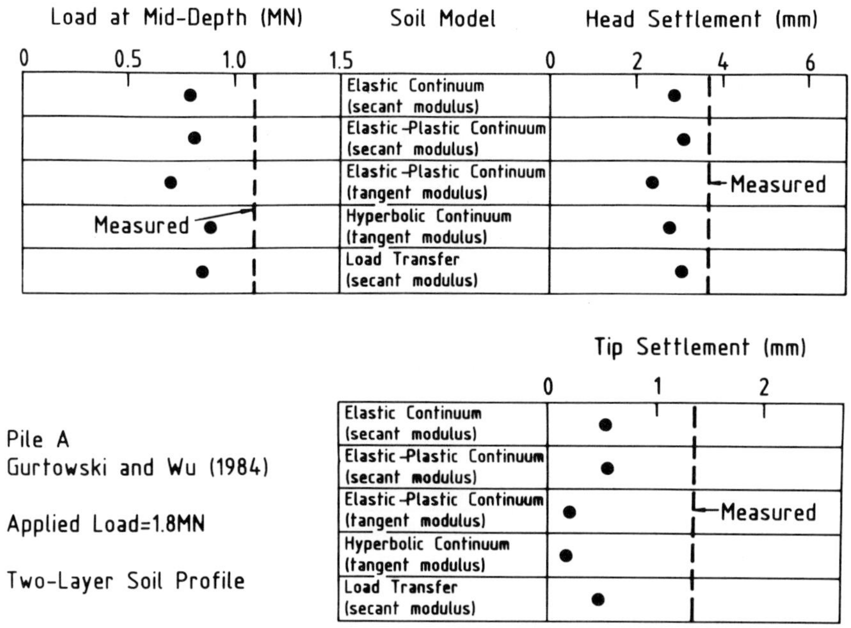

Fig. 42. Influence of soil model on predicted behaviour of single pile

for groups of nine, five and four piles, as well as two single piles. The piles were about 13 m long, 0·273 m dia. steel tubes with a 9·3 mm wall thickness.

In examining the influence of the method of analysis, only a limited number of methods were considered. The programs DEFPIG, PIGLET and GAPFIX were employed, with two DEFPIG analyses being carried out, the first a conventional analysis in which the soil modulus between the piles was assumed to be the same as that adjacent to the piles, and the second a modified analysis in which the soil between the piles was assumed to be stiffer than near the piles. The seismic cross-hole data were used to estimate the small-strain Young's modulus of the soil between the piles. For all analyses, the soil Young's modulus E_s (near the piles) was assumed to vary linearly with depth, according to the relationship $E_s = 40 + 5{\cdot}38z$ MN/m^2, this relationship being based on

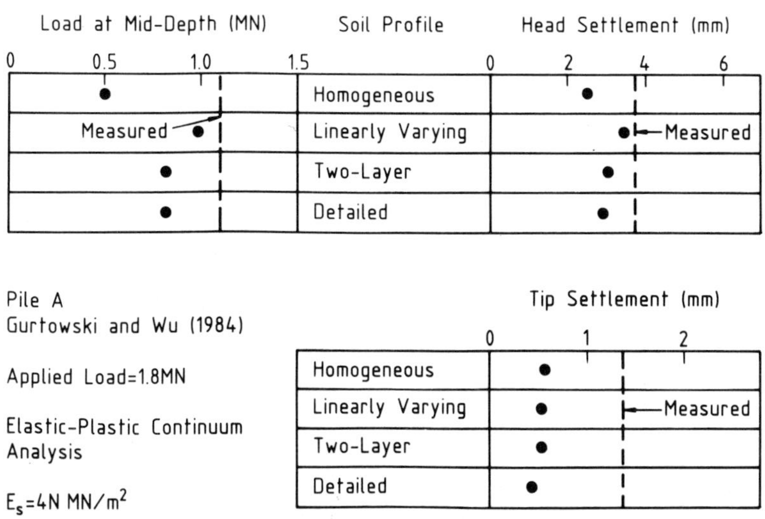

Fig. 43. Influence of soil profile idealization on predicted behaviour of single pile

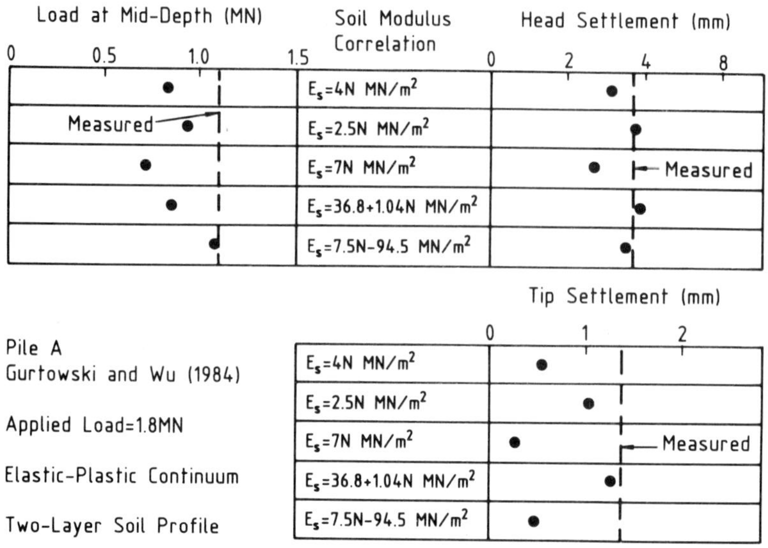

Fig. 44. Influence of soil modulus correlation on predicted behaviour of single pile

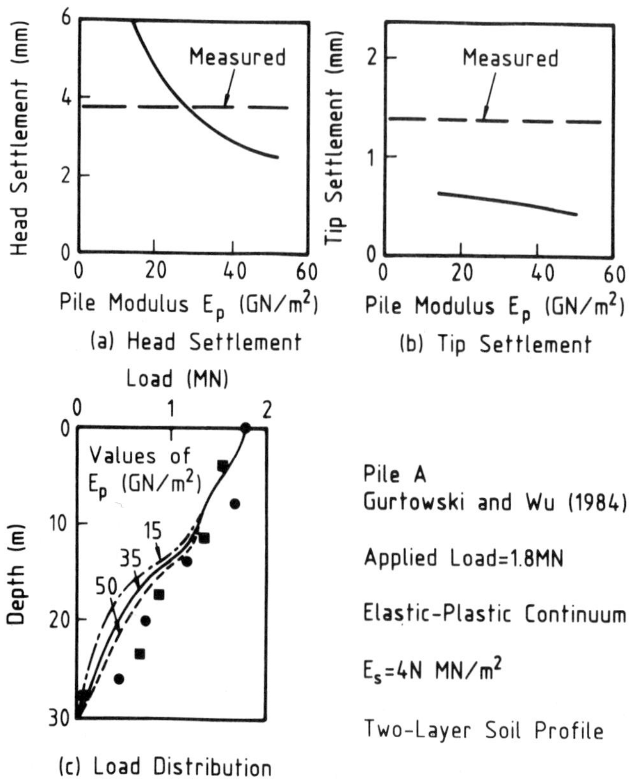

Fig. 45. Influence of pile modulus on behaviour of single pile: (a) head settlement; (b) tip settlement; (c) load distribution

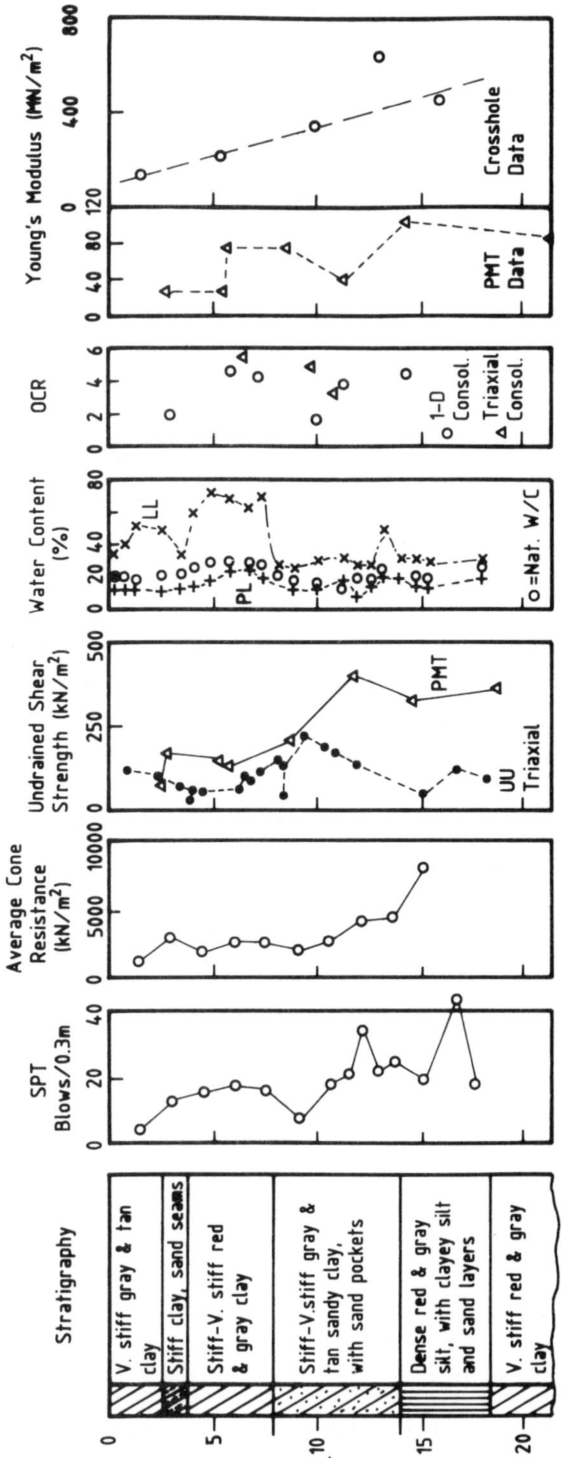

Fig. 46. Summary of geotechnical data at test site

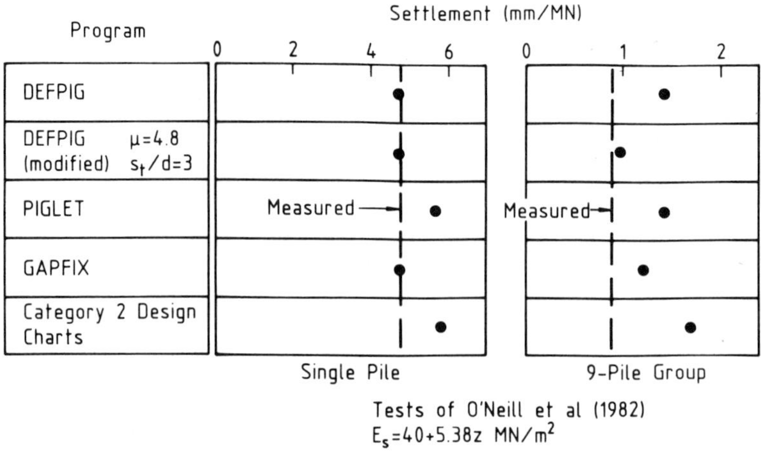

Fig. 47. Effect of analysis method on single pile and group settlements

the correlation $E_s = 750\ c_u$. The values of c_u were obtained from the pressuremeter test data, the profile being simplified to vary linearly with depth from 53 kN/m² at the surface to 147 kN/m² at a depth of 13 m.

The calculated settlements of both the single pile and the nine-pile group are shown in Fig. 47. Also shown are the estimates of settlement derived from the category 2 design charts presented in this Paper for driven piles in stiff clay. There are some differences between the settlements derived from the different analyses, but these are not great. The single pile settlements given by the DEFPIG and GAPFIX analyses are close to the measured value, but the group settlements are overestimated by all approaches. The modified DEFPIG analysis gives the most satisfactory prediction of group settlement. The category 2 design charts overestimate both the single pile and group settlements, although this is to be expected as these charts are meant for design rather than prediction and therefore tend to be conservative.

For all the group configurations tested by O'Neill *et al.*, Fig. 48 compares the measured settlement ratios with those determined from the conventional and the modified DEFPIG analyses (Poulos, 1988a). The conventional analysis considerably overestimates R_s, whereas the modified analysis gives significantly closer agreement with the measured values.

The predicted distributions of load among the piles from the four analyses agree reasonably closely, although all give a less uniform distribution than was actually measured. Fig. 49 shows a detailed comparison between the pile load distributions with depth computed by the program GAPFIX and the measured values. While there are some differences near the pile head, the overall pattern of load transfer along each of the piles appears to be reasonably well predicted by the analysis.

The foregoing comparisons suggest that, given a set of common parameters, most of the methods considered give a similar prediction of group behaviour. However, for group settlement, better agreement is obtained with the modified DEFPIG analysis in which the greater stiffness of the soil between the piles is allowed for.

To examine the influence of the idealization of the soil profile on the group settlement, four different distributions of soil modulus with depth have been used, each being based on the undrained shear strength distribution obtained from the pressuremeter tests

(*a*) the linearly varying distribution of E_s considered above

(*b*) a homogeneous profile in which $E_s = 75$ MN/m²

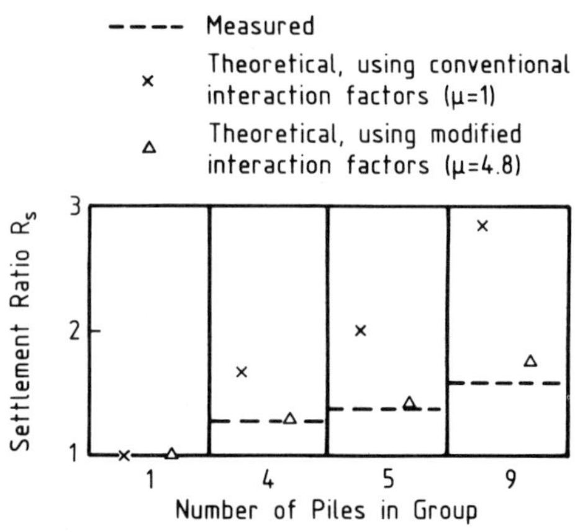

Fig. 48. Theoretical and measured group settlement behaviour (Tests of O'Neill *et al.*, 1982)

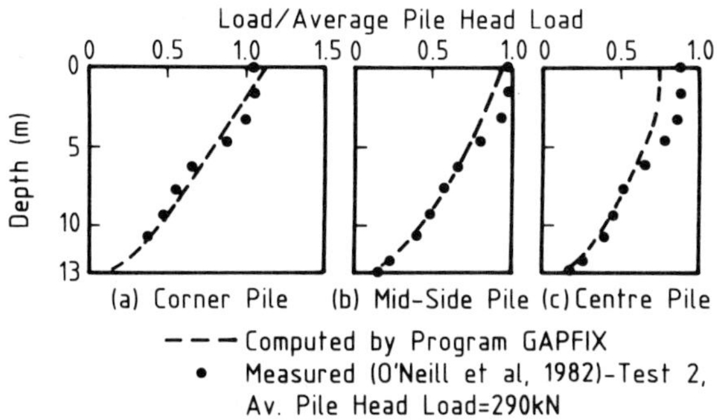

Fig. 49. Comparison between computed and measured load distribution in 9 pile group test

(*c*) a two layer profile in which the first layer extends to a depth of 8 m and has $E_s = 55$ MN/m^2, and the second extends to great depth and has $E_s = 100$ MN/m^2

(*d*) a profile which follows the detailed distribution of c_u with depth, and in which $E_s = 750\ c_u$.

The program DEFPIG was used for all calculations; no account was taken of increased soil stiffness between the piles.

Figure 50 shows the predicted settlements of the single pile and the nine-pile group. All four profiles give similar settlements, which agree well with the measured settlement for the single pile, but are about 50% too large for the group. There is also little variation among the solutions for the pile head load distribution, but in all cases the predicted load distribution is more non-uniform than that measured. It therefore appears that, in this particular case, the idealization of the soil profile is not a crucial factor in the prediction of pile group behaviour.

The influence of the method of determination of the soil Young's modulus has been investigated by using six different procedures and employing the conventional DEFPIG analysis. Fig. 51 shows the predicted and measured settlements of the single pile and the nine-pile group. It is immediately apparent that the predicted settlements are more sensitive to E_s than to either the method of analysis or the soil profile idealization. The closest predictions are given by the correlations $E_s = 750\ c_u$ and $E_s = 4N$ MN/m^2, whereas $E_s = 200\ c_u$ and $E_s = 20\ q_c$ lead to gross overestimates of both single pile and group settlements. The correlations leading to large settlements also lead to more non-uniform distributions of pile head load.

For this case study, therefore, in which the piles were significantly shorter and less compressible than in the preceding case, the two factors which appear to be crucial in accurately predicting the group behaviour are

(*a*) the magnitude of the soil Young's modulus E_s

(*b*) the greater stiffness of the soil between the piles than locally near the piles.

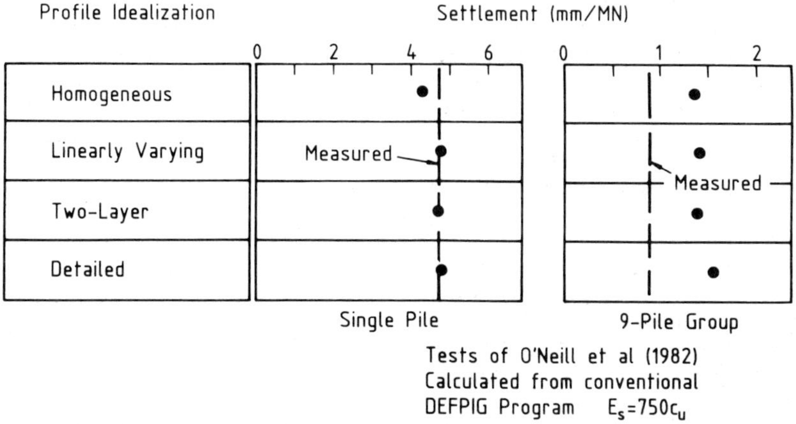

Fig. 50. Effect of soil profile idealization on single pile and group settlements

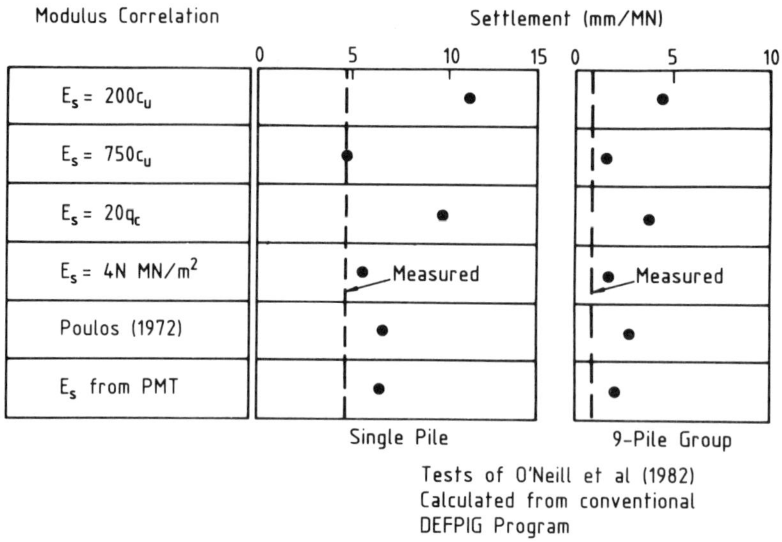

Fig. 51. Effect of modulus correlation on single pile and group settlements

The general tendency with all conventional methods of group analysis is to overestimate the interaction and to predict a settlement which is too large and a load distribution which is too non-uniform. Such tendencies may not always be as significant as they appear to be for the case considered here.

Case study 3—a class A prediction of pile behaviour

In conjunction with the Fifth Australia–New Zealand Conference on Geomechanics, held in Sydney in 1988, geotechnical engineers were invited to predict the load–settlement performance of two driven precast concrete piles at a site in Hemmant, Queensland, in Australia. After driving, each pile was left for several weeks and then tested to failure by static load testing procedures in accordance with the Australian Standard Piling Code. Sixteen engineers, including the Author, submitted 'class A' predictions based on static analyses, while four engineers undertook dynamic analyses using data from restriking tests performed four weeks after the static failure. A detailed description of the prediction exercise is given by Douglas (1989).

The piles were Balken piles, of high strength precast concrete, 275 mm square, cast in lengths of up to 12·2 m, with mechanical jointing during driving. The piles were driven using a Banut 600 piling rig with a 5 t hydraulic hammer. Each pile was instrumented so that dynamic measurements could be taken during driving. Pile 1 was driven to a depth of 30·5 m and pile 2 to 25 m. Fig. 52 shows the soil profile and summarizes the available geotechnical data at the Hemmant site. The upper 26 m was essentially clay, underlain by very dense sand. Data were obtained from in situ static cone and dilatometer tests, and from laboratory triaxial tests. Values of undrained shear strength in the clays, derived from these three sources, agreed well.

In making his predictions, the Author derived the required parameters as follows.

(*a*) The shaft resistance f_s in the clay soils was correlated with the undrained shear strength c_u, by way of the adhesion factor α which was determined from the suggestions of Semple & Rigden (1984). In the sandy soils, f_s was taken to be 0·25 times the vertical effective overburden pressure. The values of f_s thus computed were generally about 20% larger than those determined from the Schmertmann–Nottingham correlations with cone sleeve resistance.

(*b*) The end bearing resistance f_b was determined from the average cone penetration data within the vicinity of the pile tip. Values of 14 MN/m^2 and 2·45 MN/m^2 were determined for piles 1 and 2 respectively.

(*c*) The initial tangent value of Young's modulus E_s of the clays was estimated to be 25 times the cone resistance. In the underlying dense sands, a value of E_s of 350 MN/m^2 was adopted. Poisson's ratio was taken as 0·5. The hyperbolic curve fitting factor R_f was taken as 0·9 for both shaft and base elements.

(*d*) The pile modulus E_p was taken as 35 000 MN/m^2.

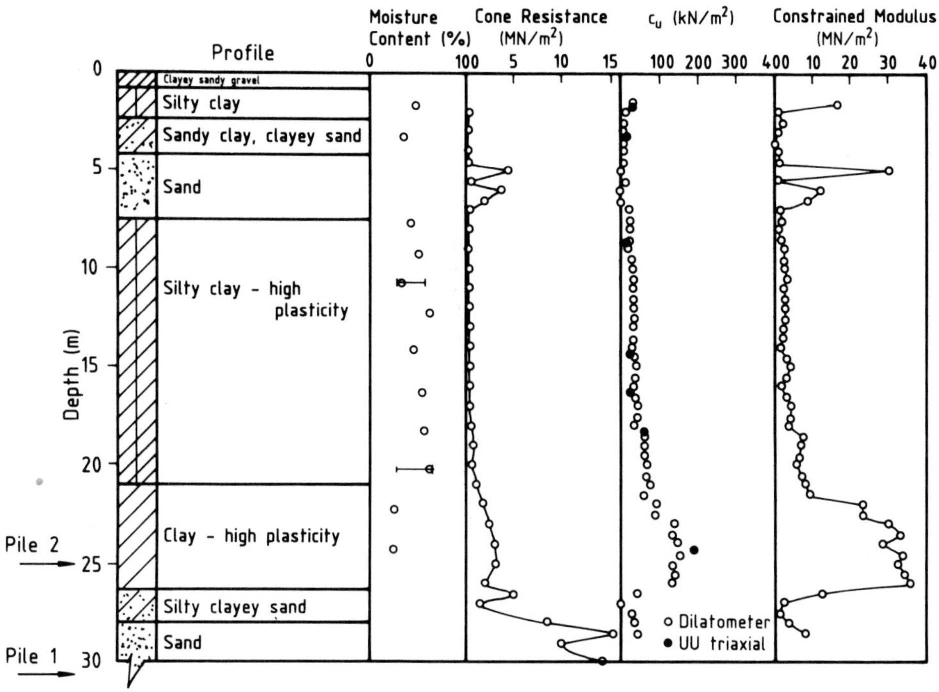

Fig. 52. Geotechnical data at Hemmant test pile site

A category 3B boundary element analysis was performed to predict the load–settlement behaviour, using the computer program PIES (Table 3).

Table 14 summarizes the Author's predictions of ultimate load, the measured values, and also the range of predictions made by the other participants. For both piles, the Author's prediction lay between the extremes, and was within ±20% of the measured load. For pile 1, all 16 predictors underestimated the capacity, presumably because of underestimation of the shaft resistance f_s. Table 14 also summarizes the predicted and measured pile head stiffness values for pile 1. It is noticeable that the range of predictions is substantially greater than for the ultimate load predictions. Twelve of the 16 predictors, including the Author, underpredicted the pile head stiffness. For pile 2, the measured stiffness was similar to

Table 14. Summary of predicted and measured performance

	Load capacity:* kN		Pile head stiffness:† kN/mm
	Pile 1	Pile 2	Pile 1
Author's predictions	2490	1270	125
Range of other participants' predictions (static analyses)	1090–3200	710–1400	49–450
Measured values	2080	1420	205

* At a deflexion of 50 mm.
† At a deflexion of 3·5 mm.

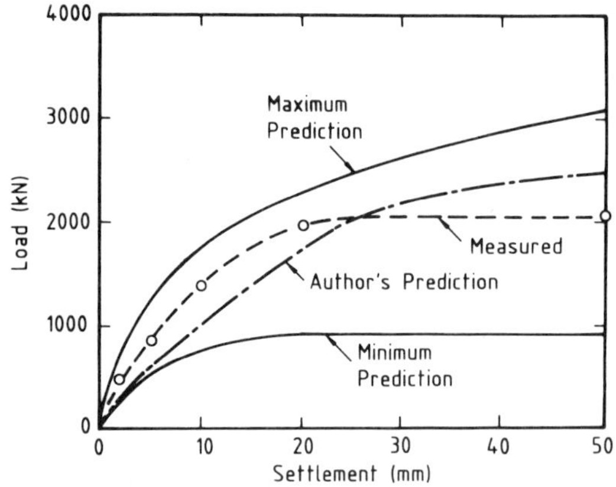

Fig. 53. Predicted and measured load–settlement behaviour pile 1

that of pile 1, and the Author and most of the predictors correctly predicted this characteristic.

Figure 53 shows the measured load–settlement curve for pile 1, the extreme predictions, and the Author's prediction. The most notable features of this figure are

(*a*) the wide spread of predicted performance
(*b*) the considerable difference between the Author's prediction and the measurements, despite the use of a site-specific category 3B analysis.

Following the revelation of the field measurements, the Author re-computed the performance of pile 1 using simple category 2 methods of calculation based on the parameters shown in Table 12. For calculation of ultimate load, the soil profile along the shaft was idealized as a 1 m zone of zero resistance, 3 m of soft clay, 4 m of loose sand, 13 m of soft clay, 5 m of stiff clay and 5·5 m of dense sand. The tip was assumed to also be in dense sand. The computed capacity in this case was 2420 kN, close to the initial predicted value. For settlement, the pile head stiffness was computed to be between 70 and 180 kN/mm, depending on whether the soil profile was taken broadly as a soft clay or a stiff clay. The average of these two values is 125 kN/mm, which corresponds to the Author's initial predicted value. Thus, for this case, rapid category 2 calculations, performed within about 10 min, give predictions which, although not in close agreement with the measurements, were no more inaccurate than the original category 3B predictions. This case study emphasizes once more that the method of calculation may play a far less significant role in pile performance prediction than does the idealization of the soil profile and the selection of the geotechnical parameters.

CONCLUSIONS

This Paper has attempted to review the analysis of piles and pile groups under axial loading and to classify and place in perspective a number of existing methods of analysis. Several of these can be placed within the framework of the boundary element method, and can be used, at least in principle, to model many of the practical features of real piling problems.

A detailed examination has been made of some of the more significant aspects of the behaviour of single piles and pile groups, as revealed by the theoretical solutions. The problems considered include piles subjected to conventional static loading, to cyclic loading, and to loadings which arise from externally-imposed soil movements. Despite the simplifications inherent in the theory, there is compelling evidence to demonstrate that the behavioural characteristics revealed by the theory are consistent with observations made from field and laboratory tests on piles.

The importance of selecting appropriate geotechnical parameters has been emphasized, especially for problems that involve conventional loading. For practical application of most of the theories, it is generally necessary to make use of empirical correlations in order to obtain values of shaft resistance, end-bearing resistance and soil Young's modulus. A number of such correlations are summarized in the Paper.

By means of two case studies, an examination has been made of the sensitivity of pile performance predictions to factors over which the geotechnical analyst has some control, in particular, the method of analysis, the idealization of the soil profile and the values of the soil parameters used in the analyses. It is concluded that, while all these factors may have an influence, the latter two are generally of greater importance than the method of analysis.

A further case study has been described to demonstrate the difficulty of accurate prior prediction of pile performance, even in relatively simple geotechnical conditions in which a considerable amount of geotechnical data is available. In this case, much depends on the experience of the individual making the prediction, and the way in which he or she interprets the available data to obtain the required geotechnical parameters for the analysis employed.

The Author firmly believes that theoretical analyses have led to a significantly increased understanding of the mechanics of pile–soil interaction and an improved appreciation of the factors which influence pile behaviour. A most important feature of a soundly-based theoretical analysis is its ability to answer 'what-if' questions, for example, what if the pile length is increased, what if the number of piles is reduced, what if the load is cyclic rather than static? Theoretical analysis can provide a sound quantitative basis for answering such questions, provided that due care is taken in modelling the problem in hand. Indeed, the major challenge in applying theoretical approaches to practical pile design remains the proper characterization, idealization and geotechnical quantification of the site.

For most practical problems that involve conventional static loading of conventional-sized piles, there appears to be little justification for using very sophisticated methods of analysis, and quite often, category 2 solutions should provide an adequate basis for design. For a practical viewpoint, more refined category 3 analyses are only likely to be justified under the following circumstances

(*a*) detailed design for major projects
(*b*) unusually long or large-diameter piles
(*c*) when detailed information is required on load transfer along the pile, or soil movements away from the piles are required
(*d*) problems that involve unconventional loading, such as cyclic loading, or loading induced by movement of the soil past the pile.

From the point of view of gaining an increased understanding of pile behaviour, there is a definite need to pursue the development of more refined category 3 (especially 3C) analyses. A number of significant aspects of pile behaviour lack a proper theoretical framework of understanding, including

(*a*) the influence of the method of installation on pile performance
(*b*) the behaviour of piles in sand
(*c*) creep deformation of piles
(*d*) the mechanics of cyclic degradation of shaft resistance
(*e*) the accumulation of permanent displacements under cyclic loading.

There is substantial scope for further research into such aspects so that those aspects of pile design which are still empirical may ultimately be replaced by more soundly-based approaches. The proper role of theoretical analysis in the development of improved pile design procedures has been well-stated by Terzaghi (1939)

> To accomplish its mission in engineering, science must be assigned the role of a partner and not that of a master.

Theory (the science to which Terzaghi refers) should not be developed in isolation from controlled experiments and field observations, and should not be applied without due recognition of the importance of proper characterization of the site geology.

ACKNOWLEDGEMENTS

The Author is indebted to his colleagues, Professor J. R. Booker, and Drs J. P. Carter, J. C. Small and P. T. Brown, for their advice and encouragement over many years, and for their constructive comments on this Paper. Valuable comments have also been provided by Professor F. H. Kulhawy and Dr M. F. Randolph, while Professor M. W. O'Neill and Mr D. J. Douglas have generously provided information on two of the case histories discussed in the Paper. A number of former postgraduate students have contributed significantly to the development of the analysis of pile foundations, including Dr N. S. Mattes, Dr T. S. Hull, Dr C. Y. Lee and Dr C. M. Hewitt, who also undertook some of the analyses of group behaviour reported herein. The Author also acknowledges the encouragement of the late Professor E. H. Davis, who first suggested to the Author that there was something to be gained from a theoretical examination of pile behaviour.

REFERENCES

American Petroleum Institute. (1984). *Recommended practice for planning, designing and constructing fixed offshore platforms* API RP2A, 15th ed, American Petroleum Institute.

Aschenbrenner, T. B. & Olsen, R. E. (1984). Prediction of settlement of single piles in clay. *Anal. Design Pile Fdns., Am. Soc. Civ. Engrs*, 41–58.

Baguelin, F. J., Bustamante, M. G. & Frank, R. A. (1986). The pressuremeter for foundations: French experience. *Proc. In-Situ 86 Conf., Am. Soc. Civ. Engrs*, 31–46.

Baguelin, F. & Frank, R. (1980). Theoretical studies of piles using the finite element method. *Proc. Int. Conf. on Numer Methods in Offshore Piling*, Instn Civ. Engrs, London, 83–91.

Balaam, N. P., Poulos, H. G. & Booker, J. R. (1975). Finite element analysis of the effects of installation on pile load–settlement behaviour. *Geotech. Engng* **6,** No. 1, 33–48.

Banerjee, P. K. (1978). Analysis of axially and laterally loaded pile groups. In *Developments in Soil Mechanics.* ed. C. Scott, Ch. 9, London: Applied Science Publishers.

Banerjee, P. K. & Davies, T. G. (1977). Analysis of pile groups embedded in Gibson soil. *Proc. 9th Int. Conf. Soil Mechs Fdn Engng*, Tokyo, **1,** 381–386.

Banerjee, P. K. & Davies, T. G. (1978). The behaviour of axially and laterally loaded single piles embedded in non-homogeneous soils. *Géotechnique* **28,** No. 3, 309–326.

Banerjee, P. K. & Driscoll, P. M. (1976). Three-dimensional analysis of raked pile groups. *Proc. Instn Civ. Engrs*, Part 2, **61,** Dec., 653–671.

Belcotec (1985). Pile foundation problems. In *Belgian Geotechnical Volume*, Ch. 1, Pub. for 1985 Golden Jubilee of ISSMFE, Belgian Soc. Soil Mechs Fdn. Engng.

Berezantzev, V. G., Khristoforov, V. & Golubkov, V. (1961). Load bearing capacity and deformation of piled foundations. *Proc. 5th Int. Conf. Soil Mech. Fdn Engng, Paris,* **2,** 11–15.

Blight, G. E. (1965). The time–rate of heave of structures on expansive clays. In *Moisture equilibria and moisture changes in soils beneath covered areas.* p. 78, Sydney: Butterworths.

Blight, G. E. (1984). Power station foundation in deep expansive soils. *Proc. 5th Int. Conf. Expansive Soils, Adelaide,* **1,** 22.

Booker, J. R. & Poulos, H. G. (1976). Analysis of creep settlement of pile foundations. *J. Geotech. Engng, Am. Soc. Civ. Engrs* **102,** No. GT1, 1–14.

Briaud, J.-L. & Tucker, L. M. (1988). Measured and predicted axial response of 98 piles. *J. Geotech. Engng, Am. Soc. Civ. Engrs* **114,** No. 9, 984–1001.

Burland, J. B. (1973). Shaft friction of piles in clay—a simple fundamental approach. *Ground Engng* **6,** No. 3, 30–42.

Bustamante, M., Frank, R. & Gianeselli, L. (1987). Le dimensionnement des fondations profondes. *Bull. Liaison Labs P. et Ch* **149,** May–June, 13–22.

Bustamante, M. & Gianeselli, L. (1982). Pile bearing capacity prediction by means of static penetrometer CPT'. *Proc. ESOPT II, Amsterdam* **2,** 493–500.

Butterfield, R. & Abdrabbo, F. M. (1983). Static working load tests on small scale pile groups in clay. In *Developments in soil mechanics and foundation engineering 1, model studies,* pp. 111–168. London: Applied Science Publishers.

Butterfield, R. & Banerjee, P. K. (1971). The elastic analysis of compressible piles and pile groups. *Géotechnique* **21,** No. 1, 43–60.

Butterfield, R. & Douglas, R. A. (1981). *Flexibility coefficients for the design of piles and pile groups.* Tech. Note 108. London: CIRIA.

Butterfield, R. & Ghosh, N. (1977). The response of single piles in clay to axial load. *Proc. 9th Int. Conf. Soil Mechs Fdn Engng, Tokyo* **1,** 451–457.

Callanan, J. F. & Kulhawy, F. H. (1985). *Evaluation of procedures for predicting foundation uplift movements.* Report to Electric Power Research Inst., No. EPRI EL-4107, Cornell University.

Cameron, D. A. & Walsh, P. F. (1984). The prediction of moisture induced foundation movements using the instability index. *Aust. Geomechs News,* No. 8, 5–11.

Chan, K. S., Karashudi, P. & Lee, S. L. (1974). Force at a point in the interior of a layered elastic half-space. *Int. J. Solids Struct.* **10,** 1179–1199.

Cheung, Y. K., Tham, L. G. & Guo, D. J. (1988). Analysis of pile group by infinite layer method. *Géotechnique* **38,** No. 3, 415–431.

Chin, J. T. (1988). Analysis of piles and pile groups embedded in a layered half-space. ME thesis, National University of Singapore.

Chow, Y. K. (1986). Analysis of vertically loaded pile groups. *Int. J. Numer. Anal. Meth. Geomechs* **10,** 59–72.

Christoulas, S. G. (1988). Dimensionnement de pieux. Quelques expériences et recherches en Grèce. *Bull. Liaison Labs. P. et. Ch.* **154,** 5–10.

Christoulas, S. & Pachakis, M. (1987). Pile settlement based on SPT results. *Bull. Res. Centre of Public Works Dept. of Greece* **3,** July–Sept, 221–226.

Coyle, H. M. & Reese, L. C. (1966). Load transfer for axially loaded piles in clay. *J. Soil Mechs Fdn Engng, Am. Soc. Civ. Engrs* **92,** No. SM2, 1–26.

D'Appolonia, D. J., D'Appolonia, E. & Brissette, R. F. (1970). Closure to settlement of spread footings on sand. *J. Soil Mechs Fdn Engng, Am. Soc. Civ. Engrs* **96,** No. SM2, 754–761.

Datta, M., Gulhati, S. K. & Rao, G. V. (1980). An appraisal of the existing practise of determining the axial load capacity of deep penetration piles in calcareous sands. *Proc. 12th Annual OTC, Houston* Paper OTC 3867, 119–130.

Decourt, L. (1982). Prediction of the bearing capacity of piles based exclusively on *N* values of the SPT. *Proc. ESOPT2, Amsterdam* **1,** 29–34.

Denver, H. (1982). Modulus of elasticity determined by SPT and CPT. *Proc. ESOPT2, Amsterdam* **1,** 35–40.

De Ruiter, J. & Beringen, F. L. (1979). Pile foundations for large North Sea structures'. *Marine Geotechnol.,* **3,** No. 3, 267–314.

Desai, C. S. (1974). Numerical design—analysis for piles in sands. *J. Geotech. Engng, Am. Soc. Civ. Engrs* **100,** No. GT6, 613–635.

Donaldson, G. W. (1967). The use of small-diameter piles in expansive soil. *Proc. 4th Regional Conf. for Africa on Soil Mechs Fdn Engng* **1,** 253.

Douglas, D. J. (1989). Prediction study for driven piles. *5th Australia–New Zealand Geomechs Conf., Aust. Geomechs,* special issue.

Dutt, R. N. & Ingram, W. B. (1984). Jackup rig siting in calcareous soils. *Proc. 16th Annual OTC, Houston* Paper OTC 4840, 541–548.

Edil, T. B. & Mochtar, I. B. (1988). Creep response of model pile in clay. *J. Geotech. Engng, Am. Soc. Civ. Engrs* **114,** No. 11, 1245–1260.

El Sharnouby, B. & Novak, M. (1985). Static and low-frequency response of pile groups. *Can. Geotech. J.* **22,** No. 1, 79–94.

Findlay, J. D. (1984). Discussion, *Piling and ground treatment,* pp. 189–190, London: Instn. Civ. Engrs. Thomas Telford.

Fleming, W. G. K. & Thorburn, S. (1984). Recent piling advances. *Piling and Ground Treatment,* pp. 1–16. London: Instn of Civil Engrs. Thomas Telford.

Fleming, W. G. K., Weltman, A. J., Randolph, M. F. & Elson, W. K. (1985). *Piling engineering.* New York: Surrey University Press, Halsted Press.

Fletcher, M. S. & Mizon, D. H. (1984). Piles in chalk for Orwell bridge. *Piling and Ground Treatment,* pp. 203–209, London: Instn. Civil Engrs. Thomas Telford.

Focht, J. A. (1967). Discussion to paper by Coyle and Reese, *J. Soil Mechs. Fdn. Engng, Am. Soc. Civ. Engrs* **93,** No. SM1, 133–138.

Frank, R. (1985). Recent developments in the prediction of pile behaviour from pressuremeter results. *Proc. Symp. from Theory to Practice on Deep Foundations, Porto Alegre, Brazil* **1,** 69–99.

Giroud, J. P., Tran-Vo-Nhiem & Obin, J. P. (1973). *Tables pour le calcul des fondations* **3,** Paris: Dunod.

Gurtowski, T. M. & Wu, M.-J. (1984). Compression load tests on concrete piles in alluvium. *Anal. and Design of Pile Foundations, Am. Soc. Civ. Engrs,* 138–153.

Ha, H. B. & O'Neill, M. W. (1983). Discussion on Theoretical *t–z* curves. *J. Geotech. Engng, Am. Soc. Civ. Engrs* **109**, No. GT10, 1353–1355.

Hewitt, C. M. (1988). *Cyclic response of offshore pile groups*, PhD thesis, University of Sydney, Australia.

Hobbs, N. B. (1977). Behaviour and design of piles in chalk—an introduction to the discussion of the papers on chalk. *Piles in Weak Rock*, pp. 149–175, London: Instn. Civil Engrs.

Holeyman, A. (1985). Dynamic non-linear skin friction of piles. *Proc. Symp. on Penetrability, Drivability of Piles, 11th Int. Conf. Soil Mechs Fdn Engng*, San Francisco.

Holloway, D. M., Clough, G. W. & Vesic, A. S. (1978). The effects of residual driving stresses on pile performance under axial loads. *Proc. 10th OTC, Houston*, 2225–2236.

Hull, T. S. (1987). The static behaviour of laterally loaded piles, PhD thesis, University of Sydney, Australia.

Hyden, A. M., Hulett, J. M., Murff, J. D. & Abbs, A. F. (1988). Design practice for grouted piles in Bass Strait calcareous soils. In *Engineering for calcareous sediments* **1**, pp. 297–304. Rotterdam: A. A. Balkema.

Imai, T. & Tonouchi, K. (1982). Correlation of *N* value with *S*-wave velocity and shear modulus. *Proc. ESOPT2, Amsterdam* **1**, 67–72.

Jardine, R. J., Potts, D. M., Fourie, A. B. & Burland, J. B. (1986). Studies of the influence of non-linear stress–strain characteristics in soil–structure interaction'. *Géotechnique* **36**, No. 3, 377–396.

Johnston, I. W. & Lam, T. S. K. (1984). Frictional characteristics of planar concrete–rock interfaces under constant normal stiffness conditions. *Proc. 4th Aust.–New Zealand Conf. on Geomechs, Perth*, **2**, 397–401.

Johnston, I. W., Carter, J. P., Novello, E. A. & Ooi, L. H. (1988). Constant normal stiffness direct shear testing of calcarenite. In *Engineering for calcareous sediments* **2**, pp. 541–553, Rotterdam: A. A. Balkema.

Kog, Y. C., Karunaratne, G. P. & Lee, S. L. (1986). Effects of negative friction on piles in layered soils. *Geotech. Engng* **17**, 211–234.

Komornik, A. (1974). Penetration testing in Israel. State-of-the-art report. *Proc. ESOPT, Stockholm* **1**, 183–192.

Kraft, L. M. & Lyons, C. G. (1974). State-of-the art: ultimate axial capacity of grouted piles. *Proc. 6th Annual OTC, Houston* Paper OTC 2081, 487–503.

Kraft, L. M., Ray, R. P. & Kagawa, T. (1981). Theoretical *t–z* curves. *J. Geotech. Engng, Am. Soc. Civ. Engrs* **107**, No. GT11, 1543–1561.

Kulhawy, F. H., O'Rourke, T. D., Stewart, J. P. & Beech, J. F. (1982). *Transmission line structure foundations for uplift/compression loading. Supplemental Volume: Load Test Summaries*, Res. Proj. 1493-1, final report to EPRI, Cornell University.

Kulhawy, F. H., Trautmann, C. H., Beech, J. F., O'Rourke, T. D., McGuire, W., Wood, W. A. & Capano, C. (1983). *Transmission line structure foundations for uplift–compression loading.* Report EL-2870, Electric Power Research Inst., Palo Alto.

Kuwabara, F. & Poulos, H. G. (1989). Downdrag forces in a group of piles, to be published in *J. Geotech. Engng, Am. Soc. Civ. Engrs.*

Lambe, T. W. (1973). Predictions in soil engineering. *Géotechnique* **23**, No. 2, 149–202.

Lee, C. Y. (1988). The behaviour of grouted piles in offshore foundations. PhD thesis, University of Sydney, Australia.

Leonards, G. A. & Darrag, A. A. (1989). Discussion to Analysis of residual stress effects in piles, H. G. Poulos. *J. Geotech. Engng, Am. Soc. Civ. Engrs* **115**, No. 4, 589–596.

Martin, R. E., Seli, J. J., Powell, G. W. & Bertoulin, M. (1987). Concrete pile design in Tidewater, Virginia. *J. Geotech. Engng Am. Soc. Civ. Engrs* **113**, No. 6, 568–585.

Matlock, H. & Foo, S. C. (1980). Axial analysis of piles using a hysteretic and degrading soil model. *Proc. Conf. Numer. Methods in Offshore Piling* pp. 165–185, London, Inst. Civil Engrs.

Mattes, N. S. & Poulos, H. G. (1971). Model tests on piles in clay. *Proc. 1st Aust.–New Zealand Conf. Geomechs, Melbourne*, 254–259.

McClelland, B. (1974). Design of deep penetration piles for ocean structures. *J. Geotech. Engng, Am. Soc. Civ. Engrs* **100**, No. GT7, 705–747.

Meyerhof, G. G. (1956). Penetration tests and bearing capacity of cohesionless soils. *J. Soil Mechs. Fdn Engng, Am. Soc. Civ. Engrs* **82**, No. SM1, 1–19.

Meyerhof, G. G. (1959). Compaction of sands and bearing capacity of piles. *J. Soil Mechs Fdn Engng, Am. Soc. Civ. Engrs* **85**, No. SM6, 1–29.

Meyerhof, G. G. (1963). Some recent research on the bearing capacity of foundations. *Can. Geotech. J.* **1**, No. 1, 16–26.

Meyerhof, G. G. (1976). Bearing capacity and settlement of pile foundations'. *J. Geotech. Engng, Am. Soc. Civ. Engrs* **102**, No. GT3, 195–228.

Milovic, D. & Stevanovic, S. (1982). Some soil parameters determined by cone penetration tests. *Proc. ESOPT2, Amsterdam*, **2**, 709–714.

Mindlin, R. D. (1936). Force at a point in the interior of a semi-infinite solid. *Physics* **7**, 195–202.

Nauroy, J.-F., Brucy, F., Le Tirant, P. & Kervadec, J.-P. (1986). Design and installation of piles in calcareous formations. *Proc. 3rd Int. Conf. Numer. Methods in Offshore Piling, Nantes*, 461–480.

Nystrom, G. A. (1984). Finite–strain axial analysis of piles in clay. *Anal. Design Pile Fdns, Am. Soc. Civ. Engrs*, 1–20.

Ohya, S., Imai, T. & Matsubara, M. (1982). Relationships between *N*-value by SPT and LLT measurement results. *Proc. ESOPT2, Amsterdam* **1**, 125–130.

O'Neill, M. W. (1983). Group action in offshore piles. *Proc. Am. Soc. Civ. Engrs Conf., Geotech. Practice in Offshore Engng, Austin*, 25–64.

O'Neill, M. W., Ghazzaly, O. I. & Ha, H. B. (1977). Analysis of three-dimensional pile groups with non-linear soil response and pile–soil–pile interaction. *Proc. 9th Annual OTC, Houston* Paper OTC 2838, 245–256.

O'Neill, M. W. & Ha, H. B. (1982). Comparative modelling of vertical pile groups. *Proc. 2nd Int. Conf. Numer. Methods in Offshore Piling, Austin*, 399–418.

O'Neill, M. W., Hawkins, R. A. & Mahar, L. J. (1981). *Field study of pile group action: final report.* Report

Nos. FHWA/RD-81/002, FHWA/RD-81/008, Federal Highway Administration, US Department of Transportation.

O'Neill, M. W., Hawkins, R. A. & Mahar, L. J. (1982). Load transfer mechanisms in piles and pile groups. *J. Geotech. Engng, Am Soc. Civ. Engrs* **108,** No. GT12, 1605–1623.

Ooi, L. H. & Carter, J. P. (1987). A constant normal stiffness direct shear device for static and cyclic loading. *Geotech. Testing, J. ASTM* **10,** 3–12.

Ooi, L. H., Boey, C. F. & Carter, J. P. (1989). Modified load transfer analysis of axially loaded pile. *Piletalk '89* Kuala Lumpur, 217–233.

Ottaviani, M. (1975). Three-dimensional finite element analysis of vertically loaded pile groups. *Géotechnique*, **25,** No. 2, 159–174.

Parry, R. H. G. (1971). A direct method of estimating settlements in sand from SPT values. *Proc. Symp. Interaction of Structs. and Fdns, Midlands SMFE Soc, Birmingham, UK*, 29–37.

Pells, P. J. N. & Turner, R. (1979). Elastic solution for the design and analysis of rock socketed piles. *Can. Geotech. J*, **16,** 481–487.

Poulos, H. G. (1968). Analysis of the settlement of pile groups. *Géotechnique* **18,** 449–471.

Poulos, H. G. (1976). Discussion to paper by Ottaviani. *Géotechnique*, **26,** No. 1, 238–239.

Poulos, H. G. (1978). *User's guide to TAPILE.* School of Civil Engineering, University of Sydney.

Poulos, H. G. (1979a). Settlement of single piles in non-homogeneous soil. *J. Geotech. Engng, Am. Soc. Civ. Engrs* **105,** No. GT5, 627–641.

Poulos, H. G. (1979b). Development of an analysis for cyclic axial loading of piles. *Proc. 3rd Int. Conf. Numer. Methods in Geomechs, Aachen* **4,** 1513–1530.

Poulos, H. G. (1980). User's guide to DEFPIG—Deformation Analysis of Pile Group. School of Civil Engineering, University of Sydney.

Poulos, H. G. (1983). Cyclic axial pile response—alternative analyses. *Am. Soc. Civ. Engrs Specialty Conf. on Geotech. Practice in Offshore Engng, Austin* 403–421.

Poulos, H. G. (1987). Analysis of residual stress effects in piles. *J. Geotech. Engng, Am. Soc. Civ. Engrs* **113,** No. 3, 216–229.

Poulos, H. G. (1988a). Modified calculation of pile group settlement interaction. *J. Geotech. Engng, Am. Soc. Civ. Engrs* **114,** No. 6, 697–706.

Poulos, H. G. (1988b). Cyclic stability diagram for axially loaded piles. *J. Geotech. Engng, Am. Soc. Civ. Engrs,* **114,** No. 8, Aug., 877–895.

Poulos, H. G. (1988c). *Marine Geotechnics.* London: Unwin Hyman.

Poulos, H. G. (1988d). The mechanics of calcareous sediments. Jaeger Memorial Lecture, *5th Australia–New Zealand Geomechs. Conf., Austr. Geomechs,* 8–41.

Poulos, H. G. (1989). *Users manual—program PIES—Pile in Expansive Soil.* Centre for Geotech. Research, University of Sydney, Australia.

Poulos, H. G. & Davis, E. H. (1968). The settlement behaviour of single axially-loaded incompressible piles and piers. *Géotechnique* **18,** No. 3, 351–371.

Poulos, H. G. & Davis, E. H. (1980). *Pile foundation analysis and design.* New York: Wiley.

Poulos, H. G. & Hewitt, C. M. (1986). Axial interaction between dissimilar piles in a group. *Proc. 3rd Int. Conf. Numer. Methods in Offshore Piling, Nantes* 253–270.

Poulos, H. G. & Hull, T. S. (1989). The role of analytical geomechanics in foundation engineering. *Proc. Am. Soc. Civ. Engrs Foundation Engng Congress, Chicago* (in publication).

Poulos, H. G. & Mattes, N. S. (1971). Displacements in a soil mass due to pile groups. *Aust. Geomechs J.* **G1,** No. 1, 18–28.

Pressley, J. S. & Poulos, H. G. (1986). Finite element analysis of mechanisms of pile group behaviour. *Int. J. Numer. Anal. Meth. Geomechs* **10,** 213–221.

Randolph, M. F. (1983). Design considerations for offshore piles. *Am. Soc. Civ. Engrs, Spec. Conf. Geotech. Practice in Offshore Engng, Austin,* 422–439.

Randolph, M. F. (1986). *RATZ–load transfer analysis of axially loaded piles.* Rep. No. Geo 86033, Perth: Dept Civil Engineering, University of Western Australia.

Randolph, M. F. (1987). *PIGLET, a computer program for the analysis and design of pile groups.* Report GEO 87036, Perth: University of Western Australia.

Randolph, M. F. (1988). Evaluation of grouted insert performance. *Proc. Int. Conf. Calcareous Sediments, Perth* **2,** Rotterdam: A. A. Balkema.

Randolph, M. F. & Wroth, C. P. (1978). Analyses of deformation of vertically loaded piles. *J. Geotech. Engng, Am. Soc. Civ. Engrs* **104,** No. GT12, 1465–1488.

Randolph, M. F. & Wroth, C. P. (1979). An analysis of the vertical deformation of pile groups. *Géotechnique* **29,** No. 4, 423–439.

Randolph, M. F., Carter, J. P. & Wroth, C. P. (1979). Driven piles in clay—the effects of installation and subsequent consolidation. *Géotechnique* **29,** No. 4, 361–393.

Robertson, P. K., Campanella, R. G. & Brown, P. T. (1985). Design of axially and laterally loaded piles using in-situ tests: a case history. *Can. Geotech. J.* **22,** 518–527.

Schmertmann, J. H. (1975). Measurement of in situ shear strength. *Spec. Conf. on In Situ Measurement of Soil Properties, Am. Soc. Civ. Engrs.* Raleigh NC **2,** 57–138.

Schmertmann, J. H. (1978). *Guidelines for cone penetration test—in performance and design.* US Dept. of Transportation, Federal Highways Administration, Washington DC.

Semple, R. M. & Rigden, W. J. (1984). Shaft capacity of driven piles in clay. *Anal. Design Pile Fdns, Am. Soc. Civ. Engrs,* 59–79.

Shioi, Y. & Fukui, J. (1982). Application of *N*-value to design of foundations in Japan. *Proc. ESOPT2, Amsterdam* **1,** 159–164.

Skempton, A. W. (1953). Discussion: Piles and pile foundations, settlement of pile foundations. *Proc. 3rd Int. Conf. Soil Mechs Fdn Engng* **3,** 172.

Skempton, A. W. (1959). Cast-in-situ bored piles in London clay. *Géotechnique* **9,** No. 4, 153–173.

Small, J. C. (1988). Finite element analysis of downdrag on piles. *Proc. Conf. Numer. Methods in Geomechanics, Innsbruck,* 1109–1113.

Stas, C. V. & Kulhawy, F. H. (1984). *Critical evaluation*

of design methods for foundations under axial uplift and compression loading. Report for EPRI, No. EL-3771, Cornell University.
Steinbrenner, W. (1934). Tafeln zur Setzungberechnung. *Die Strasse* **1,** 221.
Stewart, J. F. & Kulhawy, F. H. (1981). Interpretation of uplift load distribution data. *Proc. 10th Int. Conf. Soil Mechs Fdn Engng, Stockholm* **2,** 277–280.
Terzaghi, K. (1939). Soil mechanics—a new chapter in engineering science. *J. Instn Civ. Engrs* **12,** 106–141.
Terzaghi, K. & Peck, R. B. (1967). *Soil mechanics in engineering practice*. 2nd Ed., New York: Wiley.
Thorburn, S. & MacVicar, S. L. (1971). Pile load tests to failure in the Clyde alluvium. *Behaviour of Piles*, London: Institution of Civil Engineers. 1–7, 53–54.
Tomlinson, M. J. (1957). The adhesion of piles driven into clay soils. *Proc. 4th Int. Conf. Soil Mechs Fdn Engng, London*, **2,** 66–71.
Tomlinson, M. J. (1977). *Pile design and construction practice*. London: Cement and Concrete Association.
Valliappan, S., Lee, I. K. & Boonlualohr, P. (1974). Settlement of piles in layered soils. *Proc. 7th Biennial Conf., Aust. Road Res. Board, Adelaide* **7,** Pt 7, 144–153.
Van Der Merwe, D. H. (1964). The prediction of heave from the plasticity index and the percentage clay fraction. *Civil Engr in South Africa* **6,** No. 6, 103.
Van Impe, W. F. (1986). Evaluation of deformation and bearing capacity parameters of foundations, from static CPT-results. *4th Int. Geotech. Seminar Field Instrumentn and In-Situ Measurements Nanyang Tech. Inst., Singapore*, 51–70.
Verbrugge, J. C. (1982). The evaluation of pile cap load–settlement diagram from the CPT results. *Proc. ESOPT2, Amsterdam* **2,** 923–926.
Vesic, A. S. (1972). Expansion of cavities in infinite soil mass. *J. Soil Mechs Fdn Engng, Am. Soc. Civ. Engrs* **98,** No. SM3, 265–290.
Williams, A. F. & van der Zwaag, G. L. (1988). Analysis and evaluation of grouted section tests, *Engng for calcareous sediments* **2,** pp. 493–502, Rotterdam: Balkema.
Withiam, J. L. & Kulhawy, F. H. (1979). Analytical model for drilled shaft foundations. *Proc. 3rd Int. Conf. Numer. Methods in Geomechs, Aachen* **3,** 1115–1122.
Wright, S. J. & Reese, L. C. (1979). Design of large diameter bored piles. *Ground Engng* Nov., 17–50.
Yamashita, K., Tomono, M. & Kakurai, M. (1987). A method for estimating immediate settlement of piles and pile groups. *Soils and Fdns* **27,** No. 1, 61–76.

VOTE OF THANKS

S. F. BROWN

Ladies and Gentlemen,

About 18 months ago, I wrote to Harry Poulos in my then capacity as BGS chairman inviting him to deliver the 1989 Rankine Lecture. As a consequence, I have carried a certain responsibility over this period, which is now happily discharged. My responsibility was enhanced by having had the unusual opportunity, with respect to a distant overseas lecturer, of briefing him at first hand when I attended the Australia–New Zealand Geomechanics Conference in Sydney in August 1988.

In my invitation to Harry, I said that we assumed the subject of piling could well feature in his presentation. Within two months, I had a letter outlining the lecture much as we heard it this evening. By the time we had lunch together in Sydney eight months later, all I had to do was make encouraging noises, since it was clear to me that a balanced, well planned and extremely appropriate lecture was in prospect. I took the precaution of telling Harry how even greater men had found the Rankine Lecture occasion a difficult one, that the audience varied from the least to the greatest in British Geotechnics, and that all would wish to go away having learned something. I mentioned visual aids *en passant* remembering some past examples of slides that left something to be desired.

Towards the end of the Sydney Conference, I had the further opportunity of listening to Harry Poulos eloquently deliver the John Jaeger Memorial Address on the subject of calcareous soils. By then no doubts remained.

Against this background, therefore, and in common with many others who had been privileged to hear Harry Poulos speak on other occasions, I arrived this evening, not merely with the usual anticipation of the big event, but with the confidence that we were to hear an excellent lecture presented with clarity and authority. Ladies and Gentlemen, we have not been disappointed.

Harry, you have demonstrated vividly the use which can be made of what Terzaghi somewhat scornfully referred to as 'Science' and you have put in clear perspective the other less prudent statement he made with Ralph Peck that theoretical refinements for piles are completely out of place. Your balanced approach giving a sense of perspective on the input parameters for design through your sensitivity analyses and presenting applications of theory through case studies and a class A prediction have clearly demonstrated how theory should be used, not as a master but as an essential tool in geotechnical engineering.

Professor Poulos, the British Geotechnical Society is enormously grateful to you for preparing such an excellent and stimulating Rankine Lecture, which was full of useful practical information and will, I predict, become a standard reference when it appears in *Géotechnique* in due course.

We thank you and admire you for delivering this lecture with such EASE—eloquence, aplomb, stamina and enthusiasm.

The Rankine Lecture

The thirtieth Rankine Lecture of the British Geotechnical Society was given by Professor J. B. Burland at Imperial College on 28 March 1990. The following introduction was given by Mr S. Thorburn, OBE.

John Boscawen Burland, born 4 March 1936, was educated at Parktown Boys' High School, Johannesburg, and read civil engineering at the University of the Witwatersrand under Professor Jennings. He was awarded his Doctorate of Philosophy at the University of Cambridge in 1966. His work with Professor K. H. Roscoe at Cambridge released a latent scientific ability and enhanced his gift of analytical perception. The initial formative years of practical training in the discipline of civil engineering were spent with Ove Arup & Partners prior to his sojourn at Cambridge. John became a Senior Scientific Officer with the Building Research Station in 1966 and within six years succeeded Dr Tom Whitaker as Head of the Geotechnics Division of BRE. In 1979 he was promoted to the post of Assistant Director. In 1980 Dr Burland was appointed Professor of Soil Mechanics at Imperial College in succession to the late Professor Alan Bishop. He was elected to the Fellowship of Engineering in 1981 and was awarded his Doctorate of Science in 1983.

To capture the essence of a Rankine Lecturer in a short introduction is a pleasurable but difficult task. Having enjoyed the friendship and stimulation of the company of John Burland for more than twenty years I have been aware of his considerable talents which are given to few. His major contributions to science and engineering have been recognized by the awards of the Baker and the Kelvin medals of the Institution of Civil Engineers. He has been awarded the British Geotechnical Society Prize on four occasions and the Telford Premium of the Institution of Civil Engineers on two occasions. I hesitate to list numerous other awards.

John possesses abilities belonging to three distinct disciplines; those of an engineer, of a scientist, and of a teacher. The discipline of engineering occupies the prime position in the list since John displays a practical ability which enhances his scientific faculty. To reduce the relative importance of his scientific skills would, however, convey an incorrect impression of him. Over a period of three decades John has published more than seventy papers and his excitement on discovery of a new concept or physical relationship is infectious. John displays both the emotion of scientific discovery and a personal contentment with life—a sublime combination. Einstein said, 'All our science measured against reality is primitive and childlike—and yet it is the most precious thing we have.' John Burland possesses this awareness of scientific limitation in the real world and the engineer within him recognizes that adequacy of solution is acceptable in practice where accuracy is unattainable. His search for better ways to model physical systems and to explain mechanisms of real behaviour is unceasing.

His teaching faculty has been expressed best in the Nash Lecture given to the Ninth European Congress of Soil Mechanics and Foundation Engineering in Dublin in 1987. John concluded in his lecture, 'A key aim in teaching is to give the students "anchor points" whose security and limitations have been clearly established whether they be experimental, empirical, or theoretical.' That particular lecture has been acknowledged as a milestone in educational philosophy.

It would be remiss of me to complete this introduction without drawing your attention to the sense of timing which John possesses. Over the past twenty years I have become aware of John choosing the ideal time to make a valuable contribution to the theory and practice of civil engineering. When practitioners are struggling with subjects such as settlement of buildings on sands, earth retaining structures, piled foundations, or soil–structure interaction, the Burland influence emerges and new light is thrown on the subject. I suspect that the ancient Greeks anticipated the Burland Factor when they devised the letter β.

The life of John Burland is conveyed best by a quotation from Sir Henry Wooton (1568–1639) in his *Character of a Happy Life.*

How happy is he born and taught
That serveth not another's will
Whose armour is his honest thought
And simple truth his utmost skill

Special lectures by John Burland have become events not to be missed because of his characteristic style and interesting delivery and it is with great pleasure and anticipation that I invite Professor John Boscawen Burland to deliver the thirtieth Rankine Lecture.

Professor J. B. Burland, FEng

Burland, J. B. (1990). *Géotechnique* **40**, No. 3, 329–378

On the compressibility and shear strength of natural clays

J. B. BURLAND, FEng*

The compressibility and strength characteristics of reconstituted clays are used as a basic frame of reference for interpreting the corresponding characteristics of natural sedimentary clays. The properties of reconstituted clays are termed 'intrinsic' properties since they are inherent to the soil and independent of the natural state. The properties of a natural clay differ from its intrinsic properties due to the influence of soil structure (fabric and bonding). Thus the intrinsic properties provide a frame of reference for assessing the in situ state of a natural clay and the influence of structure on its in situ properties. A new normalizing parameter called the void index is introduced to aid in correlating the compression characteristics of various clays. The sedimentation compression curves for most, but not all, natural clays lie well above the corresponding intrinsic compression curves. A consequence of this is that such clays are more sensitive and brittle than the reconstituted material and the post-yield compression index C_c is usually much greater than the intrinsic value. This observation has important consequences for stress-path testing of soft clays. The location of the natural sedimentation curve relative to the intrinsic one is shown to depend on depositional conditions and on post-depositional processes such as leaching. The undrained strength of a normally consolidated natural sediment is shown to be primarily a function of the in situ effective stresses and of the soil structure and not of the moisture content. For overconsolidated natural clays the intrinsic compression line provides a useful means of assessing the degree of overconsolidation. Also the ratio of the intrinsic swelling index to the undisturbed swelling index (the swell sensitivity) is a valuable measure of bonding. The strength properties of two overconsolidated clays (Todi Clay and London Clay) are presented and the intact strengths are shown to be greater than the corresponding intrinsic strengths. However, both clays show brittle behaviour with the formation of shear surfaces at peak intact strength. The strength on such a shear surface drops rapidly to a well defined post-rupture strength after a few millimeters relative displacement. The post-rupture strength must be clearly distinguished from the residual strength which requires much larger relative displacements to

* Imperial College of Science, Technology and Medicine, London.

Les caractéristiques de compressibilité et de résistance des argiles reconstituées s'emploient comme base générale pour interpréter les caractéristiques correspondantes des argiles sédimentaires naturelles. Les propriétés des argiles reconstituées sont définies comme des propriétés 'intrinsèques,' parce qu'elles sont propres au sol et indepéndantes de l'état naturel. Les propriétés d'une argile naturelle diffèrent de ses propriétés intrinsèques à cause de l'influence de la structure du sol (fabrique et liage). Les propriétés intrinsèques fournissent ainsi une base générale pour évaluer l'état in situ d'une argile naturelle et l'influence de la structure sur ses propriétés in situ. Un nouveau paramètre normalisant appelé indice des vides est introduit pour aider dans la corrélation des caractéristiques de compression des argiles diverses. Pour la plupart des argiles naturelles, mais pas pour toutes, les courbes de compression de sédimentation se situent bien au-dessus des courbes de compression intrinsèques correspondantes. De telles argiles sont par par conséquent plus sensibles et fragiles que la matière reconstituée et l'indice de compression après l'écoulement C_c est normalement plus élevé que la valeur intrinsèque. Cette constatation a d'importantes conséquences pour les expériences effectuées au suget du chemin de contrainte des argiles tendres. On démontre comment l'emplacement d'une courbe de sédimentation naturelle par rapport à la courbe intrinsèque dépend des conditions de dépôt et des événements suivant le dépôt, tels que le lessivage. On démontre aussi que la résistance non-drainée d'un sédiment naturel normalement consolidé est en premier lieu une fonction des contraintes effectives in situ et de la structure du sol, et non de la teneur en eau.

Pour les argiles naturelles surconsolidées la ligne de compression intrinsèque fournit un moyen utile pour évaluer le degré de surconsolidation. Le rapport entre l'indice de gonflement intrinsèque et l'indice de gonflement non remanié (la sensibilité au gonflement) représente une indication très utile des liaisons. Les propriétés de résistance de deux argiles surconsolidées (argile de Todi et argile de Londres) sont presentées, et on démontre que les résistances intactes sont supérieures aux résistances intrinsèques correspondantes. Cependant les deux argiles se comportent de façon fragile avec la formation de surfaces de cisaillement à la résistance intacte de pic. Sur une telle surface de cisaillement la résistance décrit rapidement à une résistance bien-définie après-rupture apres quelques

develop. Evidence is given which indicates that the post-rupture strength may be relevant to many stability problems in stiff clays and may also control the in situ stresses during geological unloading. For Todi Clay and London Clay the post-rupture strengths at low confining stresses are close to the intrinsic critical state strengths. More study is required before this can be accepted as a general result for most clays.

KEYWORDS: clays; compressibility; fabric/structure of soils; sedimentation; shear strength; soil properties

millimètres de déplacement relatif. Il faut distinguer clairement entre la résistance après-rupture et la résistance résiduelle, qui ne se développe que pour des déplacements relatifs plus importants. Des données sont présentées que indiquent que la résistance après-rupture peut être importante pour beaucoup de problèmes de stabilité dans les argiles raides et peut aussi contrôler les contraintes in situ pendant le déchargement géologique. Dans les cas de l'argile de Todi et de l'argile de Londres les résistances après-rupture à des valeurs basses de contrainte avec étreinte latérale sont très près des résistances intrinsèques de l'état critique. Des études approfondies seront nécessaires pour confirmer la validité de ce résultat pair la plupart des argiles.

NOTATION

A Skempton's pore pressure coefficient
c' effective cohesion
C_c compression index
C_c^* intrinsic compression index
C_s swelling index
C_s^* intrinsic swelling index
e void ratio
e_L void ratio at liquid limit
e_{100}^* void ratio on ICL for $\sigma_v' = 100$ kPa
G_s specific gravity
I_v void index defined by equation (1)
ICL intrinsic compression line
log logarithm to the base 10
p' effective mean normal stress
p_k' capillary pressure, isotropic swelling pressure
SCL sedimentation compression line
S_u undrained strength
S_{uTC} undrained strength in triaxial compression
w percentage water content
w_L liquid limit
w_p plastic limit
θ angle of slip plane to horizontal
σ' effective normal stress
σ_a' effective axial stress
σ_h' effective horizontal stress
σ_r' effective radial stress
σ_v' effective vertical stress
σ_{ve}^* equivalent stress on the ICL corresponding to the void ratio, or void index, of the soil
σ_{vo}' effective overburden pressure
σ_{vy}' effective vertical yield stress
τ shear stress
ϕ^* intrinsic angle of shearing resistance

INTRODUCTION

Much of modern soil mechanics has developed from the results of careful, comprehensive studies of the properties of remoulded or reconstituted natural soils or artificial materials such as kaolinite or illite. These studies have been of outstanding importance—perhaps the two most notable being those of Hvorslev and Rendulic, both in the mid 1930s. It is on these, and similar later studies that the framework of what has come to be called critical state soil mechanics has been built. In recent years this phrase has become generic in its use with some of the precision of the original critical state models being lost.

The critical state framework, which was formulated so elegantly by the Cambridge soil mechanics school under the late Professor Roscoe, has provided a coherence which the subject previously lacked. It also provides a logical framework for incorporating theories of plasticity, yield and flow for the mathematical modelling of soil behaviour. Over the last twenty years critical state soil mechanics has been widely taught and increasingly applied to the solution of engineering problems.

But natural soils differ from reconstituted soils in a number of important ways. These differences stem from the influence of micro- and macrostructure. Following Mitchell (1976) the term 'structure' means the combination of 'fabric' (arrangement of particles) and interparticle 'bonding.'

When I was invited to deliver this lecture I quickly came to the conclusion that it would be both timely and appropriate to undertake a review of the basic compressibility and shear strength properties of some natural sedimentary clays and to compare these with the corresponding properties of the reconstituted material. Only results from the highest quality undisturbed samples have been used. The longer term objective of this lecture is to stimulate efforts to bring to natural soils the same unity and coherence which critical state soil mechanics in its broadest sense has brought to reconstituted soils. Significant progress has already been made in this respect (e.g. Leroueil *et al.*, 1979; Leroueil & Vaughan, 1990; Hight *et al.*, 1987; Wood, 1990).

The logical starting point is to examine the compressibility of some normally consolidated natural clays followed by their shear strength properties. The corresponding properties of some overconsolidated natural clays are then considered.

SEDIMENTATION COMPRESSION OF NATURAL CLAYS

In 1970 Skempton published an important paper on the consolidation of natural clays by gravitational compaction. Curves relating in situ void ratio e_0 to effective overburden pressure σ_{vo}' were presented for twenty deposits representing a wide range of lithologies as shown in Fig. 1. The void ratios were corrected to allow for changes in liquid and plastic limits with depth (Skempton 1944). In all cases the deposits are normally consolidated in the sense that the strata have never been under greater effective pressures than those existing at the present time. Excluded from the study were quick clays, diatomaceous clays, clays containing more than 5% organic matter as well as clays with a carbonate content of more than 25%. The average Atterberg limits for each of the deposits all plot above the A line on a plasticity chart.

The curves in Fig. 1 show the progressive changes in void ratio from recently deposited muds on the sea floor, to Quaternary clays at depths of several tens of metres to hard clays and mudstones of Pliocene and late Pleistocene age extending to about 3000 m. Each curve is termed the 'sedimentation compression curve' for the natural material—a term first used by Terzaghi (1941). Skempton drew the following conclusions from the results given in Fig. 1.

(*a*) The relationship between e_0 and log σ_{vo}' (i.e. the sedimentation compression curve) is essentially linear for any particular clay.
(*b*) At a given value of σ_{vo}' the void ratio of a normally consolidated natural clay depends on the nature and amount of clay minerals present, as indicated by the liquid limit. The higher the liquid limit the higher is the void ratio.
(*c*) A most striking observation is the converging pattern formed by the various compression curves.
(*d*) When plotted in terms of liquidity index, rather than void ratio, the results lie within a moderately narrow band. Clays with a high sensitivity lie towards the upper part of the

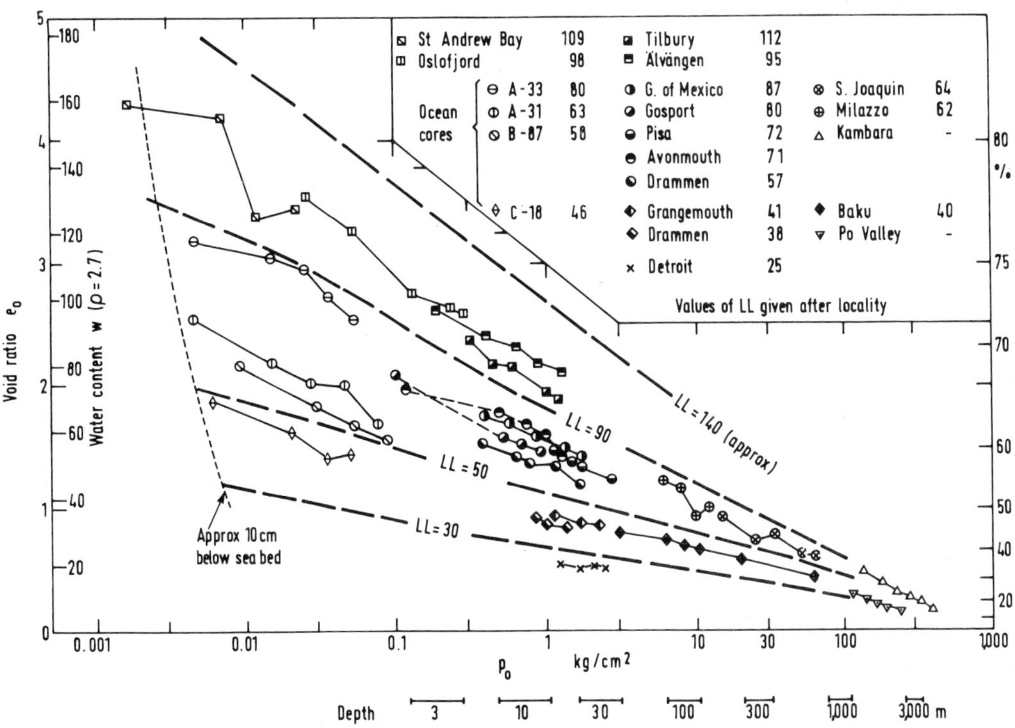

Fig. 1. Sedimentation compression curves for normally consolidated argillaceous sediments (Skempton 1970)

band while those with low sensitivity lie towards the lower part of the band.

(*e*) For sea-bed deposits the depositional water content in the uppermost 250 mm is equivalent to a liquidity index of about 1·75 while that for tidal mudflats is about 1·0.

How do these sedimentation compression curves relate to the corresponding laboratory compression curves on reconstituted material? Do these sedimentation compression curves represent the in situ compressibility associated with the loading of the stratum over a timescale associated with normal construction activities? More generally, how do the properties of these naturally sedimented clays relate to the properties of one-dimensionally consolidated reconstituted clays? Answers to these questions will help to extend our generalized understanding of the properties of reconstituted soils to natural soils.

COMPRESSIBILITY OF RECONSTITUTED CLAYS

A reconstituted clay is defined as one that has been thoroughly mixed at a water content equal to or greater than the liquid limit (w_L). Fig. 2 shows the one-dimensional compression curves for some reconstituted natural clays covering a wide range of plasticities. Values of the liquid limit and the void ratio corresponding to the liquid limit (e_L) are given for each clay. Note that, although Kleinbelt Ton and Argile Plastique have the same liquid limit, Argile Plastique has a lower specific gravity and hence a lower e_L. It appears that e_L is a more fundamental parameter than w_L. At any given value of σ_v' the void ratio is related to e_L, increasing as e_L increases. Note also the converging pattern of the various compression curves as σ_v' increases. It is evident from Fig. 2 that the compression curves are all similar in shape being slightly concave upwards. It is useful to normalize these laboratory compression curves with respect to the void ratio.

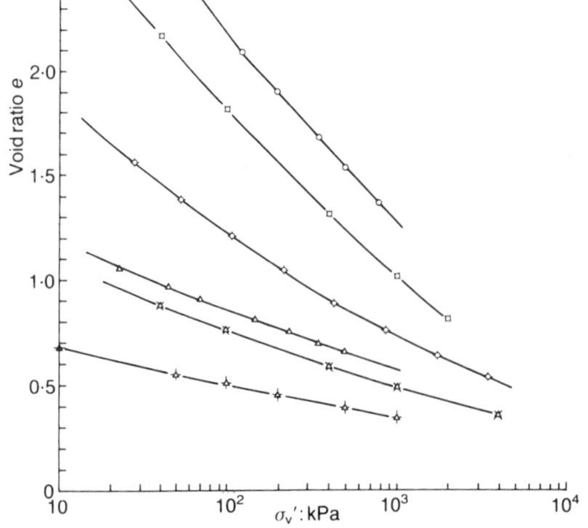

Fig. 2. One-dimensional compression curves for various reconstituted clays

Intrinsic properties

At this stage the concept of intrinsic properties of a given clay is introduced. The term 'intrinsic' is used to describe the properties of clays which have been reconstituted at a water content of between w_L and $1{\cdot}5w_L$ (preferably $1{\cdot}25w_L$) without air drying or oven drying, and then consolidated—preferably under one-dimensional conditions. Ideally the chemistry of the water should be similar to that of the pore water in the clay in its natural state. It is very important to distinguish clearly between the properties of a natural soil and its intrinsic properties. The term intrinsic has been chosen since it refers to the basic, or inherent, properties of a given soil prepared in a specified manner and which are independent of its natural state.† An asterisk is used to denote an intrinsic property (e.g. $C_c{}^*$ is the intrinsic compressibility, and ϕ^* the intrinsic angle of shearing resistance of a soil).

The compression curves plotted in Fig. 2 represent the intrinsic compression curves for the various clays since they were all reconstituted at water contents such that $w_L \leqslant w \leqslant 1{\cdot}5w_L$. Fig. 3(a) shows the intrinsic compression curve for a given clay. The quantities e^*_{100} and e^*_{1000} are the intrinsic void ratios corresponding to $\sigma_v' =$ 100 kPa and 1000 kPa respectively. The intrinsic compression index $C_c{}^*$ is defined as $e^*_{100} - e^*_{1000}$. Following Terzaghi (1925) the parameters e^*_{100} and C^*_c are called the constants of intrinsic compressibility.

Void index

The curves in Fig. 2 may be normalized by assigning fixed values to e^*_{100} and e^*_{1000}. The nor-

† Leroueil *et al.* (1985) define four states of structure: intact, destructured, remoulded and resedimented. A close examination of their definitions indicates that 'reconstituted' is a fifth important state of structure which is used here as a reference state.

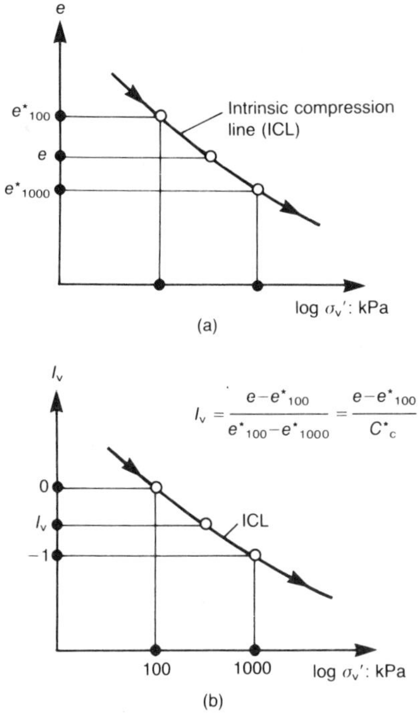

Fig. 3. The use of void index I_v to normalize intrinsic compression curve

malizing parameter chosen is defined as the void index I_v such that

$$I_v = \frac{e - e^*_{100}}{e^*_{100} - e^*_{1000}} = \frac{e - e^*_{100}}{C_c{}^*} \tag{1}$$

Thus the compression curve in Fig. 3(a) may be transformed to the normalized curve in Fig. 3(b) where the void index I_v, defined by equation (1), is the ordinate. When $e = e^*_{100}$, $I_v = 0$ and when $e = e^*_{1000}$, $I_v = -1$. The void index may be thought of as a measure of the intrinsic compactness of a sediment. When I_v is less than zero the sediment is compact and when I_v is greater than zero the sediment is loose.

Clearly there is a close analogy between void index ($= (e - e^*_{100})/C_c{}^*$) and liquidity index ($= (w - w_p)/(w_L - w_p)$). It is of the utmost importance to be clear about the difference between these two indices. The void index is defined in terms of two directly measured mechanical properties (e^*_{100} and $C_c{}^*$) derived from a one-dimensional compression test. In contrast liquidity index is defined in terms of two essentially empirical tests (the liquid limit and plastic limit tests) both of which subject the soil to extremely complex physical processes.

Intrinsic compression line

Three of the intrinsic compression curves from Fig. 2 covering a wide range of liquid limits and of pressures have been replotted in Fig. 4 in terms of void index I_v versus log σ_v'. It can be seen that a reasonably unique line is achieved which is termed the intrinsic compression line (ICL). The co-ordinates of the ICL are given in Fig. 4 and may be represented with sufficient accuracy by the cubic

$$I_v = 2{\cdot}45 - 1{\cdot}285x + 0{\cdot}015x^3 \tag{2}$$

where $x = \log \sigma_v'$ in kPa.

The intrinsic compression line may either be measured directly for a clay or, if the values of e^*_{100} and $C_c{}^*$ are known for the clay, the ICL may be constructed using Fig. 4 or equation (2). In the latter case, if it is required to plot the ICL in

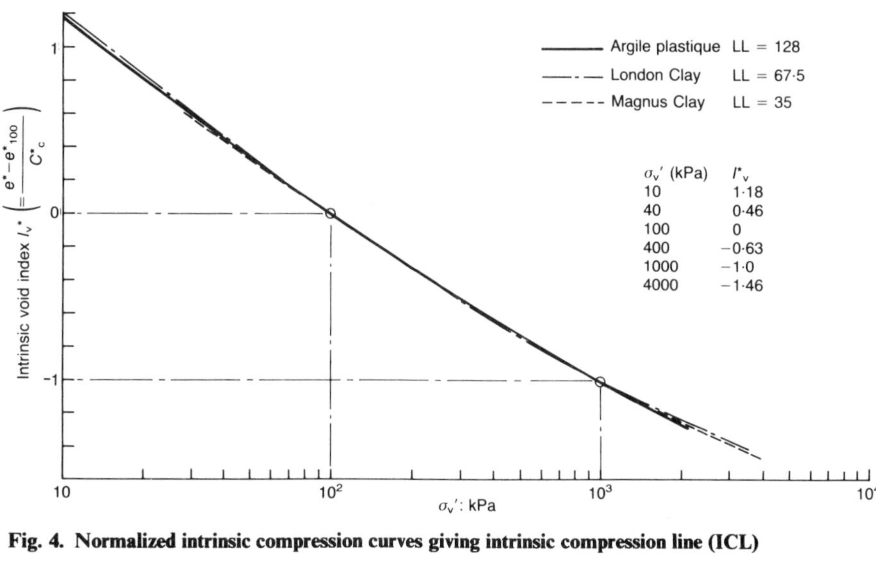

Fig. 4. Normalized intrinsic compression curves giving intrinsic compression line (ICL)

terms of e versus log σ_v', then the values of e corresponding to various values of log σ_v' may be obtained from equation (1)

$$e = I_v C_c^* + e_{100}^* \quad (3)$$

where, again, the values of I_v may be obtained from Fig. 4 or equation (2).

The available experimental evidence suggests that the ICL is insensitive to the test conditions. Fig. 5(a) shows the results of some oedometer tests on three clays in which each clay was reconstituted at various water contents (Skempton, 1944; Leonards & Ramiah, 1959). The number against each curve gives the mixing water content expressed as a proportion of the liquid limit of the clay. At pressures less than about 100 kPa the compression curves for each soil tend to diverge, but for $\sigma_v' \geqslant 100$ kPa the differences are less. Fig. 5(b) shows some results by Leonards & Ramiah (1959) in which the influence of load increment duration was investigated for two clays which were reconstituted at water contents equal to the liquid limit. Clearly there is little difference between the curves for each clay. If anything the curves for the longer duration lie slightly above those for the shorter durations. Northey (1956) obtained similar results from oedometer tests on three reconstituted New Zealand clays. Preliminary results from tests carried out at Imperial College indicate that the ICL is also insensitive to load increment ratios in excess of unity. These and other data lead to the conclusion that, provided the soil is reconstituted at a water content of between w_L and $1{\cdot}5w_L$, and provided the duration of each load increment is sufficiently long to allow primary consolidation to occur, then the ICL is well defined (i.e. it is 'robust') for pressures equal to or greater than 100 kPa.

There is much evidence to show that ageing significantly influences the compressibility of reconstituted clays. Leonards & Ramiah (1959) studied the influence of ageing on the one dimensional compression of a reconstituted residual clay and their results are given in Fig. 6. The top curve is for a standard test with a load increment ratio of one and a load duration of one day. The

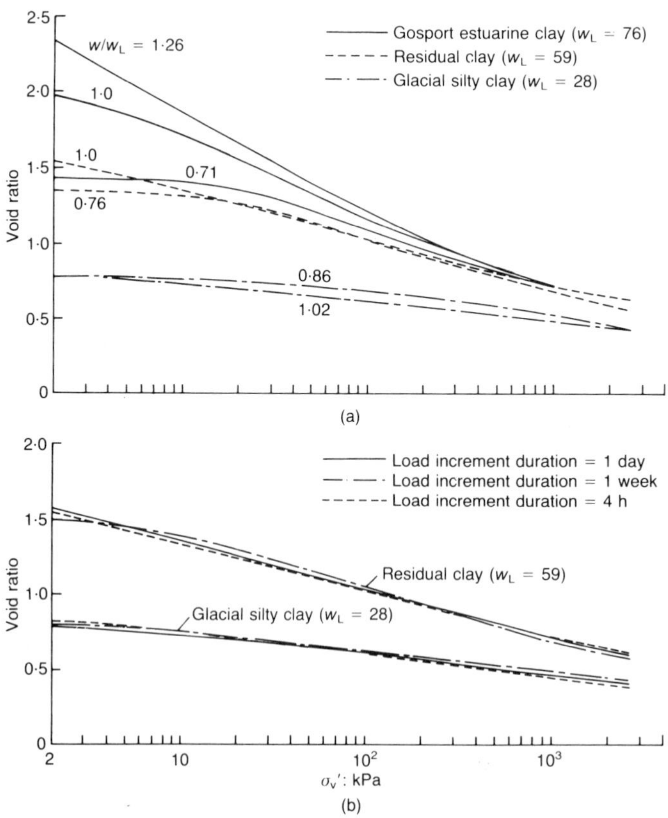

Fig. 5. Influence of (a) mixing moisture content on compression curves for reconstituted clays (load increment duration 1 day); (b) load increment duration on compression curves for reconstituted clays (initial moisture content w_L) (Leonards & Ramiah, 1959)

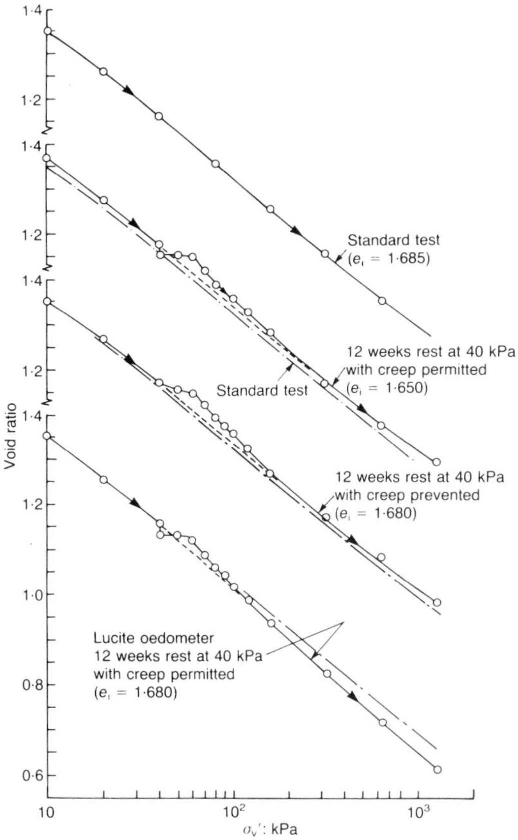

Fig. 6. Influence of ageing on compression characteristics of a reconstituted residual clay (Leonards & Ramiah, 1959)

second curve shows the effect of 12 weeks rest at 40 kPa followed by small load increments. It is evident that creep occurred during ageing but that the 'preconsolidation pressure' lies well to the right of the standard virgin compression line. The third curve shows the effect of 12 weeks ageing with creep prevented. Again the preconsolidation pressure lies well to the right of the virgin compression line. The bottom curve is a repeat of the second but using a lucite oedometer for which the side friction was known to be very small (Leonards & Girault, 1961).

These results demonstrate that the micro-fabric of a clay can develop increased resistance to compression during ageing and that this resistance does not depend on volume reduction due to creep. It can be seen from Fig. 6 that when an aged clay is loaded the structural resistance breaks down at a critical pressure and the subsequent compression curve is initially significantly steeper than the standard virgin line. Leonards and others have used the term 'quasi-preconsolidation pressure' to describe this critical pressure. It is recommended that the term 'yield stress', or more precisely 'vertical yield stress' should be used and be denoted by σ_{vy}'. The term 'preconsolidation pressure' should be reserved for situations in which the magnitude of such a pressure can be established by geological means. Similarly the term 'overconsolidation ratio' should be reserved for describing a known stress history. Where a yield stress has been observed then the ratio between it and the effective overburden pressure ($\sigma_{vy}'/\sigma_{vo}'$) could be termed the 'yield stress ratio'.

CORRELATIONS BETWEEN THE CONSTANTS OF INTRINSIC COMPRESSIBILITY AND THE ATTERBERG LIMITS

The ICL is not, at present, routinely measured, although it is easy enough to do so. Hence it is necessary to make use of empirical correlations between the Atterberg limits and the intrinsic constants of compressibility e_{100}^* and C_c^*. Skempton (1944) tabulated the results of numerous oedometer tests on reconstituted natural clays, many of them carried out at the Building Research Station. These data have been supplemented by other published results and are given in Table 1. In Fig. 7 the data are plotted on a plasticity chart and it can be seen that all except the results for Whangamarino Clay lie above the A line.

Figure 8 shows the correlation between e_L (void ratio at the liquid limit) and e_{100}^* and C_c^*. Regression analyses have been carried out and the best fit regression lines are given by the following equations

$$e_{100}^* = 0{\cdot}109 + 0{\cdot}679e_L - 0{\cdot}089e_L^2 + 0{\cdot}016e_L^3 \tag{4}$$

and

$$C_c^* = 0{\cdot}256e_L - 0{\cdot}04. \tag{5}$$

The coefficients of correlation for equations (4) and (5) are 0·991 and 0·985 respectively. These equations should of course only be used for values of e_L within the range 0·6 to 4·5 (i.e. w_L = 25 to 160). Moreover these correlations only hold for soils with Atterberg limits lying above the A line. It has been found that when the Atterberg limits lie below the A line the values of e_{100}^* and C_c^* do not fit the correlations well—an example being Whangamarino clay, which is shown as a full point in Figs 7 and 8.

The broken lines in Fig. 8 are derived from the work of Nagaraj & Srinivasa Murthy (1986) who established a relationship between the ratio e/e_L and σ_v' based on considerations of physical chem-

Table 1. **Intrinsic constants of compressibility for reconstituted natural clays**

Soil	w_L	w_p	G_s	e_L	e^*_{100}	C_c^*	Reference
Lower Cromer Till	25	13	2·65	0·663	0·503	0·154	Gens (1982)
Boulder clay	28	14	2·69	0·753	0·52	0·12	Skempton (1944)
Silty clay	28	20	2·72	0·762	0·603	0·136	Ramiah (1959)
Magnus Clay	35	17·2	2·73	0·956	0·76	0·27	Jardine (1985)
Grangemouth	35	21	2·78	0·973	0·659	0·229	This study
Ton V	36	18	2·71	0·976	0·74	0·25	Skempton (1944)
Weald clay	39	19	2·73	1·065	0·77	0·24	Skempton (1944)
Boston blue clay	39	23	2·78	1·084	0·80	0·21	Skempton (1944)
Red soil	45·3	22	2·667	1·208	0·785	0·27	Nagaraj *et al.* (1986)
River Severn alluvium	46	25	2·59	1·191	0·80	0·21	Skempton (1944)
Wiener Tegel	46·7	22	2·76	1·288	0·859	0·297	Hvorslev (1937)
Oxford clay	53	27	2·57	1·362	0·96	0·30	Skempton (1944)
Ton IV	58	26	2·85	1·653	0·97	0·32	Skempton (1944)
Residual clay	58	27	2·74	1·589	1·024	0·337	Ramiah (1959)
London Clay	62·3	24·3	2·73	1·707	1·200	0·446	Jardine (1985)
Belfast estuarine clay	67	30	2·66	1·782	1·00	0·32	Skempton (1944)
London Clay	67·5	26·5	2·71	1·829	1·227	0·494	Som (1968)
Ganges delta clay	69	28	2·77	1·911	1·22	0·42	Skempton (1944)
Gosport clay	76	29	2·67	2·029	1·20	0·48	Skempton (1944)
London Clay	77	28	2·71	2·087	1·28	0·49	Skempton (1944)
Brown London Clay	88	32	2·65	2·332	1·32	0·56	Skempton (1944)
Black cotton clay	97·3	32	2·73	2·656	1·744	0·69	Nagaraj *et al.* (1986)
Kleinbelt Ton	127	36	2·77	3·518	2·18	0·91	Hvorslev (1937)
Argile plastique	128	31	2·58	3·302	1·82	0·81	Skempton (1944)
Whangamarino clay	136	61	2·78	3.74	2·44	0·797	Newland & Allely (1956)
SAIL	159·3	46	2·826	4·443	2·769	1·05	Nagaraj *et al.* (1986)

istry. It can be seen that the two approaches give similar correlations over a wide range of e_L values but that at low and high values there are significant differences, particularly for e^*_{100}. If, for a given clay, the intrinsic constants of compressibility e^*_{100} and C_c^* have been measured then it would be appropriate to allow for small changes in e_L between samples of that soil by correcting e^*_{100} and C_c^* in direct proportion to the changes in e_L (or w_L).

The question might well be asked as to why the intrinsic constants of compressibility were not

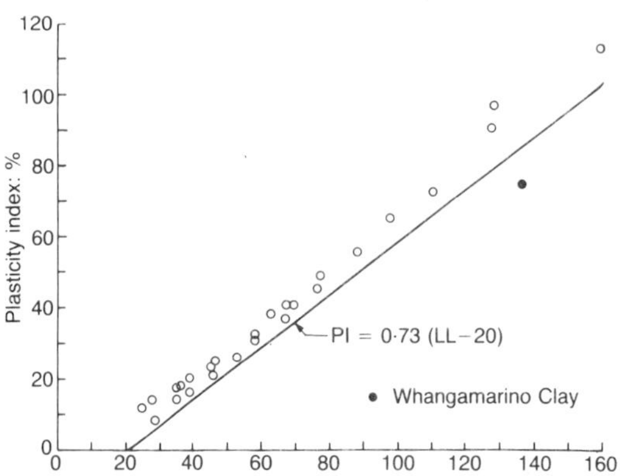

Fig. 7. **Plasticity chart for reconstituted clays in Table 1**

correlated with plasticity index, or its void ratio equivalent, instead of e_L. A statistical analysis has shown that equally good correlations are achieved at high values of plasticity index but at low values the correlations are significantly worse. This is because small errors in w_L and w_p become significant when one is subtracted from the other.

For all the data listed in Table 1 the soils were reconstituted at water contents of between w_L and $1{\cdot}5w_L$. Recently Nakase *et al.* (1988) published an independent data set for reconstituted marine clays from a number of locations in Japan. The key difference between the two data sets is that Nakase *et al.* reconstituted the soils at very high water contents to form liquid slurries. Fig. 9 shows a comparison between the results of Nakase *et al.* and equations (4) and (5). It can be seen that there is excellent agreement for C_c^*. However, the experimental values of e^*_{100} lie a little above the regression line. This is consistent with the very high mixing water contents and serves to emphasize the need to standardize these when determining the intrinsic constants of compressibility. It is encouraging that the two entirely independent sets of data are in reasonable agreement. Note that the values of C_c^* and e^*_{100} for the soil lying just below the A line in Fig. 9 tend to be displaced from the other results in the same manner as in Fig. 8.

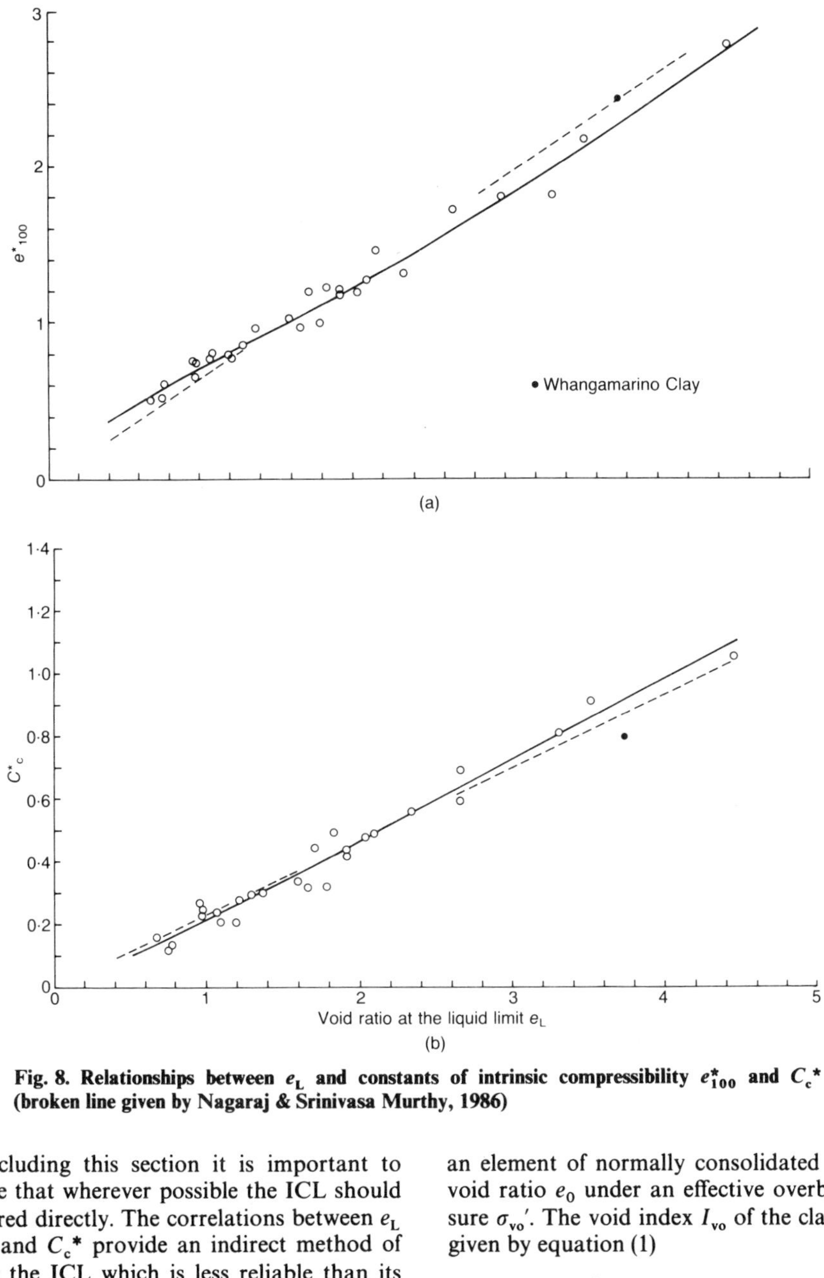

Fig. 8. Relationships between e_L and constants of intrinsic compressibility e^*_{100} and $C_c{}^*$ (broken line given by Nagaraj & Srinivasa Murthy, 1986)

In concluding this section it is important to appreciate that wherever possible the ICL should be measured directly. The correlations between e_L and e^*_{100} and $C_c{}^*$ provide an indirect method of obtaining the ICL which is less reliable than its direct experimental determination.

COMPARISON BETWEEN THE SEDIMENTATION COMPRESSION OF NATURAL CLAYS AND THE INTRINSIC COMPRESSION OF RECONSTITUTED CLAYS

Using the void index I_v as a normalizing parameter, it is possible to compare the sedimentation compression curves obtained by Skempton (see Fig. 1) with the corresponding ICL. Consider an element of normally consolidated clay with a void ratio e_0 under an effective overburden pressure σ_{vo}'. The void index I_{vo} of the clay element is given by equation (1)

$$I_{vo} = \frac{e_0 - e^*_{100}}{C_c{}^*} \tag{6}$$

The values of e^*_{100} and $C_c{}^*$ are preferably measured by means of an oedometer test on the reconstituted soil, but for the present purposes they are obtained from equations (4) and (5). Thus successive values of e_0 and σ_v' down a soil profile may be used to plot a graph of I_{vo} against $\log \sigma_v'$ to give the sedimentation compression curve which can then be compared directly with

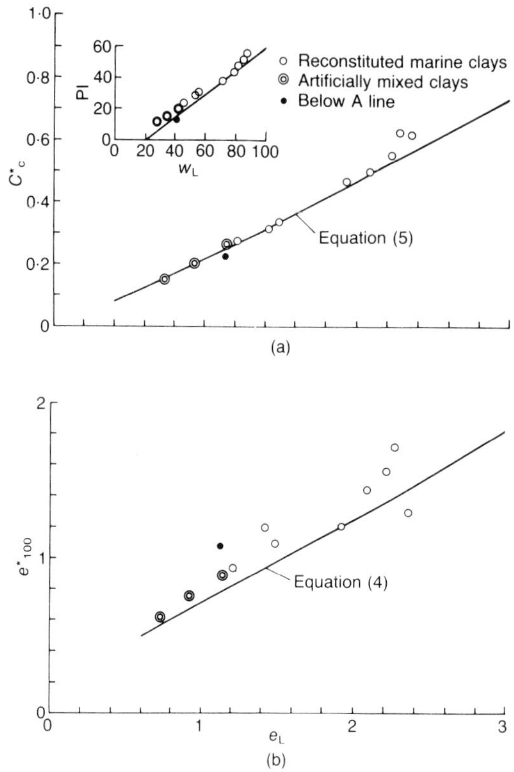

Fig. 9. Comparison of correlations from Fig. 8 with independent data set given by Nakase *et al.* (1988)

the ICL which is uniquely defined in Fig. 4 or by equation (2). Professor Skempton has kindly made his files available to the author and the detailed sedimentation compression curves have been derived for most of the profiles referred to in Fig. 1. The geology of each site has been described by Skempton (1970) and will not be repeated here.

Figure 10 shows the sedimentation compression curves for three of the Pliocene deposits plotted on axes of I_{vo} versus log σ_{vo}'. The results show marked scatter which is due to in part to errors in the determinations of water content and liquid limit but is also believed to be due to variations in depositional conditions as the profiles were being formed. The extreme variations have been removed by taking the average of successive pairs of points, thereby preserving trends but eliminating extreme fluctuations. All three curves lie well above the ICL. The results from Baku are of particular interest because of the wide range of overburden pressures. Note the 'saw-tooth' shape of the sedimentation curve which is also a feature of the other two curves.

There is no reason to anticipate a smooth sedimentation compression curve. Rates and modes of deposition are likely to vary considerably during the formation of a sedimentary soil profile and in these circumstances a wavey curve must be expected (Edge & Sills, 1989). Thus each element will retain the imprint of the conditions under which it was deposited.

Figure 11 shows the sedimentation compression curves for three British post-glacial clays from widely differing locations and having a wide range of liquid limits. The curves all lie above the ICL. The results from Shellhaven are for the lowest layer of clay at the site. The results from higher up the profile will be described later.

Figure 12 shows the sedimentation compression curves for two Scandinavian post-glacial

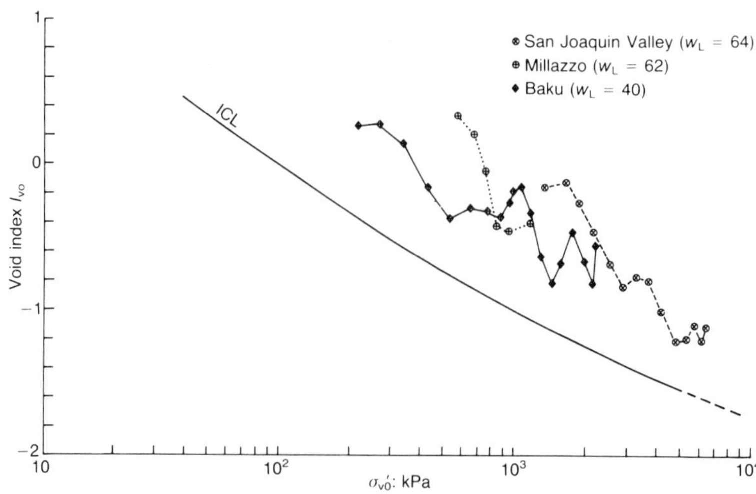

Fig. 10. Sedimentation compression curves for Pliocene and early Pleistocene clays and mudstones

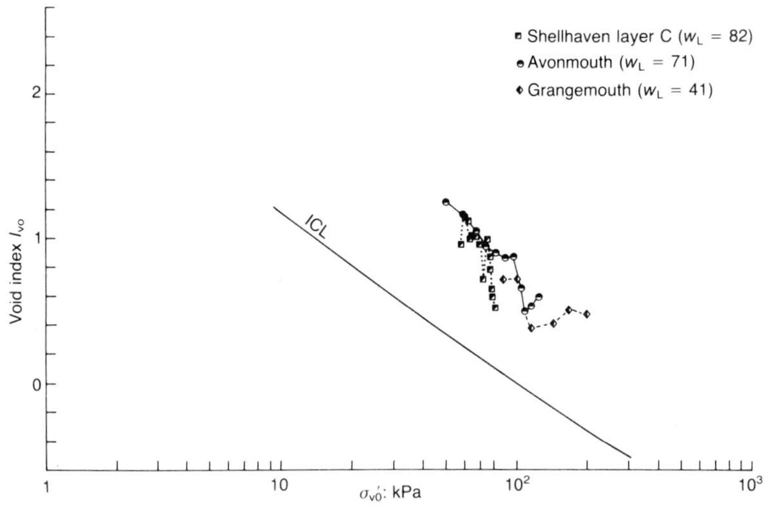

Fig. 11. Sedimentation compression curves for some British post-glacial clays

clays and once again they lie well above the ICL. The profile at Älvängen in Sweden is unusually uniform and gives relatively smooth compression curves—note the high liquid limit. The profile for Drammen was referred to by Bjerrum (1967). It consists of an upper plastic stratum (shown as circles) underlain by a lean stratum (shown as diamonds). In spite of the differences in liquid limit between these two strata it can be seen that the sedimentation compression curve is reasonably continuous. This implies that the upper plastic layer has not undergone substantially more delayed consolidation than the underlying lean layer as was suggested by Bjerrum.

The sedimentation compression line

Having considered some of the individual sedimentation compression curves the data from most of the sites considered by Skempton (1970) are assembled in Fig. 13 including the results for the shallow marine deposits. It can be seen that the various sedimentation curves all lie in a well defined continuous band when plotted on a graph of I_{v0} versus log σ_{v0}'. A regression line has been fitted to the data as shown and is called the sedimentation compression line (SCL), the co-ordinates of which are tabulated in Fig. 13. Most of the data lie within the range $I_{vo} = \pm 0{\cdot}3$ of the SCL.

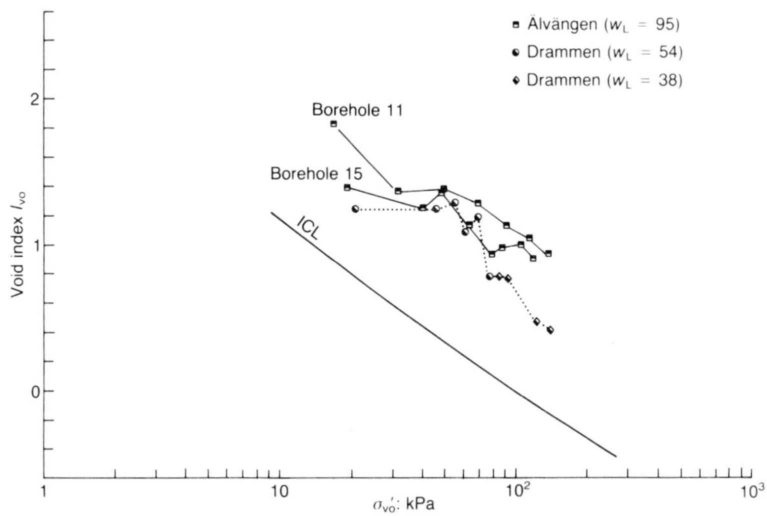

Fig. 12. Sedimentation compression curves for two Scandinavian post-glacial clays

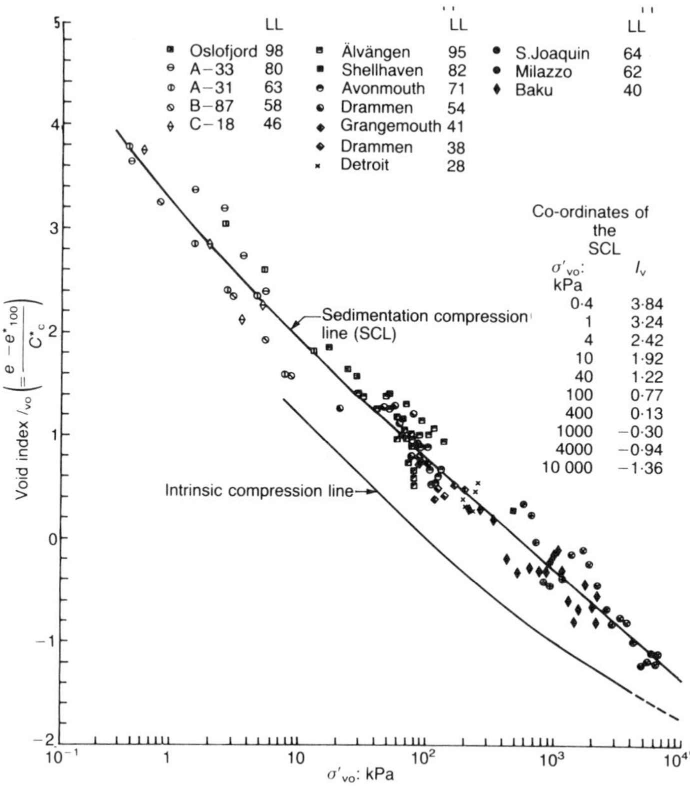

Fig. 13. Relationship between I_{vo} and log σ_{vo}' for many of the normally consolidated clays designated in Fig. 1: best-fit regression line through the data is termed sedimentation compression line (SCL)

Over the range of $\sigma_v' = 10$ kPa to 1000 kPa the ICL and the SCL can be seen from Fig. 13 to be approximately parallel. Over this region, for a given value of I_{vo}, the effective overburden pressure carried by the natural clay is approximately five times that carried by the equivalent reconstituted clay. This figure is a measure of the enhanced resistance of a naturally deposited clay over a reconstituted one and results from differences in the fabric and bonding (i.e. the structure) of the soil skeleton. The influence of the natural structure was first recognized by Terzaghi (1941) and confirmed by Skempton (1944). At pressures in excess of 1000 kPa the ICL and SCL tend to converge.

Not all normally consolidated natural clays lie close to the SCL. Fig. 14 shows the sedimentation compression curves for three such clays. The open circles are for a freshwater glacial lake clay from Sault Ste Marie, near Chicago (Wu, 1958). The reason for these data laying well above the SCL is not difficult to find. The clay is reddish in colour due to the presence of haematite which has undoubtedly given rise to cementation between

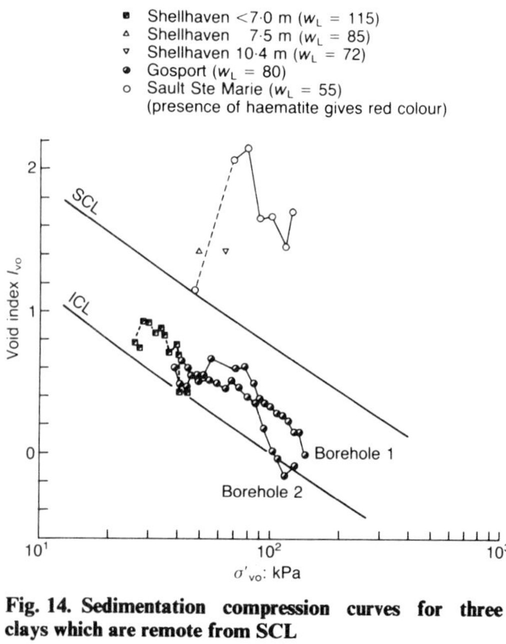

Fig. 14. Sedimentation compression curves for three clays which are remote from SCL

the particles. Wu carried out a study of the fabric by means of a polarizing microscope and found that it was essentially random. In contrast, the sedimentation compression curve for a nearby glacial lake clay at Detroit, shown as crosses in Fig. 13, lies on the SCL. The fabric of this clay, which contained no haematite, was shown to exhibit some horizontal orientation.

Also shown in Fig. 14 are the sedimentation compression curves for two British post-glacial clays—the upper clay layer at Shellhaven (Skempton & Henkel, 1953) and Gosport (Skempton, 1970). Both these clays lie well below the SCL. The reason for this is not immediately obvious but evidence will be presented later which supports the hypothesis that it is due to the deposition conditions. The deeper clays at Shellhaven lie on the SCL (see Figs 11 and 13) and the triangles in Fig. 14 are for samples from depths of 7·5 m and 10·4 m—both lie a little above the SCL. Oedometer tests were carried out on the three clays referred to in Fig. 14 and the results are of considerable interest.

Results of some oedometer tests

Figure 15(a) shows the results of four oedometer tests on undisturbed samples of Sault Ste Marie Clay from various depths. The void index I_v has been used as a normalizing parameter (in conjunction with equations (4) and (5)) so that the oedometer compression curves can be compared with the intrinsic compression line and the sedimentation compression line from Fig. 13. It can be seen that the post-yield compression curves for the three deepest samples are significantly steeper than the SCL. The curves cross the SCL from above and then flatten, converging slowly on the ICL. Note that the shallowest sample from 3·51 m depth is lightly overconsolidated due to desiccation.

The oedometer results for Sault Ste Marie Clay, which lies well above the SCL, may be contrasted with those for Shellhaven lying close to the SCL, and for Gosport lying below the SCL. Fig. 15(b) shows the results for the latter two clays. The triangular points are for Shellhaven. The full points are for a reconstituted sample of the clay and the reconstituted compression line is seen to lie very close to the one derived from equations (4) and (5) and labelled ICL. The agreement is encouraging. The post-yield compression curves for the two undisturbed samples, the initial states of which are given in Fig. 14, are steeper than the SCL crossing it from above and again converging slowly with the ICL.

The circles are for Gosport clay. The full points are for reconstituted samples and lie slightly below the ICL but the agreement is nevertheless satisfactory. The compression curves for the undisturbed samples are very different from the Sault Ste Marie and Shellhaven clays as they do not exhibit a high post-yield compressibility and the curves more or less coincide with the ICL.

In summary it appears that for normally consolidated clays whose natural states lie close to or above the SCL, the post-yield oedometer compression curve is much steeper than the SCL. It crosses the SCL and converges slowly on the ICL. In contrast, for normally consolidated clays whose natural states lie on or close to the ICL the oedometer compression curves are essentially parallel to this line.

Results from the Mississippi delta

Some work published by McClelland (1967) on the clays from the Mississippi delta provide some important clues about the factors influencing the in situ state of sedimentary clays relative to the SCL and the ICL. The continental shelf in the Gulf of Mexico off the coast of south eastern Louisiana is blanketed by clay sediments of Late Quaternary age. These clays have the Mississippi river as a common source and consist essentially of a common suite of minerals. However the depositional environments differ significantly as a result of sea level changes and changes in the course of the river.

Figure 16 shows the sedimentation compression curves for two locations remote from any of the deltas associated with the present standing-sea period. The clays are continental shelf deposits more than 15000 years old. It is evident that the data lie close to the SCL. Oedometer tests on undisturbed samples from these two boreholes give post-yield compression curves which are steeper than the SCL and which tend to converge with the ICL in accordance with the behaviour depicted in Fig. 15(b).

A borehole was also sunk through the present delta front of the river. It revealed about 85 m of recent delta deposits underlain by continental shelf deposits. The top 60 m have been deposited so rapidly over the last 400 years that they are largely unconsolidated. Since the in situ effective stresses within this top layer are not known the sedimentation compression curve cannot be constructed. However, oedometer tests on undisturbed samples from the top layer and deeper layers give interesting results as shown in Fig. 17. The open circles are for samples from the overlying rapidly deposited underconsolidated clays. It can be seen that the compression curves lie on the ICL. In contrast the compression curve for the sample from 86·6 m depth in the continental shelf deposit (closed circles) drops from the SCL down towards the ICL. The sample from 119·6 m depth

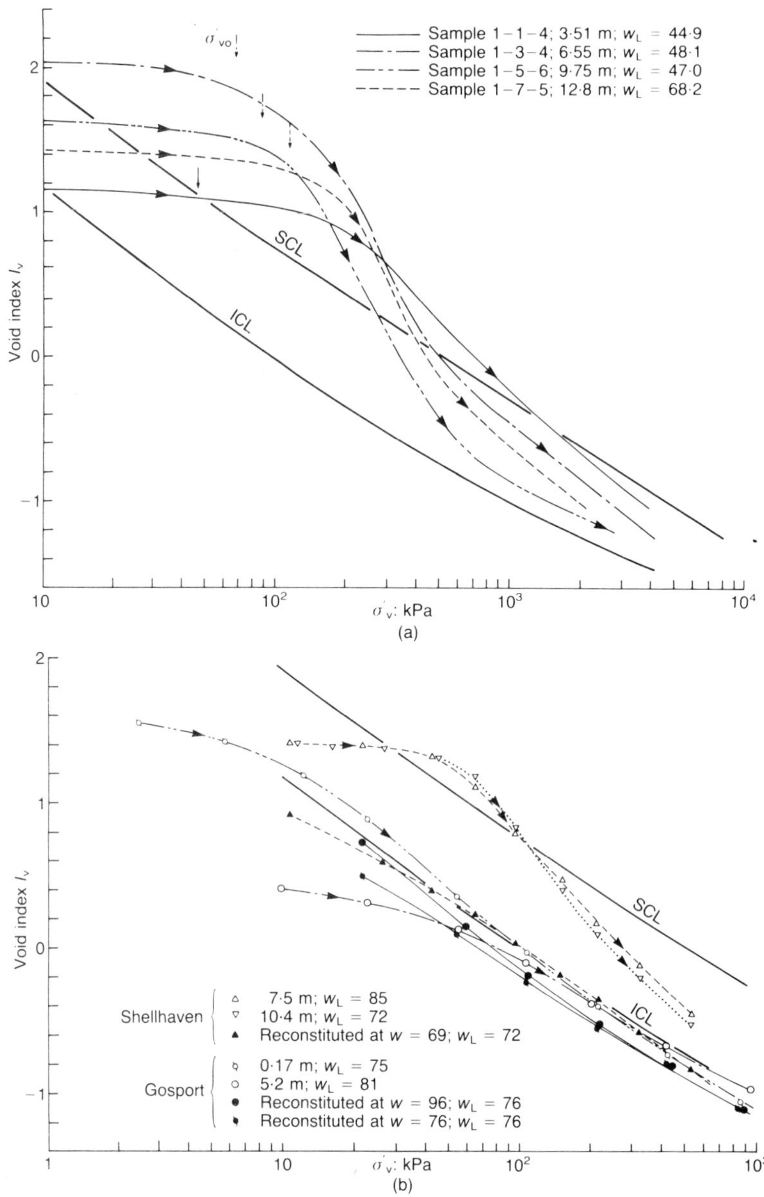

Fig. 15. Oedometer compression curves for (a) Sault Ste Marie Clay, site 1, and (b) Shellhaven and Gosport clays

has almost certainly suffered some disturbance. Nevertheless the compression curve lies well to the right of the ICL.

These results confirm that the deposition conditions profoundly affect the fabric of the sediment which is then not easily changed by subsequent increases in effective overburden pressure. The two most significant depositional factors are likely to be the rate of deposition and the stillness of the water. Slow deposition in still water leads to an open random fabric with high values of void index laying on or above the SCL. On the other hand rapid deposition from a dense suspension, possibly with significant currents, will give rise to a more oriented fabric which is consequently more compact with a lower void index.

For a soil whose state lies on or above the SCL the rate of application of load in an oedometer is sufficient to disrupt the interparticle bonding and fabric such that the compression curve is signifi-

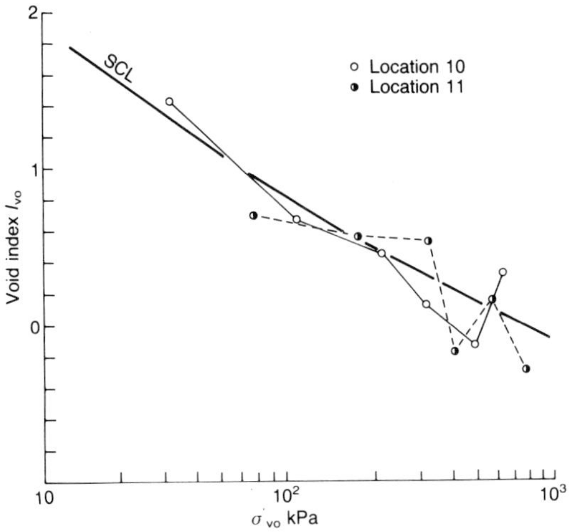

Fig. 16. Mississippi Delta: sedimentation compression curves for late Quaternary continental shelf deposits

cantly steeper than the SCL and it falls towards the ICL. However, if the state of the soil is already on the ICL due to its deposition conditions, the fabric will already be oriented and compression in an oedometer will not change things significantly.

Leroueil *et al.* (1979) have termed the post-yield disruption of the clay structure as 'destructuration'. The results given, for instance, in Figs 15 and 17 imply that this process is a gradual one and that the precise definition of a 'destructured' state is not clear. There are clear advantages in using the 'intrinsic' state as a reference state.

Results from Bothkennar test bed site

Recently the UK Science and Engineering Research Council (SERC) selected a soft clay test bed site at Bothkennar in the upper Forth Estuary, Scotland. Details of the ground conditions are given by Hawkins *et al.* (1989). It is of considerable interest to establish whether the ground conditions at this site fit the general pattern portrayed in Fig. 13.

Figure 18 is a summary of the basic properties for borehole D1 at the Bothkennar site. The clay is of medium to high plasticity, the yield stress ratio (otherwise referred to as the OCR) is about 1·7 and the undrained strength from vane tests shows a linear increase with depth with a sensitivity of about 4 to 6. These results indicate that the clay is normally consolidated although Hawkins *et al.* point out that there is some evidence to suggest that the top 1 m or so may have been removed by erosion. The sedimentation compression curve for borehole D1 is shown in Fig. 19. The curve is somewhat jagged due to significant variations in water content but it can be seen to lie very close to the SCL. The broken line is for the top 2 m which is overconsolidated due to desiccation.

High quality samples were obtained by means of a Laval sampler (La Rochelle *et al.*, 1981) and standard incremental oedometer tests were carried out on them. Fig. 20 shows the results of two oedometer tests on a sample from a depth of 6·5 m plotted as void ratio against log σ_v'. The full circles are for a sample which was reconstituted at the liquid limit to give the experimentally determined ICL. This compares very well with the broken line which was obtained from equa-

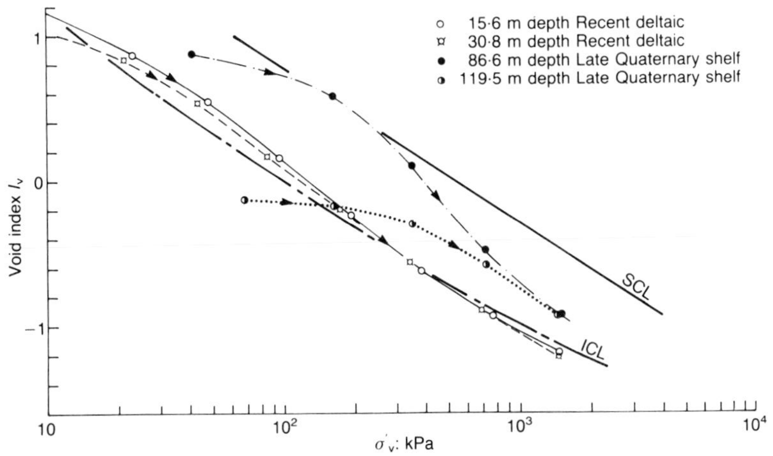

Fig. 17. Mississippi Delta: results of oedometer tests on underconsolidated deltaic deposits and underlying Quaternary shelf deposits

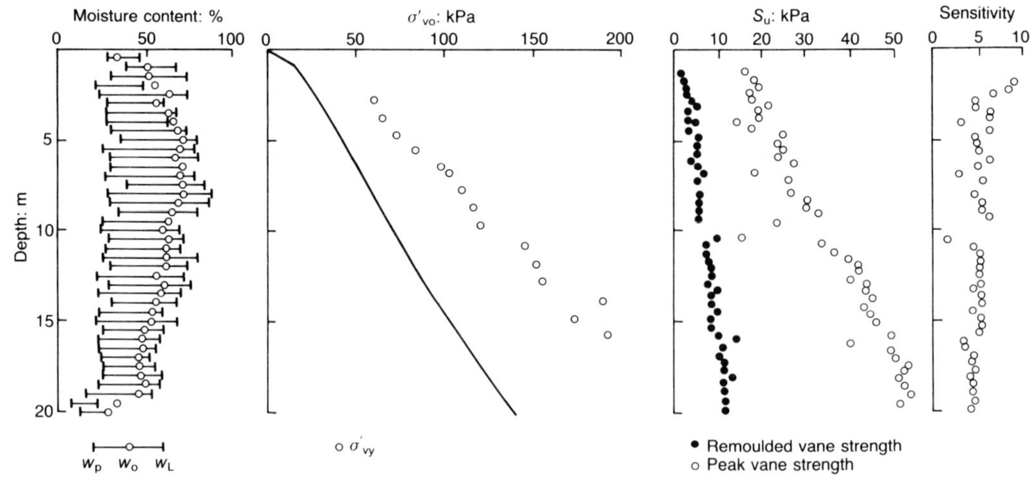

Fig. 18. Bothkennar: profile for borehole D1 (Hawkins *et al.*, 1989)

tions (4) and (5) knowing w_L and hence e_L. The open circles are for an undisturbed sample. It can be seen that the compression curve drops steeply from the SCL eventually converging with the ICL. Thus the Bothkennar test bed site appears to conform to a typical normally consolidated sensitive clay profile.

CASE RECORD FROM SURABAYA, INDONESIA

Field measurements on a land reclamation project in Indonesia provide valuable observations of in situ compressibility which may be compared with the pattern of oedometer compression curves presented in the previous sections. The project is the phase II development of Surabaya Port and Rendel Palmer and Tritton were the consulting engineers for the client—the Directorate General Sea Communications, Government of the Republic of Indonesia. The work involved the construction of a container stacking yard on land reclaimed from tidal mud flats, just to the west of the existing port of Surabaya. The site consists of about 5 m of silty sand overlying a deep soft clay layer which is underlain by stiff clay and sand. The soft clay is derived from local volcanic clays and is highly plastic with an average liquid limit of about 100. A typical profile through the soft clay as given by two boreholes is shown in Fig. 21(a).

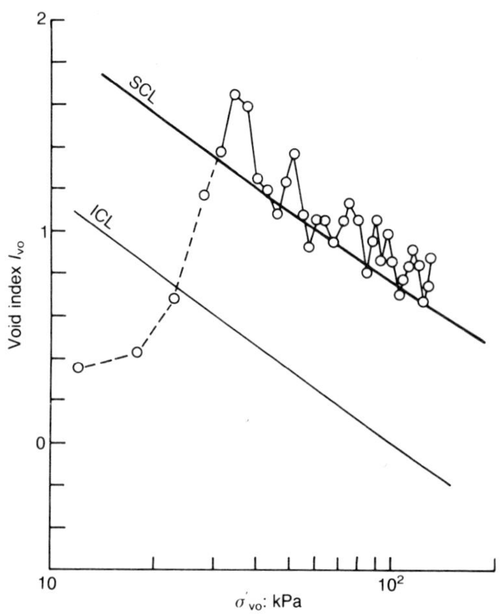

Fig. 19. Bothkennar: sedimentation compression curve for borehole D1

Accelerated consolidation by wick drains was adopted for the reclaimed area. A number of sections were instrumented by installing settlement plates at various depths and piezometers between the drains. Inclinometers were used near the slopes. The fill consisted of hydraulically placed sand. The soft clay settled considerably more than was predicted on the basis of normally consolidated behaviour using C_c values from oedometer tests on samples obtained by Shelby tubes. These values of C_c were consistent with the established correlations with I_p and hence approximate to C_c^*. Settlements were measured at various depths and the vertical effective stresses were estimated from the unit weights from the typical borehole profile and the measured pore pressures.

Figure 21(b) shows a typical instrumented section. The settlements and vertical compressions one year after completion of loading are shown in Fig. 21(c). Also shown in Fig. 21(c) are

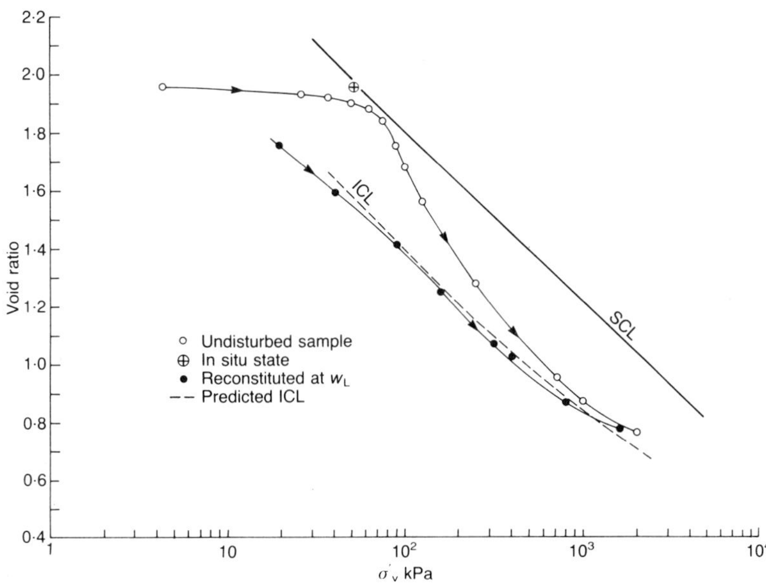

Fig. 20. Bothkennar: oedometer tests on undisturbed and reconstituted soil from 6·5 m depth (w_L = 85·4, w_p = 41·0)

the predicted settlements and compressions—the differences from the measured values are large. The measured compressions at various depths and locations can be used in conjunction with the initial void ratios to calculate the void ratios one year after completion of loading. Fig. 22 shows a plot of in situ values of I_v versus log σ_v'. The closed points and corresponding open points represent the initial and subsequent values of I_v respectively. The full lines are the in situ compres-

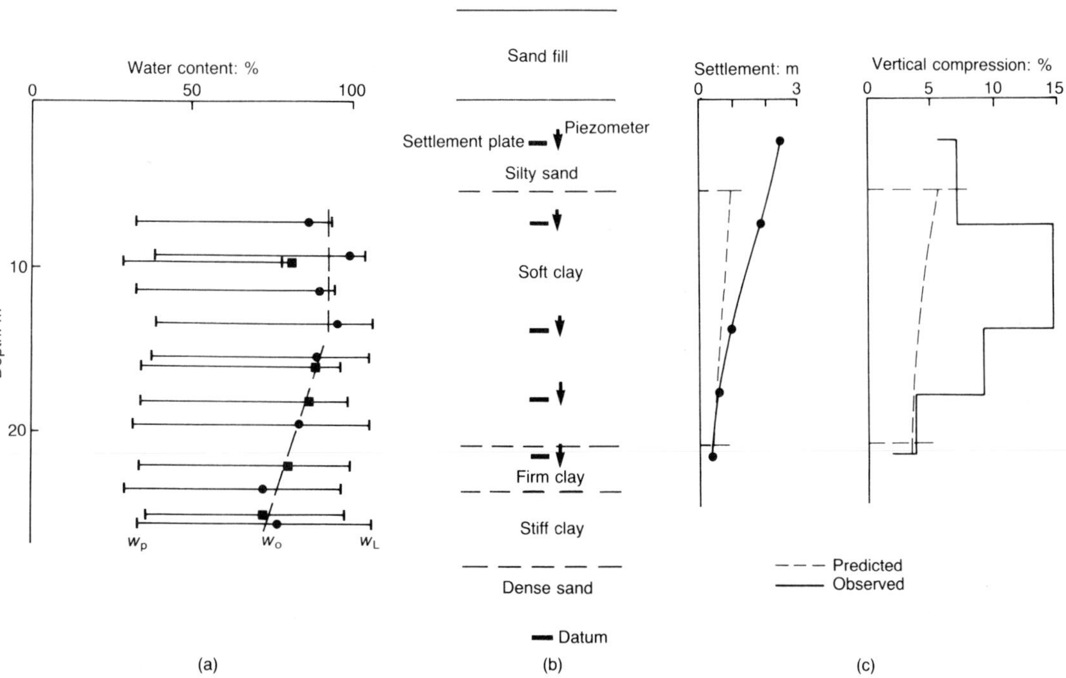

Fig. 21. Surabaya, Indonesia: (a) profile of soft clay from two adjacent boreholes; (b) typical instrumented section; (c) Observed settlements and compressions 1 year after completion of loading (band drains at 1·53 m centres)

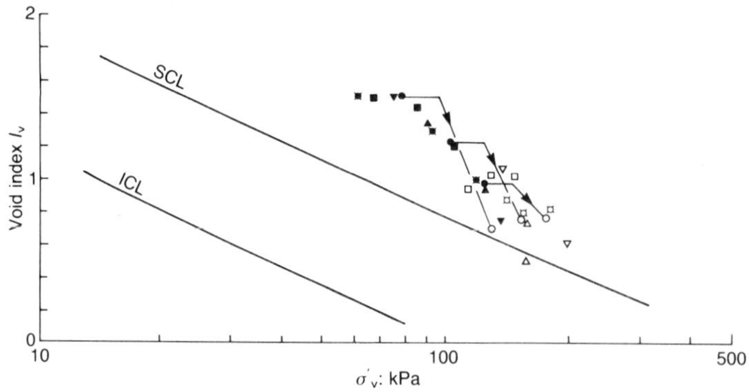

Fig. 22. Surabaya: in situ relationship between I_v and log σ_v'—closed points represent values of I_{v0} giving the sedimentation compression curve; open points give the corresponding values of I_v 1 year after completion of loading

sion curves for the section shown in Fig. 21(b). There was a threshold stress change of about 20 kPa up to which settlement was negligible.

The results plotted in Fig. 22 show that the sedimentation compression curve, as given by the full points, lies well above the SCL and is steeper than it is. The in situ compression curves resulting from the placement of the fill are very much steeper than the ICL and it is clear that they will all drop below the SCL at higher values of σ_v'. These observations are consistent with the oedometer compression curves given in the previous section.

LABORATORY SEDIMENTATION STUDIES

An interesting and important question is whether or not it is possible to reproduce the natural sedimentation compression line in the laboratory. The results of the classic studies of Bjerrum & Rosenqvist (1956) and Leonards & Altschaeffl (1964) can be used to examine this question. Bjerrum & Rosenqvist carried out a series of experiments in which a late glacial marine clay was artificially sedimented into a salt water solution over a two month period and the sediment was then left for 6 weeks. Small increments of pressure were then applied, after which the samples were left for a further three months. At this stage a number of the samples were subjected to leaching over an 18 month period in which the salt concentration was reduced from 32 g/l to 5 g/l. The whole process took about $2\frac{1}{2}$ years.

Figure 23(a) shows the equilibrium void ratios for the unleached samples (open points) and the leached samples (closed points). Clearly the process of leaching, involving the application of an hydraulic gradient across the sample, has resulted in reductions in void ratio. The leaching process resulted in a reduction of liquid limit from an average of 48·8 to 28·1. In Fig. 23(b) the results are plotted in terms of I_v so that they can be compared with the ICL and the SCL. The unleached samples lie just below the SCL. The reductions in w_L due to leaching cause the values of I_v for the leached samples to increase substantially so that the results lie well above the SCL—a characteristic of quick clays.

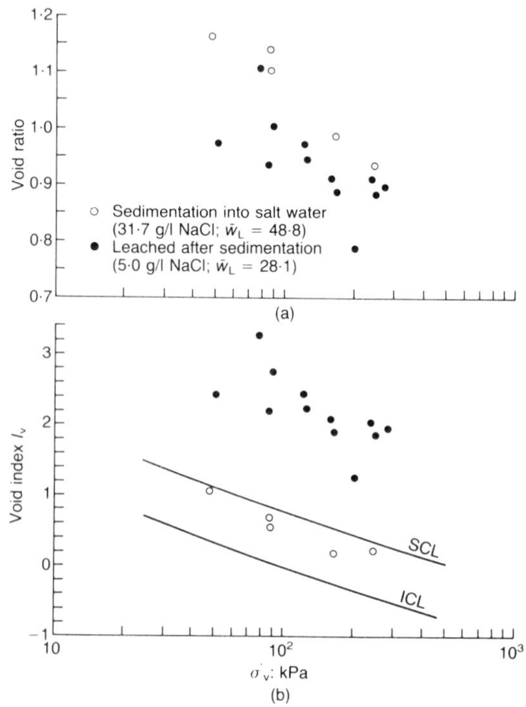

Fig. 23. Results of laboratory sedimented marine clay in terms of (a) e against log σ_v' and (b) I_v against log σ_v' (Bjerrum & Rosenqvist, 1956)

In the experiments carried out by Leonards & Altschaeffl (1964) a flocculated slurry of a residual clay was slowly loaded first by means of a hydraulic gradient and then by applied load through a plunger. The rate of change of load was controlled by syphoning oil from a counterbalancing tank. The resulting compression curve is shown in Fig. 24. When σ_v' had reached 48·7 kPa the pressure was held constant for 90 days resulting in some creep. Unloading then took place and a further rest period of 90 days was allowed. The sample was then loaded in daily increments and the results are shown by the open circles in Fig. 24.

It can be seen that the compression curve for slow loading falls steeply towards the SCL and appears to be converging with it. The compression curve for incremental loading shows a sharp yield point at $\sigma_v' = 64{\cdot}5$ kPa (giving a yield stress ratio of 1·32) after which the curve drops steeply through the SCL and converges on the ICL. The full circles are for a test on a specimen which had been sampled after unloading. The process of sampling resulted in a slightly reduced yield stress (= 60·7 kPa).

The results given in Figs 23 and 24 bear a striking resemblance to the measured compressibility of natural clays. Locat & Lefebvre (1986) describe similar tests on Grande-Baleine Clay and refer to a number of other studies on artificially sedimented clays. Contrary to the views expressed by Casagrande (1932) it can be concluded that it is possible to reproduce the behaviour of natural

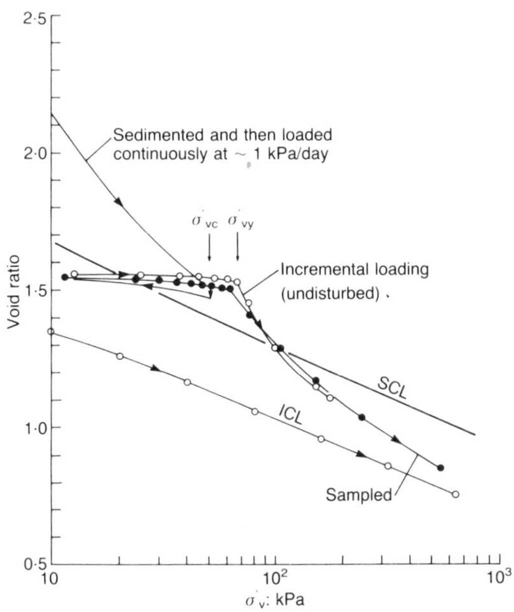

Fig. 24. Laboratory sedimented residual clay (Leonards & Altschaeffl, 1964)

clays in the laboratory but the preparation of the samples involves considerable lengths of time.

SHEAR STRENGTH OF NORMALLY CONSOLIDATED CLAYS

The discussion on the compressibility of normally consolidated natural clays was preceded by summarizing some basic properties of reconstituted clays. These properties are termed the intrinsic properties. Similarly, before examining some aspects of the shear strength of normally consolidated natural clays it is important to establish a clear picture of the intrinsic shearing behaviour of one-dimensionally consolidated reconstituted clays. For simplicity only the behaviour in triaxial compression is considered.

Intrinsic shear strength of normally consolidated clays

Figure 25(a) shows the one-dimensional intrinsic compression line for a reconstituted clay plotted on a graph of e versus σ_a'. Point 0 lies on the ICL and Fig. 25(b) shows the corresponding Mohr's circle of effective stress. The maximum shear stress is given by point A′ which lies on the K_0 effective stress path. Point A′ projects as point A in Fig. 25(a) which lies on a compression line for the average of the axial and radial stresses (σ_a' + σ_r')/2 shown as chain dotted.

A standard drained triaxial test entails increasing σ_a' with σ_r' constant. Fig. 25(c) shows the initial and failure Mohr's circles of stress for a sample initially consolidated to an axial effective stress σ_{ao}'. The Mohr's circle at failure is tangential to the intrinsic failure line and A′D′ represents the effective stress path for the test. The stress–strain and volumetric strain behaviour is shown in the adjacent diagram. It can be seen that the sample contracts and that at failure the rate of contraction is approximately zero. Thus failure corresponds to a critical state condition and in recognition of this the intrinsic angle of shearing resistance is designated ϕ_{cv}* where the asterisk denotes an intrinsic property. The stress path A′D′ plots as the path AD in Fig. 25(a) where D lies on the projection of the critical state line shown as a broken line.

The undrained behaviour of the clay is shown in Fig. 25(d). Most one-dimensionally consolidated reconstituted natural clays show brittle stress–strain behaviour with the peak undrained compressive strength being reached at very small strains, as shown by the point B″ in Fig. 25(d). Thereafter the contractant behaviour of the soil skeleton results in a falling stress–strain curve coupled with large increases in pore water pressure. The effective stress path for an undrained

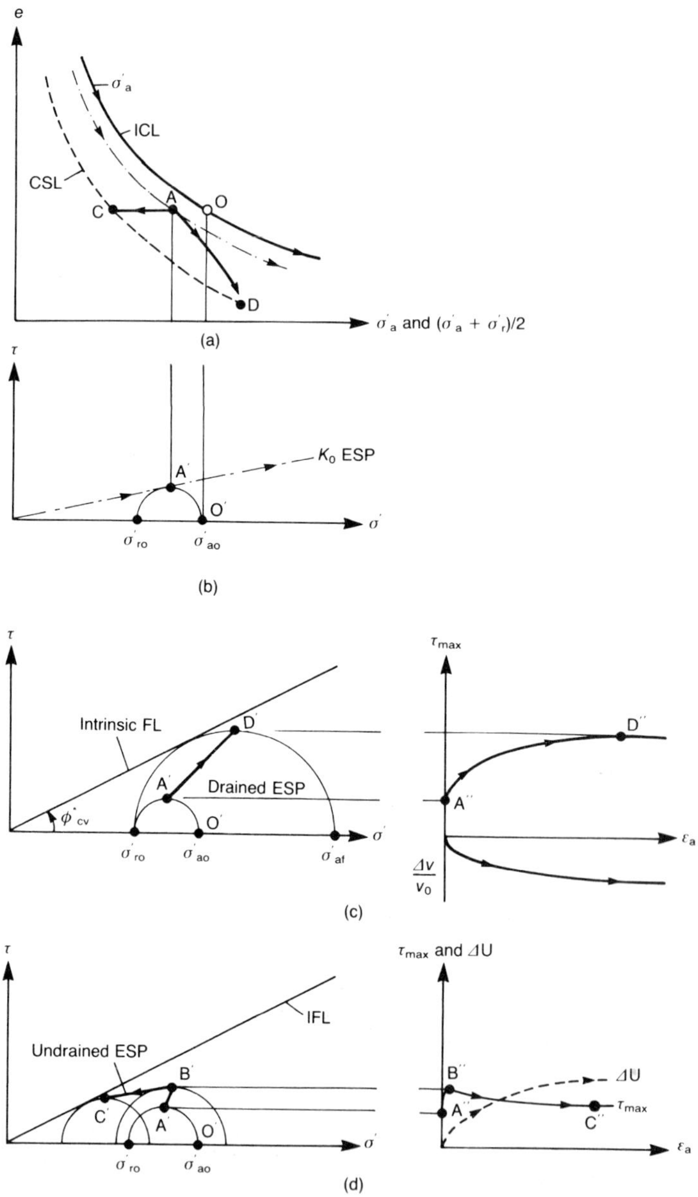

Fig. 25. Ideal behaviour of one-dimensionally consolidated reconstituted clay in triaxial compression: (a) void ratio changes; (b) K_0 stresses; (c) drained test; (d) undrained test

triaxial compression test is of the form given by A′B′C′ in Fig. 25(d) were B′ corresponds to peak strength and C′ to the critical state strength. The corresponding path in Fig. 25(a) is AC where C lies on the critical state line. The broken line CD is the projection of the intrinsic critical state line since it relates to a reconstituted soil. Note that the critical state line lies well to the left of the ICL. Although the strength of the soil decreases along the path B′C′ in Fig. 25(d) the stress ratio is actually increasing and the soil skeleton is therefore strain hardening. It can therefore be anticipated that the sample will deform in a homogeneous manner as the stresses move from B′ to C′. The significance of this will become apparent later.

It was shown previously that the effect of ageing during one-dimensional compression is to

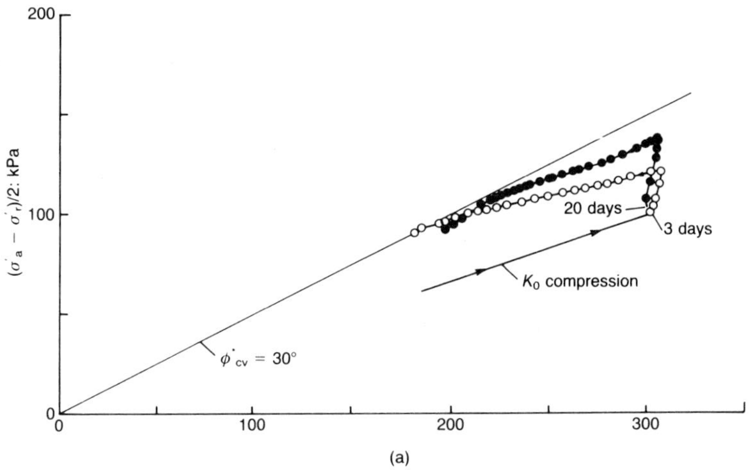

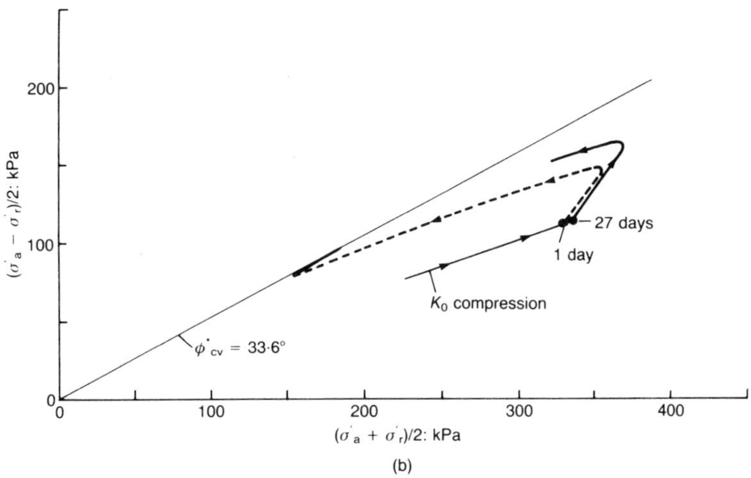

Fig. 26. Influence of ageing on undrained effective stress paths for triaxial compression tests on reconstituted soils for (a) Magnus Clay ($w_L = 35$) (Jardine, 1985), and (b) Gullfaks clayey sand (Georgiannou, 1988)

increase the vertical yield stress σ_{vy}'. Similar behaviour takes place in undrained compression. Fig. 26(a) shows the effect of ageing on reconstituted Magnus clay from the North Sea giving rise to a significant increase in peak undrained strength. There is also an increase in brittleness. Similar results are shown in Fig. 26(b) for reconstituted clayey sand from the Gullfaks field in the North Sea. In this case the volumetric strains during ageing were negligible so that the gain in strength must have been due primarily to interparticle bonding.

In the next two sections the results of undrained triaxial tests on high quality undisturbed samples of some normally consolidated clays are compared with the the framework given in Fig. 25.

Results from the Troll field in the North Sea

The Troll field is located in the Norwegian sector of the northern North Sea. Extensive site investigations have been carried out for the design of offshore gravity oil production platforms. The data presented here are for block 31/2 and high quality samples were obtained using thin walled tube samplers pushed into the ground at a steady rate. A comprehensive programme of laboratory testing was carried out jointly by Fugro-McClelland and the Norwegian Geotechnical Institute.

Figure 27 shows a typical soil profile. It consists of 23 m of a medium plasticity clay overlying low plasticity clay to a depth of about 65 m. Results of oedometer tests and anisotropically consolidated undrained (CAU) triaxial compres-

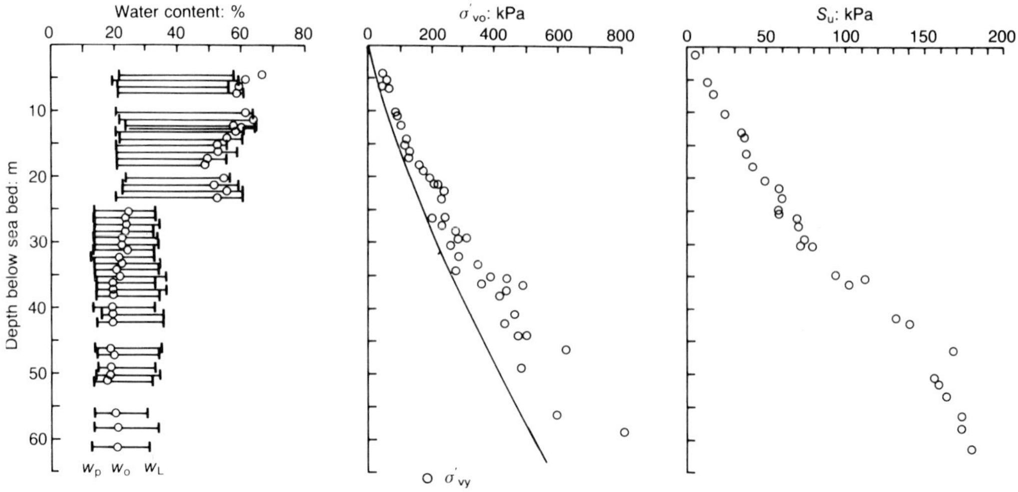

Fig. 27. Troll field, block 31/2, North Sea: soil profile

sion tests show that the soils are normally consolidated with a yield stress ratio of about 1·3. The upper clay is a glacial marine deposit laid down between 10 000 and 13 000 years BP (Sejrup *et al.*, 1989). There is some uncertainty about the mode of deposition of the lower clay but it is thought to be a glacial marine deposit or a lodgement till or a combination of both. The upper part of it was probably laid down during the retreat of the Scandinavian ice sheet about 13 000 years BP.

The sedimentation compression curves for the Troll profile (Fig. 28) are particularly interesting. The upper clay, shown by the open circles, lies a little above the SCL while the lower clay (open triangles) lies around the ICL. These results suggest that the deposition conditions for the two layers were entirely different.

The results of oedometer tests on samples from the two layers confirm the differences in the depositional environments. In Fig. 29 the open circles are for two oedometer tests on undisturbed

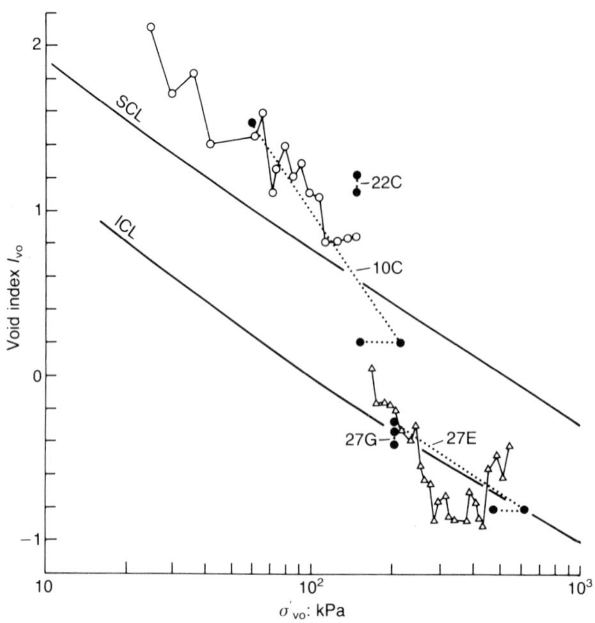

Fig. 28. Troll field: sedimentation compression curves

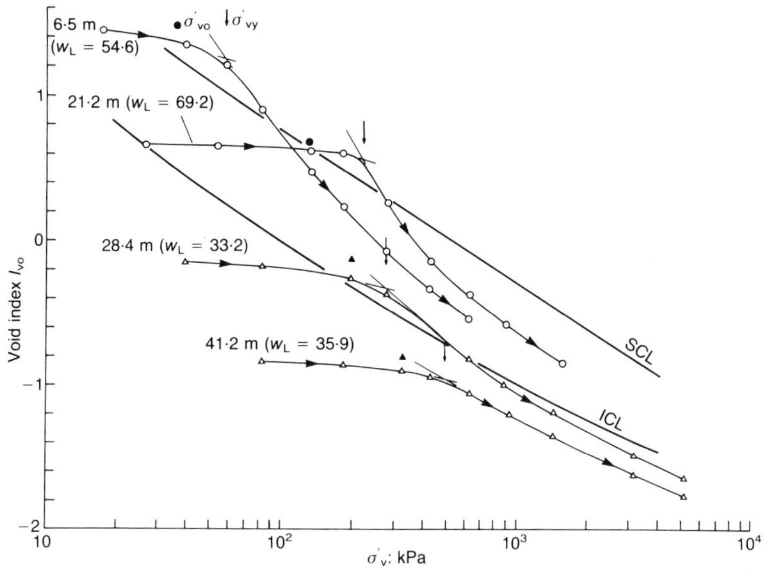

Fig. 29. Troll field: oedometer tests on upper and lower clays

samples of the upper clay. The compression curves follow the well established pattern of falling steeply through the SCL and then flattening off and converging slowly with the ICL. In contrast the compression curves for the lower clay (open triangles) remain close to the ICL.

Although a variety of types of shear test were carried out during the investigation the results from the following two types will be considered here.

(*a*) CAU triaxial compression and extension tests in which the samples were consolidated to their estimated in situ effective stress state prior to undrained shearing.
(*b*) SHANSEP tests in which the samples were compressed anisotropically to well beyond their in situ states of stress and then unloaded a little to model the apparent preconsolidation. This procedure was introduced as a method of overcoming sampling disturbance.

Figure 28 shows the void ratio changes associated with the two types of test. Tests 22C and 27G were CAU tests and it can be seen that small reductions in void ratio took place when the in situ stress state was re-established. Tests 10C and 27E were SHANSEP tests and it is evident that large reductions in void ratio took place during the consolidation phase.

The broad framework of behaviour shown in Fig. 25 may be used to assess the likely behaviour of the samples referred to in Fig. 28. Sample 22C lies above the SCL. If undrained shearing were to cause its state to reach the intrinsic critical state line the effective stresses would have to reduce enormously such that a constant void ratio path would travel to the left of the ICL. Thus the behaviour would be predicted to be very brittle and sensitive. In contrast, since sample 27G lies on the ICL its behaviour would be expected to be similar to an aged reconstituted sample with low sensitivity and little or no brittleness.

The SHANSEP procedure has caused sample 10C to move from well above the SCL to some distance below it. Thus it would be expected to be much less brittle than sample 22C. On the other hand, since sample 27E has remained on the ICL during consolidation, its behaviour would be expected to be similar to sample 27G.

The results of the undrained triaxial tests are given in Figs 30 and 31 for the upper and lower clays respectively. These figures should be studied in conjunction with Fig. 28. It can be seen from Fig. 30(a) that sample 22C shows brittle behaviour as predicted. The effective stress path (Fig. 30(b)) rises to the ultimate failure line and then travels down it towards the origin with the average effective stress reducing to about 65 kPa. If the intrinsic critical state had been reached the average effective stresses would have reduced to about 5 kPa. Clearly shearing in triaxial compression does not induce sufficient destruction of the microstructure to bring the soil to the intrinsic critical state.

As expected the SHANSEP test 10C, shown by the broken lines in Fig. 30, is very much less brittle than the CAU test. Moreover the stress path does not rise all the way to the ultimate

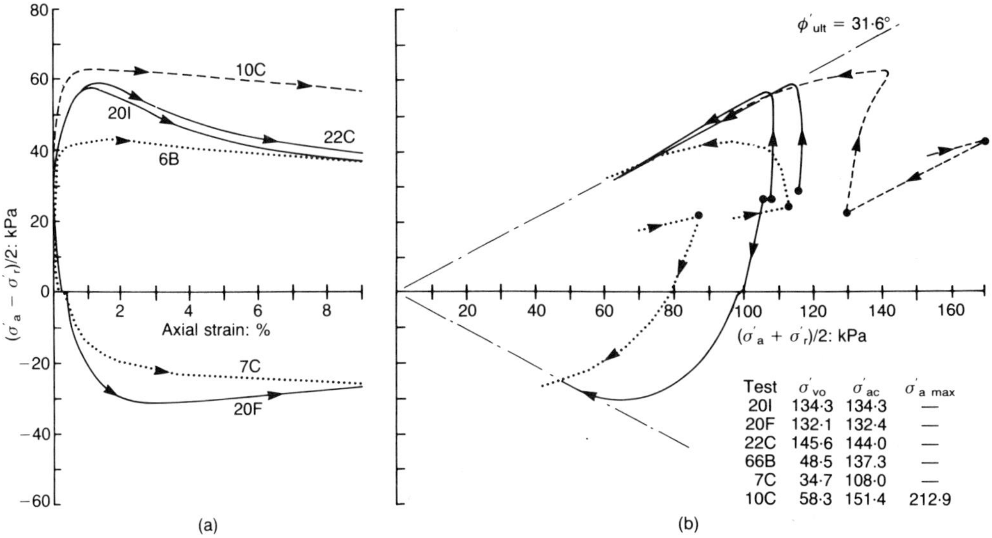

Fig. 30. Troll field: CAU triaxial tests on samples from upper clay

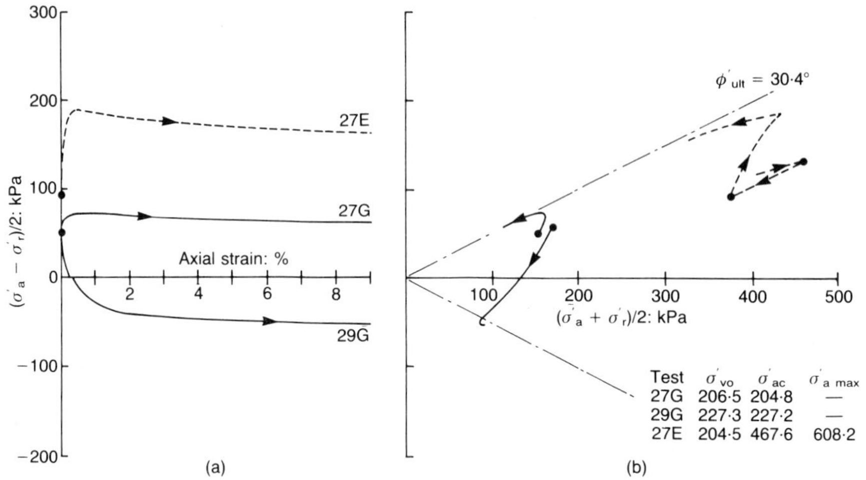

Fig. 31. Troll field: CAU triaxial tests on samples from lower clay

failure line but bends sharply to the left before reaching it—as a reconstituted soil would do. Thus, by altering the structure of the clay, the SHANSEP test procedure underestimates both the peak strength and the brittleness of a clay for which the in situ state lies on or above the SCL. It can be seen from Fig. 30 that the undrained extension tests behave in broadly the same manner. Tavenas & Leroueil, 1985, draw attention to the limitations of the SHANSEP procedure due to 'destructuration'. Smith (1990) shows that if a SHANSEP sample is allowed to 'age' under K_0 stresses the undrained strength and brittleness increase.

Figure 31 shows the undrained triaxial test results for the lower clay. It can be seen that the stress–strain and stress path behaviour of sample 27G is reasonably well modelled by SHANSEP test 27E, since the experimental curves are similar in shape with only small brittleness. It seems probable that the ultimate state of both samples closely approach the intrinsic critical state line. As no tests were carried out on reconstituted material it is not possible to be definite about this. Thus for a clay which lies close to the ICL the SHANSEP procedure provides a reasonable normalized pattern of behaviour for the natural material since the soil structure is not significantly changed during the initial consolidation.

In summary the use of the void index, ICL and SCL, in conjunction with the framework for the behaviour of reconstituted soils in Fig. 25, have been valuable in gaining an understanding of the undrained behaviour of the clays at the Troll site.

Results from three sites in Norway

Lacasse *et al.* (1985) have published the results of laboratory tests on three normally consolidated Norwegian marine clays. Two key features of the published data are:

(*a*) The tests were carried out on block samples so that sampling disturbance was reduced to a minimum. (Comparisons were also carried out with samples obtained with a fixed piston tube sampler).
(*b*) The clays from the three sites cover a wide spectrum from a sensitive clay at Onsøy, through a lean quick clay at Ellingsrud, to an extremely quick clay at Emmerstad.

The profiles for the three sites are given in Figs 32(a) to (c). The following features should be noted. The liquidity index increases significantly for Onsøy through to Emmerstad. The yield stress ratio $\sigma_{vy}'/\sigma_{vo}'$ increases for Onsøy through to Emmerstad. The vane tests show that the two quick clays, Ellingsrud and Emmerstad, have extremely high sensitivities.

The sedimentation compression curves for the three sites are plotted in Fig. 33. For the Onsøy site (open triangles) the clay in the top 4 m lies on the SCL, but at greater depths it lies a little above the SCL. The chain-dotted line is the oedometer compression curve for a sample from a depth of 9·07 m. After yield, the curve plunges steeply and drops below the SCL. The full circles represent the sedimentation compression curve for the quick clay at Ellingsrud. The sedimentation compression curve lies well above the SCL corresponding to a void index of about 3. The broken line is the oedometer compression curve for a sample from 8·05 m in depth. It is clear that the curve remains well above the SCL.

The sedimentation compression curve for the extremely quick clay at Emmerstad is given by the open circles and it can be seen that the void index is very high (about 5). As for the other quick clay site, although the oedometer compression curve falls steeply following yield, it remains well above the SCL. It appears from these results that the process of one-dimensional compression does not disrupt the structure of a lean quick clay sufficiently to cause it to compress down to or below the SCL. More drastic mechanical disturbance would be required to do this.

For each of the three sites a number of CAU triaxial tests were carried out with the estimated in situ effective stresses applied prior to shearing. Typical results are given in Figs 34(a) to (c). For the sensitive clay at Onsøy (Fig. 34(a)) the stress–strain curves for samples from the upper clay lying on the SCL show less brittleness than for the lower clay which lies above the SCL. The stress paths for triaxial compression bend to the left before reaching the ultimate failure line and travel some distance down it towards the origin. The quick clay from Ellingsrud (Fig. 34(b)) shows considerably more brittleness than for Onsøy. The stress paths rise up to the ultimate failure line before bending to the left and travelling a considerable distance down it. The stress–strain curves for the extremely quick clay at Emmerstad (Fig. 34(c)) show sharp peaks but the brittleness is no greater than for Ellingsrud. As remarked by Lacasse *et al.* (1985), the stress paths are unusual. They rise to above the ultimate failure line and as peak strength is approached the stress paths bend to the right which is indicative of dilatant behaviour. Beyond peak the paths drop down to the ultimate failure line and travel down it. This interesting behaviour might be accounted for by a soil fabric consisting of 'packets' of particles with bonded contacts. During shear up to peak the packets behave as a granular material giving rise to mildly dilatant behaviour. Once peak strength has been reached the individual packets begin to break down giving rise to contractant behaviour.

In none of the three cases do the stress paths approach the intrinsic critical state. For the two quick clays the very high values of I_{vo} would require that the critical state is very close to the origin in a stress path diagram. Thus, as for the oedometer test, the triaxial test does not appear to provide sufficient mechanical energy to break down the natural fabric and bonding of lean quick clays completely although this might be achieved by remoulding with a vane test.

Peak undrained strength

It has been shown that the critical state framework, when used in conjunction with the void index as a normalizing parameter, is helpful in accounting for the brittleness and sensitivity of natural clays, although frequently their states do not reach the intrinsic critical state in a triaxial test. However, in its present form, the critical state framework cannot be used to predict the peak undrained strength S_u of normally consolidated natural sediments.

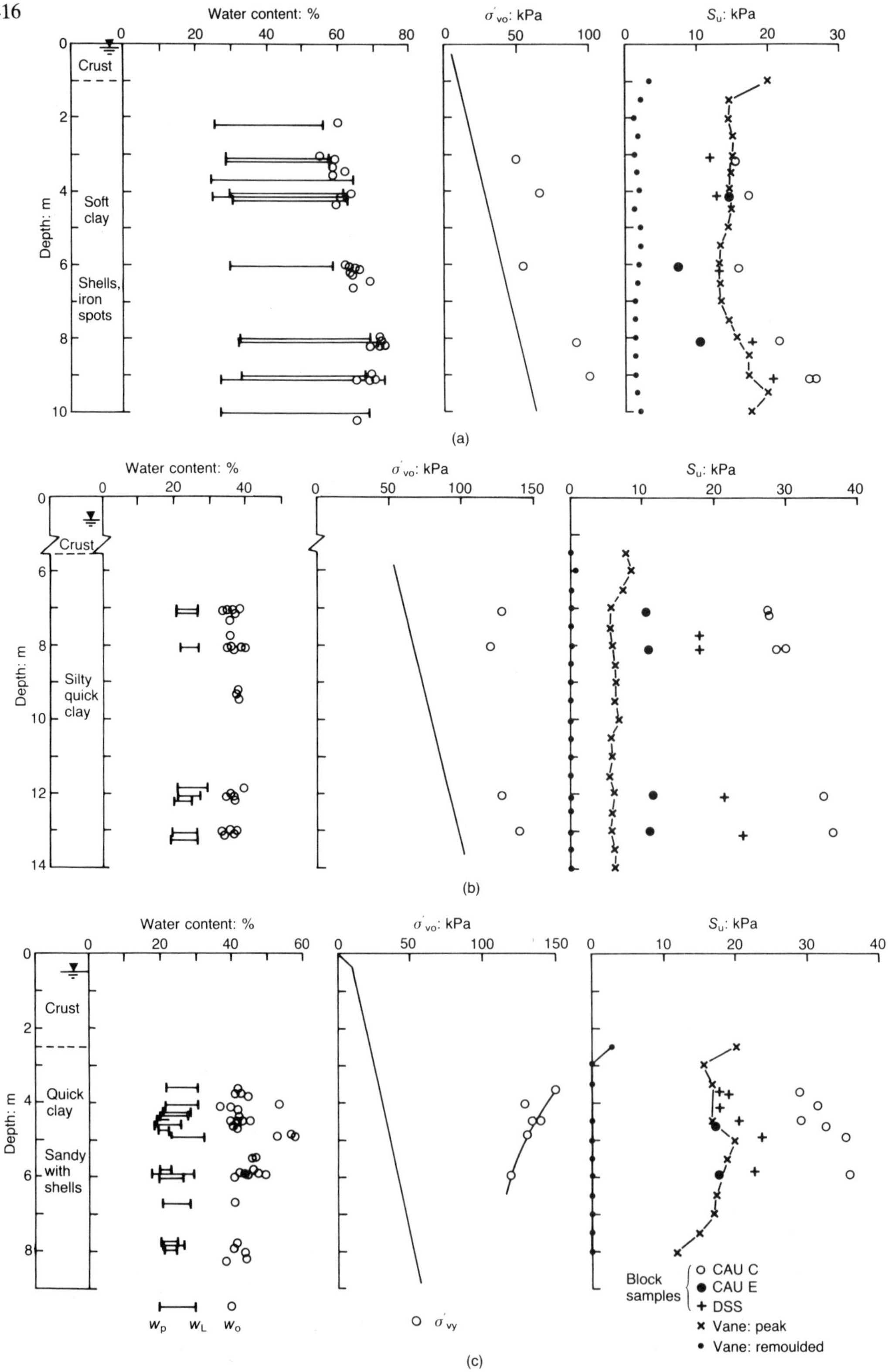

Fig. 32. Soil profiles for (a) Onsøy, (b) Ellingsrud and (c) Emmerstad, all in Norway (Lacasse *et al.*, 1985)

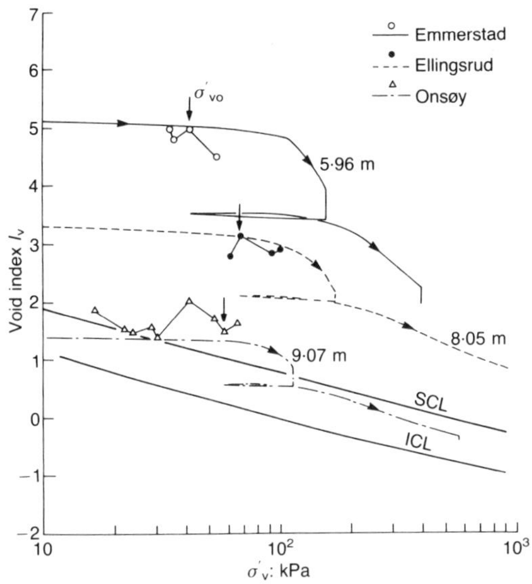

Fig. 33. Results of oedometer tests on block samples of three Norwegian sensitive clays (Lacasse *et al.*, 1985)

It is a central tenet of critical state soil mechanics that, for a given type of clay, S_u is primarily related to water content, or void ratio, and more generally to liquidity index, or void index (Wood, 1985). At the Troll site the upper clay has a much higher liquidity index and void index than the lower clay (see Fig. 27). Thus, for a given effective overburden pressure, critical state soil mechanics would predict that the upper clay would have a lower S_u than the lower clay. It can be seen from Fig. 27 that, at the junction between the two clays, there is little difference between the S_{uTC} values above and below it. The value of S_{uTC}/σ_{vo}' is about 0·4 for both clays. Expressed as a proportion of the vertical yield stress σ_{vy}', the values of S_{uTC}/σ_{vy}' are approximately 0·32 and 0·28 for the upper and lower clays respectively. These values are within the normally expected range for soft clays (Hight *et al.*, 1987).

Referring again to the three Norwegian clays in Fig. 32, critical state soil mechanics would predict that, for a given overburden pressure, the clay at Emmerstad would be weaker than at Ellingsrud, which in turn would be weaker than at Onsøy because the void indices and liquidity indices decrease in that order. In fact the reverse is the case. At an effective overburden pressure of 50 kPa the values of S_{uTC} for Emmerstad, Ellingsrud and Onsøy are approximately 35·7 kPa, 27·4 kPa and 22·0 kPa respectively. When expressed as a proportion of σ_{vy}' the corresponding values of S_{uTC}/σ_{vy}' are 0·21 to 0·31 for Emmerstad, 0·23 to 0·27 for Ellingsrud and 0·27 for Onsøy.

In summary, for the Troll and the Norwegian sites, it appears that the peak undrained strength is more directly related to soil fabric and bonding as reflected by the yield stress σ_{vy}' than it is to liquidity index or void index.

COMPRESSIBILITY OF OVERCONSOLIDATED CLAYS

Point A in Fig. 35(a) represents the in situ state of an element of overconsolidated clay in an e against log σ_v' diagram. The locations and slopes of the natural sedimentation compression curve and swelling curve are unknown. In Fig. 35(b) the void ratio has been transformed to I_v and point A plots as A′. Using this plot the position of A′ relative to the ICL and the SCL is known and this gives an immediate indication of the approximate degree of overconsolidation for the soil assuming that compression took place close to the SCL.

In this section the location of some oedometer compression curves relative to the ICL and SCL are investigated as shown in Fig. 35(c). Also the measured swelling characteristics of some natural overconsolidated clays are compared with the intrinsic swelling line (ISL) as shown in Fig. 35(d). It should be noted that the intrinsic swelling index C_s^* is defined as the slope of the ISL at an overconsolidation ratio of 10. For this study only the results from block samples are considered in order to minimize the effects of sampling disturbance.

Gault Clay

Samuels (1975), working at the Building Research Station, carried out a number of oedometer tests on block samples of heavily overconsolidated Gault Clay extracted from shafts associated with the Ely–Ouse tunnel. He also carried out oedometer tests on samples that had been reconstituted at twice the liquid limit. Fig. 36 shows some typical results for a block from 85·3 m depth. Values of e and I_v are plotted on the left- and right-hand vertical axes respectively. The intrinsic compression and swelling lines are shown as chain-dotted. The ratio between the intrinsic compression and swelling indices $C_s^*/C_c^* = 0{\cdot}398$.

For the tests on the undisturbed samples the swelling pressure was measured by adding weights to the hanger to prevent swelling following soaking of the sample. It can be seen that the swelling pressure is slightly less than the value of σ_{vo}'. The oedometer compression curve crosses the ICL and then bends down. The stresses were not sufficiently high to establish whether or not

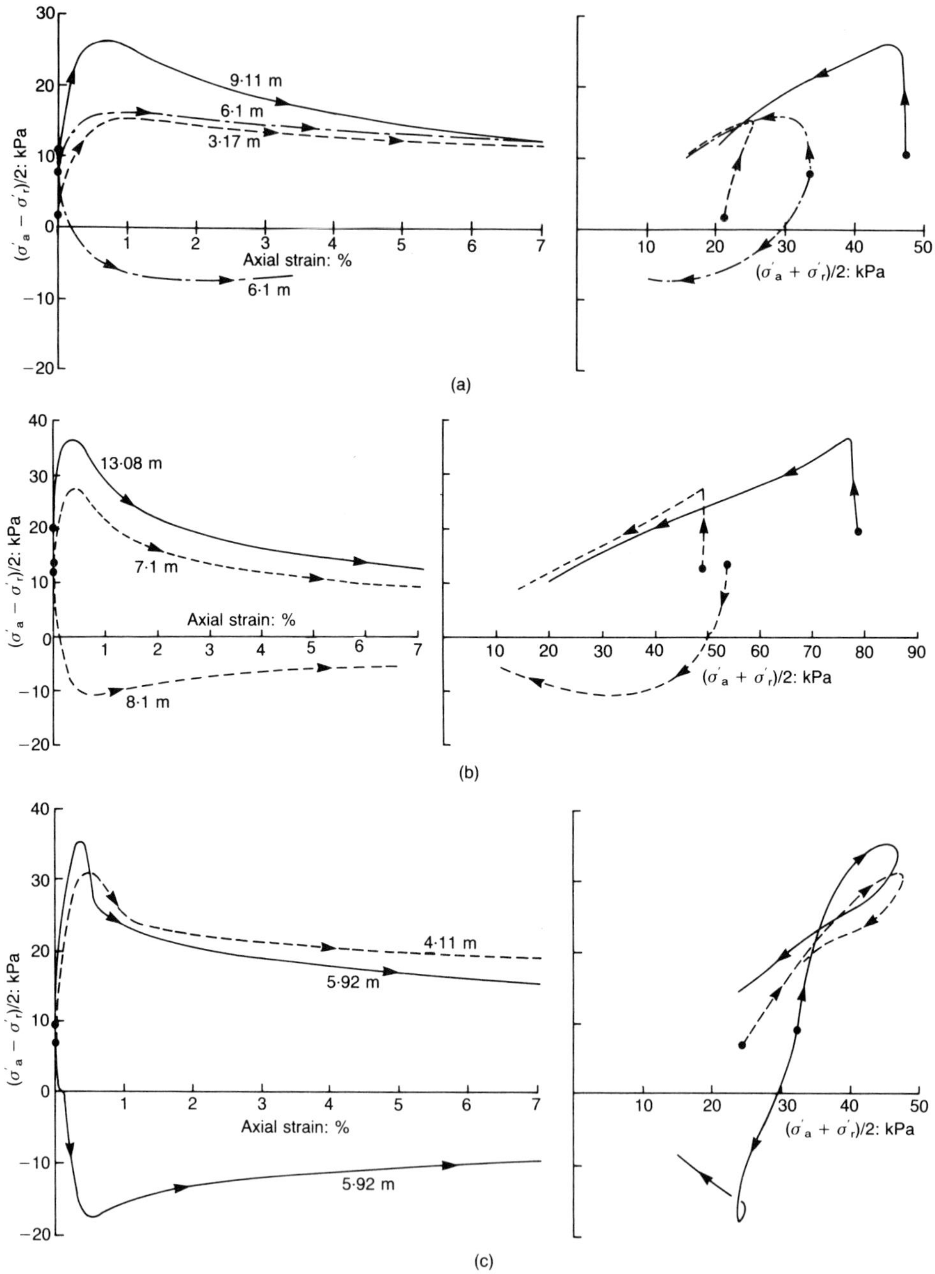

Fig. 34. Results of CAU triaxial tests on block samples of Norwegian sensitive clays from (a) Onsøy, (b) Ellingsrud and (c) Emmerstad (Lacasse *et al.* 1985)

the curve intersects the SCL but it is clear that the normal consolidation line has not been reached.

A swelling test was carried out on an identical sample. It can be seen that it is four times less expansive than the reconstituted material. The ratio C_s^*/C_s for a soil may be a sensitive indicator of fabric and interparticle bonding in the natural soil. Schmertmann (1969) defined this ratio as the 'swell sensitivity'. Note that, after loading up to 7000 kPa, the first sample had become approximately twice as expansive as the one only subjected to unloading. Thus the process of loading must have destroyed some of the bonding although the clay is still less than half as expansive as the reconstituted clay.

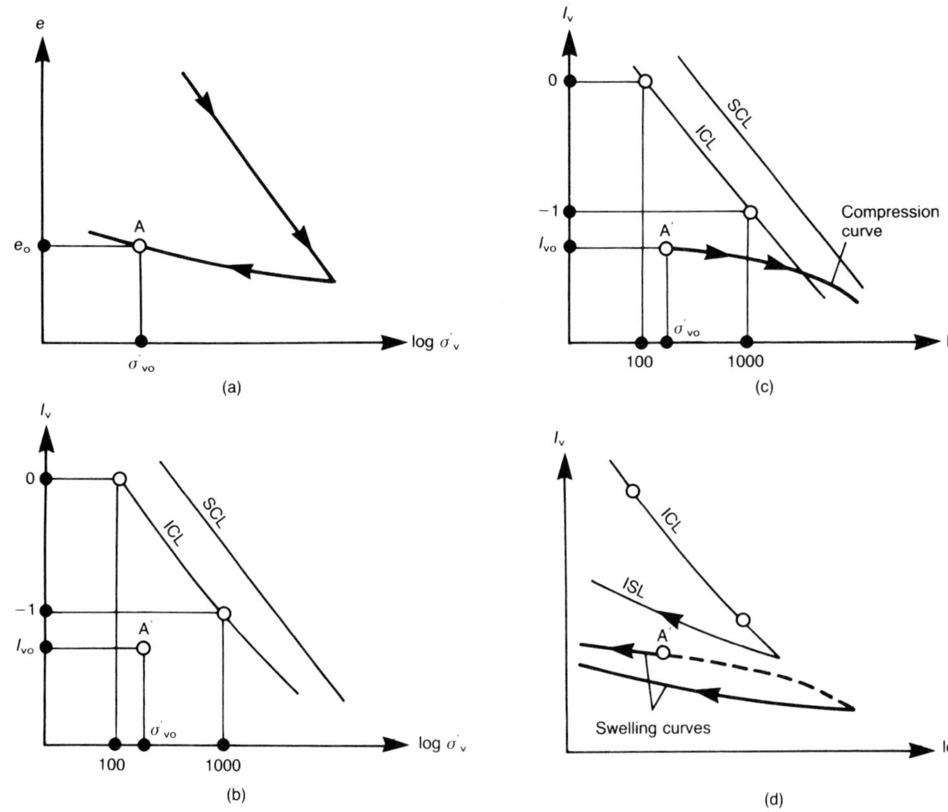

Fig. 35. Comparison of compression and swelling properties of overconsolidated clay with corresponding intrinsic properties

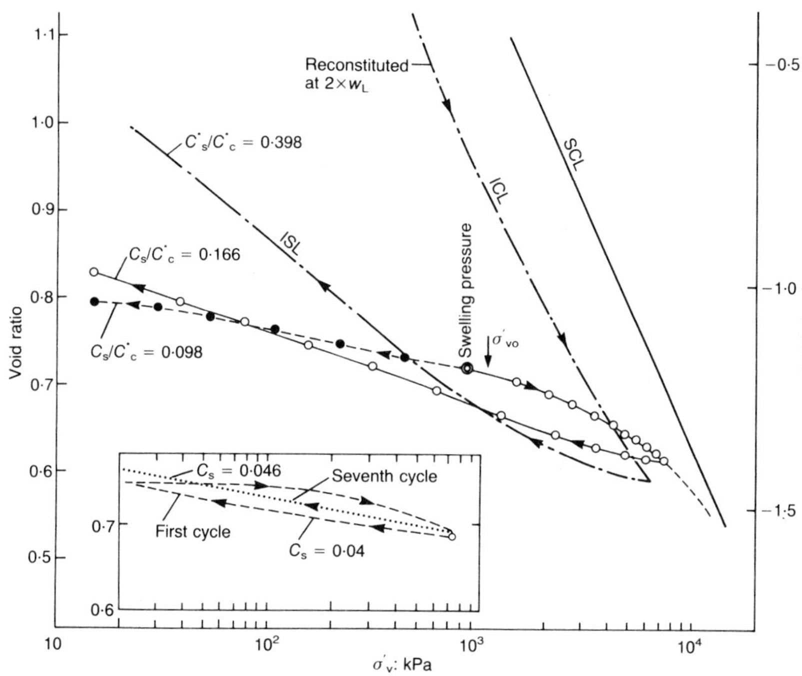

Fig. 36. Gault Clay (w_L = 79·4): oedometer tests on block sample from Ely–Ouse tunnel, shaft 10, depth = 85·3 m (Samuels, 1975)

The inset diagram in Fig. 36 shows the results of a cyclic swelling and compression oedometer test devised to investigate the susceptibility of the clay to structural breakdown. Evidently the bonding was sufficiently strong to resist this process. It should be noted that the clay had a calcium carbonate content of about 30% and this may have been the source of strong interparticle bonding.

Boom Clay

Horseman *et al.* (1987) have published the results of some high pressure oedometer tests on block samples of Boom Clay from Mol in Belgium. The tests are of interest because geologically the clay is only lightly overconsolidated but, because of the great depth from which the samples were taken (247 m), the clay is stiff. The results of a typical oedometer test are plotted in Fig. 37. The in situ state is seen to lie between the ICL and the SCL. Moreover the swelling pressure is considerably less than σ_{vo}'. Both of these observations confirm that the soil is only lightly overconsolidated.

The compression curve exhibits a reasonably well defined yield point and thereafter it drops steeply towards the ICL appearing to join the extension of it. The normally consolidated state prior to geological unloading must have been located fairly close to the SCL with a preconsolidation pressure of approximately 6 MPa giving an overconsolidation ratio of about 2·4. Horseman *et al.* (1987) state that it is difficult to reconcile the preconsolidation pressure with present geological evidence which points to a much lower preconsolidation pressure. They suggest that the yield stress may be larger than the preconsolidation pressure due to mechanisms such as creep and diagenesis. However, the fact that the yield stress lies below the SCL suggests that such mechanisms were not of major significance. Perhaps the geological evidence requires further evaluation.

Unfortunately no tests were carried out on the reconstituted material so that the value of C_s* is not known. However, it is evident that as the material is compressed the swelling index increases pointing to a progressive disruption of the natural fabric and bonding.

Todi Clay

Over the last decade a programme of fundamental research into the properties of Todi Clay has been carried out at the University of Rome under the direction of Professor G. Calabresi. Todi is an attractive hill top city to the north of Rome. It has suffered from landslip problems in recent years and the main thrust of the research has been directed towards understanding

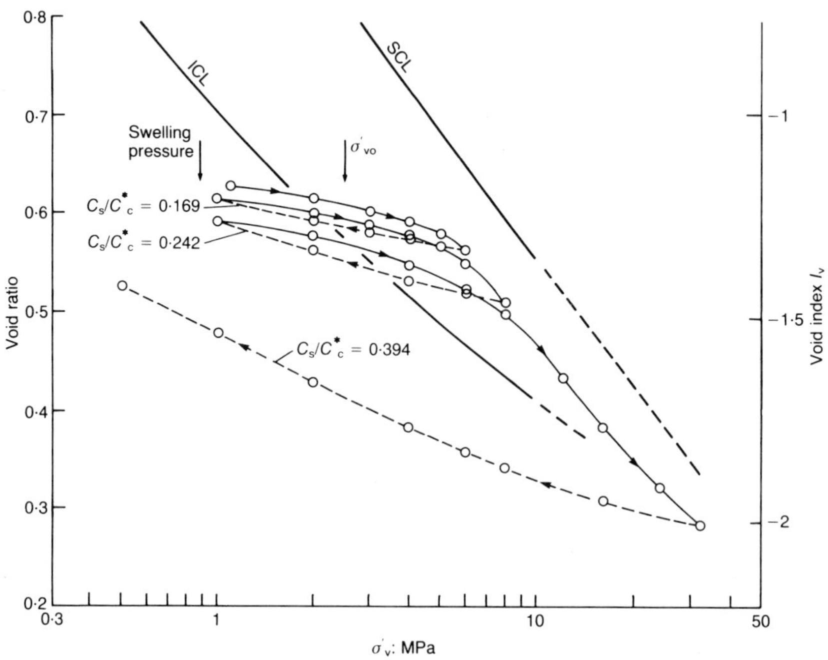

Fig. 37. Boom Clay (w_L = 65): high pressure oedometer test on block sample from Mol, depth = 247 m (Hoiseman *et al.*, 1987)

the influences of swelling and weathering on the shear strength properties of the clay (Calabresi & Scarpelli, 1985). Todi Clay is a low to medium plasticity lacustrine clay of Pleistocene age. It is overconsolidated and intensely fissured.

Block samples of the clay were extracted from vertical faces of a brick pit as and when they were required. Fig. 38 shows some oedometer compression curves for the clay. The chain-dotted line shows the experimentally determined intrinsic compression and swelling lines for the material. The SCL is shown as a full line. The open circles are for a compression test on an undisturbed sample starting from the swelling pressure.

The compression curve crosses the ICL and bends down without reaching the SCL. The subsequent swelling index is a little less than the intrinsic value. The open triangles are for a test which was allowed to swell under a very low pressure in the oedometer prior to compressing. The compression curve appears to join up with that for the sample compressed from the swelling pressure (open circles). The closed triangles are for a sample which was immersed in saturated loose sand and left to swell freely for three months prior to testing. The compression curve crosses the ICL but lies beneath the curves for the other two samples. The swelling index is about the same as for the other two samples. It can be concluded that most of the differences between the natural clay and the reconstituted clay are due to differences in the fabric. Since the swelling index C_s is insensitive to loading history and is only slightly less than the intrinsic value C_s^* it appears that interparticle bonding is not strong.

SHEAR STRENGTH OF INTACT OVERCONSOLIDATED TODI CLAY

Figure 39 shows three Mohr–Coulomb failure envelopes for Todi Clay. The broken line labelled intact strength is for intact samples (i.e. not containing fissures) which were compressed or swelled from their natural moisture content prior to shearing in drained and undrained triaxial compression. The intact failure envelope shows significant curvature for confining pressures of less than 1500 kPa. The chain-dotted line is the failure envelope for samples which were allowed to swell freely for three months after which they were reconsolidated and sheared. The free-swell failure envelope lies below the intact strength envelope. Tests on normally consolidated reconstituted samples were also carried out giving the intrinsic strength envelope. The differences in strength are due to two main factors: the void ratio at failure and the soil structure (fabric and bonding).

The influence of void ratio may be eliminated using the normalization procedure first developed by Hvorslev (1937). Fig. 40 shows the void ratio of a number of samples prior to shearing. It can be seen that the samples which were allowed to swell freely for three months (closed points) have

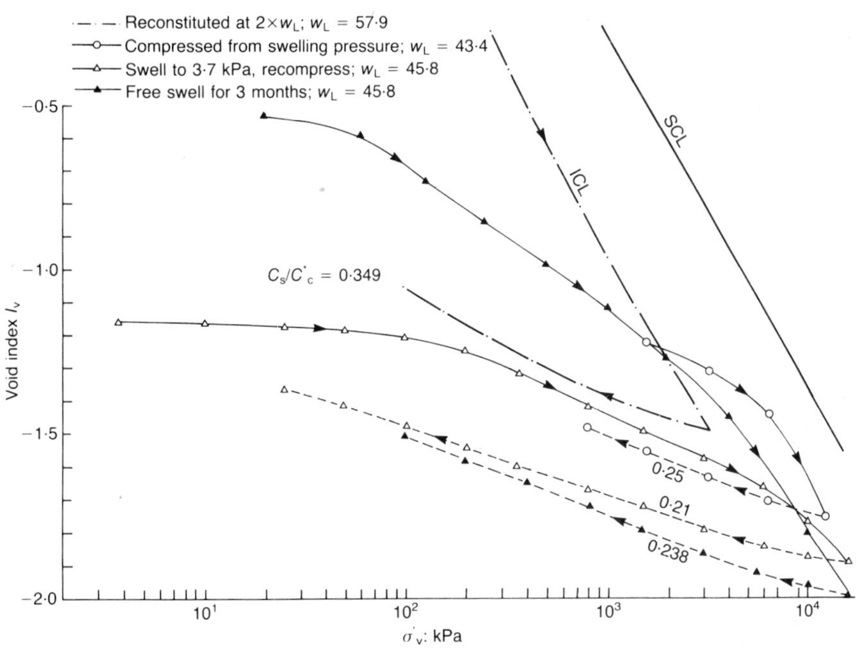

Fig. 38. Todi Clay: oedometer tests on block samples after various swelling regimes (Rampello, 1989)

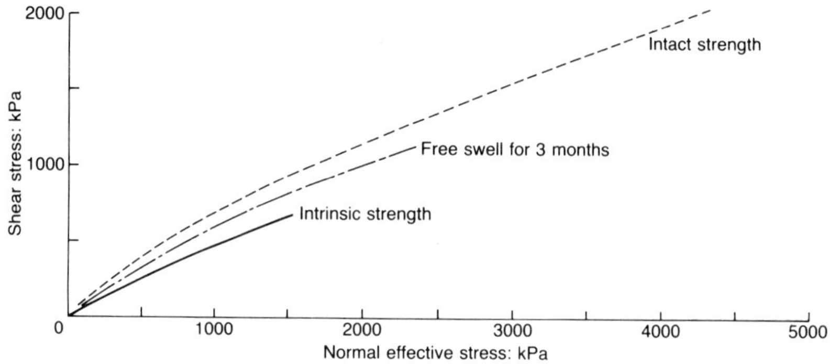

Fig. 39. Todi Clay: Mohr–Coulomb failure envelopes

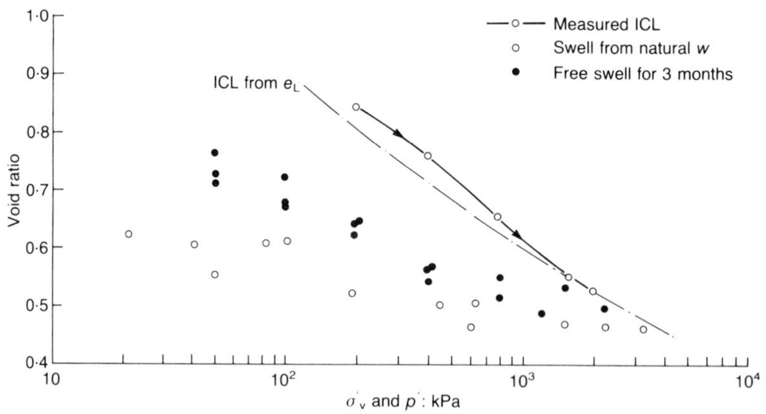

Fig. 40. Todi Clay: void ratios prior to shearing in triaxial compression

higher void ratios than the samples which were swelled or compressed from their natural moisture content. Following Hvorslev, the vertical effective pressure on the ICL corresponding to the void ratio of the soil is termed the 'equivalent intrinsic pressure' σ_{ve}^*. By dividing the strength and normal effective pressure by σ_{ve}^* the influence of differences in void ratio are eliminated.

Figure 41 shows a plot of $(\sigma_a' - \sigma_r')/2\sigma_{ve}^*$ against $(\sigma_a' + \sigma_r')/2\sigma_{ve}^*$. The dotted lines show the state boundary surfaces† for normally and overconsolidated reconstituted Todi Clay in which the initial consolidation took place under isotropic stresses. The critical state line plots as a single point in this diagram separating the Hvorslev from the Rendulic surfaces. It can be seen that the normalized failure surfaces for the intact and freely swelled Todi Clay lie a little above the intrinsic Hvorslev surface. This inherently greater strength of the natural clay is attributable to microstructural effects. It is important to note that the natural clay can exist in states well to the right of the intrinsic critical state line and Rendulic surface. This is a logical consequence of the natural SCL lying well to the right of the ICL.

Also shown in Fig. 41 are some typical undrained stress paths for the natural clay. It can be seen that the intact clay is strongly dilatant with the stress paths moving a considerable distance up the failure envelope prior to rupture. Even for stress paths lying outside the intrinsic Rendulic surface the clay is strongly dilatant. The freely swelled material shows less dilatancy.

† These surfaces are termed the Hvorslev surface for overconsolidated clays and the Roscoe surface for normally consolidated clays (Atkinson & Bransby 1978). Historically it is more appropriate to call this latter surface after Rendulic (Burland 1989).

POST-RUPTURE STRENGTH

A number of tests on intact samples of Todi Clay were carried out at Imperial College by Dr Rampello and Dr Georgiannou as a collaborative project with the University of Rome. The instru-

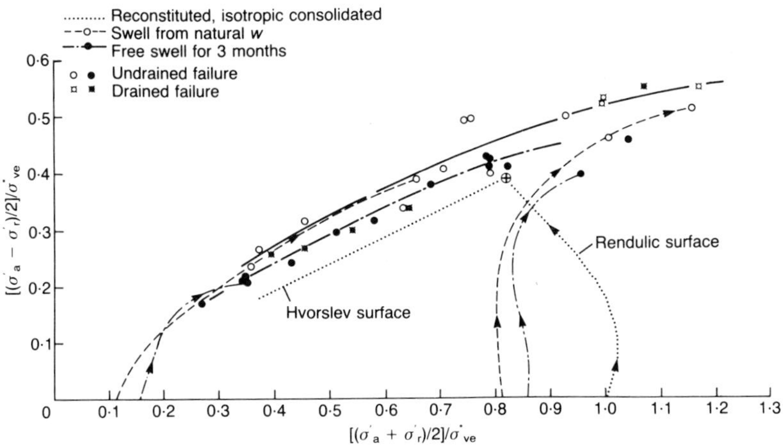

Fig. 41. Todi Clay: results of triaxial compression tests normalized by the equivalent pressure σ_{ve}^* at failure

mentation included local strain transducers (Burland & Symes, 1982; Burland, 1989) and a local pore pressure probe (Hight, 1982). For all the tests failure took place abruptly along a single slip surface as shown by Fig. 42. By good fortune the slip surface for this test passed outside both of the local strain transducers and close to the pore pressure probe. This has made possible a detailed and reliable study of the process of rupture.

In Fig. 43(a) the closed circles show the relationship between deviator force and notional overall axial strain and the open circles are for local axial strains. The excess pore pressures measured by the probe and at the base are shown by the open and closed triangles respectively. The following important observations can be made

(*a*) the local strain transducers show that the formation of the failure plane coincides with peak strength
(*b*) after peak the curve of deviator force versus notional overall strain falls steeply to a well defined plateau
(*c*) the excess pore pressure changes cease abruptly shortly after peak strength is reached
(*d*) prior to peak strength the local strains are less than the overall strains, as expected (Jardine *et al.*, 1984)
(*e*) after peak strength is reached the local axial strains decrease as a result of the unloading process; thus the post-rupture deformation consists of near-rigid body sliding on the failure plane with very slight axial extension in the surrounding clay.

Fig. 43(b) shows the relationship between maximum shear stress and overall axial strain up to rupture. Thereafter the relationship between the shear stress τ on the slip surface and the relative displacement across it is plotted.† It can be seen that the shear stress drops rapidly at first but reaches a nearly constant value after a relative displacement of about 1 mm. The closed circles show the ratio τ/σ_n' doing the same. The strength corresponding to the post-peak plateau is defined as the post-rupture strength.

Figure 44(a) shows the post-rupture failure envelope for Todi Clay. It can be seen that the envelope is bi-linear with a transition between low and high pressures at a normal effective stress of about 1500 kPa. For high stresses the envelope is defined by the parameters $c' = 0$, $\phi_{pr}' = 20{\cdot}2°$ where ϕ_{pr}' is the post-rupture angle of shearing resistance. The post-rupture failure envelope is seen to lie well below the intact failure line and a little above the residual failure line for which ϕ_r' is approximately 17° (Calabresi, 1990). The results are shown to a larger scale in Fig. 44(b) for low to intermediate stresses. For these conditions the post-rupture strength parameters are $c' = 23$ kPa and $\phi_{pr}' = 23{\cdot}7°$. The cohesive intercept may result from the fact that the failure plane is slightly wavy. The chain dotted line is the intrinsic failure line from tests on reconstituted normally consolidated samples. It is somewhat curved with $\phi_{cv}^* = 28°$ at the origin decreasing to 24° at $\sigma_n' = 600$ kPa. Over the range of $\sigma_n' = 100$ kPa to 1000 kPa the post-rupture and intrinsic failure envelopes lie very close to each other.

† Chandler (1966) and Webb (1969) give expressions for the surface area of the slip surface which is used for calculating τ and σ_n'. They also give correlations for membrane restraint and lateral restraint of the end caps.

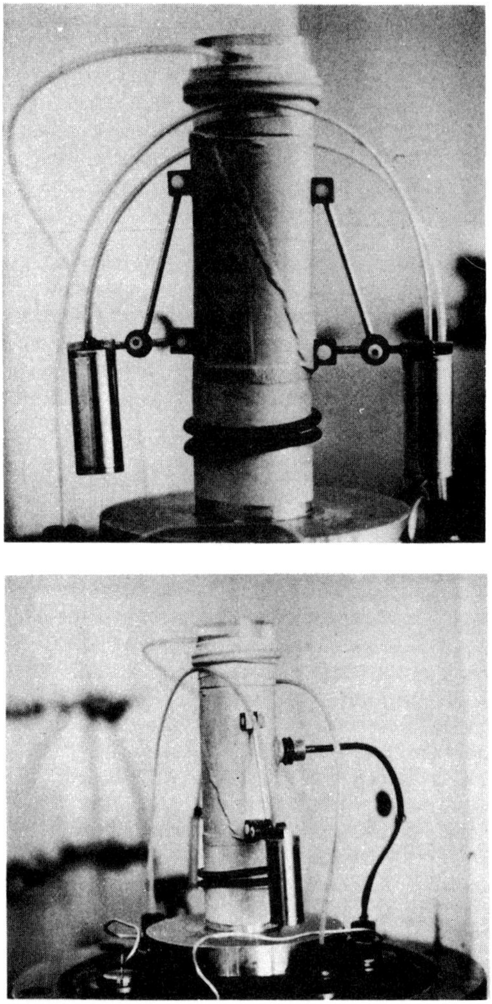

Fig. 42. Sample of Todi Clay after rupture showing local strain and pore pressure instrumentation (Rampello, 1989)

PRELIMINARY STUDY OF SLIP SURFACES IN KAOLIN

The identification of the development of rupture planes using local strain devices represents an unexpected and important application of this type of instrumentation. The results of other testing being carried out at Imperial College were studied to see if the phenomenon was apparent in other materials. Mr Ninis, working with anisotropically consolidated kaolin, has observed the formation of rupture planes even in normally consolidated samples. Fig. 45 shows the effective stress path for an undrained triaxial compression test on anisotropically consolidated aged kaolin. Peak strength occurs well below the intrinsic failure line followed by strongly contractant behaviour with a small reduction in strength. Once the stress path touches the intrinsic failure line a slip surface forms causing the stress path to descend as shown.

Figure 46(a) shows the observed relationship between the deviator force and axial strain and between the excess pore pressure and axial strain. (Note the change of scale at a strain of 1%). The full circles are for notional overall strain whereas the open circles represent the local strain. It can be seen that at a strain of about 7% the two diverge sharply with very little local axial strain taking place thereafter and with only small increases in excess pore pressures. Thus it appears that, following the formation of the slip surface, nearly rigid body sliding takes place along it. Fig. 46(b) shows an analysis of the pre- and post-rupture stresses. The ratio τ/σ_n' on the slip surface drops to a stationary value of about 0·27 after approximately 5 mm relative displacement corresponding to $\phi_{pr}' = 15{\cdot}1°$. The residual angle of friction for the kaolin is 12·4°. It would appear that the process of one-dimensional consolidation induces an aligned fabric which may be more prone than isotropically prepared samples to the formation of slip surfaces.

Figure 47 shows a thin section of a specimen of kaolin containing a slip surface. It is evident that the slip surface is very thin with only a narrow zone of orientation on either side. It can be seen that a conjugate slip surface is in the process of progressive development. This study indicates that the fixed direction of the principal stresses during shearing in the triaxial apparatus encourages the rapid development of single discrete shear surfaces. This contrasts with the shear box where the rotating stress field coupled with stress concentrations give rise to successive partial shears and Riedel shears.

ASHFORD COMMON REVISITED

The identification of a well defined post-rupture strength envelope for intact Todi Clay, together with the preliminary studies on kaolin, stimulated me to look for similar behaviour in London Clay. An obvious source of data was the classic work carried out jointly by the Building Research Station and Imperial College at Ashford Common in the early 1960s. The testing was carried out on 38 mm diameter specimens cut from block samples extracted from six different levels during the sinking of a 44 m deep shaft. The testing consisted of quick undrained triaxial compression tests and drained and undrained effective stress compression tests on vertical and horizontal specimens. Most of the effective stress testing

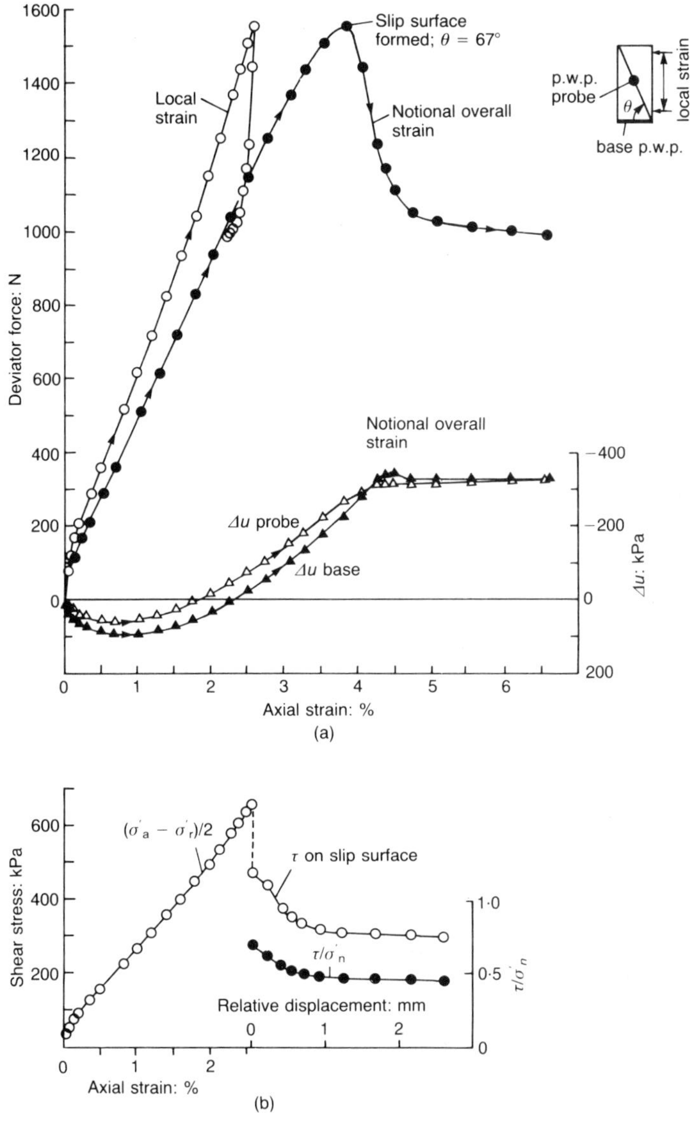

Fig. 43. Todi Clay: unconsolidated undrained triaxial test with pore pressure measurement showing post-rupture behaviour

was carried out on intact samples with those containing obvious fissures being rejected. The samples used for quick undrained testing were not selected in this way and included many containing fissures. The results have been published in two classic papers in *Géotechnique* (Ward, Marsland & Samuels, 1965; and Bishop, Webb & Lewin, 1965). Moreover, Webb's doctoral thesis contains most of the original data (Webb, 1964). These data were analysed by Wroth (1972) in his study of the elastic behaviour of overconsolidated clay. Table 2 lists the basic index properties together with the estimated in situ effective stresses. These properties have been obtained from a thorough re-analysis of all the data and differ slightly from those published by Bishop *et al.* (1965).

Intact strength

Figure 48 shows the Mohr–Coulomb failure envelopes for the intact clay at various depths. Tests were also carried out on isotropically consolidated reconstituted clay from level E giving the intrinsic failure line shown in the figure. At

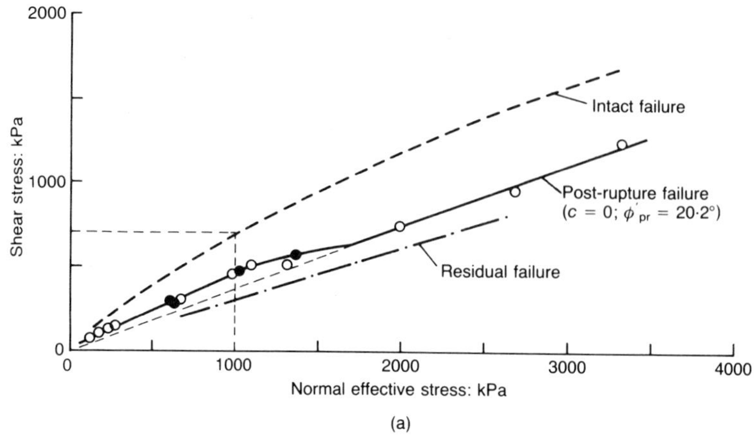

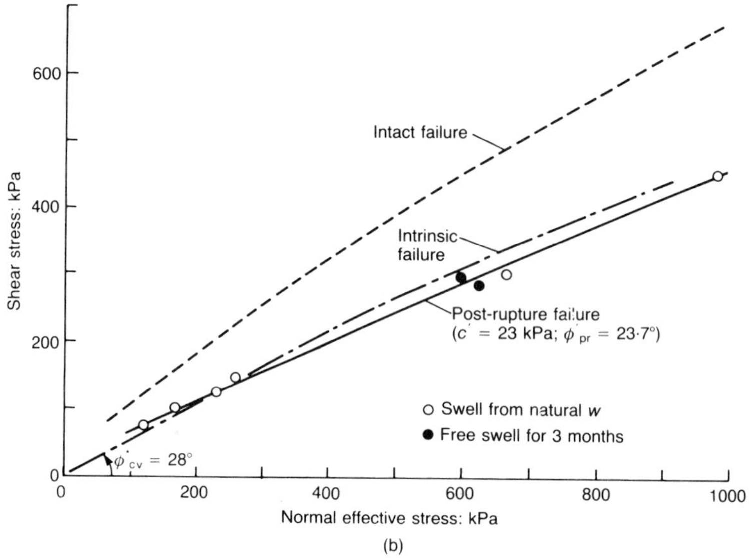

Fig. 44. Todi Clay: post-rupture failure envelope for (a) high pressures and (b) low to medium pressures compared with intact, intrinsic and residual failure lines

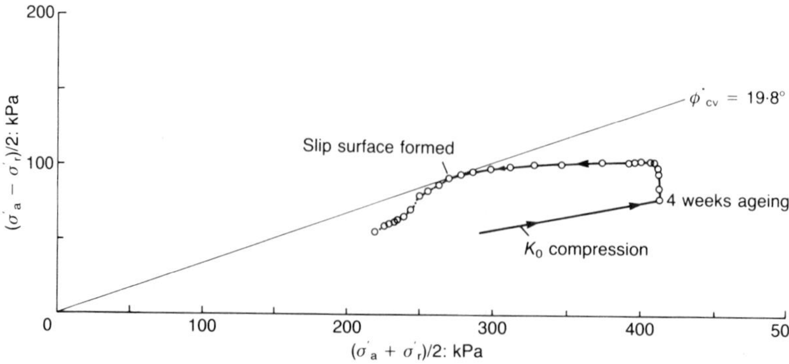

Fig. 45. Effective stress path for CAU triaxial compression test on normally consolidated aged kaolin (Ninis, 1990)

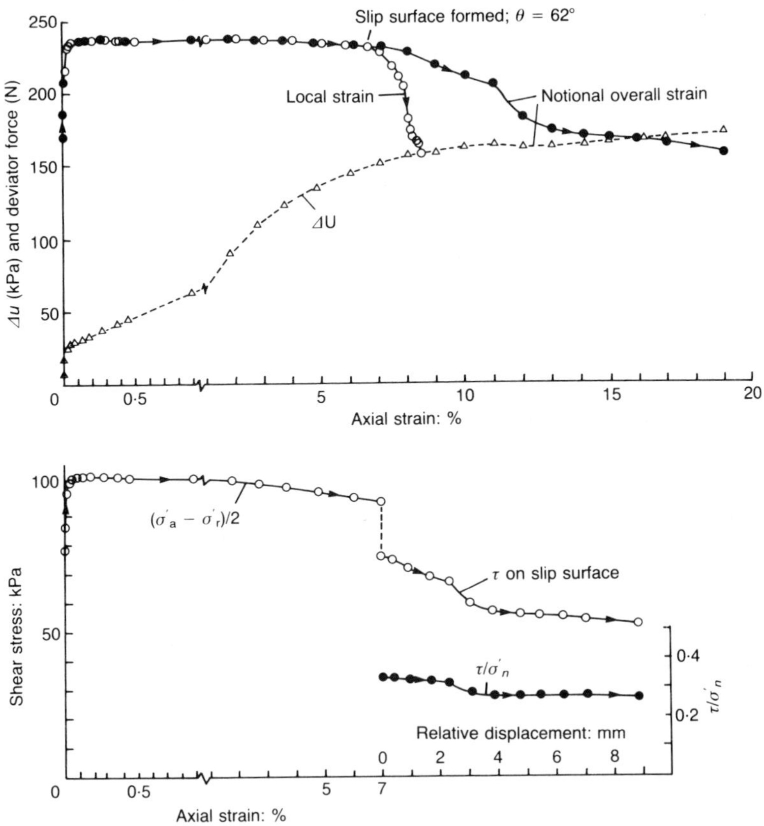

Fig. 46. Stress–strain behaviour for test given in Fig. 45 showing post-rupture behaviour

low stresses the value of $\phi_{cv}^* = 20{\cdot}1°$ and this decreases somewhat for effective stresses greater than 1000 kPa.

Figure 49(a) shows the relationship between void ratio and log p' for the samples from level E after swelling or compressing from the initial void ratio. The ICL and SCL are also shown. It can be seen that the isotropic compression curve crosses the ICL but the applied cell pressures were not sufficient to bring the clay to a state of normal consolidation. The chain dotted line represents the relationship between e and log σ_v' for a sample which was swelled isotropically to $p' = 69$ kPa and then compressed one-dimensionally in the triaxial apparatus. Fig. 49(b) shows the relationship between void ratio and the log of the maximum shear stress at failure for drained and undrained conditions. It can be seen that for shear strengths greater than about 1000 kPa the failure line is approximately parallel to the ICL. These data may be used to derive the value of σ_{ve}^* at failure for each test (remembering that σ_{ve}^* is the pressure on the ICL corresponding to the void ratio of the soil).

Table 2. Ashford Common—index properties and in situ effective stresses

Level	Depth: m	w_L: %	w_p: %	G_s	$<2\mu$: %	w_o: %	e_o	e_L	I_{vo}	σ_{vo}': kPa	p_k': kPa	K_o	σ_{ho}': kPa
A	9·1	58·9	23·8	2·74	42	22·59	0·619	1·614	−1·140	117	317	3·4	400
B	15·2	68·5	28·7	2·75	59	25·68	0·706	1·884	−1·076	179	373	2·6	469
C	20·1	70·6	28·9	2·77	53	24·82	0·688	1·956	−1.151	235	448	2·3	538
D	27·7	62·3	26·6	2·72	47	22·70	0·617	1·695	−1·191	310	524	2·0	621
E	34·8	70·0	27·0	2·77	57	23·89	0·662	1·939	−1·200	386	690	2·1	814
F	42·1	67·8	29·0	2·74	60	23·84	0·653	1·856	−1·184	455	759	2·0	911

Fig. 47. Thin section of kaolin after a CAU triaxial compression test showing slip surfaces

Figure 50 shows the intact failure surfaces for all the levels except B in normalized plots of $(\sigma_a' - \sigma_r')/2\sigma_{ve}^*$ against $(\sigma_a' + \sigma_r')/2\sigma_{ve}^*$. The following features should be noted.

(*a*) The failure surfaces for levels A to E are almost identical, suggesting that the small differences in strength in Fig. 48 are only due to differences in I_v. The failure surface for level F lies distinctly above the others indicating a structural difference, probably due to increased bonding.

(*b*) The failure surfaces lie above the intrinsic Hvorslev surface reflecting the stronger microstructure of the natural material.

(*c*) For values of $(\sigma_a' + \sigma_r')/2\sigma_{ve}^*$ greater than about 0·8 the failure surfaces bend over and become horizontal. If significantly larger effective confining pressures had been applied the surfaces would have curved downwards, eventually intersecting the isotropic stress axis. The full lines in Fig. 50 are, in fact, state boundary surfaces for the natural soil. They show very strikingly that, though for a given value of I_v the natural material can sustain somewhat larger shear stresses than the equivalent reconstituted soil, its structure will resist very much larger confining pressures than the reconstituted soil.

(*d*) The thin full lines and the broken lines are state paths for undrained and drained tests respectively. It can be seen that for values of $(\sigma_a' + \sigma_r')/2\sigma_{ve}^*$ less than about 1·3 the natural clay is dilatant with the state paths bending to the right just prior to failure. For $(\sigma_a' + \sigma_r')/2\sigma_{ve}^*$ greater than about 1·3 the clay becomes increasingly contractant.

(*e*) It must be emphasized that the state boundary surfaces in Fig. 50, are for samples that have been consolidated isotropically prior to shearing. Had the samples been consolidated one-dimensionally the state boundary surfaces would probably lie above the present ones, particularly for values of $(\sigma_a' + \sigma_r')/2\sigma_{ve}^*$ greater than about 1.

Figure 51 shows the failure points for drained and undrained tests on vertical and horizontal samples from level E. Note that the measured swelling pressure (p_k') is about 700 kPa. It can be seen that the failure line for the intact material is uniquely defined. Some typical undrained effective stress paths are shown. The significant difference in the initial slopes of the vertical and horizontal samples reflects the anisotropic stiffness properties of the clay with the horizontal stiffness being markedly greater than the vertical

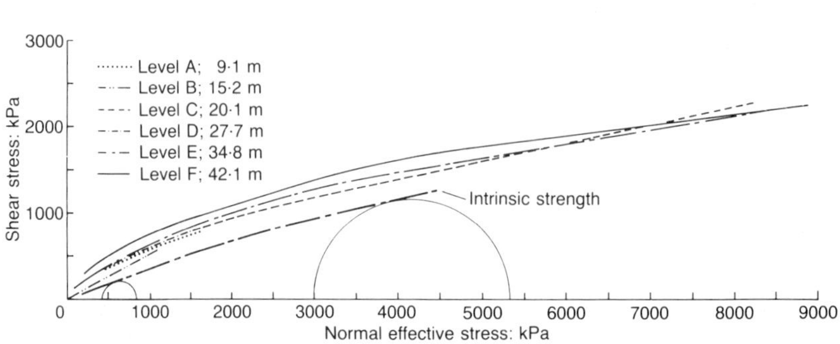

Fig. 48. London Clay at Ashford Common: Mohr–Coulomb failure envelopes for intact strength

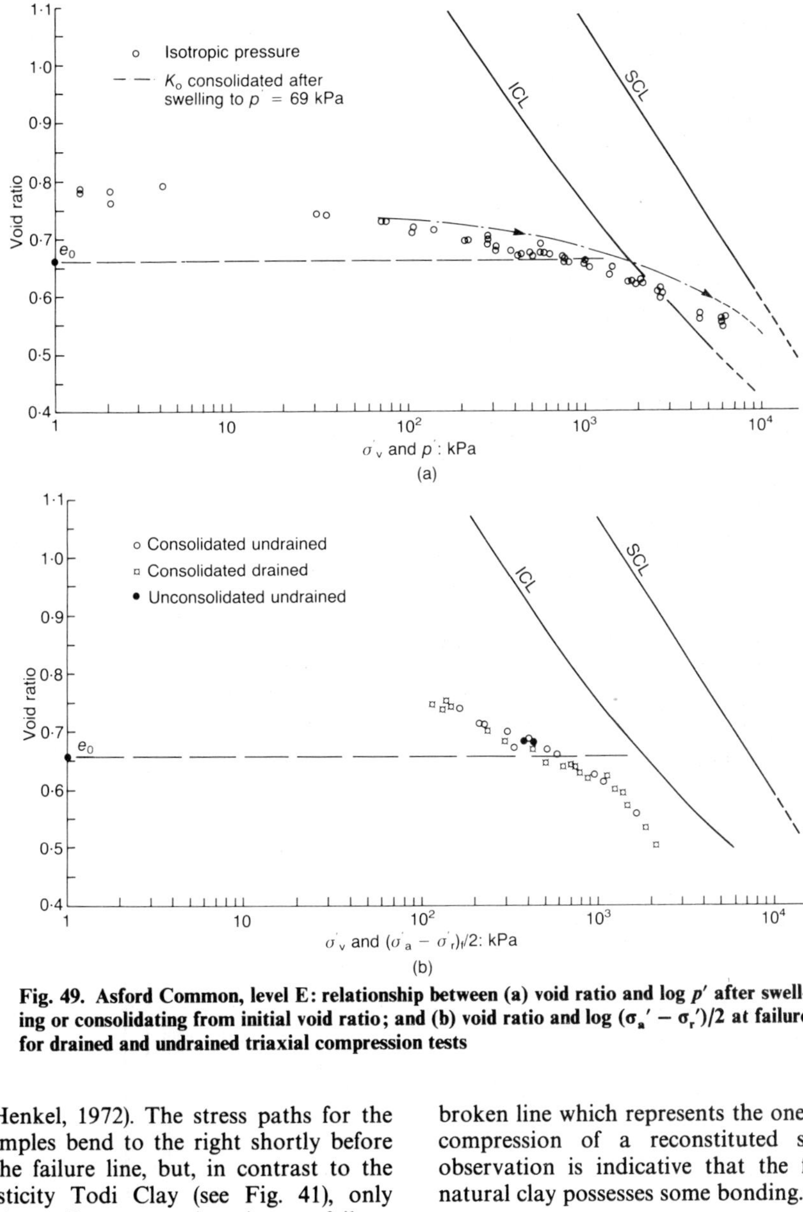

Fig. 49. Asford Common, level E: relationship between (a) void ratio and log p' after swelling or consolidating from initial void ratio; and (b) void ratio and log $(\sigma_a' - \sigma_r')/2$ at failure for drained and undrained triaxial compression tests

stiffness (Henkel, 1972). The stress paths for the vertical samples bend to the right shortly before reaching the failure line, but, in contrast to the lower plasticity Todi Clay (see Fig. 41), only travel a short distance up it prior to failure. Marsland (1977) has published the results of undrained tests on natural clays with a range of plasticities and these show very clearly that the stress paths for overconsolidated low plasticity clays tend to travel much further up the failure line before rupture than do medium to high plasticity clays.

The chain dotted line in Fig. 51 is the stress path followed by a sample which was first swelled isotropically to $p' = 69$ kPa and then compressed one-dimensionally in the triaxial apparatus. It can be seen that the stress path lies well above the broken line which represents the one-dimensional compression of a reconstituted sample. This observation is indicative that the fabric of the natural clay possesses some bonding.

Post-rupture strength

Most of the tests at low to moderately high confining pressure exhibited brittle behaviour with a well defined slip surface forming at peak strength (Bishop, Webb & Lewin, 1965). Fig. 52 shows the results for a typical undrained test and the general pattern of behaviour is strikingly similar to Todi Clay (Fig. 43) except that for the London Clay the excess pore pressures remain positive throughout the test. It can be seen from the bottom diagram that the shear stress on the

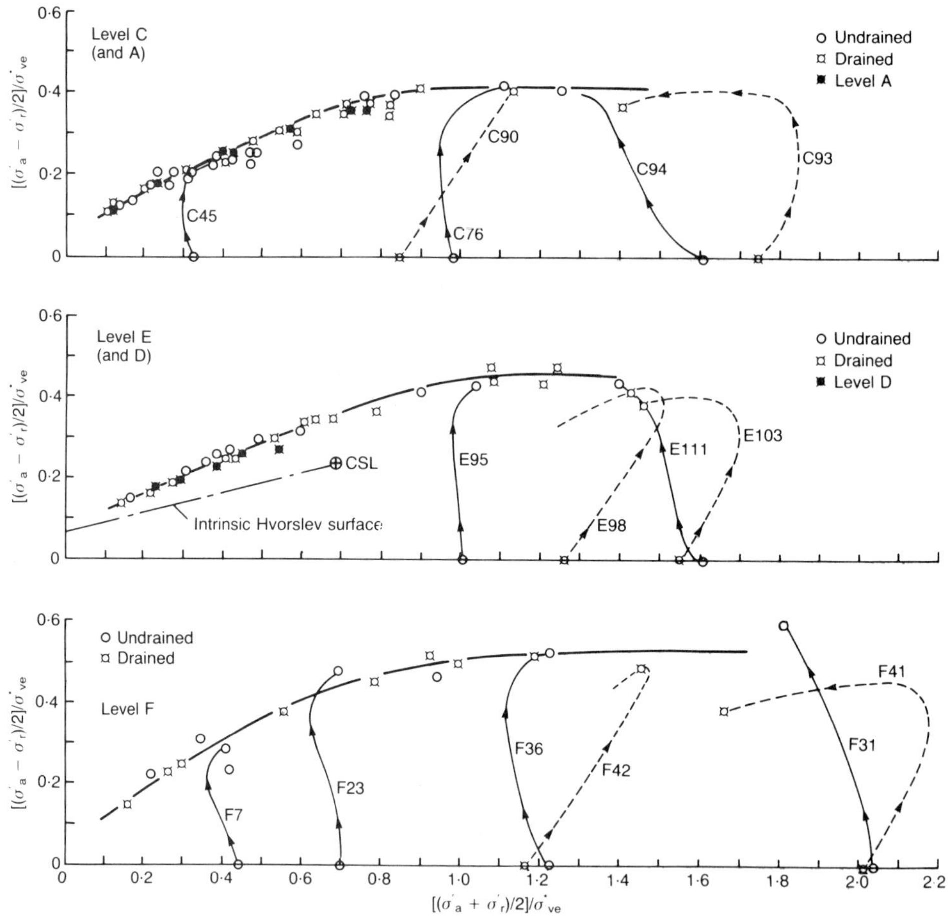

Fig. 50. Ashford Common: intact effective strength envelopes normalized by the equivalent pressure σ_{ve}* at failure

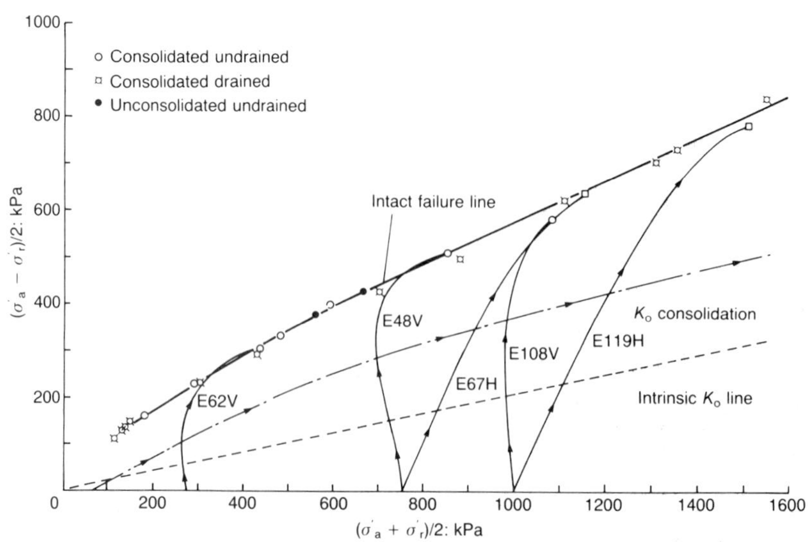

Fig. 51. Ashford Common, level E: results of consolidated drained and undrained triaxial compression tests on vertical and horizontal intact samples

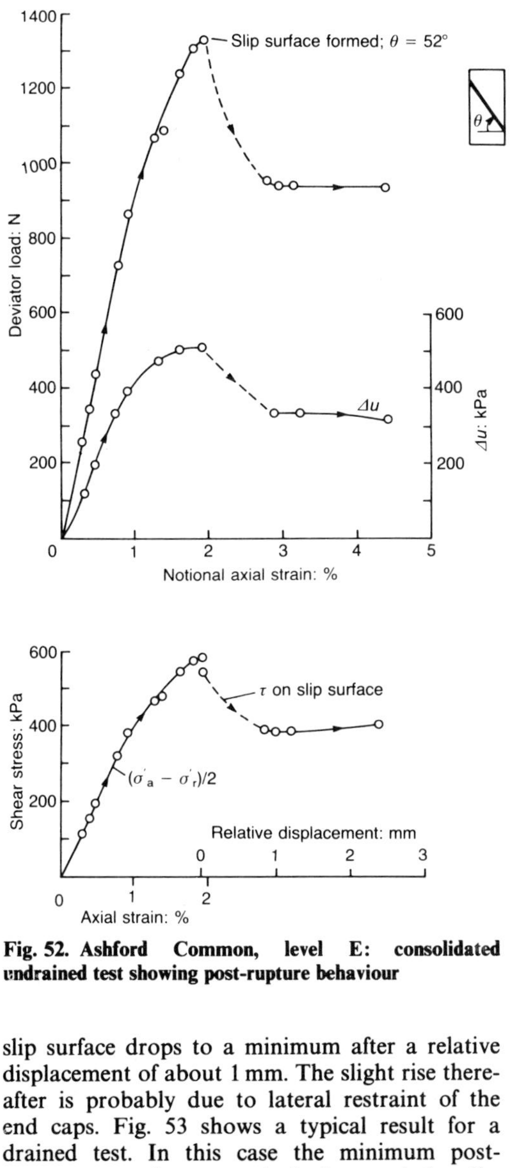

Fig. 52. Ashford Common, level E: consolidated undrained test showing post-rupture behaviour

slip surface drops to a minimum after a relative displacement of about 1 mm. The slight rise thereafter is probably due to lateral restraint of the end caps. Fig. 53 shows a typical result for a drained test. In this case the minimum post-rupture strength was reached after a relative displacement of about 3 mm. In general the overall notional strain between peak and post-rupture strength seldom exceeded 5%. Chandler (1966) has concluded that the membrane corrections are reliable up to strains of about 12%.

Figure 54(a) shows the results of all the post-rupture strength measurements for levels C and E over the full range of stresses. The full line represents the post-rupture failure envelope. This is similar in shape to that for Todi Clay (Fig. 44(a)) as it has an initially steep portion with a transition to a flatter envelope at a normal effective stress of about 2000 kPa. For high pressures the post-rupture failure line is defined by $c' = 0$, $\phi_{pr}' = 15{\cdot}2°$, a somewhat higher value than the residual angle of friction which is about 12°. The chain dotted line in Fig. 54(a) is the intrinsic failure line for the reconstituted soil. It lies below the post-rupture failure line at low stresses and above it at higher stresses. Fig. 54(b) shows a more detailed comparison between the post-rupture strength and the intrinsic strength envelopes at low and intermediate stresses. Again the general picture is strikingly similar to Todi Clay. At low stresses the intrinsic failure line is defined by $c' = 0$; $\phi_{cv}{}^* = 20{\cdot}1°$. Initially the post-rupture failure line has a slightly higher angle of friction and a cohesion intercept of about 10 kPa. However, it bends over and drops below the intrinsic failure line at about 750 kPa.

As mentioned previously, almost all the samples used for effective stress testing were initially intact. However, Webb (1964) noted that a few specimens failed on obvious pre-existing fissures (marked F). In Fig. 54(b) it can be seen that two of the samples containing fissures have values of ϕ_{pr}' close to the high pressure value of 15·2°. It is of interest to note that Skempton *et al.* (1969) obtained a post-peak angle of friction of 16° for the strength of fissures and joints in the London Clay at Wraysbury.

In the original Ashford Common publication Bishop *et al.* (1965) referred to what I have termed the post-rupture strength as the residual strength. Similarly Skempton *et al.* (1969) referred to the post-peak strength on fissures and joints as the residual strength. Subsequently attention shifted from the immediate post-peak strengths to ultimate values after large displacements. It is now generally agreed that the term residual strength refers to the ultimate steady state condition, usually after large displacements. The post-rupture strengths I refer to here are therefore not residual values. They may, however, be relevant to many stability problems such as bearing capacity, first-time slides in excavated slopes and retaining walls.

In-situ stresses

Bishop *et al.* (1965) used the laboratory measurements of the swelling pressure p_k' to estimate the in situ horizontal effective stresses at the various levels at Ashford Common (see Table 2). Figs 55(a) and (b) show the resulting Mohr's circles of in situ effective stress for level A and levels B to F respectively. Also shown in these figures are the corresponding post-rupture failure lines. (Note that the post-rupture failure line for level A lies above that for the other levels and this is consistent with its lower plasticity).

It is important to bear in mind that a number of assumptions are involved in deriving the in situ

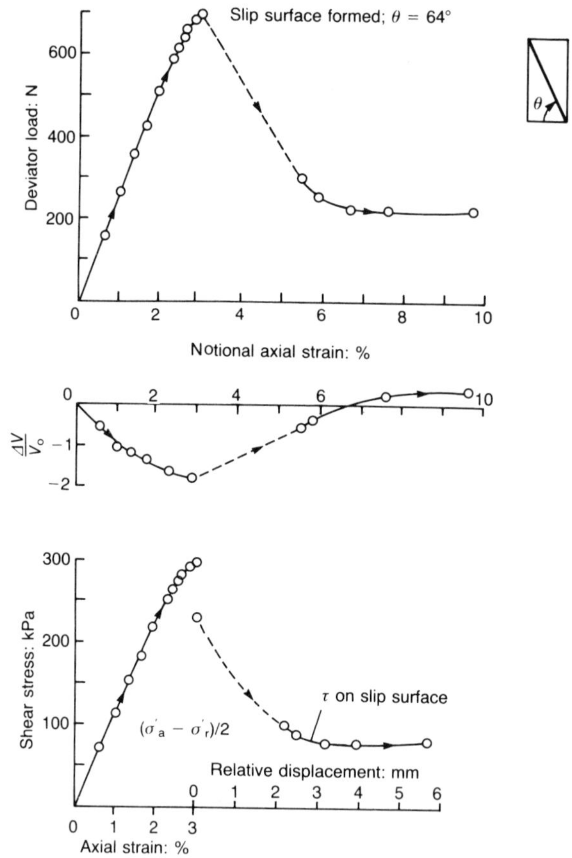

Fig. 53. Ashford Common, level E: consolidated drained test showing post-rupture behaviour

values of σ_{ho}'. Nevertheless, on the basis of the evidence given in Fig. 55, it is plausible that the post-rupture strength limits the magnitude of the horizontal effective stresses in a heavily over-consolidated fissured clay since the fissures themselves probably result from the brittle nature of the intact material.

Quick undrained tests

It is now widely accepted that the undrained strength of a stiff fissured clay is primarily a function of the volume of soil being sheared and that the presence of fissures and joints play a major role in this. At Ashford Common a large number of quick undrained triaxial tests were carried out at Imperial College and at the Building Research Station. The results of this work have not so far been properly integrated with the effective stress testing. For some of the tests it was noted that failure appeared to take place prematurely on one or more pre-existing fissures. The inclination of these failure planes was carefully measured.

Figure 56 shows histograms of the unconsolidated undrained strength of vertical samples at the six levels. The black histograms refer to tests in which failure was known to take place on an obvious fissure. Some of the low results not shown in black may also have resulted from the presence of less obvious fissures. Although the scatter is large, the results at any level can be broadly divided into two groups: those obviously affected by fissures near the lower limit of the range and those for which the samples were more or less intact giving higher strengths. An important question is: how do the quick undrained strengths affected by fissures relate to the post-peak and fissured effective strengths given in Fig. 54?

As mentioned previously a few of the undrained effective stress tests carried out by Webb (1964) showed premature failure on pre-existing fissures. Fig. 57 shows a comparison between the behaviour of two such samples (C62 and C65) with an intact sample (C50). The initial portions of the stress paths are similar in slope

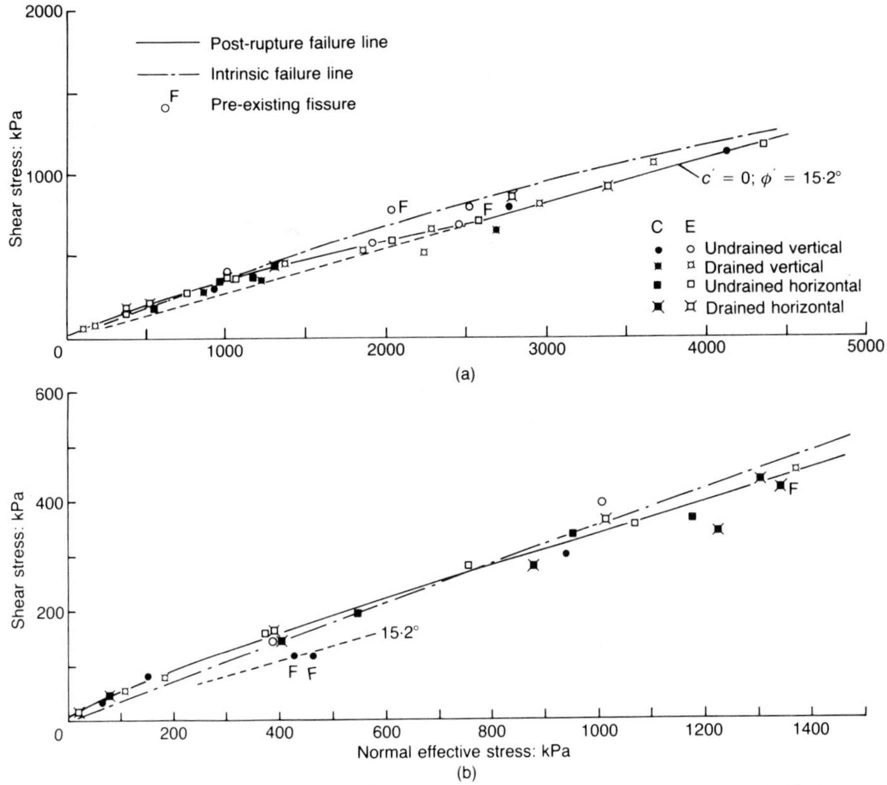

Fig. 54. Ashford Common: post-rupture failure envelopes for (a) high pressures and (b) low to medium pressures compared with the intrinsic failure line

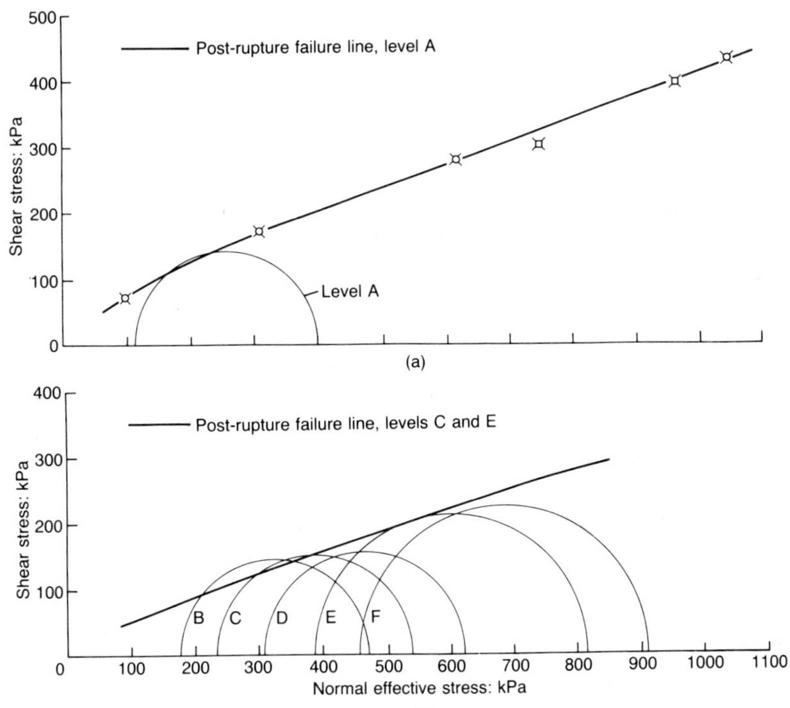

Fig. 55. Ashford Common: in situ Mohr's circles of effective stress superimposed on the post-rupture failure envelopes for (a) level A and (b) all other levels

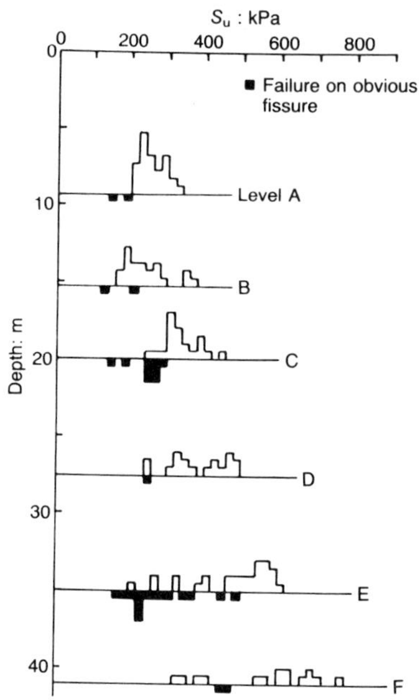

Fig. 56. Ashford Common: histograms of undrained strength from quick unconsolidated undrained triaxial compression tests on vertical samples

but premature failure on a fissure truncates the stress path and in particular eliminates most or all of the dilatant portion.

These observations assist in the analysis of the standard quick undrained tests as shown in Fig. 58. The average value of the swelling pressure p_k' is known for each level. The slope of the stress path is related to the pore pressure parameter A. Average initial values of A for the tests on the vertical and the horizontal samples are 0·67 and 0·29 respectively. The values of τ_{max} and θ for each of the tests on fissured samples are known. From the geometry of the problem it can be shown that the shear stress on the plane of the fissure at failure is

$$\tau_f = \tau_{max} \sin 2\theta \quad (7)$$

and the corresponding normal effective stress is

$$\sigma_{nf}' = \tau_{max}(1 - 2A + \cos 2\theta) + p_k' \quad (8)$$

Equations (7) and (8) have been used to calculate the values of τ_f and σ_{nf}' for all the quick undrained tests on vertical and horizontal samples which obviously failed on pre-existing fissures. The results are plotted in Fig. 58. It must be emphasized that the individual values of p_k' and A are not known, only the average values at each level. Hence some of the points may be significantly in error. Nevertheless it is clear that the broken line for $\phi' = 15{\cdot}2°$ (taken from Fig. 54) forms a reasonable lower bound to the data. The post-rupture failure line for initially intact specimens and the intrinsic failure line are also shown. Up to normal effective stresses of about 600 kPa the experimental points lie on either side of these lines. At higher stresses, particularly for level E, the experimental points tend to lie below these lines.

It can be concluded that the strengths from the quick undrained tests on samples containing fissures are consistent with the strength envelopes established from effective stress tests on samples in which slip surfaces have formed. Rate effects seem to be less important than they would be for intact strength.

OPERATIONAL STRENGTH OF STIFF FISSURED CLAYS

Figure 59 shows the well known results obtained by Marsland (1974) for London Clay at

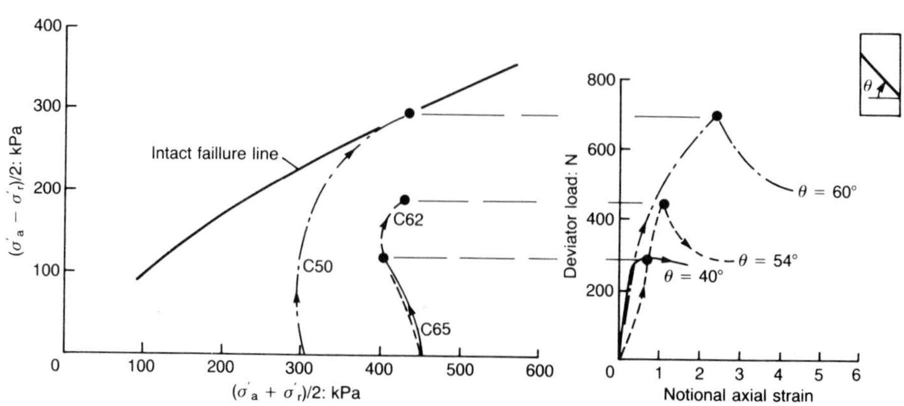

Fig. 57. Ashford Common, level C: results of two consolidated undrained tests which failed prematurely on pre-existing fissure compared with test on intact sample

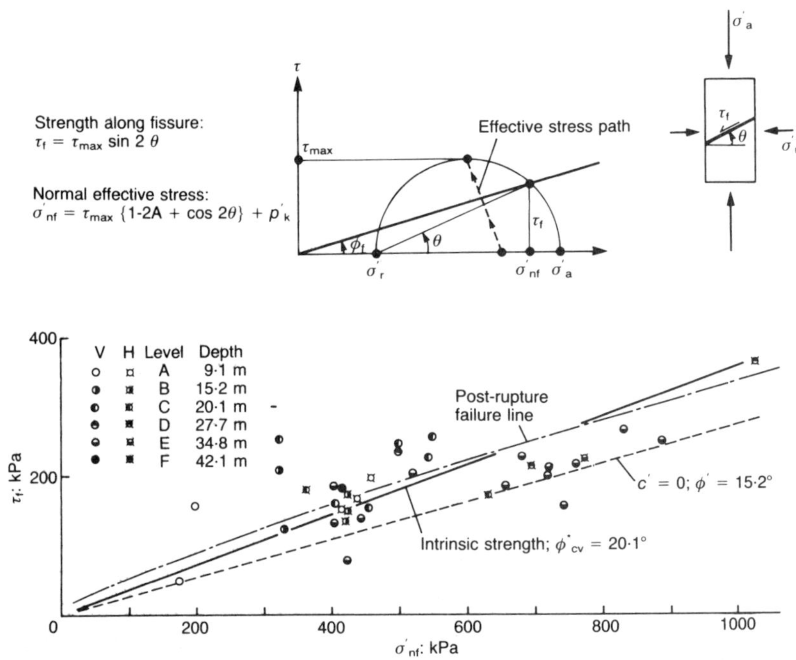

Fig. 58. Ashford Common: analysis of peak undrained strengths from quick unconsolidated undrained triaxial compression tests on vertical and horizontal samples containing pre-existing fissures of known inclination θ

Hendon in which he compared the operational undrained strengths back-analysed from large diameter in situ plate loading tests with the undrained strengths from two sizes of sample: 98 mm and 38 mm diameter. The samples were obtained using a thin wall push sampler. The operational strengths deduced from the plate loading tests show little variability and lie near to the lower limit of the scatter for both sizes of sample.

There is a striking similarity between Fig. 59, comparing the large-scale operational strength with quick undrained tests, and Fig. 56 comparing the strength of samples with fissures with those mostly without fissures. It follows that the operational effective strengths relevant to undrained bearing capacity probably lie between the post-rupture failure line and the lower limit fissure failure line shown in Fig. 58.

Does it follow that the operational strength of a clay mass is controlled entirely by the fissures? Work by Tarzi *et al.* (1982) and, more recently, by Chan & Morgenstern (1989) shows that strain-softening substantially reduces the bearing capacity of a footing. As the rate of strain-softening increases the bearing capacity tends towards that predicted by the ultimate strength. Even quite moderate rates of strain-softening reduce the bearing capacity significantly, and where strain-softening manifests as localized slip surfaces (as is the case with intact Todi Clay and London Clay) it might be anticipated that the effect is even more pronounced. Hence analysis indicates that the post-rupture strength forms a very reasonable lower bound for the bearing capacity of a footing even when the material is initially intact. It can therefore be concluded that the operational strengths deduced from the large diameter plate loading tests are controlled by two dominant factors: (i) the presence of fissures and the strength along them and (ii) strain-softening due to brittleness of the intact material. A similar combination of fissuring and brittleness may well control first-time slides in stiff fissured clays. Skempton (1977) has concluded that the operational strength parameters relevant to first-time slides in brown London Clay are given by $c' = 1$ kPa; $\phi' = 20°$. It may not be entirely coincidental that this failure envelope lies between the lower bound envelope for the fissured strength and the post-rupture strength envelope for intact samples (see Figs 54(b) and 58). So of course does the intrinsic strength from tests on reconstituted samples. Chandler (1984) has commented that the agreement with the ϕ_{cv}* value is surprising and may be fortuitous. Further study of this question is required and the concept of post-rupture strength offers a way forward.

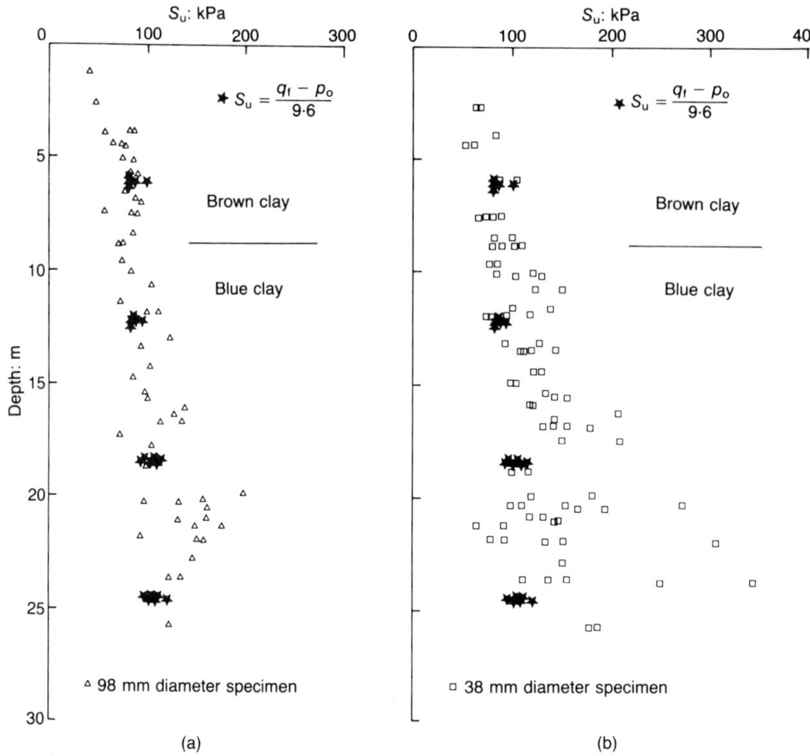

Fig. 59. London Clay, Hendon: comparison between operational undrained strengths back-analysed from 865 mm dia. plate loading tests and peak strengths from quick unconsolidated undrained triaxial compression tests on (a) 98 mm dia. specimens and (b) 38 mm dia. specimens (Marsland, 1974)

DISCUSSION AND CONCLUSIONS

This lecture has demonstrated the value of using the compressibility and strength characteristics of young reconstituted clays as a framework for interpreting the corresponding properties of natural clays. A reconstituted clay is one that has been thoroughly mixed at between 1 and 1·5 times the liquid limit and preferably consolidated one-dimensionally. The properties of such a clay are termed 'intrinsic properties' since these are inherent to the soil and independent of its natural state. An intrinsic property is denoted by an asterisk. Examples of intrinsic parameters are e^*_{100}, C_c^* and C_s^* for compression and swelling and ϕ_{cv}^* for the intrinsic critical state angle of shearing resistance. The intrinsic Hvorslev strength parameters ϕ_e^*, $\mathcal{K}^*$ and σ_{ve}^* are relevant to the study of overconsolidated clays. Numerous other intrinsic properties could of course be measured including those relating to permeability.

In the past insufficient distinction has been made between intrinsic properties and the properties of natural undisturbed soils. Natural soils differ from the corresponding reconstituted soil both with respect to fabric and bonding (both of these constituting the soil structure). The structure of a natural clay depends on many factors such as depositional conditions, ageing, cementation and leaching. These structural features profoundly affect the mechanical properties of the natural material. One objective of this Lecture has been to show that the intrinsic properties of a natural clay provide a robust frame of reference against which to assess the in situ state of the soil, its structure and the measured mechanical properties of undisturbed samples.

It has been demonstrated that the intrinsic compression line (ICL) is a valuable reference line for studying the compression characteristics of natural normally and overconsolidated sedimentary clays. The ICL is defined by the two constants of intrinsic compressibility e^*_{100} and C_c^* (see Fig. 3(a)). Provided the Atterberg limits lie above the A line, there is a good correlation between these constants of intrinsic compressibility and the void ratio at the liquid limit e_L as shown in Fig. 8. These correlations have proved useful when e^*_{100} and C_c^* have not been directly determined. It is recommended that, whenever possible, the constants should be determined experimentally. To take account of small variations in liquid limit within a given clay stratum the values of e^*_{100} and $C_c{}^*$ may be assumed to vary in direct proportion to the liquid limit. Thus it would not normally be necessary to carry out a large number of determinations of e^*_{100} and $C_c{}^*$ down a given profile.

The effects of variations in soil type, as reflected in the values of e^*_{100} and $C_c{}^*$, may largely be eliminated by replacing the void ratio e with a normalizing parameter I_v (void index) defined by equation (1) in terms of the two constants of intrinsic compressibility. The ICL forms an almost unique line in a plot of I_v against log $\sigma_v{}'$ as shown in Fig. 4. This plot has proved useful for comparing sedimentation compression curves for various soil profiles and for studying the one-dimensional compression characteristics of natural clays in a unified way.

The majority of normally consolidated natural clays have sedimentation compression curves which, when expressed in terms of I_v, lie within a narrow band well above the ICL (see Fig. 13). The regression line through this band has been termed the sedimentation compression line (SCL). Not all natural clays lie on the SCL reflecting differences in depositional and post-depositional environments. Moreover the SCL for most soils is not a smooth curve and is often 'saw-toothed', again reflecting temporal variations in depositional conditions. Thus the SCL is not a fundamental line but is nevertheless useful since it represents a norm for the majority of natural sedimentary clays.

The location of the SCL to the right of the ICL shown in Fig. 13 implies that, for a given value of I_{vo}, the effective overburden pressure carried by the natural clay is approximately five times that carried by the equivalent reconstituted clay. This is a measure of the enhanced resistance of the structure of most natural clays. For quick clays and cemented clays the enhanced resistance is many times larger than the above figure.

It has been shown that, for clays whose natural state lies above the ICL, the one-dimensional compression curve is usually significantly steeper than the ICL and tends to converge with it at high pressures (e.g. Fig. 20). This behaviour results from the progressive collapse of the natural soil structure. However, for clays whose mode of deposition is such that its in situ state lies close to the ICL, the one-dimensional compression curve will tend to lie parallel to the ICL since the structure of the natural clay is similar to that of the reconstituted material.

For overconsolidated clays the ICL and SCL provide a useful means of assessing the degree of overconsolidation of a natural clay particularly when the yield pressure $\sigma_{vy}{}'$ is not well defined. Also the ratio of the intrinsic swelling index to the natural swelling index $C_s{}^*/C_s$ (the swell sensitivity) provides an important measure of bonding in the natural soil.

The critical state framework provides a coherent model of the behaviour of reconstituted soils in terms of void index, shear stress and direct effective stress. This framework has been shown to explain qualitatively why normally consolidated natural clays lying above the ICL are more brittle and sensitive than reconstituted soils. For these clays the SHANSEP test procedure is not appropriate. It appears that, when sheared in the triaxial apparatus, most natural clays do not reach the intrinsic critical state. Much more vigorous shearing is evidently required to break down the natural structure of the clay.

In its present form the critical state framework cannot be used to predict peak undrained strength S_u of normally consolidated clays. S_u depends primarily on the structure of the clay and the in situ effective stresses and not on the void ratio or void index. It has been shown that for undisturbed natural sensitive clays $S_{uTC}/\sigma_{vy}{}' \simeq 0{\cdot}3$, although for quick clays this ratio may be somewhat lower. The yield stress $\sigma_{vy}{}'$ is a measure of the yield properties, or yield locus, of the clay.

The intact strength properties of two heavily overconsolidated undisturbed clays have been studied: a low plasticity clay from Todi, Italy, and high plasticity London Clay from Ashford Common. For both clays the intact failure surfaces lie above the intrinsic Hvorslev surfaces clearly demonstrating the enhanced strength of the natural microstructure. In the case of the Todi Clay a prolonged period of free swell does not entirely eliminate this enhanced strength.

Both these clays exhibit brittle behaviour at low and intermediate stresses with the formation of shear surfaces at peak intact strength. The strength on a shear surface drops rapidly to a reasonably steady value after only a few millimeters' relative displacement. This is termed the post-rupture strength and should be clearly distinguished from the residual strength which is reached after very much larger relative displacements.

For Todi Clay and London Clay the post-rupture failure envelopes and intrinsic critical state failure envelopes lie close together at low stresses, but at higher stresses the post-rupture strengths are less than the intrinsic critical state strengths. Further work is required to investigate the phenomenon of post-rupture strength in other intact materials and to carry out comparisons with the intrinsic critical state strength. A preliminary study on normally consolidated kaolin gives the post-rupture angle of shearing resistance ϕ_{pr}' somewhat less than ϕ_{cv}*.

In addition to studying the strengths of initially intact samples, the results of tests on samples containing existing fissures have also been examined. The results of many of the tests give strengths on the fissures close to the post-rupture strength of initially intact specimens. However, some strengths are somewhat lower and a well defined lower limit to the fissured strength has been identified for the London Clay.

It is suggested that the post-rupture strength may be relevant to many stability problems in stiff clays. For example the in situ effective stresses at Ashford Common deduced by Bishop *et al.* (1965) are consistent with the mobilization of post-rupture strengths during the process of geological unloading and the formation of fissures. It is also demonstrated that the operational undrained strength of a stiff fissured clay *en masse* is consistent with the post-rupture strength and probably results from a combination of the presence of fissures and progressive failure due to the brittle nature of the intact material. Finally the operational effective strength envelope for first time slides in brown London Clay deduced by Skempton (1977) lies between the lower bound envelope for the strength on fissures and the post-rupture strength for initially intact samples.

ACKNOWLEDGEMENTS

I would like to express my gratitude to the British Geotechnical Society for inviting me to deliver the thirtieth Rankine Lecture, which I regard as one of the highest honours not only in soil mechanics but also in civil engineering. Also I wish to thank Mr Thorburn for his kind and generous remarks.

I am grateful to the following for permission to publish experimental data in this Lecture: Statoil for the results from the Troll field; Rendel, Palmer and Tritton for the results for Surabaya; the Norwegian Geotechnical Institute for the results from the three sensitive clay sites in Norway and the Building Research Establishment for the results on Gault Clay. Dr B. McClelland made available the results of oedometer tests from the Mississippi delta, Mr D. Nash provided the oedometer test results and other soils data from Bothkennar, and Professor A. Nakaşe provided the data for Fig. 9 on the marine clays from Japan. I am indebted to Professor G. Calabresi and his colleagues for allowing the publication of the results on Todi Clay and for many stimulating discussions. Professor A. W. Skempton kindly made available his files for the data in Fig. 1 as did Dr P. I. Lewin for the data for levels B, D and F from Ashford Common. During the study described here I have benefited from discussions with many colleagues and friends too numerous to list. I am particularly grateful to Professor G. A. Leonards, Professor A. W. Skempton, Professor P. R. Vaughan, Dr R. J. Chandler, Dr R. J. Jardine, Dr D. W. Hight and Dr Suzanne Lacasse, I also wish to acknowledge the support and encouragement of all my colleagues in the Soil Mechanics Section at Imperial College.

REFERENCES

Atkinson, J. H. & Bransby, P. L. (1978). *The mechanics of soils. An introduction to critical state soil mechanics.* London: McGraw-Hill.

Bishop, A. W., Webb, D. L. & Lewin, P. I. (1965). Undisturbed samples of London Clay from the Ashford Common shaft: strength–effective stress relationship. *Géotechnique* **15**, No. 1, 1–31.

Bjerrum, L. (1967) The seventh Rankine Lecture. Engineering geology of Norwegian normally-consolidated marine clays as related to settlements of buildings. *Géotechnique* **17**, No. 2, 81–118.

Bjerrum, L. & Rosenqvist, I. Th. (1956). Some experiments with artificially sedimented clays. *Géotechnique* **6**, No. 3, 124–136.

Burland, J. B. (1989). The ninth Bjerrum Memorial Lecture: 'Small is beautiful'—the stiffness of soils at small strains. *Can. Geotech. J.* **26**, 499–516.

Burland, J. B. & Symes, M. (1982). A simple axial displacement gauge for use in the triaxial apparatus. *Géotechnique* **32**, No. 1, 62–65.

Calabresi, G. (1990). Private communication.

Calabresi, G. & Scarpelli, G. (1985). Effects of swelling caused by unloading in overconsolidated clays. *Proc. 11th Int. Conf. on Soil Mech.*, San Francisco, **2**, 411–415.

Casagrande, A. (1932). The structure of clay and its importance in foundation engineering. *J. Boston Soc. Civ. Engrs.* **19**, No. 4, 168–209.

Chan, D. H. & Morgenstern, N. R. (1989). *Bearing capacity of strain softening soil.* De Mello volume, pp. 59–68. Editora Edgard Blucher Ltda.

Chandler, R. J. (1966). The measurement of residual strength in triaxial compression. *Géotechnique* **16**, No. 3, 181–186.

Chandler, R. J. (1984). Recent European experience of landslides in over-consolidated clays and soft rocks. State-of-the-art report. *4th Int. Symp. on Landslides*, Toronto, **1**, 61–81.

Edge, M. J. & Sills, G. C. (1989). The deve opment of layered sediment beds in the laboratory as an illustration of possible field processes. *Q. J. Engng. Geol.* **22**, Part 4, 271–279.

Gens, A. (1982). *Stress–strain and strength characteristics of a low plasticity clay.* PhD thesis, University of London.

Georgiannou, V. N. (1988). *The behaviour of clayey sands under monotonic and cyclic loading.* PhD thesis, University of London.

Hawkins, A. B., Larnach, W. J., Lloyd, I. M. & Nash, D.

F. T. (1989). Selecting the location, and the initial investigation of the SERC soft clay test bed site. *Q. J. Engng. Geol.* **22,** Part 4, 281–316.

Henkel, D. J. (1972). The relevance of laboratory measured parameters in field studies. *Proc. Roscoe Memorial Symp.*, Foulis, pp. 669–675.

Hight, D. W. (1982). A simple piezometer probe for the routine measurement of pore pressure in triaxial tests on saturated soils. *Géotechnique* **32,** No. 4, 396–401.

Hight, D. W., Jardine, R. J. & Gens, A. (1987). *The behaviour of soft clays.* chapter 2, Embankment on soft clays, pp. 33–158. Athens: Public Works Research Centre of Greece.

Horseman, S. T., Winter, M. G. & Entwistle, D. C. (1987). *Geotechnical characterisation of Boom Clay in relation to disposal of radioactive waste.* Luxembourg: Office for Official Publications of the European Communities.

Hvorslev, M. J. (1937). Uber die Festigkeitseigenschaften Gestorter Bindiger Boden. Danmarks Naturvidenskabelige Samfund. *Ingeniorvidenskabelige Skrifter*, A, No. 45.

Jardine, R. J. (1985). *Investigation of pile–soil behaviour with special reference to the foundations of offshore structures.* PhD thesis, University of London.

Jardine, R. J., Symes, M. J. & Burland, J. B. (1984). The measurement of soil stiffness in the triaxial apparatus. *Géotechnique* **34,** No. 3, 323–340.

Lacasse, S., Berre, T. & Lefebvre, G. (1985). Block sampling of sensitive clays. *Proc. 11th Int. Conf. on Soil Mech.*, San Francisco, **2,** 887–892.

La Rochelle, P., Sarrailh, J., Tavenas, F., Roy, M. & Leroueil, S. (1981). Causes of sampling disturbance and design of a new sampler for sensitive soils. *Can. Geotech. J.* **18,** No. 1, 52–66.

Leonards, G. A. & Ramiah, B. K. (1959). *Time effects in the consolidation of clay.* ASTM Special technical publication No. 254, pp. 116–130. Philadelphia: ASTM.

Leonards, G. A. & Girault, P. (1961). A study of the one-dimensional consolidation test. *Proc. 5th Int. Conf. on Soil Mech*, Paris, **1,** 213–219.

Leonards, G. A. & Altschaeffl, A. G. (1964). Compressibility of Clay. *Soil Mech. Div., Proc. Am. Soc. Civ. Engrs* **90,** 133–155.

Leroueil, S., Tavenas, F., Brucy, F., La Rochelle, P. & Roy, M. (1979). Behaviour of destructured natural clays. *Proc. Am. Soc. Civ. Engrs* **105,** GT6, 759–778.

Leroueil, S., Tavenas, F. & Locat, J. (1985). Discussion: Correlations between index tests and the properties of remoulded clays. W. D. Carrier III and J. F. Beckman. *Géotechnique* **35,** No. 2, 223–226.

Leroueil, S. & Vaughan, P. R. (1990). The important and congruent effects of structure in natural soils and weak rocks. *Géotechnique* **40,** No. 3.

Locat, J. & Lefebvre, G. (1986). The origin of structuration of the Grande-Baleine marine sediments, Quebec. Canada. *Q. J. Engng. Geol.* **19,** Part 4, 365–374.

Marsland, A. (1974). Comparison of the results from static penetration tests and large in-situ plate tests in London Clay. *Proc. European Symp. Penetration Testing*, Stockholm.

Marsland, A. (1977). The evolution of the engineering design parameters for glacial clays. *Q. J. Engng. Geol.* **10,** Part 1, 1–26.

McClelland, B. (1967). Progress of consolidation in delta front and prodelta clays of the Mississippi River. In A. F. Richards (Ed). *Marine Géotechnique*, pp. 22–40. Urbana: University of Illinois Press.

Mitchell, J. K. (1976). *Fundamentals of soil behaviour.* New York: Wiley.

Nagaraj, T. S. & Srinivasa Murthy, B. R. (1986). A critical reappraisal of compression index equations. *Géotechnique* **36,** No. 1, 27–32.

Nakase, A., Kamei, T. & Kusakabe, O. (1988). Constitutive parameters estimated by plasticity index. *J. Geotech. Engng. Div., Am. Soc. Civ. Engrs* **114,** No. 7, 844–858.

Newland, P. L. & Allely, B. H. (1956). Results of some investigations of two sensitive clays. *Proc. 2nd Aust.–NZ Conf. Soil Mech.*, 39–45.

Ninis, N. (1990). Private communication.

Northey, R. D. (1956). Rapid consolidation tests for routine investigations. *Proc. 2nd Aust.–NZ Conf. Soil Mech.*, 20–26.

Ramiah, B. K. (1959). *Time effects in the consolidation properties of clays.* PhD thesis, Purdue University.

Rampello, S. (1989). *Effetti del rigonfiamento sul comportamento meccanico di argille fortemente sovraconsolidate.* Doctoral thesis, University of Rome.

Rendulic, L. (1937). Ein Grundgesetz der Tonmechanik und sein Experementeller Beweis. *Der Bauingeneur* **18.**

Samuels, S. G. (1975). Some properties of the Gault Clay from the Ely–Ouse Essex water tunnel. *Géotechnique* **25,** No. 2, 239–264.

Schmertmann, J. H. (1969). Swell sensitivity. *Géotechnique* **19,** No. 4, 530–533.

Sejrup, H. P., Nagy, J. & Brigham-Grette, J. (1989). Foraminiferal stratigraphy and amino acid geochronology of Quaternary sediments in the Norwegian Channel, northern North Sea. *Norsk Geologisk Tidsskrift* **69,** No. 2, 111–124.

Skempton, A. W. (1944). Notes on the compressibility of clays. *Q. J. Geol. Soc.*, **100,** 119–135.

Skempton, A. W. (1970). The consolidation of clays by gravitational compaction. *Q. J. Geol. Soc.* **125,** 373–411.

Skempton, A. W. (1977). Slope stability of cuttings in brown London Clay. *Proc. 9th Int. Conf. on Soil Mech.*, Tokyo, **3,** 261–270.

Skempton, A. W. & Henkel, D. J. (1953). The postglacial clays of the Thames Estuary at Tilbury and Shellhaven. *Proc. 3rd Int. Conf. Soil Mech.*, Zurich, **1,** 302–308.

Skempton, A. W., Schuster, R. L. & Petley, D. J. (1969). Joints and fissures in London Clay at Wraysbury and Edgware. *Géotechnique* **19,** No. 2, 205–217.

Smith, P. R. (1990). Personal communication.

Som, N. N. (1968). *The effects of stress path on the deformation and consolidation of London Clay.* PhD thesis, University of London.

Tarzi, A. I., Kalteziotis, N. A. & Menzies, B. K. (1982). Finite element analysis of strip footings on strain-softening clay. *Proc. Int. Symp. on Num. Models in Geomechanics*, Rotterdam, 194–200.

Tavenas, F. & Leroueil, S. (1985). Discussion on Theme

Lecture 2. *Proc. 11th Int. Conf. on Soil Mech.*, San Francisco, **5**, 2693–2694.

Terzaghi, K. (1925). *Erdbaumechanik auf bodenphysikalischer Grundlage.* Vienna: Deuticke.

Terzaghi, K. (1941). Undisturbed clay samples and undisturbed clays. *J. Boston Soc. Civ. Engrs* **28**, No. 3, 45–65.

Ward, W. H., Marsland, A. & Samuels, S. G. (1965). Properties of the London Clay at the Ashford Common shaft: in-situ and undrained strength tests. *Géotechnique* **15**, No. 4, 321–344.

Webb, D. L. (1964). *The mechanical properties of undisturbed samples of London Clay and Pierre shale.* PhD thesis, University of London.

Webb, D. L. (1969). Residual strength in conventional triaxial tests. *Proc. 7th Int. Conf. on Soil Mech.*, Mexico City, **1**, 433–441.

Wood, D. M. (1985). Index properties and consolidation history. *Proc. 11th Int. Conf. on Soil Mech.*, San Francisco, **2**, 703–706.

Wood, D. M. (1990). *Soil behaviour and critical state soil mechanics.* New York: Cambridge University Press.

Wroth, C. P. (1972). Some aspects of the elastic behaviour of overconsolidated clay. *Proc. Roscoe Memorial Symp.*, Foulis, pp. 347–361.

Wu, T. H. (1958). Geotechnical properties of glacial lake clays. *Proc. Am. Soc. Civ. Engrs* **84**, SM3, paper 1732.

VOTE OF THANKS

DR R. J. MAIR

We have been privileged to hear a superbly delivered 30th Rankine Lecture by Professor Burland, displaying the combination of skills of a distinguished engineer, scientist and teacher referred to by Mr Thorburn in his introduction. John Burland has more than lived up to his reputation for clarity of thought and ability to reduce apparently complex geotechnical problems to a simple framework. In his outstanding Nash Lecture at the Dublin Conference in 1987, John referred to Terzaghi's aim to 'maintain that vital balance between idealization and reality'. In this Rankine Lecture John Burland himself has made a most valuable contribution to that balance between idealization and reality by clarifying the factors affecting the compressibility and shear strength of natural clays.

John Burland's career uniquely qualifies him to address the behaviour of natural clays. At Cambridge he was closely associated with the valuable framework of critical state soil mechanics describing idealized soil behaviour in a new and fundamental way. At Imperial College he and his colleagues have been concerned with the reality and complexities of behaviour of natural soils. In between Cambridge and Imperial College, John had a distinguished period at the Building Research Establishment, where he was primarily concerned with a wide range of field measurements—these have enabled him to identify the strength and weaknesses of idealized soil behaviour and to appreciate the complexities of real soils.

He has introduced the important concept of what he terms the intrinsic properties of clays: the properties of reconstituted clays. He has emphasized the importance of the combination of testing good quality undisturbed samples and performing tests on reconstituted material. He has introduced new definitions: the intrinsic compression line , the sedimentation compression line and the void index. It is my belief that these definitions will be widely referred to in the future by practitioners and research workers alike.

By introducing these definitions, John Burland has elegantly distinguished between the behaviour of natural soft clays and reconstituted clays. He has demonstrated with characteristic clarity the difference in their behaviour due to structure—the combination of fabric and bonding. By drawing on examples from the Mississippi Delta and the North Sea, he has highlighted the importance of the deposition conditions on a clay's subsequent behaviour.

He has also covered overconsolidated clays and introduced the concept of post-rupture strength. He has provided clear insight into the complex behaviour associated with formation of rupture surfaces—it must be gratifying for him to see the electrolevel device he originally proposed for measurement of local strains lead to an improved understanding of soil behaviour. Practitioners are constantly faced with the problem of selection of appropriate design parameters for stiff overconsolidated clays, and it is likely that the concepts of post-rupture failure line and intrinsic failure line presented by John Burland this evening will become of significant practical value in the years to come.

In this Lecture we have seen the results of a thorough re-analysis of data from a wide variety of sources from across the world. The meticulous way in which this has evidently been done, and the enthusiasm with which the results have been presented, are both hallmarks of John Burland's style. It is with the greatest pleasure that I propose a warm vote of thanks to Professor Burland for an excellent and memorable 30th Rankine Lecture.